Les Houches, Août 1972
Cours de l'Ecole d'été de Physique théorique·
Organe d'intérêt commun de l'U.S.M.G.
et I.N.P.G. subventionné par l'OTAN et
le Commissariat à l'Energie Atomique

BLACK HOLES
LES ASTRES OCCLUS

edited by **C. DeWitt**
Faculté des Sciences, Grenoble
Dept. of Astronomy, University of Texas, Austin, et
B. S. DeWitt
Dept. of Physics, University of Texas, Austin

GORDON AND BREACH SCIENCE PUBLISHERS
New York London Paris

Gordon and Breach Science Publishers, Inc.
One Park Avenue
New York, N.Y. 10016

Editorial office for the United Kingdom

Gordon and Breach Science Publishers Ltd.
42 William IV Street
London W.C.2

Editorial office for France

Gordon & Breach
7–9 rue Emile Dubois
Paris 14e

Library of Congress catalog card number Applied for. ISBN 0 677 15610 3. Printed in Greated Britain.

CONTENTS

Contributors	vi
Preface (English)	vii
Préface (Français)	ix
List of Participants	xi
S. W. HAWKING: The Event Horizon	1
B. CARTER: Black Hole Equilibrium States	57
J. M. BARDEEN: Timelike and Null Geodesics in the Kerr Metric	215
J. M. BARDEEN: Rapidly Rotating Stars, Disks, and Black Holes	241
H. GURSKY: Observations of Galactic X-ray Sources	291
I. D. NOVIKOV and K. S. THORNE: Black Hole Astrophysics	343
R. RUFFINI: On the Energetics of Black Holes	451
Table of Contents	547

CONTRIBUTORS

J. M. BARDEEN
Yale University, New Haven, Conn.

B. CARTER
Institute of Astronomy, Cambridge

H. GURSKY
American Science and Engineering, Cambridge, Mass.

S. W. HAWKING
Institute of Astronomy, Cambridge

I. D. NOVIKOV
Institute of Applied Mathematics, Moscow

R. RUFFINI
Princeton University, Princeton, N.J.

K. S. THORNE
California Institute of Technology, Pasadena, Calif.

PREFACE

The story of the phenomenal transformation of general relativity within little more than a decade, from a quiet backwater of research, harboring a handful of theorists, to a booming outpost attracting increasing numbers of highly talented young people as well as heavy investment in experiments, is by now familiar. The amazing thing about this revolution is that the physics is entirely classical. But it is classical physics with a new twist. The student who embraces it must learn to chart unfamiliar conceptual waters, where even the firmament of classical fixed stars (energy conservation, causality, thermal equilibrium and entropy) has become strangely distorted. That this revolution is yet far from having run its course is clear from its classical nature. The quantum revolution is yet to come.

No single object or concept epitomizes more completely the present stage of the revolution than the Black Hole. This volume, based on the lectures given at the 23rd session of the Summer School of Les Houches, contains nearly everything that is currently (1972) known about black holes, and much that will remain permanently useful to workers in the field. The contents, which are deliberately pedagogical, begin with Hawking's masterful presentation of the fundamentals: the definition of a black hole, event horizons, trapped surfaces, singularity theorems, the area theorem, final state theorems. This is followed by Carter's beautiful elaboration, with proofs, of the properties of Kerr-Newman and other axisymmetric black holes: separability theorems, uniqueness theorems, variational principles. Next comes a very detailed study by Bardeen of the properties of timelike and null geodesics in the Kerr metric, of rapidly rotating stars, and of extreme relativistic disks, illustrated by explicit computer results.

After this thorough theoretical introduction, the facts are brought on stage. The data currently most likely to tell us whether black holes really do exist in our universe are the X-ray data from the UHURU satellite. The article by Gursky outlines the present status of X-ray astronomy, emphasizing the difficulties of interpreting the data as well as the knowledge gained so far about the various species of X-ray sources in the sky. Gursky presents the sober evidence placing Cygnus X-1 in the running as a prime candidate for black hole status. Then comes a superb attempt at a grand synthesis. Starting with a pedagogical review of the astrophysical processes relevant to black holes, Novikov and Thorne, in one of the longest articles in the volume, deal successively with: accretion of interstellar gas onto an isolated black hole, with accompanying emission of optical and ultraviolet radiation; accretion of gas onto a black hole from a companion star in a close binary system, with accompanying emission of X-rays, ultraviolet and optical radiation; and accretion of gas onto a supermassive black hole in the nucleus of a galaxy, with accompanying emission of ultraviolet, optical, infrared and radio radiation. For the first time, explicit models are built that incorporate the full general relativistic effects of the Kerr metric.

In the final chapter of this volume, Ruffini marshals most of the known results on mass limits for neutron stars, radiation (both electromagnetic and gravitational) emitted by single objects falling onto black holes, and general theory of the energetics of black holes, including possible energy-extraction mechanisms.

Warmest thanks are due to the authors for the very many hours of preparation and hard work that have made, first, the summer lectures and, finally, this volume possible.

Cécile DeWitt
Bryce S. DeWitt

PRÉFACE

L'histoire de la transformation prodigieuse de la Relativité Générale pendant ces dix dernières années est chose connue; d'une baie tranquille où quelques théoriciens poursuivaient leurs recherches, elle est passée aux avant postes, en pleine effervescence, qui attirent un nombre croissant de jeunes talents, ainsi que de crédits importants destinés aux recherches expérimentales. Et ce qui est étonnant, c'est que cette explosion reste dans le cadre de la physique classique – classique il est vrai quant à ses principes mais si nouvelle quant aux domaines qu'elle explore que les principes classiques (conservation de l'énergie, causalité, équilibre thermique et entropie), ces étoiles fixes du physicien, prennent un aspect étrangement nouveau avec lequel le navigateur doit se familiariser pour déterminer la route à suivre sur ces eaux inconnues. Et nous n'en sommes qu'au début du voyage, les phénomènes quantiques n'ont pas encore été abordés.

Plus que tout autre objet ou concept, les Astres occlus† sont l'épitomé de cette évolution. Ce volume a pour base les cours donnés à la 23ème session de L'Ecole des Houches; ils contiennent presque tout ce qui est connu à l'heure actuelle (1972) sur les occlus; beaucoup en restera utile à ceux qui travaillent dans ce domaine. Les exposés sont franchement pédagogiques; ils commencent par la présentation, de main de maître, par Hawking des notions et théorèmes fondamentaux: définition d'un occlus, de l'horizon des évènements, et, des surfaces occlusives, théoremes de singularité, de surface, et, d'états finaux. Cette présentation est suivie de la belle élaboration de Carter sur les propriétés des occlus Kerr-Newman et autres occlus à symmétrie axiale, ainsi que sur les théorèmes qui s'y rapportent: théorèmes de séparabilité, unicité, principes de variation. Ensuite vient une étude très détaillée faite par Bardeen sur les propriétés des géodesiques temporelles et isotropes dans la métrique de Kerr, sur les propriétés des étoiles à rotation rapide et des disques extrèmement relativistes; cette étude est illustrée par des résultats numériques obtenus par ordinateurs.

Après cette introduction théorique approfondie, les faits sont presentés sur la scène. Les données actuelles les plus susceptibles de nous dire si les occlus existent réellement dans notre univers sont celles des rayons X obtenues par le satellite UHURU. L'article de Gursky schématise l'état actuel de l'astronomie de rayons X, et souligne les difficultés d'interprétation des données expérimentales tout en précisant les connaissances acquises sur les différentes sources de rayons X

---

† Après de nombreuses discussions avec les spécialistes de la question, l'expression "Astres occlus" a semblé la plus adéquate compte tenu des propriétés de cet objet et de ses dénominations dans d'autres langues. "Trapped surface" devient ainsi surface occlusive (qui produit l'occlusion).

dans le ciel. Gursky présente sobrement les indications suggérant la possibilité que Cygnus X-1 soit un occlus.

Vient ensuite, un essai de grande synthese: commençant par une revue pédagogique des processus astrophysiques se rapportant aux occlus, Novikov et Thorne, dans l'article le plus long de ce livre, traite successivement: la capture, par un occlus isolé, de gaz interstellaire, avec émission de rayonnement ultraviolet et optique; la capture, par un occlus d'un système binaire, de gaz provenant de l'autre composante du système, avec émission de rayonnement X, ultraviolet et optique; la capture de gaz par un occlus supermassif dans le noyau d'une galaxie, avec émission de rayonnement ultraviolet, optique et radio. Pour la première fois, des modèles ont été construits qui tiennent compte de tous les effets de Relativité Générale présents dans la métrique de Kerr.

Pour terminer, Ruffini rassemble et classe presque tous les résultats connus sur les masses limites des étoiles de neutrons, le rayonnement électromagnétique et gravitationnel émis par un objet tombant dans un occlus, et la théorie générale de l'énergétique des occlus y compris les mécanismes possibles d'extraction de l'énergie.

Les remerciements les plus chaleureux sont dûs aux auteurs pour les très longues heures de préparation et le lourd labeur, grâce auxquels la session, puis ce volume a pu être réalisé.

Cécile DeWitt
Bryce S. DeWitt

LIST OF PARTICIPANTS

ABRAMOWICZ, Marek	Institut Astronomie, Varsovie
ANILE, Angelo	University Observatory, Oxford
BEKENSTEIN, Jacob D.	Center for Relativity Theory, University of Texas, Austin
BLAND, Roger W.	Dept. of Natural Philosophy, University of Glasgow
BREUER, Reinhard	Int. für Theoretische Physik II, Wurzburg
CADEZ, Andrej	Fakult. Naravoslovje Technologijo, Ljubliana
CUNNINGHAM	Dept. Astronomy, University of Washington, Seattle
CHRISTENSEN, Steven M.	Centre for Relativity Theory, University of Texas
CHRZANOWSKI, Paul	Dept. Physics and Astronomy, University of Maryland
DEMARET, Jacques	Inst. Astrophysique, University of Liege
DEMIANSKI, Marek	Inst. Physique théorique, Universitet Warszawskieg, Warszawa
DENARDO, Gallieno	Inst. di Fisica Teorica, University of Trieste
DYER, Charles C.	Dept. of Astronomy, University of Toronto, Toronto
GIESSWEIN, Michael	Inst. Physique théorique, Vienne
GODDARD, Andrew J.	University Observatory, Oxford
HAJICEK, Petr	Inst. Physique Théorique, Berne
HARRISON, Bertrand K.	Dept. of Physics and Astronomy, Brigham Young University, Provo, Utah
HU, Bie-Lok	Jadwin Hall, Dept. Physics, Princeton University, Princeton
HUGHES, Henry G.	1834 Marshall, Houston, Texas 77006, USA
HUGHSTON, Lane P.	1206 W. Louisiana, MacKinney, Texas 75069, USA
KALLMAN, Cerl-Gustav	Inst. for Pysik Abo Akademi, Porthansg 3-5 20500 ABO 50, Finlande
MAGNON, Anne M.	Dept. de Mathématiques, Université de Clermont
MEINHARDT, Roberto	Inst. Physique théorique, Hamburg
PERJES, Zoltan	Central Research Inst. for Physics, Budapest
PINEAULT, Serge	Dept. of Astronomy, University of Toronto, Toronto
PRESS, William H.	California Inst. of Technology, Pasadena
QUINTANA, Hernan	Inst. of Theor. Astronomy, Cambridge

RAKAVY, Gideon	Hebrew University, Jerusalem
RASBAND, S. Neil	Dept. of Physics, Brigham Young University, Provo, Utah
ROSS, Dennis K.	Physics Dept., Iowa State University, Ames
SCHNEIDER, Jean	Dept. d'Astronomie Fondamentale, Observatoire de Meudon
SILVESTRO, Giovanni	Ist. di Fisica Generale, Università di Torino, Turin
SMARR, Larry	Center for Relativity Theory, University of Texas, Austin
SOBOUTI, Yousef	Physics Dept. University of Pahlavi, Shiraz
SOMMERS, Paul D.	Center for Relativity Theory, University of Texas, Austin
STREUBEL, Michael	Inst. Physique théorique, Hamburg
TEITELBOIM, Claudio	Joseph Henry Laboratories, Princeton University, Princeton
TEUKOLSKY, Saul A.	California Inst. of Technology, Pasadena
TOD, Kenneth P.	University Observatory, Oxford
TREVES, Aldo Renato	Ist. di Fisica dell'Università di Milano, Milan
TSIANG, Elaine	Center for Relativity Theory, University of Texas, Austin
VANDERMOLEN, Jean-Claude	Inst. d'Astronomie et Géophysique, G. Lemaire Kardinaal Mercierlaan, Louvain
VAN NIEUWHENHUIZEN, P.	Lab. Physique théorique et Hautes Energies, Université Paris, Orsay
WALKER, Martin A.	Inst. Max Planck Physik und Astrophysik, Munchen
WILKINS, Daniel C.	Dept. of Physics, Stanford University, Stanford
WILL, Clifford M.	Dept. of Physics, California Inst. of Technology, Pasadena
WOODHOUSE, Nicholas	University of London, King's College, London

The Event Horizon

Stephen W. Hawking

Institute of Astronomy
Cambridge, Great Britain

Contents

Introduction 5
1 Spherically Symmetric Collapse 6
2 Nonspherical Collapse 11
3 Conformal Infinity 13
4 Causality Relations 16
5 The Focusing Effect 20
6 Predictability 24
7 Black Holes 29
8 The Final State of Black Holes 37
9 Applications 44
A Energy Limits 44
B Perturbations of Black Holes 46
C Time Symmetric Black Holes 51
References 54

Introduction

We know from observations during eclipses and radio measurements of quasars passing behind the sun that light is deflected by gravitational fields. One would therefore imagine that if there were a sufficient amount of matter in a certain region of space, it would produce such a strong gravitational field that light from the region would not be able to escape to infinity but would be "dragged back". However one cannot really talk about things being dragged back in general relativity since there are not in general any well defined frames of reference against which to measure their progress. To overcome this difficulty one can use the following idea of Roger Penrose. Imagine that the matter is transparent and consider a flash of light emitted at some point near the centre of the region. As time passes, a wave front will spread out from the point (Figure 1). At first this wave front will be nearly spherical and its area will be proportional

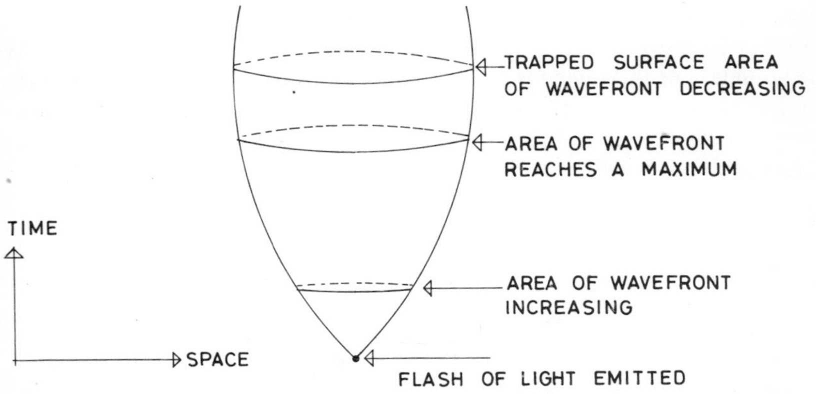

Figure 1. The wavefront from a flash of light being focused and dragged back by a strong gravitational field.

to the square of the time since the flash was emitted. However the gravitational attraction of the matter through which the light is passing will deflect neighbouring rays towards each other and so reduce the rate at which they are diverging from each other. In other words, the light is being focused by the gravitational effect of the matter. If there is a sufficient amount of matter, the divergence of neighbouring rays will be reduced to zero and then turned into a convergence. The area of the wave front will reach a maximum and start to decrease. The effect of passing through any more matter is further to step up the rate of decrease of the area of the wave front. The wave front therefore will not expand and reach infinity since, if it were to do so, its area would have to become arbitrarily large. Instead, it is "trapped" by the gravitational field of the matter in the region.

We shall take this existence of a wave front which is moving outward yet decreasing in area as our criterion that light is being "dragged back". In fact it does not matter whether or not the wave front originated at a single point. All that is important is that it should be a *closed* (i.e. compact) surface, that it should be *outgoing* and that at each point of the wavefront neighbouring rays should be *converging* on each other. In more technical language, such a wave front is a compact space like 2-surface [without edges] such that the family of outgoing future-directed null geodesics orthogonal to it is converging at each point of the surface. I shall call this an *outer trapped surface* (or simply, a trapped surface). This differs from Penrose's definition (Penrose, 1965a) in that he required the *ingoing* future-directed null geodesics orthogonal to the surface to be converging as well. The behaviour of the ingoing null geodesics is of importance in proving the occurrence of a spacetime singularity in the trapped region. However, in this course we are primarily interested in what can be seen by observers at a safe distance. Modulo certain reservations which will be discussed in section 2, the existence of a closed outgoing wave front (or null hypersurface) which is decreasing in area implies that information about what happens behind the wavefront cannot reach such observers. In other words, there is a region of spacetime from which it is not possible to escape to infinity. This is a *black hole.* The boundary of this region is formed by a wavefront or null hypersurface which just does not escape to infinity; its rays are asymptotically parallel and its area is asymptotically constant. This is the *event horizon.*

To show how event horizon and black holes can occur I shall now discuss the one situation that we can treat exactly, spherical symmetry.

1 Spherically Symmetric Collapse

Consider a non-rotating star. After its formation from an interstellar gas cloud, there will be a long period (10^9–10^{11} years) in which it will be in an almost stationary state burning hydrogen into helium. During this period the star will be supported against its own gravity by thermal pressure and will be spherically symmetric. The metric outside the star will be the Schwarzschild solution—the only empty spherically symmetric solution

$$ds^2 = \left(1 - \frac{2M}{r}\right) dt^2 - \left(1 - \frac{2M}{r}\right)^{-1} dr^2 - r^2 (d\theta^2 + \sin^2 \theta d\phi^2) \qquad (1.1)$$

This is the form of the metric for r greater than some value r_0 corresponding to the surface of the star. For $r < r_0$ the metric has some different forms depending on the distribution of density in the star. The details do not concern us here.

When the star has exhausted its nuclear field, it begins to lose its thermal energy and to contract. If the mass M is less than about 1.5–$2M_\odot$, this contraction can be halted by degeneracy pressure of electrons or neutrons resulting in a white dwarf or neutron star respectively. If, on the other hand, M is greater than this limit, contraction cannot be halted. During this spherical contraction the metric outside the star remains of the form (1.1) since this is the only spherically symmetric empty solution. There is an apparent difficulty when the surface of the star gets down to the Schwarzschild radius $r = 2M$ since the metric (1.1) is singular there. This however is simply because the coordinate system goes wrong here. If one introduces an advanced time coordinate v defined by

$$v = t + r + 2M \log (r - 2M) \tag{1.2}$$

the metric takes the Eddington-Finkelstein form

$$ds^2 = \left(1 - \frac{2M}{r}\right) dv^2 - 2dvdr - r^2(d\theta^2 + \sin^2\theta d\phi^2) \tag{1.3}$$

This metric is perfectly regular at $r = 2M$ but still has a singularity of infinite curvature at $r = 0$ which cannot be removed by coordinate transformation. The orientation of the light-cones in this metric is shown in Figure 2. At large values of r they are like the light-cones in Minkowski space and they allow a particle or photon following a nonspacelike (i.e., timelike or null) curve to move outwards or inwards. As r decreases the light-cones tilt over until for $r < 2M$ all nonspacelike curves necessarily move inwards and hit the singularity at $r = 0$. At $r = 2M$ all nonspacelike curves except one move inwards. The exception is the null geodesic r, θ, ϕ constant which neither moves inwards nor outwards. From this behaviour it follows that light emitted from points with $r > 2M$ can escape to infinity whereas that from $r \leqslant 2M$ cannot. In particular the singularity at $r = 0$ cannot be seen by observers who remain outside $r = 2M$. This is an important feature about which I shall have more to say later.

The metric (1.3) holds only outside the surface of the star which will be represented by a timelike surface which crosses $r = 2M$ and hits the singularity at $r = 0$. Inside the star the metric will be different but the details again do not matter. One can analyse the important qualitative features by considering the behaviour of a series of flashes of light emitted from the centre of the star which again is taken to be transparent. In the early stage of the collapse when the density is still low, the divergence of the outgoing light rays or null geodesics will not be reduced much by the focusing effect of the matter. The wavefront will therefore continue to increase in area and will reach infinity. As the collapse continues and the density increases, the focusing effect will get bigger until there will be a critical wavefront whose rays emerge from the surface of the star with zero divergence. Outside the star the area of this wavefront will remain constant and it will be the surface $r = 2M$ in the metric (1.3). Wavefronts corresponding to flashes of light emitted after this critical time will be focused so much by the matter that their rays will begin to *converge* and their area to decrease. They will then form *trapped surfaces.* Their area will continue to decrease, reaching zero when they hit the singularity at $r = 0$.

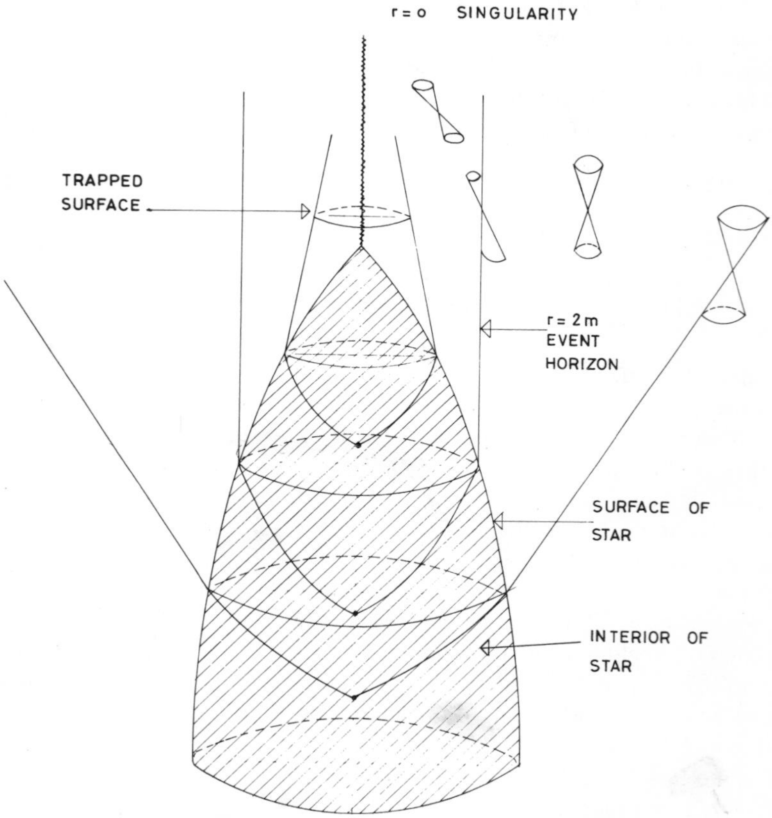

Figure 2. The collapse of a spherical star leading to the formation of trapped surfaces, event horizon and spacetime singularity.

The critical wavefront which just avoids being converged is the *event horizon*, the boundary of the region of space-time from which it is not possible to escape to infinity along a future directed nonspacelike curve. It is worth noting certain properties of the event horizon for future reference.

(1) The event horizon is a null hypersurface which is generated by null geodesic segments which have no future end-points but which do have past end-points (at the point of emission of the flash).

(2) The divergence of these null geodesic generators is positive during the collapse phase and is zero in the final time independent state. It is never negative.
(3) The area of a 2-dimensional cross-section of the horizon increases monotonically from zero to a final value of $16\pi M^2$.

We shall see that the event horizon in the general case without spherical symmetry will also have these properties with a couple of small modifications. The first modification is that in general the null geodesic generators will not all have their past end-points at the same point but will have them on some caustic or crossing surface. The second modification is that if the collapsing star is rotating, the final areas of the event horizon will be

$$8\pi[M^2 + (M^4 - L^2)^{\frac{1}{2}}] \tag{1.4}$$

where L is the final angular momentum of the black hole, i.e., that part of the original angular momentum of the star that is not carried away by gravitational radiation during the collapse. This formula (1.4) will play an important role later on.

In the example we have been considering the event horizon has another property in the time-independent region outside the star. It is the boundary of the part of spacetime containing trapped surfaces. This is not true however in the time-dependent region inside the star. There has in the past been some confusion between the event horizon and the boundary of the region containing trapped surfaces, so it is worth spending a little time to clarify the distinction. Let us introduce a family of spacelike surfaces $S(\tau)$ labelled by a parameter τ which we shall interpret as some sort of time coordinate. In the example we are considering τ could be chosen to be $v - r$ but the exact form is not important. Given a particular surface $S(\tau)$, one can find whether there are any trapped surfaces which lie in $S(\tau)$. The boundary of the region of $S(\tau)$ containing trapped surfaces lying in $S(\tau)$ will be called the *apparent horizon* in $S(\tau)$. This is not necessarily the same as the intersection of the event horizon with $S(\tau)$ which is the boundary of the region of $S(\tau)$ from which it is not possible to escape to infinity. To see the differences consider a situation which is similar to the previous example of a collapsing spherical star of mass M but where there is also a thin spherical shell of matter of mass δM which collapses from infinity at some later time and hits the singularity at $r = 0$. (Figure 3). Between the surface of the star and the shell the metric is of the form (1.3) while outside the shell it is of the form (1.3) with M replaced by $M + \delta M$. The apparent horizon in $S(\tau_1)$, the boundary of the trapped surfaces in $S(\tau_1)$, will be at $r = 2M$. It will remain at $r = 2M$ until the surface $S(\tau_2)$ when it will suddenly jump out to $r = 2(M + \delta M)$. On the other hand, the event horizon, the boundary of the points from which it is not possible to escape to infinity, will intersect $S(\tau_1)$ just outside $r = 2M$. It will move out continuously reach $r = 2(M + \delta M)$ at the surface $S(\tau_2)$. Thereafter it will remain at this radius provided no more shells of matter fall in from infinity.

The apparent horizon has the practical advantage that one can locate it on a given surface $S(\tau)$ knowing the solution only on that surface. On the other hand one has to know the solution at all times to locate the event horizon. However, the event horizon has the mathematical advantage of being a null hypersurface

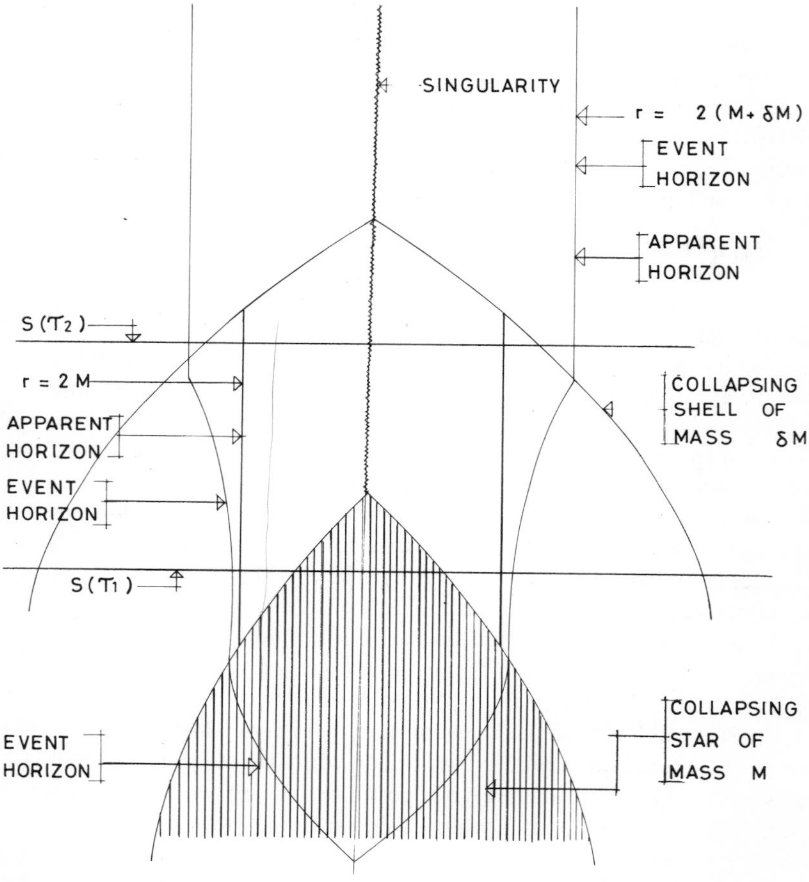

Figure 3. The collapse of a star followed by the collapse of a thin shell of matter. The apparent horizon moves outwards discontinuously but the event horizon moves in a continuous manner.

with nice properties like the area always increasing whereas the apparent horizon is not in general null and can move discontinuously. In this course I shall therefore concentrate on the event horizon. I shall show that it will always coincide with or be outside the apparent horizon. During periods when the solution is nearly time independent and nothing is just about to fall into the black hole, the two horizons will nearly coincide and their areas will be almost equal. If the black hole now undergoes some interaction and settles down to another almost stationary state, the area of the event horizon will have increased. Thus the area of the apparent horizon will also have increased. I shall show how the area increase can be used to measure the amounts of energy and angular momentum which fell into the black hole.

2 Nonspherical Collapse

No real star is exactly spherical; they all are rotating a bit and have magnetic fields. One must therefore ask whether their collapse will show the same features as the spherical case we discussed before. One would not expect this necessarily to be the case if the departure from spherical symmetry were too large. For example a rapidly rotating star would not collapse to within $r = 2M$ but would form a thin rotating disc, maintaining itself by centrifugal force against the gravitational attraction. However one might hope that the picture would be qualitatively similar to the spherical case for departures from spherical symmetry that are small initially. One can divide this question of stability under small perturbations of the initial conditions into three parts.

(1) Is the occurrence of a singularity a stable feature?
(2) Is the form of the singularity stable?
(3) Is the fact that the singularity cannot be seen from infinity stable?

The Einstein equations being a well behaved system of differential equations have the property of local stability. The solution at nonsingular points depends continuously on the initial data (see Hawking and Ellis, 1973. I shall refer to this as HE). In other words, given a compact nonsingular region V in the Cauchy development of an initial surface S, one can find a perturbation of the initial data on S which is sufficiently small that the solution on V changes by less than a given amount. One can apply this result to show that small initial departure from spherical symmetry will not affect the fact that the wavefronts corresponding to flashes of light emitted from the centre of the star will be focused and made to start to reconverge. It follows from a theorem of Penrose and myself (Hawking and Penrose, 1970) that the existence of such a reconverging wavefront implies the occurrence of a space time singularity provided that certain other reasonable conditions like positive energy density and causality are satisfied. Thus the answer to question (1) is "yes"; the occurrence of a singularity is a stable feature of gravitational collapse.

As the local stability result holds only at non-singular points it cannot be used to answer question (2): is the form of the singularity stable? In fact the answer is "no". For example adding a small amount of electric charge to the star changes the singularity from that in the Schwarzschild solution to that in the Reissner-Nordström solution which is completely different. It is reasonable to expect that a small departure from spherical symmetry would also completely change the singularity. This makes it very difficult to study singularities since one does not know what a "generic" singularity would look like. The work of Liftshitz, Belinsky and Khalatnikov suggests that it is probably very complicated. Fortunately we do not have to worry about this in this course provided we have an affirmative answer to question (3): is the fact that the singularity cannot be seen from infinity stable? One cannot use the local stability result to answer this since it applies only to the behaviour of perturbations over a finite interval of time. The question of whether the singularities can be seen from infinity depends on the behaviour of the solution at arbitrarily large times and at such times the perturbations might have grown large. In fact this question which is absolutely fundamental to the whole study of black holes has not yet been properly answered. However there are grounds for optimism. The first of these is that linearized perturbation studies of spherical collapse by Regge and Wheeler (1957), Doroshkevich, Zeldovich and Novikov (1965), Price (1972) and others have shown that all perturbations except one die away with time. The one exception corresponds to a rotational perturbation which changes the Schwarzschild solution into a linearized Kerr solution. In this the singularities are also hidden from infinity. These perturbation calculations do not completely answer the stability question since they are only first order: one would need to show that the perturbations of the second and higher orders also died away and that the perturbation series converged.

The second ground for believing that the singularities are hidden is that Penrose and Gibbons have tried and failed to devise situations in which they are not. The idea was to try and obtain a contradiction with the result that the area of the event horizon increases which is a consequence of the assumption that the singularities are hidden. However they failed. Of course their failure does not prove anything but it does strengthen my personal conviction that the singularities in gravitational collapse will not be visible from infinity. One has to be slightly careful how one states this because one can always devise situations where there are naked singularities of a sort. For example, if one has pressure-free matter (dust), one can arrange the flow-lines to intersect on caustics which will be three dimensional surfaces of infinite density. However such singularities are really trivial in the sense that the addition of a small amount of pressure or a slight variation in the initial conditions would remove them. I believe that if one starts from a non-singular, asymptotically flat initial surface there will not be any non-trivial singularities which can be seen from infinity.

If there are non-trivial singularities which are naked, i.e. which can be seen from infinity, we may as well all give up. One cannot predict the future in the presence of a space-time singularity since the Einstein equations and all the known laws of physics break down there. This does not matter so much if the singularities are all safely hidden inside black holes but if they are not we could be in for a shock every time a star in the galaxy collapsed. People working in General Relativity have a strong vested interest in believing that singularities are hidden.

In order to investigate this in more detail one needs precise notions of infinity and of causality relations. These will be introduced in the next two sections.

3 Conformal Infinity

What can be seen from infinity is determined by the light-cone structure of spacetime. This is unchanged by a conformal transformation of the metric, i.e., $g_{ab} \to \Omega^2 g_{ab}$ where Ω is some suitably smooth positive function of position. It is therefore helpful to make a conformal transformation which squashes everything up near infinity and brings infinity up to a finite distance. To see how this can be done consider Minkowski space:

$$ds^2 = dt^2 - dr^2 - r^2(d\theta^2 + \sin^2\theta\phi^2) \tag{3.1}$$

Introduce retarded and advanced time coordinates, $w = t - r, v = t + r$. The metric then takes the form

$$ds^2 = dvdw - r^2(d\theta^2 + \sin^2\theta d\phi^2) \tag{3.2}$$

Now introduce new coordinates p and q defined by $\tan p = v$, $\tan q = w$, $p - q \geqslant 0$. The metric then becomes

$$ds^2 = \sec^2 p \sec^2 q [dpdq - \tfrac{1}{4}\sin^2(p - q)(d\theta^2 + \sin^2\theta d\phi^2)] \tag{3.3}$$

This is of the form $ds^2 = \Omega^{-2}d\tilde{s}^2$ where $d\tilde{s}^2$ is the metric within the square brackets. In new coordinates $t' = \frac{1}{2}(p + q), r' = \frac{1}{2}(p - q)$ the conformal metric $d\tilde{s}^2$ becomes

$$d\tilde{s}^2 = dt'^2 - dr'^2 - \tfrac{1}{4}\sin^2 2r'(d\theta^2 + \sin^2\theta d\phi^2) \tag{3.4}$$

This is the metric of the Einstein universe, the static spacetime where space sections are 3-spheres. Minkowski space is conformal to the region bounded by the null surface $t' - r' = -\pi/2$ [this can be regarded as the future light-cone of the point $r' = 0, t' = -(\pi/2)$] and the null surface $t' + r' = \pi/2$ (the past light cone of $r' = 0, t' = \pi/2$) (Figure 4). Following Penrose (1963, 1965b) these null surfaces will be denoted by $\mathscr{I}^-$ and $\mathscr{I}^+$ respectively. The point $r' = 0, t' = \pm\pi/2$ will be denoted by $i^\pm$ and the points $r' = \pi/2, t' = 0$ will be denoted by i^0. (It is a point because $\sin^2 2r'$ is zero there.) Penrose originally used capital I's for these points but this would cause confusion with the symbol for the timelike future which will be introduced in the next section.

All timelike geodesics in Minkowski space start at i^- which represents past timelike infinity and end at i^+ which represents future timelike infinity. Spacelike geodesics start and end at i^0 which represents spacelike infinity. Null geodesics, on the other hand, start at some point on the null surface $\mathscr{I}^-$ and end at some point on $\mathscr{I}^+$. These surfaces represent past and future null infinity respectively (Figure 5).

When one says that space time is asymptotically flat one means that near infinity it is like Minkowski space in some sense. One would therefore expect the conformal structure of its infinity to be similar to that of Minkowski space. In fact it turns out that the conformal metric is singular in general at the points corresponding to i^- i^+ i^0. However it is regular on the null surfaces $\mathscr{I}^-$ $\mathscr{I}^+$. This led Penrose (1963, 1965b) to adopt this feature as a *definition* of asymptotic

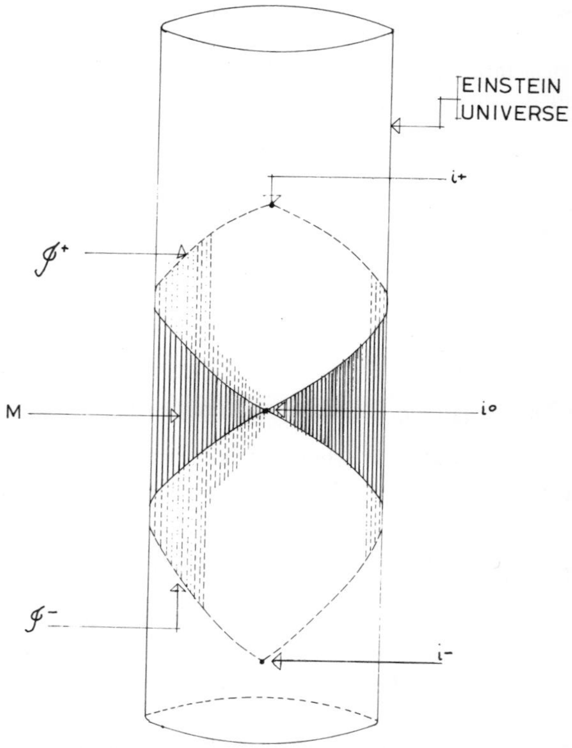

Figure 4. Minkowski space M conformally imbedded in the Instein Static Universe. The Conformal boundary is formed by the two null surfaces $\mathscr{I}^+$, $\mathscr{I}^-$ and the points ι^+ ι° and ι^-.

flatness. A manifold M with a metric g_{ab} is said to be *asymptotically simple* if there exists a manifold $\tilde{M}$ with a metric $\tilde{g}_{ab}$ such that

(1) M can be imbedded in $\tilde{M}$ as a manifold with boundary ∂M
(2) On M, $\tilde{g}_{ab} = \Omega^2 g_{ab}$
(3) On ∂M, $\Omega = 0$, $\Omega_{;a} \neq 0$
(4) Every null geodesic in M has past and future end points on ∂M.
(5) The Einstein equations hold in M which is empty or contains only an electromagnetic field near ∂M (Penrose did not actually include this last condition in the definition but it is useful really only if this condition holds)

Condition (3) implies that the conformal boundary ∂M is at infinity from the point of view of someone in the manifold M. Penrose showed that conditions (4) and (5) implied that ∂M consisted of two disjoint null hypersurfaces, labelled $\mathscr{I}^-$ and $\mathscr{I}^+$ which each had topology $R^1 \times S^2$. An example of an asymptotically simple space would be a solution containing a bounded object such as a star which did not undergo gravitational collapse. However the definition is too strong to apply to solutions containing black holes because condition (4) requires that every null geodesic should escape to infinity in both directions. To overcome this

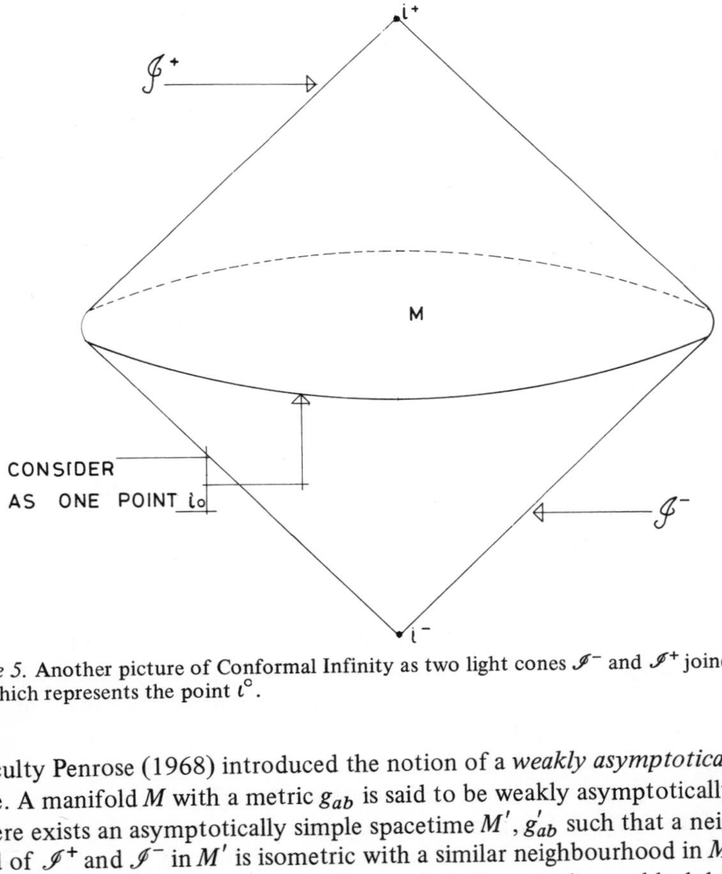

Figure 5. Another picture of Conformal Infinity as two light cones $\mathscr{I}^-$ and $\mathscr{I}^+$ joined by a rim which represents the point i^0.

difficulty Penrose (1968) introduced the notion of a *weakly asymptotically simple* space. A manifold M with a metric g_{ab} is said to be weakly asymptotically simple if there exists an asymptotically simple spacetime M', g'_{ab} such that a neighbourhood of $\mathscr{I}^+$ and $\mathscr{I}^-$ in M' is isometric with a similar neighbourhood in M. This will be the definition of asymptotic flatness I shall use to discuss black holes. Since condition (4) no longer holds for the whole of M there can be points from which it is not possible to reach future null infinity $\mathscr{I}^+$ along a future directed timelike or null curve. In other words these points are not in the past of $\mathscr{I}^+$. The boundary of these points, the event horizon, is the boundary of the past of $\mathscr{I}^+$. I shall discuss properties of such boundaries in the next section.

Exercise

Show that the Schwarzschild solution is weakly asymptotically simple.

4 Causality Relations

I shall assume that one can define a consistent distinction between past and future at each point of spacetime. This is a physically reasonable assumption. Even if it did not hold in the actual spacetime manifold M, there would be a covering manifold in which it did hold (Markus 1955).

Given a point p, I shall denote by $I^+(p)$ the *timelike or chronological future* of p, i.e. the set of all points which can be reached from p by future directed timelike curves. Similarly $I^-(p)$ will denote the past of p. Many of the definitions I shall give will have duals in which future is replaced by past and plus by minus. I shall regard such duals as self-evident. Note that p itself is not contained in $I^+(p)$ unless there is a timelike curve from p which returns to p. Let q be a point in $I^+(p)$ and let $\lambda(v)$ be a future directed timelike curve from p to q. The condition that $\lambda(v)$ is timelike is an inequality:

$$g_{ab}\frac{dx^a}{dv}\frac{dx^b}{dv}>0$$

where $\frac{dx^a}{dv}$ is the tangent vector to $\lambda(v)$. One can deform the curve $\lambda(v)$ slightly without violating the inequality to obtain a future directed timelike curve from p to any point in a small neighbourhood of q. Thus $I^+(p)$ is an open set.

The *causal future* of p, $J^+(p)$, is defined as the union of p with the set of points that can be reached from p by future directed nonspacelike, i.e., timelike or null curves. If one considers only a small neighbourhood of p, then $I^+(p)$ is the interior of the future light cone of p and $J^+(p)$ is $I^+(p)$ with the addition of the future light cone itself including the vertex. Note that the boundary of $I^+(p)$, which I shall denote by $\dot{I}^+(p)$, is the same as $\dot{J}^+(p)$, the boundary of $J^+(p)$, and is generated by null geodesic segments with past end points at p.

When one is dealing with regions larger than a small neighbourhood, there is the possibility that some of the null geodesics through p may reintersect each other and the forms of $I^+(p)$ and $J^+(p)$ may be more complicated. To see the general relationship between them consider a future directed curve from a point p to some point $q \in J^+(p)$. If this curve is not a null geodesic from p, one can deform it slightly to obtain a timelike curve from p to q. From this one can deduce the following:

(a) If q is contained in $J^+(p)$ and r is contained in $I^+(q)$, then r is contained in $I^+(p)$. The same is true if q is in $I^+(p)$ and r is in $J^+(q)$.

(b) The set $E^+(p)$, defined as $J^+(p) - I^+(p)$, is contained in (not necessarily equal to) the set of points lying on future directed null geodesics from p.

(c) $\dot{I}^+(p)$ equals $\dot{J}^+(p)$. It is not necessarily the same as $E^+(p)$.

A simple example of a space in which $E^+(p)$ does not contain the whole of the future directed null geodesics from p is provided by a 2-dimensional cylinder with the time direction along the axis of the cylinder and the space direction round the circumference (Figure 6). The null geodesics from the point p meet

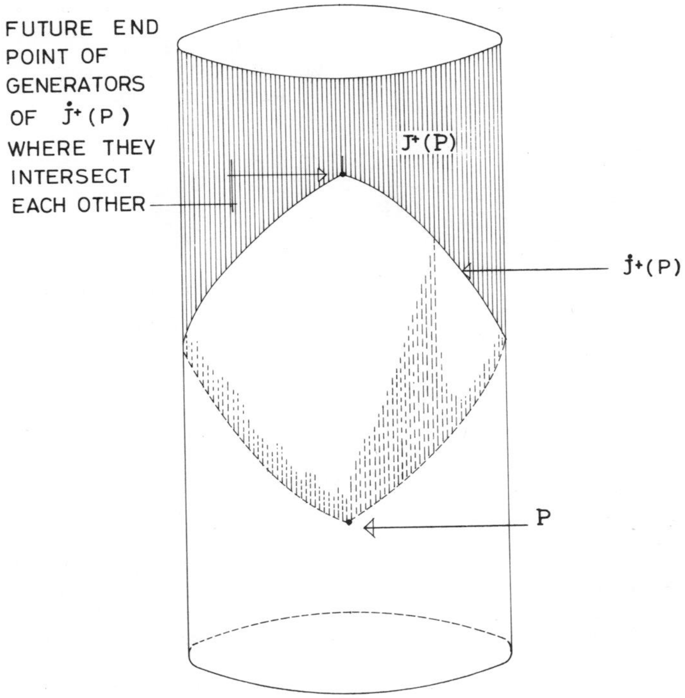

Figure 6. A space in which the future directed null geodesics from a point P have future endpoints as generators of $\dot{J}^+(P)$.

up again at the point q. After this they enter $I^+(p)$. An example in which $E^+(p)$ does not form all of $\dot{I}^+(p)$ is 2-dimensional Minkowski space with a point r removed (Figure 7). The null geodesic in $\dot{I}^+(p)$ beyond r does not pass through p and is not in $J^+(p)$.

The definitions of timelike and causal futures can be extended from points to sets: for a set S, $I^+(S)$ is defined to be the union of $I^+(p)$ for all $p \in S$. Similarly for $J^+(S)$. They will have the same properties (a), (b) and (c) as the futures of points, Suppose there were two points q, r on the boundary $\dot{I}^+(S)$ of the future of a set S with a future directed timelike curve λ from q to r.

One could deform λ slightly to give a timelike curve from a point x in $I^+(S)$ near q to a point y in $M - I^+(S)$ near r. This would be a contradiction since $I^+(x)$ is contained in $I^+(S)$. Thus one has

(d) $\dot{I}^+(S)$ does not contain any pair of points with timelike separation. In other words, the boundary $\dot{I}^+(S)$ is null or spacelike at each point.

Consider a point $q \in \dot{I}^+(S)$. One can introduce normal coordinates x^1, x^2, x^3, x^4 (x^4 timelike) in a small neighbourhood of q. Each timelike curve x^i = constant

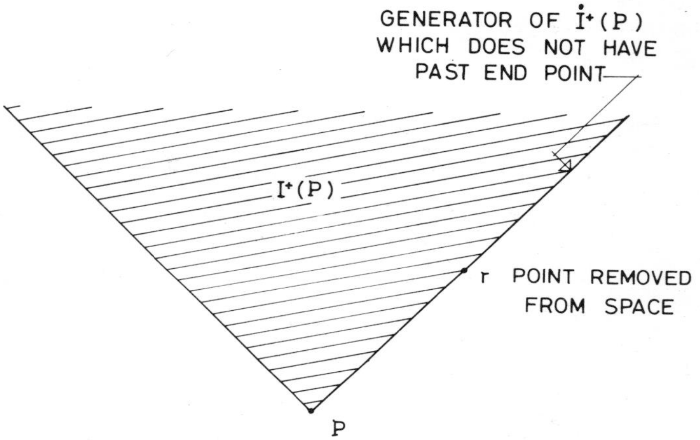

Figure 7. The point r has been removed from two dimensional Minkowski space.

(i = 1, 2, 3) will intersect $\dot{I}^+(S)$ once and once only. These curves will give a continuous map of a small region of $\dot{I}^+(S)$ to the 3-plane $x^4 = 0$. Thus

(e) $I^+(s)$ is a manifold (not necessarily a differentiable one).

Now consider a point q in $\dot{I}^+(S)$ but not in S itself, or its topological closure $\bar{S}$. One can thus find a small convex neighbourhood U of q which does not intersect $\bar{S}$. In U one can find a sequence $\{y_n\}$ of points in $I^+(S)$ which converge to the point q (Figure 8). From each y_n there will be a past directed timelike curve λ_n to S. The intersections of the $\{\lambda_n\}$ with the boundary $\dot{U}$ of U must have some limit point z since i$\dot{U}$ is compact. Any neighbourhood of z will intersect an infinite number of the $\{\lambda_n\}$. Thus z will be in $\bar{I}^+(S)$. The point z cannot be spacelike separated from q since, if it were, it would not be near timelike curves from points y_n near q. It cannot be timelike separated from q since if it were one could deform one of the λ_n passing near z to give a timelike curve from S to q which would then have to be in the interior of $I^+(S)$ and not on boundary. Thus z must lie on a past directed null geodesic segment γ from q. Each point of γ between q and z will be in $\dot{I}^+(S)$. One can now repeat the construction at z and obtain a past directed null geodesic segment μ from z which lies in $\dot{I}^+(S)$. If the direction of μ were differed from that of γ one could join points of μ to points of γ by timelike curves. This would contradict property (d) which says that no two points of $\dot{I}^+(S)$ have timelike separation. Thus μ will be a continuation of γ. One can continue extending γ to the past in $\dot{I}^+(S)$ unless and until it intersects S.

If there are two past directed null geodesic segments γ_1 and γ_2 lying in $\dot{I}^+(S)$ from a point $q \in \dot{I}^+(S)$, there can be no future directed such segment from q since if there were, it would be in a different direction to and be timelike separated from, either γ_1 or γ_2. One therefore has

(f) $\dot{I}^+(S)$ (and also $\dot{J}^+(S)$) is generated by null geodesic segments which have

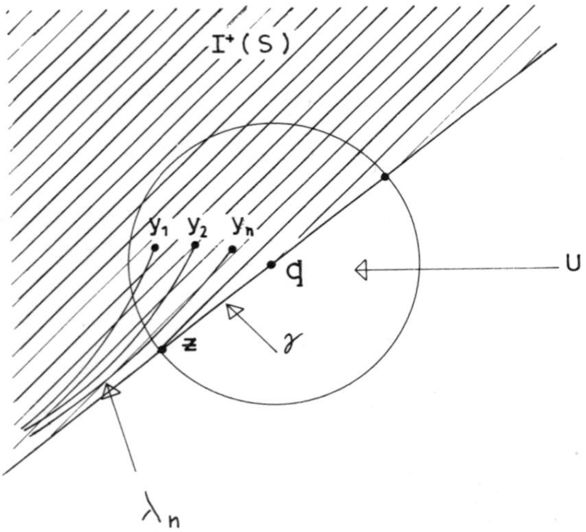

Figure 8. The points y_n converge to the point q in the boundary of $I^+(S)$. From each y_n there is past directed timelike curve λ_n to S. These curves converge to the past directed null geodesic segment γ through q.

future end-points where they intersect each other but which can have past end-points only if and when they intersect S.

The example of 2-dimensional Minkowski space with a point removed shows that there can be null geodesic generators which do not intersect S and which do not have past end-points in the space.

The region of spacetime from which one can escape to infinity along a future directed nonspacelike curve is $J^-(\mathscr{I}^+)$ the causal past of future null infinity. Thus $\dot{J}^-(\mathscr{I}^+)$ is the *event horizon*, the boundary of the region from which one cannot escape to infinity (Figure 9). Interchanging future and past in the results above, one sees that the event horizon is a manifold which is generated by null geodesic segments which may have past end-points but which could have future end-points only if they intersected $\mathscr{I}^+$. Suppose there were some generator γ of $\dot{J}^-(\mathscr{I}^+)$ which intersected $\mathscr{I}^+$ at some point q. Let λ be the generator of the null surface $\mathscr{I}^+$ which passes through q. Since the direction of λ would be different from that of γ, one could join points on λ to the future of q by timelike curves to points on γ the past of q. This would contradict the assumption that γ was in $\dot{J}^-(\mathscr{I}^+)$. Thus the null geodesic generators of the event horizon *have no future-end-points.* This is one of the fundamental properties of the event horizon. The other fundamental property, that neighbouring generators are never converging, will be described in Section 6.

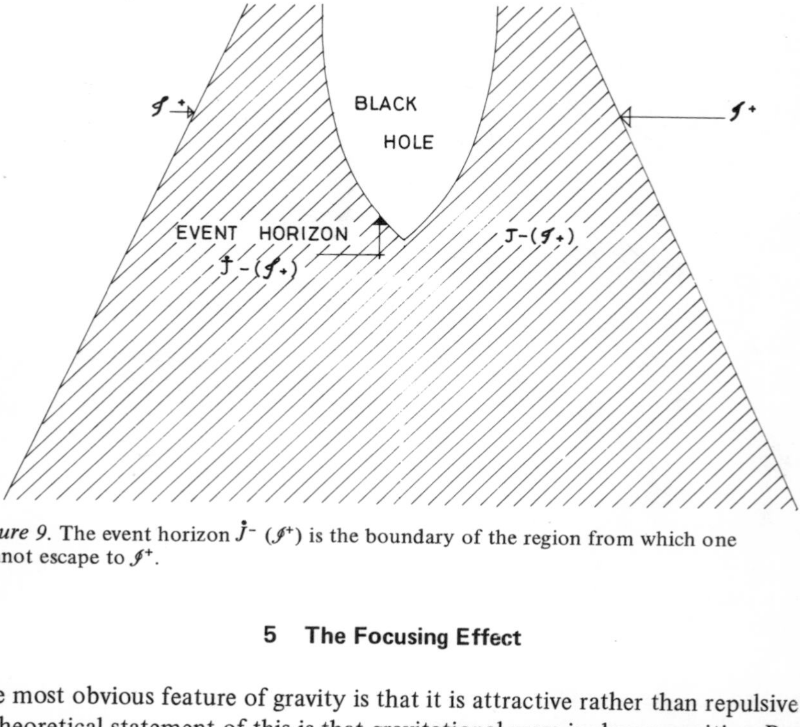

Figure 9. The event horizon $\dot{J}^-(\mathscr{I}^+)$ is the boundary of the region from which one cannot escape to $\mathscr{I}^+$.

5 The Focusing Effect

The most obvious feature of gravity is that it is attractive rather than repulsive. A theoretical statement of this is that gravitational mass is always positive. By the principle of equivalence the positive character of gravitational mass is related to the positive definiteness of energy density which in turn is normally considered to be a consequence of local quantum mechanics. There are possible modifications to this positive definiteness in the very strong fields near singularities. However these will not worry us if, as we shall assume, the singularities are safely hidden behind an event horizon. We shall be concerned, in this course, only with the region outside and including the event horizon.

The fact that gravity is always attractive means that a gravitational field always has a net focusing (i.e., converging) effect on light rays. To describe this effect in more detail, consider a family of null geodesics. Let $l^a = dx^a/dv$ denote the null tangent vectors to these geodesics where v is some parameter along the geodesic. At each point one can introduce a pair of unit spacelike vectors a^a and b^a which are orthogonal to each other and to l^a. It turns out to be more convenient to work with the complex conjugate vectors

$$\sqrt{2}m^a = a^a + ib^a, \qquad \sqrt{2}\bar{m}^a = a^a - ib^a$$

These are actually null vectors in the sense that $m^a m_a = \bar{m}^a \bar{m}_a = 0$, they are

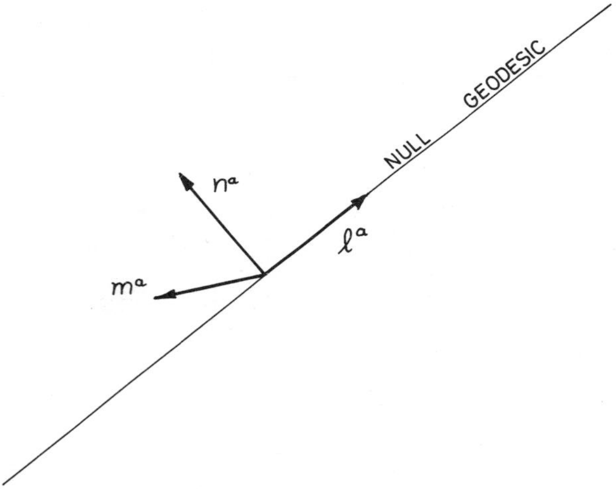

Figure 10. The null vector l^a lies along the null geodesic. The null vector n^a is such that $l^a n_a = 1$. The null vector m^a is complex combination of two spacelike vectors orthogonal to l^a, n^a and to each other.

orthogonal to l^a, $l^a m_a = l^a \bar{m}_a = 0$ and they satisfy $m^a \bar{m}_a = -1$. These conditions determine m^a up to a spatial rotation

$$m^a \rightarrow m^a e^{i\phi} \tag{5.1}$$

and up to the addition of a complex multiple of l^a

$$m^a \rightarrow m^a + cl^a \tag{5.2}$$

where c is a complex number. This is called a null rotation. Given m^a there is a unique real null vector n^a such that $l^a n_a = 1$, $n^a m_a = n^a \bar{m}_a = 0$. The vectors $(1^a, n^a, m^a, \bar{m}^a)$ form what is called a null tetrad or vierbein (Figure 10).

Using this null tetrad one can express the fact that the curves of the family are geodesics as

$$l_{a;b} m^a l^b = 0 \tag{5.3}$$

where semi-colon indicates covariant derivative. One can also define complex quantities ρ and σ as

$$\rho = l_{a;b}m^a\bar{m}^b, \qquad \sigma = l_{a;b}m^a m^b \tag{5.4}$$

The imaginary part of ρ measures the twist or rate of rotation of neighbouring null geodesics. It is zero if and only if the null geodesics lie in 3-dimensional null hypersurfaces. This will always be the case in what follows so I shall henceforth take ρ to be real. The real part of ρ measures the average rate of convergence of

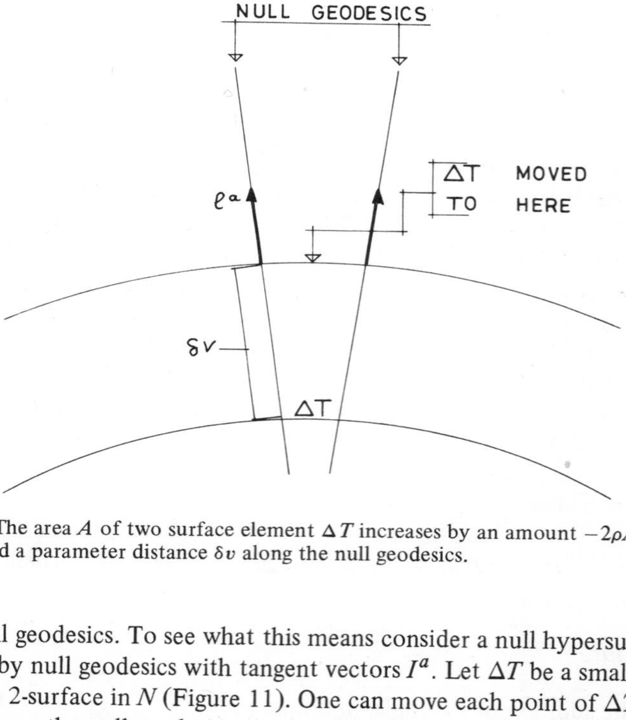

Figure 11. The area A of two surface element ΔT increases by an amount $-2\rho A\delta v$ when ΔT is moved a parameter distance δv along the null geodesics.

nearby null geodesics. To see what this means consider a null hypersurface N generated by null geodesics with tangent vectors l^a. Let ΔT be a small element of a spacelike 2-surface in N (Figure 11). One can move each point of ΔT a parameter distance δv up the null geodesics. As one does so the area of ΔT changes by an amount

$$\delta A = -2A\rho\delta v \tag{5.5}$$

The quantity σ measures the rate of distortion or shear of the null geodesics, that is, the difference between the rates of convergence of neighbouring geodesics in the two spacelike directions orthogonal to l^a. The effect of shear is to make a small 2-surface which was spatially circular, become elliptical as it is moved up the null geodesic.

The rate of change of the quantities ρ and σ along the null geodesics is given by two of the Newman-Penrose (1962) equations

$$\frac{d\rho}{dv} = \rho^2 + \sigma\bar{\sigma} + (\epsilon + \bar{\epsilon})\rho + \phi_{00} \tag{5.6}$$

$$\frac{d\sigma}{dv} = 2\rho\sigma + (3\epsilon - \bar{\epsilon})\sigma + \psi_0 \tag{5.7}$$

where

$$\epsilon = \tfrac{1}{2}(l_{a;b}n^a l^b + \bar{m}_{a;b}m^a l^b)$$

$$\phi_{00} = \tfrac{1}{2}R_{ab}l^a l^b$$

$$\psi_0 = C_{abcd}l^a m^b l^c m^d$$

(Note that my definitions of the Ricci and Weyl tensors have the opposite sign to those of Newman and Penrose.)

The imaginary part of ϵ is the rate of spatial rotation of the vectors m^a and $\bar{m}^a$ relative to a parallelly transported frame as one moves along the null geodesics. In what follows m^a will always be chosen so that $\epsilon - \bar{\epsilon} = 0$. The real part of ϵ measures the rate at which the tangent vector l^a changes in magnitude compared to a parallelly transported vector as one moves along the null geodesics. It is zero if $l^a = dx^a/dv$ where v is an affine parameter. It is convenient however in some situations to choose v not to be an affine parameter.

The Ricci tensor term ϕ_{00} in equation (5.6) represents the focusing effect of the matter. By the Einstein equations

$$R_{ab} - \tfrac{1}{2}g_{ab}R = 8\pi T_{ab} \tag{5.8}$$

it is equal to $4\pi T_{ab}l^a l^b$. The local energy density of matter (i.e. non-gravitational) fields measured by an observer with velocity vector v^a is $T_{ab}v^a v^b$. It seems reasonable from local quantum mechanics to assume that this is always non-negative. It then follows from continuity that $T_{ab}w^a w^b \geqslant 0$ for any null vector w^a. I shall call this the *weak energy condition*, (Penrose 1965a, Hawking and Penrose 1970, HE) and shall assume it in what follows. With this assumption one can see from equation (5.6) that the effect of matter is always to increase the average convergence ρ, i.e., to focus the null geodesic.

The Weyl tensor term ψ_0 can be throught of as representing, in a sense, the gravitational radiation crossing the null hypersurface N. One can see from equation (5.7) that it has the effect of inducing shear in the null geodesic. This shear then induces convergence by equation (5.6). Thus both matter and pure gravitational fields have a focusing effect on null geodesics.

To see the significance of this, consider the boundary $\dot{I}^+(S)$ of the future of a set S. As I showed in the last section this will be generated by null geodesic segments. Suppose that the convergence of neighbouring segments has some positive value ρ_0 at a point $q \in \dot{I}^+(S)$ on a generator γ. Then choosing v to be an affine parameter, one can see from equation (5.6) that ρ will increase and become infinite at a point r on the null geodesic γ within an affine distance of $1/\rho_0$ to the future of q. The point r will be a *focal point* where neighbouring null geodesics intersect. We saw in the last section that the generators of $\dot{I}^+(S)$ have future end points where they intersect other generators. Strictly speaking, this was shown only for generators which intersect each other at a finite angle but it is true also for neighbouring generators which intersect at infinitesimal angles (see HE for proof). Thus the generator γ through q will have an end-point at or before the point r. (It may be before r because γ may intersect some other generator at a finite angle.) In other words, once the generators of $\dot{I}^+(S)$ start converging, they are destined to have future end-points within a finite affine distance. They may not, however, attain this distance because they may run into a singularity first.

The importance of this result will be seen in the next section.

6 Predictability

A 3-dimensional spacelike surface S without edges will be said to be a *partial Cauchy surface* if it does not intersect any nonspacelike curve more than once. Given suitable data on such a surface one can solve the Cauchy problem and predict the solution on a region denoted by the $D^+(S)$ and called the *future Cauchy development of S*. This can be defined as the set of all points q such that every past directed nonspacelike curve from q intersects S if continued far enough. Note that this definition is not the same as the one used in Penrose (1968) and Hawking and Penrose (1970) where nonspacelike is replaced by timelike. However the difference affects only whether points on the boundary of $D^+(S)$ are considered to be in $D^+(S)$ or not.

When one is dealing with the gravitational collapse of a local object such as a star or even a galaxy, it is reasonable to neglect the curvature of the universe and the "big-bang" singularity 10^{10} years ago and to consider spacetime to be asymptotically flat and initially nonsingular. As I said earlier in section 4, I shall take asymptotically flat to mean that the spacetime manifold M and metric g_{ab} are weakly asymptotically simple. This means that there are well-defined past and future null infinities $\mathscr{I}^-$ and $\mathscr{I}^+$. The assumption that we are implicitly making in this Summer School that one can predict the future, at least in the region far away from the collapsing object, can now be expressed as the assumption that there is a partial Cauchy surface S such that points near $\mathscr{I}^+$ lie in $D^+(S)$ (Figure 12). ($\mathscr{I}^+$ cannot lie actually in $D^+(S)$ since its null geodesic generators do not intersect S. However the solution on $D^+(S)$ determines the conformal structure of $\mathscr{I}^+$ by continuity.) I shall say that a weakly asymptotically simple spacetime M, g_{ab} which admits such a partial Cauchy surface S is (*future*) *asymptotically predictable*. This definition, and a slightly stronger version which I shall introduce shortly, will form the basis of my course. Asymptotic predictability implies that every past directed nonspacelike curve from points near $\mathscr{I}^+$ continues back to S and does not run into a singularity on the way. One can think of this as a precise statement to the effect that there are no singularities to the future of S which are naked, i.e., visible from $\mathscr{I}^+$.

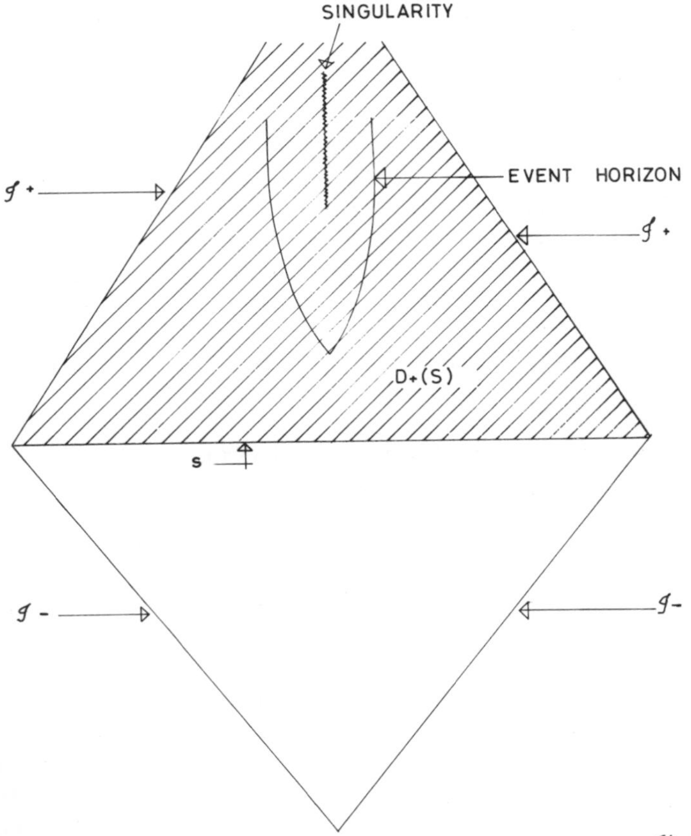

Figure 12. A space with a partial Cauchy surface S such that the points near $\mathscr{I}^+$ are contained in the future Cauchy development $D^+(S)$.

Asymptotic predictability implies that the future Cauchy development $D^+(S)$ contains $J^+(S) \cap J^-(\mathscr{I}^+)$, i.e., it contains all points to the future of S which are outside the event horizon. Suppose there were a point p on the event horizon to the future of S which was not contained in $D^+(S)$. Then there would be a past directed nonspacelike curve λ (in fact a null geodesic) from p which did not [illegible]rsect S but ran into some sort of singularity instead. This singularity would [illegible] naked" in that the slightest variation of the metric could result in it being visible from $\mathscr{I}^+$. Since we are assuming that the non-existence of naked singularities is a stable property, we would wish to rule out such an unstable situation. One can also argue that the metric of spacetime is some classical limit of an underlying quantum reality. This would mean that the metric could not be defined so exactly as to distinguish between nearly naked singularities and those which are actually naked. These considerations motivate a slightly stronger version of asymptotic predictability. I shall say that a weakly asymptotically simple spacetime M, g_{ab} is *strongly (future) asymptotically predictable* if there is a partial Cauchy surface S such that

(a) $\mathscr{I}^+$ lies in the boundary of $D^+(S)$,
(b) $J^+(S) \cap \dot{J}^-(\mathscr{I}^+)$ is contained in $D^+(S)$.

Suppose that at some time after the initial surface S, a star starts collapsing and gives rise to a trapped surface T in $D^+(S)$. Recall that a trapped surface is defined to be a compact spacelike 2-surface such that the future directed outgoing null geodesics orthogonal to it have positive convergence ρ. This definition assumes that one can define which direction is outgoing. I shall assume that the 2-surface is orientable and shall require that the initial surface S has the property:

(α) S is simply connected.

Physically, one is interested only in black holes which develop from non-singular situations. In such cases the partial Cauchy surface S can be chosen to be R^3 and so will be simply connected. It is however convenient to frame the definitions so that they can be applied also to spaces like the Schwarzschild and Kerr solutions which are not initially non-singular but which may approximate the form of initially non-singular solutions at late times. In these solutions also one can find partial Cauchy surfaces S which are simply connected.

Given a compact orientable spacelike 2-surface T in the future Cauchy development $D^+(S)$ one can define which direction is outwards. To do this one uses the fact that on any manifold M with a metric g_{ab} of Lorentz signature one can find a vector field X^a which is everywhere nonzero and timelike. Using the integral curves of this vector field, one can map the 2-surface T onto a 2-surface $\hat{T}$ in S. Since S is simply connected, this 2-surface $\hat{T}$ separates S into two regions. One can label the region which contains the part of S near infinity in the asymptotically flat space as the outer region and the other as the inner region. The side of $\hat{T}$ facing the outer region is then the outer side and carrying this up the integral curves of the vector field X^a one can define which is the outgoing direction on T.

Now suppose that one could escape from a point on T to infinity, i.e., suppose that T intersected $J^-(\mathscr{I}^+)$ (Figure 13). Then there would be some point $q \in \mathscr{I}^+$ which was in $J^+(T)$. Proceeding to the past along the null geodesic generator λ of $\mathscr{I}^+$ through q one would eventually leave $J^+(T)$. Thus λ must contain a point r of $\dot{J}^+(T)$. The null geodesic generator γ of $\dot{J}^+(T)$ through r would enter the physical manifold M. If it did not have a past end point it would inter partial Cauchy surface S. This is impossible since it lies in the boundary of the future of T and T is to the future of S. Thus it would have to have a past end point which, from section 4, would have to be on T. It would have to intersect T orthogonally as otherwise one could join points of T to points of γ by timelike curves. However the outgoing null geodesic orthogonal to T are

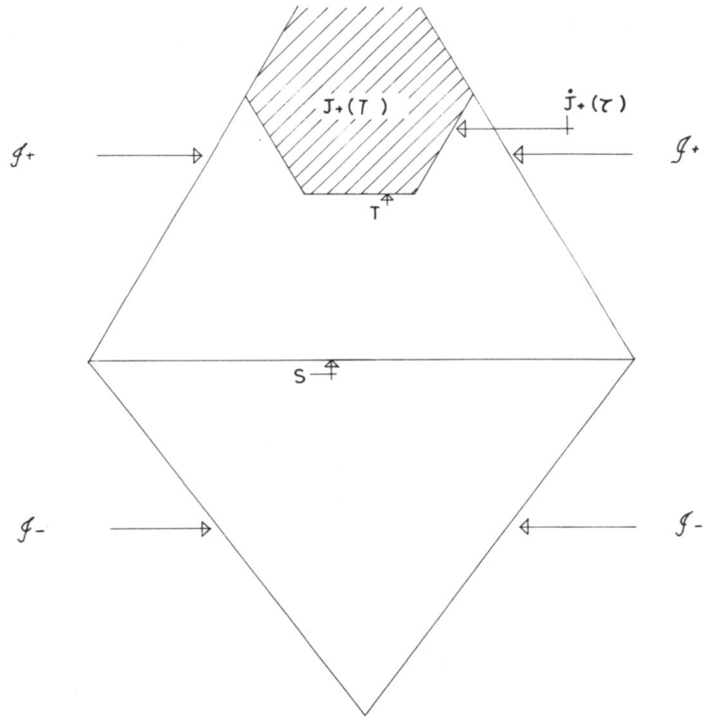

Figure 13. If a trapped surface T intersected $J^-(\mathscr{I}^+)$, there would be a null geodesic generator of $\dot{J}^+(T)$ from T to $\mathscr{I}^+$. This would be impossible as all null geodesics orthogonal to T contain a conjugate point within a finite affine distance of T.

converging because T is a trapped surface. As we saw in the last section, this implies that neighbouring null geodesics would intersect γ within a finite affine distance. This means that the generator γ of $\dot{J}^+(T)$ would have a future end point and would not remain in $\dot{J}^+(T)$ all the way out to $\mathscr{I}^+$. This establishes a contradiction which shows that the supposition that T intersects $J^-(\mathscr{I}^+)$ must be false. In other words, every point on or inside a trapped surface really is trapped: one cannot escape to $\mathscr{I}^+$ along a future directed nonspacelike curve.

The same applies to a compact orientable 2-surface T which is *marginally* which is such that the outgoing future directed null geodesics orthogonal to T have zero convergence ρ at T. For suppose T intersected $J^-(\mathscr{I}^+)$, then $\dot{J}^+(T)$ would intersect $\mathscr{I}^+$. The area of this intersection would be infinite since it is at infinity. However the generators of $\dot{J}^+(T)$ start off with zero convergence and therefore cannot ever be diverging. Thus the area of $\dot{J}^+(T) \cap \mathscr{I}^+$ could not be greater than that of T. This shows that marginally trapped surfaces in $D^+(S)$ cannot intersect $J^-(\mathscr{I}^+)$.

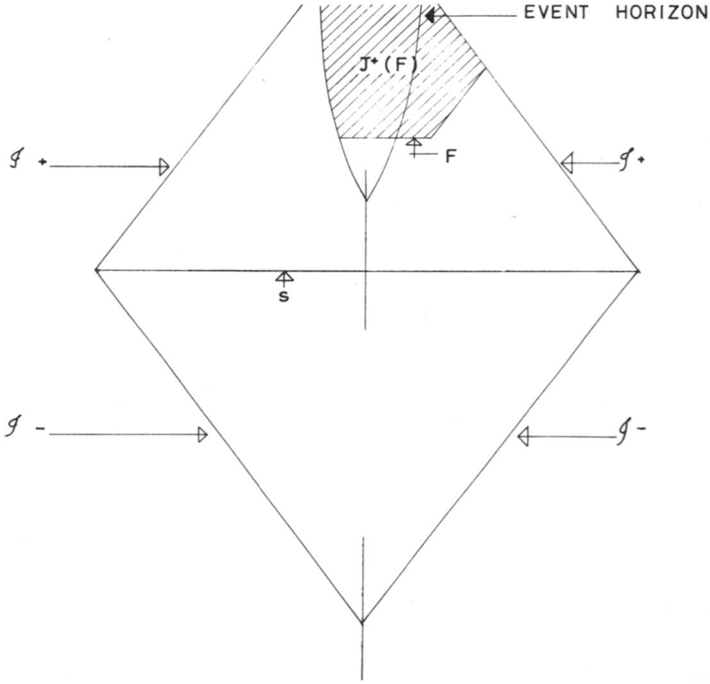

Figure 14. If the null geodesics orthogonal to a two surface F in the event horizon were converging, one could deform F outwards slightly and obtain a contradiction similar to that in Figure 13.

What has been shown in that a trapped surface implies *either* a breakdown of asymptotic predictability (i.e., the occurrence of naked singularities) *or* the existence of an event horizon. I shall assume that the first alternative does not occur and shall concentrate on the second. As was shown is section 4, the event horizon will be generated by null geodesic segments which have no future end-points. If one assumed that these generators were geodesically complete in future directions it would follow that the convergence of neighbouring generators could not be positive anywhere on the horizon since, if it were, neighbouring generators would intersect and have future end-points within a finite affine distance. In examples such as the Kerr solution, the generators *are* geodesically the future direction but there does not seem to be any a priori reason why this should always be the case. I shall now show, however, that asymptotic predictability itself without any assumption of completeness of the horizon is sufficient to prove that ρ is non-positive.

Consider a spacelike 2-surface F lying in the event horizon to the future of S. The null geodesic generators of the horizon will intersect F orthogonally. Suppose their convergence ρ was positive at some point $p \in F$. In a small neighbourhood of p one could deform the 2-surface F slightly outwards into $J^-(\mathscr{I}^+)$ so that the convergence ρ of the outgoing null geodesics orthogonal to F was still positive (Figure 14). This would lead to a contradiction similar to the one we have just considered. The null geodesics in $J^-(\mathscr{I}^+)$ which are orthogonal to F would intersect each other within a finite affine distance and hence could not be generators of $\dot{J}^+(F)$ all the way out to $\mathscr{I}^+$, which being at infinity is at an infinite affine distance.

This shows that the convergence ρ of neighbouring generators of the event horizon cannot be positive anywhere to the future of S. Together with the result that the generators of the event horizon do not have future endpoints, this implies that the area of a two dimensional cross section of the horizon must increase with time. This will be discussed further in the next section.

7 Black Holes

In order to describe the formation and evolution of black holes, one needs a suitable time coordinate. The usual coordinate t in the Schwarzschild and Kerr solutions is no good because all the surfaces of constant t intersect the horizon at the same place (see Carter's lectures). What one wants is a coordinate τ such that the surfaces of constant τ cover the future Cauchy development $D^+(S)$. By the assumption of strong future asymptotic predictability the event horizon to the future of S will be contained in $D^+(S)$ and so will be covered by the surfaces of constant τ. I shall denote the surface $\tau = \tau_0$ by $S(\tau_0)$ with $S(0) = S$. Near infinity the surfaces $S(\tau)$ for $\tau > 0$ could be chosen to be asymptotically flat spacelike surfaces like S which approached spacelike infinity i^0 and which were such that $\mathscr{I}^+$ lay in the boundary of $D^+(S(\tau))$ for each $\tau \geqslant 0$. However it is somewhat more convenient to choose the surface $S(\tau)$ for $\tau > 0$ so that they intersect $\mathscr{I}^+$ (Figure 15). This means that asymptotically they tend to null surfaces of constant retarded time. The advantage of such a choice of surfaces $S(\tau)$ is that the gravitational radiation emitted during the formation and interaction of black holes will escape to $\mathscr{I}^+$ and will not intersect the surfaces $S(\tau)$ for τ sufficiently large. When the solution settles down to a nearly stationary state, one can relate the properties of the event horizon at the time τ to the values of the mass and angular tum measured on the intersection of $\mathscr{I}^+$ and $S(\tau)$. There is no unique surface $S(\tau)$ and of the correspondence between points on the horizon and points on $\mathscr{I}^+$ at the same values of τ. This arbitrariness does not matter provided one relates the properties of the event horizon to the mass and angular momentum measured on $\mathscr{I}^+$ only during periods when the system is nearly stationary. I shall be concerned with relations between initial and final quasi-stationary states.

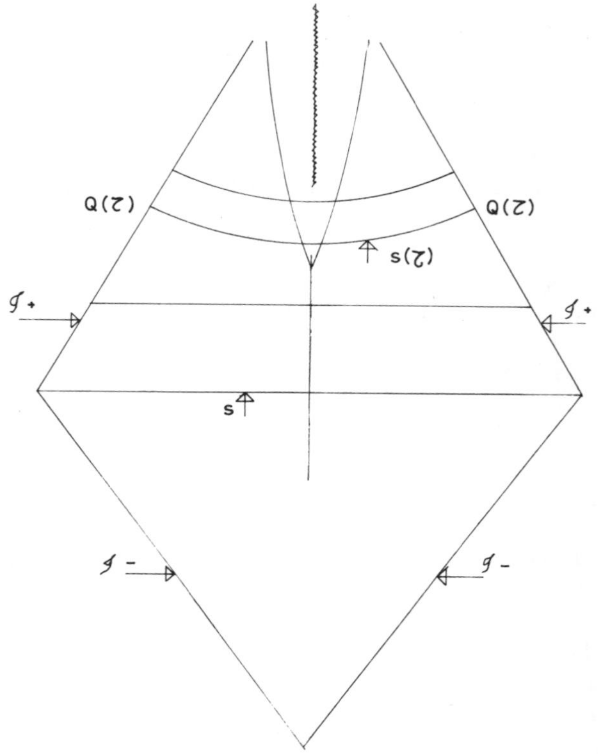

Figure 15. The surface $S(\tau)$ of constant τ intersect $\mathscr{I}^+$ in the two-spheres $Q(\tau)$.

It turns out that one can always find such a time coordinate τ if the solution is strongly asymptotically predictable, i.e., if there exists a partial Cauchy surface S such that

(a) $\mathscr{I}^+$ lies in the boundary of $D^+(S)$,
(b) $J^+(S) \cap \dot{J}^-(\mathscr{I}^+)$ lies in $D^+(S)$.

More precisely, one can find a function $\tau \geqslant 0$ on $D^+(S)$ such that the surfaces $S(\tau)$ of constant τ are spacelike surfaces without edges in M and satisfy

(i) $S(0) = S$,
(ii) $S(\tau_2)$ lies to the future of $S(\tau_1)$ for $\tau_2 > \tau_1$,

(iii) Each $S(\tau)$ for $\tau > 0$ intersects $\mathscr{I}^+$ in a 2-sphere $Q(\tau)$. The $\{Q(\tau)\}$ for $\tau > 0$ cover $\mathscr{I}^+$,
(iv) Every future directed nonspacelike curve from any point in the region of $D^+(S)$ between S and $S(\tau)$ intersects either $\mathscr{I}^+$ or $S(\tau)$ if continued far enough.
(v) $S(\tau)$ minus the boundary 2-sphere $Q(\tau)$ is topologically equivalent to S.

The point that one can find such a time function τ is somewhat technical so I shall just give an outline here. Full details are in HE. It is based on an idea of Geroch (1968). One first chooses a volume measure $d\mu$ on M so that the total volume of M in this measure is finite. In the case of a weakly asymptotically simple space such as I am considering, this volume measure could be that defined by the conformal metric $\tilde{g}_{ab}$ which is regular on $\mathscr{I}^-$ and $\mathscr{I}^+$. For a point $p \in D^+(S)$ one can then define a quantity $f(p)$ which is the volume of $J^+(p) \cap D^+(S)$ evaluated in the measure $d\mu$. Now choose a family $\{Q(\tau)\}$, $\tau > 0$ of 2-spheres which cover $\mathscr{I}^+$ and which are such that $Q(\tau_2)$ lies to the future of $Q(\tau_1)$ for $\tau_2 > \tau_1$. Then, given $p \in D^+(S)$ one can define a quantity $h(p, \tau)$ as the volume in the measure of $d\mu$ of $D^+(S) \cap [J^-(p) - J^-(Q(\tau))]$. The functions $f(p)$ and $h(p, \tau)$ are continuous in p and τ. The surface $S(\tau)$ can now be defined as the set of points p for which $h(p, \tau) = \tau f(p)$. Properties (i)–(v) can easily be verified.

With the time function τ one can describe the evolution of black holes. Suppose that a star collapses and gives rise to a trapped surface T. As was shown in the last section, the assumption of strong asymptotic predictability implies that one cannot escape from T to $\mathscr{I}^+$. There must thus be an event horizon $\dot{J}^-(\mathscr{I}^+)$ to the future of S. Also by the assumption of strong asymptotic predictability, $J^+(S) \cap \dot{J}^-(\mathscr{I}^+)$ will be contained in $D^+(S)$. For sufficiently large τ, the surface $S(\tau)$ will intersect the horizon and the set $B(\tau)$ defined as $S(\tau) - J^-(\mathscr{I}^+)$ will be nonempty. I shall define a *black hole* on the surface $S(\tau)$ to be a connected component of $B(\tau)$. In other words, it is a connected region of the surface $S(\tau)$ from which one cannot escape to $\mathscr{I}^+$. As τ increases, black holes may grow or merge together and new black holes may be formed by further stars collapsing but a black hole, once formed, *cannot disappear, nor can it bifurcate.* To see that it cannot disappear is easy. Consider a black hole $B_1(\tau_1)$ on a surface $S(\tau_1)$. Let p be a point of $B_1(\tau_1)$. By property (iv), every future directed nonspacelike curve λ from p will intersect either $\mathscr{I}^+$ or $S(\tau_2)$ for any $\tau_2 > \tau_1$. The former is impossible since p is not in $J^-(\mathscr{I}^+)$. This also implies that λ must intersect $S(\tau_2)$ at some point q which is not in $J^-(\mathscr{I}^+)$. Thus q must be contained in some black hole $B_2(\tau_2)$ on the surface $S(\tau_2)$ which will be said to be *descended from* the black hole $B_1(\tau_1)$. Since black holes can merge together, $B_2(\tau_2)$ may be descended from more than one black hole on the surface $S(\tau_1)$. Alternatively, a black hole on $S(\tau_2)$ may not be descended from any on $S(\tau_1)$ but have formed between τ_1 and τ_2 (Figure 16). ...ck hole cannot bifurcate can be expressed by saying that ...re than one descendant on a later surface $S(\tau_2)$. This follows from the fact that any future directed nonspacelike curve from a point $p \in B_1(\tau_1)$ can be continuously deformed through a sequence of such curves into any other future directed nonspacelike curve from p. Since all these curves will intersect $S(\tau_2)$, their intersection with $S(\tau_2)$ will form a continuous curve in $S(\tau_2)$. Thus $J^+(p) \cap S(\tau_2)$ will be connected. Similarly $J^+(B_1(\tau_1)) \cap S(\tau_2)$ will be connected. It must be contained in $B(\tau_2)$ and so will be contained in only one connected component of $B(\tau_2)$. There will thus be only one black hole on $S(\tau_2)$ which is descended from $B_1(\tau_1)$.

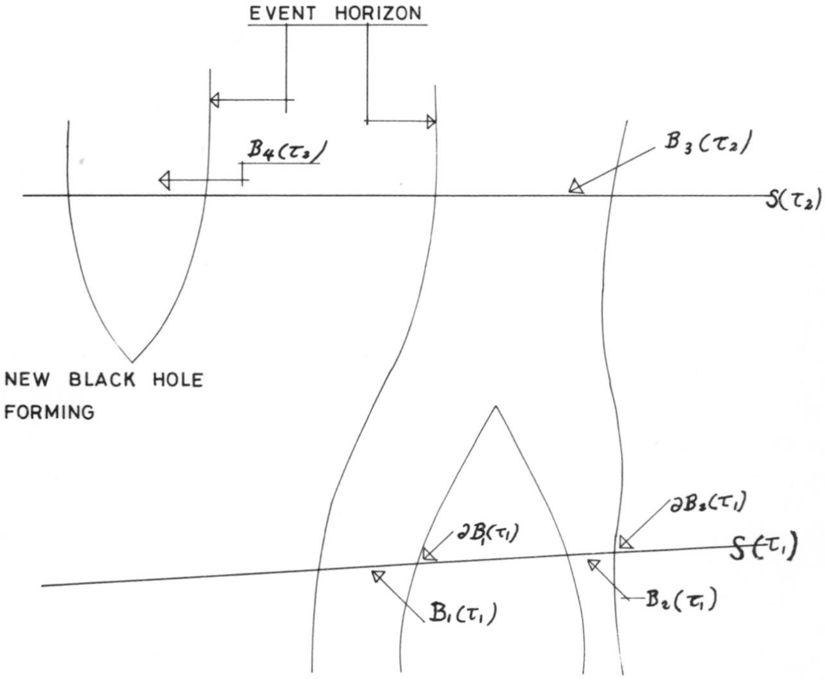

Figure 16. The two black holes $B_1(\tau_1)$ and $B_z(\tau_1)$ on the surface $S(\tau_1)$ merge to form the black hole $B_3(\tau_2)$ on the surface $S(\tau_2)$. A new black hole $B_4(\tau_2)$ is formed between $S(\tau_1)$ and $S(\tau_2)$.

The boundary $\partial B_1(\tau_1)$ in $S(\tau_1)$ of a black hole $B_1(\tau_1)$ is formed by part of the intersection of the event horizon with the surface $S(\tau_1)$. Since we are assuming that the initial surface S is simply connected, it follows from property (v) that each of the surfaces $S(\tau)$ is also simply connected. This implies that the boundary $\partial B_1(\tau_1)$ is connected. For suppose that $\partial B_1(\tau_1)$ consisted to two components $\partial_1 B_1(\tau_1)$ and $\partial_2 B_1(\tau_1)$. One could join a point $q_1 \in \partial_1 B_1(\tau_1)$ to $q_2 \in \partial_2 B_1(\tau_1)$ by a curve μ lying in $B_1(\tau_1)$ and a curve λ ly
Joining μ and λ, one would obtain a closed curve in $S(\tau_1$

deformed to zero in $S(\tau_1)$ since it crossed the closed surface $\partial_1 B_1(\tau_1)$ only once. This would contradict the fact that $S(\tau_1)$ is simply connected.

If the black holes are formed by collapses in a space which is nonsingular initially, the surface S can be chosen to have a topology of Euclidean 3-space R^3. By property (v) each surface $S(\tau)$ minus the bounding 2-sphere $Q(\tau)$ on $\mathscr{I}^+$ will also have this topology. It then follows that the boundary $\partial B_1(\tau)$ of a black hole $B_1(\tau)$ will be compact and that the topology of $S(\tau) \cap \bar{J}^-(\mathscr{I}^+)$, the space outside and including the horizon, will have the topology of R^3 minus a number of open sets with compact closure. As I said earlier, it is sometimes convenient to consider black hole solutions which are not initially nonsingular but which may outside the event horizon approximate the behaviour of initially nonsingular solutions at large times. If they are to do this it is not necessary that the surfaces $S(\tau) - Q(\tau)$ have the topology R^3 (indeed they do not in the Schwarzschild and Kerr solutions), but they should have the same topology outside the event horizon. One can ensure this by requiring that the initial surface S has the property.

(β) $S \cap J^-(\mathscr{I}^+)$ has the topology of R^3 minus a finite number of open sets with a compact closure.

It is easy to show that if S has the property (β) then each surface $S(\tau) - Q(\tau)$ has the property (β) also.

I showed earlier that the null geodesic generator of the event horizon did not have any future endpoints and had negative or zero convergence ρ. It follows from this that the area of the boundary $\partial B_1(\tau)$ of a black hole $B_1(\tau)$ cannot decrease with increasing τ. If two black holes $B_1(\tau_1)$ and $B_2(\tau_2)$ on a surface $S(\tau_1)$ later collide and merge to form a black hole $B_3(\tau_2)$ on the surface $S(\tau_2)$, the area of $\partial B_3(\tau_2)$ must be at least as great as the sum of the areas of the boundaries $\partial B_1(\tau_1)$ and $\partial B_2(\tau_1)$ of the original black holes. In fact it must be strictly greater because $\partial B_3(\tau_2)$ will contain two disjoint closed sets corresponding to generators which intersected $\partial B_1(\tau_1)$ and $\partial B_2(\tau_1)$ respectively. Since $\partial B_3(\tau_2)$ is connected, it must also contain an open set corresponding to generators which had past endpoints between $S(\tau_1)$ and $S(\tau_2)$.

The area of the boundaries of black holes has strong analogies to the concept of entropy in thermodynamics: it never decreases and it is additive. We shall see later that the area will remain constant only if the black hole is in a stationary state. When the black hole interacts with anything else the area will always increase. Under favourable circumstances one can arrange that the increase is arbitrarily small. This corresponds to using nearly reversible transformations in thermodynamics. I shall show later how the area of a black hole in a stationary state is related to its mass and angular momentum. The fact that the area cannot decrease will impose certain inequalities on the change of the mass and angular momentum of the black hole as a result of interaction.

I shall denote by $T(\tau)$ the region of the surface $S(\tau)$ that contains trapped or marginally trapped surfaces lying in $S(\tau)$. I shall call the boundary $\partial T(\tau)$ of $T(\tau)$, the *apparent horizon* in the surface $S(\tau)$. In the last section it was shown that trapped or marginally trapped surfaces cannot intersect $J^-(\mathscr{I}^+)$. Thus $T(\tau)$

must be contained in $B(\tau)$ and the apparent horizon must lie behind or coincide with the event horizon. The apparent horizon $\partial T(\tau)$ will be a *marginally trapped surface.* That is, it is a spacelike 2-surface such that the convergence ρ of the outgoing null geodesics orthogonal to it is zero. As τ increases, these null geodesics may be focused by matter or gravitational radiation and the position of the apparent horizon will move outwards on the surface $S(\tau)$ at or faster than the speed of light. As the example of the spherical collapsing shell shows, it can move outwards discontinuously. When the solution is in a quasi-stationary state, the apparent horizon will lie just inside the event horizon and the area of $\partial T(\tau)$ will be nearly equal to that of $\partial B(\tau)$. In the transition from one quasi-stationary state to another the area of $\partial B(\tau)$ will increase and so the area of $\partial T(\tau)$ must be greater in the final state than in the initial one. I have not been able to show, however, that the area of $\partial T(\tau)$ increases monotonically though I believe it probably does.

It is interesting to see the behaviour of the event and apparent horizon in the case of two black holes which collide and merge together. Suppose two stars a long way apart collapse to form black holes $B_1(\tau)$ and $B_2(\tau)$ which have settled down to a quasi-stationary state by the surface $S(\tau_1)$ (Figure 17). Just inside the two components $\partial B_1(\tau_1)$ and $\partial B_2(\tau_1)$ of the event horizon there will be two components $\partial T_1(\tau_1)$ and $\partial T_2(\tau_1)$ of the apparent horizon. The 2-surfaces $\partial T_1(\tau_1)$ and $\partial T_2(\tau_1)$ will be smooth but the 2-surfaces $\partial B_1(\tau_1)$ and $\partial B_2(\tau_1)$ will each have a slight cusp on the side facing the other. As the black holes approach each other, these cusps will become more pronounced and will join up to give a single component $\partial B_3(\tau)$ of the event horizon. The apparent horizon $\partial T_1(\tau)$ and $\partial T_2(\tau)$ on the other hand, will not join up. As they approach each other there will be some surfaces $S(\tau_2)$ on which there will be a third component $\partial T_3(\tau_2)$ which surrounds both $\partial T_1(\tau_2)$ and $\partial T_2(\tau_2)$.

I shall now show that each component of the apparent horizon $\partial T(\tau)$ must have the topology of a 2-sphere. I originally developed this proof for the event horizon in the stationary situations considered in the next section but I am grateful to G. W. Gibbons for pointing out that it can be applied to the apparent horizon at any time. The idea is to show that if a connected component $\partial T_1(\tau)$ of the apparent horizon had any topology other than that of a 2-sphere, one could deform it to give a trapped or marginally trapped surface just outside $\partial T(\tau)$. This would be a contradiction of the fact that the apparent horizon is the outer boundary of such surfaces.

Let u^a be the unit timelike vector field orthogonal to the surface $S(\tau)$. Let l^a and n^a be respectively the future directed outgoing and ingoing null vector fields, orthogonal to $\partial T_1(\tau)$ and normalized so that

$$l^a u_a = 2^{-\frac{1}{2}}, n^a u_a = 2^{-\frac{1}{2}}, l^a n_a = 1$$

The complex null vectors m^a and $\bar{m}^a$ will then lie in the 2-surface $\partial T_1(\tau)$. The vector $\omega^a = 2^{-\frac{1}{2}}(l^a - n^a)$ will be the unit outward spacelike vector in $S(\tau)$

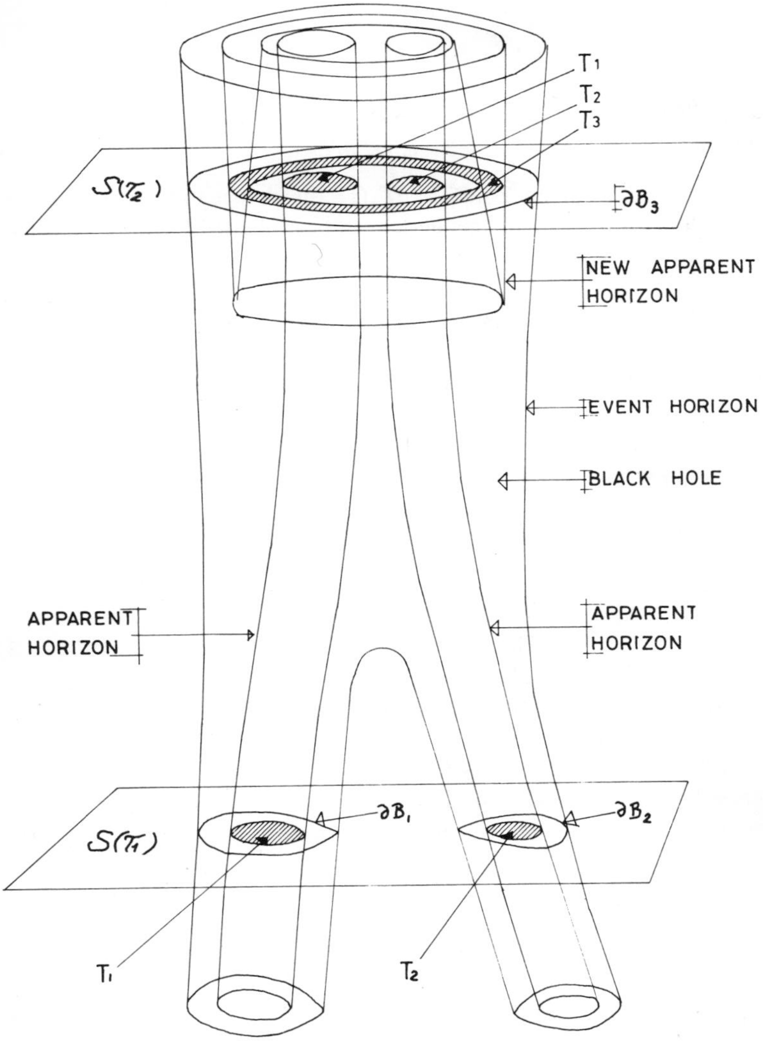

Figure 17. The collision of two Black Holes. The event horizons ∂B_1 and ∂B_2 merge to form the event horizon ∂B_3. The apparent horizons ∂T_2 do not merge but are enveloped by a new apparent horizon ∂T_3.

orthogonal to $\partial T_1(\tau)$. Suppose one now moves each point of $\partial T_1(\tau)$ a parameter distance h outwards along the vector field $y^a = \omega^a e^f$ where f is some function on $\partial T_1(\tau)$. To maintain the orthogonality of l^a and n^a to the 2-surface requires

$$\kappa - \tau - \delta f + \bar{\alpha} + \beta = 0 \tag{7.1}$$

$$\nu - \pi + \bar{\delta} f + \alpha + \bar{\beta} = 0 \tag{7.2}$$

where

$$\kappa = l_{a;b}m^a l^b, \qquad \tau = l_{a;b}m^a n^b$$

$$\nu = -n_{a;b}\bar{m}^a n^b, \qquad \pi = -n_{a;b}\bar{m}^a l^b$$

$$\delta f = m^a f_{;a} \qquad \text{and} \qquad \bar{\alpha} + \beta = l_{a;b}n^a m^b$$

Under this movement of the 2-surface $\partial T_1(\tau)$, the change in the convergence ρ at the outgoing orthogonal null geodesics can be evaluated from the Newman-Penrose equations:

$$\frac{d\rho}{du} = 2^{-\frac{1}{2}}e^f\,[\sigma\bar{\sigma} + \phi_{00} + (\kappa - \tau)(\bar{\kappa} - \bar{\tau}) + \rho(\rho + \epsilon + \bar{\epsilon} - \bar{\mu} - \gamma - \bar{\gamma})$$

$$+ \eth\delta f - \bar{\eth}(\bar{\alpha} + \beta) + \sigma\lambda + \psi_2 + 2\Lambda] \tag{7.3}$$

where

$$\lambda = -n_{a;b}\bar{m}^a\bar{m}^b,\; \mu = -n_{a;b}\bar{m}^a m^b,\; \gamma = -\tfrac{1}{2}(n_{a;b}l^a n^b - \bar{m}_{a;b}m^a n^b),$$

$$\psi_2 = -\tfrac{1}{2}C_{abcd}(l^a n^b l^c n^d - l^a u^b m^c \bar{m}^d),$$

$$\Lambda = \frac{-R}{24} \qquad \text{and} \qquad \bar{\eth} = \bar{\delta} - (\alpha - \bar{\beta})$$

where $\alpha - \bar{\beta} = \bar{m}_{a;b}m^a\bar{m}^b$. The first three terms on the right of equation (7.3) are non-negative. The term $\eth\delta f$ is the Laplacian of f in the 2-surface. One can choose f so that the sum of the last five terms on the right of equation (7.3) is constant over the 2-surface. The sign of this constant value will be determined by that of the integral of $(\sigma\lambda + \psi_2 + 2\Lambda)$ over the 2-surface ($\eth(\bar{\alpha} + \beta)$ being a divergence, has zero integral). This integral can be evaluated from another Newman-Penrose equation which can be written as

$$\eth(\alpha + \bar{\beta}) - \bar{\eth}(\bar{\alpha} + \beta) + \eth(\alpha - \beta) + \bar{\eth}(\bar{\alpha} - \bar{\beta})$$

$$= -2\sigma\lambda - 2\psi_2 + 2\Lambda + 2\phi_{11} \tag{7.4}$$

where

$$\phi_{11} = \tfrac{1}{4}R_{ab}(l^a n^b + m^a\bar{m}^b)$$

When integrated over the 2-surface the terms in $\bar{\alpha} + \beta$ disappear but there is in general a contribution from the $\bar{\alpha} - \beta$ terms because the vector field m^a will have singularities on the 2-surface. The contribution from these singularities is determined by the Euler number χ of the 2-surface. Thus

$$\int(-\sigma\lambda - \psi_2 + \phi_{11} + \Lambda)dA = 2\pi\chi \tag{7.5}$$

(The real part of the equation is, in fact the Gauss-Bonnet theorem). Therefore

$$-\int(\sigma\lambda + \psi_2 + 2\Lambda)dA = 2\pi\chi - \int(\phi_{11} + 3\Lambda)dA \tag{7.6}$$

Any reasonable form of matter will obey the Dominant Energy condition (Hawking 1971):– $T^{00} \geqslant |T^{ab}|$ in any orthonormal tetrad. This and the Einstein equations imply that $\phi_{11} + 3\Lambda \geqslant 0$. The Euler number χ is + 2 for a sphere, 0 for a torus and negative for any other compact orientiable 2-surface. ($\partial T_1(\tau)$ has to be orientable as it is a boundary.) Suppose $\partial T_1(\tau)$ was not a sphere. Then one could choose f so that the right hand side of equation (7.3) was everywhere positive or zero. This would mean that there would be a trapped or marginally trapped surface just outside $\partial T(\tau)$, which is supposed to be the outer boundary of such surfaces. Thus each component of the apparent horizon has the topology of a 2-sphere.

In the next section I shall show that the event horizon will coincide with the apparent horizon in the final stationary state of the solution. Thus each connected component $\partial B_1(\tau)$ of the event horizon will have spherical topology at late times. It might, however, have some other topology during the earlier, time dependent, phase of the solution.

8 The Final State of Black Holes

During the formation of a black hole in a stellar collapse, the solution will change rapidly with time. Gravitational radiation will propagate out to $\mathscr{I}^+$ and across the event horizon into the black hole. By the conservation law for asymptotically flat space (Bondi *et al.* 1962, Penrose 1963), the energy of the gravitational radiation reaching $\mathscr{I}^+$ will reduce the mass of the system as measured from $\mathscr{I}^+$. The radiation crossing the event horizon will cause the area of the horizon to increase. The amount of energy that can be radiated to $\mathscr{I}^+$ or down the black hole is presumably bounded by the original rest mass of the star. Thus one might expect that the area of the horizon and the mass measured on $\mathscr{I}^+$ might eventually tend to constant values and the solution outside the horizon settle down to a stationary state. Although we cannot at the moment describe in detail the time dependent formation phase, it seems that we probably can find all these final stationary states. In this section therefore I shall consider stationary black hole solutions in the expectation that outside the horizon they will approximate to time dependent solutions at late times.

More precisely, I shall consider spacetimes M, g_{ab} which satisfy

(1) M, g_{ab} is strongly asymptotically predictable.
(2) M, g_{ab} is stationary, i.e., there exists a one parameter isometry group $\phi_t: M \to M$ whose Killing vector K^a is timelike near $\mathscr{I}^-$ and $\mathscr{I}^+$. (Note that is may be spacelike near the black hole.)

Since these stationary spaces are not necessarily nonsingular initially, the partial Cauchy surface S may not have the topology R^3. In fact, in most cases it will be $R^1 \times S^2$. However, one wants these spaces to approximate physical initially nonsingular solutions in the region outside and including the horizon at late times, i.e., on $S(\tau) \cap \bar{J}^{-}(\mathscr{I}^{+})$ for large τ. Thus $S(\tau) \cap \bar{J}^{-}(\mathscr{I}^{+})$ must have the same topology as it would have in an initially nonsingular solution. One can ensure this by requiring the property

(β)$S \cap \bar{J}^{-}(\mathscr{I}^{+})$ has the topology of R^3 minus a finite number of open sets with compact closure.

It is also convenient (but not essential) to require

(α)S is simply connected.

Finally, one is interested only in black holes that one could fall into from infinity. Thus it is reasonable to require

(γ). There is some τ_0 such that for $\tau \geqslant \tau_0$, $S(\tau) \cap J^{-}(\mathscr{I}^{+})$ is contained in $J^{+}(\mathscr{I}^{-})$.

I shall call a space satisfying (1), (α), (β), (γ) a *regular predictable* space. If, in addition, (2) is satisfied, I shall call it a *stationary regular predictable space.* I shall show that in such a space the convergence ρ and shear σ of the generators of the horizon are zero. It then follows that the Ricci tensor term $\phi_{00} = 4\pi T_{ab} l^a l^b$ and Weyl tensor term $\psi_0 = C_{abcd} l^a m^b l^c m^d$ must be zero on the horizon. One can interpret this as saying that no matter or gravitational radiation is crossing the horizon.

The fact that ρ is zero implies that each connected component $\partial B_i(\tau)$ of the event horizon is a marginally trapped surface. Since, there are no trapped or marginally trapped surfaces outside the event horizon $\partial B_i(\tau)$ must coincide with a component $\partial T_i(\tau)$ of the apparent horizon. Thus all stationary black holes are topologically spherical; there are no toroidal ones. There could be several components $\partial B_i(\tau)$ of the event horizon corresponding to black holes which maintain themselves at constant distances from each other. This is possible in the limiting case of non-rotating black holes carrying electric charge equal to their mass (Hartle and Hawking 1972): the electric repulsion just cancels the gravitational repulsion. It seems probable but has not yet been proved that these solutions are the only stationary regular predictable spaces containing more than one black hole.

Assuming there is only one black hole, the question of the final state has two branches according as to whether or not the solution is static. A stationary solution is said to be *static* if the Killing vector K^a is hypersurface orthogonal, i.e., if the twist $\omega^a = \frac{1}{2}\eta^{abcd} K_b K_{a;d}$ is zero. In a static regular predictable space which is empty or contains only an electromagnetic field one can apply Israel's theorem (Israel, 1968) to show that the space must be the Schwarzschild or Reissner-Nordström solution.

If the solution is not static but only stationary, I shall show (modulo one point) that the black hole must be rotating. I shall prove that a stationary regular predictable space containing a rotating black hole must be axisymmetric. One can then appeal to Carter's theorem (see his lectures) to show that such spaces, if empty, can depend only on two parameters; the mass and angular momentum.

One two parameter family is known, the Kerr solutions for $a^2 \leqslant m^2$ (the Kerr solutions for $a^2 > m^2$ contain naked singularities). It seems unlikely that there are any others. Thus it appears that the final state of a black hole is a Kerr solution. In the case where the collapsing star carries a net electric charge one would expect it to be a Newman-Kerr solution.

I shall only give outlines of the results mentioned above. The full gory details will be found in HE.

To show that the convergence and shear of the generators of the event horizon are zero, consider a compact spacelike 2-surface F lying in the horizon. Under the time translation ϕ_t the surface F will be moved into another 2-surface $\phi_t(F)$ in the event horizon. Assuming that $\phi_t(F)$ lies to the future of F on the event horizon for $t > 0$, one can compare their areas by moving each element of F up the generators of the horizon to $\phi_t(F)$. I showed earlier that the generators had no future end-points and did not have positive convergence ρ. If any of them had past end-points or negative convergence between F and $\phi_t(F)$, the area of $\phi_t(F)$ would be greater than that of F. But the area of $\phi_t(F)$ must be the same as that of F since ϕ_t is an isometry. Thus the generators of the event horizon cannot have any past end-points and must have zero convergence ρ. From the Newman-Penrose equations

$$\frac{d\rho}{dv} = \rho^2 + \sigma\bar{\sigma} + (\epsilon + \bar{\epsilon})\rho + \phi_{00}$$

$$\frac{d\sigma}{dv} = 2\rho\sigma + (3\epsilon - \bar{\epsilon})\sigma + \psi_0$$

it follows that the shear σ, the Ricci tensor term ϕ_{00} and the Weyl tensor term ψ_0 are zero on the horizon.

The only complication in this proof comes from the fact that the Killing vector K^a which represents infinitesimal time translations, may be spacelike on and near the horizon. (I shall have more to say about this later.) This means that for an arbitrary 2-surface F in the horizon these may be some points of $\phi_t(F)$ for $t > 0$ which lie to the past of F. However one can construct a 2-surface F for which $\phi_t(F)$ lies wholly to the future of F in the following way. Choose a compact spacelike 2-sphere C on $\mathscr{I}^-$. The Killing vector K^a will be directed along the null geodesic generators on $\mathscr{I}^-$. Thus $\phi_t(C)$ will lie to the future of C for $t > 0$. The intersection of $\dot{J}^+(C)$, the boundary of the future of C, with the event horizon will define a 2-surface F with the required properties.

If the solution is static, one can apply Israel's theorem. If the solution is only stationary but not static one can apply a generalization of the Lichnerowicz theorem (cf. Carter) to show that the Killing vector K^a is spacelike in a non-zero region (called the *ergosphere*) part of which lies outside the horizon. The non-trivial part of this generalization consists of showing that a certain surface integral over the horizon would be zero if K^a were not spacelike there. Details are given in HE.

There are now two possibilities: either the ergosphere intersects the horizon or it does not. The horizon is mapped into itself by the time translation. ϕ_t. In the former case the Killing vector K^a will be spacelike on part of the horizon and so some null geodesic generators will be mapped into other ones. The generators form a 2-dimensional space Q which is topologically a 2-sphere, and which has a metric corresponding to the constant separation of the generators. The time translation ϕ_t which moves generators into generators can be regarded as an isometry group on Q. Thus its action corresponds to rotating Q about an axis. One can interpret this as follows. A point of Q represents a generator of the horizon. As one moves along a generator one is moving relative to the stationary frame defined by the integral curves of K^a, i.e., relative to infinity. Thus the horizon would be *rotating* with respect to infinity. I shall show that such a rotating black hole must be axisymmetric.

The other possibility is that the ergosphere might be disjoint from the horizon. Hajicek (1972) has shown that in general the ergosphere must intersect the horizon if the region outside the horizon is null geodesically complete in both the future and the past directions. However, these stationary spaces approximate to physical solutions only at late times. There is thus no physically compelling reason why they should not contain geodesics in the exterior region which are incomplete in the past direction. I shall therefore give an alternative intuitive argument to show that the ergosphere must intersect the horizon.

When there is an ergosphere one can extract energy from the solution by the Penrose process (Penrose 1969). This consists of sending a particle with energy $E_1 = P_1^a K_a$ from infinity into the ergosphere. It then splits into two particles with energies E_2 and E_3. By local conservation $E_1 = E_2 + E_3$. Since the Killing vector K^a is spacelike in the ergosphere, one can choose the momentum p_2^a of the second particle such that E_2 is negative. Thus E_3 is greater than E_1. The particle 3 can escape to infinity where its total energy (the rest mass + kinetic energy) will be greater than that of the original particle 1. Thus one has extracted energy. Particle 2, having negative energy, must remain in the region where K^a is spacelike. Suppose that the ergosphere did not intersect the horizon. Then particle 2 would have to remain outside the horizon. One could repeat the process and extract more energy. As one did so the solution would presumably change gradually. However the ergosphere could not disappear because there has to be somewhere for the negative energy particles to exist. If the ergosphere remained disjoint from the horizon one could extract an arbitrarily large amount of energy. This does not seem reasonable physically. On the other hand, if the ergosphere moved so that it intersected the horizon, the solution would have to become axisymmetric. At the moment the ergosphere touched the horizon one would have a stationary, non static, axisymmetric black hole solution. This could not be a Kerr solution because in a non static Kerr solution the ergosphere actually intersects and does not merely touch the horizon. However it appears from the results of Carter that the Kerr solutions are the only stationary axisymmetric black hole solutions. Thus it seems that one ends up with a contradiction if one supposes that the ergosphere is disjoint from the horizon. I shall therefore assume that any stationary, non static, black hole is rotating.

My original proof (Hawking 1972) that a stationary rotating black hole must be axisymmetric had the great advantage of simplicity. However it involved the assumption that as well as the *future* event horizon $\dot{J}^-(\mathscr{I}^+)$ there was a *past* event horizon $\dot{J}^+(\mathscr{I}^-)$ and that the two horizons intersected in a compact spacelike 2-surface. Penrose pointed out that there is no necessity for this assumption to hold. These stationary spaces represent physical solutions only at

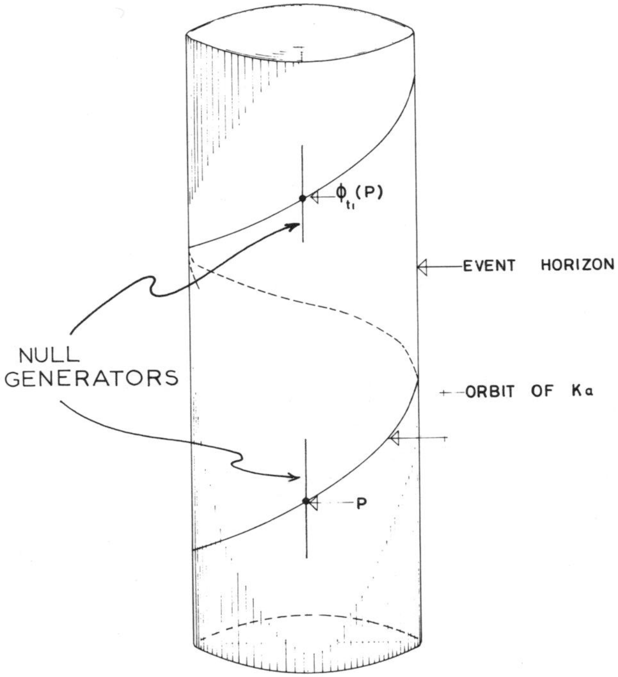

Figure 18. The time translation ϕ_{t_1} moves a point P on the horizon along the orbit of K^a to the point $\phi_{t_1}(P)$ on the same generator of the horizon.

large times. There would be a past horizon if the solution were time-symmetric. By the Papapetrou theorem (see Carter) time-symmetry is a *consequence* of stationary and axial-symmetry. It should not be assumed to prove axial symmetry. I therefore developed another proof of axial symmetry which depends only o future horizon. Unfortunately, this proof is rather long and messy. I shall t give an intuitive picture of it here and shall give the full details in HE

Consider a rotating black hole. Let t_1 be the period of rotation This means that for a point p on a generator λ of the horizon $\phi_{t_1}($ on λ (Figure 18). One can choose a parameter v on λ so that l^a

$$l_{a;b}l^b = 2\epsilon l_a$$

where ϵ is constant on λ and so that difference between the values of v at p and at $\phi_{t_1}(p)$ is t_1. This fixes the scaling of l^a. One can now form the vector field

$$\tilde{K}^a = \frac{t_1}{2\pi}(l^a - K^a)$$

on the horizon. The orbits of $\tilde{K}^a$ will be closed spacelike curves in the horizon. The aim will be to show that they correspond to rotations of the solution about

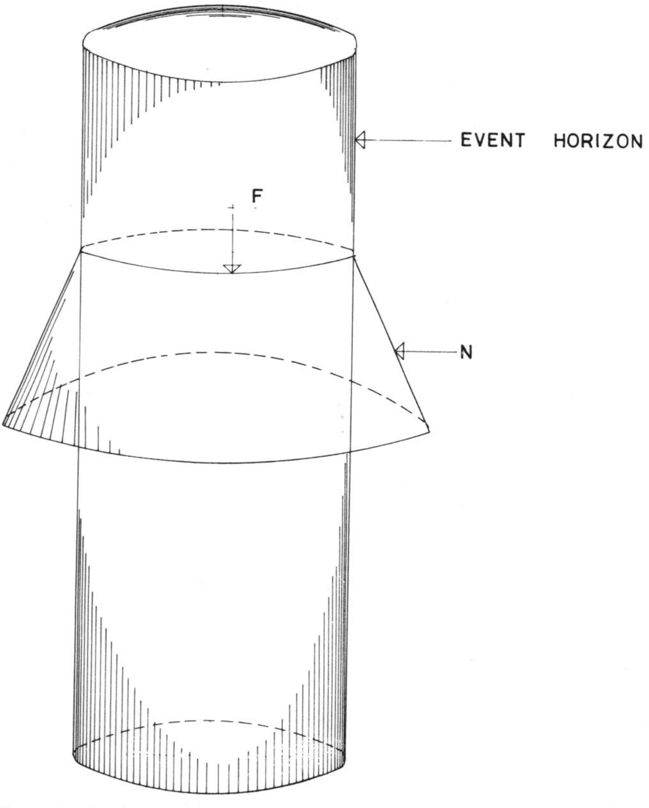

Figure 19. The event horizon and the null surface N intersected in the spacelike surface F.

an axis of symmetry. Choose a spacelike 2-surface F in the horizon tangent to $\tilde{K}^a$. Let N be the null surface generated by the ingoing null geodesics orthogonal to F (Figure 19). The idea of the proof is to consider the Cauchy problem for the region to the past of both the horizon and N. The Cauchy data for the empty space Einstein equations in this situation consists of ψ_0 on the horizon $\psi_4 = C_{abcd} n^a \bar{m}^b n^c \bar{m}^d$ on N where n^a is the null vector tangent to N and ρ, $\mu = -n_{a;b}\bar{m}^a m^b$

and $\psi_2 = \frac{1}{2}C_{abcd}(l^a n^b l^c n^d - l^a n^b m^c m^d)$ on the 2-surface F. If there are other fields present (e.g., an electromagnetic field) one has to give additional data for them. I shall consider only the empty case but similar arguments hold in the presence of any fields obeying well-behaved hyperbolic equations.

By the stationarity of the horizon, ρ and ψ_0 are zero and one can show from the Newman-Penrose equations that ψ_2 is constant along the generators of the horizon. Thus the only non-trivial Cauchy data are that on the null surface N. The idea now is to show that these Cauchy data are unchanged if one moves N by moving each point of the 2-surface F an equal parameter distance down the generators of the horizon. If this is the case, it follows from the uniqueness of the Cauchy problem that the solution admits a killing vector $\hat{K}^a$ which coincides with l^a on the horizon. Then $\tilde{K}^a$ defined as $t_1/2\pi(\hat{K}^a - K^a)$ will also be a Killing vector. Since the oribits of $\tilde{K}^a$ are closed curves on the horizon, they will be closed everywhere and so will correspond to rotations about an axis of symmetry.

To show that the data on N are unchanged on moving each point of F down the generators of the horizon, I assume that the solution is analytic though this is almost certainly not necessary. The data on N can then be represented by their partial derivatives at F in the direction along N. From the Newman-Penrose equations one can evaluate the derivatives along a generator λ of the horizon of these and certain other quantities. If one takes them in a certain order one obtains equations of the form

$$\frac{dx}{dv} = ax + b$$

where x is the quantity in question and a and b are constant along λ.

Now moving F a parameter distance t_1 (the period of rotation of the black hole) to the past along the generators of the horizon is the same as moving F by the time translation ϕ_{-t_1}. Since ϕ_{-t_1} is an isometry, the quantity x will be unchanged under it. Thus x must be periodic along the generator λ with period t_1. This is possible only if x is constant along λ and equal to $-(b/a)$. One then uses this to calculate the derivative along λ of another quantity and shows that it is constant by a similar argument. Proceeding by induction one shows that all the derivatives at the horizon of the Cauchy data on N are constant along the generators of the horizon.

The first quantity x that one considers is $\bar{\alpha} + \beta$. The Newman-Penrose equation for this is

$$\frac{d}{dt}(\bar{\alpha} + \beta) = \delta(\epsilon + \bar{\epsilon}) + \psi_1.$$

By construction $\delta(\epsilon + \bar{\epsilon})$ is constant along the generators of the horizon and by another Newman-Penrose equation, $\psi_1 = 0$ on the horizon. Therefore in order for $\bar{\alpha} + \beta$ to be periodic it has to be constant along the generators and $\delta(\epsilon + \bar{\epsilon})$ has to be zero. This means that $\epsilon + \bar{\epsilon}$ must be constant over the whole horizon.

In the next section we shall see $\epsilon + \bar{\epsilon}$ can be interpreted as the restoring force or effective surface gravity of the black hole.

One now applies similar arguments to show that $(\bar{\alpha} - \beta)$, μ, λ, ψ_3 and ψ_4 are constant along the generators of the horizon. One then repeats the arguments to show that the first and higher derivatives of all quantities along the vector n^a are constant along the generators. This completes the proof.

It turns out that if ϵ is nonzero (as it is in general) the solution is completely determined by a knowledge of ψ_2 on each generator. I shall use this fact in one of the applications in the next section. It holds true even if the space outside the black hole is not empty but contains, say, a ring of matter (in which case the space would not be a Kerr solution).

The proof of axial symmetry implies that a rotating black hole cannot be exactly stationary unless all distant matter and all fields are arranged axisymmetrically. In real life this will never be the case. Thus a rotating black hole can never be exactly stationary, it must be slowing down. However, calculations by Press (1972), Hawkin and Hartle (1972), and Hartle (1972) have shown that the rate of slowing down is very small in most cases. I shall discuss this further in the next section.

9 Applications

In this final section I shall outline some of the ways in which the theory described so far can be used to obtain quantitative results, which is what most people want. I shall discuss three applications:

(A) The limits that can be placed from the area theorem on the amount of energy that can be extracted from black holes.
(B) The change in the mass and angular momentum of a nearly stationary black hole produced by small perturbations.
(C) Time symmetric black holes. (These are not very realistic but they provide some concrete examples.)

A Energy Limits

In view of the last section it seems reasonable to assume that a black hole settles down to a Kerr solution or, if carrying an electric charge, to a Newman-Kerr solution. The area of the event horizon of such a solution is

$$A = 4\pi[2M^2 - e^2 + 2(M^4 - M^2e^2 - L^2)^{\frac{1}{2}}] \tag{9.1}$$

where M is the mass, e the electric charge and L the angular momentum of the black hole. (All in units such that $G = c = 1$.) Now suppose that the black hole, having settled down by the surface $S(\tau_1)$ to a nearly stationary state with parameters $M_1, e_1 L_1$, now undergoes some interaction with external particles or fields and then settles down again by the surface $S(\tau_2)$ to a nearly stationary state with parameters M_2, e_2, L_2. Since the area of the horizon cannot decrease

$$A_2 \geqslant A_1 \tag{9.2}$$

where A_1 and A_2 are given by equation (9.1) with the appropriate values of M, e and L. In fact (9.2) is a strict inequality if there is any disturbance at the horizon. It puts an upper limit on $M_1 - M_2$, which represents the amount of energy extracted from the black hole by the interaction. To see what this limit is, it is convenient to express equation (9.1) in the form:

$$M^2 = \frac{A}{16\pi} + \frac{4\pi L^2}{A} + \frac{\pi e^4}{A} + \frac{e^2}{2} \tag{9.3}$$

The first term on the right can be regarded as the "irreducible" part of M^2, the part that is irretrievably lost down the black hole. The second term can be regarded as the contribution of the rotational energy of the black hole and the third and fourth terms as the contribution of the electrostatic energy. Christodoulou (1970) has shown that one can extract an arbitrarily large fraction of the rotational energy by the Penrose process of sending a particle from infinity into the ergosphere where it splits into two particles one of which returns to infinity with more than the original energy while the other falls through the horizon and reduces the mass and angular momentum of the black hole. Similarly, using charged particles, one can extract an arbitrarily large fraction of the electrostatic energy.

Note that it is M^2 and not M which has an irreducible part. This distinction does not matter when there is only one black hole but it means that one can extract energy, other than rotational or electrostatic energy, by allowing black holes to collide and merge. Consider two black holes $B_1(\tau)$ and $B_2(\tau)$ a long way apart which have settled down to nearly stationary states. One can neglect the interaction between them and regard the solution near each as a Kerr solution with the parameters M_1, e_1, L_1 and M_2, e_2 and L_2 respectively. The areas A_1 and A_2 of $\partial B_1(\tau)$ and $\partial B_2(\tau)$ will be given by equation (9.1). Suppose that at some later time the two black holes come together and merge to form a single black hole $B_3(\tau)$ which settles down to a nearly stationary state with parameters M_3, e_3 and L_3. During the collision process a certain amount of gravitational and possibly electromagnetic radiation will be emitted to infinity. The energy of this radiation will be $M_1 + M_2 - M_3$. This is limited by the requirement that the area A_3 of $\partial B_3(\tau)$ must be greater than the sum of A_1 and A_2. The fraction $\epsilon = (M_1 + M_2)^{-1}(M_1 + M_2 - M_3)$ of the total mass that can be radiated is always less than $1 - 2^{-\frac{3}{2}}$, i.e., about 65%. If the black holes are uncharged or carry the same sign of charge, the fraction is less than a half, i.e., 50%. If the black holes are also non-rotating the fraction is less than $1 - 2^{-\frac{1}{2}}$, i.e., about 29%.

By the conservation of charge $e_3 = e_1 + e_2$. Angular momentum, on the other hand, can be carried away by the radiation. This cannot happen, however, if the situation is axisymmetric, i.e., if the rotation axes of the black holes are aligned along their direction of approach to each other. Then $L_3 = L_1 + L_2$. One can see from equation (9.3) that M_3 can be smaller, i.e., there can be more energy radiated, if the rotations of the black holes are in opposite directions than if they are in the same direction. This suggests that there may be an orientation dependent force between black holes analogous to that between magnetic dipoles. Unlike the electromagnetic case, the force is repulsive if the orientations are the same and attractive if they are opposite. Even in the limiting case when $L_1 = M_1^2$ and $L_2 = M_2^2$, there is still energy available to be radiated. Thus it seems that the force can never be sufficiently repulsive to prevent the black holes colliding.

B Perturbations of Black Holes

To perform dynamic calculations about black holes seems to require the use of a computor in general. However there are a number of situations that can be treated as small perturbations of stationary black holes, i.e., Kerr solutions. The general idea in these calculations is to solve the linearized equations for a perturbation field (scalar, electromagnetic or gravitational) in a Kerr background and to try to find the radiation emitted to infinity and the rate of change of the mass and angular momentum of the black hole. In the case of the scalar and electromagnetic field these latter can be evaluated by integrating the appropriate components of the energy-momentum tensor of the field over the horizon. For gravitational perturbations, however, there is no well defined local energy-momentum tensor. Instead I shall show how one can determine the change in the mass and angular momentum of the black hole by calculating the change in the area of the horizon and the quantity ψ_2 on the horizon. It turns out that these depend only on the Ricci tensor terms $\phi_{00} = 4\pi T_{ab}l^a l^b$ and $\phi_{01} = 4\pi T_{ab}l^a m^b$ and the Weyl tensor term ψ_0 on the horizon. This is fortunate because it seems that the full equation for gravitational perturbations in a Kerr background are not solvable by separation of functions but Teukolsky (1972) has obtained decoupled separable equations for the quantities ψ_0 and ψ_4.

The mass, the magnitude of the angular momentum and its orientation make up four parameters in all. However, in many uses there are constraints which make it sufficient to calculate the change in only one function of these four parameters. The simplest such function is the area of the horizon which is given by equation (9.1). The rate of charge of this area can be calculated from the Newman-Penrose equations

$$\frac{d\rho}{dv} = \rho^2 + \sigma\bar{\sigma} + 2\epsilon\rho + \phi_{00} \tag{9.4}$$

$$\frac{d\sigma}{dv} = 2\rho\sigma + 2\epsilon\sigma + \psi_0 \tag{9.5}$$

Choose a spacelike surface S which intersects the event horizon of the background Kerr solutions in $J^+(\mathscr{I}^-)$ and is tangent to the rotation Killing vector $\tilde{K}^a$. Then one can define a family $S(t)$ of such surfaces by moving S under the time translation ϕ_t, i.e., by moving each point of S a parameter distance t along orbits of the Killing vector K^a of the unperturbed metric. This defines a time coordinate t on the horizon. It is convenient to choose the parameter v along the generators of the horizon to be equal to t. Then in the unperturbed Kerr metric

$$\epsilon = \frac{y}{4M(M^2 + y)}, \tag{9.6}$$

where

$$y = (M^4 - L^2)^{\frac{1}{2}}$$

There are two kinds of perturbations one can consider, those in which there is some matter fields like the scalar or electromagnetic field on the horizon with energy momentum tensor T_{ab} and those in which the perturbations at the horizon are purely gravitational and are produced by matter at a distance from the black hole. Consider first a matter field perturbation where the field is proportional to a small parameter λ. The energy momentum tensor and so the perturbation in the metric and in ψ_0 will be proportional to λ^2. Thus ρ and σ will be proportional to λ^2 and to order λ^2 the equation (9.4) becomes

$$\frac{d\rho}{dt} = 2\epsilon\rho + 4\pi T_{ab} l^a l^b \tag{9.7}$$

where ϵ is given by (9.6). Suppose that the perturbation field is turned off after some time t_1. The black hole will then settle down to a stationary state with $\rho = 0$.

Thus the solution of (9.7) for ρ is

$$\rho = -4\pi \int_t^\infty \exp\{2\epsilon(t - t')\} T_{ab} l^a l^b dt'. \tag{9.8}$$

The rate of increase of area of the horizon is

$$\frac{dA}{dt} = -2 \int \rho dA$$

where the integral is taken over the two surface $\partial B(t)$ which is the intersection of the event horizon with the surface $S(t)$. Substituting from equation (9.8) and performing a partial integration with respect to time one finds that the total area increase of the horizon is

$$\delta A = \frac{4\pi}{\epsilon} \int T_{ab} l^a d\Sigma^b \tag{9.9}$$

where $d\Sigma^b = l^b dAdt$ is the 3-surface element of the event horizon. The null vector l^a tangent to the horizon can be expressed in terms of K^a and $\tilde{K}^a$ the Killing vectors of the background Kerr metric which correspond to time translations and spatial rotations respectively.

$$l^a = K^a + \omega\tilde{K}^a + 0(\lambda^2), \tag{9.10}$$

where

$$\omega = \frac{L}{2M(M^2 + y)} \tag{9.11}$$

is the angular velocity of the black hole. The vectors $T_{ab}K^a$ and $-T_{ab}\tilde{K}^a$ represent the flow of energy and angular momentum respectively in the matter fields. They are conserved in the background Kerr metric and their fluxes across the horizon give change of mass and angular momentum of the black hole.

Thus

$$\delta A = \frac{4\pi}{\epsilon}\,[\delta M - \omega\delta L]. \tag{9.12}$$

This is just the change needed to preserve the formula (9.1) for the area of the horizon of Kerr solution. It is therefore consistent with the idea that the perturbation changes the black hole from one Kerr solution to one with slightly different parameters.

The case of purely gravitational perturbations is rather more interesting because one does not have an energy momentum tensor from which to compute the fluxes of energy and angular momentum into the black hole. Instead one can use the area increase as a measure of a certain combination of them. One takes the gravitational perturbation field to be proportional to a small parameter λ. Then from equations (9.4), (9.5) σ will be proportional to λ and ρ to λ^2

$$\frac{d\rho}{dt} = \sigma\bar{\sigma} + 2\epsilon\rho, \tag{9.13}$$

$$\frac{d\sigma}{dt} = 2\epsilon\sigma + \psi_0. \tag{9.14}$$

From (9.14)

$$\sigma = -\int_t^\infty \exp\{2\epsilon(t - t')\}\psi_0 dt' \tag{9.15}$$

and

$$\delta A = \frac{1}{\epsilon}\int \sigma\bar{\sigma}dAdt. \tag{9.16}$$

One can apply this formula in at least two situations. First there are stationary gravitational perturbations induced by distant matter which is stationary or nearly stationary. In such perturbations there will be no radiation at infinity and the energy of the sources of the perturbation will be nearly constant. Thus there can be no energy flow into or out of the black hole and its mass must remain constant. From equation (9.1) it then follows that the increase in the area A of the horizon must be accompanied by a decrease in the angular momentum of the black hole. In other words, the effect of stationary perturbation is to slow down the rotation of the black hole. What is happening is that the rotational energy part of M^2 in equation (9.3) is being dissipated into the irreducible part of M^2 represented by A.

There is a strong analogy between this process and ordinary tidal friction in a shallow sea covering a rotating planet. A nearly stationary external body such as a moon will raise tides in the sea. As the planet rotates, the shape of a fluid element will change and so the fluid will be shearing. There will be dissipation of energy at a rate proportional to the coefficient of viscosity times the square of the shear. This energy must come from the rotational energy of the planet. Thus the planet will slow down.

Similarly one can regard the perturbation field of a stationary external object as tidally distorting the horizon of the black hole (Figure 20) with consequent shearing as the black hole rotates and dissipation of rotational energy at a rate proportional to the square of the shear. The dimensionless analogue of the viscosity in this case is of order unity. Hartle (1972) has calculated the rate of slowing down of a slowly rotating black hole caused by a stationary object of mass M' at coordinates r and θ. For r/M large he finds

$$\frac{dL}{dt} = -2/5\,\frac{L}{M}\left(\frac{M'}{M}\right)^2 \sin^2\theta \left(\frac{M}{r}\right)^6$$

Because of the last factor, this seems too small ever to be of astrophysical significance. The situation might be different, however, for a rapidly rotating black hole with L nearly equal to M^2. In this case the quantity ϵ which acts as a restoring force in equations (9.13) and (9.14) is very small. In a sense the black hole is rotating with nearly break up velocity so centrifugal force almost balances gravity and a small object can raise a large tide on the horizon. For maximum effect, the object should be orbiting the black hole near the horizon with nearly the same angular velocity as that of the black hole. Under these circumstances the black hole would lose energy and angular momentum at a significant rate to the object. The object would also be losing energy and angular momentum in radiation to infinity. It is possible that the rates would balance to give what is called a *floating orbit*. To find out whether this could happen, it would be sufficient to calculate the rate of increase of the area of the horizon and the rate of radiation of energy and angular momentum to infinity since an object in a circular orbit can gain or lose energy and angular momentum only in a certain ratio.

For other problems it would be helpful to be able to calculate separately the rate of change of the mass and the three components of angular momentum. In the last section we saw that a stationary black hole solution is in general determined by a knowledge of the quantity ψ_2 on a 2-dimensional section of the

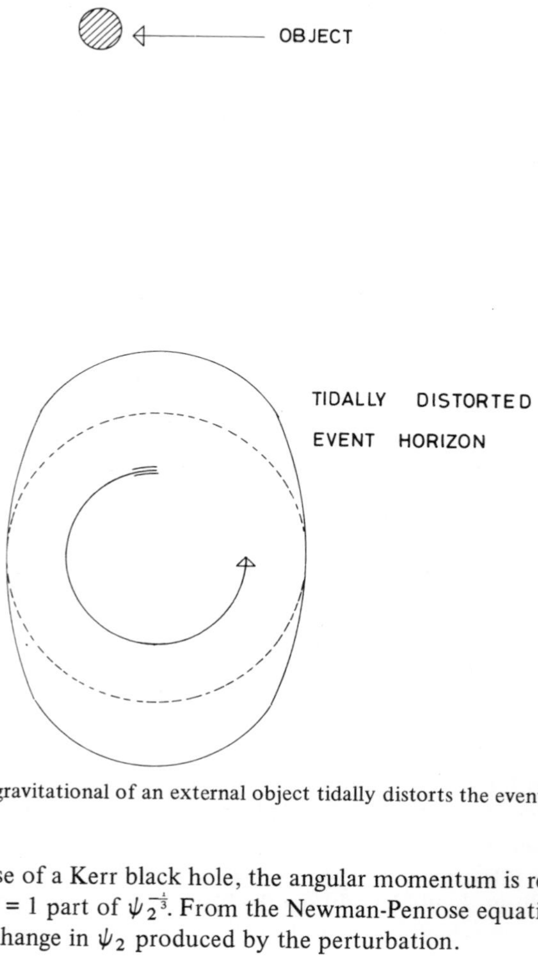

Figure 20. The gravitational of an external object tidally distorts the event horizon.

horizon. In the case of a Kerr black hole, the angular momentum is represented by the imaginary $l = 1$ part of $\psi_2^{-\frac{1}{3}}$. From the Newman-Penrose equations one can calculate the change in ψ_2 produced by the perturbation.

$$\frac{d}{dt}[(\psi_2 + \sigma\lambda - \mu\rho - \Phi_{11} - \Lambda)dA]$$

$$= \bar{\eth}\bar{\eth}\sigma - \bar{\eth}\eth\rho + \bar{\eth}[(\bar{\alpha} + \beta)\rho + \Phi_{01}] - \eth[(\alpha + \bar{\beta})\rho + \Phi_{10}]$$

where $\bar{\eth}$ acting on a spin weight s quantity is $\bar{\delta} + s(\alpha - \bar{\beta})$. Further details will be given elsewhere.

C Time Symmetric Black Holes

The last application I shall describe is largely based on the work of G. W. Gibbons. Some of it is about to be published (Gibbons 1972) and more will be in his Ph.D. thesis.

To calculate the evolution of a section of the Einstein equation one requires initial data on a partial Cauchy surface S. The Cauchy data on a spacelike surface can be represented by two symmetric 3-dimensional tensor fields h_{ij} and χ_{ij}. The negative definite tensor h_{ij} is the first fundamental form or induced metric of the 3-surface S imbedded in the 4-dimensional space-time manifold M. It is equal to $g_{ij} - u_j u_j$ where u_i is the unit timelike vector orthogonal to S. The tensor χ_{ij} is the second fundamental form or extrinsic curvature of S imbedded in M. It is equal to $u_{k;l} h_i^k h_i^l$. The fields h_{ij} and χ_{ij} have to obey the constraint equations:

$$\chi_{ab\|c} h^{bc} - \chi_{bc\|a} h^{bc} = 8\pi T^{bd} h_{ab} u_d$$

$$\tfrac{1}{2}[{}^{(3)}R + (\chi_{ab} h^{ab})^2 - \chi_{ab}\chi_{cd} h^{ac} h^{bd}] = 8\pi T^{ab} u_a u_b$$

where $\|$ indicates covariant differentiation with respect to the 3-dimensional metric h_{ij} in the surfaces. The constraint equations are non-linear and difficult to solve in general. However the problem is much simpler if the solution is time symmetric. The solution is said to be *time-symmetric* about the surface S if there is an isometry which leaves the surface S pointwise fixed but reverses the direction of time, i.e., it moves a point to the future of S on a timelike geodesic orthogonal to S to the point on the some geodesic an equal distance to the past of S. The time symmetry isometry maps χ_{ij} to $-\chi_{ij}$ since it reverses the direction of the normal u_i to S. Thus $\chi_{ij} = 0$. The first constraint is trivially satisfied and the second one becomes in the empty case

$${}^{(3)}R = 0$$

The convergence of the outgoing null geodesics orthogonal to a 2-surface F in S is

$$\rho = 2^{-\frac{1}{2}} m^i \bar{m}^j (u_{i;j} + w_{i;j})$$

where w_i is the unit spacelike vector in S orthogonal to F. The first term is zero because $\chi_{ij} = 0$. Thus if F is a marginally trapped surface, the convergence of its normals in S must be zero. This means that it is an extremal surface, i.e., its area is unchanged to first order under a small deformation. In fact F must be a minimal surface if it is an apparent horizon, i.e., if it is the outer boundary of a region containing closed trapped surfaces. Conversely any minimal 2-surface in S is an apparent horizon.

One can write down an explicit family of solutions of the remaining constraint equation by taking the metric h_{ij} on S to be $V^4\eta_{ij}$ where η_{ij} is the three dimensional flat metric and V satisfies the Laplace's equation in this metric

$$\nabla^2 V = 0.$$

I shall consider solutions of the form $V = 1 + \Sigma M_i/2r_i$ representing the field of a number of point masses M_i where the distance from the ith mass is r_i.

The solution with only one mass is the Schwarzschild solution expressed in isotropic coordinance. The minimal surface, which in this case is both the apparent and event horizon, is at $r = \frac{1}{2}M$ and has area $16\pi M^2$. Now consider the case of two equal mass points M_1 and M_2. If they are far apart the minimal surfaces around each will be almost at $r_1 = \frac{1}{2}M$ and $r_2 = \frac{1}{2}M$ and their areas will be nearly $16\pi M^2$. Each surface will however be slightly distorted by the field of the other points and their areas will be slightly greater than $16\pi M^2$. As the solution evolves the two black holes containing these two apparent horizons will fall towards each other and will merge to form a single black hole which will settle down to a Schwarzschild solution with mass M'. The energy of the gravitational radiation emitted in this process will be the initial mass $2M$ of the system minus the final mass M'. This is limited by the fact that the area $16\pi M'^2$ of the event horizon of the final black hole must be greater than the sum of the areas of the event horizons around the two original black holes. The area of these event horizons must be greater than those of the corresponding apparent horizons since these are minimal surfaces. Thus the upper limit on the fraction ϵ of the initial mass that can be radiated is somewhat less than $1 - 2^{-\frac{1}{2}}$. If the two mass points are moved nearer to each other in the initial surface S the minimal surfaces around them become more distorted and their area increases. Thus the upper limit on the fraction of energy that can be radiated becomes less. This is what one would expect since the available energy of each black hole is reduced by the negative gravitational potential of the other. In fact to first order, the reduction in the upper limit on ϵ just corresponds to the Newtonian gravitational interaction energy of the two point masses. When the two mass points are moved close to each other the area of the minimal surface around each becomes greater than $32\pi M^2$. This seems to indicate that the amount of energy that could be radiated would be negative which would be a contradiction. However before the two mass ponts are close enough for this to happen, it seems that a third minimal surface will be formed which surrounds them both and has area less than $64\pi M^2$ (Figure 21).

Gibbons (1972) has shown that any minimal surface in a conformally flat initial surface must have an area greater than

$$2\pi \left(\frac{1}{2\pi} \int V_{\|a} dA^a \right)^2$$

where the integral is taken over the minimal surface. The expression in the brackets represents the contribution to the total mass on the initial surface arising from points within the minimal surface. The solution that evolves from the initial surface will eventually settle down to a Schwarzschild solution with an event horizon of area $16\pi M'^2$. Since this area must be greater than the area of the event horizon on the initial surface which in turn must be greater than the area of the minimal surface, the difference between the initial mass M and the final mass M' must be less than $(1 - 2^{-3/2})M$. This means that a single distorted black hole on a surface of time symmetry cannot radiate more than 65% of its initial mass M in relaxing to a spherical black hole.

The black holes that have been considered so far in this subsection are non-rotating. This is because the condition that the solution be invariant under $t \to -t$ rules out any rotation. However, one can include rotation in a simple way if the solution is invariant under the simultaneous transformation $t \to -t$, $\varphi \to -\varphi$. I shall call such a solution (t, φ) symmetric. To obtain such a solution the initial data must be of the form

$$\chi_{ab} = \frac{2J_{(a}\tilde{K}_{b)}}{\tilde{K}_c\tilde{K}^c}$$

where J_a is an axisymmetric vector field orthogonal to the Killing vector $\tilde{K}^b$

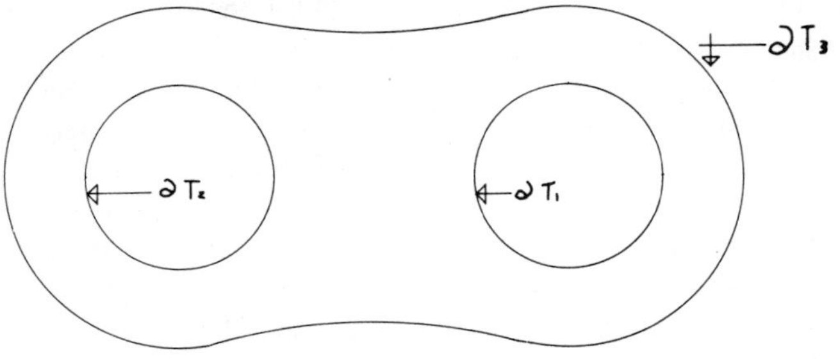

Figure 21. The two apparent horizons ∂T_1 and ∂T_2 are surrounded by another apparent horizon ∂T_3.

which corresponds to rotations about the axis of symmetry. The first constraint equation then becomes

$$J^a_{\|a} = 8\pi T_{ab}u^a \tilde{K}^b.$$

One can integrate this equation to obtain the total angular momentum within a given 2 – surface

$$L = -\frac{1}{8\pi}\int J_a\, dA^a.$$

In the empty case, to which I shall now restrict myself, the angular momentum will arise from singularities of the field J^a. The solution will be asymtotically predictable and will represent black holes if those singularities are contained within apparent horizons. From the form of χ_{ab} it follows that the apparent horizons in the initial surface of (t, φ) symmetry are minimal 2-surfaces. Note that this is the case only in a surface of time symmetry or (t, φ) symmetry. It is not true in later space-like surfaces.

In the empty case the second constraint equation becomes

$$^{(3)}R = \frac{2J_a J^a}{\tilde{K}_a \tilde{K}^a}.$$

This equation can be solved by a technique of Lichnerowicz. Choose a spatial metric h_{ab}. Then choose a spatial vector field J_a which is axisymmetric, orthogonal to $\tilde{K}^a$ and which satisfies $J^a_{\|a} = 0$ in the metric h_{ab}. One then makes a conformal transformation $\tilde{h}_{ab} = V^4 h_{ab}$. The first constraint equation will remain satisfied if J_a transforms as $\tilde{J}_a = V^{-2} J_a$. The second constraint equation will be satisfied if V is chosen so that

$$V^{(3)}R - 8V_{\|ab}h^{ab} = 2V^{-7}\frac{J_a J^a}{\tilde{K}_b \tilde{K}^b}$$

where the covariant derivatives are with respect to the metric h_{ab}. This equation is non-linear so one cannot write down explicit solutions even in the case where the metric h_{ab} is chosen to be flat. However one can note certain qualitative features. One of these is that the addition of angular momentum tends to increase the total mass of the solution. Thus it seems that the rotational energy of black holes is positive as one would expect. Calculations by Gibbons in the case of two black holes indicate that the ratio of the area of the apparent horizons to the square of the total mass is bigger when the angular momenta of the black holes are in opposite directions than when they are in the same direction. This indicates that there is less energy available to the radiated in the former case than in the latter which is consistent with the idea that there is a spin dependent force between black holes which is attractive in the case of opposite angular momenta and repulsive in the other case.

The calculations of Gibbons indicate that when the black holes are far apart the force is proportional to the inverse fourth power of the separation which is what one would expect from the analogy with magnetic dipoles.

References

H. Bondi, M. G. J. Van der Burg and A. W. K. Metzner, *Proc. Roy. Soc.* (London) **A269**, 21 (1962).
D. Christodoulou, *Phys. Rev. Letters* **25**, 1596 (1970).
A. G. Doroshkevich, Ya, B. Zel'dovich and I. D. Novikov, *Sov. Phys.* JETP **22**, 122 (1966).
R. P. Geroch, *J. Math. Phys.* **9**, 1739 (1968).
G. W. Gibbons, *Comm. in Math. Phys.* **27**, 87 (1972).
J. B. Hartle, 1972 (preprint).
J. B. Hartle and S. W. Hawking, *Comm. in Math. Phys.* **26**, 87 (1972).

S. W. Hawking and R. Penrose, *Proc. Roy. Soc.* **A314**, 529 (1970).
S. W. Hawking, *Comm. in Math. Phys.* **18**, 301 (1970).
S. W. Hawking and J. Hartle, *Comm. in Math. Phys.* **27**, 283 (1972).
S. W. Hawking, *Comm. in Math. Phys.* **25**, 152 (1972).
S. W. Hawking and G. F. R. Ellis, *The Large Scale Structure of Space-Time* (Cambridge University Press, Cambridge, to be published).
W. Israel, *Phys. Rev.* **164**, 1776 (1967).
L. Markus, *Ann. Math.* **62**, 411 (1955).
E. T. Newman and R. Penrose, *J. Math. Phys.* **3**, 566 (1962).
R. Penrose, *Phys. Rev. Letters* **10**, 66 (1963).
R. Penrose, *Phys. Rev. Letters* **14**, 57 (1965a).
R. Penrose, *Proc. Roy. Soc.* **A284**, 159 (1965b).
R. Penrose, *Battelle Rencontres,* 1967 *Lectures in Mathematics and Physics,* edited by C. DeWitt and J. A. Wheeler (W. A. Benjamin, Inc., New York 1968).
R. Penrose, *Riv. Nuovo Cimento* **1** (Num. spec.) 252 (1969).
W. H. Press, *AP.J.* **175**, 243 (1972).
R. Price, *Phys. Rev.* **D5**, 2419 (1972).
T. Regge and J. A. Wheeler, *Phys. Rev.* **108**, 1063 (1957).
S. Teukolsky, 1972 (Caltech preprint).

Black Hole Equilibrium States

Brandon Carter

Institute of Theoretical Astronomy,
Cambridge, Great Britain

Contents

Part I Analytic and Geometric Properties of the Kerr Solutions

1 Introduction 61
2 Spheres and Pseudo-Spheres 62
3 Derivation of the Spherical Vacuum Solutions 68
4 Maximal Extensions of the Spherical Solutions 74
5 Derivation of the Kerr Solution and its Generalizations . . . 89
6 Maximal Extensions of the Generalized Kerr Solutions . . . 103
7 The Domains of Outer Communication 112
8 Integration of the Geodesic Equations 117

Part II General Theory of Stationary Black Hole States

1 Introduction 125
2 The Domain of Outer Communications and the Global Horizon . . 133
3 Axisymmetry and the Canonical Killing Vectors 136
4 Ergosurfaces, Rotosurfaces and the Horizon 140
5 Properties of Killing Horizons 146
6 Stationarity, Staticity, and the Hawking-Lichnerowicz Theorem . 151
7 Stationary-Axisymmetry, Circularity, and the Papapetrou Theorem . 159
8 The Four Laws of Black Hole Mechanics 166
9 Generalized Smarr Formula and the General Mass Variation Formula . 177
10 Boundary Conditions for the Vacuum Black Hole Problem . . 185
11 Differential Equation Systems for the Vacuum Black Hole Problem . 197
12 The Pure Vacuum No-Hair Theorem 205
13 Unsolved Problems 210

PART I Analytic and Geometric Properties of the Kerr Solutions

1 Introduction

The Kerr solutions (Kerr 1963) and their electromagnetic generalizations (Newman *et al.* 1965) form a 4-parameter family of asymptotically flat solutions of the source-free Einstein-Maxwell equations, the parameters being most conveniently taken to be the asymptotically defined mass M, electric charge Q, and magnetic monopole charge P, together with a rotation parameter a, which is such that (in units of the form we shall use throughout, where the speed of light c and Newton's constant G are set equal to unity) the asymptotically defined angular momentum J is given by

$$J = Ma$$

The parameters may range over all real values subject to the restriction

$$M^2 \geqslant a^2 + P^2 + Q^2$$

which must be satisfied if the solution is to represent the exterior to a *hole* rather than *naked singularity*. It turns out (Carter 1968a) that the solutions all have the same gyromagnetic ratios as those predicted by the simple Dirac theory of the electron, as the asymptotic magnetic and electric dipole moments cannot be specified independently of the angular momentum but are given, in terms of the same rotation parameter, as Qa and Pa respectively.

The solutions are all geometrically unaltered by variations of P and Q provided that the sum $P^2 + Q^2$ remains constant, and since it is in any case widely believed that magnetic monopoles do not exist in nature, attention in most studies is restricted to the 3-parameter subfamily in which P is zero. This 3-parameter subfamily, and specially the 2-parameter pure vacuum subfamily in which Q is also zero, has come recently to be regarded as being at least potentially of great astrophysical interest, since the Kerr solutions do not merely represent the *only known* stationary source-free black hole exterior solutions: they are also widely believed (for reasons which will be presented in Part II of this course) to be the *only possible* such solutions.

We shall devote the whole of Part I of the present lecture course to the derivation and geometric investigation of these Kerr solutions. In a strictly logical approach, Part II of this course (which will consist of a general examination of stationary black hole states with or without external sources) should come first, but it is probably advisable for a reader who is not already familiar with the subject to start with Part I, since the significance of the reasoning to be presented in Part II will be more easily appreciated if one has in mind the concrete examples described in Part I. For the same reason many readers may find it easier to appreciate the accompanying lecture course of Hawking if they have first become familiar with the examples described here, although only the final stages of Hawking's course actually depend on the results presented here, the bulk of it being logically antecedent to both Part II and Part I of the present course. This present lecture course is intended to serve both logically and pedagogically as an introduction to the immediately following course of Bardeen and hence also (but less directly) to the subsequent courses in this volume.

2 Spheres and Pseudo-Spheres

A space time manifold $\mathscr{M}$ is said to be *spherically symmetric* if it is invariant under an action of the rotation group SO(3) whose surfaces of transitivity are 2-dimensional. The metric on any one of these 2-surfaces must have the form

$$ds_{\odot}^2 = r^2\,(d\theta^2 + \sin^2\theta d\varphi^2) \tag{2.1}$$

in terms of a suitably chosen azimuthal co-ordinate θ running from 0 to π, and a periodic ignorable co-ordinate φ defined modulo 2π, where the scale factor r is the *radius of curvature* of the 2-sphere. In what follows we shall frequently find it convenient to use the equivalent alternative form

$$ds_{\odot}^2 = r^2\left\{\frac{d\mu^2}{1-\mu^2} + (1-\mu^2)d\varphi^2\right\} \tag{2.2}$$

expressed in terms the customary alternative azimuthal co-ordinate

$$\mu = \cos\theta \tag{2.3}$$

running from -1 to $+1$. See Figure 2.1.

A space-time manifold $\mathscr{M}$ is said to be *pseudo-spherically* symmetric if it is invariant under an action of the 3-dimensional Lorentz group whose surfaces of transitivity are timelike and 2-dimensional. The metric on one of these 2-surfaces can be expressed in a form analogous to (2) as

$$ds_{\times}^2 = -s^2\left\{\frac{dx^2}{1-x^2} - (1-x^2)dt^2\right\} \tag{2.4}$$

where s is the radius of curvature and where the co-ordinate t is ignorable.

The two simple and familiar metric forms $ds_{\odot}^2$ and $ds_{\times}^2$ illustrate a feature which will crop up repeatedly in the present course, that is to say they both have removable *co-ordinate singularities*. The spherical form (2.2) is obviously singular at $\mu = \pm 1$; moreover we cannot remove this singularity simply by returning to the form (2.1) in which the infinity is eliminated only at the price of introducing an equally undesirable vanishing determinant which will of course lead to an infinity in the inverse metric tensor. We can cure the singularity and show explicitly that the space is well behaved at the poles (as we know it must be by the homogeneity)

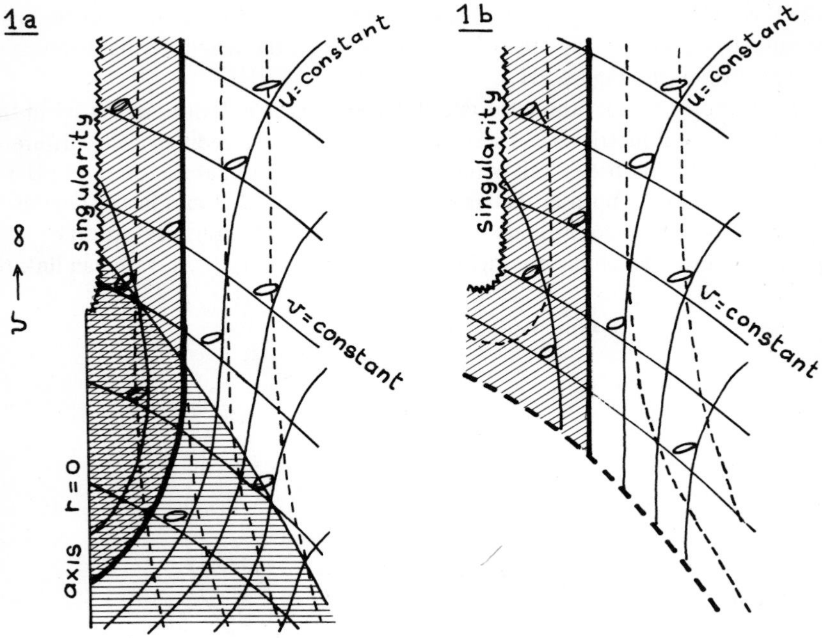

Figure 1. Figure 1a shows a plan of a timelike 2-section, in which the spherical co-ordinates θ, φ are held constant, of the space time manifold of a spherical collapsing star which occupies the part of space time marked by horizontal shading. Locuses on which r is constant are marked by dotted lines and null lines are marked by continuous lines. The horizon $\mathscr{H}^+$ is indicated by a heavy continuous line and the region outside the domain of outer communications is marked by diagonal shading. Figure 1b shows the extrapolation back under the group action of the pseudo-stationary empty outer region of Figure 1a. The past boundary marked by a heavy dotted line would become the horizon $\mathscr{H}^-$ in an extended manifold. (*See Part II, Section 1*)

only by giving up the use of the ignorable co-ordinate φ. Thus for example we can cure the singularity at the north pole $\mu = 1$ by introducing Cartesian type co-ordinates x, y defined by

$$x = 2r\frac{\sqrt{1-\mu^2}}{1+\mu}\sin\varphi \tag{2.5}$$

$$y = 2r\frac{\sqrt{1-\mu^2}}{1+\mu}\cos\varphi \tag{2.6}$$

to obtain the conformally flat form

$$ds_{\odot}^2 = \frac{dx^2 + dy^2}{1 + \dfrac{x^2+y^2}{4r^2}} \tag{2.7}$$

which is well behaved everywhere except at the south pole. It is of course impossible to obtain a form which is well behaved on the whole of the 2-sphere at the same time (see Figure 2.1.).

Let us now consider the algebraically analogous singularities at $x = \pm 1$ in the pseudo-spherical metric ds_x^2. These singularities are of a geometrically different nature since the metric is well behaved with the same signature, on *both sides* of the singularities i.e. both in the static regions with $|x| > 1$ and in the regions where the ignorable co-ordinate is spacelike $|x| < 1$. (In contrast with the spherical metric which has the right signature only for $|\mu| < 1$.) We can link the

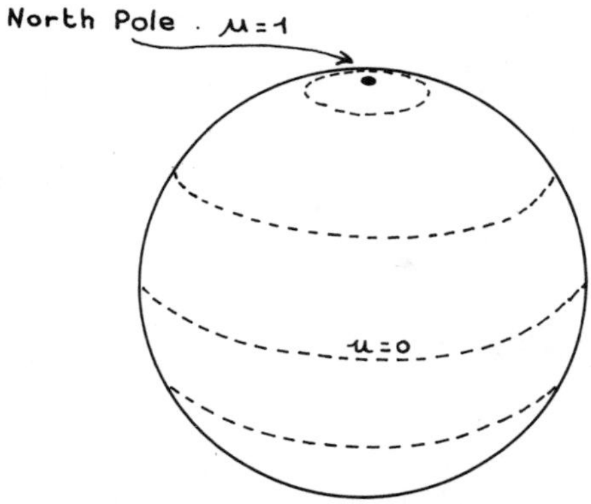

Figure 2.1. The 2-sphere. The light dotted lines represent trajectories of the Killing vector $\partial/\partial\varphi$.

well behaved domains $x > 1$ and $-1 < x < 1$ by a co-ordinate patch extending right across the divisions $x = \pm 1$, by introducing new co-ordinates

$$\tau = \frac{-x}{|1 - x^2|^{\frac{1}{2}}} e^{-t} \tag{2.8}$$

$$\lambda = |1 - x^2|^{\frac{1}{2}} e^t \tag{2.9}$$

which leads to the form

$$ds_x^2 = s^2 \left\{ \frac{d\lambda^2}{\lambda^2} - \lambda^2 d\tau^2 \right\} \tag{2.10}$$

in which the new co-ordinate τ is ignoreable. This form is well behaved over the whole of the region $\lambda > 0$ including the loci $\lambda\tau = \pm 1$ which correspond respectively to the divisions $-x = \pm 1$ in the original system (2.4). Moreover it can easily be seen that these loci are in fact *null lines.* The situation can be visualized most easily in terms of null-co-ordinates u, v which we introduce by the defining relations

$$u = \tau + \frac{1}{\lambda} \tag{2.11}$$

$$v = \tau - \frac{1}{\lambda} \tag{2.12}$$

which leads to the form

$$ds_x^2 = -4s^2 \frac{du\,dv}{(u - v)^2} \tag{2.13}$$

(whose u and v are restricted by the condition $u - v > 0$) in which the divisions $x = \pm 1$ are represented by the lines $u = 0$ and $v = 0$ respectively.

The relationship between the co-ordinate patches (2.13) (which is equivalent to (2.10) with $\lambda > 0$) and the co-ordinate patches of the form (2.7) is shown in the *conformal diagram* of Figure 2.2a. This is the first example of a technique, which we shall employ frequently, of representing the geometry of a timelike 2-surface on the plane of the paper, in a manner which takes advantage of the fact that any such metric is conformly flat and thus can be expressed in the canonical null form $ds^2 = C du dv$ where C is a conformal factor. The method consists simply of identifying the null co-ordinates u and v with ordinary Cartesian co-ordinates on the paper which conventionally are placed diagonally so that timelike directions lie in a cone of angles within 45 degrees of the vertical. There is nothing unique about such a representation since there is a wide choice of canonical null form preserving transformations in which the co-ordinates u, v are replaced respectively by new co-ordinates u^*, v^* of the form $u^* = u^*(u)$, $v^* = v^*(v)$ and in which the conformal factor C is replaced by $C^* = C(du^*/du)(dv^*/dv)$. This freedom can be used to arrange for the co-ordinate range u^*, v^* to be finite even if the original co-ordinate range is not (e.g. by taking $u^* = \tanh u$, $v^* = \tanh v$) thereby making it possible to represent an infinite timelike 2-manifold in its entirety on a finite piece of paper.

The co-ordinate patch of Figure 2a with the spacially homogeneous form (2.10) is in fact still incomplete, as would have been expected from the fact that starting from the time symmetric form (2.4) one could have extended into the past instead of the future by replacing t by $-t$ in the transformation equations (2.8) and (2.9). We can obtain a new form which covers both extensions by setting

$$u = \tan \frac{U}{2} \tag{2.14}$$

$$v = -\cot \frac{V}{2} \tag{2.15}$$

which leads to the new null form

$$ds_{\times}^2 = -s^2 \left\{1 + \tan^2 \frac{U - V}{2}\right\} dUdV \tag{2.16}$$

In terms of these co-ordinates (which can both range from $-\infty$ to ∞ subject to the restriction $-\pi < U - V < \pi$) the entire co-ordinate range of u and v subject to $u + v < 0$, i.e. the entire range of λ, τ subject to $\lambda > 0$, is covered by the range

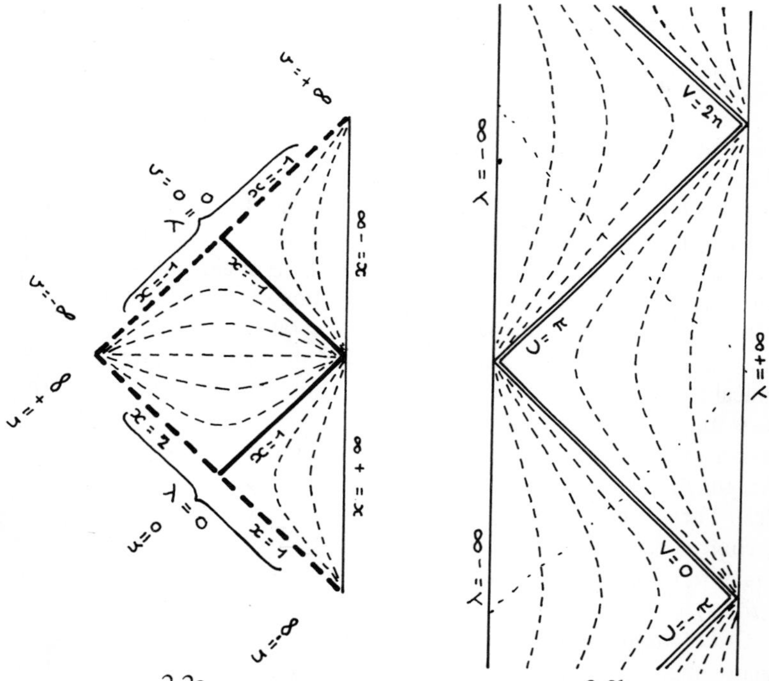

2.2a 2.2b

Figure 2.2 Conformal diagram of the 2-dimensional pseudo-sphere. In Figure 2a the light dotted lines represent trajectories of the Killing vector $\partial/\partial t$ and the heavy lines represent the corresponding non-degenerate Killing horizons. In the extended diagram of Figure 2b the light dotted lines represent trajectories of the Killing vector $\partial/\partial\tau$, and the double lines represent the corresponding degenerate Killing horizons.

$0 < V < 2\pi$, and $-\pi < U < \pi$. The situation is illustrated in Figure 2b. The manifold in this figure is geodesically complete and hence maximal in the sense that no further extension can be made. The null form (2.16) covering this maximal extension can be converted to an equivalent static form by introducing co-ordinates X, T defined by

$$X = \tan \frac{U - V}{2} \tag{2.17}$$

$$T = \frac{U + V}{2} \tag{2.18}$$

which gives the maximally extended static form

$$ds_{\times}^2 = +s^2 \left\{ \frac{dX^2}{1 + X^2} - (1 + X^2) dT^2 \right\} \tag{2.19}$$

in which X and T can range from $-\infty$ to ∞ without restriction.

Looking back from the vantage point of this complete manifold we can see clearly what has been happening. The timelike Killing vector whose trajectories are the static curves x = constant, $|x| > 0$ in the system (2.4), becomes null on the surfaces on which U or V are multiples of π, these surfaces representing past and future *event horizons* for observers who are fixed in that x remains constant relative to this system. [The past and future event horizon (cf. Rindler 1966) of an observer being respectively the boundary of the past of his worldline, i.e. the set of events he will ultimately be able to know of, and the boundary of the future of his world line i.e. of the set of events which he could in principle have influenced.] Thus for the Killing vector whose trajectories are the static lines x = constant, $|x| > 0$ in the system (2.4), the corresponding event horizons are the lines on which U and V are multiples of π. In the extension (2.10) in which the ignorable co-ordinate t has been sacrificed, there is a new manifest symmetry generated by the Killing vector whose trajectories are the lines λ = constant, and for the corresponding static observers the event horizons coincide with alternate horizons of the previous set, specifically the lines on which $U + \pi$ and V are multiples of 2π. In the maximal extension in which the ignorable co-ordinate τ has in its turn been replaced by T, there is a third non-equivalent Killing vector field, whose static trajectories are the lines X = constant, and in this case the corresponding observers have no event horizons.

There is a fairly close analogy between the removable co-ordinate singularities associated with rotation axes, as exemplified by the case of the ordinary 2-sphere discussed earlier, and the removable co-ordinate singularities associated with Killing horizons. Both arise from the inevitable bad behaviour of an *ignorable* co-ordinate which is used to make manifest symmetry group action generated by a Killing vector. The former case arises in the case of a spacelike Killing vector generating a rotation group action when it becomes zero on a rotation axis, while the latter arises in the case of a static Killing vector (and also under appropriate conditions as we shall see later a stationary Killing vector) when it becomes null on a Killing horizon. Killing horizons can be classified as degenerate or non-degenerate according to whether the gradient of the square of the Killing vector is zero or not. In the case of a *non-degenerate Killing horizon* (as exemplified by the horizons on which the Killing vector whose trajectories are x = constant) the relevant Killing vector must change from being timelike on one side to being spacelike on the other. In the case of a *degenerate Killing horizon*, the relevant

Killing vector *may be* (and in the sample of the Killing vector whose trajectories are λ = constant, actually *is*) timelike on both sides of the horizon. We can see in the example of the pseudo-sphere another phenomenon that will later be shown to be true in general, namely that non-degenerate Killing horizons of a given Killing vector field cross each other at what we shall call a Boyer-Kruskal axis, on which the Killing vector is actually zero, this axis being very closely analogous to a rotation axis. On the other hand degenerate Killing horizons (which have no exact analogue in the case of axisymmetry) continue unimpeded over an unbounded range.

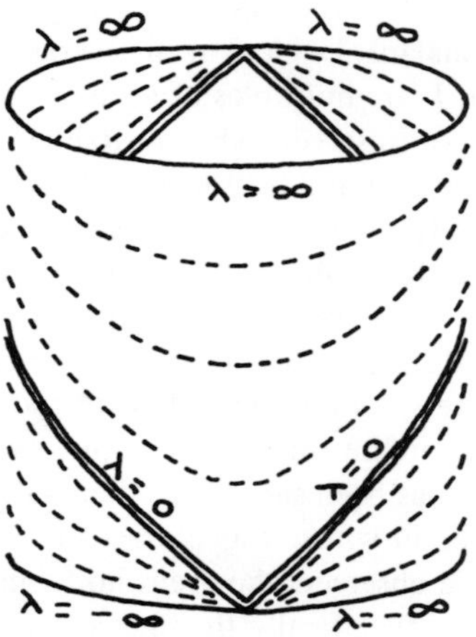

Figure 2.3. Sketch of conformal diagram of 2-dimensional pseudo-sphere with the canonical global topology obtained by identifying points with the same value of X but with values of T differing by 2π in the maximally extended covering manifold of Figure 2.2b.

Since T is ignorable, we could construct a new reduced manifold from the manifold of Figure 2.2b by identifying points with the same values of X but for which the values of T differ by any arbitrarily chosen fixed p period; the locus of points at unit spacelike distance from the origin in 3-dimensional Minkowski space is a manifold of the form obtained in this way where the period of T has the canonical value 2π (see Figure 2.3).

3 Derivation of the Spherical Vacuum Solutions

In this section we shall run rapidly through the steps by which the spherically symmetric vacuum (both pure vacuum and source free Einstein Maxwell) solutions are derived. Thus we start from the condition that the space time manifold under consideration is invariant under an action of SO(3) which is transitive over 2-surfaces. Since the rotation group action is necessarily invertible in the sense that associated with any point there is a discrete subgroup action—a rotation by 180°—which reverses the senses of the tangent vectors to the surfaces of transitivity, it follows automatically (cf. e.g. Schmidt 1967, Carter 1969) that the group action is *orthogonally transitive* in the sense that the surfaces of transitivity must themselves be orthogonal to another family of 2-surfaces. Therefore by requiring that they be constant on these 2-surfaces, we can extend the polar co-ordinate θ, φ or equivalently μ, φ from the individual 2-surfaces to the 4-dimensional manifold $\mathscr{M}$. It follows further that apart from an overall conformal factor r^2 (where r is the local radius of the 2-spheres) the space-time will locally have the form of a direct product of the unit 2-sphere with a certain *time-like 2-space*, with metric dl^2 say, so that the overall metric will have the canonical form

$$ds^2 = r^2 \left\{ dl^2 + \frac{d\mu^2}{1-\mu^2} + (1-\mu^2)\, d\varphi^2 \right\}$$

Since we are only considering stationary spaces in the present course, we can introduce an ignorable co-ordinate t on the timelike 2-spaces (in fact by the well known theorem of Birkhoff, the stationary assumption involves no loss of generality at all in so far as spherically symmetric *pure vacuum* or *electromagnetic vacuum* solutions are concerned). Moreover except in the special case (which we shall consider later since it is not as irrelevant as is often assumed) when r is constant, we may take the radius r itself as a co-ordinate on the time-like 2-space. The t co-ordinate can then be fixed uniquely by the requirement that it be orthogonal to the r co-ordinate. This leads to the canonical spherical metric form

$$\boxed{ds^2 = r^2 \left\{ \frac{d\mu^2}{1-\mu^2} + (1-\mu^2) d\varphi^2 + \frac{dr^2}{\Delta_r} - \frac{\Delta_r dt^2}{Z_r^2} \right\}} \tag{3.2}$$

where Δ_r and Z_r are two arbitrary functions of r only, and where the factors Δ_r have been distributed in such a manner as to cancel out of the expression for the metric determinant g, so that the ubiquitous volume density weight factor $\sqrt{-g}$ which appears in so many expressions takes the very simple form

$$\sqrt{-g} = \frac{r^4}{Z_r} \tag{3.3}$$

The easiest way that I know of solving Einstein's equations in a case like this is the method described by Misner (1963), which is based on the application of Cartan type calculus to the differential forms of the canonical tetrad. In this method the metric is expressed in terms of 4 differential forms $\omega^{(i)} = \omega^{(i)}_a\, dx^a$, $i = 1, 2, 3, 0$, which are orthonormal, so that the metric takes the form

$$ds^2 = g_{(i)(j)} \omega^{(i)} \omega^{(j)} \tag{3.4}$$

where the $g_{(i)(j)}$ are scalar constants which are the components of the standard Minkowski metric tensor in a flat co-ordinate system. Instead of working with a large number of Christofel symbols one works with what will (specially if the original metric form is fairly simple) be a comparatively small number of connection forms, $\gamma^{(i)}{}_{(j)} = \gamma^{(i)}{}_{(j)}\, dx^a$ defined (but *not* in practice computed) by the equations

$$\omega^{(i)}{}_{a;b} = \gamma^{(i)}{}_{(j)b}\,\omega^{(j)}{}_a \tag{3.5}$$

These forms will automatically be symmetric in the sense that if the labelling indices are lowered by contractions with the Minkowski scalors i.e. setting $\gamma_{(i)(j)} = g_{(i)(k)}\gamma^{(k)}{}_{(j)}$, then we have they will satisfy

$$\gamma_{(i)(j)} = \gamma_{(j)(i)} \tag{3.6}$$

Moreover the antisymmetrized part of the equation (3.5) can be expressed in terms of Cartan form language as

$$d\omega^{(i)} = -\gamma^{(i)}{}_{(j)} \wedge \omega^{(j)} \tag{3.7}$$

and it can easily be seen that the two expression (3.6) and (3.7) together (which can easily be worked out without using covariant differentiation) can be used to determine the forms $\gamma^{(i)}{}_{(j)}$ instead of the computationally more awkward defining relation (3.4). Furthermore the tetrad components $R_{(i)(j)(k)(l)}$ of the Riemann tensor can also be read out without the use of covariant differentiation from the expression

$$\theta^{(i)}{}_{(j)} = \tfrac{1}{2}R^{(i)}{}_{(j)(k)(l)}\,\omega^{(k)} \wedge \omega^{(l)} \tag{3.8}$$

where the curvature two-forms $\theta^{(i)}{}_{(j)} = \theta^{(i)}{}_{(j)ab}\, dx^a \wedge dx^b$ are given (as can easily be checked by differentiating (1) and using the defining relation $\omega^{(i)}{}_{a[b;c]} = \tfrac{1}{2}R_{abcd}\,\omega^{(i)d}$ by

$$\theta^{(i)}{}_{(j)} = d\gamma^{(i)}{}_{(j)} + \gamma^{(i)}{}_{(k)} \wedge \gamma^{(k)}{}_{(j)} \tag{3.9}$$

The tetrad components of the Ricci tensor are obtained simply by contracting

$$R_{(i)(j)} = R^{(k)}{}_{(i)(j)(k)} \tag{3.10}$$

and if required the co-ordinate form is given by

$$R_{ab} = R_{(i)(j)}\,\omega^{(i)}{}_a\,\omega^{(j)}{}_b. \tag{3.11}$$

In the present case the obvious tetrad of forms consists of

$$\left.\begin{aligned}
\omega^{(1)} &= \frac{r}{\sqrt{\Delta_r}}\, dr \\
\omega^{(2)} &= \frac{r}{\sqrt{1-\mu^2}}\, d\mu \\
\omega^{(3)} &= r\sqrt{1-\mu^2}\, d\varphi \\
\omega^{(0)} &= \frac{r\sqrt{\Delta_r}}{Z_r}\, dt
\end{aligned}\right\} \tag{3.12}$$

and a straightforward calculation on the lines described above shows that the only solution of the pure Einstein vacuum equations

$$R_{ij} = 0 \tag{3.13}$$

are given (after use of co-ordinate scale change freedom to achieve a standard normalization) by

$$Z_r = r^2 \tag{3.14}$$

$$\Delta_r = r^2 - 2Mr \tag{3.15}$$

where M is a constant. The corresponding values of the curvature forms may be tabulated as

$$\left.\begin{array}{ll} \theta^{(1)}{}_{(2)} = -\dfrac{M}{r^3}\,\omega^{(1)} \wedge \omega^{(2)} & \theta^{(3)}{}_{(0)} = -\dfrac{M}{r^3}\,\omega^{(3)} \wedge \omega^{(0)} \\ \theta^{(3)}{}_{(1)} = \dfrac{M}{r^3}\,\omega^{(1)} \wedge \omega^{(3)} & \theta^{(3)}{}_{(2)} = 2\dfrac{M}{r^3}\,\omega^{(2)} \wedge \omega^{(3)} \\ \theta^{(0)}{}_{(1)} = 2\dfrac{M}{r^3}\,\omega^{(1)} \wedge \omega^{(0)} & \theta^{(0)}{}_{(2)} = \dfrac{M}{r^3}\,\omega^{(2)} \wedge \omega^{(0)} \end{array}\right\} \tag{3.16}$$

The comparative conciseness of this array, from which, if desired, all 20 of the ordinary Riemann tensor components can be read out, shows clearly the advantage of the Cartan formulation. If one is interested in the Petrof classification, it can be seen directly from the above array that the tetrad $\omega^{(i)}$ is in fact a canonical tetrad and that the Riemann tensor, which in the vacuum case is the same as the Weyl conformal tensor, is of type D. (See Pirani 1964, 1962; Ehlers and Kundt 1962.

The familiar Schwarzschild metric itself is given explicitly by

$$ds^2 = r^2 \left\{ \frac{dr^2}{r^2 - 2Mr} + \frac{d\mu^2}{1 - \mu^2} - (1 - \mu^2)d\varphi^2 \right\} - \frac{(r^2 - 2Mr)}{r^2}\,dt^2 \tag{3.17}$$

To obtain the *electromagnetic* vacuum solutions we must first find the most general forms of the electromagnetic field consistent with spherical symmetry. We start from the well known fact that there are no spherically symmetric vector fields on the 2-sphere, and only one unique (apart from a scale factor) spherically symmetric 2-form on the 2-sphere, which takes a very simple form in terms of the co-ordinate system (2.2), namely $d\mu \wedge d\varphi$. Since any cross components of the Maxwell field between directions orthogonal to and in the 2-spheres of transitivity would define vector fields in the surfaces of transitivity it follows that the most general spherical Maxwell field is a linear combination with co-efficients depending only on r, of $dr \wedge dt$ and $d\mu \wedge d\varphi$, and hence can be expressed in terms of the canonical tetrad in the form

$$F = 2E_r\omega^{(1)} \wedge \omega^{(0)} + 2B_r\omega^{(2)} \wedge \omega^{(3)} \tag{3.18}$$

where E_r and B_r are functions of r only. The dual field form $*F$ is then given simply by

$$*F = 2B_r\omega^{(1)} \wedge \omega^{(0)} + 2E_r\omega^{(2)} \wedge \omega^{(3)} \tag{3.19}$$

and in terms of these expressions the source free Maxwell equations simply take the form $dF = 0$, $d*F = 0$ and thus they too can be worked out by the Cartan method without recourse to covariant differentiation. The electromagnetic energy tensor will be given by

$$8\pi T_{ab} = (E_r^2 + B_r^2)\{\omega_a^{(0)}\omega_b^{(0)} + \omega_a^{(3)}\omega_b^{(3)} + \omega_a^{(2)}\omega_b^{(2)} - \omega_a^{(1)}\omega_b^{(1)}\} \tag{3.20}$$

and hence the Einstein-Maxwell equations

$$R_{ab} = 8\pi T_{ab} \tag{3.21}$$

can easily be worked out with the Ricci tensor evaluated in the way described above.

The solutions are given by

$$E_r = \frac{Q}{r^2} \qquad B_r = \frac{P}{r^2} \tag{3.22}$$

(where Q and P are constants which correspond respectively to electric and magnetic monopole charges) with

$$Z_r = r^2 \tag{3.23}$$

$$\Delta_r = r^2 - 2Mr + Q^2 + P^2 \tag{3.24}$$

In presenting the information giving the curvature, it is worthwhile to make a distinction now between the Riemann tensor, and the Weyl conformal tensor since the Ricci tensor which may be read out by substituting (3.20) in (3.21) and using (3.22), is no longer zero. The Weyl forms $\Omega_{(i)(j)} = \Omega_{(i)(j)ab}\,\omega_a^{(i)}\omega_b^{(j)}$ defined in terms of the tetrad components $C^{(i)}{}_{(j)(k)(l)}$ by

$$\Omega^{(i)}{}_{(j)} = \tfrac{1}{2}C^{(i)}{}_{(j)(k)(l)}\omega^{(k)} \wedge \omega^{(l)} \tag{3.25}$$

are related to the curvature forms $\theta^{(i)}{}_{(j)}$ by

$$\Omega^{(i)(j)} = \theta^{(i)(j)} + R^{[(i)}{}_{(k)}\omega^{(j)]} \wedge \omega^{(k)} + \tfrac{1}{6}R\omega^{(i)} \wedge \omega^{(j)} \tag{3.26}$$

where R is the Ricci scalar (which is of course zero in the present case). The Weyl form may be tabulated as

$$\left.\begin{aligned}
\Omega^{(1)}{}_{(2)} &= -\frac{Mr - Q^2 - P^2}{r^4}\,\omega^{(1)} \wedge \omega^{(2)} & \Omega^{(0)}{}_{(3)} &= -\frac{Mr - Q^2 - P^2}{r^4}\,\omega^{(3)} \wedge \omega^{(0)}\\
\Omega^{(3)}{}_{(1)} &= \frac{Mr - Q^2 - P^2}{r^4}\,\omega^{(3)} \wedge \omega^{(1)} & \Omega^{(3)}{}_{(2)} &= 2\frac{Mr - Q^2 - P^2}{r^4}\,\omega^{(3)} \wedge \omega^{(3)}\\
\Omega^{(0)}{}_{(1)} &= 2\,\frac{Mr - Q^2 - P^2}{r^4}\,\omega^{(0)} \wedge \omega^{(1)} & \Omega^{(0)}{}_{(2)} &= \frac{Mr - Q^2 - P^2}{r^4}\,\omega^{(0)} \wedge \omega^{(2)}
\end{aligned}\right\} \quad (3$$

Again the form of this array makes it immediately clear to a connoisseur that the solution is of Petrov type D. The explicit form of the electromagnetic field is

$$F = \frac{2Q}{r^2}\, dr \wedge dt + 2P\, d\mu \wedge d\varphi \tag{3.28}$$

which may be derived via the relation

$$F = 2\, dA \tag{3.29}$$

from a vector potential A given by

$$A = \frac{Q}{r}\, dt - P\mu\, d\varphi \tag{3.30}$$

The explicit form of the metric itself is

$$ds^2 = r^2 \left\{ \frac{dr^2}{r^2 - 2Mr + Q^2 + P^2} + \frac{d\mu^2}{1-\mu^2} + (1-\mu^2)\, d\varphi^2 \right\} - \frac{r^2 - 2Mr + Q^2 + P^2}{r^2}\, dt \tag{3.31}$$

this being the solution of Riessner and Nordstrom, which of course includes the Schwarzschild solution in the limit when Q and P are set equal to zero.

Our search for spherical solutions is not quite complete at this point because by working with the spherical radius r as a co-ordinate, we have excluded the special case where r is a constant. Of course a solution with this property cannot be asymptotically flat, unlike the solutions which we have obtained so far, and therefore it might be thought not to have much physical interest. However as we shall see in the next section, it is impossible to have a full understanding of the global structures of the solutions we have obtained so far, without considering this special case, which arises naturally in a certain physically interesting limit.

To deal with this special case we must alter the canonical metric form (3.2) by replacing the co-ordinate r by a new co-ordinate, λ say, *except* in the conformal factor r^2 outside the large brackets which remains formally as before, and is now to be held constant.

Using the same methods as before, we find that there are no pure vacuum solutions of this form, but that there do exist source free electromagnetic solutions, which can be expressed as follows: The electromagnetic field takes the form

$$F = 2Q\, d\lambda \wedge dt + 2P\, d\mu \wedge d\varphi \tag{3.32}$$

which can be derived from a vector potential

$$A = Q\lambda\, dt - P\mu\, d\varphi \tag{3.33}$$

and the metric is given by

$$ds^2 = (Q^2 + P^2) \left\{ \frac{d\lambda^2}{\lambda^2} + \frac{d\mu^2}{1-\mu^2} + (1-\mu^2)\, d\varphi - \lambda^2\, dt^2 \right\} \tag{3.34}$$

This is the solution of Robinson and Bertotti. This metric is in fact almost as symmetric as possible, since it can easily be seen that it is the direct product of a 2-sphere whose radius is the square root of $Q^2 + P^2$ with a pseudo sphere (cf the form (2.10)) of the same radius. Its maximal symmetry group therefore has not four parameters as in the previous solutions (three for sphericity and one for stationarity) but six.

4 Maximal Extensions of the Spherical Solutions

It is relatively easy to analyse the global structures of the solutions which we have just derived, since, as indicated by (3.1), all of them are equivalent, modulo a conformal factor r^2, to the direct product of a 2-sphere with a timelike 2-surface. Thus the problem boils down to an analysis of the timelike 2-surfaces with metric $ds_\perp^2 = r^2 dl^2$, whose structures can be represented by the simple conformal diagrams whose use was described in section (1).

The simplest case is of course simply flat space to which the Schwarzschild form (3.17) reduces when the mass parameter is set equal to zero. The corresponding timelike 2-dimensional metric is simply $dr^2 - dt^2$ with the restriction $r > 0$, which, via the transformation $u = r - t$, $v = r + t$, is equivalent to the flat null form $2\, du\, dv$, with the restriction $u + v > 0$. The corresponding conformal diagram is given in Figure 4.1, in which the null boundaries $\mathscr{I}^+$ and $\mathscr{I}^-$ which play a key role in the Penrose definition of asymptotic flatness are marked. To qualify as asymptotically flat in the Penrose sense—which he refers to as the condition of weak asymptotic simplicity—a spacetime manifold must be conformally equivalent to an extended manifold-with-boundary with well behaved null boundary horizons $\mathscr{I}^+$ and $\mathscr{I}^-$ isomorphic to those of flat space. (Asymptotic flatness is discussed in more detail in the accompanying course of Hawking.)

The Schwarzschild solution is of course asymptotically flat in this sense, and indeed when M is negative the conformal diagram of the metric

$$ds_\perp^2 = \frac{r^2\, dr^2}{r^2 - 2Mr} - \frac{r^2 - 2Mr}{r^2}\, dt^2 \tag{4.1}$$

is topologically identical to the flat space diagram of Figure 4.1, although there is the important geometric difference that whereas in the former case the boundary $r = 0$ represented only a trivial co-ordinate degeneracy at the spherical centre, in the latter case, as can be seen from a glance at the array (3.16) it represents geometric singularity of the Riemann tensor. In the physically more interesting case where M is positive the Schwarzschild conformal diagram still agrees with the flat space diagram for large values of r, in accordance with the Penrose criterion, but it has an entirely different behaviour going in the other direction due to the fact that the metric form (3.17) becomes singular not only at $r = 0$ but also at $r = 2M$. The fact that the Riemann components, as exhibited in the array (3.16), are perfectly well behaved there suggests that this may not be a true geometric singularity but merely a removable co-ordinate singularity of the Killing horizon type which we have already come across in section 2.

It is not difficult to verify that this is indeed the case. By introducing just one null co-ordinate v, defined by

$$v = t + \{r + 2M \log |r - 2M|\} \tag{4.2}$$

we obtain the form

$$ds_{\perp}^2 = 2\, dv\, dr - \frac{r^2 - 2Mr}{r^2} dv^2 \tag{4.3}$$

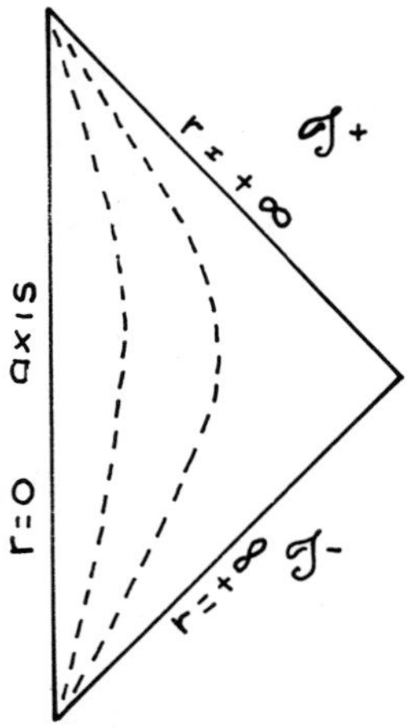

Figure 4.1. Conformal diagram of timelike 2-section with constant spherical co-ordinates θ, φ of Minkowski space.

This metric form is well behaved over the whole co-ordinate range $0 < r < \infty$ $-\infty < v < \infty$. The removability of the co-ordinate singularity at $r = 2M$ was first pointed out by Lemaitre (1933). The type of transformation we have used here in which the manifest stationary symmetry is preserved is due to Finkelstein (1958). The conformal diagram of this Finkelstein extension is shown in Figure 4.2.

The Finkelstein extension is not of course maximal since we could equally well have extended in the past direction, by introducing the alternative null co-ordinate

$$u = t - \{r + 2M \log |r - 2M|\} \tag{4.4}$$

and thereby obtaining a symmetric but distinct metric extension:

$$ds_{\perp}^2 = -2\, du\, dr - \frac{r^2 - 2Mr}{r^2} du^2 \tag{4.5}$$

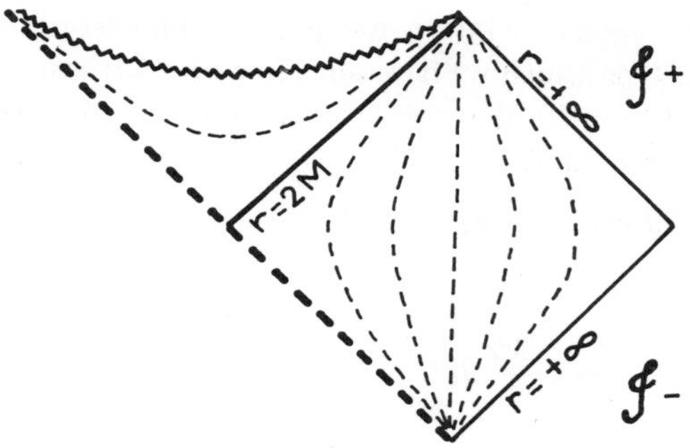

Figure 4.2. Conformal diagram of timelike 2-section with constant spherical co-ordinates θ, φ of Finkelstein extension of Schwarzschild manifold. In this figure and in all the conformal diagrams of section 2 and section 4 the convention is employed that the Killing vector trajectories (which in this case are the curves on which r is constant) are marked by light dotted lines, while Killing horizons are marked by heavy lines, except in that degenerate Killing horizons are marked by double lines. Curvature singularities are indicated by zig-zag lines.

In fact we can make a sequence of such extensions alternating between outgoing and ingoing null lines. After four such Finkelstein extensions, we can arrange to get back to our original starting point, in the manner shown in Figure 4.3. None of the four patches covers the central crossover point in the diagram, but our

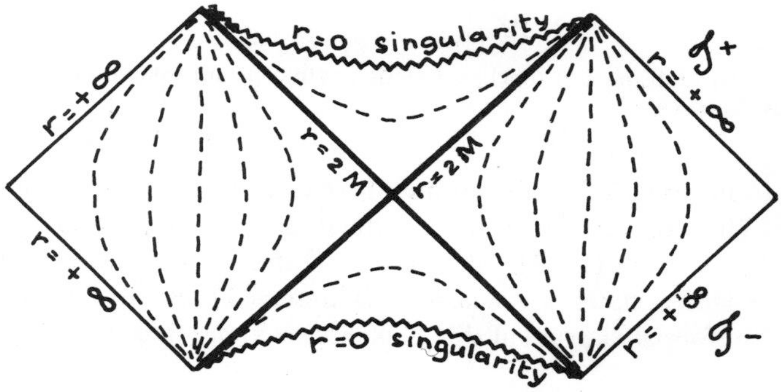

Figure 4.3. Conformal diagram of timelike 2-section with constant spherical co-ordinates θ, φ of Krushal's maximally extended Schwarzschild's solution.

experience with the analogous Killing horizon crossover points in the pseudo-sphere, suggests that just as the extension from (2.4) to (2.10) could be augmented to the maximal extension (2.19), so also here there should be a further extension to cover the crossover point. This is indeed the case, but unfortunately this time in dropping the manifest symmetry represented by the ignorability of the co-ordinate t we cannot expect to obtain any new compensating manifest symmetry as we did in the previous case, and so the new extended system will not be very elegant.

To construct the required extension we first introduce the null co-ordinates u, v simultaneously to obtain the null form

$$ds_\perp^2 = -\frac{r^2 - 2Mr}{r^2}\, du\, dv \tag{4.6}$$

where now r is given implicitly as the solution of

$$r + 2M \log |r - 2M| = \tfrac{1}{2}(u + v) \tag{4.7}$$

At this point we are even worse off than with the original Finkelstein extension, since we have restored the co-ordinate degeneracy at $r = 2M$. However we can now easily make an extension both ways at once by setting

$$u = -4M \ln |U| \tag{4.8}$$

$$v = 4M \ln |V| \tag{4.9}$$

which gives

$$ds_\perp^2 = \frac{e^{-r/2M}\, dU\, dV}{r} \tag{4.10}$$

where r is given implicitly as a function of U and V by

$$(r - 2M)\, e^{r/2M} = -UV \tag{4.11}$$

and where if required, the original time co-ordinate t can be recovered using the equation

$$e^{t/2} = -\frac{V}{U} \tag{4.12}$$

This form is well behaved in the whole of the UV plane. The original static patch is determined by $U < 0$, $V > 0$ and the original Finkelstein extension is determined by $V > 0$. It is fairly obvious that this extension is maximal since most geodesics either can be extended into the asymptotically flat region or to the geometrically singular limit. However to prove strictly that it is maximal we should check that there are no incomplete geodesics lying on or tending to the Killing horizons $U = 0$ and $V = 0$. We shall cover this point later on when we discuss geodesics. [Checking that all geodesics are either complete or tend to a curvature singularity is a standard way of proving that an extension is maximal, but it is not the only way; for example a compact manifold must clearly be inextensible even though it may—as was pointed out by Misner (1963)—contain incomplete geodesics.] This maximal extension was first published—in a somewhat different co-ordinate system—by Kruskal (1960), surprisingly recently when one considers that the Schwarzschild solution, was discovered in 1916.

For studying the stationary exterior field of a collapsed star, that is to say the classical (if it is not premature to use such a word) black hole situation, the Finkelstein extension is sufficient, as is indicated by the conformal diagram given in Figure 4.4. The full Kruskal extension has the feature, which seems unlikely to be relevant except in a rather exotic situation, of possessing *two distinct asymptotically flat parts*, which are connected by wha' has come to be known as a bridge. The nature of the bridge can be under' ›od by considering the geometry of the three dimensional space sections, e.g. the locus $U = V$ (which coincides with $t = 0$ in the original co-ordinate system) whose geometry is suggested by Figure 4.5 which is meant to illustrate an imbedding in 3-dimensional

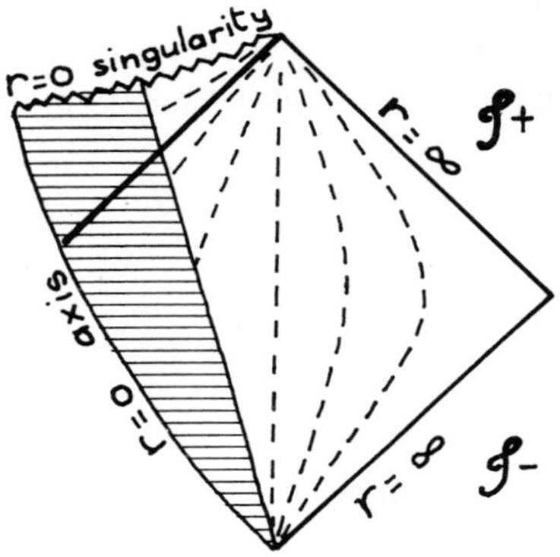

Figure 4.4. Conformal diagram of timelike 2-section with constant spherical co-ordinates θ, φ of maximally extended space-time of a collapsing spherically symmetric star.

flat space of the 2-dimensional circumferential section $\theta = \pi$ of the locus $U = V$. These three sections have spherical cross sections whose radius diminishes to a minimum value $r = 2M$ (at the throat of the bridge) and then increases again without bound. It can be seen from the conformal diagram of Figure 4.3 that an observer on a timelike trajectory cannot in fact cross this throat, since after crossing the horizon $U = 0$, the co-ordinate r must inevitably continue to decrease, until after a finite proper time the observer hits the geometric singularity at $r = 0$; this means in fact not only that such an observer is unable to reach the region of expanding r on the other side of the throat, but also that he is unable to return or even send a signal to the regions $r > 2M$ on the side from which he came. It is this latter phenomenon which is apparent even on the restricted Finkelstein extension (Figure 4.2), and in the dynamically realistic collapse diagram (Figure 4.4) which justifies the description of the null hypersurface at $r = 2M$ as a *horizon.* Thus technically, the hypersurface $U = 0$ is the past event horizon of $\mathscr{I}^+$, and together with the hypersurface $V = 0$ bounding the future of $\mathscr{I}^-$, it forms the boundary of the *domain of outer communications* which we shall denote by $\langle\!\langle \mathscr{I} \rangle\!\rangle$ meaning the region which can both receive signals from, and send them back to, asymptotically large distances which in the present case is the connected region with $r > 2M$ specified by $U < 0$, $V > 0$. The horizon $U = 0$ has the *infinite red shift* property (which is quite generally characteristic of a past event horizon of $\mathscr{I}^+$) that the light emitted by any physical object which crosses it is spread out over an infinite time as seen by a stationary observer in the asymptotically flat region, and thus not only gets infinitely red shifted as the body approaches the horizon (which of course it can reach in a finite proper time from its own point of view) but also progressively fades out. Thus the body, which might for example be the collapsing star represented on Figure 4.4, will

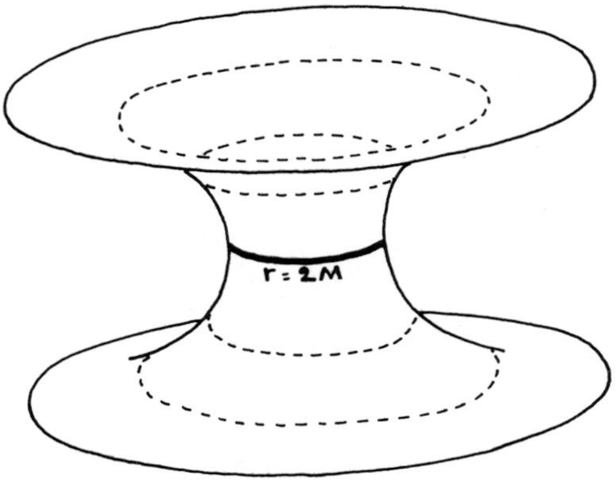

Figure 4.5. Sketch of part of a space-like equatorial 2-section ($\cos\theta = 0$) of one of the t = constant hypersurfaces through the throat of Kruskal's maximally extended Schwarzschild manifold. Trajectories of the Killing vector $\partial/\partial\varphi$ are indicated by light dotted lines.

appear to become not only redder but also blacker until with a characteristic time of the order of that required for light to cross the Schwarzschild radius ($\sim 10^{-4}$ seconds in the case of a collapsing star of typical mass) it effectively fades out of view altogether. It is for this reason that the region inside the horizon is referred to as a black hole.

Of course in the more exotic situation when the full Kruskal extension is present, the region of the past of $V = 0$ is by no means invisible from outside, and it would in fact be possible to see right back to the singularity (unless of course, like the big bang of cosmology theory, it were hidden in some opaque cloud of particles). The region to the past of $V = 0$ is often referred to as a *white* hole, although with less justification than the application of the description black to the future of $U = 0$. Without worrying about the question of colour, it is obviously reasonable in all cases to describe the regions outside the domain of outer communications $\langle\!\langle \mathscr{I} \rangle\!\rangle$ as *holes.*

Let us now move on to see how the situation is modified in the electromagnetic case. In the Reissner-Nordstrom solutions the timelike 2-sections have the metric form

$$ds_{\perp}^2 = \left(\frac{r^2}{r^2 - 2Mr + Q^2 + P^2}\right) dr^2 - \left(\frac{r^2 - 2Mr + Q^2 + P^2}{r^2}\right) dt^2 \tag{4.13}$$

When the charge is large, more precisely when $Q^2 + P^2 > M^2$, this metric is well behaved over the whole range from the singularity at $r = 0$ (which from the array (3.26) can be seen to be a genuine geometric curvature singularity) to the asymptotic limit $r \to \infty$, and the conformal diagram is the same as in the negative M

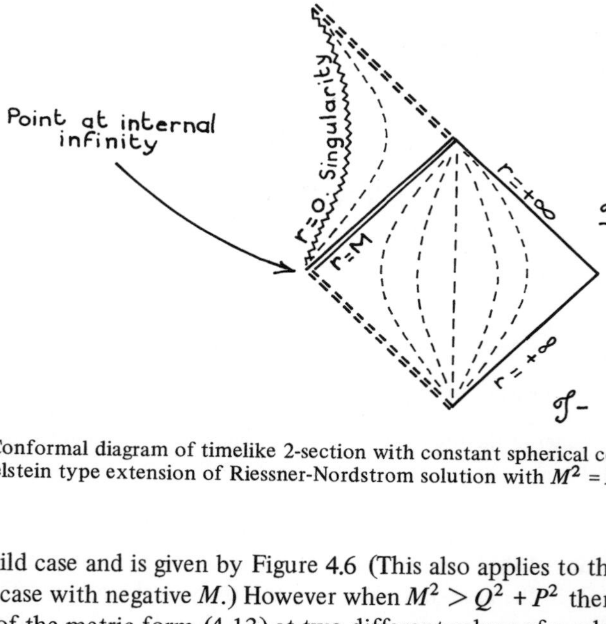

Figure 4.7. Conformal diagram of timelike 2-section with constant spherical co-ordinates θ, φ of Finkelstein type extension of Riessner-Nordstrom solution with $M^2 = P^2 + Q^2$.

Schwarzschild case and is given by Figure 4.6 (This also applies to the Reissner-Nordstrom case with negative M.) However when $M^2 > Q^2 + P^2$ there is a singularity of the metric form (4.13) at two different values of r, which we shall label r_+ and r_-, and which are given by

$$r_{\pm} = M \pm \sqrt{M^2 - Q^2 - P^2} \tag{4.14}$$

We can use this notation to write the metric in the more convenient form

$$ds_{\perp}^2 = \frac{r^2}{(r - r_+)(r - r_-)} dr^2 - \frac{(r - r_+)(r - r_-)}{r^2} dt^2 \tag{4.15}$$

As is suggested by the regularity of the Weyl curvature array (3.26), these singularities are removable in the manner with which we are beginning to become familiar, as can easily be seen by making the obvious Finkelstein type transformation

$$v = t + r + \frac{r_+^2}{2M} \ln |r - r_+| - \frac{r_-^2}{2M} \ln |r - r_-| \tag{4.16}$$

which leads to the form

$$ds_\perp^2 = 2\, dv\, dr - \frac{(r - r_-)(r - r_+)}{r^2} dv^2 \tag{4.17}$$

which is clearly well behaved in the whole range $0 < r < \infty$. In this system there is an inner static domain $0 < r < r_-$ in addition to the outer static domain as is shown in the conformal diagrams of Figure 4.7. In the limit as the charges Q and P tend to zero the inner horizon $r = r_-$ collapses down on to the curvature singularity $r = 0$ and the inner domain is squeezed out of existence, so that the manifold goes over continuously to the Schwarzschild Finkelstein extension. The way in which this limiting process takes place is illustrated in the conformal diagram of Figure 4.8 in which the freedom to adjust the conformal factor is used to represent the Schwarzschild Finkelstein extension in a manner which approaches the zero charge limit of the Reissner-Nordstrom Finkelstein extension.

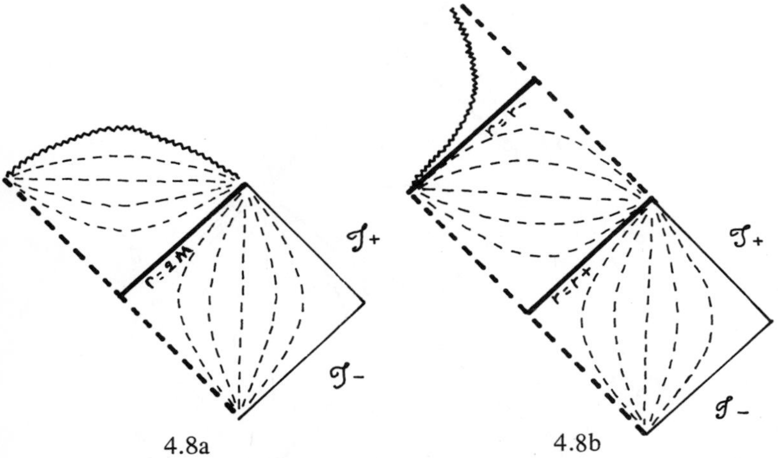

4.8a 4.8b

Figure 4.8. Figure 4.8a is a modified version of Figure 4.2 and Figure 4.8b is a modified version of Figure 4.7 showing how the Reissner-Nordstrom solution approaches the Schwarzschild solution as $P^2 + Q^2 \to 0$.

As in the Schwarzschild case, we can make a Finkelstein type extension not only in the forward direction but also in the backward, by introducing the outgoing null co-ordinate

$$u = t - r - \frac{r_+^2}{2M} \ln |r - r_+| + \frac{r_-^2}{2M} \ln |r - r_-| \tag{4.18}$$

which leads to the form

$$ds_{\perp}^2 = -2\,du\,dr - \frac{(r - r_+)(r - r_-)\,du^2}{r^2} \tag{4.19}$$

By patching together forward and backward going extensions of this kind we can build up what is in fact a maximal extension in the manner illustrated in the

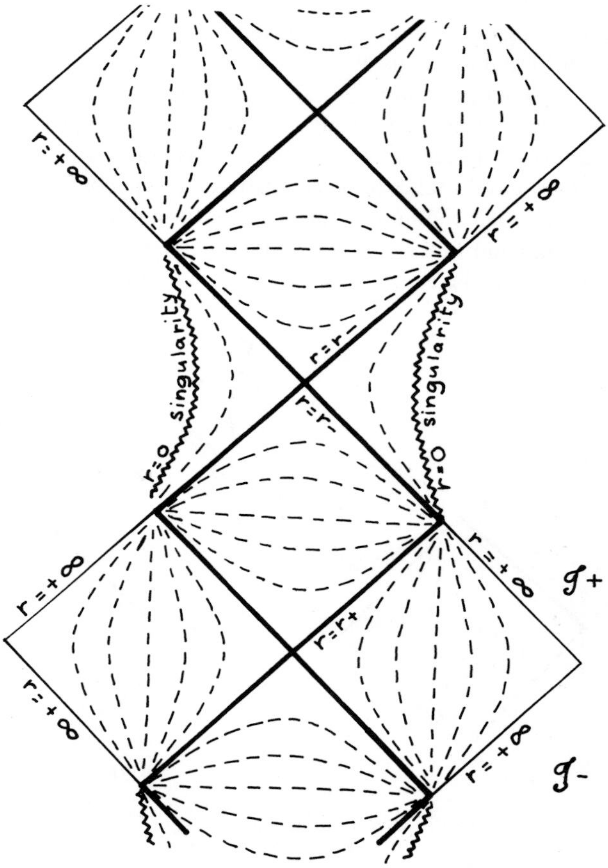

Figure 4.9. Conformal diagram of timelike 2-section with constant spherical co-ordinates θ, φ of maximally extended Riessner-Nordstrom solution with $M^2 > P^2 + Q^2$.

conformal diagram of Figure 4.9 to obtain an infinite double chain of asymptotically flat universes linked together in pairs by wormholes. This manifold was first described by Graves and Brill (1960). As in the Schwarzschild case it is possible to construct Kruskal type co-ordinate extensions to cover the crossover points of the horizons. It is possible to construct a co-ordinate system covering the whole manifold at once, but it is simpler to construct local patches adapted to the kinds of horizon $r = r_+$ and $r = r_-$ separately. The notation we are using enables us to describe both kinds of patch at once as follows. Starting as in the previous case from the double null form.

$$ds_\perp^2 = -\frac{(r - r_+)(r - r_-)}{r^2}\, du\, dv \tag{4.20}$$

where r is given implicitly as a function of u and v by

$$r + \frac{r_+^2}{2M} \ln |r - r_+| - \frac{r_-^2}{2M} \ln |r - r_-| = \tfrac{1}{2}(v - u) \tag{4.21}$$

we introduce new co-ordinates U^+, V^+ or U^-, V^- depending on whether we want to remove the singularity at $r = r_+$ or $r = r_-$, defined by

$$\left.\begin{aligned} u &= \mp \frac{r_\pm^2}{M} \ln |-U^\pm| \\ v &= \pm \frac{r_\pm^2}{M} \ln |+V^\pm| \end{aligned}\right\} \tag{4.22}$$

which leads to the form

$$ds_\perp^2 = -\left(\frac{r_\pm}{M}\right)^2 \exp\left(-\frac{2Mr}{r_\pm^2}\right) |r - r_\mp|^{(r_\pm^2 + r_\mp^2)/r_\pm^2} \frac{dU^\pm\, dV^\pm}{r^2} \tag{4.23}$$

where r is given implicitly as a function of $U^\pm$, $V^\pm$ by

$$|r - r_\pm| \exp\left(\pm\frac{2Mr}{r_\pm^2}\right) |r - r_\mp|^{-r_\mp^2/r_\pm^2} = -U^\pm V^\pm \tag{4.24}$$

and the co-ordinate t, if required, is given by

$$\exp\left(\frac{2M}{r_\pm^2}\, t\right) = -\frac{V^\pm}{U^\pm} \tag{4.25}$$

It can be seen that these forms are well behaved in the neighbourhood of $r = r_\pm$ respectively, (although not in the neighbourhood of $r = r_\mp$).

In the Graves and Brill manifold the time symmetric 3-dimensional space sections (t = constant) fall into two classes, as illustrated in Figure 4.10 which shows imbeddings in 3-dimensional flat space of 2-dimensional sections. The first kind, exemplified by the locuses $U^+ + V^+ = 0$, are topologically similar to the Kruskal bridges connecting two asympototically flat regions except that the throat is somewhat narrower, and can be (if the charges are larger) very much longer. The second kind, as exemplified by the locus $U^- + V^- = 0$, represents a sort of tube connecting two curvature singularities $r = 0$; only part of the tube is illustrated since the flat space imbedding breaks down before the curvature singularities are reached.

Let us pause for a moment to consider the implications of what we have seen so far for a gravitational collapse situation. We notice that a material sphere can only undergo catastrophic gravitational collapse if the condition $Q^2 + P^2 < M^2$ (or simply $Q^2 < M^2$ provided we take for granted the physical condition that there are no material magnetic monopoles) is satisfied since otherwise electromagnetic repulsion would predominate over gravitational attraction, and that it

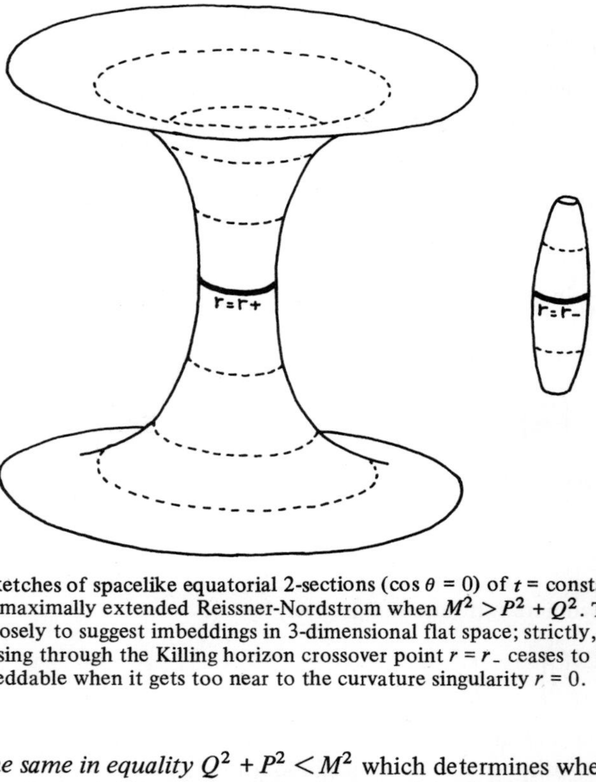

Figure 4.10. Sketches of spacelike equatorial 2-sections ($\cos\theta = 0$) of t = constant hypersurfaces in the maximally extended Reissner-Nordstrom when $M^2 > P^2 + Q^2$. The sketches are intended loosely to suggest imbeddings in 3-dimensional flat space; strictly, however, the section passing through the Killing horizon crossover point $r = r_-$ ceases to be symmetrically imbeddable when it gets too near to the curvature singularity $r = 0$.

is *precisely the same in equality* $Q^2 + P^2 < M^2$ which determines whether a horizon will be formed. This suggests what Penrose (1969) has termed the *cosmic censorship hypothesis* according to which, in any situation arising from the gravitational collapse of an astrophysical object (such as the central part of a star after a supernova explosion) starting from a well behaved initial situation, the singularities which result are hidden from outside by an event horizon i.e. that *naked singularities* which can both be approached from and seen from outside (as in the case of the singularity at $r = 0$ in the Reissner-Nordstrom solutions with $Q^2 + P^2 > M^2$) *cannot arise naturally* from a well behaved initial situation. This hypothesis does not exclude the visibility of *preexisting* singularities (such as that of the big bang in cosmology theory not to mention the so called white holes) which cannot be reached in the future by a timelike trajectory and in this sense are not completely naked.

Let us now move on to consider the special case of the Reissner-Nordstrom solutions for which the critical condition $Q^2 + P^2 = M^2$ is satisfied, i.e. which are poised between the normal hole situation $Q^2 + P^2 < M^2$ and the naked singularity situation $Q^2 + P^2 > M^2$. According to the cosmic censorship hypothesis a critical case such as this should represent a physically unattainable limit, but one which is approachable and therefore of considerable interest in the same way as for example the extreme relativistic limit is of great interest in particle scattering theory.

The metric in this limiting case simply takes the form

$$ds_{\perp}^2 = \frac{(r-M)^2}{r^2}\, dr^2 - \frac{r^2}{(r-M)^2}\, dt^2 \tag{4.26}$$

This metric form is static *both* for $r > M$ *and* for $r < M$ but it is none the less singular at $r = m$. Its extension was first discussed by myself (1966). The co-ordinate singularity is in fact the first example in the present section of a Killing horizon which is *degenerate* in the sense described in section 2, as would be expected when it is thought of as the limiting case in which the two non-degenerate Killing horizons $r = r_+$ and $r = r_-$ have coalesced.

As usual the Finkelstein type extension can be carried out without difficulty by introducing the null co-ordinate

$$v = t + (r - M) + 2M \ln |r - M| - \frac{M^2}{r - M} \tag{4.27}$$

which leads to the metric form

$$ds_{\perp}^2 = 2\, dr\, dv - \frac{(r-M)^2}{r^2}\, dv^2 \tag{4.28}$$

The corresponding co-ordinate patch which is well behaved over the whole range $0 < r < \infty$, $-\infty < v < \infty$, is illustrated in Figure 4.10, in which we have also shown the way in which this limit is approached by the Finkelstein extensions of the ordinary Riessner-Nordstrom metrics as $Q^2 + P^2$ approaches M^2 from below.

As usual we can also make the symmetric past extension, by introducing a co-ordinate

$$u = t - (r - M) - 2M \ln |r - M| + \frac{M^2}{r - M} \tag{4.29}$$

thus obtaining the form

$$ds_{\perp}^2 = -2\, du\, dr - \frac{(r-M)^2}{r^2}\, dr^2 \tag{4.30}$$

This time however there is no way of carrying out a further Kruskal type extension, because the transformation (4.27) contains not only the by now familiar logarithmic singularity but also a first order pole singularity. However it turns out that a Kruskal type extension is quite unnecessary, since the Finkelstein extensions can be fitted together in the manner illustrated in Figure 4.1.2 to form an extended manifold which is in fact maximal since (as we shall verify later) all geodesics either intersect the curvature singularity or can be completed. The situation is rather different from those which we have come across so far, because

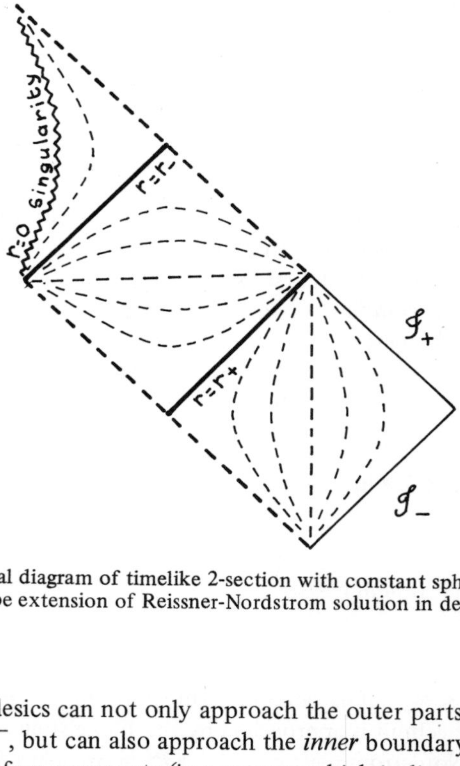

Figure 4.11. Conformal diagram of timelike 2-section with constant spherical co-ordinates θ, φ of Finkelstein-type extension of Reissner-Nordstrom solution in degenerate case when $M^2 = P^2 + Q^2$.

the completed geodesics can not only approach the outer parts of the diagram labelled $\mathscr{I}^+$ and $\mathscr{I}^-$, but can also approach the *inner* boundary points labelled x, each of which therefore represents (in a manner which is disguised by the conformal factor) infinitely distant limit in the inward direction.

The nature of the limits represented by the points x, (which take the place of the missing Kruskal crossover axes) can best be understood by considering the time symmetric space sections, t = constant. In this case the analogue of Figure 4.10 is given by Figure 4.13 which shows that instead of having a minimum (the throat) or a maximum of r in the respective cases $r > M, r < M$, the space sections extend indefinitely, both approaching asymptotically the same infinite spherical cylinder.

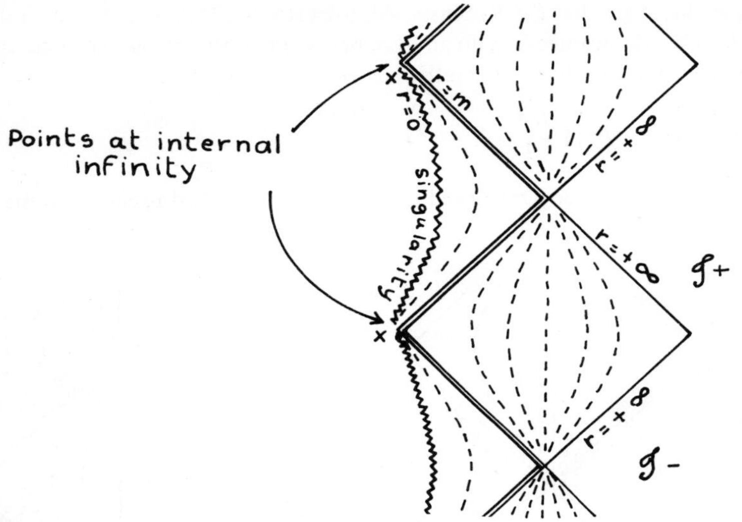

Figure 4.12. Conformal diagram of timelike 2-section with constant spherical co-ordinates θ, φ of maximal extension of Reissner-Nordstrom solution in degenerate case when $M^2 = P^2 + Q^2$.

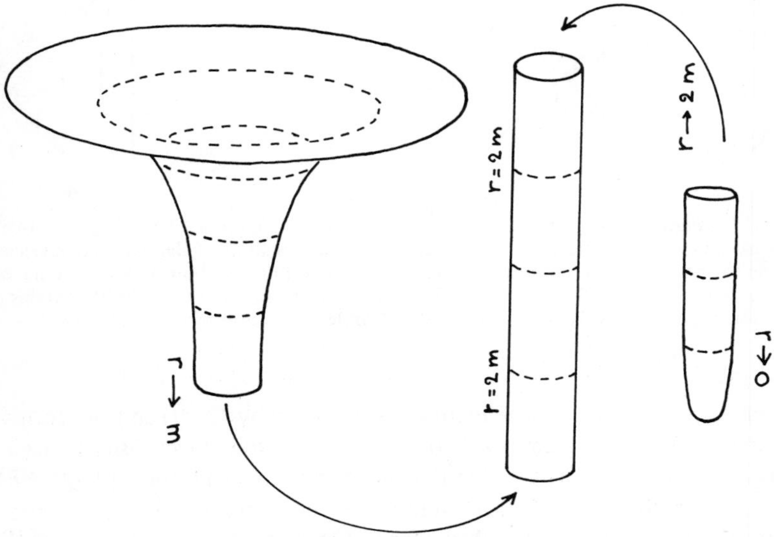

Figure 4.13. Sketches of spacelike equatorial 2-sections ($\cos \theta = 0$) of t = constant hyper surfaces in the maximally extended Reissner-Nordstrom solution in the case $M^2 = P^2 + Q^2$ and also in a Robertson-Bertotti universe, indicating how the latter represents an asymptotic limit of the former.

It is at this stage that the Robinson-Bertotti solution (3.34) enters into the discussion. We do not need to do any further work to find its global structure, since its time sections have the metric

$$ds_{\perp}^2 = (Q^2 + P^2)\left(\frac{d\lambda^2}{\lambda^2} - \lambda^2\, d\tau^2\right) \tag{4.31}$$

which, as we have already remarked is the metric of a pseudo-sphere, so that the

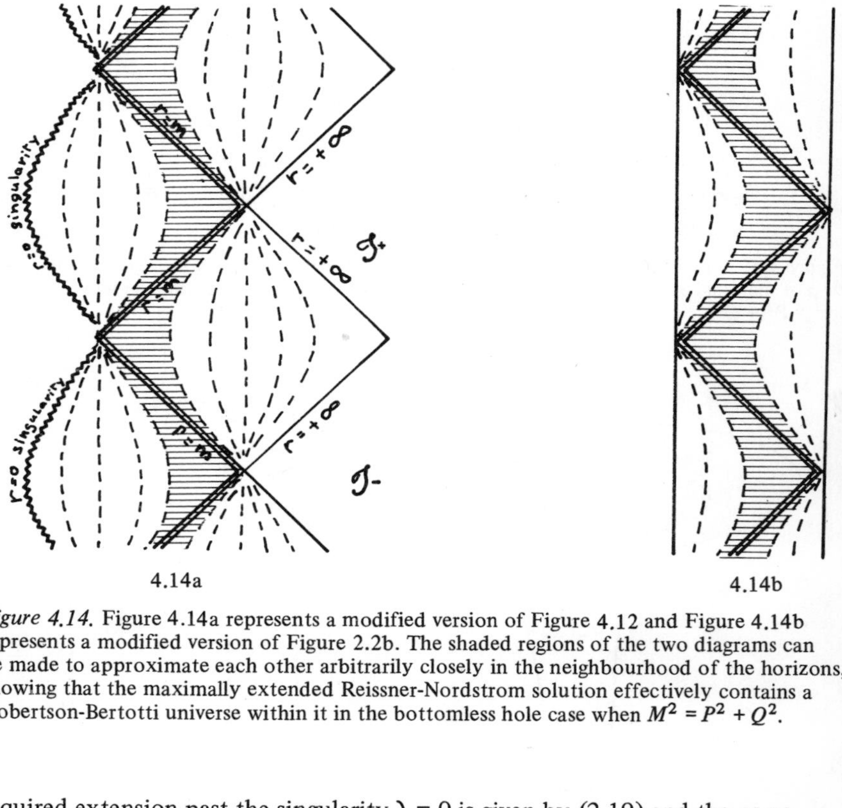

Figure 4.14. Figure 4.14a represents a modified version of Figure 4.12 and Figure 4.14b represents a modified version of Figure 2.2b. The shaded regions of the two diagrams can be made to approximate each other arbitrarily closely in the neighbourhood of the horizons, showing that the maximally extended Reissner-Nordstrom solution effectively contains a Robertson-Bertotti universe within it in the bottomless hole case when $M^2 = P^2 + Q^2$.

required extension past the singularity $\lambda = 0$ is given by (2.19) and the corresponding conformal diagram is that of Figure 2.2b. Now it can easily be seen by setting $r = \lambda + M$, $t = M^2\tau$, that the Reissner-Nordstrom solution with $Q^2 + P^2 = M^2$ approaches the Robertson-Bertotti solution in the asymptotic limit as $r \to \infty$. The limiting spherical cylinder illustrated in Figure 4.13 is in fact the same as a τ = constant cross section of a Robinson-Bertotti universe. The conformal diagrams of the $Q^2 + P^2 = M^2$ Reissner-Nordstrom solution, and of the Robinson-Bertotti solution are shown side by side for comparison in Figure 4.14. The shaded regions

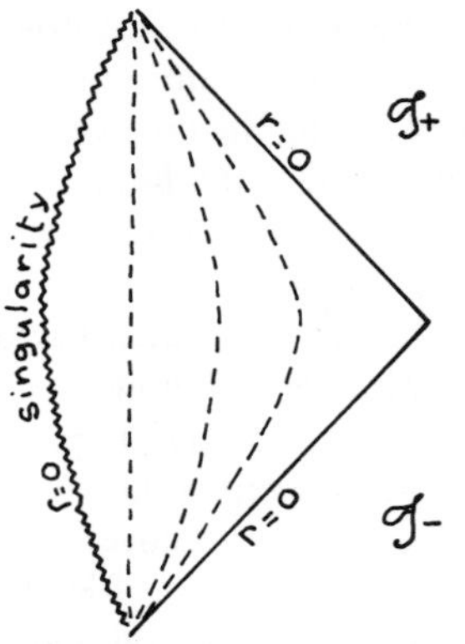

Figure 4.6. Conformal diagram of timelike 2-section with constant spherical co-ordinates θ, φ in the maximally extended Reissner-Nordstrom solution in the naked singularity case when $M^2 < P^2 + Q^2$.

can be made to coincide as closely as one pleases by adjusting the conformal factor so that they correspond to a sufficiently small range of the co-ordinate $\lambda = r - M$.

5 Derivation of the Kerr Solution and its Generalizations

Having now looked fairly thoroughly at the spherical vacuum solutions, we have clearly arrived at a stage where it would be interesting to see how the various phenomena—horizons, naked singularities etc.—which we have come across would be modified in more general non-spherical situations, particularly when angular momentum, the most obvious source of deviation from spherical symmetry, is present.

Since, as we have seen, the derivation of the spherical Schwarzschild solution is very easy (it was achieved in 1916 within a year of the publication of Einstein's Theory in 1915) one might have guessed that it would be comparatively not too difficult to derive a rotating (and hence non-static but still stationary) generalization in a fairly straightforward manner, starting from some suitably simple and natural canonical metric form. Moreover, as I plan to make clear in this section, one would have guessed correctly. Nevertheless, after forty years, repeated attempts to find a canonical form leading to a natural vacuum generalization of the Schwarzschild solution had turned up nothing (or more precisely nothing which was asymptotically flat) except the Weyl solutions, which are unfortunately static, and therefore useless in so far as showing the effect of angular momentum is concerned. Moreover the first (and so far the only) non-static pure vacuum generalization of the Schwarzschild solution was found at last by Kerr in 1963 using a method which is by no means straightforward, and which arose as a bi-product of the sophisticated Petrov-Pirani approach to gravitational radiation theory which was developed during the nineteen fifties. In consequence of this history it is still widely believed that the Kerr solution can only be derived using advanced modern techniques. There is however an elementary approach of the old fashioned kind which was rather surprisingly overlooked by the searchers in the nineteen twenties and thirties, and which I actually found myself, with the aid of hindsight, in 1967. This approach now seems so obvious that I am sure it will be clear to anyone who follows it that despite the apparent messiness of the form in which it is customarily presented, no non-static generalization of the Schwarzschild solution which may be discovered in the future can possibly be simpler in its algebraic structure than that of Kerr.

This approach starts from the observation that for practical computational purposes one of the most useful, indeed almost certainly *the* most useful, algebraic property which the Schwarzschild solution possesses as a consequence of spherical symmetry, and which one might hope to return in a simple non-spherical generalization, is that of separability of its Dalembertian wave equation and the associated integrability of its geodesic equations.

Now it is physically evident from the correspondence principle of quantum mechanics, (and it follows mathematically from standard Hamilton Jacobi theory) that integrability of the geodesics as well, obviously, as separability of the Dalembertian wave equation $\psi^{;a}{}_{;a} = 0$, will follow if the slightly more general Klein-Gordon wave equation $\psi^{;a}{}_{;a} - m^2\psi = 0$ (where m^2 is a freely chosen constant which may be interpreted as a squared test particle mass) is separable.

Separability is of course something which depends not only on the geometry but on a particular choice of co-ordinates x^2 ($a = 0, 1, 2, 3$), and in terms of these co-ordinates it depends less directly on the form of the ordinary covariant tensor g_{ab} defined by

$$ds^2 = g_{ab}\, dx^a\, dx^b \tag{5.1}$$

than on the form of the contravariant metric tensor g^{ab} defined in terms of the inverse co-form

$$\left(\frac{\partial}{\partial s}\right)^2 = g^{ab}\,\frac{\partial}{\partial x^a}\,\frac{\partial}{\partial x^b} \tag{5.2}$$

In terms of the contravariant metric components and of the determinant

$$g = \det(g_{ab}) = \{\det(g^{ab})\}^{-1} \tag{5.3}$$

the Klein Gordon equation can be expressed in terms of simple partial derivatives in the form

$$\psi^{-1}\frac{\partial}{\partial x^a}\sqrt{-g}\, g^{ab}\,\frac{\partial}{\partial x^b}\,\psi - m^2\sqrt{-g} = 0 \tag{5.4}$$

The standard kind of separability takes place if substitution of the product expression

$$\psi = \prod_i \psi_i \tag{5.5}$$

where each function $\psi_i (i = 0, 1, 2, 3)$ is a function of just the single variable x^i, causes the left hand side of (5.4) to split up into four independent single variable ordinary differential equations, expressed in terms of four independent freely chosen constants of which $\bar{m}^2$ is one.

To see how this works out in the spherical case, we note that the inverse metric corresponding to our general spherical canonical form (3.2) is

$$\left(\frac{\partial}{\partial s}\right)^2 = \frac{1}{r^2}\left\{(1-\mu^2)\left(\frac{\partial}{\partial \mu}\right)^2 + \frac{1}{1-\mu^2}\left(\frac{\partial}{\partial \varphi}\right)^2 + \Delta_r\left(\frac{\partial}{\partial r}\right)^2 - \frac{Z_r^2}{\Delta_r}\left(\frac{\partial}{\partial t}\right)^2\right\} \tag{5.6}$$

and hence using (3.3), the Klein Gordon equation takes the form

$$\frac{r^2}{Z_r}\left\{\psi^{-1}\frac{\partial}{\partial \mu}(1-\mu^2)\frac{\partial \psi}{\partial \mu} + \psi^{-1}\frac{1}{1-\mu^2}\frac{\partial^2}{\partial \varphi^2}\psi - m^2 r^2\right\}$$

$$+\psi^{-1}\frac{\partial}{\partial r}\frac{r^2\Delta_r}{Z_r}\frac{\partial \psi}{\partial r} - \psi^{-1}\frac{rZ_r}{\Delta_r}\frac{\partial^2 \psi}{\partial t^2} = 0 \tag{5.7}$$

which has solutions of the form

$$\psi = R_r P_n^l(\mu) e^{in\varphi} e^{i\omega t} \tag{5.8}$$

where l, n, ω are separation constants (of which l and n must be integers if the solution is to be regular) and $P_n^l(\mu)$ is a solution of the (l, n) associated Legendre equation (and is thus an associated Legendre-polynomial in the regular case) and where R_r is a solution of the equation

$$R_r^{-1}\frac{d}{dr}\frac{r^2\Delta_r}{Z_r}\frac{dR_r}{dr} + \frac{\omega^2 r^2 Z_r}{\Delta_r} + \frac{r^2}{Z_r}[l(l+1) - m^2 r^2] = 0 \tag{5.9}$$

What we want to do now is find the simplest possible non-static generalization of the canonical coform (5.6), including in our criteria for simplicity not only the most obvious requirement of all, namely that the manifest symmetry property of stationarity and axisymmetry represented by the ignorability of the co-ordinates t and φ be retained but also the requirement that the separability property of the corresponding Klein Gordon equation be retained. There is a third obvious simplicity property of the coform (5.6) which can be retained without prejudice to the other two, (and which greatly simplifies the computation of the Riemann tensor etc.) namely the fact that it determines a natural canonical orthonormal tetrad, such that two of the tetrad vectors contribute in the separation only to the terms independent of r, while the other two contribute only to the terms independent of μ.

This leads us to try the canonical coform

$$\left(\frac{\partial}{\partial s}\right)^2 = \frac{1}{Z}\left\{\Delta_\mu\left(\frac{\partial}{\partial\mu}\right)^2 + \frac{1}{\Delta_\mu}\left[Z_\mu\frac{\partial}{\partial t} + Q_\mu\frac{\partial}{\partial\varphi}\right]^2\right\}$$
$$+\frac{1}{Z}\left\{\Delta_r\left(\frac{\partial}{\partial r}\right)^2 - \frac{1}{\Delta_r}\left[Z_r\frac{\partial}{\partial t} + Q_r\frac{\partial}{\partial\varphi}\right]^2\right\} \qquad (5.10)$$

where Δ_μ, Z_μ, Q_μ are functions of μ only, and where Δ_r, Z_r, Q_r are functions of r only, and the form of the conformal factor Z remains to be determined.

It is to be observed that the factors Δ_μ are redundant, since the one in the first term could be eliminated by renormalizing μ as a function of itself while the one in the second term could be eliminated by a proportional readjustment of Z_μ and Δ_μ. The same applies to the factor Δ_r. These factors have been included explicitly however firstly because they are suggested by the canonical spherical coform (5.6) (to which (5.10) reduces when one sets $\Delta_\mu = 1 - \mu^2$ with $\mu = 1$, $Q_\mu = Q_r = 0$ and $Z = r^2$) but also for the more compelling reason that the freedom of adjustment of Δ_μ and Δ_r can be used to achieve considerable simplification of the form of Z required to achieve separability. In any case they are arranged so as to cancel out of the determinant which is simply given by

$$\sqrt{-g} = \frac{Z^2}{|Z_r Q_\mu - Z_\mu Q_r|} \qquad (5.11)$$

Let us now investigate the conditions which must be imposed on Z to achieve separability. Using (5.10) and (5.11) we obtain the Klein Gordon equation in the form

$$\psi^{-1}\left\{\frac{\partial}{\partial\mu}\frac{\sqrt{-g}}{Z}\Delta_\mu\frac{\partial}{\partial\mu} + \frac{\partial}{\partial r}\frac{\sqrt{-g}}{Z}\Delta_r\frac{\partial}{\partial r} - \frac{\sqrt{-g}}{Z}m^2 Z\right\}\psi$$
$$+\psi^{-1}\left\{\frac{\partial}{\partial\varphi}Q_\mu + \frac{\partial}{\partial t}Z_\mu\right\}\frac{1}{\Delta_\mu}\left(\frac{\sqrt{-g}}{Z}\right)\left(Q_\mu\frac{\partial}{\partial\varphi} + Z_\mu\frac{\partial}{\partial t}\right)\psi$$
$$-\psi^{-1}\left(\frac{\partial}{\partial\varphi}Q_r + \frac{\partial}{\partial t}Z_r\right)\frac{1}{\Delta_r}\left(\frac{\sqrt{-g}}{Z}\right)\left(Q_r\frac{\partial}{\partial\varphi} + Z_r\frac{\partial}{\partial t}\right)\psi = 0 \qquad (5.12)$$

It is clear that if these terms are to separate, the factor $Z^{-1}\sqrt{-g}$ which occurs in each one must depend on r and μ only as a product of single variable functions which can be absorbed into Δ_r and Δ_μ (using our freedom to rescale these functions) so as to reduce the factor $Z^{-1}\sqrt{-g}$ to unity. Thus we are led to choose the definition

$$Z = Z_r Q_\mu - Z_\mu Q_r \tag{5.13}$$

for the conformal factor. This is still not quite sufficient for separability except in the case of the pure Dalembertian wave equation, since there remains the mass term which now takes the form $m^2 Z$. In this expression we have not only made no provision for cross terms between the non-ignorable co-ordinates, i.e. terms proportional to $(\partial/\partial r)(\partial/\partial\mu)$, which would obviously destroy the separability, but we have also, in accordance with our principle of maximum simplicity, excluded all other cross terms *except* those directly between the ignorable co-ordinates, i.e. those proportional to $(\partial/\partial t)(\partial/\partial\varphi)$ whose presence is essential if we are to have non-zero angular momentum. To achieve complete separability this term also must split up into the sum of two parts each depending on only one variable, i.e. Z must have the algebraic form

$$Z = U_\mu + U_\lambda \tag{5.14}$$

where U_μ depends only on μ and U_λ only on λ. From the expression (5.13) we see that this requirement will be satisfied if and only if

$$\frac{dZ_r}{dr}\frac{dQ_\mu}{d\mu} - \frac{dZ_\mu}{d\mu}\frac{dQ_r}{dr} = 0 \tag{5.15}$$

There are basically two ways in which this can be satisfied: a more general case in which either both Z_r and Z_μ or both Q_r and Q_μ are constants, and a more special case in which at least one of the four functions is zero, or can be reduced to zero by a form preserving co-ordinate transformation in which φ and t are replaced by linear combinations of themselves. In the more general case we can take it without loss of algebraic generality that it is the Q's which are constants. Thus replacing Q_r, Q_μ by constant C_r, C_μ respectively we obtain the basic separable canonical form

$$\left(\frac{\partial}{\partial s}\right)^2 = \frac{1}{[C_\mu Z_r - C_r Z_\mu]}\left\{\Delta_\mu\left(\frac{\partial}{\partial\mu}\right)^2 + \frac{1}{\Delta_\mu}\left[C_\mu\frac{\partial}{\partial\varphi} + Z_\mu\frac{\partial}{\partial t}\right]^2\right\}$$
$$+\frac{1}{[C_\mu Z_r - C_r Z_\mu]}\left\{\Delta_r\frac{\partial}{\partial r} - \frac{1}{\Delta_r}\left[C_r\frac{\partial}{\partial\varphi} + Z_r\frac{\partial}{\partial t}\right]^2\right\} \tag{5.16}$$

The original expression (5.10) from which we started had algebraic symmetry not only between r and μ (apart from a sign change) but also between t and φ. In the coform (5.16), in which the symmetry between t and φ has been lost, we have chosen to set Q_r and Q_μ rather than Z_r and Z_μ constant in order that it should include the spherical coform (5.6), in which t and φ have their usual quite distinct interpretations in the appropriate limit (that is to say when $C_\mu = 1$, $C_r = 0, Z_\mu = 0, Z_r = r^2, \Delta_\mu = 1 - \mu^2$).

In the alternative way of satisfying (5.15) it can be arranged either that Z_μ is zero, and Q_r a constant or that Q_μ is zero and Z_r a constant. I have studied the resulting canonical forms (Carter 1968) and found that there are no vacuum solutions except those which are in fact special cases of (5.16) and therefore we shall not consider these alternative possibilities any further here.

The steps by which we have arrived at the canonical separable coform 5.16 are so simple that it is really quite surprising that it was not found by any researcher of the nineteen thirties, such as for example Eisenhart or Robertson who both worked on separability of wave equations in curved spacetime (cf. Robertson 1927, Eisenhart 1933). The corresponding metric determinant is given by

$$\sqrt{-g} = |C_\mu Z_r - C_r Z_\mu| \tag{5.17}$$

and the covariant metric form is given by

$$ds^2 = [C_\mu Z_r - C_r Z_\mu]\left\{\frac{dr^2}{\Delta_r} + \frac{d\mu^2}{\Delta_\mu}\right\} + \frac{\Delta_\mu [C_r dt - Z_r d\varphi]^2 - \Delta_r [C_\mu dt - Z_\mu d\varphi]^2}{[C_\mu Z_r - C_r Z_\mu]} \tag{5.18}$$

It is very easy to derive the vacuum solutions corresponding to this canonical form, using the same method as in the spherical case, working in terms of the natural canonical orthogonal tetrad of forms,

$$\omega^{(1)} = \left\{\frac{C_\mu Z_r - C_r Z_\mu}{\Delta_r}\right\}^{1/2} dr \tag{5.19}$$

$$\omega^{(2)} = \left\{\frac{C_\mu Z_r - C_r Z_\mu}{\Delta_\mu}\right\} d\mu \tag{5.20}$$

$$\omega^{(3)} = \left\{\frac{\Delta_\mu}{C_\mu Z_r - C_r Z_\mu}\right\}^{1/2} [C_\mu\, dt - Z_\mu\, d\varphi] \tag{5.21}$$

$$\omega^{(0)} = \left\{\frac{\Delta_r}{C_\mu Z_r - C_r Z_\mu}\right\} [C_\mu\, dt - Z_\mu\, d\varphi] \tag{5.22}$$

The complete set of solutions for the four unknown variable functions Z_r, Z_μ, Δ_r, Δ_μ in terms of the constants C_μ, C_r turns out to be remarkably simple: the variables functions are all required to be quadratic polynomials, whose coefficients are subject to a few linear restraints. These quadratic functions can be stated in a very compact form if we temporarily exclude the special cases which arrive when either C_μ or C_r is zero (in the same way that it was convenient to give the general spherical solutions in a form in which r was assumed to be a variable, dealing separately with the special Robinson Bertotti case). When C_r and C_μ are non-zero they may both be normalized to unity by a change of scale of ψ and t. We shall impose this normalization only for C_μ which is unity in the spherical canonical form, but in order to be able to go over smoothly to the spherical limit we shall retain C_r as a freely renormalizable parameter which—with a certain prescience—we shall relable a. Thus setting

$$C_\mu = 1 \tag{5.23}$$

$$C_r = a \tag{5.24}$$

we can express the general solution (excluding special limiting cases where C_μ or C_r is zero) in the concise form

$$Z_r = r^2, \qquad Z_\mu = -a\mu^2 \tag{5.25}$$

$$\Delta_r = hr^2 - 2Mr + pa^2 \tag{5.26}$$

$$\Delta_\mu = -h\mu^2 - 2q\mu + p \tag{5.27}$$

where h, M, q and p are arbitrary parameters. (This would look more symmetric between r and μ if we chose to use the available co-ordinate freedom to adjust a to unity.) The complete set of solutions, including the special limiting cases, can be obtained from these by first making linear co-ordinate transformations in which r and $a\mu$ are replaced respectively by $cr + d$ and $a\mu + f$, and then allowing the parameters c, d, a, f to vary freely over *all* values in the resulting somewhat more complicated expressions including the values $c = 0$ and $a = 0$ for which the original transformation would have been singular. The solutions, (5.25), (5.26), (5.27) represent a vast class of vacuum metrics most of which are unacceptable for our present purposes for global geometric reasons, examples being the Taub-N.U.T. space (cf. Misner, 1963) which is included in the family as one of the special limiting cases and the more general family discovered by Newman and Demianski (1966).

In order to understand the global geometry of these solutions, we note that the canonical metric form (5.18) has singularities whenever Δ_r and Δ_μ are zero which are very similar respectively to the Killing horizon type and symmetry axis type co-ordinate singularities with which we are already familiar in the spherical case. Moreover while we can conceive that r should be able to vary across a region where Δ_r vanishes in an extended metric, it is clear that if the metric is to retain the correct signature μ must be *absolutely restricted to the range in which* Δ_μ *is non-negative.* Now since we wish the co-ordinate to represent an azimuthal variation between extending from a south polar to a north polar symmetry axis, it is clear that we must require that the quadratic function Δ_μ be positive in a restricted range, whose limits will we hope—if things work out—turn out to be the symmetry south and north polar symmetry axes. To achieve these conditions it is clear that we need $h > 0$ and also $hp + q^2 > 0$. Now we know that if the symmetry axis is to be well behaved the overall coefficient of $d\varphi^2$ in the metric form (5.18) must be zero that is to say Z_μ must vanish for the same values of μ as Δ_μ. Now the values of Z_μ and Z_r may be adjusted to the extent of the addition of *the same* arbitrary constant to both of them (by replacing t by a constant coefficient linear combination of t and φ) but it is clear that even with these adjustments the zeros of Z_μ will always occur for equal and opposite values of μ, and can therefore match the zeros of Δ_μ only if q is zero. Thus we see that the restrictions

$$q = 0 \tag{5.28}$$

$$h > 0 \tag{5.29}$$

$$p > 0 \tag{5.30}$$

are necessary for regular angular behaviour of the solutions. When the two latter conditions are satisfied we can make scale changes of μ and r so as to obtain $h = p = 1$ thereby ensuring that μ varies over the conventional co-ordinate range $-1 < \mu < 1$. In doing so we use up our co-ordinate freedom to adjust a which thereafter becomes a geometrically well determined parameter). The adjustment of t and φ necessary to ensure that the zeros of Z_μ coincide with those of Δ_μ leads us to replace the forms (5.25) of the solution (which were previously adjusted for maximum algebraic simplicity) by

$$Z_r = r^2 + a^2 \tag{5.31}$$

$$Z_\mu = a(1 - \mu^2) \tag{5.32}$$

while the other conditions we have imposed cause the expressions (5.26), (5.27) to reduce to the standard expressions

$$\Delta_r = r^2 - 2Mr + a^2 \tag{5.33}$$

$$\Delta_\mu = 1 - \mu^2 \tag{5.34}$$

On substituting these expressions together with (5.23) and (5.24) back into the canonical form (5.18) and making the substitution

$$\mu = \cos\theta \tag{5.35}$$

we obtain the solution of Kerr (1963) in the standard co-ordinate system introduced by Boyer and Lindquist (1966), which takes the explicit form

$$ds^2 = (r^2 + a^2\cos^2\theta)\left\{\frac{dr^2}{r^2 - 2Mr + a^2} + d\theta^2\right\} + \frac{\sin^2\theta\,[a\,dt - (r^2 + a^2)\,d\varphi]^2 - (r^2 - 2Mr + a^2)[dt - a(1 - \mu^2)\,d\varphi]^2}{r^2 + a^2\cos^2\theta} \tag{5.36}$$

where we have retained the grouping of the terms to make manifest the canonical tetrad with respect to which the separability (which of course is not affected by the replacement of μ by a function of itself) takes place. In terms of this tetrad as given explicitly by (5.19), (5.20), (5.21), (5.22) and of the curvature forms defined in section (3), the Weyl tensor, which in this case is equal to the Riemann tensor, may be presented in the form

$$
\begin{aligned}
\Omega^{(1)}{}_{(2)} &= -I_1 \omega^{(1)} \wedge \omega^{(2)} - I_2 \omega^{(0)} \wedge \omega^{(3)} \\
\Omega^{(0)}{}_{(3)} &= -I_1 \omega^{(0)} \wedge \omega^{(3)} + I_2 \omega^{(1)} \wedge \omega^{(2)} \\
\Omega^{(0)}{}_{(1)} &= 2I_1 \omega^{(1)} \wedge \omega^{(0)} - 2I_2 \omega^{(2)} \wedge \omega^{(3)} \\
\Omega^{(3)}{}_{(2)} &= 2I_1 \omega^{(2)} \wedge \omega^{(3)} + 2I_2 \omega^{(1)} \wedge \omega^{(0)} \\
\Omega^{(0)}{}_{(2)} &= I_1 \omega^{(2)} \wedge \omega^{(0)} - I_2 \omega^{(1)} \wedge \omega^{(3)} \\
\Omega^{(3)}{}_{(1)} &= I_1 \omega^{(1)} \wedge \omega^{(3)} + I_2 \omega^{(2)} \wedge \omega^{(0)}
\end{aligned}
\tag{5.37}
$$

where

$$I_1 = Mr \frac{(r^2 - 3a^2\mu^2)}{(r^2 + a^2\mu^2)^3} \tag{5.38}$$

$$I_2 = Ma\mu \frac{(3r^2 - a^2\mu^2)}{(r^2 + a^2\mu^2)^3} \tag{5.39}$$

It is obvious (to an expert) from this array that the canonical separation tetrad is also a canonical Petrov tetrad, and that the Weyl tensor is of Petrov type D. It was by searching for vacuum solutions with type D Weyl tensors that Kerr originally found this metric.

Let us now move on to consider the electromagnetic generalization of the method we have just applied. The obvious thing to do is to seek a generalization of the spherical canonical form (3.32) to canonical form of the electromagnetic potential A in a separable background metric of the canonical form (5.18) which will be such that not only the ordinary Klein Gordon equation but also its electromagnetic generalization is separable, thus ensuring (as a consequence of Hamilton-Jacobi theory, or from a physical point of view by the correspondence principle) that not only geodesics but also charged particle orbits will be integrable. In terms of an electromagnetic field potential

$$A = A_a \, dx^a \tag{5.40}$$

the electromagnetic Klein-Gordon equation takes the form

$$\psi^{-1} \left(\frac{\partial}{\partial x^a} - ieA_a \right) \sqrt{-g}\, g^{ab} \left(\frac{\partial}{\partial x^a} - ieA_a \right) \psi - m^2 \psi = 0 \tag{5.41}$$

In order not to introduce unnecessary cross terms we shall start off by requiring that A contain only the same two components as were necessary in the spherical case, i.e. that it has the form

$$A = A_3\, d\varphi + A_0\, dt \tag{5.42}$$

where A_3 and A_0 are scalars, independent of φ and t, whose functional dependenc on μ and r remains to be determined. On substituting this expression into (3.41), using the canonical co-form (5.18) we obtain

$$\begin{aligned}\psi^{-1}&\left\{\frac{\partial}{\partial\mu}\Delta_\mu\frac{\partial}{\partial\mu}+\frac{\partial}{\partial r}\Delta_r\frac{\partial}{\partial r}-m^2(C_\mu Z_r - C_r Z_\mu)\right\}\psi\\ &+\psi^{-1}\left\{\frac{\partial}{\partial t}Z_\mu+\frac{\partial}{\partial\varphi}C_\mu+ieX_\mu\right\}\frac{1}{\Delta_\mu}\left\{\frac{\partial}{\partial t}Z_\mu+\frac{\partial}{\partial\varphi}C_\mu+ieX_\mu\right\}\psi\\ &-\psi^{-1}\left\{\frac{\partial}{\partial t}Z_r+\frac{\partial}{\partial\varphi}C_r-ieX_r\right\}\frac{1}{\Delta_r}\left\{\frac{\partial}{\partial t}Z_r+\frac{\partial}{\partial\varphi}C_r-ieX_r\right\}\psi=0\end{aligned} \tag{5.43}$$

where we have set

$$\left.\begin{aligned}A_0Z_r + A_3C_r &= X_r\\ A_0Z_\mu + A_3C_\mu &= -X_\mu\end{aligned}\right\} \tag{5.44}$$

In order for the equation to separate we clearly need to choose A_0 and A_3 in such a way that X_μ and X_r are respectively functions of μ and r only. This can be done simply by taking arbitrary functions X_r, X_μ of r and μ respectively and solving (5.44) for A_0 and A_3. Thus we obtain the expression

$$\boxed{A = \frac{X_r(C_\mu\, dt - Z_\mu\, d\varphi) + X_\mu(C_r\, dt - Z_r\, d\varphi)}{C_\mu Z_r - C_r Z_\mu}} \tag{5.45}$$

for the canonical separable vector potential associated with the form (5.18). We can now go on to solve the source free Einstein-Maxwell equations for the forms (5.45) and (5.18) in conjunction, using the method described in section 3. As in the pure vacuum case the functions $\Delta_\mu, \Delta_r, Z_r, Z_\mu$ turn out to just quadratic polynomials, and the new functions X_r, X_μ are even simpler—they are linear. Again there is a large class of solutions most of which have undesirable global behaviour, and from which a small subclass can be selected by requiring that μ should be a well behaved azimuthal angle co-ordinate. We shall not repeat this selection procedure but merely present the final form of the well behaved subclass, using the same normalization conditions as in our presentation of the vacuum Kerr solutions. Thus the solutions for which C_r or C_μ are non-zero are given directly, setting

$$C_\mu = 1 \tag{5.46}$$

$$C_r = a \tag{5.47}$$

by

$$X_r = Qr \tag{5.48}$$

$$X_\mu = P\mu \tag{5.49}$$

$$Z_r = r^2 + a^2 \tag{5.50}$$

$$Z_\mu = a(1 - \mu^2) \tag{5.51}$$

$$\Delta_r = r^2 - 2Mr + a^2 + Q^2 + P^2 \tag{5.52}$$

$$\Delta_\mu = 1 - \mu^2 \tag{5.53}$$

where in addition to the parameters M and a which by comparison with the Lenz-Thirring solution can be seen to represent mass and angular momentum per unit respectively, we now have two more parameters Q and P which represent electric and magnetic monopole moments. This solution was first obtained by Newman and his co-workers (1965) using a guessing method based on an algebraic trick. Written out explicitly, and again setting $\mu = \cos\theta$, this Kerr-Newman solution takes the standard form

$$ds^2 = (r^2 + a^2\cos^2\theta)\left\{\frac{dr^2}{r^2 - 2Mr + a^2 + Q^2 + P^2} + d\theta^2\right\}$$
$$+ \frac{\sin^2\theta[a\,dt - (r^2 + a^2)\,d\varphi]^2 - (r^2 - 2Mr + a^2 + Q^2 + P^2)[dt - a\sin^2\theta\,d\varphi]^2}{r^2 + a^2\cos^2\theta} \tag{5.54}$$

with

$$A = \frac{Qr[dt - a\sin^2\theta\,d\varphi] + P\cos\theta\,[a\,dt - (r^2 + a^2)\,d\varphi]}{r^2 + a^2\cos^2\theta} \tag{5.55}$$

[These solutions can be obtained from the general solution [A] (Carter 1968) by making the restrictions $q = 0, h > 0, p > 0$ and setting $r = \lambda$, $a\cos\theta = \mu$, $\varphi = a\psi$, $t = \chi + a^2\psi$ where $a^2 = p$.] The metric (but not of course the field) is invariant under a duality rotation in which a and b are altered in such a way as to preserve the value of the sum of their squares. The Ricci tensor for this metric can be expressed in terms of the canonical tetrad as

$$R_{ab} = \frac{Q^2 + P^2}{r^2 + a^2\mu^2}\,[\omega_a^{(0)}\omega_b^{(0)} + \omega_a^{(3)}\omega_b^{(3)} + \omega_a^{(2)}\omega_b^{(2)} - \omega_a^{(1)}\omega_b^{(1)}] \tag{5.56}$$

and the Weyl tensor has the same basic form (5.37) as in the vacuum case, but this time with the more general expressions

$$I_1 = \frac{Mr[r^2 - 3a^2\mu^2] - (Q^2 + P^2)(r^2 - a^2\mu^2)}{(r^2 + a^2\mu^2)^3} \tag{5.57}$$

$$I_2 = \frac{Ma\mu[3r^2 - a^2\mu^2] - 2(Q^2 + P^2)a\mu r}{(r^2 + a^2\mu^2)^3} \tag{5.58}$$

for the coefficients.

We now come to the special case—the generalization of the Robinson-Bertotti solution—which has been left out so far. These special cases can be obtained from the general form by making the linear co-ordinate transformation $r \to cr + k$ and then taking the singular limit when c tends to zero. Again we select a subset from a much wider class by the requirement that μ should behave as a regular azimuthal co-ordinate. The solutions could be expressed algebraically in terms of the general canonical form (5.18) but it would be geometrically misleading to do so since t and φ would come the wrong way round. Instead we express the solution directly in an alternative and in this case geometrically more enlightening canonical form as follows. [These solutions can be obtained from the form [$\hat{B}(-)$] (Carter 1968) in the subcase $q = 0$, $h > 0$, $k^2 > Q^2 + P^2$ by adjusting m to zero, n to unity, setting $a^2 = k^2 - Q^2 - P^2$ and replacing μ by $a\mu$, ψ by t and $a\chi$ by $(a^2 + k^2)\varphi$]

$$ds^2 = (a^2\mu^2 + k^2)\left\{\frac{d\lambda^2}{\Delta_\lambda} + \frac{d\mu^2}{\Delta_\mu} - \Delta_\lambda\, d\tau^2\right\}$$
$$+ \frac{\Delta_\mu}{a^2\mu^2 + k^2}\,[(a^2 + k^2)\, d\varphi - 2ak\lambda\, d\tau]^2 \tag{5.59}$$

with

$$A = Q\lambda\, d\tau + P_\mu\, \frac{[(a^2 + k^2)\, d\varphi - 2ak\lambda\, d\tau]}{a^2\mu^2 + k^2} \tag{5.60}$$

where

$$\left.\begin{aligned}\Delta_\mu &= 1 - \mu^2\\ \Delta_\lambda &= \lambda^2 + n\end{aligned}\right\} \tag{5.61}$$

where Q, P, a, n are independent parameters, of which the last, n, can be adjusted to zero by suitable but non-trivial form preserving co-ordinate transformations using the fact that the solution has a 4-parameter isometry group (of which only 2 degrees of freedom—those corresponding to the ignorability of t and φ—are manifest) under which the hypersurfaces on which μ is constant are homogeneous and partially isotropic. The parameter k is not independent of the others, but must satisfy

$$k^2 = a^2 + Q^2 + P^2 \tag{5.62}$$

[These solutions are not fundamentally new since they could have been obtained from Taub-N.U.T. space (in the case where the "mass" term is zero) by analytic continuation in complex co-ordinate planes—the parameter k turning up as the analogue of the standard N.U.T. parameter l.]

In a more explicit form, having chosen $n = 0$, and having set $\mu = 0$, these solutions can be expressed as

$$ds^2 = [a^2(1 + \cos^2\theta) + Q^2 + P^2] \left\{\frac{d\lambda^2}{\lambda^2} + d\theta^2 - \lambda^2\, d\tau^2\right\}$$
$$+ \frac{\sin^2\theta\,[(2a^2 + Q^2 + P^2)\, d\varphi - 2a(a^2 + Q^2 + P^2)^{1/2}\, \lambda\, d\tau]^2}{[a^2(1 + \cos^2\theta) + Q^2 + P^2]} \tag{5.63}$$

with

$$A = Q\lambda\, d\tau + P \cos\theta \frac{[(2a^2 + Q^2 + P^2)\, d\varphi - 2a(a^2 + Q^2 + P^2)^{1/2}\, \lambda\, d\tau]}{[a^2(1 + \cos^2\theta) + Q^2 + P^2]} \tag{5.64}$$

which clearly reduces to the Robinson-Bertotti metric when a is set equal to zero.

The derivation of the source-free solutions of the separable form (5.18) with (5.45) worked out so well, and the separability property is so valuable for subsequent applications, that I was tempted to go on and search for more general solutions, in which for example, a perfect fluid is present. Unfortunately the separability conditions are in fact very restrictive—so much so that it is rather remarkable that there are any vacuum solutions at all—and nothing that was of any obvious physical interest turned up (not that I would by any means claim to have exhausted the subject). However, there was one generalization which came out at once, namely to solutions with no material sources but in which a cosmological Λ term is present. Although I don't think there is much physical justification for believing a non-zero Λ term, it is perhaps worth quoting the result as a geometrical curiosity. When a Λ term is present the solutions are still polynomials, X_r and X_μ being linear and Z_r and Z_μ being quadratic, as before, but now Δ_r and Δ_μ are no longer quadratic but quartic. As before, the solutions which I quote below are selected out of a much wider class by the requirement that μ should behave as a regular and azimuthal angle co-ordinate, which ensures that the solutions tend asymptotically this time not to Minkowski space as before but to de-Sitter space. The pure vacuum solutions are given by

$$ds^2 = (r^2 + a^2 \cos\theta)\left\{\frac{dr^2}{\Delta_r} + \frac{d\theta^2}{1 - \frac{1}{3}a^2\Lambda\cos^2\theta}\right\}$$
$$+ \sin^2\theta\left(\frac{1 - \frac{1}{3}a^2\Lambda\cos^2\theta}{r^2 + a^2\cos^2\theta}\right)\left[\frac{a\,dt - (r^2 + a^2)\,d\varphi}{1 - \frac{1}{3}a^2\Lambda}\right]^2$$
$$- \frac{\Delta_r}{r^2 + a^2\cos^2\theta}\left[\frac{dt - a\sin^2\theta\,d\varphi}{1 - \frac{1}{3}a^2\Lambda}\right]^2 \tag{5.65}$$

where

$$\Delta_r = \tfrac{1}{3}\Lambda(r^4 + a^2r^2) + r^2 - 2Mr + a^2 \tag{5.66}$$

A straightforward electromagnetic generalization exists, but I shall not bother to write it out here. [This solution can be obtained from the general solution [A] (Carter 1968) by setting $q = 0, p = a^2, h = 1 + \tfrac{1}{3}a^2\Lambda$ with the restriction $1 - \tfrac{1}{3}a^2\Lambda > 0$, and setting $r = \lambda$, $a \cos\theta = \mu$, $\varphi = a(1 - \tfrac{1}{3}a^2\Lambda)\psi$ and $t = (1 - \tfrac{1}{3}a^2\Lambda)(\chi + a^2\psi)$].

This solution has the property that it can be expressed in the form

$$ds^2 = ds_0^2 + 2Mr\left[\frac{d\tilde{t} - a\sin^2\theta\, d\tilde{\varphi}}{(1 - \tfrac{1}{3}a^2\Lambda)(r^2 + a^2\cos^2\theta)} - \frac{dr}{(r^2 + a^2)(1 + \tfrac{1}{3}\Lambda r^2)}\right]^2 \tag{5.67}$$

where

$$ds_0^2 = (r^2 + a^2\cos^2\theta)\left\{\frac{dr^2}{(a^2 + r^2)(1 + \tfrac{1}{3}\Lambda r^2)} + \frac{d\theta^2}{1 - \tfrac{1}{3}a^2\Lambda\cos^2\theta}\right\}$$
$$+ \sin^2\theta\left(\frac{1 - \tfrac{1}{3}a^2\Lambda\cos^2\theta}{r^2 + a^2\cos^2\theta}\right)\left[\frac{a\,dt - (r^2 + a^2)\,d\varphi}{(1 - \tfrac{1}{3}\Lambda a^2)}\right]^2 \tag{5.68}$$

where the new time and angle co-ordinates $\tilde{t}$ and $\tilde{\varphi}$ are defined by

$$d\tilde{t} = dt + \frac{2Mr\,dr}{(1 + \tfrac{1}{3}\Lambda r^2)\,\Delta_r} \tag{5.69}$$

$$d\tilde{\varphi} = d\varphi + \left(\frac{a}{r^2 + a^2}\right)\frac{2Mr\,dr}{(1 + \tfrac{1}{3}\Lambda r^2)\,\Delta_r} \tag{5.70}$$

The form whose square appears in the second term of (5.67) is in fact a null form, either with respect to the full metric ds^2 or with respect to the metric ds_0^2 to which ds^2 reduces when m is set equal to zero. The existence of the expression (5.67) establishes the claim that the solution is asymptotically de-Sitter (or asymptotically flat when $\Lambda = 0$) since the metric ds_0^2 to which ds^2 tends in the limit is in fact exactly de-Sitter space (or exactly Minkowski space when $\Lambda = 0$) albeit in a somewhat twisted co-ordinate system. The co-ordinate system may be untwisted, and (5.68) reduced to a familiar expression for de-Sitter space (or flat space when $\Lambda = 0$) by introducing new co-ordinates $\hat{r}, \hat{\theta}, \hat{\varphi}, \hat{t}$ defined by

$$(1 - \tfrac{1}{3}a^2\Lambda)\hat{r}^2 = r^2 + a^2\sin^2\theta - \tfrac{1}{3}a^2\Lambda r^2\cos^2\theta \tag{5.71}$$

$$\hat{r}\cos\hat{\theta} = r\cos\theta \tag{5.72}$$

$$\hat{\varphi} = \tilde{\varphi} + \frac{\tfrac{1}{3}a\Lambda t}{1 - \tfrac{1}{3}a^2\Lambda^2} \tag{5.73}$$

$$\hat{t} = \frac{\tilde{t}}{1 - \frac{1}{3}a^2\Lambda} \tag{5.74}$$

which leads to

$$ds_0^2 = \frac{d\hat{r}^2}{1 + \frac{1}{3}\Lambda\hat{r}^2} + \hat{r}^2(d\hat{\theta}^2 + \sin^2\hat{\theta}\, d\hat{\varphi}^2) - (1 + \tfrac{1}{3}\Lambda\hat{r}^2)\, d\hat{t}^2 \tag{5.75}$$

The fact that the original Kerr metric (with $\Lambda = 0$) can be expressed as flat-space-plus-squared-null-form was discovered by Kerr and Schild (1965). This property, with its cosmological term generalization to de-Sitter-space-plus-squared-null-form (which also applies to the electromagnetic generalizations) distinguishes the asymptotically well behaved solutions selected above from the much wider class of separable solutions (including Taub-N.U.T. space) which have been rejected.

6 Maximal Extensions of the Generalized Kerr Solutions

It turns out that despite their much greater complexity, the generalized Kerr solutions have maximally extended manifolds which are closely analogous to those of the spherical special cases as described in section 4. The construction of the appropriate maximally extended manifolds has been discussed in detail by Boyer and Lindquist (1967) and Carter (1968). The present section will consist primarily of a somewhat abbreviated description of these extension constructions.

The metric forms (5.36) and (5.54) are of course singular on the locus $r = 0$, $\mu \equiv \cos\theta = 0$ where the factor Z is zero, and where the curvature array (5.37) is clearly singular itself. However, when $M^2 > Q^2 + P^2 + a^2$, these forms are also singular on the loci where $r = r_+$ or $r = r_-$ with the definitions

$$r_\pm = M \pm \sqrt{M^2 - a^2 - Q^2 - P^2} \tag{6.1}$$

i.e. where Δ_r is zero. Since the curvature array (5.36) is well behaved where Δ_r is zero, one might have guessed that the metric form singularities at $r = r_+$ and $r = r_-$ should be removable by co-ordinate transformations, as is in fact shown to be the case by the existence of the transformation (5.69), (5.70) to the form (5.67) which is perfectly well behaved where $r = r_\pm$. It will be convenient to make a slightly different transformation, exactly analogous to the Finkelstein transformation described in section 4, introducing an ingoing null co-ordinate v, and a corresponding angle co-ordinate $\tilde{\varphi}$ by

$$\left.\begin{aligned} dv &= dt + \frac{r^2 + a^2}{\Delta_r}\, dr \\ d\tilde{\varphi} &= d\varphi + \frac{a}{\Delta_r}\, dr \end{aligned}\right\} \tag{6.2}$$

so as to obtain from (5.54) the form

$$\begin{aligned} ds^2 = {} & 2\,dr\,dv - 2a\sin^2\theta\,dr\,d\tilde{\varphi} + Z\,d\theta^2 \\ & + Z^{-1}[(r^2+a^2)^2 - \Delta_r a^2\sin^2\theta]\sin^2\theta\,d\tilde{\varphi}^2 \\ & - 2aZ^{-1}(2Mr - Q^2 - P^2)\sin^2\theta\,d\tilde{\varphi}\,dv \\ & - [1 - Z^{-1}(2Mr - Q^2 - P^2)]\,dv^2 \end{aligned} \tag{6.3}$$

with

$$Z = r^2 + a^2\cos^2\theta \tag{6.4}$$

the corresponding electromagnetic potential form being

$$\tilde{A} = Z^{-1}[Qr + Pa\cos\theta)\,dv - Z^{-1}[Qra\sin^2\theta + P(r^2+a^2)\cos\theta]\,d\tilde{\varphi} \tag{6.5}$$

where the new potential $\tilde{A}$ differs from A by a gauge transformation. It was in this form that both the pure vacuum solution (Kerr 1963) and its electromagnetic generalization (Newman *et al.* 1965) were originally discovered (the curves on which the co-ordinates v, $\tilde{\varphi}$ and θ are all held constant being the integral trajectories of one of the two degenerate Debever (1958) eigenvector fields, namely $\omega^{(0)} - \omega^{(1)}$, the Weyl tensor). The transformation (6.2), from this original Kerr-Newman form (6.3) to the standard symmetric form (5.36) was first discovered by Boyer and Lindquist (1967). The Kerr-Newman form (6.3) is well behaved over the whole of a co-ordinate patch whose topology is that of the product of a 2-plane and a 2-sphere, with co-ordinates v, r running from $-\infty$ to ∞ and with spherical type co-ordinates θ running from 0 to π and φ periodic with period 2π, except for the usual rotation axis co-ordinate singularity at $\theta = 0$ and $\theta = \pi$ where φ ceases to be well defined, and except also for a *ring singularity* on the locus $r = 0$, $\theta = \pi/2$ where the function Z is zero. The rotation axis singularity is easily removable, for for example by introducing Cartesian-type co-ordinates x, y, z defined by

$$x + iy = (r + ia)\,e^{i\varphi}\sin\theta \qquad z = r\cos\theta \tag{6.6}$$

in terms of which the form (5.67) is transformed (in the case $\Lambda = 0$ under consideration here) to the Cartesian Kerr-Schildt form

$$\begin{aligned} ds^2 = {} & dx^2 + dy^2 + dz^2 - dt^2 \\ & + \frac{2Mr - Q^2 - P^2}{r^4 + a^2 z^2}\left\{\frac{r(x\,dx + y\,dy) - a(x\,dy - y\,dx)}{r^2 + a^2} + \frac{z\,dz}{r} + dt\right\}^2 \end{aligned} \tag{6.7}$$

The ring singularity is irremovable, since as we have already remarked the curvature array (5.37) is singular when Z is zero.

Two dimensional cross sections at constant values of v and $\tilde{\varphi}$ are illustrated in Figures 6.1 to 6.5. The ring singularity (which is irremovable since as we have already remarked, the curvature components in the array (5.37) are unbounded in the limit as Z tends to zero) can be thought of as representing the frame of an

Alice-type looking glass separating the positive r part of the patch (which is asymptotically flat in the limit $r \to \infty$) from the negative r part of the patch (which can easily be seen to be asymptotically flat also in the limit $r \to -\infty$).

It can be seen directly from the form (6.3) that when they exist (i.e. when the inequality $M^2 > Q^2 + P^2 + a^2$ is satisfied) the locuses $r = r_+$ and $r = r_-$ are null

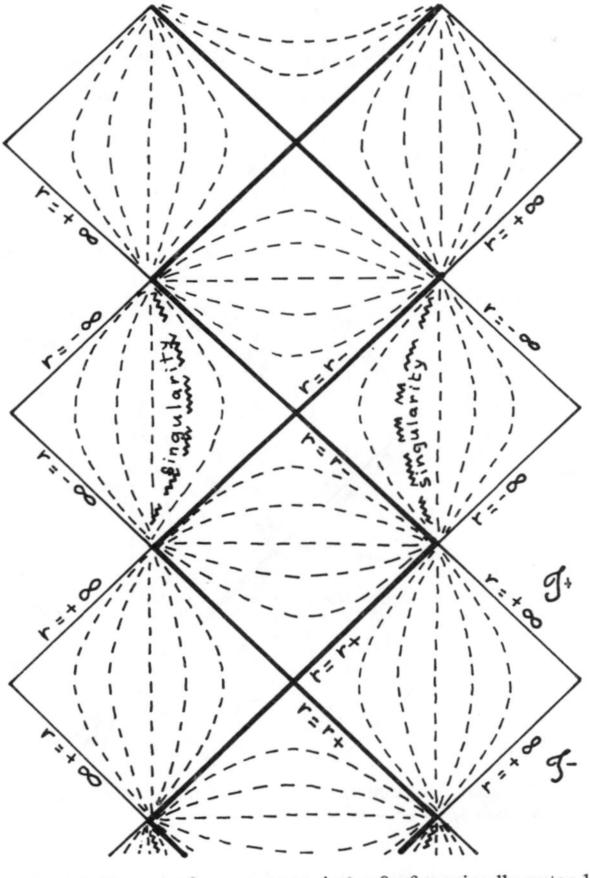

Figure 6.1. Conformal diagram of symmetry axis $\theta = 0$ of maximally extended Kerr or Kerr-Newman solution when $M^2 > a^2 + P^2 + Q^2$. In all the diagrams of this section the locus $\theta = 0$, where the axis passes through (without intersecting) the ring singularity, is marked by a broken zig-zag line.

hypersurfaces which can be crossed in the ingoing (decreasing r) direction only by future directed timelike lines. The surfaces on which r has a constant value between these limits (i.e. $r_- < r < r_+$) are spacelike and can also be crossed only in the sense of decreasing r by a future directed timelike line. Of course starting from the t, φ reversal symmetric standard form 5.54, we could also have made an analogous extension by introducing outgoing null co-ordinates $u, \overrightarrow{\varphi}$ defined by

$$\left.\begin{aligned} du &= dt - \frac{r^2 + a^2}{\Delta_r}\, dr \\ d\overrightarrow{\varphi} &= d\varphi - \frac{a}{\Delta_r}\, dr \end{aligned}\right\} \tag{6.8}$$

The metric form in the resulting co-ordinate system is identical to (6.4) except that u is replaced by $-v$ and $\overleftarrow{\varphi}$ by $-\overrightarrow{\varphi}$. This metric form is well behaved on a co-

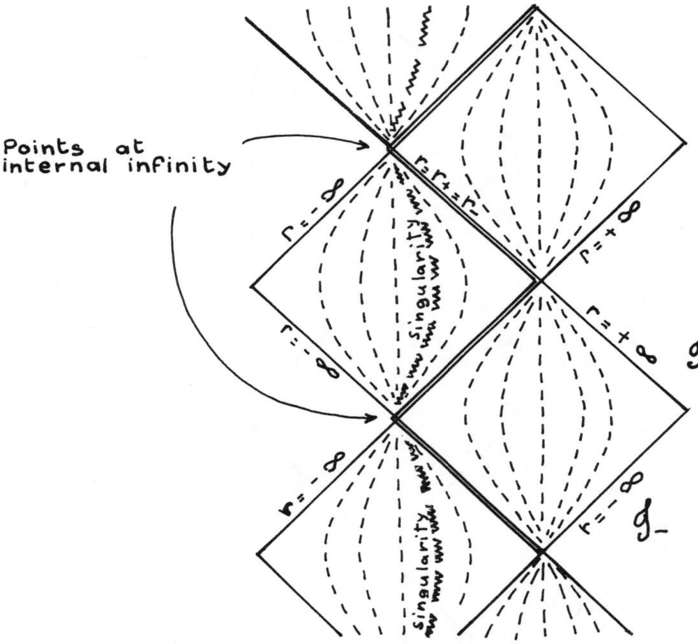

Figure 6.2. Conformal diagram of symmetry axis $\theta = 0$ of maximally extended Kerr or Kerr-Newman solution when $M^2 = a^2 + P^2 + Q^2$ $(a^2 > 0)$.

ordinate patch with the topology of the product of a 2-plane and a 2-sphere, with co-ordinates u, r running from $-\infty$ to ∞ and with spherical type co-ordinates θ running from 0 to π and φ periodic with periodic 2π, except as before for the standard rotation axis co-ordinate singularity at $\theta = 0, \theta = \pi$, and the ring singularity at $r = 0, \theta = \pi/2$. Just as was done in the special case of the Reissner-Nordstrom solutions in section 4, so also in this more general case, it is possible to build a maximally extended manifold by combining ingoing null co-ordinate patches $(r, \theta, \overleftarrow{\varphi}, v)$ and outgoing null co-ordinate patches $(r, \theta, \overrightarrow{\varphi}, u)$ in the criss-cross pattern illustrated in Figures 6.1, 6.2 and 6.4 (which are analogous to 4.9,

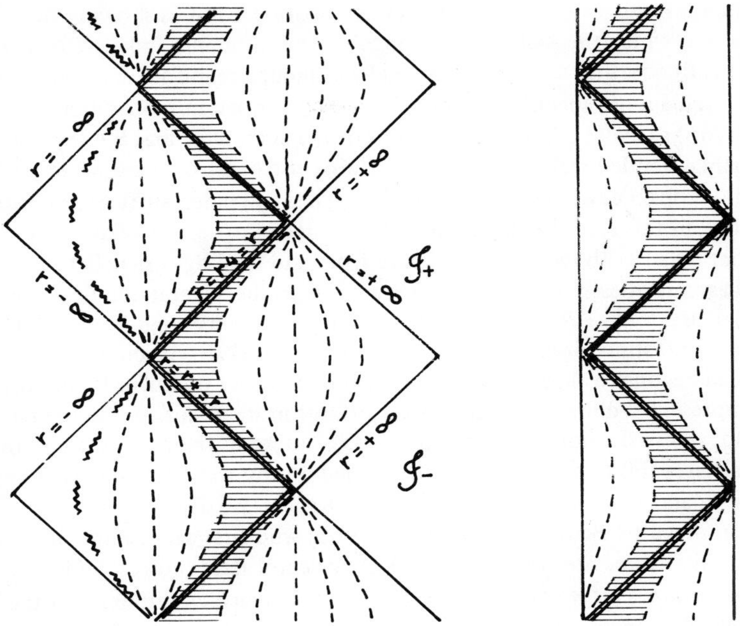

6.3a 6.3b

Figure 6.3. Figure 6.3a is a modified version of Figure 6.2 and Figure 6.3b is a conformal diagram of the maximally extended homogeneous universe whose metric is given by equation (5.63). The shaded regions of the two diagrams can approximate each other arbitrarily closely in the neighbourhood of the horizons, showing that the maximally extended Kerr or Kerr-Newman solution effectively contains a homogeneous universe within it in the bottomless hole case when $M^2 = P^2 + Q^2 + a^2$.

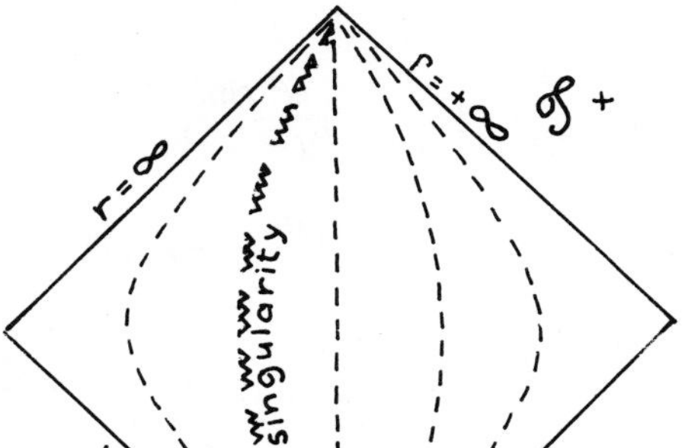

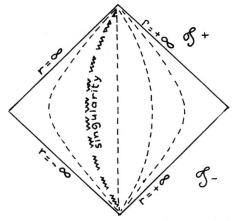

Figure 6.4. Conformal diagram of symmetry axis $\theta = 0$ of maximally extended Kerr or Kerr-Newman solution in the naked singularity case when $M^2 < a^2 + P^2 + Q^2$.

4.12 and 4.6 respectively). These figures can be thought of as showing the basic pattern which the 4-dimensional ingoing and outgoing Finkelstein-type extensions fit together to form a maximally extended manifold, and they can also be given a more specific interpretation (as described in more detail by Carter 1966) as exact conformal diagrams of the symmetry axis $\theta = 0$ of the maximally extended manifold. As in the Reissner-Nordstrom case, when there are no horizons, i.e. when $M^2 < Q^2 + P^2 + a^2$, a single patch suffices as illustrated in Figure 6.4; in this case the singularity where Z tends to zero is naked in the sense that it can both receive light signals from and send light signals to the same original asymptotically flat limit $\mathscr{I}$ where $r \to \infty$. In the general non-degenerate case when $M^2 > Q^2 + P^2 + a^2$, i.e. when the horizons $r = r_+$ and $r = r_-$ both exist and are distinct, it is necessary to construct Kruskal type co-ordinate patches analogous to those of 4.9 to cover the crossover points in Figure 6.1, which represent 2-dimensional spacelike surfaces in the 4-dimensional extended manifold, where the horizons $r = r_-$ and $r = r_+$ intersect themselves. As in the spherical cases this construction can be carried out by first introducing the co-ordinates u and v simultaneously in place of r and t. We also take advantage of the fact noticed by Boyer and Lindquist (1968) who first carried out this construction, that there are two particular Killing vector fields corresponding to the two operators $(r_\pm^2 + a^2)\,\partial/\partial t + a\,\partial/\partial\varphi$ in the standard co-ordinates of the form (5.64), which coincide everywhere with the null generator of the horizons $r = r_\pm$ respectively. This makes it possible to define new ignorable angle co-ordinates $\varphi^\pm$ given by

$$2\,d\varphi^\pm = d\vec{\varphi} - d\overleftarrow{\varphi} + a(r_\pm^2 + a^2)^{-1}\,(du - dv) \tag{6.9}$$

which are constant on the null generators of the horizons at $r = r_\pm$ respectively. We thus obtain two symmetric double null co-ordinate systems, analogous to (4.20) (and adapted respectively to the horizons $r = r_\pm$) given by

$$\begin{aligned} ds^2 = {} & \frac{\Delta_r}{Z}\left(\frac{Z}{r^2 + a^2} + \frac{Z_\pm}{r_\pm^2 + a^2}\right)\frac{(r^2 - r_\pm^2)a^2\sin^2\theta}{(r^2 + a^2)(r_\pm^2 + a^2)}\frac{(du^2 + dv^2)}{4} \\ & + \frac{\Delta_r}{Z}\left[\frac{Z^2}{(r^2 + a^2)^2} + \frac{Z_\pm^2}{(r^2 + a^2)^2}\right]\frac{du\,dv}{2} + Z\,d\theta^2 \\ & - \frac{\Delta_r a \sin^2\theta}{Z}\left[a\sin^2\theta\,d\varphi^\pm - \frac{Z_\pm}{r_\pm^2 + a^2}(du - dv)\,d\varphi^\pm\right]^2 \\ & + \frac{\sin^2\theta}{Z}\left[a\,\frac{(r_\pm^2 - r^2)}{r_\pm^2 + a^2}\frac{(du - dv)}{2} - (r^2 + a^2)\,d\varphi^\pm\right]^2 \end{aligned} \tag{6.10}$$

with the obvious abbreviation $Z_\pm = r_\pm^2 + a^2\cos^2\theta$ and where r is defined implicitly as a function of u and v by

$$F(r) = u + v \tag{6.11}$$

where

$$F(r) = 2r + \kappa_+^{-1} \ln |r - r_+| + \kappa_-^{-1} \ln |r - r_-| \tag{6.12}$$

with the constants κ_+ and κ_- defined by

$$\kappa_\pm = \tfrac{1}{2}(r_\pm^2 + a^2)^{-1}(r_\pm - r_\mp) \tag{6.13}$$

As in section 4, we now introduce new co-ordinates U^+ and V^+ or U^- and V^- (depending on whether we wish to remove the co-ordinate singularity at $r = r_+$ or $r = r_-$) defined by

$$\left.\begin{aligned} u &= -\kappa_\pm^{-1} \ln |U^\pm| \\ v &= +\kappa_\pm^{-1} \ln |V^\pm| \end{aligned}\right\} \tag{6.14}$$

which leads directly to the forms

$$\begin{aligned} ds^2 = {} & Z^{-1}\left(\frac{Z}{r^2 + a^2} + \frac{Z_\pm}{r_\pm^2 + a^2}\right)\frac{(r - r_\mp)(r + r_\pm)a \sin^2\theta}{(r^2 + a^2)(r_\pm^2 + a^2)}\kappa_\pm^{-2}G_\pm^2(r) \\ & \times \frac{(U^{\pm 2}dV^{\pm 2} + V^{\pm 2}dU^{\pm 2})}{4} + Z^{-1}\left(\frac{Z^2}{(r^2 + a^2)^2} + \frac{Z_\pm^2}{(r_\pm^2 + a^2)^2}\right) \\ & \times (r - r_\mp)\kappa_\pm^{-2}G_\pm(r)\frac{dU^\pm dV^\pm}{2} + Z\, d\theta^2 - \frac{a^2 \sin^2\theta}{Z} \\ & \times \left[\Delta_r a \sin^2\theta\, d\varphi^\pm - \frac{Z_\pm^2}{r_\pm^2 + a^2}(r - r_\mp)G_\pm(r)\kappa_\pm^{-1}(V^\pm dU^\pm - U^\pm dV^\pm)\right]d\varphi^\pm \\ & + \frac{\sin^2\theta}{Z}\left[(r^2 + a^2)\, d\varphi^\pm + a\frac{(r + r_\pm)}{r_\pm^2 + a^2}\kappa_\pm^{-1}G_\pm(r)\frac{(V^\pm dU^\pm - U^\pm dV^\pm)}{2}\right]^2 \end{aligned} \tag{6.15}$$

[which is the generalization of (4.23)] where r is now determined implicitly as a function of $U^\pm$ and $V^\pm$ by

$$U^\pm V^\pm = (r - r_\pm)G_\pm^{-1}(r) \tag{6.16}$$

where $G_\pm(r)$ is defined by

$$G_\pm(r) = e^{-2\kappa_\pm r}|r - r_\mp|^{|\kappa_\pm/\kappa_\mp|} \tag{6.17}$$

The corresponding electromagnetic field potential forms are

$$\begin{aligned} A^\pm = {} & \frac{[Q(rr_\pm - a^2\cos^2\theta) - Pa\cos\theta(r + r_\pm)]}{Z(r_\pm^2 + a^2)}\frac{(V^\pm dU^\pm - U^\pm dV^\pm)}{2} \\ & + \frac{[Qra\sin^2\theta + P\cos\theta(r^2 + a^2)]}{Z}d\varphi^\pm \end{aligned} \tag{6.18}$$

where the new potentials $A^{\pm}$ differ from A by gauge transformations. Since the relations (6.16) and (6.17) ensure that r is an analytic function of U^{+} and V^{+} in the neighbourhood of the horizons $r = r_{+}$ and an analytic function of U^{-}, V^{-} in the neighbourhood of the horizons $r = r_{-}$, it can easily be checked that the forms (6.15) and (6.18) are well behaved in the neighbourhood of the horizons $r = r_{\pm}$ respectively except for the usual co-ordinate singularities on the rotation axes $\theta = 0$, $\theta = \pi$ which could easily be removed by changing to a Cartesian type system. Since it is rather complicated the form (6.15) is of very little practical use, but its *existence* is important in order to establish that the extended manifold includes the *intersections* of the two null hypersurfaces $U^{\pm} = 0$ and $V^{\pm} = 0$ which make up the locuses $r = r_{\pm}$. The fitting together of the co-ordinate patches covering the maximally extended manifold of Figure 6.1 is described in more explicit detail by Boyer and Lindquist (1967) and Carter (1968).

In the naked-singularity case, i.e. when $M^2 < Q^2 + P^2 + a^2$, the patches covered by the original standard metric form (5.54) and by the Kerr–Newman form (6.3) are equivalent, and so a single patch gives the maximally extended manifold as illustrated in Figure 6.4. In the limiting intermediate case $M^2 = Q^2 + P^2 + a^2$ where the horizons $r = r_{+}$ and $r = r_{-}$ have coalesced to form a single degenerate horizon at $r = M$, the Kerr–Newman patches of the form (6.3) are sufficient to build the whole maximally extended manifold illustrated in Figure 6.2, as is described in more explicit detail by Carter (1968). As in the degenerate spherical case of Figure 4.12, so also in Figure 6.2 there are *internal* boundary points which represent limits which can only be reached by curves of infinite affine length (in addition to the boundary points representing the ordinary external asymptotically flat limit, i.e. the boundary hypersurfaces congruent to $\mathscr{I}$). As in the spherical case, so also in this more general case, it can easily be seen that the degenerate limit manifold illustrated by Figure 6.2 approximates a homogeneous universe whose metric is given by the special solution (5.63), in the limit as the degenerate horizon $r = M$ is approached, in the manner shown by Figure 6.3 (which is analogous to Figure 4.14). This limiting case will be given special attention in the following course of Bardeen.

The rigorous proof that these manifolds (as illustrated by Figures 6.1 to 6.4) really are maximal (in the sense that they are not submanifolds of more extended solution manifolds) depends (as in the spherical case) on the demonstration that all geodesics can either be extended to arbitrary affine length or else approach the curvature singularities, i.e. they are such that Z tends to zero along them. The method of integrating the geodesics is described in the following section. The explicit demonstration that the only incomplete geodesics are those which approach the curvature singularity is decribed by Carter (1968) and will not be repeated in this course.

We conclude this section by remarking that the procedures described here can also be applied in a straightforward manner to the asymptotically de Sitter Λ-term generalization (5.65) of the Kerr solutions. In this case when the rotation parameter a is small compared with M, and when M is small compared with the radius parameter $\Lambda^{-1/2}$ of the asymptotic de Sitter universe, there will be four zeros of Δ_r, of which two, which may be denoted by $r = r_+$, $r = r_-$, will be analogues of the zeros $r = r_\pm$ in the ordinary Kerr solutions, while the other two, which may be denoted by $r = r_{++} \gg r_+$ and $r = r_{--} \ll r_-$ would exist even in the ordinary de Sitter universe. The conformal diagram for the symmetry axis of the ordinary de Sitter universe has already been given by Figure 2.3. It is easy to guess that the

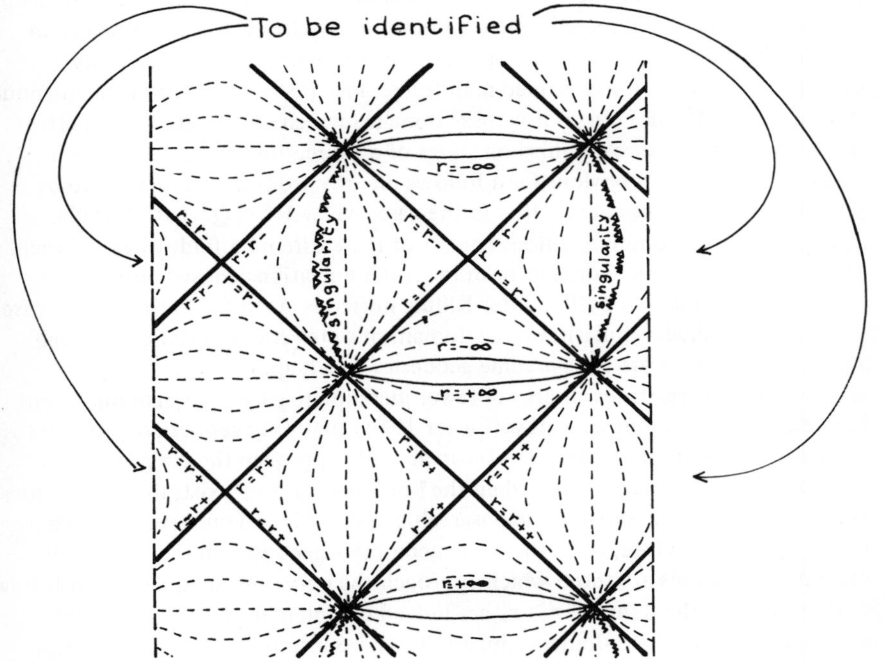

Figure 6.5. Conformal diagram of the symmetry axis $\theta = 0$ of the Kerr-de Sitter solution when $\Lambda^{-4} \gg M^2 \gg a^2$. The opposite sides of the diagram are to be identified [compare Figure 2.3 which is also the conformal diagram of the symmetry axis $\theta = 0$ for the pure de Sitter universe].

corresponding conformal diagram of the symmetry axis of the simplest maximally extended Kerr–de Sitter space will be as shown by Figure 6.5 which shows a sequence of de Sitter universes in which antipodal points are connected by Kruskal type throats. It is a straightforward (albeit lengthy) exercise, using the techniques described in this section, to construct appropriate Finkelstein type and Kruskal type co-ordinate patches to cover the manifold whose structure is shown by Figure 6.5 and then to integrate the geodesic equations and use them to prove that this manifold is indeed maximal. Since the Kerr–de Sitter solution is of more geometrical than physical interest, we shall not discuss it further in this course.

7 The Domains of Outer Communication

The only parts of the manifolds described in the previous sections which are strictly relevant to studies of black holes are the *domains of outer communications* (i.e. the parts which can be connected to one of the asymptotically flat regions by both future and past directed timelike lines), more particularly the domains of communications which are non-singular in the sense of having geodesically complete closures in the extended manifolds, since (as is discussed in the introduction to Part II of this course) it is these domains which are believed to represent the possible asymptotic equilibrium states of a source-free exterior field of a black hole. It is fairly obvious (and follows directly from the lemma 2.1 to be given in Part II of this course) that in the cases when $M^2 < Q^2 + P^2 + a^2$ the domains of outer communications consist of the entire manifold, whereas when $M^2 > Q^2 + P^2 + a^2$ the domains of outer communications consist only of the regions $r > r_+$ bounded by the outer Killing horizons at $r = r_+$. In the former case —the case of naked singularities—the domain of outer communications is itself singular, since there are always some geodesics which approach the curvature singularities. (Nevertheless there are fewer incomplete geodesics than one might have expected since one can show (Carter 1968) that only geodesics confined to the equatorial symmetry plane $\mu = 0$ can actually approach the curvature singularities.) In the latter case, when the horizons at $r = r_+$ exist, the singularities will always be hidden outside the observable region $r > r_+$ (since we always have $r_+ \geqslant r_- > 0$); therefore, since the only incomplete geodesics in the maximally extended manifolds are those which approach the curvature singularities, it follows that closures of the domains $r > r_+$ are indeed complete in the sense that any geodesic in one of these domains must either have infinite affine length or else have an end point within the domain $r > r_+$ or on the boundary $r = r_+$ in the extended manifold.

The remainder of our discussion will therefore be limited to these black hole exterior domains $r > r_+$ in the cases when the inequality $M^2 \geqslant a^2 + P^2 + Q^2$ is satisfied. An important property of these domains (without which they could not be taken seriously as the basis of a physical theory) is that they satisfy the causality condition that there are no compact (topologically circular) causal (i.e. timelike or null) curves within them. This follows from the fact that the hypersurfaces on which t is constant form a well-behaved congruence of *spacelike* hypersurfaces within the domains $r > r_+$ [which implies that t must increase monotonically along any timelike or null curve which remains in one of these domains]. The fact that these hypersurfaces are spacelike when $r > r_+$ follows from the fact that the two-dimensional metric on the surfaces on which t and φ are *both* constant is positive definite whenever Δ_r is positive, (i.e. both when $r > r_+$ and when $r < r_-$) and from the fact that the coefficient given by

$$X = \frac{(r^2 + a^2)^2 - \Delta_r a^2 \sin^2\theta}{r^2 + a^2 \cos^2\theta} \sin^2\theta \tag{7.1}$$

(which represents the squared magnitude of the Killing vector corresponding to the operator $\partial/\partial\varphi$ in r, θ, t, φ co-ordinates) turns out to be strictly positive also throughout the domains $r = r_-$ except of course on the rotation axis where it is zero. The coefficient X does become negative in a subregion [indicated by double shading in Figures 7.1 to 7.7] in the neighbourhood of the curvature singularity, thereby giving rise to causality violation in the inner regions of the extended manifold. In the naked singularity case, when $M^2 < a^2 + P^2 + Q^2$, it can easily be shown (Carter 1968, Carter 1972) that the entire extended manifold is a single *vicious set* in the sense that any event can be connected to any other by both a

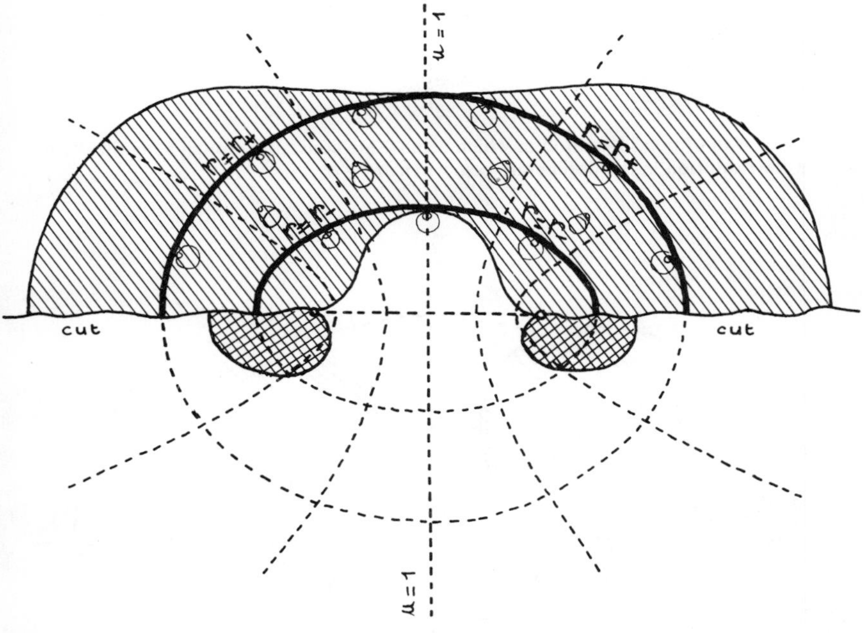

Figure 7.1. Plan of a polar 2-section on which v and $\tilde{\varphi}$ are constant through maximally extended Kerr solution with $M^2 > a^2$. The ring singularity is treated as a branch point and only half of the 2-section (corresponding roughly to $\cos\theta > 0$) bounded by cuts is shown—the other half should be regarded as being superimposed on the first half in the plane of the paper. The same comments apply to Figures 7.2 and 7.3. In all the diagrams of this section dotted lines are used to represent locuses on which r or θ is constant, and the positions of the Killing horizons are marked by a heavy line except for degenerate horizons which are marked by a double line. The regions in which V is negative are indicated by single shading and the regions where X is negative are marked by double shading. Some projected null cones are marked.

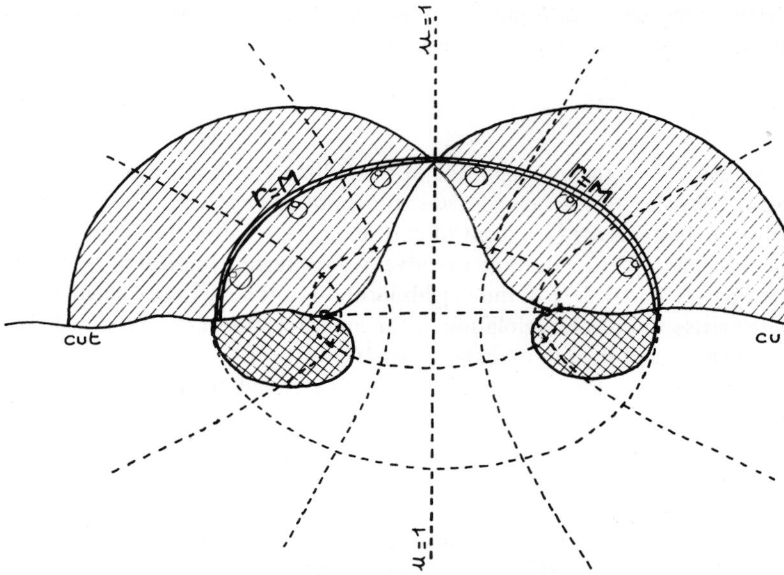

Figure 7.2. Plan of polar 2-section in which v and $\tilde{\varphi}$ are constant through a maximally extended Kerr solution in the degenerate case when $M^2 = a^2$.

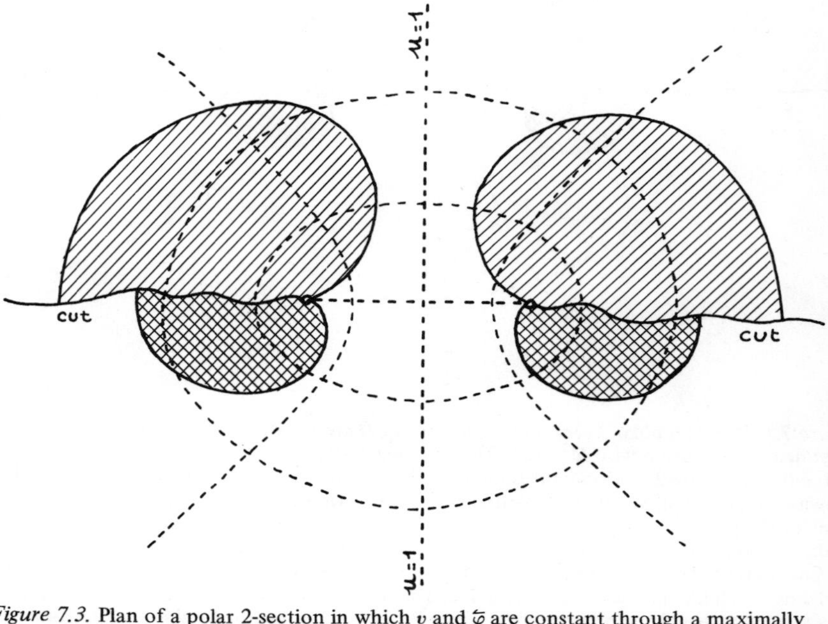

Figure 7.3. Plan of a polar 2-section in which v and $\tilde{\varphi}$ are constant through a maximally extended Kerr solution when $M^2 < a^2$

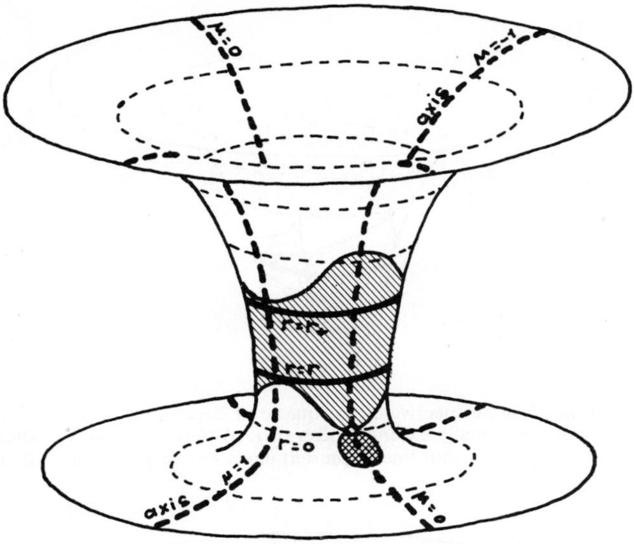

Figure 7.4. Rough sketch in perspective of a polar 2-section in which v and $\tilde{\varphi}$ are constant through a maximally extended Kerr solution when $M^2 > a^2$ (this is an alternative representation of the same section as is shown in Figure 7.1).

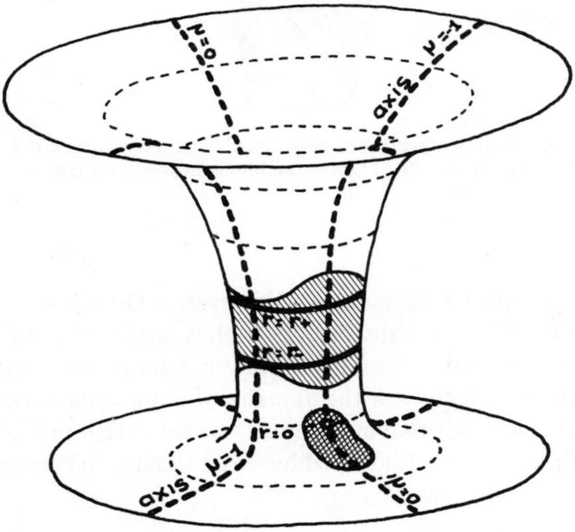

Figure 7.5. Rough sketch in perspective of a polar 2-section in which v and $\tilde{\varphi}$ are constant through a maximally extended Kerr-Newman solution when $M^2 > a^2 + P^2 + Q^2$.

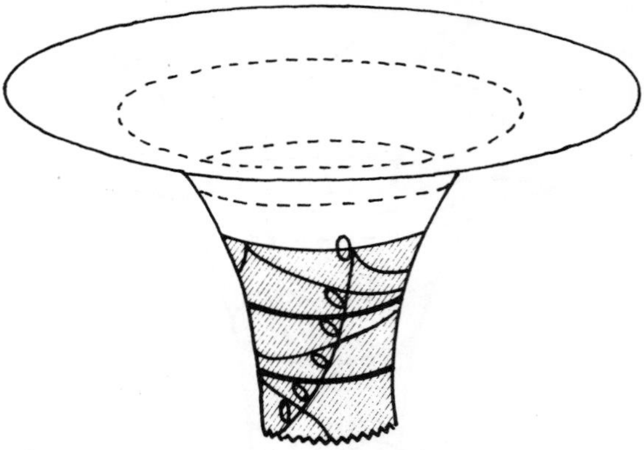

Figure 7.6. Rough sketch in perspective of an equatorial 2-section in which v is constant and $\cos\theta = 0$ through a maximally extended Kerr or Kerr-Newman solution when $M^2 > a^2 + P^2 + Q^2$. The continuous lines represent envelopes of projected null cones.

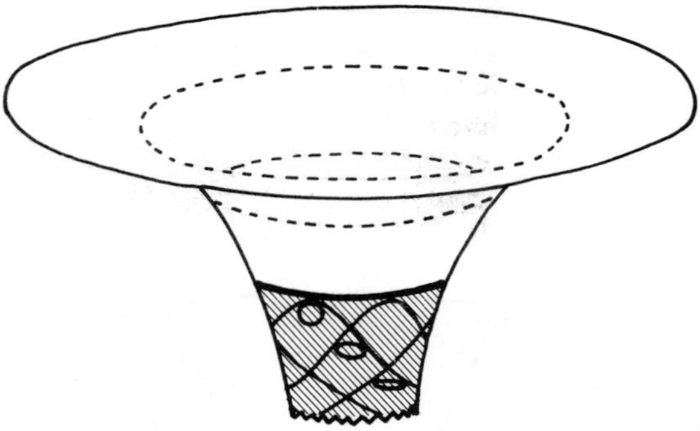

Figure 7.7. Rough sketch in perspective of equatorial 2-section in which v is constant and $\cos\theta = 0$ through maximally extended Schwarzschild solution. This represents the static limit of Figure 7.6.

future and a past directed timelike curve. However, in the black hole cases, i.e. when $M^2 \leqslant a^2 + P^2 + Q^2$, the causality violation is restricted to the domains $r < r_-$, and so does not affect the domains of outer communications.

Another noteworthy feature of the domains of outer communications (although its significance is often exaggerated) is that except in the spherical case they overlap with the regions [indicated by single shading in Figures 7.1 to 7.6] in which the coefficient V given by

$$V = 1 - \frac{2Mr - P^2 - Q^2}{r^2 + a^2 \cos^2\theta} \tag{7.2}$$

is negative. This coefficient V is the negative squared magnitude of the time Killing vector corresponding to the operator $\partial/\partial t$ in r, θ, φ, t co-ordinates; it is the same as the metric component g_{00} in a notation system where the co-ordinates x^1, x^2, x^3, x^0 are taken to be r, θ, φ, t. We shall refer to the outer boundary of this region as the *outer ergosurface* since it bounds the region within which the energy E of an uncharged particle orbit (as defined by equation (8.22) of the next section) can be negative. It is sometimes misleadingly described as an "infinite red shift surface" on the grounds that the light emitted by a particle which is static (in the sense that its co-ordinates remain fixed) will be redshifted at large asymptotic distances by an amount which tends to infinity as the position of the static particle approaches the ergosurface. However, the real significance of the region $V < 0$ bounded by the ergosurface is that physical particles which are static in this sense may not exist at all within it (if their world lines are to be timelike). In practice, moreover, one would not expect static particles to exist even in the neighbourhood of the ergosurface, and therefore the infinite red shift phenomenon would never be observed. On the other hand a genuine physically observable infinite red shift will take place whenever a particle crosses the actual horizon at $r = r_+$, since it is obvious that the light emitted by such a particle in the finite proper time-interval preceding the moment at which it crosses the horizon will emerge at large distances spread during an unbounded time interval.

8 Integration of the Geodesic Equations

The geodesic equation, and also the orbit equations

$$mv^a{}_{;b}v^b = eF^a{}_bv^b \tag{8.1}$$

for a particle of mass m charge e moving in a free orbit with unit tangent vector v^a can both be obtained from the same simple Lagrangian

$$L = \tfrac{1}{2}g_{ab}\dot{x}^a\dot{x}^b - eA_a\dot{x}^a \tag{8.2}$$

where a dot denotes differentiation with respect to an affine parameter, λ say. In the case of a particle orbit, λ must be related to the proper time τ along the orbit by

$$\tau = m\lambda \tag{8.3}$$

which is equivalent to imposing the normalization condition

$$g_{ab}\dot{x}^a\dot{x}^b = -m^2 \tag{8.4}$$

which means that we shall have

$$\dot{x}^a = mv^a \tag{8.5}$$

The geodesic equations are obtained from the Lagrangian (8.2) when e is set equal to zero.

Introducing the momentum co-vector

$$p_a = g_{ab}\dot{x}^b - eA_a \tag{8.6}$$

we see that the Hamiltonian corresponding to the Lagrangian (8.2) has the very simple form

$$H = \tfrac{1}{2}g^{ab}(p_a + eA_a)(p_b + eA_b) \tag{8.7}$$

Since the affine parameter λ does not appear explicitly in the Hamiltonian, the Hamiltonian itself will automatically be a constant of the motion, and by the normalization condition (8.4) it is clear that this constant is given simply by

$$H = -\tfrac{1}{2}m^2 \tag{8.8}$$

When one of the co-ordinates, x^0 say, is ignorable (i.e. does not appear explicitly in the Hamiltonian) the corresponding momentum p_0 will commute with the Hamiltonian (in the sense that its Poisson bracket with the Hamiltonian will be zero) and hence it will also be a constant of the motion. In the case of the metric form (5.18) with the electromagnetic field potential (5.45) there are two independent ignorable co-ordinates, namely t and φ and hence the corresponding momenta p_t and p_φ will both be conserved, giving two constants of the motion in addition to the constant given by (8.7) which would always exist.

To obtain a complete set of explicit first integrals of the motion we need to have a *fourth* constant of the motion. In a general stationary-axisymmetric space-time an independent fourth integral would not exist, but the condition that the charged Klein Gordon equation be separable (which we imposed in our derivation of the Kerr solutions) automatically guarantees the weaker condition that the Hamilton–Jacobi equation corresponding to the Hamiltonian (8.6) will be separable in the present case (Carter 1968a, Carter 1968b), thus ensuring the existence of the required fourth integral. We shall not use the Hamilton–Jacobi formalism in the present section, but instead shall show how the fourth constant may be obtained directly by inspection of the Hamiltonian using the following lemma:

LEMMA Let the Hamiltonian have the form

$$H = \frac{1}{2}\frac{H_r + H_\mu}{U_r + U_\mu} \tag{8.9}$$

where U_r and U_μ are single variable functions of the co-ordinates r and μ respectively and where H_r is independent of the momentum p_μ and of all the co-ordinate functions other than r, and H_μ is independent of the momentum p_r and of all the co-ordinate functions other than μ. Then the quantity

$$K = \frac{U_r H_\mu - U_\mu H_r}{U_r + U_\mu} \tag{8.10}$$

commutes with H and hence is a constant of the motion.

Proof Since the stipulated conditions clearly ensure that H_r commutes with H_μ, and since H_μ naturally commutes with itself, we obtain

$$[H_r, H] = \tfrac{1}{2}(H_r + H_\mu)\left[H_r, \frac{1}{U_r + U_\mu}\right] \tag{8.11}$$

Similarly since U_r clearly commutes with H_μ and with $U_r + U_\mu$ we obtain

$$[U_r, H] = \frac{1}{2(U_r + U_\mu)}[U_r, H_r] \tag{8.12}$$

Working out the Poisson brackets on the right hand sides we find

$$\begin{aligned}\left[H_r, \frac{1}{U_r + U_\mu}\right] &= -\frac{1}{(U_r + U_\mu)^2}\frac{\partial H_r}{\partial p_r}\frac{dU_r}{dr} \\ &= \frac{1}{(U_r + U_\mu)^2}[U_r, H_r]\end{aligned} \tag{8.13}$$

Thus we can eliminate the right hand sides of (8.11) and (8.12) to obtain

$$[H_r, H] = 2H[U_r, H] \tag{8.14}$$

It follows immediately that the quantity

$$K = 2U_r H - H_r \tag{8.15}$$

commutes with H and hence is a constant of the motion. It is clear from (9.8) that this quantity K can also be expressed in the form

$$K = H_\mu - 2U_\mu H \tag{8.16}$$

and also in the more symmetric form (8.10). This completes the proof.

Now the Hamiltonian corresponding to our canonical separable metric form (5.18) with the canonical vector potential (5.45) can be read off directly from the inverse metric (5.16) as

$$\begin{aligned}H = &\frac{\Delta_\mu p_\mu^2 + \Delta_\mu^{-1}[C_\mu p_\varphi + Z_\mu p_t - eX_\mu]^2}{2(C_\mu Z_r - C_r Z_\mu)} \\ &+ \frac{\Delta_r p_\mu^2 - \Delta_r^{-1}[C_r p_\varphi + Z_\mu p_t + eX_r]^2}{2(C_\mu Z_r - C_r Z_\mu)}\end{aligned} \tag{8.17}$$

which clearly has the form (8.9) with

$$H_r = \Delta_r p_r^2 - \Delta_r^{-1}[C_r p_\varphi + Z_r p_t + eX_r]^2 \tag{8.18}$$

$$H_\mu = \Delta_\mu p_\mu^2 + \Delta_\mu^{-1}[C_\mu p_\mu + Z_\mu p_t - eX_\mu]^2 \tag{8.19}$$

$$U_r = C_\mu Z_r \tag{8.20}$$

$$U_\mu = -C_r Z_\mu \tag{8.21}$$

It follows at once from the lemma that there will indeed be a fourth constant of the motion K given equivalently by (8.10) or (8.16) in terms of the expressions (8.18), (8.19), (8.20), (8.21), in addition to the constant H given by (8.17) and the energy and polar angular momentum constants E and L_z given by

$$E = -p_t \tag{8.22}$$

$$L_z = p_\varphi \tag{8.23}$$

In terms of the actual source free solutions (5.46) to (5.53) we shall have

$$H = \frac{(1-\mu^2)p_\mu^2 + (1-\mu^2)^{-1}[p_\varphi + a(1-\mu^2)p_t - eP\mu]^2}{2(r^2 + a^2\mu^2)} + \frac{\Delta_r p_r^2 - \Delta_r^{-1}[ap_\varphi + (r^2+a^2)p_t + eQr]^2}{2(r^2+a^2\mu^2)} \tag{8.24}$$

and the corresponding expression for K will be

$$K = \frac{a^2(1-\mu^2)\{\Delta_r p_r^2 - \Delta_r^{-1}[ap_\varphi + (r^2+a^2)p_t + eQr]^2\}}{r^2+a^2\mu^2} + \frac{(r^2+a^2)\{(1-\mu^2)p_\mu^2 + (1-\mu^2)^{-1}[p_\varphi + a(1-\mu^2)p_t - eP\mu]^2\}}{r^2+a^2\mu^2} \tag{8.25}$$

where Δ_r is given by (5.52).

Transforming from momentum to velocity co-ordinates and setting $\mu = \cos\theta$ we obtain the four constants of the motion explicitly as two linear combinations

$$E = \frac{(r^2 - 2Mr + a^2\cos^2\theta + P^2 + Q^2)\dot{t}}{r^2 + a^2\cos^2\theta} + \frac{(2Mr - P^2 - Q^2)a\sin^2\theta\dot{\varphi} + e(Qr + aP\mu)}{r^2+a^2\cos^2\theta} \tag{8.26}$$

$$L_z = \frac{[(r^2+a^2)^2 - \Delta_r a^2\sin^2\theta]\sin^2\theta\dot{\varphi}}{r^2+a^2\cos^2\theta} + \frac{-(2Mr - P^2 - Q^2)a\sin^2\theta\dot{t} + e[aQr\sin^2\theta + P(r^2+a^2)\cos\theta]}{r^2+a^2\cos^2\theta} \tag{8.27}$$

and two quadratic combinations

$$m^2 = -(r^2 + a^2 \cos^2 \theta)\left(\frac{\dot{r}^2}{\Delta_r} + \dot{\theta}^2\right)$$
$$- \frac{\sin^2 \theta}{r^2 + a^2 \cos^2 \theta}[(r^2 + a^2)\dot{\varphi} - a\dot{t}]^2 + \frac{\Delta_r}{r^2 + a^2 \cos^2 \theta}[a \sin^2 \theta\dot{\varphi} - \dot{t}]^2 \tag{8.28}$$

$$\mathrm{K} = a^2 \sin^2 \theta \left\{\frac{r^2 + a^2 \cos^2 \theta}{\Delta_r}\dot{r}^2 - \frac{\Delta_r}{r^2 + a^2 \cos^2 \theta}[a \sin^2 \theta\dot{\varphi} - \dot{t}]^2\right\}$$
$$+ (r^2 + a^2)\left\{(r^2 + a^2 \cos^2 \theta)\dot{\theta}^2 + \frac{\sin^2 \theta}{r^2 + a^2 \cos^2 \theta}[(r^2 + a^2)\dot{\varphi} - a\dot{t}]^2\right\} \tag{8.29}$$

Introducing the functions

$$R(r) = [E(r^2 + a^2) - aL_z - eQr]^2 - \Delta_r(m^2r^2 + \mathrm{K}) \tag{8.30}$$

$$\Theta(\theta) = \mathrm{K} - m^2a^2 \cos^2 \theta - \sin^{-2} \theta\,[Ea \sin^2 \theta - L_z + eP \cos \theta]^2 \tag{8.31}$$

$$= C + 2aePE \cos \theta + [a^2(E^2 - m^2) - (L_z^2 + e^2P^2) \sin^{-2} \theta] \cos^2 \theta \tag{8.32}$$

where C is a constant given in terms of K by

$$C = \mathrm{K} - (L_z - aE)^2 \tag{8.33}$$

we can express the non-ignorable co-ordinate velocities by the equations

$$Z^2\dot{r}^2 = R(r) \tag{8.34}$$

$$Z^2\dot{\theta}^2 = \Theta(\theta) \tag{8.35}$$

It is clear from the form of (8.35) and (8.31) that K must always be positive if $\dot{\theta}$ is to be real. The ignorable co-ordinate velocities can be expressed directly as

$$Z\dot{\varphi} = L_z\left[\frac{1}{\sin^2 \theta} - \frac{a^2}{\Delta_r}\right] - \frac{aE(2Mr - P^2 - Q^2)}{\Delta_r} \tag{8.36}$$

$$Z\dot{t} = E\left[\frac{(r^2 + a^2)^2}{\Delta_r} - a^2 \sin^2 \theta\right] - aL_z\frac{(2Mr - P^2 - Q^2)}{\Delta_r} \tag{8.37}$$

Hence we can obtain the final fully integrated form of the orbit equations as

$$\int^\theta \frac{d\theta}{\sqrt{\Theta}} = \int^r \frac{dr}{\sqrt{R}} \tag{8.38}$$

$$\lambda = \int^\theta \frac{a^2 \cos^2 \theta\, d\theta}{\sqrt{\Theta}} + \int^r \frac{r^2\, dr}{\sqrt{R}} \tag{8.39}$$

$$\varphi = +\int^r \frac{[L_z + aE(2Mr - P^2 - Q^2)]\,dr}{\Delta_r\sqrt{R}} + \int^\theta \frac{L_z\,d\vartheta}{\sin^2\theta\sqrt{\Theta}} \tag{8.40}$$

$$t = +\int^r \frac{[E(r^2 + a^2)^2 - aL_z(2Mr - P^2 - Q^2)]\,dr}{\Delta_r\sqrt{R}} - \int^\theta \frac{a^2E\sin^2\theta\,d\theta}{\sqrt{\Theta}} \tag{8.41}$$

Applications of these equations are discussed in the following course by Bardeen.

We conclude by remarking that the linear constants of the motion E and L_z can be expressed in the form

$$E = -k^a p_a \tag{8.42}$$

and

$$L_z = m^a p_a \tag{8.43}$$

in terms of the Killing vector components k^a and m^a defined by

$$\frac{\partial}{\partial t} = k^a \frac{\partial}{\partial x^a} \tag{8.44}$$

$$\frac{\partial}{\partial \varphi} = m^a \frac{\partial}{\partial x^a} \tag{8.45}$$

It can easily be seen that the Killing equations

$$k_{(a;b)} = 0 \tag{8.46}$$

$$m_{(\alpha;b)} = 0 \tag{8.47}$$

are both necessary and sufficient for E and L_z as defined by (8.42) and (8.43) to be constant along all geodesics. In order for them to be constant along *charged* particle orbits as well it is further necessary and sufficient that the invariance conditions

$$A^a k_{a;b} + A_{b;a}k^a = 0 \tag{8.48}$$

$$A^a m_{a;b} + A_{b;a}m^a = 0 \tag{8.49}$$

be satisfied, as they are in the present case.

Now the quadratic constant of the motion (8.29) can analogously be expressed in the form

$$\mathrm{K} = a_{ab}\dot{x}^a\dot{x}^b \tag{8.50}$$

where a_{ab} is a symmetric tensor given by

$$a_{ab}\,dx^a\,dx^b = (r^2 + a^2\cos^2\theta)\left[\frac{a^2\sin^2\theta\,dr^2}{\Delta_r} + (r^2 + a^2)\,d\theta^2\right] + \frac{(r^2 + a^2)\sin^2\theta\,[(r^2 + a^2)\,d\varphi - a\,dt]^2}{r^2 + a^2\cos^2\theta} - \frac{\Delta_r a^2\sin^2\theta\,[a\sin^2\theta\,d\varphi - dt]^2}{r^2 + a^2\cos^2\theta} \tag{8.51}$$

[This expression (8.50) for K having a similar form to the expression (8.4) for the constant m^2.] It is a standard result (see Eisenhart 1926) that the equations

$$a_{(ab;c)} = 0 \tag{8.52}$$

are both necessary and sufficient for an expression of the form (8.50) to be constant along all geodesics. Since the equation (8.52) is analogous to the Killing equations (8.46) which are necessary and sufficient for the analogous linear expression to be constant along geodesics, Penrose and Walker (1969) in a recent discussion have introduced the term Killing tensor to describe a symmetric tensor satisfying the equations (8.52). Personally I would prefer to call it a Stackel tensor since Stackel (a contemporary of Killing) actually made the first studies of the kind of Hamilton–Jacobi separability which gives rise to the existence of such a tensor (Stackel 1893).

It is to be noted that the existence of the Stackel–Killing tensor is not in itself sufficient to ensure that the quadratic expression (8.50) is constant along charged particle orbits as well as geodesics. It is easy to see that in general a necessary and sufficient condition for the quadratic expression $a_{ab}\dot{x}^a\dot{x}^b$ to be constant along charged particle orbits is that in addition to (8.52) the electromagnetic field should be such that

$$a_{a(b}F_{c)}{}^a = 0 \tag{8.53}$$

It is clear that in any space the metric tensor g_{ab} will satisfy the Stackel–Killing equations (8.52) and also that it will satisfy the condition (8.53) for *arbitrary* F_{ab}, thus giving rise to the constant of motion (8.4). Also when there are Killing vectors present their symmetrized products also form Stackel–Killing tensors, so that in the Kerr metrics there is actually a five parameter family of Stackel–Killing tensors of which five linearly independent members may be taken to be g_{ab}, the tensor a_{ab} defined by (8.51), together with $k_{(a}k_{b)}$, $k_{(a}m_{b)}$ and $m_{(a}m_{b)}$. These give rise to five linearly independent constants of which of course only four are *algebraically* independent.

The approach to the Kerr metrics which we have adopted here is to *start* with the separability properties as a fundamental postulate, and then to derive other basic algebraic properties such that the fact that the Weyl tensor is of Petrov type D. The more traditional approach has been to start from the Petrov type D property and then to *derive* the separability properties. The investigations initiated by Penrose and Walker (1969) are aimed at deriving the separability properties directly from the Petrov type D property without actually going through the explicit solution of the field equations. [It is to be emphasized however that the direct derivation of the existence of the Stackel–Killing tensor can only be an intermediate stage in this programme, and is by no means sufficient to establish the full separability properties of the Kerr metrics: there is at present no known way of deducing the separability of the *wave equation*—as opposed merely to the separability of the Hamilton–Jacobi equation—from the Stackel–Killing tensor.] Some important recent work by Teukolsky (1972) has shown that the Kerr metric possess even stronger separability properties than those we have used in the present course, by which *higher spin* wave equations (as well as the ordinary scalar wave equation) can be at least partially separated by combining the separation methods described at the beginning of section 5 with Petrov-type analysis. This discovery is likely to prove to be of very great value in future perturbation analyses of the Kerr solutions.

Bibliography

Bertotti, B. (1959) *Phys. Rev.* **116,** 1331.
Boyer, R. H. and Price, T. G. (1965) *Proc. Camb. Phil. Soc.* **61,** 531.
Boyer, R. H. and Lindquist, R. W. (1967) *J. Math. Phys.* **8,** 265.
Carter, B. (1966a) *Phys. Rev.* **141,** 1242.
Carter, B. (1966b) *Physics Letters* **21,** 423.
Carter, B. (1968a) *Phys. Rev.* **174,** 1559.
Carter, B. (1968b) *Commun. Math. Phys.* **10,** 280.
Carter, B. (1969) *J. Math. Phys.* **10,** 70.
Debever, R. (1969) *Bull. Akad. Roy. de Belgique* **55,** 8.
Eisenhart, L. P. (1926) *Riemmannian Geometry* (Princeton)
Finhelstein, D. (1958) *Phys. Rev.* **110,** 965.
Graves, J. C. and Brill, D. R. (1960) *Phys. Rev.* **120,** 1507.
Goldberg, J. N. and Sachs, R. K. (1962) *Acta Phys. Polon.* **13.**
Hughston, L. P., Penrose, R., Sommers, P. and Walker, M. (1972) *Commun Math. Phys.* **27,** 303.
Kerr, R. P. (1963) *Phys. Rev. Letters* **11,** 238.
Kerr, R. P. and Schild, A. (1964) Am. Math. Soc. Symposium.
Kundt, W. and Trumper, M. (1962) *Akad. Wiss. Lit. Mainz* **12.**
Lemaitre, G. E. (1933) *Am. Soc. Scient. Bruxelles,* **A53,** 51.
Misner, C. (1963) *J. Math. Phys.* **4,** 924.
Robinson, I. (1959) *Bull. Akad. Pol.* **7,** 351.
Schwarzschild, K. (1916) *Berl. Ber.* **189.**
Stachel (1893) *Compt. Rendus Acad. Sci.* **116,** 1284.
Teukolski, S. A. (1972) preprint Caltech.
Walker, M. and Penrose, R. (1970) *Commun. Math. Phys.* **18,** 265.

PART II General Theory of Stationary Black Hole States

1 Introduction

The necessity of facing up to the problem of catastrophic gravitational collapse in astrophysics was first appreciated by Chandrasekhar in 1931 when he discovered the upper mass limit for a spherical equilibrium configuration of a cold sphere of ordinary degenerate matter in Newtonian gravitational theory. Subsequent work by Chandrasekhar, Landau, Oppenheimer and others established fairly clearly by the end of the nineteen thirties that the upper mass limit would not be greatly affected by taking into account the existence of exotic forms of high density matter, such as a degenerate neutron fluid, nor by the corrections resulting from the use of Einstein's theory of gravity, according to which an object above this limit must ultimately disappear from sight within its Schwarzschild horizon—which occurs when the circumferential radius r is diminished to twice the conserved mass M (where here, as throughout this work, we use units in which Newton's constant G and the speed of light c are set equal to unity)—with the subsequent formation of a singularity.

It is true of course that an arbitrarily large mass can exist in quasi-equilibrium in a sufficiently extended differentially rotating disc-like configuration. However, it would be rather surprising if nature contrived to avoid the formation of at least some large stars with sufficiently small angular momentum for its effects to be negligible even after contraction to the Schwarzschild radius. Moreover there are many diverse effects (including viscosity, magnetic fields, and gravitational radiation) which would all tend to transfer angular momentum outwards, except in the special case of uniformly rotating axially symmetric configurations, for which there is in any case an upper mass limit, somewhat larger than the spherical upper mass limit, but of the same order of magnitude. (See for example the work of Ostriker *et al.*) Hence although its onset may in some cases be postponed or prevented, the basic issue of catastrophic collapse cannot be avoided merely by taking into account natural deviations from spherical symmetry.

Despite these considerations there was very little work in the field until the mid nineteen-sixties. The first signs of renewed interest were a pioneering attempt by Regge and Wheeler (1957) to determine the stability of the Schwarzschild horizon, and an equally important pioneering work by Lifshitz and Khalatnikov (1961) in which it was suggested that the ultimate singularity which occurs in the spherical collapse case might be avoided in more general situations. These two works forshadowed the subsequent subdivision of the field of gravitational collapse investigations into two main branches, the first—with which we shall be concerned here—being concerned primarily with horizons and astronomically observable effects, and the second being primarily concerned with the fundamental physical question of the nature of the singularities and the breakdown of the classical Einstein theory. Another event which was of key importance in stimulating interest in the field was the accidental discovery by Kerr (1963) of a rotating generalization of the spherical Schwarzschild solution of Einstein's vacuum equations—accidental in the sense that Kerr's investigations were not directly motivated by the astrophysical collapse problem.

However, the event which laid the foundations of the modern mathematical theory of gravitational collapse was the publication by Penrose (1965) of the now famous singularity theorem which conclusively refuted the suggestion that singularities are merely a consequence of the spherical idealization used in the early work. The actual conclusion of the Penrose theorem was rather restricted, being essentially negative in nature, and although its range of application has been very greatly extended by the subsequent work of Stephen Hawking, there is so far very little positive information about the nature of the singularities whose existence is predicted. The real importance of the Penrose theorem lay rather in the wealth of new techniques and concepts which were introduced in its proof, particularly the notion of a trapped surface and the idea of treating boundary horizons as dynamic entities in their own right. These developments lead directly to the realization that the outcome of a gravitational collapse—in so far as it can be followed up in terms of the classical Einstein theory—would have to be either (a) the formation of a well behaved event horizon separating a well behaved *domain of outer communication* (within which light signals can not only be received from but also sent to arbitrarily large distances) from hidden regions (for which the term *black holes* was subsequently introduced by Wheeler) within which the singularities would be located, and from which no light could escape to large distances, or (b) the formation of what have come to be known as *naked singularities,* i.e. singularities which can be approached arbitrarily closely by time-like curves or light rays which subsequently escape to infinity. If the latter situatio were indeed to occur it would mean that the classical Einstein theory would be essentially useless for predicting the subsequent astronomically observable phenomena, so that the development of a more sophisticated gravitational theory would be an urgent practical necessity. However there are a number of features of the situation as we understand it at present which encourage belief in the conjecture which Penrose has termed the *cosmic censorship hypothesis*, which postulates that naked singularities do not in fact occur, i.e. that the only possible outcome of a gravitational collapse is the formation of black holes. It is probably fair to say that the verification or refutation of this conjecture is the most important unresolved mathematical problem in General Relativity theory today. If the cosmic censorship conjecture is correct then there will be no theoretical reason why Einstein's theory of gravity should not be adequate for all foreseeable astronomical purposes, except for dealing with primordial singularities (sometimes referred to as white holes) such as the cosmological big bang. (It will still of course be possible that the theory may fail empirically by conflict with observation.)

Having arrived at the idea of a black hole as a possible outcome, if not an inevitable outcome, of a gravitational collapse, it is natural to conjecture that it should settle down asymptotically in time towards a stationary final equilibrium state—any oscillations being damped out by gravitational radiation—as indicated in Figure 1. It was after all the non-existence of such a stationary final equilibrium state for an ordinary massive star which led to the introduction of the alternative black hole concept in the first place. With these considerations in mind, and the recently discovered Kerr solution as an example, several workers including myself, and notably Bob Boyer, began a systematic study of the properties of stationary black hole states during the years 1965–1966. Too much concentration on the stationary equilibrium states could not of course have been justified if the settling down process took place over astronomically long timescales, but it has subsequently become fairly clear from the dynamical perturbation investigations of the Schwarzschild solution by Doroshkeviteh, Zel'dovich and Novikov (1966), Vishveshwara (1970), Price (1972) and others that (as was suspected, by dimensional considerations, from the outset) any dynamic variations of a black hole can be expected to die away over timescales of the order of the time for light to cross a distance equal to the Schwarzschild radius. These timescales are in fact extremely short by astronomical timescales—of the order of milliseconds for the collapse of an ordinary massive star, and at most hours or days for the collapse of an entire galactic nucleus.

This early work served to clarify many of the elementary properties of the black hole event horizons in the stationary case, and also their relationship with other geometrically defined features such as what has subsequently come to be known as the ergosurface, which are described in detail in sections 2 to 5. This phase of elementary investigation was almost complete by the time of Boyer's tragic death in the summer of 1966, although the results were not actually published until somewhat later, (Boyer (1969), Carter (1969), Vishveshwara (1968)). The results described in section 8 were overlooked in this earlier period and were obtained quite recently (by Hartle, Hawking and myself) while the rigorous proofs in section 9 were actually worked out at Les Houches by Bardeen, Hawking and myself.

An analogous first phase of investigation of the elementary properties of black hole even horizons in the more general *dynamic* case has been carried out more recently almost exclusively by Stephen Hawking, using some of the techniques developed by Penrose, Geroch and himself for proving singularity theorems; the most important of these elementary properties is that the area of the intersection of a constant time hypersurface with the event horizon (which is of course constant in the stationary case) can only increase but never decrease with time in the dynamic case (Hawking 1971). An off-shoot of this work has been a much more sophisticated second phase of investigations of stationary black hole event

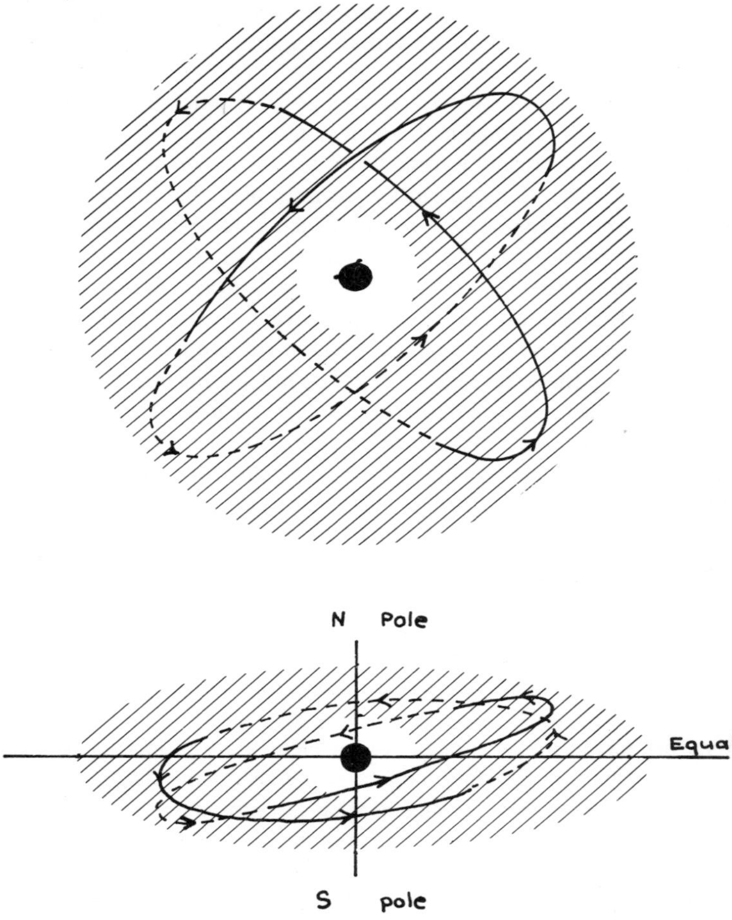

Figure 2. Equatorial and polar sections of a conceivable axially and equatorially symmetric galactic nucleus are illustrated. The nucleus is supposed to consist of a stellar system (whose intersections with the equatorial and polar planes are represented by the shaded regions) with a massive central black hole. (The hole will be non-Kerr-like due to the distorting influence of the stellar mass distribution.) Projections onto the equatorial and polar planes of two stellar orbits (related by a 90 degree axisymmetry rotation) are marked, using dotted lines where the orbits lie behind the relevant planes. These would be typical orbits in a system which is approximately stationary (in that mass redistribution as the stars fall into the hole takes place over timescales long compared with the orbital periods) but in which the circularity condition is violated due to a net inflow of matter at high latitudes compensated by a net outflow near the equator. (In order to obtain this circularity violation it is necessary not only that the orbits be eccentric but also that they should have a consistent average tilt.) The Papapetrou theorem is sufficiently powerful to ensure that the spacetime will be stationary-circular in the empty region in the neighbourhood of the hole despite the non-circularity resulting from the quasi-convective matter flow in the region occupied by the stars. (*See section 7.*)

horizons, in which (using the techniques originally developed by Newman and Penrose for dealing with gravitational radiation problems) Hawking has shown that many of the simplifying assumptions which were taken for granted in the earlier work on stationary event horizons, particularly that of axisymmetry, can in fact be rigorously justified. I shall use what I shall refer to as Hawking's *strong rigidity theorem* as the starting point of section 3 of this course. The actual derivation is discussed by Stephen Hawking himself in the accompanying lecture course, but the full technical details of the proof are beyond the scope of the present proceedings (see Hawking (1972), Hawking and Ellis (1973)). The main content of sections 6 and 7 will be the demonstration that this theorem can be combined with suitable generalizations (which I shall present in full detail) of earlier theorems, originally derived outside the context of the black hole problem, by Lichnerowicz (1945) and Papapetrou (1966), to justify the principle simplifying conditions namely staticity and circularity, which are involved at various stages in section 2 and in the subsequent sections.

The existence of the kinds of the results described in sections 2 to 7 were in most cases at least conjectured before 1967, even though many of the detailed formulae and rigorous proofs which are now available, particularly those due to Hawking, are of more recent origin. However in that year a new result was announced by Israel which took me, and as far as I know everyone else in the field, by surprise, despite the fact that there were already many hints which might have suggested it to us. This was the theorem which could be taken, subject to some fairly obvious assumptions which have since been fully justified, as implying that the Schwarzschild solution is the only possible pure vacuum exterior solution for a stationary black hole which is non-rotating (in the sense made precise in section 4). Until then I think that we had all thought of a black hole even in the stationary limit, as potentially a fairly complicated object. I know that I had always imagined it should have many internal degrees of freedom representing the vestigial multipole structure of the star or other object from whose collapse it had arisen. Yet here was a theorem which implies that if rotational effects were excluded, there could be no internal degrees of freedom at all, apart from the mass itself. My immediate reaction—and as far as I can remember this was how Israel himself saw the situation—was to suppose that this meant that only stars starting off with artificially restricted multipole structures, and in the non-rotating case only exactly spherical stars, could form black holes with well behaved event horizons, so that the natural outcome of a gravitational collapse under physically realistic conditions would always be the formation of a naked singularity. In short it seemed that the cosmic censorship conjecture was utterly erroneous, and that our work on black hole theory had been, from a physical point of view, rather a waste of time. This state of alarm did not last very long, however, since in the discussions that followed, in which Roger Penrose took the leading part, it soon became clear that there was an almost diametrically opposite and far more plausible alternative interpretation which could be made. This was that far from being an essentially unstable phenomenon, the formation of a black hole is actually a stabilizing effect, which enables an object collapsing from one of a very wide range of initial configurations, not only to form a well behaved event horizon, but also to settle down (as far as the part of space-time outside the event horizon is concerned) towards one of a very restricted range of possible stationary equilibrium states, the excess multipoles being lost in the form of gravitational radiation. Thus on this interpretation, an uncharged non-rotating body (if it collapses in a part of space sufficiently far from external perturbing influence) should give rise ultimately to a Schwarzschild black hole. The subsequent investigation of the decay of perturbations in a Schwarzschild background space by Vishveshwara, Price, and others, provides strong support for this way of viewing the situation, which is now almost universally accepted as correct, at least for cases where the initial deviations from spherical symmetry are not too large.

Having reached this point of view, it was natural to conjecture that the only possible pure vacuum exterior solutions for a *rotating black hole* were those which were already known, that is to say the Kerr solutions, which until then we had thought of as merely simple examples from a potentially much wider class. The idea that the multipole moments could settle down to preordained values by gravitational radiation made it seem particularly natural to conjecture that in the pure vacuum case the only multipole moments which would remain as independent degrees of freedom would be those for which there were no corresponding degrees of freedom in the asymptotic gravitational radiation field, that is to say the monopole and dipole moments, i.e. the moments corresponding to mass and angular momentum (which cannot be radiated away directly, although they can of course be varied indirectly in consequence of the non-linearity of the field equations). These are in fact just the two degrees of freedom which actually are possessed by the Kerr solutions. It is not yet known for certain whether the Kerr solutions are absolutely the only possible stationary black hole exterior solutions, but it can be established conclusively (using appropriate global conditions, and with the aid of Israel's theorem and the strong rigidity theorem of Stephen Hawking, which in conjunction make it possible to take axisymmetry for granted) that all solutions must indeed fall into discrete classes within which there are at most two degrees of freedom, as the intuitive physical reasoning described above would suggest. (It is this property of having no degrees of freedom except those corresponding to multipoles which can not be directly radiated away which Wheeler has described by the statement that "a black hole has no hair".) It can also be shown that in the unlikely event that there does exist any other family than the Kerr family, it must be somewhat anomalous, in so much as the range of variation of the angular momentum within the family cannot include zero.

In addition to the strong rigidity theorem, Stephen Hawking has also derived a very valuable theorem which he will describe in the accompanying lecture course, stating that the constant time sections of stationary black hole boundaries must be topologically spherical. Using these two theorems as the starting point and making the further highly plausible but not yet rigorously justified assumption that there cannot be more than one topologically spherical black hole component in an asymptotically flat pure vacuum stationary exterior space, I shall give a fairly complete derivation of the basic no-hair theorem whose content I have just described. I shall also demonstrate, subject to the further *assumption* of axisymmetry that the Schwarzschild black holes are the only ones with stationary pure vacuum exteriors and with zero angular momentum; I hope that the treatment given here will also be useful as an introduction to the study of the full scale Israel theorem, in which axisymmetry is not assumed. Israel's original version of the theorem (1967) contained an alternative simplifying assumption which has since been removed, at considerable cost in technical complexity, by Muller Zum Hagen, Robinson, and Siefert (1972). Both these versions of the theorem are incomplete in that they assume various boundary conditions on what is presumed to be the surface of the black hole without reference to a proper global definition of the black hole event horizon; it should however be clear from the present treatment that these boundary conditions can in fact be justified without too much difficulty by means of the lemma described in section 4.

The net effect of the mathematical results which I have just outlined is to make it virtually certain that for practical applications a stationary pure vacuum black hole can be taken to be a Kerr black hole. However it is by no means certain that the whole range of Kerr black hole states is physically attainable. The recent work by Chandrasekhar (1969), (1970), and Bardeen (1971) on equilibrium configurations for a self gravitating liquid in General relativity, suggests by analogy the likelihood that although the Kerr solutions are almost certainly stable for sufficiently small values of the dimensionless parameter J^2/M^4, (where M is the mass and J the angular momentum in natural units) there may be an eigenvalue of this parameter (below the cosmic censorship limit $J^2/M^4 = 1$) above which they would become unstable. The existence of such an eigenvalue might be revealed in a purely stationary analysis by the presence of a first order perturbation solution bifurcating from the Kerr solutions at the corresponding value of J^2/M^4. I have shown in the proof of the no-hair theorem that there is no axisymmetric first order bifurcating solution, but the theorem of Hawking, which excludes the existence of *exact* non-axisymmetric perturbations, does not rule out the possibility of first order non-axisymmetric perturbations. This suggests that if there is an instability in the higher angular momentum Kerr black holes it is likely that it will lead to the formation of *non*-axisymmetric deformations. If the cosmic censorship hypothesis is correct the deformed black hole would presumably get rid of its excess angular momentum by gravitational radiation, and ultimately settle down towards one of the stable Kerr black holes below the critical eigenvalue. The recent developments in dynamic perturbation theory, in which very diverse techniques have been introduced by groups such as Chandrasekhar and Friedman, Hawking and Hartle, and Press and Teukolsky, encourage the hope that the question of the stability of the upper part of the Kerr black hole sequence will be solved conclusively, one way or the other, in the not too distant future. Although it appears that the most exciting future development in black hole theory will be concerned with dynamic aspects, there remains a great deal to be done in stationary black hole theory, particularly in relation with non-vacuum black holes. To start with there is the question of how the results would be affected by the presence of an electromagnetic field. I have been able to allow for the presence of such a field in all the results described in sections 2 to 11, but although Israel was able by an impressive tour de force (Israel 1968) to generalize his theorem to cover the electromagnetic case, I have not yet been able to do the same for the no-hair-theorem (given in section 12). One would expect (as it is discussed in section 13) that a stationary black hole solution of the source-free Einstein–Maxwell equations would have just two extra degrees of freedom, (in addition to the mass and angular momentum) which would correspond to the conserved electric and magnetic monopole charges. In particular this means that an uncharged black hole should have no magnetic dipole moment.) While no one has yet been able to exclude the existence of a solution family branching off from the Kerr–Newman solution family (which has just the four degrees of freedom one would expect), it is not so difficult (as has been shown by Wald, and also independently by Ipser) to prove that no such bifurcating solution family can start off from the pure vacuum Kerr subfamily.

For practical astrophysical applications the question of electromagnetic generalizations is probably not of very great importance, since magnetic monopoles appear not to exist, and due to the fact that charges can be carried by electrons and positrons (which are very light compared with the baryons which normally give the principal contribution to the mass), it is very easy for electric charges to be neutralized by ionization or pair creation processes, so that in an actual collapse situation it is hard to see how any excess charge carried down into the black hole could ever be more than a small fraction of a per cent (in natural units) of the mass. A potentially more important generalization is to the case where the black hole is distorted from the simple Kerr form by the presence of some sort of fluid. In the case of a black hole formed from the collapse of an ordinary star such effects would probably be negligible, since in order to produce significant deviation effects it is necessary that the perturbing mass be not too small compared with the mass of the hole, and also that it be within a distance not too large compared with the Schwarzschild radius of the hole. In other words the perturbing material must be condensed to a density not too far below that determined by the dimensions of the hole. In the case of an ordinary star sized black hole this could only arise if the perturbing material itself consisted of a neutron star or another black hole in a close binary configuration, and due to its lack of overall axisymmetry such a system would lost energy rapidly by emission of gravitational radiation, and could survive for at most a matter of seconds. However it seems much more likely that a supermassive black hole of the kind whose existence in galactic nuclei was originally suggested by Lynden-Bell (1969) could deviate significantly from the Kerr form, at least during an initial active period, lasting perhaps for a million years or so after its formation. The sort of object I have in mind would be an axisymmetric system consisting of a central black hole of something like 10^9 solar masses, surrounded at a distance of a few Schwarzschild radii, by an equally massive cloud consisting of 10^9 ordinary individual stellar mass Kerr black holes. To minimize the rate of collisions (which would in any case be extremely small in terms of timescales comparable with the several hour long orbital periods) one could make the further not implausible supposition that the small Kerr black holes move in fairly coherent orbits in a Saturn-like disc configuration. Being out of phase, the gravitational radiation contributions of the individual Kerr black holes would almost completely cancel out, and as far as calculations of the large scale properties of the central non-Kerr black hole were concerned, the cloud of small black holes could be treated approximately as a (suitably non-isotropic) fluid. Except in the final sections 10 to 13 which are specialized to the vacuum cases, all the results which follow will be valid for general stationary but non-Kerr black hole configurations of the kind which I have just described. A more detailed treatment of many aspects of non-vacuum black hole theory is contained in the following course given by Jim Bardeen.

2 The Domain of Outer Communications and the Global Horizon

Throughout this course we shall be dealing with a space-time manifold $\mathcal{M}$ which is *asymptotically flat* and *pseudo-stationary*. By *asymptotically flat* we mean that $\mathcal{M}$ is weakly asymptotically simple, in the sense of Penrose (1967), and by *pseudo-stationary* we mean that $\mathcal{M}$ is invariant under a one-parameter isometry group action π^s:R(1) x $\mathcal{M} \to \mathcal{M}$ of the one-parameter translation group R(1), and that the trajectories of this action π^s (i.e. the sets of points $\{x \in \mathcal{M}: x = \pi^s(t, x_0), t \in \mathrm{R}(1)\}$ for fixed $x_0 \in \mathcal{M}$) are *timelike* curves at least at sufficiently large asymptotic distances. If the trajectories of π^s were timelike curves everywhere, $\mathcal{M}$ would be said to be stationary in the strict sense. The maximal connected asymptotically flat subdomain $\mathcal{S}$ of $\mathcal{M}$ that is stationary in this strict sense will be referred to as the *outer stationary domain* of $\mathcal{M}$.

Our attention in this course will be almost exclusively restricted to the *domain of outer communications* $\langle\!\langle \mathcal{I} \rangle\!\rangle$ of $\mathcal{M}$ which is defined as the set of points from which there exist both future and past directed timelike curves extending to arbitrarily large asymptotic distances. In the terminology of Penrose (1967), $\langle\!\langle \mathcal{I} \rangle\!\rangle$ is the intersection of the chronological past of $\mathcal{I}^+$ with the chronological future of $\mathcal{I}^-$. It is evident that $\langle\!\langle \mathcal{I} \rangle\!\rangle$ (like $\mathcal{S}$) is an open subset of $\mathcal{M}$, and can therefore be regarded as an asymptotically flat space-time manifold in its own right. Clearly the boundary $\langle\!\langle \mathcal{I} \rangle\!\rangle^\bullet$ of $\langle\!\langle \mathcal{I} \rangle\!\rangle$ in $\mathcal{M}$, which we shall denote more briefly by $\mathcal{H}$, can be thought of as a union of the form $\mathcal{H} = \mathcal{H}^+ \cup \mathcal{H}^-$, where the subset of points $\mathscr{H}^+$ forms part of the boundary of the chronological past of $\mathscr{I}^+$, (i.e. the boundary of the subdomain of $\mathscr{M}$ from which there are future directed timelike curves extending to arbitrarily large asymptotic distances) and where similarly the subset $\mathscr{H}^-$ forms part of the boundary of the chronological future of $\mathscr{I}^-$. Using the lemmas of Penrose and Hawking (Penrose (1965), Hawking (1966)) it is easy to see that $\mathscr{H}^+$ must be a *global past horizon* in the sense (Carter 1971) that it is an achronal hypersurface (i.e. a not necessarily differentiable hypersurface with the property that no two points on it can be connected by a timelike curve) such that from any point on it there is a null geodesic which can be extended without bound towards the future (but not necessarily towards the past) lying entirely in $\mathscr{H}^+$. Similarly $\mathscr{H}^-$ is a future global horizon.

These definitions and properties of $\langle\!\langle \mathscr{I} \rangle\!\rangle$, $\mathscr{H}^+$ and $\mathscr{H}^-$ do not depend in any way on the pseudo-stationary property. In a dynamic collapse starting from well behaved intial conditions $\mathscr{H}^-$ would not exist, and $\mathscr{H}^+$ could be regarded as being either the *past* boundary of the *black hole* (i.e. the region of space-time outside $\langle\!\langle \mathscr{I} \rangle\!\rangle$ or equivalently as the *future* boundary of the domain of outer communications (see Figure 1). The example of spherical collapse shows however that one cannot expect that the corresponding pseudo-stationary limit space will, when extrapolated back in time by the group action, have well behaved initial conditions, although one can at least demand that the closure of the domain of outer communications should be geodesically complete (in the sense that any inextensible geodesic of finite length will have an end point on the boundary) if a horizon $\mathscr{H}^-$ is allowed for in $\mathscr{M}$. The region lying to the future of $\mathscr{H}^+$ can still be quite appropri ately described as a *black hole* since, by its definition, no light from within it can escape to an external observer. The analogous region to the past of $\mathscr{H}^-$ (if it exists), which has the property that it can never be reached *by* any signal or probe from outside, has been referred to as a white hole–a rather misleading term, since light of any colour might in principle emerge from it. Personally I should prefer to refer to any such region (i.e. a region lying outside the future of $\mathscr{I}^-$) as a *primordial hole.* As far as the present course is concerned it will not be necessary to worry about these distinctions, and I shall simply refer to the whole of the region outside the domain of outer communications as a *hole*† without further qualification. Since this work is motivated by the problem of gravitational collapse, our attention will be entirely devoted to the domain of outer communications $\langle\!\langle \mathscr{I} \rangle\!\rangle$ and its future boundary, the horizon $\mathscr{H}^+$. It will at no stage be of any particular importance whether the horizon $\mathscr{H}^-$ exists, nor what its properties may be if it does.

Since the domain of outer communications and its bounding horizon $\mathscr{H}^+$ are globally defined subsets, it is not always easy to ascertain their position in relation

† A consensus at this school has agreed that the most appropriate French translation of the term black hole is "piège noir". For more general purposes it has been suggested that the term hole should be translated as "poche"; thus primordial hole would be translated as "poche primordiale".

to locally defined structures on $\mathscr{M}$. Under favourable conditions however it is possible to determine the position of $\mathscr{H}^+$ locally, in relation to the symmetry group structure, as will be shown in the next section. In establishing such relationships, I have found the following lemma, which gives an alternative global characterization of the domain of outer communications, extremely useful.

LEMMA 2 $\ll\mathscr{I}\gg$ is the maximal connected asymptotically flat domain of $\mathscr{M}$ with the property that the trajectory $\pi^s(x_0)$ of the pseudo-stationary action π^s through any point x_0 in the domain will, if extended sufficiently far in the forward direction, enter and remain in the chronological future of x_0 (i.e. in the set of points which can be reached by a future directed time-like curve from x_0).

Proof The proof consists of two parts: Part I in which it is shown that any connected asymptotically flat domain $\mathscr{D}$ of $\mathscr{M}$, with the property that the trajectory $\pi^s(x_0)$ of π^s through any point $x_0 \in \mathscr{D}$ will enter the chronological future of x_0 if extended sufficiently far in the forward direction, must lie within $\ll\mathscr{I}\gg$; and Part II in which it is shown that the trajectory of π^s through any point x_0 of $\ll\mathscr{I}\gg$ will enter and remain in the chronological future of x_0 if extended sufficiently far in the forward direction.

Part I: The required result follows at once from the fact that any point $x_0 \in \mathscr{D}$ can be connected to some point y' on the trajectory $\pi^s(y)$ of the action π^s through any point $y \in \mathscr{D}$ by a future directed timelike curve λ in $\mathscr{D}$. The existence of λ is established as follows. Since $\mathscr{D}$ is connected there certainly exists some curve γ from x_0 to y. By the defining property of $\mathscr{D}$, there must exist some point x_0' on the trajectory $\pi^s(x_0)$ such that x_0 lies in the chronological future of x_0'. Therefore any point x_1 on γ sufficiently close to x_0 will lie on some future directed timelike curve λ_1 from x_0'. Since γ is compact, it is possible by repeating this process to obtain a finite sequence of points x_i on γ starting with x_0 and ending with y, such that each point x_i lies on a future directed timelike curve λ_i which starts from some point x_{i-1}' on the trajectory $\pi^s(x_{i-1})$ of π^s through the preceding point x_{i-1} of the sequence. By transporting the timelike segments λ_i suitably under the group action π^s one can clearly construct a sequence of image segments which connect up end to end to form a future directed timelike curve λ from x_0 to some point on $\pi^s(y)$.

Part II: It is evident that a possible subdomain $\mathscr{D}$, with the properties specified above, is the outer stationary domain $\mathscr{S}$. It is clear from the definition of $\ll\mathscr{I}\gg$ that any point x_0 in $\ll\mathscr{I}\gg$ can be connected to some point x_1 in $\mathscr{S}$ by a future directed timelike curve λ_1, and that x_0 can be reached from some other point x_2 in $\mathscr{S}$ by a future directed timelike curve λ_3. Moreover by the properties of $\mathscr{D}$ discussed above, there must exist a future directed timelike curve λ_2 from x_1 to some point x_2' in the trajectory $\pi^s(x_2)$ of π^s through x_2. Therefore by transporting λ_3 suitably under the group action π^s, it is possible to construct an image segment λ_3' which links up with λ_1 and λ_2 to form a timelike curve λ from x_0 to some point x_0' on $\pi^s(x_0)$ via x_1 and x_2'. This establishes that $\pi^s(x_0)$ enters the chronological future of x_0. $\pi^s(x_0)$ may of course leave the future of x_0 after remaining within it only for a limited group parameter distance, but in this case (by repetition of the preceding argument) it will re-enter after a finite distance, and at the second entry must remain in the chronological future of x_0 for at least twice as long (as measured by the group parameter) as on the first occasion. It is easy to see that after at most a finite number of departures and re-entries, a point will be reached beyond which will never leave the chronological future of x_0. This completes the proof of the lemma.

In most of the work of this course, we shall need to employ the following postulate:

CAUSALITY AXIOM There are no compact (i.e. to be more explicit, topologically circular) causal (i.e. everywhere timelike or null) curves in $\mathscr{M}$.

This is almost the weakest possible global causality condition. Its necessity for any reasonably well behaved physical situation can hardly be question. The lemma which has just been established does not itself depend on the causality condition, but most of its applications do. An example is the following immediate corollary:

COROLLARY If the causality axiom holds in $\mathscr{M}$, then any degenerate trajectory of π^s (i.e. any fixed point of the action) and also any topologically circular trajectory of π^s, must lie outside the domain of outer communications $\ll\mathscr{I}\gg$.

Fixed points and circular trajectories *can* however lie on the boundary $\mathscr{H}$ of $\ll\mathscr{I}\gg$–indeed this is where one would normally expect them to turn up, as is shown by the example of the Kerr black holes. The causality axiom is satisfied in a Kerr black hole solution if, in defining $\mathscr{M}$, we exclude the region within the inner horizon at $r = r_-$. In this case there are in fact circular trajectories of the pseudo-stationary action on the Boyer–Kruskal axis $\mathscr{H}^+ \cap \mathscr{H}^-$, and also fixed points of the pseudo-stationary action where $\mathscr{H}^+ \cap \mathscr{H}^-$ intersects the rotation axis. In the more special case of a Schwarzschild black hole, $\mathscr{H}^+ \cap \mathscr{H}^-$ consists entirely of fixed points of the pseudo-stationary action.

The lemma which we have just proved can also be used, again in conjunction with the causality axiom, to show that the domain of outer communications of a pseudo-stationary space-time manifold is a fibre bundle over a well behaved three-dimensional base manifold, the fibres being the trajectories of the group action (Carter 1972b).

3 Axisymmetry and the Canonical Killing Vectors

Whenever it is convenient to do so, I shall suppose that the space-time manifold $\mathscr{M}$ under consideration is axisymmetric, i.e. that it is invariant under an action $\pi^A: SO(2) \times \mathscr{M} \to \mathscr{M}$ of the one-parameter rotation group SO(2), in addition to being invariant under the pseudo-stationary action π^s of R(1). In fact apart from the discussion of static ergosurfaces in the present section, the only stage at which

I shall *not* assume axisymmetry will be in the derivation of the Hawking-Lichnerowicz theorem (which would not in any way be simplified by assuming axisymmetry).

The original justification, (when the work described here was first undertaken) for postulating axisymmetry was simply that it is difficult to imagine any realistic astrophysical system to which the theory might be applied which could, as a good approximation be treated as pseudo-stationary without being able, as an equally good approximation, to be treated as axisymmetric (the idea being that any non-axisymmetric bulge would probably need to rotate to support itself, thereby radiating gravitationally). It is now possible to justify the axisymmetry condition as a mathematical necessity, at least in the case of *rotating* black holes, by appeal to Hawking's strong-rigidity theorem according to which, subject to very weak and general assumptions, which are described by Hawking in the accompanying notes, there exists an isometry group action $\pi^{s\dagger}$ of R(1) on $\mathscr{M}$ with the property that the null geodesic generators of $\mathscr{H}^+$ are trajectories of $\pi^{s\dagger}$. The black hole is said to be *non-rotating* if $\pi^{s\dagger}$ is the same (up to a scale factor) as π^s, and otherwise it is described as *rotating*. On account of the restrictions imposed by asymptotic flatness, it may be taken for granted that the axisymmetry action π^A whose existence I have postulated commutes with the pseudo-stationary action π^s (see Carter 1970; I should point out that in this reference I ought to have stated that the vector fields referred to in the statement of Proposition 5 and in the proof of Proposition 6 must be complete). Hawking has shown directly that the action $\pi^{s\dagger}$ predicted by his theorem commutes with π^s and that in the rotating case the two together generate a two parameter Abelian isometry group action π^{sA} with a subgroup action π^A of SO(2). In the non-rotating case it is possible to establish axisymmetry—and indeed spherical symmetry—by a rather different method provided that $\ll\mathscr{I}\gg$ is a pure or electromagnetic vacuum, by using first the generalized Hawking-Lichnerowicz theorem to be described in section 6, and then the theorems of Israel (1967), (1968). In the case of a non-rotating hole with external matter present, axisymmetry is *not* a mathematical necessity, but even in this case it seems unlikely that there would be any natural astrophysical applications which would not be axisymmetric in practice.

Even when $\mathscr{M}$ has a many parameter isometry group, resulting from axial or even spherical symmetry, there will still be only one unique pseudo-stationary subgroup action π^s (up to a scale factor), since any other subgroup action will have spacelike trajectories at large distances. We shall denote the Killing vector generator of π^s by

$$\frac{\partial}{\partial t} = k^a \frac{\partial}{\partial x^a} \tag{3.1}$$

where the x^a ($a = 0, 1, 2, 3$) are general local co-ordinates, and where t is a group parameter along the trajectories. (At a later stage we shall impose further restrictions in order that t shall be well defined as a canonical co-ordinate function.) We shall fix the scale factor by imposing the standard normalization condition, which consists of the requirement that the squared magnitude scalar

$$V = -k^a k_a \tag{3.2}$$

should satisfy

$$V \to 1 \tag{3.3}$$

in the asymptotic limit at large distances. Subject to this requirement, the vector k^a is uniquely determined. It will of course satisfy the Killing equations

$$k_{a;b} = k_{[a;b]} \tag{3.4}$$

Except in the spherically symmetric case, in which a choice must be exercised, the axisymmetry action π^A will also be uniquely determined, (up to a scale factor) since other one-parameter subgroup actions will in general have non-compact trajectories. We shall denote its generator by

$$\frac{\partial}{\partial \varphi} = m^a \frac{\partial}{\partial x^a} \tag{3.5}$$

where φ is a group parameter along the trajectories. The vector field m^a will be zero on a necessarily timelike two dimensional *rotation axis.* We shall fix the scale factor by imposing the standard normalization condition, which consists of the requirement that the squared magnitude scalar

$$X = m^a m_a \tag{3.6}$$

should satisfy

$$\frac{X_{,a} X^{,a}}{4X} \to 1 \tag{3.7}$$

in the limit on the rotation axis. This ensures that the group parameter φ has the standard periodicity 2π. Subject to this requirement m^a is uniquely determined in the spherical case. It will of course satisfy the Killing equations

$$m_{a;b} = m_{[a;b]} \tag{3.8}$$

We have already remarked that the actions π^s and π^A must commute, meaning that together they generate a 2-parameter action $\pi^{sA} = \pi^s \oplus \pi^A$ of the group R(1) x SO(2), defined by

$$\pi^{sA}(t, \varphi, x) = \pi^s(t, \pi^A(\varphi, x)) = \pi^A(\varphi, \pi^s(t, x))$$

for any $t \in R(1)$, $\varphi \in SO(2)$, $x \in \mathcal{M}$. It follows that the generators k^a and m^a satisfy the local commutation condition

$$m^a{}_{;b} k^b = k^a{}_{;b} m^b \tag{3.9}$$

The surfaces of transitivity of the action π^{sA} i.e. the sets of points

$$\{x \in \mathcal{M}: x = \pi^{sA}(t, \varphi, x_0), t \in R(1), \varphi \in SO(2)\}$$

will clearly be timelike in $\mathcal{S}$. We shall define the *outer stationary axisymmetric domain* $\mathcal{W}$ of $\mathcal{M}$ as the maximal asymptotically flat sub-domain of $\mathcal{M}$ in which these surfaces of transitivity are timelike, or equivalently in which the Killing bivector

$$\rho_{ab} = 2k_{[a}m_{b]} \tag{3.10}$$

is timelike. Thus $\mathcal{S}$ is the maximal connected asymptotically flat region in which $V > 0$ and $\mathcal{W}$ is the maximal connected asymptotically flat region in which $\sigma > 0$, where we introduce the notation

$$\sigma = -\tfrac{1}{2}\rho_{ab}\rho^{ab} = VX + W^2 \tag{3.11}$$

where

$$W = k^a m_a \tag{3.12}$$

We note that

$$W = X = \sigma = 0 \tag{3.13}$$

on the rotation axis, where m^a is zero. By a trivial application of the lemma of the previous section it is clear that $\mathcal{S}$ must lie within $\ll\mathcal{I}\gg$. By considering the locally intrinsically flat geometry of one of the cylindrical timelike 2-surfaces of transitivity of the action π^{sA} in $\mathcal{W}$, it is clear that any trajectory of the action through a point x_0 in the 2-cylinder must ultimately enter the chronological future of x_0 defined in relation to the intrinsic geometry on the 2-cylinder, and must hence, *a fortiori*, enter the chronological future of x_0 in the 4-dimensional space-time geometry of $\mathcal{M}$. Therefore by application of the lemma of the previous section we deduce that $\mathcal{W}$ must also lie in $\ll\mathcal{I}\gg$. Since we have already remarked that $\mathcal{S}$ must lie within $\mathcal{W}$ we see therefore that we must have

$$\mathcal{S} \subseteq \mathcal{W} \subseteq \ll\mathcal{I}\gg \tag{3.14}$$

whenever the relevant domains are defined.

We conclude this section by noting that while $\mathcal{S}$ is characterized by $V > 0$, and $\mathcal{W}$ is characterized by $\sigma > 0$, the *whole* of $\mathcal{M}$ except for the rotation axis on which $X = W = \sigma = 0$ must be characterized by

$$X > 0. \tag{3.15}$$

if the causality axiom holds.

4 Ergosurfaces, Rotosurfaces and the Horizon

In any space-time manifold with a pseudo-stationary isometry group action generated by a Killing vector k^a, the energy E of a particle moving on a timelike geodesic orbit is a constant given by $E = -mk^a v_a$ where m is the rest mass of the particle and v^a is the future oriented unit tangent vector of the orbit. It is evident that this energy must always be positive within the outer stationary domain $\mathscr{S}$. More generally E must always have a definite sign (depending on the time-orientation of k^a) in any region in which $V = -k^a k_a$ is positive, while on the other hand it may have either sign at any point where k^a is spacelike. We shall refer to any locus on which V is zero, i.e. any boundary separating regions where k^a is timelike from regions where it is spacelike, as *an ergosurface*, and more particularly, we shall refer to the boundary $\mathscr{S}^\bullet$ of $\mathscr{S}$ as *the outer ergosurface.*

When $\mathscr{M}$ is axisymmetric as well as pseudo-stationary we shall in an analogous manner refer to a locus on which σ is zero as a *rotosurface*, and more particularly we shall refer to the boundary $\mathscr{W}^\bullet$ of the outer stationary-axisymmetric domain $\mathscr{W}$ as the *outer rotosurface.*

Another way of thinking of these surfaces is to regard $\mathscr{S}^\bullet$ as a *staticity* limit, i.e. the boundary of the outer connected region within which it is kinematically possible for a timelike particle orbit to satisfy the condition of *staticity* i.e. for its tangent vector v^a to be parallel to k^a; similarly $\mathscr{W}^\bullet$ can be thought of as a *circularity* limit i.e. the boundary of the outer connected region within which it is kinematically possible for a timelike orbit to satisfy the condition of *circularity,* i.e. for its tangent vector to be a linear combination of k^a and m^a so that the orbit represents a uniform circular motion. [The ergosurface $\mathscr{S}^\bullet$ has sometimes been referred to in the literature as an "infinite red-shift surface", but this is misleading since the only physical particles whose observed light will suffer a genuine infinite red shift are those which cross the black hole horizon $\mathscr{H}^+$ itself.]

In consequence of (3.14) we see that the rotosurface $\mathscr{W}^\bullet$—when it is defined—must always lie outside or on the hole boundary $\mathscr{H}$ and similarly that the ergosurface $\mathscr{S}^\bullet$ must always lie outside or on both $\mathscr{W}^\bullet$ and $\mathscr{H}$. The main objective of this section is to show that there are natural conditions under which either $\mathscr{S}^\bullet$ or $\mathscr{W}^\bullet$ (or both) must actually coincide with $\mathscr{H}$. This will be of great value when we come on to the study of black hole uniqueness problems, since (being globally defined) the hole boundary may in general be difficult to locate without integrating the null geodesic equations. When $\mathscr{H}$ can be identified with $\mathscr{S}^\bullet$ or $\mathscr{W}^\bullet$ whose positions are locally determined in terms of the symmetry group structure, what would be an extremely intractable integro-partial differential problem reduces to a relatively straightforward (albeit non-linear) partial differential equation boundary problem.

The situations under which this simplification occurs, arise when either the pseudo stationary action π^S or the pseudo-stationary axisymmetry action $\pi^S \oplus \pi^A$

is *orthogonally transitive* meaning that the surface of transitivity (which will have dimensionality p equal respectively to 1 or 2 everywhere except on loci of degeneracy) are orthogonal to a family of surfaces of the conjugate dimension (i.e. of dimension $n - p$ which will be respectively 3 or 2). In the former case, i.e. when π^S is orthogonally transitive, the geometry is said to be *static*. In the latter case, i.e. when π^{SA} is orthogonally transitive, I shall describe the geometry as *circular* for reasons which will be made clear in section 7. By Frobenius theorem the necessary and sufficient condition for the geometry to be static in $\mathscr{S}$ (i.e. for the Killing vector k^a to be orthogonal to a family of hypersurfaces) is

$$k_{[a;b}k_{c]} = 0 \tag{4.1}$$

and similarly the necessary and sufficient condition for the geometry to be circular in $\mathscr{W}$, (i.e. for the Killing bivector ρ_{ab} to be orthogonal to a family of 2-surfaces) is

$$\begin{aligned} k_{[a;b}\rho_{cd]} &= 0 \\ m_{[a;b}\rho_{cd]} &= 0 \end{aligned} \tag{4.2}$$

It will be made clear in sections 6 and 7 that these conditions of staticity and circularity are not imposed gratuitously but that as a consequence of the Hawking strong rigidity theorem one or other can be expected to hold in the exterior of a stationary black hole under practically any naturally occurring conditions.

The purpose of the present section is to show that subject to suitable global conditions, the ergosurface (or staticity limit) $\mathscr{S}^\bullet$ coincides with $\mathscr{H}$ in the static case, while the rotosurface (or circularity limit) $\mathscr{W}^\bullet$ coincides with $\mathscr{H}$ in the circular case. The demonstration depends on the two following lemmas, of which the first was originally given (independently) by Vishveshwara (1968) and Carter (1969) and the second was originally given by Carter (1969).

LEMMA 4.1 If the staticity condition (4.1) is satisfied in $\mathscr{S}$, then the boundary $\mathscr{S}^\bullet$ consists of null hypersurface segments except at points of degeneracy of the action π^S (i.e. where k^a is zero).

LEMMA 4.2 If the circularity conditions (4.2) is satisfied in $\mathscr{W}$, then the boundary $\mathscr{W}^\bullet$ consists of null hypersurface segments except at points of degeneracy of the action $\pi^S \oplus \pi^A$ (i.e. where ρ^{ab} is zero).

Proof of Lemma 4.1 We start by using the Killing antisymmetry condition 3.3 to convert the Frobenius orthogonality condition 4.1 to the form

$$2k_{a;[b}k_{c]} = k_a k_{[b;c]} \tag{4.3}$$

from which, on contracting with k^a and using the definition (3.2) of V, we obtain

$$V_{,[b}k_{c]} = Vk_{[b;c]} \tag{4.4}$$

This tells us immediately that on the locus $\mathscr{S}^\bullet$ on which V is zero, the gradient $V_{,b}$ is parallel to k^a and hence null there, (except at points of degeneracy where k^a is actually zero). In the *non-degenerate* case, i.e. when the gradient $V_{,b}$ is non-zero, so that $V_{,b}$ determines the direction of the normal to the hypersurface on which $V = 0$, it follows immediately that this hypersurface is a null hypersurface as required. In the degenerate case where $V_{,b}$ is zero on the locus $V = 0$ more care is required since V might be zero not just on a hypersurface but over a domain of finite measure. However by more detailed consideration of the situation (cf. Carter (1969)) it can easily be verified that it will still be true that the actual boundary $\mathscr{S}^\bullet$ must be a null hypersurface.

Proof of Lemma 4.2 By analogy with the previous case, we use the Killing antisymmetry conditions (3.4) and (3.8) to convert the Frobenius orthogonality conditions (4.2) to the form

$$\left.\begin{aligned} k_{a;[b}\rho_{cd]} &= -k_a k_{[b;c}m_{d]} + m_a k_{[b;c}k_{d]} \\ m_{a;[b}\rho_{cd]} &= m_a m_{[b;c}m_{d]} - k_a m_{[b;c}m_{d]} \end{aligned}\right\} \tag{4.5}$$

from which we can directly obtain

$$2\rho_{ae;[b}\rho_{cd]} = \rho_{ae}\rho_{[cd;b]} \tag{4.6}$$

Contracting this with the Killing bivector ρ_{ab} we obtain

$$\sigma_{,[b}\rho_{cd]} = \sigma\rho_{[cd;b]} \tag{4.7}$$

This equation is analogous to (4.4), and tells us immediately that in the *non degenerate* case, i.e. when the gradient $\sigma_{,b}$ is non-zero on the boundary $\mathscr{W}^\bullet$ where σ is zero, the normal to $\mathscr{W}^\bullet$ which is parallel to $\sigma_{,b}$ lies in the plane of the Killing bivector ρ_{ab}. This is only possible (since the normal must also be orthogonal to this bivector) if the normal is null, i.e. if $\mathscr{W}^\bullet$ is a null hypersurface. As before it is still possible with rather more care (see Carter 1969) to deduce that the boundary $\mathscr{W}^\bullet$ is a null hypersurface even in the degenerate case where $\sigma_{,b}$ is zero, (except on the lower dimensional surfaces of degeneracy of the group action, where the Killing bivector ρ_{ab} is itself zero). This completes the proof.

The two preceding lemmas may be used respectively in conjunction with the lemma of section 2, to prove the two following theorems, which are of central importance in stationary black hole theory: (these results have been given in a somewhat more general mathematical context by Carter 1972b).

THEOREM 4.1 (Static Ergosurface Theorem) Let $\mathscr{M}$ be a pseudo-stationary asymptotically flat space-time manifold with a *simply-connected* domain of outer communications $\ll\mathscr{I}\gg$. Then if the (chronological) *causality axiom* and the *staticity condition* (4.1) are satisfied in $\ll\mathscr{I}\gg$ it follows that $\ll\mathscr{I}\gg = \mathscr{S}$, where $\mathscr{S}$ is the outer stationary domain and hence that *the outer ergosurface* $\mathscr{S}^\bullet$ *coincides with the hole boundary* $\mathscr{H}$.

THEOREM 4.2 (Rotosurface Theorem) Let $\mathscr{M}$ be a pseudo-stationary axisymmetric asymptotically flat space-time manifold with a *simply connected* domain of outer communications $\langle\!\langle\mathscr{I}\rangle\!\rangle$. Then if the *causality axiom* and the *circularity condition* (4.2) are satisfied in $\langle\!\langle\mathscr{I}\rangle\!\rangle$ it follows that $\langle\!\langle\mathscr{I}\rangle\!\rangle = \mathscr{W}$ where $\mathscr{W}$ is the outer stationary-axisymmetry domain, and hence that the *outer rotosurface* $\mathscr{W}^{\bullet}$ *coincides with the hole boundary* $\mathscr{H}$.

Proof of Theorem 4.1 We have already noted that $\mathscr{S}$ must always be entirely contained within $\langle\!\langle\mathscr{I}\rangle\!\rangle$. Let $\mathscr{D}$ be a connected component of the complement of $\mathscr{S}$ in $\langle\!\langle\mathscr{I}\rangle\!\rangle$. Since $\mathscr{S}$ is connected (by definition) so also is the complement of $\mathscr{D}$ in $\langle\!\langle\mathscr{I}\rangle\!\rangle$. Hence by the condition that $\langle\!\langle\mathscr{I}\rangle\!\rangle$ is *simply* connected, it follows (from elementary homotopy theory) that the boundary $\mathscr{D}^{\bullet}$ of $\mathscr{D}$, as restricted to $\langle\!\langle\mathscr{I}\rangle\!\rangle$, is *connected.*

Now (as we remarked at the end of section 2) it follows from the causality axiom by the corollary to Lemma 2 that the action π^s can never be degenerate (i.e. k^a can never be zero) in $\langle\!\langle\mathscr{I}\rangle\!\rangle$, and hence it follows from Lemma 4.1 that the boundary $\mathscr{D}^{\bullet}$ of $\mathscr{D}$, as restricted to $\langle\!\langle\mathscr{I}\rangle\!\rangle$, must consist entirely of one connected null hypersurface. It follows that the outgoing normal (from $\mathscr{D}$) of this hypersurface must therefore be everywhere future directed or everywhere past directed; in the former case no future directed timelike line in $\langle\!\langle\mathscr{I}\rangle\!\rangle$ could ever enter $\mathscr{D}$ from $\mathscr{S}$, and in the latter case no future directed timelike line in $\langle\!\langle\mathscr{I}\rangle\!\rangle$ could ever enter $\mathscr{S}$ from $\mathscr{D}$. Neither alternative is compatible with the condition that $\mathscr{D}$ lies in $\langle\!\langle\mathscr{I}\rangle\!\rangle$, unless $\mathscr{D}$ is empty. This establishes the required result.

Proof of Theorem 4.2 By analogy with the previous case we note that $\mathscr{W}$ must lie entirely within $\langle\!\langle\mathscr{I}\rangle\!\rangle$, and we choose $\mathscr{D}$ to be a connected component of the complement of $\mathscr{W}$ in $\langle\!\langle\mathscr{I}\rangle\!\rangle$. As before we see that the boundary $\mathscr{D}^{\bullet}$ of $\mathscr{D}$ as restricted to $\langle\!\langle\mathscr{I}\rangle\!\rangle$, is connected.

We now use the causality axiom to establish that the action $\pi^s \oplus \pi^A$ is nowhere degenerate (i.e. the Killing bivector ρ_{ab} is nowhere zero) in $\langle\!\langle\mathscr{I}\rangle\!\rangle$ except on the rotation axis where m^a is zero. This follows from the fact that k^a must be parallel to m^a at any point of degeneracy, so that the trajectories of the action π^s would be circles. Any such circular trajectory of π^s must lie outside $\langle\!\langle\mathscr{I}\rangle\!\rangle$ if the causality axiom holds, by the corollary to Lemma 2.

We can now use Lemma 4.2 to deduce that the boundary $\mathscr{D}^{\bullet}$ of $\mathscr{D}$ as restricted to $\langle\!\langle\mathscr{I}\rangle\!\rangle$ must consist of a null hypersurface everywhere, except perhaps at points on the rotation axis where m^a is zero. We can go on to deduce that the outgoing normal to this boundary must be everywhere future directed or everywhere past directed as before, despite the possibility of degeneracy on the rotation axis, since the rotation axis must be a *timelike* 2-surface everywhere (by local geometrical considerations) and as such could never form the boundary of a null hypersurface. Hence as in the previous case we deduce that $\mathscr{D}$ must be empty as required. This completes the proof.

The mathematically important qualification that the domain of outer communications should be simply connected does not restrict the practical application of these theorems, since Stephen Hawking has given arguments (see his discussion in the accompanying lecture course) which indicate that in any reasonably well behaved gravitational collapse situation the domain of outer communications of the resulting final equilibrium state should have the topology of the product of the Euclidean line R(1) with a 3-space which has the form of a Euclidean R(3) from which a number of three dimensional balls (solid spheres) have been removed, and which is therefore *necessarily* simply connected. Except in artificially contrived situations one would expect further that the 3-space would have the form of a Euclidean 3-space from which *only one* 3-dimensional ball has been removed, i.e. that it would have the topology S(2) x R(1) where S(2) is the 2-sphere, so that the domain of outer communications as a whole would have the topology S(2) x R(2). This latter stronger condition will be imposed in the theorems of sections 10 to 13.

We shall conclude this section by proving an important corollary to Theorem 4.2, which is as follows:

COROLLARY TO THEOREM 4.2 (Rigidity Theorem) Under the conditions of Theorem 4.2, it is possible to choose the normalization of the null tangent vector l^a of $\mathscr{H}^+$ in such a way that it has the form

$$l^a = k^a + \Omega^H m^a \tag{4.8}$$

where the scalar Ω^H is a *constant* over any connected component of $\mathscr{H}^+$.

Proof It is evident from (4.2) that l^a is a linear combination of k^a and m^a on the horizon, and hence can be expressed in the form (4.8) where Ω^H is some scalar which represents the local *angular velocity* of the horizon. The non trivial part of the proof is the demonstration that Ω^H is constant, i.e. that the rotation which it determines is *rigid*, which is a consequence of the commutation condition (3.8), which was not required for the proof of the basic Theorem 4.2. Since any vector in the horizon including m^a in particular, is orthogonal to l^a we find on contracting (4.8) with m^a that Ω^H must be given by

$$X\Omega^H = -W \tag{4.9}$$

where W and X are defined by (3.12) and (3.6). (On the rotation axis, where W and X are both zero, Ω^H must be defined by a limiting process). Using the relations

$$\left.\begin{aligned} X_{,a} &= 2m^b m_{b;a} \\ W_{,a} &= 2k^b m_{b;a} \end{aligned}\right\} \tag{4.10}$$

(of which the latter is a consequence of the commutation condition (3.8)) we find by differentiation of (4.9) that the gradient of Ω^H is given by

$$X^2 \Omega^H_{,a} = 2[Wm^b - Xk^b]m_{b;a} \tag{4.11}$$

It follows, from the second of the Frobenius equations (4.2), that we must have

$$X^2\Omega^H_{,[a}\rho_{bc]} = -4\sigma m_{[a;b}m_{c]} \tag{4.12}$$

where σ is defined by (3.11). Since we have already established that σ is zero on the horizon, this gives simply

$$\Omega^H_{,[a}\rho_{bc]} = 0 \tag{4.13}$$

This implies that the gradient of the coefficient Ω^H determined by (4.9) lies in the surface of transitivity of the group action π^{sA} at the horizon, which is impossible (since Ω^H is clearly a group invariant quantity) unless the gradient is actually zero on the horizon, i.e.

$$\Omega^H_{,a} = 0 \tag{4.14}$$

which is the required result. This completes the proof.

Having established that Ω^H as defined by (4.8) is constant on the horizon, we can use this constant in (4.8) to define a vector field l^a which will satisfy the Killing equations

$$l_{a;b} = l_{[a;b]} \tag{4.15}$$

everywhere in $\mathscr{M}$. Our demonstration of the existence of this Killing vector field l^a that is null on the horizon has been based on the assumption that either the staticity condition or the axisymmetric circularity condition is satisfied. The *strong* rigidity theorem of Hawking, to which we have already referred, establishes the existence of this Killing vector field without assuming either staticity or axisymmetry, subject only to very weak and general assumptions. If the hole is *rotating,* i.e. if l^a does not coincide with k^a then Hawking has shown, as an immediate corollary to his basic theorem, that there must exist an axisymmetry action π^A generated by a Killing vector field m^a such that l^a has the form (4.8). It then follows from the generalized Papapetrou theorem which will be described in section 7 that the circularity condition which was postulated in Theorem 4.2 must necessarily hold. On the other hand if the hole is non-rotating, i.e. if the Killing vector field l^a that is null on the horizon coincides with the pseudo-stationary Killing vector field k^a, it follows from the generalized Hawking–Lichnerowicz theorem which will be described in section 6 that the staticity condition which was postulated in Theorem 4.1 must necessarily hold (at least provided one is prepared to accept the as yet not rigorously proved supposition that the ergosurface which coincides with the horizon in the non-rotating case is the same as *the outer* ergosurface).

The net effect of the results of this section is to establish that $\mathscr{H}^+$ must be a *Killing horizon* (in the sense defined by Carter 1969) that is to say it is *a null hypersurface whose null tangent vector coincides (when suitably normalized) with the representative of some (fixed) Killing vector field*, this field being the l^a which has just been constructed. This makes it possible to carry out a precise analysis of the boundary conditions on $\mathscr{H}$ in the manner which will be described in the next section.

5 Properties of Killing Horizons

The horizon Killing vector l^a, which generates the action $\pi^{s\dagger}$ referred to in section 3, can be thought of as being in a certain sense complementary to the pseudo stationary Killing vector k^a. We shall consistently use a dagger to denote the complementary analogue, defined in terms of l^a, of a quantity originally defined in terms of k^a. Thus in particular we define

$$V^\dagger = -l^a l_a \tag{5.1}$$

$$W^\dagger = l^a m_a \tag{5.2}$$

The quantity X is self complementary, as also is σ which can be expressed in the form

$$\sigma = V^\dagger X + W^{\dagger 2} \tag{5.3}$$

Since l^a is orthogonal to any vector in the horizon, including m^a and l^a itself, it is immediately evident that $V^\dagger$ and $W^\dagger$ and hence also (as was established directly in Theorem 4.2) σ, are zero on the horizon, i.e. we have

$$V^\dagger = W^\dagger = \sigma = 0 \tag{5.4}$$

on $\mathscr{H}$, which is closely analogous to the condition (3.13) satisfied on the rotation axis. Whereas k^a is timelike at large asymptotic distances, (i.e. near $\mathscr{I}$) becoming spacelike between the ergosurface where $V = 0$ and the horizon (except in the non-rotating case where it is timelike right up to the horizon), on the other hand l^a is timelike just outside the horizon, becoming spacelike outside the *co-ergosurface,* where $V^\dagger = 0$ (except in the non-rotating case where l^a is timelike out to arbitrarily large distances). [We remark that an analogous co-ergosurface can be defined for any rigidly rotating body as the cylindrical surface at which a co-rotating frame is moving at the speed of light; the surface so defined is of importance in the Gold theory of pulsars.]

Being orthogonal to the horizon, l^a must satisfy the Frobenius orthogonality condition

$$l_{[a;b}l_{c]} = 0 \tag{5.5}$$

on $\mathscr{H}$. It is a well known consequence of this condition that the null tangent vector of any null hypersurface must satisfy the geodesic equation. To see this one uses the fact that the squared magnitude $-V^\dagger$ of l^a is a constant (namely zero) on the horizon, so that its gradient must be orthogonal to the horizon, and hence satisfies

$$V^{\dagger}_{,[a}l_{b]} = 0 \tag{5.6}$$

on the horizon. Contracting (5.5) with l^c and using (5.6) together with the condition $V^{\dagger} = 0$, one immediately obtains the geodesic equation

$$l_{[a}l_{b];c}l^c = 0 \tag{5.7}$$

This is equivalent to the condition that there exists a scalar κ which is in fact the positive root of the equation

$$\kappa^2 = \tfrac{1}{2}l_{a;b}l^{b;a} \tag{5.8}$$

such that

$$l^a{}_{;b}l^b = \kappa l^a \tag{5.9}$$

on $\mathscr{H}^+$. (On $\mathscr{H}^-$, l^a will satisfy an equation of the same form, except that κ is replaced by $-\kappa$). The scalar κ measures the deviation from affine parametrization of the null geodesic. If v is a group parameter on one of the null geodesic generators of $\mathscr{H}^+$, i.e. a parameter such that

$$l^a = \frac{dx^a}{dv} \tag{5.10}$$

where $x^a = x^a(v)$ is the equation of the geodesic, and if v^a is a renormalized tangent vector given by

$$v^a = \frac{dx^a}{dw} = \frac{dv}{dw}\,l^a \tag{5.11}$$

where w is an affine parameter, i.e. a function of v chosen so that v^a will satisfy the simple affine geodesic equation

$$v^a{}_{;b}v^b = 0, \tag{5.12}$$

then it follows that they will be related by

$$\frac{d}{dv}\left(\ln\frac{dw}{dv}\right) = \kappa \tag{5.13}$$

[The expression (5.8) for κ is obtained by contracting (5.5) with $l^{b;a}$ and using the Killing equation (4.15) with the defining relation (5.9).]

The scalar κ defined at each point on the horizon in this way, will turn up repeatedly in different contexts throughout the remainder of this course. (It also plays an important role in the accompanying course of Stephen Hawking; in terms of the Newman–Penrose notation used by Hawking, it is given by $\kappa = \epsilon + \bar{\epsilon}$ where ϵ is one of the standard spin coefficients for l^a, defined with respect to a null-tetrad which is Lie propagated by the field l^a.)

This scalar was first discussed explicitly by Boyer in some work which was written up posthumously by Ehlers and Stachel (Boyer 1968). To start with, Boyer pointed out that the condition $\kappa = 0$ is a criterion for *degeneracy* of the horizon, in the sense in which we have previously used this description, as meaning that the gradient of σ, or in the non-rotating case, of V, is zero. To see this we use (4.15) to convert (5.9) to the form

$$V^{\dagger}_{,a} = 2\kappa l_a \tag{5.14}$$

which makes it clear that the vanishing of κ is necessary and sufficient for the gradient of $V^{\dagger}$ to be zero on the horizon, and hence for the gradient of V to be zero on the horizon in the non-rotating case, and (since $W^{\dagger}$ is always zero on the horizon) for the gradient of σ to be zero on the horizon in the rotating case.

Next, Boyer pointed out that since κ is clearly constant on each null geodesic of the horizon, i.e.

$$\kappa_{,a} l^a = 0, \tag{5.15}$$

(this being evident from the fact that l^a, in terms of which κ is completely determined by (5.8) or (5.9), is a Killing vector) it follows that (5.13) can be integrated explicitly to give the affine parameter w (with a suitable choice of scale and origin) directly in terms of the group parameter v by

$$w = e^{\kappa v} \tag{5.16}$$

except in the degenerate case, i.e. when κ is zero, in which case it will be possible simply to take $w = v$. This shows that in the non-degenerate case, the affine parameter w varies only from 0 to ∞ as the group parameter ranges from $-\infty$ to ∞, and hence that the horizon $\mathscr{H}^+$ is incomplete in the past. Moreover the vector l^a itself tends to zero, as measured against the affinely parameterized tangent vector v^a by (5.11) in the limit as w tends to 0, and hence unless the space-time manifold $\mathscr{M}$ itself is incomplete along the null geodesics of $\mathscr{H}^+$, there must exist a *fixed point* of the subgroup action $\pi^{s\dagger}$ generated by l^a at the end point, with limiting parameter value $w = 0$, of each null geodesic of $\mathscr{H}^+$, and also a continuatio[n] of each null geodesic beyond the past boundary of $\mathscr{H}^+$ to negative values of the affine parameter w. By continuity, such fixed points cannot be isolated in space-time, but must clearly form spacelike 2-sections of the null hypersurface formed by continuing the null geodesics of $\mathscr{H}^+$ past their endpoints as subgroup trajectories. (The null geodesics cannot intersect each other, since their spacelike separation must remain invariant under the subgroup action.) Such a 2-surface, being invariant under the group action, will obviously determine two null hypersurfaces which intersect on it, forming the boundaries of its past and future, these null hypersurfaces being themselves also invariant under the subgroup action. Furthermore since in the present case the spacelike 2-surface is pointwise invariant under the subgroup action, it is clear that each individual null geodesic of the two null hypersurfaces will be invariant under the subgroup action, so that *both* null hypersurfaces are Killing horizons on which $V^\dagger$ is zero. [Both Boyer himself in his original unpublished work, and also Ehlers and Stachel in their rather different edited version (Boyer 1968) made rather heavier work than necessary of the proof of this last point.] Since $\mathscr{H}^+$, from which we started this construction process, forms the part of one of these null hypersurfaces lying to the future of the fixed point 2-surface, it can be seen that the part of *the other* null hypersurface lying to *the past* of the fixed point 2-surface must also lie on the boundary of $\langle\!\langle \mathscr{I} \rangle\!\rangle$, and hence must form part of $\mathscr{H}^-$. We could have made a similar construction starting from $\mathscr{H}^-$. We thus arrive at the following conclusion:

THEOREM 5.1 (Boyer's Theorem) Let the conditions of Theorem 4.1 or Theorem 4.2 be satisfied. Then if the horizon $\mathscr{H}$ is non-degenerate, and if the closure $\overline{\langle\!\langle \mathscr{I} \rangle\!\rangle}$ of $\langle\!\langle \mathscr{I} \rangle\!\rangle$ is geodesically complete, then both $\mathscr{H}^-$ and $\mathscr{H}^+$ exist and they intersect on a spacelike 2-surface (on which the Killing vector l^a is zero) which contains a future endpoint for every null geodesic of $\mathscr{H}^-$, and a past endpoint for every null geodesic of $\mathscr{H}^+$. On the other hand if $\mathscr{H}$ is degenerate, then $\mathscr{H}^+$ contains null geodesics which can be extended to arbitrary affine distance not only towards the future (as always) but also towards the past.

Since a null geodesic on $\mathscr{H}^+$ cannot approach $\mathscr{I}$ when it is extended towards the *past* (since it lies on the boundary of the *past* of $\mathscr{I}$) it follows that *in the degenerate case* $\langle\!\langle \mathscr{I} \rangle\!\rangle$ *possesses an internal infinity, or in other words the hole is bottomless,* in the manner we became familiar with in Part I of this course, in the limiting cases with $a^2 + Q^2 + P^2 = M^2$ of the generalized Kerr solution black holes. This fact lends support to the conjecture, which is associated with the cosmic censorship hypothesis, that degenerate black holes represent physically unattainable limits, in much the same way that the speed of light represents an unattainable limit for the speed of a massive particle, or (to introduce an analogy which we shall see later can be pushed considerably further) in the same way that the absolute zero of temperature represents an unattainable limit in thermodynamics.

We shall continue this section with the description of some further important boundary condition properties of Killing horizons. Substituting l^a into the Raychaudhuri identity

$$(u^a{}_{;a})_{;b}u^b = (u^a{}_{;b}u^b)_{;a} - u^a{}_{;b}u^b{}_{;a} - R_{ab}u^a u^b \tag{5.17}$$

which is satisfied by any vector field u^a whatsoever, where R_{ab} is the Ricci tensor, and using (4.15), we obtain the identity

$$V^\dagger{}_{;a}{}^{;a} = 2l^a{}_{;b}l^b{}_{;a} + 2R_{ab}l^a l^b \tag{5.18}$$

which is satisfied by any Killing vector field. Using the further identity

$$V^\dagger l^a{}_{;b}l^b{}_{;a} = \tfrac{1}{2}V^\dagger_{;a}V^{\dagger;a} - 2\omega^\dagger_a\omega^{\dagger a} \tag{5.19}$$

which also follows directly from the Killing equation, where we have introduced the rotation vector $\omega^{\dagger a}$ of the field given by

$$\omega_a^\dagger = \tfrac{1}{2}\epsilon_{abcd} l^b l^{d;c} \qquad (5.20)$$

(where ϵ_{abcd} is the alternating tensor) (5.19) can be expressed as

$$V^{\dagger\,;a}{}_{;a} = \frac{V^\dagger_{;a}V^{\dagger;a} - 4\omega_a^\dagger\omega^{\dagger a}}{V^\dagger} + 2R_{ab}l^a l^b \qquad (5.21)$$

Evaluating the various terms on the horizon, we deduce directly from (4.15) and (5.9) that the Laplacian of $V^\dagger$ takes the value

$$V^{\dagger\,;a}{}_{;a} = 4\kappa^2 \qquad (5.22)$$

on $\mathscr{H}$, and hence from (5.8) and (5.18) that

$$R_{ab}l^a l^b = 0 \qquad (5.23)$$

on $\mathscr{H}$. Since, by (5.5) the rotation vector $\omega_a^\dagger$ is zero on $\mathscr{H}$ it is clear in the non-degenerate case, and can be verified to be true even in the degenerate limit case, that the ratio $\omega_a^\dagger\omega^{\dagger a}/V^\dagger$ tends to zero on $\mathscr{H}$; it therefore follows that

$$\frac{V^\dagger_{;a}V^{\dagger;a}}{4V^\dagger} \to \kappa^2 \qquad (5.24)$$

in the limit on $\mathscr{H}$. This last equation gives rise to another interpretation of κ. The acceleration a^a of an observer rigidly co-rotating with the hole, i.e. following one of the trajectories of the field l^a, is given by

$$V^\dagger a^a = l^a{}_{;b}l^b \qquad (5.25)$$

Introducing the convention (which will be employed frequently in the following sections) of using a bar to denote quantities associated with orbits which have been renormalized by multiplication by the time dilatation factor of the orbit (which in this case is $V^{\dagger 1/2}$), we define

$$\bar{a}^a = V^{\dagger 1/2} a^a \qquad (5.26)$$

In terms of this renormalized acceleration vector, we see that (5.24) has the form

$$\bar{a}^a\bar{a}_a \to \kappa^2 \qquad (5.27)$$

on $\mathscr{H}$, i.e. κ represents the limiting magnitude of the renormalized acceleration of a co-rotating observer.

Since ordinary matter will always give positive contributions to $R_{ab}l^a l^b$, the equation (5.23) implies that the immediate neighbourhood of the hole boundary must always be empty, except possibly for the presence of an electromagnetic field, which can be contrived so as to give zero contribution to the term $R_{ab}l^a l^b$. The restrictions required to satisfy this requirement are rather stringent however, and place well defined boundary conditions on the field. Introducing the Maxwell energy tensor T_F^{ab}, which will be proportional to the Ricci tensor under these conditions, by

$$T_F^{ab} = \frac{1}{4\pi}\,[F^{ac}F_c^b - \tfrac{1}{4}F^{cd}F_{cd}g^{ab}] \qquad (5.28)$$

where F_{ab} is the electromagnetic field tensor, we obtain

$$T_{Fab}l^a l^b = \frac{1}{8\pi}\,[\bar{E}_a^\dagger \bar{E}^{\dagger a} + \bar{B}_a^\dagger \bar{B}^{\dagger a}] \qquad (5.29)$$

where we have introduced the co-rotating electric and magnetic field vectors $\bar{E}_a^\dagger$ and $\bar{B}_a^\dagger$ defined by

$$\bar{E}_a^\dagger = F_{ab}l^b \qquad (5.30)$$

$$\bar{B}_a^\dagger = \tfrac{1}{2}\epsilon_{abcd}F^{cd}l^b \qquad (5.31)$$

Since they are both orthogonal to l^a, neither can be timelike within the closure of $\ll\mathscr{I}\gg$, so that (5.29) is a sum of non-negative terms. Thus $R_{ab}l^a l^b$ and $T_{Fab}l^a l^b$ can only be zero on $\mathscr{H}$ if $E_a^\dagger$ and $B_a^\dagger$ are both null on $\mathscr{H}$, and hence (by the orthogonality) parallel to l_a on $\mathscr{H}$. In other words the electromagnetic field must satisfy the boundary conditions

$$\bar{E}_{[a}^\dagger l_{b]} = 0 \qquad (5.32)$$

$$\bar{B}_{[a}^\dagger l_{b]} = 0 \qquad (5.33)$$

on $\mathscr{H}$.

6 Stationarity, Staticity, and the Hawking–Lichnerowicz Theorem

The terms stationary and static are used widely in different physical contexts. Generally speaking a system is said to be *stationary* if it is invariant under a continuous transformation group which maps earlier events into later ones. In General Relativity this condition can be more precisely defined in terms of the requirement that space-time be invariant under the action of a continuous one-parameter isometry group generated by a Killing vector field which is either timelike everywhere (in which case the space-time is said to be stationary in the *strict* sense) or else timelike at least somewhere, e.g. in sufficiently distant asymptotically flat region, (in which case I shall refer to the space-time as *pseudo-stationary*). It is not quite so easy to give a definition of what is meant by the statement that a general physical system is *static*, without referring to the specific context, but broadly speaking a system is said to be *static* if it is not only stationary in the strict sense, but also such that there is no motion (i.e. no material flow or current) relative to the stationary reference system. How this is to be interpreted will of course depend on which kinds of flow or current are relevant. For example a river through which water is flowing at a steady rate could be said to be stationary, but not static; on the other hand a stagnant pond might be describable as static in so far as only liquid flow was being taken into account, while being non-static from the point of view of a more detailed analysis in which perhaps the presence of a steady downward heat flux might need to be considered. In general relativity the requirement that a flow represented by a current vector field be static means that the vector field must be parallel to a timelike Killing vector.

The most fundamental generally defined flow vector in General Relativity is the timelike eigenvector of the energy-momentum tensor, which by Einstein's equations is the same as the timelike eigenvector of the Ricci Tensor. Thus we are led to the idea of a static Ricci tensor as one which satisfies the equation.

$$k^c R_{c[a} k_{b]} = 0 \tag{6.1}$$

where the timelike eigenvector k^a is a Killing vector. When electromagnetism is present another fundamental flow vector is the electric current vector j^a. The electric current vector is said to be static if it satisfies

$$j_{[a} k_{b]} = 0 \tag{6.2}$$

i.e. if it only has a component (representing a fixed electric charge density) parallel to the Killing vector.

Now although the basic physical idea of what is meant by a system being static depends essentially on the behaviour of current flows, the concept can be often extended by means of the field equations, to apply to the fields of which the relevant currents are the sources. In particular the definition of the term static can be extended to apply both to the *space-time metric tensor* g_{ab} and to the *electromagnetic field tensor* F_{ab}. The space-time metric tensor (or simply the space-time itself) is said to be static if the Killing vector trajectories are everywhere orthogonal to a family of spacelike hypersurfaces, i.e. (by Frobenius theorem) if

$$k_{[a;b} k_{c]} = 0 \tag{6.3}$$

and the electromagnetic field is said to be static if it satisfies

$$F_{[ab} k_{c]} = 0 \tag{6.4}$$

The justification for these definitions is that, by the field equations, they are not only *sufficient* for the stacity condition (6.1) and (6.2) to be satisfied by the corresponding source vectors, but also, at least under suitable global conditions, *necessary*.

To prove that (6.3) and (6.4) are respectively sufficient conditions for (6.1) and (6.2) to hold it is merely necessary to differentiate and use the appropriate field equations locally. Using the Killing equations

$$k_{a;b} = k_{[a;b]} \tag{6.5}$$

together with the definition of the Riemann tensor, by which

$$k_{a;[b;c]} = \tfrac{1}{2}R^{d}_{abc}k_d , \tag{6.6}$$

and using the Riemann tensor symmetries, we obtain the equation

$$k_{a;b;c} = R^{d}_{abc}k_d \tag{6.7}$$

for the second derivatives of the Killing vector, and hence, on contracting, the condition

$$k_{a;c}{}^{;c} = R_{ac}k^c \tag{6.8}$$

for the Dalembertian of the Killing vector. Thus again using (6.5), we obtain the identity

$$\{k_{[a;b}k_{c]}\}^{;c} = \tfrac{2}{3}k^c R_{c[a}k_{b]} \tag{6.9}$$

from which it is evident that the metric staticity condition (6.3) is *sufficient* for the Ricci staticity condition (6.1) to hold.

To obtain (2) from (4) we must use the antisymmetry condition

$$F_{ab} = F_{[ab]} \tag{6.10}$$

together with the condition that F_{ab} is invariant under the action generated by k_a, i.e.

$$\mathscr{L}_k [F_{ab}] \equiv F_{ab;c}k^c - 2F_{c[a}k^c{}_{;b]} = 0 \tag{6.11}$$

This, in conjunction with the Killing equation (6.5), leads to the identity

$$\{F_{[ab}k_{c]}\}^{;c} = \tfrac{2}{3}k_{[a}F_{b]c}{}^{;c} \tag{6.12}$$

from which, using second of the Maxwell equations

$$F_{[ab;c]} = 0 \tag{6.13}$$

$$F^{ab}{}_{;b} = 4\pi j^a \tag{6.14}$$

it is evident that the electromagnetic field staticity condition (6.4) is a sufficient condition for the current staticity condition (6.2) to hold.

If the field staticity condition (6.4) is satisfied, then it follows automatically that the Maxwell energy tensor T_F^{ab} defined by 4.12 will satisfy the staticity condition

$$k_a T_F^{a[b}k^{c]} = 0 \tag{6.15}$$

If the Einstein equations

$$R_{ab} - \tfrac{1}{2}Rg_{ab} = 8\pi T_{ab} \tag{6.16}$$

hold, with the total energy tensor T_{ab} given by

$$T^{ab} = T_F^{ab} + T_M^{ab} \tag{6.17}$$

where T_M^{ab} is the matter contribution, then the Ricci staticity condition (6.1) will be satisfied at the same time as the field staticity condition 6.4 if and only if the analogous *matter staticity condition*

$$k_a T_M^{a[b} k^{c]} = 0 \tag{6.18}$$

is satisfied. If the matter tensor is expressed in the canonical form

$$T_M^{ab} = \rho u^a u^b + p^{ab} \tag{6.19}$$

where the velocity u^a and the pressure tensor p^{ab} are required to satisfy the normalization and orthogonality conditions

$$u_a u^a = -1 \qquad p^{ab} u_b = 0 \tag{6.20}$$

and where the eigenvalue ρ represents the mass density, the staticity condition (6.18) can be seen to be equivalent to

$$u_{[a} k_{b]} = 0 \tag{6.21}$$

The basic idea of theorems of the Lichnerowicz type is to establish converses to these results, i.e. to find global conditions under which the basic flow staticity conditions (6.2) and (6.21) are *sufficient* as well as necessary for the metric and field staticity conditions to hold. It is a well known result of classical electromagnetic theory that (6.2) is a sufficient condition for (6.4) to hold in asymptotically source free flat space. The first examination of this question in General Relativity theory was made by Lichnerowicz (1939). It was shown by Lichnerowicz and Choquet-Bruhat (see Lichnerowicz (1955)) that in a space-time which is asymptotically Minkowskian and topologically Euclidean, (6.2) is a consequence of (6.1) and hence, in the non-electromagnetic case, of (2.1). When (6.31) has been established it is easy to show under the same conditions, that (6.4) is a consequence of (6.2), but the demonstration that (6.3) itself is a consequence of (6.21) and (6.2) involves certain technical difficulties which I did not succeed in resolving until a few days before the beginning of this school. I have not had time to undertake a thorough search of the literature, but as far as I know this section contains the first published account of the full electromagnetic generalization of the Lichnerowicz theorem.

The global topology conditions originally assumed by Lichnerowicz and Choquet-Bruhat were such as to explicitly exclude the presence of a central black hole. However Hawking has recently shown that the original theorem can be extended to cover the case of a *non-rotating* black hole, (i.e. one whose horizon is an ergosurface) provided it is assumed that the horizon is in fact the *outer-ergosurface*. In the present section I shall show that Hawking's result can also be extended to the case where an electromagnetic field is present, at least in the non-degenerate case. (At the time of writing I have not had time to verify that this result still holds in the degenerate limit.)

Formally the results which will be established in this section may be stated as follows:

THEOREM 6.1 (Generalized Lichnerowicz Theorem If $\mathscr{M}$ is topologically Minkowskian and both asymptotically flat, and asymptotically source free, if it is stationary in the *strict* sense and if the electromagnetic and material current staticity conditions (6.2) and (6.21) are satisfied everywhere, then the electromagnetic field staticity condition (6.4) and the metric staticity condition (6.3) will be satisfied everywhere.

THEOREM 6.2 (Generalized Hawking–Lichnerowicz Theorem) If $\mathscr{M}$ is asymptotically flat and asymptotically source free, and the asymptotic magnetic monopole moment is zero, if the domain of outer communications $\langle\!\langle \mathscr{I} \rangle\!\rangle$ is stationary in the *strict* sense (which implies by Lemma 2 that $\mathscr{H}$ is the outer ergosurface) and *simply* connected, and if the electromagnetic and material current staticity conditions (6.2) and (6.21) are satisfied everywhere, then the electromagnetic field staticity condition (6.4) and the metric staticity condition (6.3) will be satisfied in $\langle\!\langle \mathscr{I} \rangle\!\rangle$, subject (provisionally) to the requirement that the horizon be non-degenerate.

Proof of Theorem 6.1 We start by introducing the rotation vector ω^a of the stationary Killing vector k^a by the definition

$$\omega_a = \tfrac{1}{2}\epsilon_{abcd}k^b k^{d;c} \tag{6.22}$$

It is clear that the metric staticity condition (6.3) that we wish to establish, has the form $\omega_a = 0$. By differentiation of (6.22), is easy to verify using the identity (6.7) that ω_a must satisfy

$$\omega_{[a;b]} = \tfrac{1}{2}\epsilon_{abcd}k^c R^{de}k_e \tag{6.23}$$

We introduce electric and magnetic field vectors $\bar{E}^a$ and $\bar{B}^a$ defined by

$$\bar{E}_a = -F_{ab}k^b \tag{6.24}$$

$$\bar{B}_a = \tfrac{1}{2}\epsilon_{abcd}k^b F^{dc}, \tag{6.25}$$

in terms of which the electromagnetic field tensor can be given by

$$VF_{ab} = -2k_{[a}\bar{E}_{b]} + \epsilon_{abcd}k^c\bar{B}^d. \tag{6.26}$$

It is clear that the field staticity condition (6.4) that we wish to establish has the form $\bar{B}_a = 0$.

Evaluating (6.23) using the Einstein equations, we obtain

$$\omega_{[a;b]} = -2\bar{E}_{[a}\bar{B}_{b]} - 4\pi\epsilon_{abcd}k^c T_M^{de}k_e \tag{6.27}$$

From Maxwell's equations (6.13) and the field invariance condition (6.11) it is clear that we shall always have

$$\bar{E}_{[a;b]} = 0 \tag{6.28}$$

When the electric current staticity condition (6.4) is satisfied, we shall also have

$$\bar{B}_{[a;b]} = 0 \tag{6.29}$$

Since we are postulating that $\ll\mathscr{I}\gg$ (which is the same as $\mathscr{M}$ under the conditions of Theorem 6.1) is simply connected, these two equations imply the existence of globally well behaved scalars Φ and Ψ such that

$$\bar{E}_a = \Phi_{,a} \tag{6.30}$$

$$\bar{B}_a = \Psi_{,a} \tag{6.31}$$

When the matter staticity condition ((6.18) which follows from (6.21)) is also true, (6.37) reduces to

$$U_{[a;b]} = 0 \tag{6.32}$$

where we have used the abbreviation

$$U_a = \tfrac{1}{2}\omega_a + \Phi\bar{B}_a - \Psi\bar{E}_a \tag{6.33}$$

As in the case of (6.21) and (6.29), this equation implies the existence of a globally well behaved scalar U such that

$$U_a = U_{,a} \tag{6.34}$$

Now by their definition the scalars Φ, Ψ and U must satisfy the divergence identity

$$\left\{\frac{(U + 2\Phi\Psi)\omega^a}{V^2} + \frac{2\Psi\bar{B}^a}{V}\right\}_{;a} = 2\,\frac{\omega^a\omega_a}{V^2} + 2\,\frac{\bar{B}^a\bar{B}_a}{V} + (U + 2\Phi\Psi)\left\{\frac{\omega_a}{V^2}\right\}^{;a} + 2\Psi\left\{\left(\frac{\bar{B}_a}{V}\right)^{;a} + 2\bar{E}^a\omega_a\right\} \tag{6.35}$$

Moreover it follows directly from the definition (6.22) that ω_a must satisfy the identity

$$\left\{\frac{\omega_a}{V^2}\right\}^{;a} = 0 \tag{6.36}$$

and it follows from the Maxwell equations (6.13) that $\bar{B}^a$ must satisfy

$$\left(\frac{\bar{B}_a}{V}\right)^{;a} + 2\bar{E}^a\omega_a = 0 \tag{6.37}$$

Hence the right hand side of (6.35) reduces to a sum of two terms which are non-negative in $\ll\mathscr{I}\gg$, since under the postulated conditions V is positive and k^a (to which ω^a and $\bar{B}^a$ are orthogonal by their definitions) is timelike in $\ll\mathscr{I}\gg$.

We can convert the identity to which (6.35) reduces, from a scalar divergence equation to a vector divergence equation by multiplication by k^a, using the group invariance conditions. In this way we obtain

$$\left\{\frac{(U+2\Phi\Psi)\omega^{[a}k^{b]}}{V^2}+\frac{2\Psi\bar{B}^{[a}k^{b]}}{V}\right\}_{;a}=\left\{\frac{\omega^a\omega_a}{V^2}+\frac{\bar{B}^a\bar{B}_a}{V}\right\}k^b \tag{6.39}$$

We now construct an asymptotically flat spacelike hypersurface Σ properly imbedded in $\langle\!\langle\mathscr{I}\rangle\!\rangle$ so that its only edge consists of a 2-surface where it intersects $\mathscr{H}^+$ (under the conditions of Theorem 4.1, Σ will have no edge at all). According to Stoke's theorem, the integral of the divergence of any antisymmetric quantity V^{ab} over a hypersurface with metric normal element $d\Sigma^a$ must satisfy

$$\int_\Sigma V^{ab}{}_{;b}\,d\Sigma_a=\oint_{\partial\Sigma} V^{ab}\,dS_{ab} \tag{6.40}$$

where dS_{ab} is the 2-dimensional antisymmetric normal element to the edge $\partial\Sigma$ of Σ. Applying this to 6.39 we obtain

$$\begin{aligned}\int_\Sigma\left\{\frac{\omega^a\omega_a}{V^2}+\frac{\bar{B}^a\bar{B}_a}{V}\right\}k^a\,d\Sigma_a&=\oint_\infty dS_{ab}\left\{\frac{(U+2\Phi\Psi)\omega^{[a}k^{b]}}{V^2}+2\Psi\frac{\bar{B}^{[a}k^{b]}}{V}\right\}\\&\quad-\oint_H dS_{ab}\left\{\frac{(U+2\Phi\Psi)\omega^{[a}k^{b]}}{V^2}+2\Psi\frac{\bar{B}^{[a}k^{b]}}{V}\right\}\end{aligned} \tag{6.41}$$

where $\oint_\infty dS_{ab}$ indicates the limit of large distance of the integral over a topologically spherical 2-surface in Σ, and $\oint_H dS_{ab}$ indicates the integral over the edge of Σ where it meets $\mathscr{H}^+$.

The product $-k^a\,d\Sigma_a$ is always positive since k^a and $d\Sigma^a$ are both timelike in $\langle\!\langle\mathscr{I}\rangle\!\rangle$, and as we have already remarked, the terms $\omega^a\omega_a/V^2$ and $\bar{B}^a\bar{B}_a/V$ are non-negative in $\langle\!\langle\mathscr{I}\rangle\!\rangle$. Hence we shall obtain the required result, i.e. that the spacelike vectors ω^a and B^a must be zero everywhere in $\langle\!\langle\mathscr{I}\rangle\!\rangle$ if we can establish that the surface integral terms on the right hand side of (6.41) must be zero.

Now by the asymptotic boundary conditions, ω_a must diminish like the inverse cube, and $\bar{E}_a$ and $\bar{B}_a$ like the inverse square of the radial distance, which makes it possible to choose U, Φ and Ψ in such a way that their asymptotic limits are all zero. With this choice, U will diminish like the inverse square, and Φ and Ψ like the inverse first power of the radial distance, which will lead at once to the condition

$$\oint_\infty dS_{ab}\left\{\frac{(U+2\Phi\Psi)}{V^2}\,\omega^{[a}k^{b]}+\frac{2\Psi\bar{B}^{[a}k^{b]}}{V}\right\}=0 \tag{6.42}$$

(since by (3.3) the asymptotic limit of V is 1). This is sufficient to establish Theorem 6.1.

When a black hole is present, the situation is rather more complicated. The integration on $\mathscr{H}^+$ is rather more delicate than the integration in the asymptotic limit, since the quantity V in the denominators is zero. However the analysis is not too difficult if the degenerate case, when the gradient of V is zero, is excluded.

Since, under the conditions of Theorem 6.2, the horizon Killing vector l^a discussed in the previous section is the same as k^a, it follows from the boundary conditions (5.32) and (5.33) that $\bar{E}_a$ and $\bar{B}_a$ will be orthogonal to $\mathscr{H}^+$, and hence that the scalars Φ and Ψ must be *constant* on $\mathscr{H}^+$, which makes it possible to choose Φ and Ψ to satisfy

$$\Phi = \Psi = 0 \tag{6.43}$$

on $\mathscr{H}^+$. With this choice, it is clear at least in the non-degenerate case that Φ/V and Ψ/V will have finite limits on $\mathscr{H}^+$. Since under the conditions assumed for Theorem 6.2 k^a is null on $\mathscr{H}^+$ and hence orthogonal to $\mathscr{H}^+$, it follows that the Frobenius orthogonality condition $\omega_a = 0$ will hold at least on $\mathscr{H}^+$, which implies that subject to the choice (6.43) we shall have $U_a = 0$ on the horizon, which will make it possible to choose

$$U = 0 \tag{6.44}$$

on $\mathscr{H}^+$. With this choice it is clear (at least in the non-degenerate case) that not only U/V but even U/V^2 will have a finite limit on $\mathscr{H}^+$, so that we shall obtain

$$\oint_H dS_{ab} \left\{ \frac{(U + 2\Phi\Psi)\omega^{[a}k^{b]}}{V^2} + \frac{2\Psi\bar{B}^{[a}k^{b]}}{V} \right\} = 0 \tag{6.45}$$

This is still not sufficient to establish Theorem 6.2 since the choice $\Phi = \Psi = U = 0$ on $\mathscr{H}^+$ is not compatible with the choice of Φ, Ψ and U in such a way that their asymptotic large distance limits are zero, which was necessary in our previous derivation of (6.42). In order to obtain 6.42 at the same time as (6.45) we must use the additional assumption that the asymptotic magnetic monopole moment is zero, which means that $\bar{B}_a$ must diminish as the cube (not just the square) of the asymptotic radial distance. In this case it is clear that 6.42 will be true independently of the choice of U, Ψ or Φ. This enables us to deduce that

$$\bar{B}_a = 0 \tag{6.46}$$

$$\omega_a = 0 \tag{6.47}$$

everywhere in $\ll\mathscr{I}\gg$ as required, thus completing the proof.

COROLLARY TO THEOREMS 6.1 AND 6.2 Under the conditions of Theorem 6.1 and 6.2 the field tensor can be derived by the standard equation

$$F_{ab} = 2A_{[a;b]} \tag{6.48}$$

from a well behaved vector potential A_a in $\ll\mathscr{I}\gg$ which satisfies both the invariance condition

$$\underset{k}{\mathscr{L}}[A_a] \equiv A_{a;b}k^b + A_b m^b_{;a} = 0 \tag{6.49}$$

and also the *vector potential staticity condition*

$$A_{[a}k_{b]} = 0 \tag{6.50}$$

Proof It can easily be checked that the vector A_a given in terms of the scalar potential Φ, (whose existence has already been established) by

$$VA_a = \Phi k_a \tag{6.51}$$

satisfies all the conditions (6.48), (6.49), (6.50). (It is to be noted that it is possible to choose Φ *either* so that A_a has a finite limit on $\mathscr{H}$, or so that A_a tends asymptotically to zero at large radial distance, but not both at once.)

The Frobenius staticity condition implies (taking into account that $\ll\mathscr{I}\gg$ is required to be simply connected) that there exist a globally well behaved time co-ordinate function t on $\ll\mathscr{I}\gg$ defined up to an additive constant, in terms of which the metric $ds^2 = g_{ab}\, dx^a\, dx^b$ can be expressed in the form

$$ds^2 = g_{\mu\nu}\, dx^\mu\, dx^\nu - V\, dt^2 \tag{6.52}$$

where μ and ν run only from 1 to 3. The electromagnetic potential form $A = A_a\, dx^a$ will take the form

$$A = \Phi\, dt \tag{6.53}$$

7 Stationary-Axisymmetry, Circularity, and the Papapetrou Theorem

Just as a flow vector is said to be static if it is everywhere parallel to the stationary Killing vector k^a, so more generally it will be said to be *circular* or equivalently *non-convective* if it is everywhere parallel to a general linear combination of the two independent Killing vectors k^a and m^a, i.e. if each integral curve of the flow field lies in one of the surfaces of transitivity of the 2-parameter group generated by k^a and m^a, and therefore coincides with an integral curve of one of Killing vector fields generating the group. (If furthermore the flow vector coincides with a linear combination of k^a and m^a with everywhere *constant* coefficients, i.e. if it coincides with a *fixed* Killing vector field generator of the group, then the flow is said to be *rigid.*) Thus the *circularity* condition requires that each flow trajectory be invariant under some one-parameter transformation sub-group, which is consistent with steady circular motion relative to a non-rotating frame at infinity, whereas the stronger *staticity* condition is incompatible with any motion at all relative to such a frame. (A reasonably realistic example of a stationary-axisymmetric but *non-circular* flow would be the case of a stationary-axisymmetric star containing two convective zones, one in the northern hemisphere and one in the southern, with fluid rising towards the surface at the equator, and descending towards the centre at the poles.)

As in the previous section the basic flow vectors we shall wish to consider are the timelike eigenvector of the energy momentum tensor, or equivalently, of the Ricci tensor, and the electric current vector. When the Ricci tensor eigenvector is circular the Ricci tensor will be referred to as *invertible* (meaning that it is invariant under the tangent space isometry transformation in which directions in the plane of the Killing vectors are reversed while orthogonal directions are left unchanged). The condition for this, i.e. the necessary and sufficient condition for the timelike eigenvector of the Ricci tensor to be invertible, is that the equations

$$\begin{aligned} k^a R_{d[a} k_b m_{c]} &= 0 \\ m^d R_{d[a} k_b m_{c]} &= 0 \end{aligned} \qquad (7.1)$$

be satisfied. This circularity condition will play a role in the present section analogous to that of the Ricci staticity condition (6.1) in the previous section. The corresponding circularity condition for the electric current vector, i.e. the analogue of (6.2), is simply

$$j_{[a} k_b m_{c]} = 0 \qquad (7.2)$$

We can pursue the analogy with staticity by defining corresponding circularity conditions for the space-time metric tensor and the electromagnetic field tensor. The circularity condition for the metric is the condition that there should exist a family of 2-surfaces everywhere orthogonal to the surfaces of transitivity of the two-parameter group action, i.e. everywhere orthogonal to the plane of the Killing vectors k^a and m^a. This *orthogonal transitivity* condition is the precise analogue of the (one-parameter) orthogonal transitivity condition for metric staticity. By Frobenius theorem the necessary and sufficient condition for orthogonal transivity is

$$\begin{aligned} k_{[a;b} k_c m_{d]} &= 0 \\ m_{[a;b} k_c m_{d]} &= 0 \end{aligned} \qquad (7.3)$$

these equations jointly being the analogue of (6.3). The circularity condition for the electromagnetic field tensor is that it should satisfy the equations

$$\begin{aligned} F_{ab} k^a m^b &= 0 \\ F_{[ab} k_c m_{d]} &= 0 \end{aligned} \qquad (7.4)$$

these being jointly the analogue of (6.4).

As for the staticity conditions of previous section, so also here we justify the description of the conditions (7.3) and (7.4) as circularity conditions by showing not only that they are sufficient for the respective source flow circularity conditions (7.1) and (7.2) to hold, but also, under global conditions much weaker than those which were necessary in the previous case, that they are necessary. To prove that (7.3) is a sufficient condition for (7.1) to hold we simply use Killing equations

$$m_{a;b} = m_{[a;b]} \tag{7.5}$$

and the consequent identity

$$\{m_{[a;b}m_{c]}\}^{;c} = \tfrac{2}{3}k^c R_{c[a}k_{b]} \tag{7.6}$$

together with the analogous equations (6.5) and (6.6) for k^a, and the commutation condition

$$m^a{}_{;b}k^b - k^a{}_{;b}m^b = 0 \tag{7.7}$$

to derive the identities

$$\begin{aligned} \{k_{[a;b}k_c m_{d]}\}^{;d} &= -\tfrac{1}{2}k^d R_{d[a}k_b m_{c]} \\ \{m_{[a;b}k_c m_{d]}\}^{;d} &= -\tfrac{1}{2}m^d R_{d[a}k_b m_{c]} \end{aligned} \tag{7.8}$$

It is immediately clear from these identities that the Ricci invertibility condition, (7.1), is a consequence of the orthogonal transitivity condition (7.3).

To prove that (7.4) is sufficient for (7.2) to hold we use the group invariance condition

$$\underset{m}{\mathscr{L}}[F_{ab}] \equiv F_{ab;c}m^c - 2F_{c[a}m^c{}_{;b]} = 0 \tag{7.9}$$

together with the analogous equation (6.11) and the commutation condition (7.7) to derive the identities

$$\begin{aligned} \{F_{ab}k^a m^b\}_{;c} &= 3F_{[ab;c]}k^a m^b \\ \{F_{[ab}k_c m_{d]}\}^{;d} &= \tfrac{1}{4}k_{[a}m_b F^{;d}_{c]d} \end{aligned} \tag{7.10}$$

from which, using the Maxwell equations (6.13) and (6.14), it is clear that the current circularity condition (7.2) will hold whenever the field circularity conditions (7.4) are satisfied.

Continuing the analogy with the previous section, we check that it follows automatically from the electromagnetic field, circularity conditions that the Maxwell energy tensor T_F^{ab} satisfies the staticity condition

$$\left.\begin{aligned} k_a T_F^{a[b}k^c m^{d]} &= 0 \\ m_a T_F^{a[b}k^c m^{d]} &= 0 \end{aligned}\right\} \tag{7.11}$$

from which it follows, if the Einstein equations hold, that the Ricci circularity condition (7.1) will hold at the same time as the field staticity condition (7.4) if and only if the analogous *matter circularity condition*

$$\left.\begin{aligned} k_a T_M^{a[b}k^c m^{d]} &= 0 \\ m_a T_M^{a[b}k^c m^{d]} &= 0 \end{aligned}\right\} \tag{7.12}$$

are satisfied, i.e. if and only if

$$u^{[a}k^{b}m^{c]} = 0 \tag{7.13}$$

The basic converse theorem is due to Papapetrou (see Papapetrou 1966, Kundt and Trumper 1966, Carter 1969). The electromagnetic generalization (Carter 1969) is quite straightforward, and does not require any special trick such as was needed for the electromagnetic generalization of the Lichnerowicz theorem. Moreover the topological and boundary conditions required for the Papapetrou theorem are very much weaker than for the Lichnerowicz theorem, so that the original theorem can be applied directly to the case where a hole is present.

The general result is as follows:

THEOREM 7 (Generalized Papapetrou Theorem) If $\mathscr{M}$ is pseudo-stationary, asymptotically flat and asymptotically source free, and if the electromagnetic and material current circularity conditions (7.2) and (7.13) are satisfied in a *connected* subdomain $\mathscr{D}$ of $\mathscr{M}$ which intersects the rotation axes (or is otherwise known to contain points at which (7.4) and (7.3) are satisfied) then the electromagnetic field circularity condition (7.4) and the metric circularity condition (7.3) will be satisfied everywhere in $\mathscr{D}$.

Proof We prove the basic Papapetrou theorem by introducing a second twist vector ψ^a, analogous to the vector ω^a introduced (by equation 6.22) in the previous section, defined by

$$\psi_a = \tfrac{1}{2}\epsilon_{abcd}m^b m^{c;d} \tag{7.14}$$

It is evident that the required metric circularity condition 7.3 will be satisfied if and only if the twist scalars $m^c\omega_c$ and $k^c\psi_c$ are both zero. Now the identities (7.7) can be converted to the equivalent dual forms

$$\left.\begin{aligned}(m^c\omega_c)_{;a} &= \epsilon_{abcd}k^b m^c R^{dl}k_l\\ (k^c\psi_c)_{;a} &= \epsilon_{abcd}k^b m^c R^{dl}m_l\end{aligned}\right\} \tag{7.15}$$

respectively. Hence it is clear that the Ricci circularity condition (7.1) will be satisfied in the domain $\mathscr{D}$ if and only if the *twist scalars* are *constant* i.e.

$$\left.\begin{aligned}(m^c\omega_c)_{,a} &= 0\\ (k^c\psi_c)_{,a} &= 0\end{aligned}\right\} \tag{7.16}$$

in $\mathscr{D}$. Since m^c, and hence also ψ^c, are zero on the rotation axis, the twist scalars are also zero on the rotation axis, and hence by conditions of the theorem, at some points of $\mathscr{D}$. Thus (7.16) implies that we shall have

$$\left.\begin{aligned}m^c\omega_c &= 0\\ k^c\psi_c &= 0\end{aligned}\right\} \tag{7.17}$$

at all points of $\mathscr{D}$. This completes the proof of the basic theorem establishing that (7.3) is a consequence of (7.1), and hence in the non-electromagnetic case of (7.13).

To cover the electromagnetic case, we proceed from the fact that using the Maxwell equations (6.13) and (6.14), the identities (7.10) can be reduced to the form

$$\left.\begin{aligned}(\bar{E}_b m^b)_{,a} &= 0\\ (\bar{B}_b m^b)_{,a} &= 4\pi\epsilon_{abcd}k^b m^c j^d\end{aligned}\right\} \qquad (7.18)$$

where $\bar{E}_b$ and $\bar{B}_b$ are as defined by (6.24) and (6.25). It is clear therefore that the correct statisticity condition (7.4) is sufficient to ensure that the scalar $\bar{B}_b m^b$ is constant in $\mathscr{D}$, while $\bar{E}_b m^b$ will be constant in any case, i.e. we shall have

$$\left.\begin{aligned}(\bar{E}_b m^b)_{,a} &= 0\\ (\bar{B}_b m^b)_{,a} &= 0\end{aligned}\right\} \qquad (7.19)$$

in $\mathscr{D}$. Now the required electromagnetic field circularity condition (7.4) is clearly equivalent to the requirement that the scalars $\bar{E}_b m^b$ and $\bar{B}_b m^b$ should be zero. These scalars will obviously be zero on the rotation axis where m^b is zero and hence at some points of $\mathscr{D}$, and therefore it follows from (7.19) that we shall have

$$\left.\begin{aligned}\bar{E}_b m^b &= 0\\ \bar{B}_b m^b &= 0\end{aligned}\right\} \qquad (7.19)$$

which is equivalent to (7.4), at all points of $\mathscr{D}$. We have already remarked that this is sufficient for (7.11) and hence, subject to (7.13), for (7.1) to hold. Therefore by the first part of the proof (7.3) will still hold in the electromagnetic case. This completes the proof of the theorem.

COROLLARY TO THEOREM 7 If the conditions of Theorem 7.1 are satisfied, and in addition the domain $\mathscr{D}$ is *simply* connected, then (except on the rotation axis) the electromagnetic field tensor F_{ab} is derivable, via (6.48) from a vector potential A_a which satisfies the group invariance condition

$$\underset{m}{\mathscr{L}}\,[A_a] \equiv A_{a;b}m^b + A_b m^b{}_{;a} = 0 \qquad (7.20)$$

and its analogue (6.49), and also the *electromagnetic potential circularity condition*

$$A_{[a}k_b m_{c]} = 0 \qquad (7.21)$$

(The vector potential will be well behaved on the rotation axis only if the magnetic flux

$$4\pi P = \tfrac{1}{2} \oint_S \epsilon^{abcd} F_{ab} dS_{cd} \qquad (7.22)$$

over any compact 2-surface S is zero, i.e. only if there are no magnetic monopoles.)

Proof It is clear that the electromagnetic field circularity condition (7.4) implies that (except on the rotation axis) the field can be expressed in the form

$$F_{ab} = 2e_{[a}k_{b]} + 2f_{[a}m_{b]} \tag{7.23}$$

where e_a and f_a are vectors which satisfy the orthogonality conditions

$$\left.\begin{aligned} e_a k^a = e_a m^a = 0 \\ f_a k^a = f_a m^a = 0 \end{aligned}\right\} \tag{7.24}$$

Moreover by the group invariance (7.8) and (6.11) e_a and f_a can be chosen to satisfy the corresponding group invariance conditions

$$\left.\begin{aligned} e_{a;b}k^b + e_a k^b_{;a} = 0 \qquad e_{a;b}m^b + e_a m^b_{;a} = 0 \\ f_{a;b}k^b + f_a k^b_{;a} = 0 \qquad f_{a;b}m^b + f_a m^b_{;a} = 0 \end{aligned}\right\} \tag{7.25}$$

Now it can be verified by a little algebra that the Maxwell equation (6.13) implies that the vectors Φ_a and B_a defined by

$$\left.\begin{aligned} \Phi_a = Ve_a - Wf_a \\ B_a = Xf_a + We_a \end{aligned}\right\} \tag{7.26}$$

will satisfy

$$\left.\begin{aligned} \Phi_{[a;b]} = 0 \\ B_{[a;b]} = 0 \end{aligned}\right\} \tag{7.27}$$

It follows from the simple connectivity condition that there will exist scalars Φ and B everywhere on $\mathscr{D}$ (including the rotation axis) such that

$$\left.\begin{aligned} \Phi_a = \Phi_{,a} \\ B_a = B_{,a} \end{aligned}\right\} \tag{7.28}$$

It can now be checked that the vector A_a given by

$$A_a = -\frac{X\Phi + WB}{\sigma} k_a + \frac{W\Phi + VB}{\sigma} m_a \tag{7.29}$$

will satisfy the conditions (6.49), (7.20), (7.21). (However it follows from the conditions (3.13) that this vector potential will be singular on the rotation axis unless B can be chosen to be zero on the rotation axis, and it can be verified by Stokes' theorem that this will be possible only if the no magnetic monopole condition is satisfied.) This completes the proof of the corollary.

Except where the Killing bivector $\rho_{ab} = 2k_{[a}m_{b]}$ is null or degenerate (and hence, by Theorem 4.2, everywhere in $\ll\mathscr{I}\gg$ except on the rotation axis) the Frobenius orthogonality condition 4.3 implies that it is possible to choose locally well behaved functions t and φ that are constant on the 2-surfaces orthogonal to the surfaces of transitivity of the action π^{sA}, where t is also constant on the trajectories of the action π^A generated by m^a, and φ is also constant on the trajectories of the action π^s generated by k^a. In other words t and φ can be chosen so as to satisfy

$$t_{,[a}k_b m_{c]} = 0 \tag{7.30}$$

$$\varphi_{,[a}k_b m_{c]} = 0 \tag{7.31}$$

$$t_{,a}m^a = 0 \tag{7.32}$$

$$\varphi_{,a}k^a = 0 \tag{7.33}$$

They will be determined uniquely up to an arbitrary additive constant if we also impose the standard normalization conditions

$$t_{,a}k^a = 1 \tag{7.34}$$

$$\varphi_{,a}m^a = 1 \tag{7.35}$$

Under these conditions the metric form $ds^2 = g_{ab}\, dx^a\, dx^b$ on $\ll\mathscr{I}\gg$ can be expressed (except on the rotation axis) in the form

$$ds^2 = g_{\alpha\beta}\, dx^\alpha\, dx^\beta + X\, d\varphi^2 + 2W\, d\varphi\, dt - V\, dt^2 \tag{7.36}$$

with X, W, V as defined in section 3, where α, β run from 1 to 2, and where the co-ordinates x^α are constant on the surfaces of transitivity of the action π^{sA}. For many purposes it will be convenient to work not with φ but with a complementary angle co-ordinate function $\varphi^\dagger$ defined analogously to φ except for the requirement that it be constant on the trajectories of the action $\pi^{s\dagger}$ instead of π^s, i.e. $\varphi^\dagger$ is defined by the requirements

$$\varphi^\dagger_{,[a}l_b m_{c]} = 0 \tag{7.37}$$

$$\varphi^\dagger_{,a}l^a = 0 \tag{7.38}$$

$$\varphi^\dagger_{,a}m^a = 1 \tag{7.39}$$

which determines it uniquely up to an additive constant, which may be chosen in such a way that $\varphi^\dagger$ is related to φ by

$$\varphi^\dagger = \varphi - \Omega^H t \tag{7.40}$$

In terms of $\varphi^\dagger$ the metric form (7.36) may be rewritten in the complementary form

$$ds^2 = g_{\alpha\beta}\, dx^\alpha\, dx^\beta + X\, d\varphi^{\dagger 2} + 2W^\dagger\, d\varphi^\dagger\, dt - V^\dagger\, dt^2 \tag{7.41}$$

It can easily be seen that if $\ll\mathscr{I}\gg$ is *simply* connected t can be taken to be a globally well behaved function on $\ll\mathscr{I}\gg$, and φ and $\varphi^\dagger$ can be taken to be well behaved angle co-ordinates, defined modulo 2π, on $\ll\mathscr{I}\gg$.

In terms of such a co-ordinate system, the electromagnetic field form $A = A_a\, dx^a$

can be expressed in terms of the scalars Φ and B introduced in the proof of the corollary to theorem 7, as

$$A = \Phi\, dt + B\, d\varphi \tag{7.42}$$

For many purposes it is convenient to use the complementary expression

$$A = \Phi^{\dagger}\, dt + B\, d\varphi^{\dagger} \tag{7.43}$$

where

$$\Phi^{\dagger} = \Phi + \Omega^{H} B. \tag{7.44}$$

8 The Four Laws of Black Hole Mechanics

For many purposes it is useful to analyse the neighbourhood of the Killing Horizon $\mathscr{H}^+$ in terms of a canonical null co-ordinate system constructed as follows. One first chooses a null hypersurface cutting across the Killing horizon $\mathscr{H}^+$. One can then construct a local null co-ordinate function v which is defined by the requirement that it should be zero on the chosen null hypersurface, and that it should satisfy $v_{,a} l^a = 1$, so that it will be constant on a family of null hypersurfaces congruent to the original one. By requiring that the other co-ordinate functions, r^*, x^2, x^3 should be constant on the trajectories of the action $\pi^{s\dagger}$ generated by l^a we ensure that v will be an ignorable co-ordinate. We choose the co-ordinates x^2, x^3 so as to be constant on each of the null geodesic generators of the null hyper-surfaces on which v is constant. The co-ordinate r^* is then specified uniquely by the requirement that it be zero on the Killing horizon $\mathscr{H}^+$ and that it varies along the null geodesics, on which x^2 and x^3 are required to be constant, in such a way that the metric takes the standard form

$$ds^2 = -V^{\dagger}\, dv^2 + 2\, dv\, dr^* + 2l_i\, dv\, dx^i + g_{ij}\, dx^i\, dx^j \tag{8.1}$$

where i, j run over 2, 3 and where $V^{\dagger}$, l_2 and l_3 are functions of r^*, x^2, x^3 only. It is clear that in addition to the familiar condition $V^{\dagger} = 0$, we shall also have

$$l_i = 0 \tag{8.2}$$

on the Killing horizon $\mathscr{H}^+$ at $r^* = 0$, and it follows further from (5.14) that in terms of such a co-ordinate system we shall have

$$\frac{\partial V^{\dagger}}{\partial r^*} = 2\kappa \tag{8.3}$$

on $\mathscr{H}^+$. Now it is evident that when the conditions of the Hawking-Lichnerowicz Theorem 6.2 or the Papapetrou Theorem 7 are satisfied, there must exist a local co-ordinate transformation of the form

$$v = t + \tfrac{1}{2} F(r^*, x^i) \tag{8.4}$$

relating the form (8.1) *either* to the form (6.52) with $x^1 = r^*$ (in the non-rotating case) *or* to the form (7.41) with $x^1 = r^*$, $\varphi^\dagger = x^3$ (in the rotating case), where F is a function of r^*, x^2, x^3 only. It is clear that this requires

$$V^\dagger \frac{\partial F}{\partial r^*} = 2 \tag{8.5}$$

$$V^\dagger \frac{\partial F}{\partial x^2} = l_2 \tag{8.6}$$

$$V^\dagger \frac{\partial F}{\partial x^3} = l_3 - W^\dagger \tag{8.7}$$

(provided we adopt the convection that $W^\dagger$ is zero in the non-rotating case). Since $V^\dagger$ is a well behaved function which tends to zero on $\mathscr{H}^+$, it follows from (8.3) that in the non-degenerate case (i.e. when κ is non-zero) we must have

$$F(r^*, x^i) = \frac{\ln r^*}{\kappa} + G(r^*, x^i) \tag{8.8}$$

if (8.5) is to be satisfied, where $G(r^*, x^i)$ is a function of r^* and x^i which is well behaved in the limit as $r^* \to 0$ on $\mathscr{H}^+$. Now since l_2 and l_3 are all well behaved functions which tend to zero on $\mathscr{H}^+$, (8.6) and (8.7) imply that the quantities $V^\dagger(\partial F/\partial x^i)$ must also be well behaved functions which tend to zero on $\mathscr{H}^+$. This will clearly be compatible with (8.4) only if the scalar κ satisfies

$$\frac{\partial \kappa}{\partial x^i} = 0. \tag{8.9}$$

Thus we arrive at the following conclusion

THEOREM 8 Under the conditions of Theorem 6.2 or Theorem 7, the quantity κ defined on the horizon $\mathscr{H}^+$by (5.9) is *constant* over $\mathscr{H}^+$. It is an immediate consequence of this theorem that a connected component of $\mathscr{H}^+$is either degenerate everywhere or not at all.

The result contained in Theorem 8 was first noticed, in the particular context of the Kerr solutions, by Boyer and Lindquist (1967) when they discovered that transformation from the original co-ordinate system (which had the form (8.1)) in which the solution was discovered by Kerr (1963) to the now standard Boyer-Lindquist co-ordinate system (which has the form 7.41). I cam across the more general result given by Theorem 8 in the course of an examination of the necessary and sufficient boundary conditions required for the black hole uniqueness problem (Carter 1971, Carter 1972). This result was discovered independently under even more general conditions by Hawking (1972) as a lemma in the proof of the strong rigidity theorem. However this result has recently acquired a much greater significance (which is the reason why I am giving it special attention at the present stage) from the work of Hartle and Hawking (1972), to be described in the accompanying course by Hawking. As a result it is now clear that Theorem 8 (which Hawking and I had previously regarded as a minor lemma) deserves to be dignified as the *zeroth law of black hole mechanics* for reasons which will be explained in this section.

Before proceeding it is worth remarking that the fact that the limit (5.24) must be constant on the horizon is closely analogous to the fact that the limit (3.7) used in normalizing the axisymmetry Killing vector is constant on the rotation axis. The analogy can be seen more clearly by considering the fixed point axis $\mathscr{H}^+ \cap \mathscr{H}^-$ predicted by Boyer's Theorem 5—such an axis bearing the same relation to a Lorentz rotation as the rotation axis does to an ordinary space rotation.

The main content of this section will be to describe the extension to black holes of very general heuristic argument originally due to Thorne (1969) and Zeldovich (Zledovich and Novikov (1971)) relating variations in the equilibrium mass of an isolated self gravitating system to corresponding variations in angular momentum, chemical composition and entropy. Before doing so I shall rapidly run through the basic argument, including its generalization (Carter 1972) to include the effect of variation of electric charge.

The generalized Thorne-Zeldovich formula is derived as follows. We consider a reversible change in the equilibrium state of the system, which to start with we shall think of as a rotating star, assumed to be both stationary and axisymmetric, which interacts with a freely falling particle of rest mass m, charge e, and unit velocity vector v^a which is sent in from infinity. The energy

$$E = -k^a p_a \tag{8.6}$$

and angular momentum

$$L_z = m^a p_a \tag{8.7}$$

of the particle are conserved during the free motion, where the momentum vector p_a is defined by

$$p_a = mv_a - eA_a \tag{8.8}$$

where A_a is the electromagnetic vector potential. Following Thorne and Zeldovich, we suppose that the particle interacts with the matter of the star at some point, transferring some of its matter and momentum in the process, and that it is then ejected back to infinity. We suppose that the material motion of the star is purely circular, in the sense of section 7, so that at any point the unit velocity vector u^a of the material is a linear combination of the Killing vectors, i.e.

$$u^a = -(\bar{u}_c \bar{u}^c)^{-1/2} \bar{u}^c \tag{8.9}$$

where the renormalized flow vector $\bar{u}^a$ has the form

$$\bar{u}^a = k^a + \Omega m^a \tag{8.10}$$

and where the quantity Ω defined by this equation is the local angular velocity at the point under consideration. (In the case of a star with *rigid* motion Ω will be independent of position and the renormalized flow vector $\bar{u}^a$ will itself be a Killing vector.) [We shall consistently use a bar to denote any quantity which has been renormalized by multiplication by the *time dilatation* factor $(-\bar{u}_a\bar{u}^a)^{1/2}$.] In the local rest frame, the energy δU transferred to the material of the star will be

$$\begin{aligned} \delta U &= u_a\, d(mv^a) \\ &= u^a(dp_a + A_a\, de) \\ &= (-\bar{u}_a\bar{u}^a)^{-1/2}\{-dE + \Omega\, dL_2 + \bar{u}^a A_a\, de\} \end{aligned} \tag{8.11}$$

Provided A_a is chosen so as to tend to zero in the limit of large distances, the contributions to the charge in total mass M, angular momentum J and charge Q of the star will be given by

$$\delta M = -dE \tag{8.12}$$

$$\delta J = -dL_z \tag{8.13}$$

$$\delta Q = -de \tag{8.14}$$

and hence we obtain

$$\delta M - \Omega\delta J - (\bar{u}^a A_a)\delta Q = (-\bar{u}_a\bar{u}^a)^{1/2}\delta U \tag{8.15}$$

It the star is initially in local thermal and chemical equilibrium with a well defined temperature Θ and well defined chemical potentials $\mu^{(i)}$ associated with various kinds of exchanged particles which are conserved in the interaction process, and in the particular case of a *thermodynamically reversible* process (which will only be possible, even in principle, if hysteresis effects can be neglected) the local energy transfer will be given by

$$\delta U = \mu^{(i)}\delta N_i + \Theta\delta S \tag{8.16}$$

where δN_i are the numbers of the various kinds of conserved particles which are transferred, and δS is the entropy transferred. Thus introducing the renormalized *effective temperature* $\bar{\Theta}$ and effective chemical potentials $\bar{\mu}^{(i)}$ defined by

$$\bar{\Theta} = (-\bar{u}^a\bar{u}_a)^{1/2}\Theta \tag{8.17}$$

$$\bar{\mu}^{(i)} = (-\bar{u}^a\bar{u}_a)^{1/2}\mu^{(i)} \tag{8.18}$$

we are led to the basic formula

$$\delta M = \Omega\delta J + \bar{u}^a A_a\delta Q + \bar{\mu}^{(i)}\delta N_{(i)} + \bar{\Theta}\,\delta S \tag{8.19}$$

for the change in mass of the star. Of course after this process the star will no longer be exactly in mechanical equilibrium. However the formula (8.19) should still be valid for the change in mass after the star has settled down to a new equilibrium (by radiation and other damping mechanisms) in the *small perturbation limit* provided that it can be argued that the energy corrections due to the initial departure from mechanical equilibrium are of second order. If this is the case, we may evaluate the total first order change dM in the mass between the mechanical equilibrium states due to a sequence of such transfer operations by integrating (8.19) in the form

$$dM = \int \Omega\delta J + \int \bar{u}^a A_a \delta Q + \int \bar{u}^{(i)}\delta N_{(i)} + \int \bar{\Theta}\, \delta S \tag{8.20}$$

In order for it to be practically useful it is necessary that this formula should be able to be interpreted as a space integral of locally well defined quantities. With any metric normal element $d\Sigma_a$ associated with an element $d\Sigma$ of a space-like hypersurface Σ, there will be associated flux elements of, charge, particle numbers, and entropy given by

$$dQ = j^a\, d\Sigma_a \tag{8.21}$$

$$dN_{(i)} = n^a_{(i)}\, d\Sigma_a \tag{8.22}$$

$$dS = s^a\, d\Sigma_a \tag{8.23}$$

where j^a, $n^a_{(i)}$ and s^a are the current vectors of charge particle numbers and entropy, which satisfy the conservation laws

$$j^a_{;a} = 0 \tag{8.24}$$

$$n^a_{(i);a} = 0 \tag{8.25}$$

and in the case of entropy, the semi-conservation law

$$s^a_{;a} \geqslant 0 \tag{8.26}$$

the latter being a strict equality for the reversible processes under consideration here. There is also a well defined angular momentum flux element given by

$$dJ = T^{ab} m_b\, d\Sigma_a \tag{8.27}$$

associated with an angular momentum current $T^{ab}m_b$ which satisfies the conservation law

$$[T^{ab} m_b]_{;a} = 0 \tag{8.28}$$

in consequence of the conservation law

$$T^{ab}{}_{;b} = 0 \tag{8.29}$$

of the total energy momentum tensor T^{ab} and of the Killing equations (7.5) satisfied by m^a. [It is not possible to give an equally meaningful local definition of conserved mass since the time symmetry generated by k^a is necessarily violated during any alteration process.] With these definitions, and the interpretation $\delta \equiv d(d)$ where the (first d refers to the alteration and the second d refers to the differential element in the integration) the terms in the formula (8.20) become well defined space integrals. In the non-electromagnetic case such an interpretation will be perfectly valid, but in the electromagnetic case it must be born in mind that the angular momentum transferred in an interaction is not well localized at the point of exchange where Ω is measured, since only part of it goes into the matter, the rest being located elsewhere in the electromagnetic field. Thus the value of Ω to be associated with a contribution $\delta J = d(T^{ab}m_b\, d\Sigma_a)$ in the angular momentum integral in (8.20) is not the locally measured value but some weighted space average of Ω over nearby and to a lesser extent more distant regions. Thus the straightforward interpretation will only be correct either when there is no electromagnetic field (the case originally treated by Thorne and Zeldovich) or when there is no differential rotation, at least of the electromagnetically interacting parts of the system. These are also conditions under which one could expect that the deviations from mechanical equilibrium caused by the variation process will be of second order in the perturbation, as required for the above formula to be applicable. In particular one can be confident that these deviations will be of second order in the particular case when the star is not only in crude mechanical equilibrium but also in equilibrium with respect to all relevant perturbation processes in the sense that there can be no energy release by internal transfer processes; clearly from (8.20) this requires (1) that the star be in thermal equilibrium in the sense that the effective temperature $\bar{\Theta}$ is constant, (2) that it be in chemical equilibrium in the sense that the effective Gibbs potentials $\bar{\mu}^{(i)}$ are constant, (3) that it be in rotational equilibrium (i.e. rigid) in the sense that Ω is constant, and (4) that it be in electrical equilibrium in the sense that the comoving electrical potential

$$\Phi^S = \bar{u}^a A_a \tag{8.30}$$

be constant within the star. It is to be noted that this last condition is equivalent, subject to rigidity, to the requirement that the locally measured electric field E_a in the star be zero, since E_a is given by

$$E_a = (-\bar{u}_c \bar{u}^c)^{-1/2} \bar{E}_a \tag{8.31}$$

where

$$\bar{E}_a = \Phi^S_{\ ,a} \tag{8.32}$$

in the rigid case. When *all* these equilibrium conditions are satisfied, (8.20) can be integrated explicitly to give the change in mass in a reversible variation directly in terms of the *total* changes in angular momentum, charge, particle numbers and entropy in the form

$$dM = \Omega\, dJ + \Phi^S\, dQ + \bar{\mu}^{(i)}\, dN_{(i)} + \bar{\Theta}\, dS \tag{8.33}$$

Under conditions when there are effective thermodynamic restraints which allow the star to exist in equilibrium with variable Ω, Φ^S, $\bar{\mu}^{(i)}$, $\bar{\Theta}$ more care is needed to verify that the energy deviations from the mechanical equilibrium value after the alteration process is really of second order. In fact non-uniform temperature $\bar{\Theta}$ and chemical potentials $\bar{\mu}^{(i)}$ cause no difficulties. In the non-electromagnetic case non-uniform angular velocity Ω (which requires zero viscosity) causes no difficulty either since it is always possible consistently with the corresponding restraint (that of local conservation of angular momentum in the individual matter rings) to minimize the energy by appropriate expansions and displacements of the matter rings. However in the electromagnetic case, when the zero viscosity condition is extended to apply to the electric current, there will be additional restraints (corresponding to conservation of magnetic flux through the matter rings) which may be incompatible with the adjustments of the current rings which would be required for energy minimization. Thus we are again led back to the requirement that the electromagnetically interacting parts of the system should be rigid if the formula (8.20) is to be applicable. Thus the most general practically applicable formula which we can obtain by this line of reasoning has the form

$$dM = \Omega_F \int \delta J_F + \int \Omega \, \delta J_M + \int \Phi^s \, \delta Q + \int \bar{\mu}^{(i)} \, \delta N_{(i)} + \int \bar{\Theta} \, \delta S \tag{8.34}$$

where we have separated the angular momentum contributions of the electromagnetic field, given by the flux element

$$dJ_F = T_F^{ab} m_b \, d\Sigma_a \tag{8.35}$$

from those of the matter field, given by

$$dJ_M = T_M^{ab} m_b \, d\Sigma_a \tag{8.36}$$

in terms of the separate electromagnetic and matter energy tensors defined by (5.28) and (6.19), and where we have been obliged to assume that there is a well defined uniform angular velocity Ω_F associated with the electromagnetic contributions.

Let us now consider the extension of this formula to cover variations, in the case where instead of or in addition to the material system, there is a central black hole. It is obvious (Carter 1972) that in the neighbourhood of the hole we should set

$$\bar{u}^a = l^a \tag{8.37}$$

where l^a is the horizon Killing vector field determined by the rigid angular velocity Ω^H of the hole as described in section 4. There is no alternative to this choice of $\bar{u}^a$ in the black hole limit, since *any* vector of the form (8.7) must approach the null tangent vector l^a of the horizon in the limit if it is to remain timelike up to the horizon. The potential corresponding to Φ^S as defined by (8.30) in the case of a star will be

$$\Phi^\dagger = l^a A_a \tag{8.38}$$

which is clearly the same as the quantity introduced by (7.43). The electric field vector $\bar{E}_a^\dagger$ introduced by (5.30) is clearly given by

$$\bar{E}_a^\dagger = \Phi^\dagger_{,a} \tag{8.39}$$

and it therefore follows from the boundary conditions (5.32) that is *constant* on the horizon. We shall denote the value of this constant by Φ^H i.e. we set

$$\Phi^\dagger = \Phi^H \tag{8.40}$$

on $\mathscr{H}^+$, where Φ^H is uniquely defined by the gauge condition, which we have been using throughout this section, that A^a (and hence also $\Phi^\dagger$) tends to zero in the large distance asymptotic limit. Thus we see that a black hole is analogous to an ordinary body (with finite viscosity and electrical conductivity) in rigid electrical equilibrium.

In extrapolating the formula (8.34) to a black hole, it is clear that the entropy and particle number contributions associated with the hole will be zero, since the time-dilation factor $(-\bar{u}^a\bar{u}_a)^{1/2}$ in the definitions (8.17) and (8.18) of the effective temperature and chemical potentials, tends to zero on the horizon. In other words it is clear that the effective temperature $\bar{\Theta}^H$ and the effective chemical potentials $\bar{u}^{(i)H}$ of the hole must all be taken to be zero i.e.

$$\bar{\Theta}^H = 0 \tag{8.41}$$

$$\bar{\mu}^{(i)H} = 0 \tag{8.42}$$

This is an expression of the so called "transcendence" of ordinary particle conservation laws and the second law of thermodynamics by the black hole.

It is to be noted that though particles which go down a black hole are neither extractable nor even externally detectable, such a process is not necessarily irreversible from a thermodynamic point of view since it is always possible to send in a corresponding number of antiparticles. Thus by considering the black hole limit of Thorne-Zledovich type interaction processes, we are lead to the heuristic deduction that a *reversible* variation of the equilibrium state of a system consisting of an axisymmetric black hole with surrounding matter rings should satisfy

$$dM = \Omega^H(dJ_H + \int \delta J_F) + \int \Omega\, dd_M + \Phi^H\, dQ_H + \int \Phi^S\, \delta Q \tag{8.43}$$
$$+ \int \bar{\Theta}\, \delta S + \int \bar{\mu}^{(i)}\, \delta N_{(i)}$$

where J_H and Q_H are the angular momentum and charge of the hole itself, and where we have supposed that the angular velocity of the electromagneticly interacting parts of the system is the same as that of the hole.

In the particular case of the Kerr-Newman vacuum solutions Φ^H and Ω^H as defined here are known functions of M, J and of Q (which in this case is the same as Q_H) given by

$$\Omega^H = JM^{-1}(2Mr_+ - Q^2)^{-1} \tag{8.44}$$

$$\Phi^H = Qr_+\,(2Mr_+ - Q^2)^{-1} \tag{8.45}$$

where

$$r_+ = M + (M^2 - J^2M^{-2} - Q^2)^{1/2}. \tag{8.46}$$

With these values, (8.43) reduces in this case to the exact differential of the mass formula of Christodoulou and Ruffini (1970, 1971) which takes the form

$$M = \{(M_0 + \tfrac{1}{4}Q^2M_0^{-1})^2 + \tfrac{1}{4}J^2M^{-2}M_0^{-2}\}^{1/2} \tag{8.47}$$

where M_0 is an integration constant, given by the expression

$$M_0^2 = \frac{1}{16\pi}\mathscr{A} = \frac{1}{4}\{2M^2 - Q^2 + 2(M^4 - M^2Q^2 - J^2)^{1/2}\} \tag{8.48}$$

where $\mathscr{A}$ is the surface area of the black hole (i.e. the integral of surface area over a spacelike 2-dimensional section of $\mathscr{H}^+$).

Although the basic variation formula for a (8.34) for a star was derived by considering *reversible* processes, we can immediately deduce that it will hold for *any* process if we assume that the equilibrium configuration is a well defined function of the distribution of J, Q, S and $N^{(i)}$ over the rotating matter rings, which will be the case under a wide range of natural conditions. Now in the case when a central black hole is present our experience with the Kerr solutions leads to the *generalized no-hair conjecture* according to which the system as a whole should have just *two additional degrees of freedom in the non-electromagnetic case, and three in the electromagnetic case* (leaving a mathematically conceivable magnetic monopole moment out of account) *in addition to the degrees of freedom* (determining the distribution of J, Q, S, $N^{(i)}$) *associated directly with the external matter rings.*

Now the formula (8.43) derived by considering reversible variations, involves just one additional degree of freedom, namely J_H, associated with the hole in the non-electromagnetic case, and just two, namely J_H and Q_H in the electromagnetic case. However it has been shown by Hawking (1971) (in the manner which he describes in the accompanying course) that the result discovered by Christodoulou and Ruffini in the special case of the Kerr-Newman black holes must be true in general, i.e. the surface area $\mathscr{A}$ of the hole must *always remain constant in any transformation which is reverisble.* Taking $\mathscr{A}$ itself as the additional degree of freedom, it therefore follows from the generalized no hair conjecture that the mass variation in a completely general (not necessarily reversible) change between neighbouring black hole equilibrium states should be given by

$$\begin{aligned} dM = \mathscr{T} d\mathscr{A} + \Omega_H(dJ_H + \int \delta J_F) + \int \Omega\, dJ_M \\ + \Phi^H dQ_H + \int \Phi^S \delta Q + \int \bar{\Theta}\, \delta S + \int \bar{\mu}^{(i)} \delta N_{(i)} \end{aligned} \tag{8.49}$$

where the form of the coefficient $\mathscr{T}$ (which has the dimensions of surface tension) remains to be determined. Since the $\mathscr{T} d\mathscr{A}$ contribution (unlike all the other terms) can only be produced by a non-reversible transformation there is no hope of evaluating it by considering a Thorne-Zeldovich type process. In the particular case of the Kerr-Newman solutions, where M is a known function of J, Q and $\mathscr{A}$ only, it is of course possible to write down a differential formula of the form (8.49) and read off the value of the co-efficient $\mathscr{T}$ empirically, as has been done by Beckenstein (1972), but this does not give any insight into the general form of the coefficient $\mathscr{T}$. However a simple and elegant solution to the problem of evaluating $\mathscr{T}$ has come out of the recent work of Hartle and Hawking (1972). By applying the Newman-Penrose equations to the perturbation effect on the horizon of a test field which represents uncharged matter falling from infinity through the horizon $\mathscr{H}^+$, without interacting with any external matter which may be present (so that there should be no contribution either to dQ_H or to the variations δJ, δQ, $\delta N_{(i)}$, δS in the matter) and using a limit in which there should be no gravitational or electromagnetic radiation to infinity, Hawking and Hartle obtained the formula

$$dM = \frac{\kappa}{8\pi} d\,\mathscr{A} + \Omega^H\, dJ \tag{8.50}$$

(where κ is the constant whose existence was established by Theorem 8) in the manner described by Hawking in his accompanying course. If the generalized no hair conjecture is correct, the coefficient of $d\mathscr{A}$ should be the same for a general variation as in the special kind of variation considered by Hawking and Hartle, and therefore we are led from (8.50) to the conclusion that the coefficient $\mathscr{T}$ in the general mass variation formula (8.49) must be given by

$$\mathscr{T} = \frac{\kappa}{8\pi} \tag{8.51}$$

Since both the generalized Thorne-Zeldovich argument and the Hartle-Hawking discussion (not to mention the generalized no-hair conjecture itself), on which the present derivation of the general black hole mass variation formula (8.49) is based, are essentially heuristic, it is obviously desirable to have a mathematical proof of the validity of the formula (8.49). Such a proof has in fact been constructed, during the course of the present summer school, by Bardeen, Hawking and myself in collaboration. The existence of this proof, which is described in the next section, provides strong evidence in favour of the generalized no-hair conjecture.

Hawking (1971) showed not only that the surface area $\mathscr{A}$ of a black hole must remain constant in any reversible variation, but also that it must increase in any irreversible transformation. This result immediately suggested to many people an analogy between the role played by the surface area $\mathscr{A}$ in black hole mechanics and the role played by entropy in what is traditionally known as thermodynamics (although the term thermal equilibrium mechanics would be more appropriate). The results of this section show that this analogy can be carried very much further, with the locally defined scalar κ playing the role analogous to that of temperature. Thus we are led to formulate the following *four laws of black hole equilibrium mechanics*, which are closely analogous to the four standard laws of thermodynamics.

The zeroth law of black hole mechanics will obviously be the result proved in Theorem 8, that the scalar κ is constant over the horizon $\mathscr{H}^+$.

The *first law of black hole mechanics* will be the mass variation formula (8.49) whose heuristic derivation has just been described, and which will be proved rigorously in the next section.

The *second law of black hole mechanics* will of course be the rule

$$d\mathscr{A} \geqslant 0 \tag{8.52}$$

whose derivation and applications are described by Hawking in the accompanying course.

Continuing the analogy, I suggest that the *third law of black hole mechanics* should be the statement that it is impossible by any procedure, no matter how idealized, to reduce the constant κ of a black hole to zero by a finite sequence of operations. In short, degenerate black hole states represent physically unattainable limits.

Unlike the other three laws which are based on rigorous proofs, this third law is still essentially conjectural. However the evidence provided by our knowledge of the extreme (bottomless) Kerr and Reissner-Nordstrom black hole spaces (as described in Part I of the present source and in the accompanying course of Bardeen) provides strong evidence in favour of this conjecture, which would appear to be a fairly direct consequence of the cosmic censorship hypothesis. Conversely by showing that all degenerated black hole equilibrium states are essentially bottomless, Boyer's Theorem 5 suggests directly that degenerate black hole states are unattainable and this in turn provides support for the cosmic censorship hypothesis. Thus it seems that the third law as stated above is virtually equivalent to the cosmic censorship hypothesis in the sense that they will stand or fall together.

The four laws collected together above are clearly of fundamental importance in their own right. Although they correspond closely to the classical laws of thermodynamics, it is to be emphasized that this is only an analogy whose significance should not be exaggerated. Although they are analogous, $\mathscr{T}$ and $\mathscr{A}$ play a quite distinct role from the temperature and entropy with which they should not be confused (and which enter separately into the first law equation (8.43)). The real effective temperature $\bar{\Theta}^H$ of a black hole is well defined and unambiguously zero, as also are its chemical potentials $\bar{\mu}^{(i)H}$. The ordinary particle conservation laws, and the ordinary second law of thermodynamics are unquestionably transcended by a black hole, in the sense that particles and entropy can be lost without trace from an external point of view. It is not possible to mitigate this trandescension by somehow relating the amount of entropy, (or the number of particles) which have gone in to the subsequent increase in surface area $\mathscr{A}$.

9 Generalized Smarr Formula and the General Mass Variation Formula

The results described in this section were worked out by Bardeen, Hawking and myself at Les Houches. This work originated as a search for a general black hole mass formula of the kind discovered empirically by Smarr (1972) in the particular case of the Kerr and Kerr-Newman solutions. This work leads us on naturally to a generalization of the previous mass variation formulae of Hartle and Sharp (1967) Bardeen (1970) and Carter (1972) so as to obtain a rigorous derivation of the first law of black hole dynamics as given by the formula (8.49).

It follows from the asymptotic boundary conditions that the total mass M and angular momentum J of a general stationary axisymmetric system can be expressed in the Komar form

$$M = -\frac{1}{4\pi}\oint_{\infty} k^{a;b}\, dS_{ab} \tag{9.1}$$

and

$$J = \frac{1}{8\pi}\oint m^{a;b}\, dS_{ab} \tag{9.2}$$

where the integrals are taken over a spacelike 2-sphere with metric normal element dS_{ab}, surrounding the system at large distance. The fact that $k^{a;b}$ and $m^{a;b}$ are antisymmetric makes it possible to apply the generalized Stokes theorem to obtain

$$M = -\frac{1}{4\pi}\int k^{a;b}{}_{;b}\, d\Sigma_a - \frac{1}{4\pi}\oint_H k^{a;b}\, dS_{ab} \tag{9.3}$$

and

$$J = \frac{1}{8\pi}\int m^{a;b}{}_{;b}\, d\Sigma_a + \frac{1}{8\pi}\oint_H m^{a;b}\, dS_{ab} \tag{9.4}$$

where the suffix is used to denote a boundary integral over a 2-sphere on the surface of the central black hole (if there is one) and $d\Sigma_a$ is a metric normal element of a spacelike hypersurface Σ extending from the boundary of the hole out to infinity.

Using the standard identities

$$k^{a;b}{}_{;b} = -R^a{}_b k^b \tag{9.5}$$

$$m^{a;b}{}_{;b} = -R^a{}_b m^b \tag{9.6}$$

(derived in sections 6 and 7) which hold for any Killing vectors, and making the obvious definitions

$$M_H = -\frac{1}{4\pi}\oint_H k^{a;b}\,dS_{ab} \tag{9.7}$$

and

$$J_H = \frac{1}{8\pi}\oint_H m^{a;b}\,dS_{ab} \tag{9.8}$$

we can relate the boundary integrals to the Ricci tensor R_{ab} in the intervening space by

$$M = \frac{1}{4\pi}\int R^a_b k^b\,d\Sigma_a + M_H \tag{9.9}$$

and

$$J = -\frac{1}{8\pi}\int R^a_b m^b\,d\Sigma_a + J_H \tag{9.10}$$

These formulae differ from the standard formulae for an ordinary star only through the presence of the black hole boundary terms. For the Kerr solutions on the other hand, these boundary terms will be the only ones that remain. The neat mass formula recently discovered by Larry Smarr for the Kerr solution prompts us to examine these terms more closely. Introducing the angular velocity Ω^H of the hole, and the rigidly co-rotating Killing vector

$$l^a = k^a + \Omega^H m^a \tag{9.11}$$

we obtain the formula

$$\tfrac{1}{2}M_H = \Omega^H J_H - \frac{1}{8\pi}\oint_H l^{a;b}\,dS_{ab} \tag{9.12}$$

Introducing a second null vector n^a orthogonal to the 2-sphere on the horizon, with the normalization condition $l^a n_a = 1$, we can express the normal element in the form $dS_{ab} = l_{[a}n_{b]}\,dS$ where dS is the element of surface area, and noting that the gravitation constant κ discussed in the previous section can be expressed in the form

$$\kappa = l^a{}_{;b}l^b n_a \tag{9.13}$$

we thus obtain

$$\tfrac{1}{2}M_H = \Omega_H J_H + \frac{\kappa}{8\pi}\mathscr{A} \tag{9.14}$$

We can use this expression to eliminate M_H from the mass formula (9), which leads to the basic generalized Smarr formula

$$\tfrac{1}{2}M = \frac{1}{8\pi}\int R^a_b k^b \, d\Sigma_a + \Omega^H J_H + \mathscr{T}\mathscr{A} , \tag{9.15}$$

where $\mathscr{T}$ is given by the expression (8.45), which reduces to the original Smarr formula, as obtained for the Kerr solutions, in the pure vacuum case when the Ricci terms on the right hand side are zero.

The expression is not merely elegant. It is also extremely convenient as a starting point for the rigorous derivation of the general mass variation formula. Before starting on the variational calculation, we shall carry out a further decomposition of the unperturbed mass formula, in order to separate the electromagnetic field contributions in the manner suggested by the original Smarr formula for the charged Kerr-Newman solutions. Thus we split up the total energy momentum tensor T^{ab} appearing in the Einstein equations

$$R^{ab} - \tfrac{1}{2}Rg^{ab} = 8\pi \, T^{ab} \tag{9.16}$$

in the form

$$T^{ab} = T^{ab}_M + T^{ab}_F \tag{9.17}$$

where T^{ab}_M is given by (6.19) and T^{ab}_F is given by (5.28).

The field is assumed to have the form $F_{ab} = 2A_{[a;b]}$ where the electromagnetic potential satisfies the group invariance conditions

$$A_{a;b}k^b + A_b k^b{}_{;a} = 0$$

$$A_{a;b}m^b + A_b m^b{}_{;a} = 0 \tag{9.18}$$

The electric current vector j^a is defined as usual by

$$F^{ab}{}_{;b} = 4\pi j^a \tag{9.19}$$

We can decompose the angular momentum into matter, field, and hole contributions, the two former being defined by

$$J_M = \int T^a_{Mb} m^b \, d\Sigma_a \tag{9.20}$$

and

$$\begin{aligned} J_F &= \int T^a_{Fb} m^b \, d\Sigma_a \\ &= \int m^c A_c j^a \, d\Sigma_a + \frac{1}{4\pi}\oint_H m^c A_c F^{ab} \, dS_{ab} \end{aligned} \tag{9.21}$$

where in deriving the last formula we have used the asymptotic boundary conditions to eliminate a surface integral contribution. (We have also used the fact—which will enable us to drop out many angular momentum contributions—that Σ can be chosen in such a way that $m^a \, d\Sigma_a = 0$.) In terms of these the total angular momentum takes the form

$$J = J_M + J_F + J_H \tag{9.22}$$

Before giving the corresponding subdivision of the mass term, we introduce the total charge Q of the system, defined analogously to the mass and angular momentum by a boundary integral at infinity as

$$Q = -\frac{1}{4\pi} \oint_\infty F^{ab} \, dS_{ab} \tag{9.23}$$

Using (9.19) we can write this as

$$Q = -\int j^a \, d\Sigma_a + Q_H \tag{9.24}$$

with the obvious definition

$$Q_H = -\frac{1}{4\pi} \oint_H F^{ab} \, dS_{ab} \tag{9.25}$$

It was shown in the previous section that

$$\Phi^\dagger = l^a A_a \tag{9.26}$$

is constant on the boundary of the hole. Using this to evaluate the boundary integrals

$$\oint_H l^c A_c F^{ab} \, dS_{ab} = -4\pi \Phi^H Q_H \tag{9.27}$$

and (with a little more work)

$$\oint_H A_c F^{ca} l^b \, dS_{ab} = 2\Phi^H Q_H \tag{9.28}$$

we obtain the basic generalization of the electromagnetic Smarr formula in the form

$$\begin{aligned} \tfrac{1}{2}M = & \int [T^a_{Mb} k^b - \tfrac{1}{2} T_M k^a] \, d\Sigma_a - \Omega^H J_M \\ & -\tfrac{1}{2} \int l^c A_c j^a \, d\Sigma_a + \int A_b j^{[b} l^{a]} \, d\Sigma_a \\ & + \Omega^H J + \mathscr{T}\mathscr{A} + \tfrac{1}{2} \Phi^H Q_H \end{aligned} \tag{9.29}$$

In the source free case, when the electric current and the matter contributions are zero, so that only the last three terms are left, this reduces to the original Smarr formula as given for the Kerr-Newman solutions.

Although the expression (9.29) shows the connection with the original Smarr formula most clearly, it is rather more convenient for starting the variational calculation to recast it in a form in which the Lagrangian densities R and $F_{cd}F^{cd}$ for the gravitational and electromagnetic fields are brought into evidence. Thus (again using (9.28)) we choose to set it out as

$$M = \int [T^a_{Mb} k^b - l^c A_c j^a] \, d\Sigma_a + \frac{1}{16\pi} \int (R + F^d_c F^c_d) k^a \, d\Sigma_a$$
$$+ \tfrac{1}{2}M + \Omega^H J_F + \Omega^H J_H + \Phi^H Q_H + \mathscr{T}\mathscr{A} \qquad (9.30)$$

where the term $\frac{1}{2}M$ on the right hand side is to be interpreted as being given directly by (9.7).

Up to this point we have been able to work with a fully general matter tensor (subject only to group invariance) but in order to carry out the variational calculation we shall now specialize to the case when the circularity conditions discussed in section 7 are satisfied, so that the vector $\bar{u}^a$ introduced in the previous section is well defined.

We now begin the actual variational calculation. In any variational calculation there is a certain freedom of choice in the way in which one identify parts of the manifold before and after the variation. When matter is present a naturally convenient choice is to identify particular particle world lines, i.e. trajectories of u^a or equivalently l^a before and after the variation. In the present cases however an even more important consideration, to which we shall give priority, is the preservation of the invariance group properties. Thus we shall require that the Killing vectors (in their natural contravariant form) be left invariant by the variation, i.e.

$$dk^a = 0 \qquad dm^a = 0 \qquad (9.31)$$

Unfortunately (except in the restrictive case when the angular velocities are left invariant by the variation, i.e. when $d\Omega = 0$) this will not be compatible with preservation of the particle world lines taken as a whole. This is not a very serious problem however, since we are only interested in quantities evaluated on a particular spacelike hypersurface Σ or on its boundary, and we can, and therefore shall, require that the points at which particle world lines cross Σ be left invariant. In particular we require that the boundary points on which particular null generators of the horizon meet the boundary of Σ remain the same, even though the canonical null tangent vector itself will have a variation given by

$$dl^a = m^a \, d\Omega^H \qquad (9.32)$$

Introducing the metric variation tensor h_{ab} given by

$$dg_{ab} = h_{ab}; \qquad dg^{ab} = -h^{ab} \qquad (9.33)$$

we can express the variation of the covariant form of l_a as

$$dl_a = h_{ab} l^b + m_a \, d\Omega^H \tag{9.34}$$

Since l_a remains normal to the horizon, which itself remains invariant, this vector must also satisfy the restriction

$$l_{[a} \, dl_{b]} = 0 \tag{9.35}$$

Using this, together with the group invariance condition

$$l^b (dl_a)_{;b} + l^b{}_{;a} \, dl_b = 0 \tag{9.36}$$

it is easy to verify that the differential of the black hole surface gravity constant κ is given by

$$\begin{aligned} d\kappa &= -\tfrac{1}{2}(dl_a)^{;a} - m_{a;b} l^a n^b \, d\Omega^H \\ &= \tfrac{1}{2} h_{a;b}{}^{;b} l^a - m_{a;b} l^a n^b \, d\Omega^H \end{aligned} \tag{9.37}$$

The corresponding formula for the differential of the black hole surface potential ϕ^H is simply

$$d\overset{H}{\phi} = l^a \, dA_a + m^a A_a \, d\Omega^H \tag{9.38}$$

We can now proceed with the evaluation of the variations of the integrals in (9.30), noting that for any integrand

$$d[\quad d\Sigma_a] = (-g)^{-1/2} \, d[(-g)^{1/2} \quad] \, d\Sigma_a \tag{9.39}$$

and that

$$d(-g)^{1/2} = \tfrac{1}{2}(-g^{1/2}) h^a_a \tag{9.40}$$

It is evident that

$$d\left\{ \frac{1}{16\pi} F^d{}_c F^c{}_d (-g)^{1/2} \right\} = \left\{ \tfrac{1}{2} T^{cd}_F h_{cd} + \frac{1}{4\pi} F^{cd} (dA_d)_{;c} \right\} (-g)^{1/2} \tag{9.41}$$

It is well known (see any good standard textbook, such as Landau and Lifshits, whose sign conventions I am following here) that

$$d\left\{ \frac{1}{16\pi} R(-g)^{1/2} \right\} = -\left\{ \frac{1}{16\pi} (R^{cd} - \tfrac{1}{2} R g^{cd}) h_{cd} + \frac{1}{8\pi} h_c{}^{[c;b]}{}_b \right\} (-g)^{1/2} \tag{9.41}$$

It is also true (see Carter 1972) that in a perfectly elastic variation of a solid or perfect fluid

$$\begin{aligned} d\{T^a_{Mb} k^b (-g)^{1/2}\} \, d\Sigma_a &= -\Omega d\{T^a_{Mb} m^b (-g)^{1/2}\} \, d\Sigma_a \\ &\quad - l^a \, d\{\rho(-g)^{1/2}\}|_\Omega \, d\Sigma_a \end{aligned} \tag{9.42}$$

where $d\rho|_\Omega$ denotes the variation in ρ calculated as it would be if $d\Omega$ were held equal to zero. (It is not true that $\partial\rho/\partial\Omega$ is itself zero, as would be the case in the

Newtonian limit, but the contribution to which it gives rise cancels out.) In a purely elastic variation, in which entropy and particle numbers are conserved, the variation will be given by $d\rho|_\Omega = (\partial\rho/\partial g_{ab})h_{ab}$ where

$$\frac{\partial \rho}{\partial g_{ab}} = -\tfrac{1}{2}(p^{ab} + \rho g^{ab} + \rho u^a u^b) \tag{9.43}$$

(see Carter 1972). In a perfectly elastic case the entropy and particle number flux vectors can have no components transverse to the flow since it is necessary to be able to regard the system as being in local thermal and chemical equilibrium, i.e. we must have

$$S^a = Su^a \tag{9.44}$$

$$n_{(i)}^a = n_{(i)}u^a \tag{9.55}$$

Under these conditions it is easy to allow for more general variations in which the local thermal and chemical equilibrium is altered, i.e. variations which do not preserve entropy and particle numbers, provided that ρ is a well defined function not only of the geometry but also of the entropy and particle number density scalars s and $n_{(i)}$ defined by (9.44) and (9.45) so that the local temperature Θ and Gibbs chemical potentials $\mu^{(i)}$ are well defined by

$$\frac{\partial \rho}{\partial s} = \Theta \tag{9.46}$$

$$\frac{\partial p}{\partial n_{(i)}} = \mu^{(i)} \tag{9.47}$$

It is to be noted that these quantities are not all independent, being related by the identity

$$\rho + p = \Theta s + \mu^{(i)} n_{(i)} \tag{9.48}$$

where we define

$$p = \tfrac{1}{3}p_a^a \tag{9.49}$$

When these more general variations are allowed, (9.43) will still be valid provided we substitute

$$d\rho|_\Omega = -\tfrac{1}{2}(p^{ab} + \rho u^a u^b + \rho g^{ab}) + \Theta\, đs + \mu^{(i)}\, đn_{(i)} \tag{9.50}$$

where $đs$ and $đn_{(i)}$ are not the total variations in s and the $n_{(i)}$ but only the contributions due to non-conservation of entropy and particle numbers in the local matter rings, i.e. they are given by

$$d(su^a\, d\Sigma_a) = đs u^a\, d\Sigma_a \tag{9.51}$$

$$d(n_{(i)}u^a\, d\Sigma_a) = đn_{(i)} u^a\, d\Sigma_a \tag{9.52}$$

Under these conditions (9.42) becomes

$$d\{T^a_{Mb}k^b(-g)^{1/2}\}\,d\Sigma_a = \Omega d\{T^a_{Mb}m^b(-g)^{1/2}\}\,d\Sigma_a$$
$$+\,l^aT^{cd}_Mh_{cd}(-g)^{1/2}\,d\Sigma_a - \{\bar{\Theta}\,ds + \bar{\mu}^{(i)}\,dn_{(i)}\}\,u^a(-g)^{1/2}\,d\Sigma_a \qquad (9.53)$$

where we have introduced the effective temperature and chemical potentials $\bar{\Theta}$ and $\bar{\mu}^{(i)}$ defined by (8.17) and (8.18).

Having thus worked out the contributions to the variational integrands, we are now ready to perform the actual integration. Using the group invariance condition

$$h_{ab;c}k^c + h_{ac}k^c_{;b} + h_{bc}k^c_{;a} = 0 \qquad (9.54)$$

to cast the integrand $k^ah_c{}^{[c;b]}{}_{;b}$ as a divergence in the form

$$k^ah_c{}^{[c;b]}{}_{;b} = \{k^ah_c{}^{[c;b]} - k^bh_c{}^{[c;\ ]}\}_{;b} \qquad (9.55)$$

we obtain

$$\frac{1}{8\pi}\int k^ah^{[c;b]}_{c;b}\,d\Sigma_a = \frac{1}{4\pi}\oint_{\infty} k^ah_c^{[c;b]}\,dS_{ab} - \frac{1}{4\pi}\oint_{H} k^ah_c^{[c;b]}\,dS_{ab}$$
$$= \tfrac{1}{2}dM + \mathscr{A}d\,\mathscr{T} + J_H\,d\Omega^H \qquad (9.56)$$

where the standard contribution $\frac{1}{2}dM$ from the surface integral at infinity is easily obtained from the a symptotic boundary conditions, and where the hole terms $\mathscr{A}d\,\mathscr{T} + J_H\,d\Omega^H$ are obtained with the aid of the formula (9.36). In a similar manner we can cast the integrand $k^aF^{cd}(dA_d)_{;c}$ as a divergence in the form

$$k^aF^{cd}(dA_d)_{;c} = 2\{F^{c[a}k^{b]}\,dA_c\}_{;b} + 4\pi k^aj^c\,dA_c$$

and hence we obtain

$$\frac{1}{4\pi}\int k^aF^{cd}(dA_d)_{;c}\,d\Sigma_c = -Q\,d\Phi^H$$
$$-(J_F - \int j^am^cA_c\,d\Sigma_a)\,d\Omega^H + \int l^aj^c\,dA_c\,d\Sigma_a \qquad (9.57)$$

with the aid of (9.37). Finally the variation of the second term in (9.30) gives simply

$$d\{\int l^cA_cj^a\,d\Sigma_a\} = \int l^cj^a\,dA_c\,d\Sigma_a$$
$$+\,d\Omega_H\int j^am^cA_c\,d\Sigma_a - \int l^cA_c\,\delta Q \qquad (9.58)$$

where we have introduced the abbreviation

$$\delta Q = d[-j^a\,d\Sigma_a] \qquad (9.59)$$

Introducing the analogous abbreviations

$$\delta N_{(i)} = d[-n^a_{(i)}\, d\Sigma_a] \tag{9.60}$$

$$\delta S = d[-s^a\, d\Sigma_a] \tag{9.61}$$

$$\delta J_M = d[-T^a_{Mb} m^b\, d\Sigma_a] \tag{9.62}$$

combining (9.40), (9.41), (9.51) with the Einstein equations, and finally using (9.54), (9.55), (9.56) to simplify the residual integrals, we obtain the variation of (9.30) in the form

$$\begin{aligned} dM = {} & \Omega^H(dJ_H + dJ_F) + \int \Omega\delta J_M + \int \bar{\Theta}\, \delta S \\ & + \int \bar{\mu}^{(i)}\, \delta N_{(i)} + \Phi^H\, dQ_H + \int l^c A_c\, \delta Q \\ & + \frac{\kappa}{8\pi} d\mathscr{A} + 2\int j^{[c}l^{a]}\, dA_c\, d\Sigma_a \end{aligned} \tag{9.63}$$

This variation formula, which represents our final expression of the first law of black hole mechanics, reduces to the formula (8.49) when the electromagnetic rigidity condition

$$j^{[c}l^{a]} = 0 \tag{9.64}$$

is satisfied so that the last term drops out; this last term also drops out for any variation which satisfies the perfect conductivity condition that the magnetic flux (cf. equation (7.22)) through any comoving circuit be conserved.

10 Boundary Conditions for the Vacuum Black Hole Problem

From this point onward we shall restrict our attention to the case of black hole spaces in which the pure vacuum or source free Einstein-Maxwell equations are satisfied, excluding the degenerate limit case due to lack to time and space. (Further details of the degenerate limit case are given by Bardeen.) As has been explained in the preceding sections it has been established with certainty in the rotating case, and it is virtually certain in the non-rotating case also, that such a space will satisfy the conditions of Theorem 7 and Theorem 4.2 and it is also virtually certain that the horizon will be connected, with topologically spherical space sections. These various conditions can be summed up as follows.

Condition 10 $\mathscr{M}$ is time orientable asymptotically flat, pseudo-stationary and axisymmetric. The causality axiom is satisfied in $\mathscr{M}$. The domain of outer communications $\ll \mathscr{I} \gg$ of $\mathscr{M}$ is topologically the product of the Euclidean 2-plane and the 2-sphere, and the horizon $\mathscr{H}^+$ is non-degenerate and is topologically the product of a Euclidean line and a 2-sphere. The Einstein equations

$$R^{ab} = 0 \tag{10.1}$$

or the Einstein-Maxwell equations

$$R^{ab} = 8\pi T_F^{ab}$$

$$j^a = 0 \qquad (10.2)$$

are satisfied in $\mathscr{M}$.

It follows from Condition 10, by Theorem 4.2 and Theorem 7, as we have seen in section 7, that the metric on the union $\ll \mathscr{I} \gg \cup \mathscr{H}^+$ can be expressed in the form

$$ds^2 = ds_{\text{II}}^2 + X\,d\varphi^2 + 2W\,d\varphi\,dt - V\,dt^2 \qquad (10.3)$$

where $ds_{\text{II}}^2 = g_{\alpha\beta}\,dx^\alpha\,dx^\beta$ ($\alpha, \beta = 1, 2$) is the metric projected orthonally onto a 2-dimensional surface, $\overline{\mathscr{B}}$ say, which intersects each surface of transitivity within $\ll \mathscr{I} \gg \cup \mathscr{H}^+$ of the action π^{sA} once. The form (10.3) will be well behaved except on $\mathscr{H}^+$ and on the rotation axis, which form an edge to $\overline{\mathscr{B}}$. Since the topological conditions are such that the rotation axis will have two branches (extending respectively from the north and south poles of the hole to infinity) $\overline{\mathscr{B}}$ will have the topology of a square in Euclidean 2-space possessing three edges, but from which the remaining edge has been excluded, the included edges corresponding in order, to the intersection of $\overline{\mathscr{B}}$ with the southern rotation axis, the horizon $\mathscr{H}^+$, and the northern rotation axis.

The purpose of this section is to show that the system of Einstein equations and global conditions required by Condition 10 can be reduced to a comparativel simple set of partial differential equations with boundary conditions for V, W, X and the scalars B, Φ defined by (7.28) as functions on $\overline{\mathscr{B}}$.

We start by considering the metric form ds_{II}^2 on $\mathscr{B}$. In deriving the general form (10.3) or (7.36) for the metric and the form (7.42) for the field potential we have in effect made use of and satisfied only the Einstein-Maxwell equations involving the cross components in the Ricci-tensor between directions lying in and orthogonal to the surfaces of transitivity, but we have made no use of the equations for the components of the Ricci tensor lying wholly in or wholly orthogonal to the surfaces of transitivity. Now it is true generally that the source-free Einstein-Maxwell equations require that the Ricci tensor as a whole be trace-free; and in the case of a field satisfying the circularity conditions (7.4), it is easy to see further that the energy momentum tensor is such that the projections of the Ricci tensor into the orthogonal to the surfaces of transitivity must be trace free separately. Now it is well known from the work of Papapetrou and others that the condition that the projection of the Ricci tensor into the surfaces of transitivity be trace free, i.e. (in terms of the co-ordinate system of the form (10.3))

$$XR_{tt} - 2WR_{t\varphi} + VR_{\varphi\varphi} = 0 \qquad (10.4)$$

is equivalent to the condition that the scalar ρ defined, (whenever it is real) as the non-negative root of the equation

$$\rho^2 = \sigma \tag{10.5}$$

must be *harmonic* i.e. its Laplacian, defined in terms of the two dimensional metric ds^2_{II}, is zero. We can use this fact to show that ρ *has no critical points in the interior of* $\overline{\mathscr{B}}$ (i.e. no points where its gradient vanishes) and hence that can ρ be used as a globally well behaved co-ordinate in $\ll \mathscr{I} \gg$ except on the axi-symmetry axis (Figure 10.1).

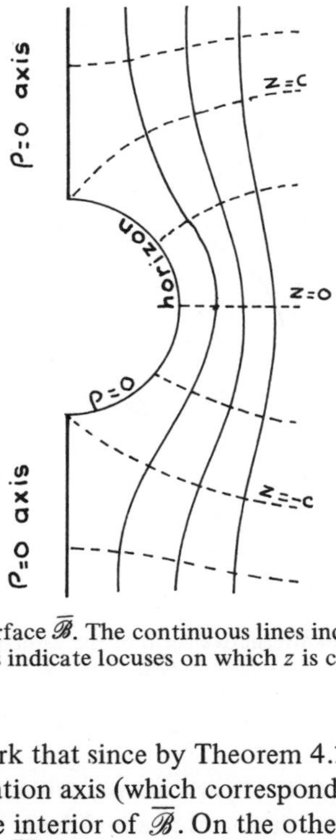

Figure 10.1. Plan of the 2-surface $\overline{\mathscr{B}}$. The continuous lines indicate locuses on which ρ is constant and the dotted lines indicate locuses on which z is constant.

To prove this we remark that since by Theorem 4.2, σ is strictly positive in $\ll \mathscr{I} \gg$ except on the rotation axis (which corresponds to the boundary of $\overline{\mathscr{B}}$), ρ is *strictly positive* in the interior of $\overline{\mathscr{B}}$. On the other hand ρ *is zero on the whole of the boundary of* $\overline{\mathscr{B}}$, since it is immediately clear that ρ is zero both on the rotation axis where X and W are both zero (because m^a is zero) and on the horizon $\mathscr{H}^+$, since, by Theorem 4.2, $\mathscr{H}^+$ lies on the rotosurface where σ is zero. Furthermore the asymptotic flatness boundary conditions at infinity ensure that ρ behaves in the limit at large distances like an ordinary cylindrical radial co-ordinate, and hence that it has no critical points at large distances. Under such well defined boundary conditions as these, ordinary Morse theory tells us that provided there are no *degenerate* critical points, (i.e. points where higher than first deviations of ρ are zero) the number of maxima plus the number of minima minus the number of saddle points of ρ in the interior of $\bar{\mathscr{B}}$ is an invariant of the differential topology. By considering the special case of a cylindrical co-ordinate in ordinary flat space it is obvious that in the present case this invariant is *zero*. Since a harmonic function can have no maxima or minima, it follows that in the present case there can be no saddle points either, provided there are no degenerate critical points.

Now the more specialized harmonic Morse theory and Heinz (1949) enables us to exclude degenerate critical points as well. The critical points of a harmonic function can be classified with a positive integral index number which is the order of the highest partial derivative which vanishes at the point under consideration, this index number being unity for a non-degenerate critical point. According

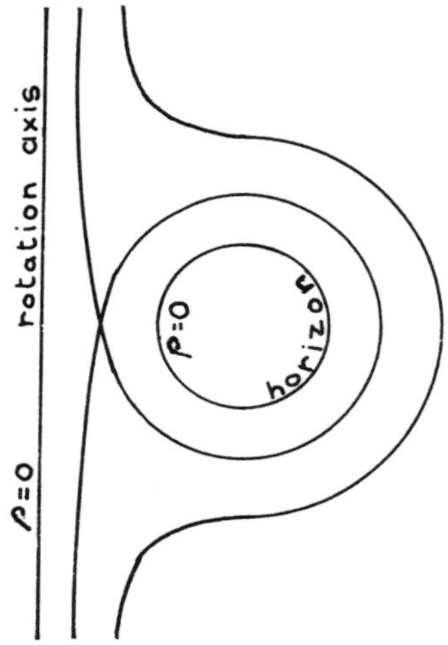

Figure 10.2. Plan of hypothetical alternative form of the 2-surface $\bar{\mathscr{B}}$ corresponding to the mathematically conceivable (but physically impossible) case of a topologically toroidal as opposed to a spherical hole. The continuous lines indicate locuses on which ρ is constant. In this case there is necessarily just one non-degenerate critical point.

to this theory, if degenerate critical parts are present, the sum of the indices of the critical points will be equal to the Morse invariant. In the present case this tells us unambiguously that *there are no critical points of ρ at all* degenerate or otherwise in $\bar{\mathscr{B}}$. (In the case where the topology of $\bar{\mathscr{B}}$ corresponded to that of a toroidal black hole, this theory would tell us equally unambiguously that there would be *one* non-degenerate critical point (Figure 10.2); in the case of two distinct toroidal black holes there would be an ambiguity, since there might be either two non-degenerate critical points or one degenerate critical point of index two.)

Having established that ρ has no critical points in the interior of $\mathscr{B}$ it follows that the curves on which it is constant have no intersections, and that they have everywhere the same topology, namely that at the Euclidean line. Hence not only can we take ρ itself to be a globally well defined co-ordinate on $\overline{\mathscr{B}}$, but we can also choose a globally well behaved scalar z without critical points, and constant on curves orthogonal to those on which z is constant, as a second globally well behaved co-ordinate. In terms of such a co-ordinate system, the metric ds_{II}^2 on $\overline{\mathscr{B}}$ will have the form

$$ds_{\text{II}}^2 = \Sigma(d\rho^2 + Zdz^2) \tag{10.6}$$

where Σ and Z are strictly positive functions in the interior of $\overline{\mathscr{B}}$. Now it follows at once by further application of the harmonicity condition that Z is a function of of z only, and hence that by a variable rescaling of the function z we can arrange to set Z equal to unity thus reducing the metric to

$$ds_{\text{II}}^2 = \Sigma\,(d\rho^2 + dz^2) \tag{10.7}$$

In this form the co-ordinate z is uniquely defined to an additive constant, and this constant may itself be specified uniquely by the requirement that z take equal and opposite values,

$$z = \pm c \tag{10.8}$$

say, where c is a positive constant, in the limit as the junction of the symmetry axis and the horizon is approached.

Thus we deduce finally that $\ll \mathscr{I} \gg$ may be covered globally apart from the degeneracy on the symmetry axis, by a Weyl-Papapetrou co-ordinate system ρ, z, φ, t in which the metric takes the canonical form

$$ds^2 = \Sigma\,(d\rho^2 + dz^2) + X\,d\varphi^2 + 2W\,d\varphi\,dt - V\,dt^2 \tag{10.9}$$

while the electromagnetic field will have the form

$$F = 2(B_{,\rho}\,d\rho + B_{,z}\,dz) \wedge d\varphi + 2(\Phi_{,\rho}\,d\rho + \Phi_{,z}\,dz) \wedge dt \tag{10.10}$$

where the co-ordinate φ is defined modulo 2π, the co-ordinate ρ ranges over the positive half of the real line, and where t and z range over the entire real line, and where these co-ordinates are defined uniquely apart from the possibility of adding a (clearly ignorable) constant to t or φ. The same applies to the complementary forms

$$ds^2 = \Sigma(d\rho^2 + dz^2) + X\,d\varphi^{\dagger 2} + 2W^\dagger\,d\varphi^\dagger\,dt - V^\dagger\,dt^2 \tag{10.11}$$

$$F = 2(B_{,\rho}\,d\rho + B_{,z}) \wedge d\varphi^\dagger + 2(\Phi^\dagger_{,\rho}\,d\rho + \Phi^\dagger_{,z}\,dz) \wedge dt \tag{10.12}$$

It is well known (see e.g. Ernst (1969) that for a metric of the simple form (10.9) with an electromagnetic field of the form (10.10), the source free Einstein-Maxwell field equations for the metric and potential components in the surfaces of transitivity, i.e. for the variables V, W, X and B, Φ decouple from the remaining field equations which are either redundant or serve to determine

Σ on $\overline{\mathscr{B}}$ uniquely, (up to a constant multiplicative factor fixed by the boundary conditions), by an explicit quadrature, as a function of V, W, X, B, Φ.

In this section we shall give boundary conditions on these five variables V, W, X, B, Φ and $\overline{\mathscr{B}}$ which we shall show to be both necessary and (subject to the field equations) sufficient for the axisymmetry axis and the horizon $\mathscr{H}^+$ in $\mathscr{M}$ to be well behaved.

We start by considering the more familiar case of the axisymmetry axis, on which m^a and hence also X and W, are zero, *leaving out of consideration for the time being the intersection* (on which V will be zero also) *of the axis with the horizon*. Since the Killing vector squared magnitude X must increase in proportion to the square of the orthogonal spacelike distance from the axis, the same must be true also of σ, and it follows that its square root ρ can be used as well behaved co-ordinate, with finite non-vanishing gradient not only in the interior of $\overline{\mathscr{B}}$ but also in the part of the boundary corresponding to the axisymmetry axis. It follows that the coefficient Σ in the metric form, must be finite on the rotation axis (in fact it must satisfy $\Sigma = V^{-1}$ there) and that the orthogonal co-ordinate z on $\overline{\mathscr{B}}$ is also well behaved there.

Now (by considering the affect of a rotation by an angle π about the axis) it is evident that in order to be well behaved on the axis the scalars V, W, X, Φ, B must all be *even* functions of ρ, and hence can be expressed as well behaved functions not only of ρ and z but also of σ and z. Thus we see at once that necessary boundary conditions on X, W, V are

$$V = V(\sigma, z) \tag{10.13}$$

$$W = \sigma W_1(\sigma, z) \tag{10.14}$$

$$X = \sigma X_1(\sigma, z) \tag{10.15}$$

where the functions W_1, X_1 and V are well behaved functions of σ and z, the two last being strictly positive on the axis where σ is zero. We can deduce corresponding conditions for Φ and B by considering the requirements for the regularity of (10.10). The form dt in $\ll \mathscr{I} \gg$ is well behaved, but the form $d\varphi$ is singular on the axis, and therefore B must be correspondingly restricted in order to ensure that $(B_{,\rho}\, d\rho + B_{,z}\, dz) \wedge d\varphi$ is well behaved. Now since we have $B_{,\rho}\, d\rho = B_{,\sigma}\, d\sigma$ and since the form $d\sigma$ tends to zero on the axis in such a way that $d\sigma \wedge d\varphi$ is well behaved, it follows that the restriction applies only to the partial derivative $B_{,z}$,which must itself tend to zero on the axis. Thus we obtain the conditions

$$\Phi = \Phi(\sigma, z) \tag{10.16}$$

$$B = B^A + \sigma B_1(\sigma, z) \tag{10.17}$$

where Φ and B_1 are well behaved functions of σ and z, and B^A is a constant, which will be the same on both the north and south branches of the axis only if the magnetic flux defined by (7.22) is zero.

The conditions (10.13), (10.14), (10.15), (10.16), (10.17) are not only necessary for the symmetry axis to be well behaved, but they are also sufficient, *provided* that the *Einstein-Maxwell field equations* are *satisfied*, since it follows from the conditions (10.16) and (10.17) that the ρ, z component of the energy-momentum tensor, and hence also at the Ricci tensor, in terms of the co-ordinate system of the metric form (10.9), must tend to zero on the symmetry axis. Now by analysing the explicit form of the Ricci tensor, using (10.13), (10.14), (10.15) it can be checked that satisfying the equation $R_{\rho z} = 0$ on the axis *automatically* ensures that the scalar function Σ in the metric satisfied the boundary condition

$$\Sigma = \epsilon^{-2}[X_1(\sigma, z) + \sigma \Sigma_1(\sigma, z)] \qquad (10.18)$$

on the axis where ϵ is a non-zero multiplicative constant of integration, Σ_1, is a well behaved function of σ and z, and X_1 is the same function as was introduced in equation (10.15). It is now easy to see that the condition (10.13), (10.14), (10.15) together with (10.18) are sufficient for the cylindrical co-ordinate degeneracy of the form (10.7) on the symmetry axis $\rho = 0$ to be removable in the usual way be replacing ρ and φ by Cartesian type co-ordinates x, y defined by

$$\left.\begin{aligned} x &= \rho \cos \varphi \\ y &= \rho \sin \varphi \end{aligned}\right\} \qquad (10.19)$$

provided that the constant ϵ *has the value unity*, which will in fact be the case under the assumed global topological conditions, since both disconnected components of the axisymmetry axis extend to asymptotically large distances, where the asymptotic flatness conditions ensure that ϵ is indeed unity. Under these conditions the use of (10.13), (10.14), (10.15), (10.18) in conjunction with the transformation (10.19) reduces the cylindrical co-ordinate form (10.9) of the metric to the Cartesian form

$$\begin{aligned} ds^2 = X_1[dx^2 + dy^2 + dz^2] + \Sigma_1[(x\,dx + y\,dy)^2 + (x^2 + y^2)\,dz^2] \\ + 2W_1(x\,dy - y\,dx)dt - V\,dt^2 \end{aligned} \qquad (10.20)$$

which can easily be seen to be well behaved on the axisymmetry axis where $x = y = 0$. Similarly the use of (10.16) and (10.17) in conjuction with the transformation (10.19) reduces the cylindrical co-ordinate form (7.41) of the vector potential to the form

$$A = \Phi\,dt + B_1(y\,dx - x\,dy) + B^A\,d\varphi \qquad (10.21)$$

Since B^A is a constant on each branch of the axis, the final singular term can be removed by a canonical form preserving gauge transformation whenever the magnetic monopole (7.22) is zero, and in any case, the electromagnetic field F will have the well behaved Cartesian form

$$F = 2\,d\Phi \wedge dt + 2dB_1 \wedge (y\,dx - x\,dy) + 4B_1\,dy \wedge dx \qquad (10.22)$$

We now move on to consider the analogous boundary conditions required by the regularity of the horizon $\mathscr{H}^+$, again excluding for the time being the junction of the horizon with the axis. For analysing the horizon, on which as we have seen, σ and $V^\dagger$, and therefore also $W^\dagger$ are zero, it is convenient to work with the complementary form (10.11) of the metric instead of (10.9). In consequence the non-degeneracy condition the gradient of $V^\dagger$, and therefore also (since we are excluding the rotation axis where X is zero) the gradient of σ in $\mathscr{M}$ are non-zero on $\mathscr{H}^+$, and it follows this time that σ can be used as a well behaved co-ordinate with finite non-vanishing gradient not only in the interior of $\overline{\mathscr{B}}$ but also on the part of the boundary of $\overline{\mathscr{B}}$ corresponding to the horizon. Thus the square root, ρ, of σ will have a singular gradient on the horizon boundary of $\overline{\mathscr{B}}$ (in contrast with the situation on the axis boundary where ρ, but not σ, is well behaved as a co-ordinate). Now using (5.24) we can easily see that we must have

$$\Sigma^{-1} = \rho_{,\alpha}\rho^{,\alpha} = \kappa X \tag{10.23}$$

on $\mathscr{H}^+$. Since Σ thus tends to a finite limit on the horizon boundary of $\overline{\mathscr{B}}$, it follows that z is a well behaved co-ordinate on the horizon boundary. Hence we deduce that $V^\dagger$, $W^\dagger$, X, $\Phi^\dagger$, B will be well behaved functions of σ and z on the horizon boundary, and therefore that in the neighbourhood of the horizon boundary they will have the form

$$V^\dagger = \sigma V_1^\dagger(\sigma, z) \tag{10.24}$$

$$W^\dagger = \sigma W_1^\dagger(\sigma, z) \tag{10.25}$$

$$X = X(\sigma, z) \tag{10.26}$$

where the functions $W_1^\dagger$, $V_1^\dagger$ and X are well behaved functions of σ and z, the two last being strictly positive, and

$$\Phi^\dagger = \Phi^H + \sigma\Phi_1^\dagger\,(\sigma, z) \tag{10.27}$$

$$B = B(\sigma, z) \tag{10.28}$$

where $\Phi_1^\dagger$ and B are well behaved functions, and Φ^H is the *constant* previously introduced by (8.40).

As before it can be seen these conditions are also *sufficient* for the regularity of the horizon when the Einstein-Maxwell field equations are satisfied since as in the previous section the conditions (10.27), (10.28) imply that the Ricci component $R_{\rho z}$ in the form (10.11) must tend to zero on the horizon, and hence by (10.24), (10.25), (10.26) that the scalar functions Σ must have the form

$$\Sigma = \kappa^{-2}\,[V_1(\sigma, z) + \sigma\,\Sigma_1(\sigma, z)] \tag{10.29}$$

where Σ_1 is a well behaved function of σ and z, and κ is a strictly positive multiplicative *constant* or integration, which is clearly, by (10.23) the same as the constant κ with which we are already familiar. Thus the boundary conditions (10.24) to (10.28) are sufficient to ensure that the condition (whose necessity was shown in Theorem 8) that κ be constant will be satisfied. We can therefore remove the co-ordinate degeneracy on the horizon by a Finkelstein (1958) type co-ordinate transformation in which t and ρ are replaced in the form (10.11) by σ and a new ignoreable co-ordinate v defined by

$$v = t + \kappa^{-1} \ln |\rho| \tag{10.30}$$

which leads to the form

$$\begin{aligned} ds^2 = V_1 [\kappa^{-2} dz^2 - (\sigma\, dv - \tfrac{1}{2}\kappa^{-1}\, d\sigma) dv] + X\, d\varphi^{\dagger 2} \\ + \kappa^{-2} \Sigma_1 [\sigma\, dz^2 + \tfrac{1}{4}\, d\sigma^2] + 2W_1 (\sigma\, dv - \tfrac{1}{2}\kappa^{-1}\, d\sigma)\, d\varphi^{\dagger} \end{aligned} \tag{10.31}$$

which, can easily be seen to be well behaved on the horizon where σ is zero, since V_1 and X are strictly positive there. The corresponding forms for the vector potential A and the electromagnetic field F are

$$A = B\, d\varphi^{\dagger} + \Phi_1^{\dagger} [\sigma\, dv - \tfrac{1}{2}\kappa^{-1}\, d\sigma] + \Phi^H \kappa^{-1} \rho^{-1}\, d\rho \tag{10.32}$$

which is non-singular except for the final term, which could be removed by a gauge change, and

$$F = 2dB \wedge d\varphi^{\dagger} + 2d\Phi_1^{\dagger} \wedge [\sigma\, dv - \tfrac{1}{2}\kappa^{-1}\, d\sigma] + 2\Phi_1^{\dagger}\, d\sigma \wedge dv \tag{10.33}$$

which is well behaved in any case.

The condition (10.24) to (10.28) are not only sufficient for the future bounding horizon $\mathscr{H}^+$of $\ll \mathscr{I} \gg$ to be well behaved, as we have just shown (still leaving aside the junction of the horizon with the axisymmetry axis) but they are also sufficient for there to exist a Kruskal type co-ordinate extension of $\ll \mathscr{I} \gg$ to cover both a corresponding past boundary horizon $\mathscr{H}^-$ of $\ll \mathscr{I} \gg$ and a crossover axis $\mathscr{H}^+ \cap \mathscr{H}^-$ on which l^a is zero, where the past and future horizons bounding $\ll \mathscr{I} \gg$ meet (whether or not the past boundary horizon and the crossover axis were included in $\mathscr{M}$ as it was originally specified). To see this we introduce co-ordinates w^+ and w^- in plane of t and ρ, by the equation

$$w^{\pm} = \rho\, e^{\pm \kappa t} \tag{10.34}$$

and thus obtain transform (10.11) to

$$\begin{aligned} ds^2 = \kappa^{-2} V_1 [dw^+ dw^- + dz^2] + X\, d\varphi^{\dagger 2} \\ + \kappa^{-2} \Sigma_1 [\tfrac{1}{4}(w^- dw^+ + w^+ dw^-)^2 + w^+ w^- dz^2] \\ + \kappa^{-1} W_1 [w^- dw^+ - w^+ dw^-]\, d\varphi^{\dagger} \end{aligned} \tag{10.35}$$

which is well behaved both in the neighbourhood of the future bounding horizon, $\mathscr{H}^+$ which is represented by $w^+ = 0$, and on the past bounding horizon $\mathscr{H}^-$ represented by $w^- = 0$, including the crossover where they both meet. The corresponding values for the vector potential and field are

$$A = B\, d\varphi^{\dagger} + \tfrac{1}{2}\kappa^{-1} \Phi_1 [w^- dw^+ - w^+ dw^-] + \Phi^H\, dt \tag{10.37}$$

which is well behaved in the neighbourhood of the horizons, except for the final singular term which is removable by a gauge transformation, and

$$F = 2dB \wedge d\varphi^{\dagger} + \kappa^{-1} d\Phi_1 \wedge [w^- dw^+ - w^+ dw^-] + 2\kappa^{-1}\Phi_1 \, dw^- \wedge dw^+ \quad (10.39)$$

which is well behaved near the horizons in any case.

We now come finally to the question of regularity on the north and south poles of the black holes i.e. on the junctions of the two rotation axis components

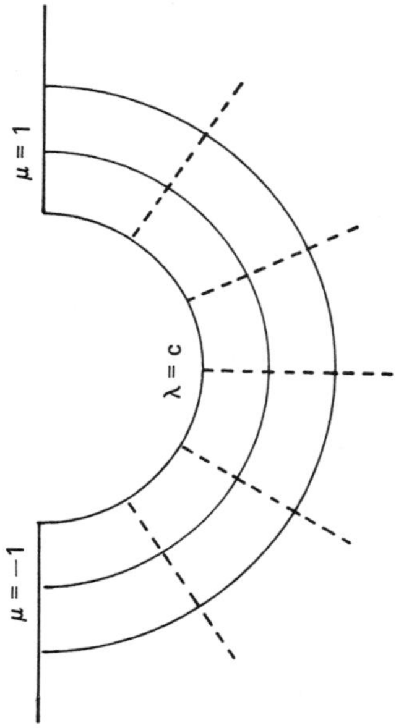

Figure 10.3. Plan of the 2-surface $\mathscr{B}$. The continuous lines indicate locuses on which λ is constant and the dotted lines indicate locuses on which μ is constant.

with the horizon. It is clear from symmetry considerations that the axis must intersect the horizons orthogonally, and therefore it will obviously be convenient to discuss the junction in terms of a new orthogonal co-ordinate system chosen transverse to the ρ, z system, in such a way that one co-ordinate is constant on $\mathscr{H}^+$ while the other is constant on the rotation axis. The simplest way to construct such a system is by introducing ellipsoidal polar type co-ordinates (Figure 10.3), μ running from -1 to $+1$ and λ running from c to ∞ (where $\pm c$ are the values of z on the junctions and where c is *strictly* greater than zero since as we have seen, z is a well behaved co-ordinate function of the horizon), these co-ordinates being defined on $\overline{\mathscr{B}}$ and hence on $\ll\mathscr{I}\gg \cup \mathscr{H}^+$ in terms of ρ and z by

$$\rho^2 = \sigma = (\lambda^2 - c^2)(1 - \mu^2) \tag{10.40}$$

$$z = \lambda\mu \tag{10.41}$$

In this system the two axisymmetry axis components are characterized by $\mu = \pm 1$ respectively, and the horizon boundary is characterized by $\lambda = c$. Corresponding to the conditions that ρ is well behaved on the axis while σ is well behaved on the horizon, we shall have the conditions that $(1 - \mu^2)^{1/2}$ is well behaved on the axis (and hence everywhere except on the equator where $\mu = 0$) and λ is well behaved on the horizon (and hence everywhere without exception).

It follows that we can replace the separate axis and horizon boundary condition that the co-ordinate functions X, etc., be well behaved as functions of σ and z, by the joint boundary conditions that they be well behaved as functions of λ and μ everywhere. Thus the necessary boundary conditions (10.13) to (10.17) and (10.24) and (10.25) can be replaced (noting that the differences between V, W, Φ and $V^\dagger$, $W^\dagger$, $\Phi^\dagger$ tend to zero on the axisymmetry axis) by the single set of necessary boundary conditions

$$X = (1 - \mu^2)\hat{X}(\mu, \lambda) \tag{10.42}$$

$$W^\dagger = (\lambda^2 - c^2)(1 - \mu^2)\hat{W}(\mu, \lambda) \tag{10.43}$$

$$V^\dagger = (\lambda^2 - c^2)\hat{V}(\mu, \lambda) \tag{10.44}$$

$$B = B^A + (1 - \mu^2)\hat{B}(\mu, \lambda) \tag{10.45}$$

$$\Phi^\dagger = \Phi^H + (\lambda^2 - c^2)\Phi(\mu, \lambda) \tag{10.46}$$

where $\hat{X}$, $\hat{W}$, $\hat{V}$, $\hat{B}$, $\hat{\Phi}$ are well behaved functions of μ, λ everywhere including both the axis where $\mu = \pm 1$ and on the horizon where $\lambda = c$ and where B^A and Φ^A are the gauge constants which have already been introduced (and where $\hat{X}$ and $\hat{V}$ are strictly positive in the neighbourhood of both the axis and the horizon).
positive in the neighbourhood of both the axis and the horizon).

It follows that the metric on $\overline{\mathscr{B}}$ takes the form

$$ds_{\text{II}}^2 = \Xi \left\{ \frac{d\lambda^2}{\lambda^2 - c^2} + \frac{d\mu^2}{1 - \mu^2} \right\} \tag{10.47}$$

where the function Ξ, which is related to the function Σ by

$$\Xi = (\lambda^2 - c^2\mu^2)\,\Sigma, \tag{10.48}$$

is well behaved and non-zero everywhere including the north and south poles, provided that the field equations are satisfied, since (in consequence of the identities $X_1 = (\lambda^2 - c^2)^{-1}\hat{X}$ and $V_1^\dagger = (1 - \mu^2)^{-1}\hat{V}$) the consequential boundary conditions (10.18) and (10.24) for Σ imply that we can write

$$\Xi = \mu^2 [\hat{X} + (1 - \mu^2)\Xi_1] \tag{10.49}$$

and

$$\Xi = \lambda^2 \kappa^{-2} [\hat{V} + (\lambda^2 - c^2)\Xi_1^\dagger] \tag{10.50}$$

for suitably chosen functions Ξ_1, and $\Xi_1^\dagger$ which will be well behaved functions of λ and μ on the axis $\mu = \pm 1$ and the horizon $\lambda = c$ respectively, and hence both well behaved also on the poles where the axis and horizon meet.

We can now introduce co-ordinates $\hat{x}, \hat{y}$ (analogous to the previous cartesian co-ordinates x, y) and $\hat{w}^+, \hat{w}^-$ (analogous to the previous Kruskal type null co-ordinates w^+, w^-) simultaneously by the definitions

$$\left.\begin{aligned} \hat{x} &= \sqrt{1-\mu^2}\,\sin\varphi^\dagger \\ \hat{y} &= \sqrt{1-\mu^2}\,\cos\varphi^\dagger \\ \hat{w}^\pm &= \sqrt{\lambda^2 - c^2}\, e^{\pm\kappa t} \end{aligned}\right\} \tag{10.51}$$

which transforms (10.11) to the form

$$\begin{aligned} ds^2 = \Xi &\left\{\frac{d\hat{w}^+ d\hat{w}^-}{\hat{w}^+\hat{w}^- + c^2} + \frac{d\hat{x}^2 + d\hat{y}^2}{1 - (\hat{x}^2 + \hat{y}^2)}\right\} \\ &+ \kappa^{-1}\hat{W}[\hat{w}^- d\hat{w}^+ - \hat{w}^+ d\hat{w}^-]\,[\hat{x}\,dy - \hat{y}\,d\hat{x}] \\ &+ \tfrac{1}{4}\kappa^2\,\Xi_1^\dagger(\hat{w}^- d\hat{w}^+ - \hat{w}^+ d\hat{w}^-)^2 - \Xi_1\,[\hat{x}\,d\hat{y} - \hat{y}\,d\hat{x}]^2 \end{aligned} \tag{10.52}$$

which is well behaved everywhere in both the northern hemisphere (i.e. where $z > 0$) and in the southern hemisphere (i.e. where $z < 0$) including the horizon on which $\hat{w}^+ = 0$, (and it can also be extended over a past bounding horizon of $\ll\mathscr{I}\gg$ on which $\hat{w}^- = 0$, and over a Kruskal crossover axis $\hat{w}^+ = \hat{w}^- = 0$) although it is singular on the equator $z = 0$ since we have $\hat{x}^2 = \hat{y}^2 = 1$ there.

The field in these co-ordinates, is derived from the vector potential

$$\begin{aligned} A = \hat{B}[\hat{x}\,d\hat{y} - \hat{y}\,d\hat{x}] &+ \kappa^{-1}\,d\hat{\Phi}\wedge[\hat{w}^+ d\hat{w}^- - \hat{w}^- d\hat{w}^+] \\ &+ B^A\,d\varphi^\dagger + \Phi^H\,dt \end{aligned} \tag{10.53}$$

and therefore has the form

$$\begin{aligned} F = 2\,d\hat{B}\wedge[\hat{x}\,d\hat{y} - \hat{y}\,d\hat{x}] &+ \kappa^{-1}\,d\Phi\wedge[\hat{w}^+ d\hat{w}^- - \hat{w}^- d\hat{w}^+] \\ &+ 4\hat{B}\,d\hat{x}\wedge d\hat{y} + 2\kappa^{-1}\hat{\Phi}\,d\hat{w}^+\wedge d\hat{w}^- \end{aligned} \tag{10.54}$$

which is also well behaved in each hemisphere, although not on the equator.

Since we have already covered the equator by the Kruskal type co-ordinate patch (10.35) (which is well behaved everywhere except at the poles), this completes the verification that (subject to the field equations) *the necessary conditions* (10.42) *to* (10.46) *are also completely sufficient for the whole of* $\ll \mathscr{I} \gg$ *and its future boundary horizon* $\mathscr{H}^+$ *to be regular*, and also that they imply the existence of a symmetric past boundary horizon $\mathscr{H}^-$, with a regular topologically two-spherical intersection $\mathscr{H}^+ \cap \mathscr{H}^-$ in a suitable extension of $\ll \mathscr{I} \gg \cup \mathscr{H}^+$ (although not necessarily in $\mathscr{M}$).

Before proceeding, it is worth commenting on the precise interpretation of the parameter c, which puzzled me when I originally worked out these results. The solution to this problem (as was pointed out to me by Bardeen) is provided by the generalized Smarr formula (9.29). It follows directly from (10.23) using the metric form (10.9) that the surface area $\mathscr{A}$ of the hole is given in terms of c by

$$\mathscr{A} = \frac{4\pi c}{\kappa} \tag{10.55}$$

Hence using the source free form of (9.29) we obtain

$$c = M - 2\Omega^H J - \Phi^H Q \tag{10.56}$$

11 Differential Equation Systems for the Vacuum Black Hole Problem

It is well known, from the work of Papapetrou and others, that the source free Einstein-Maxwell equations for the form (10.9) in conjunction with (7.41) can be reduced to a set of four independent equations for the four unknowns V, W, Φ, B and also that the asymptotic boundary conditions reduce to a corresponding set of four conditions on these four unknowns, in terms of the four corresponding conserved asymptotic quantities, namely the mass M, the angular momentum J, the electric charge Q, and the magnetic monopole moment P. What we have shown here so far is that the remaining regularity conditions for a system satisfying the conditions (10) can also be reduced to a set of four boundary conditions, namely (10.42), (10.43), (10.45), (10.46) for the four unknowns V, W, Φ, B only (since the fifth condition (10.44) for X is not independent but clearly a consequence of (10.42) and (10.43).

The tranditional form of the field equations, in terms of V, W, Φ, B is very convenient for studying properties at large asymptotic distances, but unfortunately these equations become singular on the ergosurface where V is zero. We can get over this difficulty by noticing that the metric form (10.9) and the electromagnetic field (7.41) are algebraicly invariant under a change in which V, W, Φ, B are replaced by $-X$, W, B, Φ respectively. By making this interchange we get over the singularity difficulty, since as has already been remarked, the causality condition ensures that X is never zero except of course in an entirely predictable way on the rotation axis.

The equations we obtain may be expressed conveniently in terms of the background metric

$$d\hat{s}_{\rm II}^2 = \frac{d\lambda^2}{\lambda^2 - c^2} + \frac{d\mu^2}{1 - \mu^2} \tag{11.1}$$

on $\overline{\mathscr{B}}$ as two Maxwell equations

$$\nabla \left\{ \frac{X\nabla\phi - W\nabla B}{\rho} \right\} = 0 \tag{11.2}$$

$$\nabla \left\{ \rho \frac{\nabla B}{X} + \frac{W}{X} \frac{[X\nabla\phi - W\nabla B]}{\rho} \right\} = 0 \tag{11.3}$$

and two Einstein equations

$$\nabla \left\{ \frac{X\nabla W - W\nabla X}{\rho} + 4B \frac{[X\nabla\phi - W\nabla B]}{\rho} \right\} = 0 \tag{11.4}$$

$$\nabla \left\{ \frac{\rho\nabla X}{X} \right\} + \frac{|X\nabla W - W\nabla X|^2}{\rho X^2} + 2 \frac{|X\nabla\phi - W\nabla B|^2}{\rho X} + 2\rho \frac{|\nabla B|^2}{X} = 0 \tag{11.5}$$

Noting that the co-ordinates λ, μ can be related to co-ordinates r and θ which behave asymptotically like the traditional Schwarzschild spherical co-ordinates by the transformation $\lambda = r - M$ and $\mu = \cos\theta$, we can easily express the standard Papapetrou (1948) type boundary conditions in the terms of the requirement that W, B, Φ and $\lambda^{-2}X$ are well behaved functions of μ and λ^{-1} in the limit as $\lambda^{-1} \to 0$, and that they satisfy.

$$\Phi = Q\lambda^{-1} + 0(\lambda^{-2}) \tag{11.6}$$

$$B = -P\mu + 0(\lambda^{-1}) \tag{11.7}$$

$$W = -2J\lambda^{-1} + 0(\lambda^{-2}) \tag{11.8}$$

$$\lambda^{-2}X = (1 + \mu^2)[1 + 2M\lambda^{-1}] + 0(\lambda^{-2}) \tag{11.9}$$

as $\lambda^{-1} \to 0$ (In imposing these conditions we have fixed the gauge of ϕ and B.)

The boundary conditions derived in the previous section can be expressed as

$$\frac{\partial\Phi}{\partial\mu} = 0(1) \qquad \frac{\partial\Phi}{\partial\lambda} = 0(1) \tag{11.10}$$

$$\frac{\partial B}{\partial\mu} = 0(1) \qquad \frac{\partial B}{\partial\lambda} = 0(1 - \mu^2) \tag{11.11}$$

$$W = 0(1 - \mu^2) \tag{11.12}$$

$$X = 0(1-\mu^2)\frac{1}{X}\frac{\partial X}{\partial\mu} = 1 + 0(1-\mu^2) \tag{11.13}$$

as $\mu \to \pm 1$, and

$$\frac{\partial\Phi}{\partial\lambda} = 0(1) \qquad \frac{\partial\Phi}{\partial\mu} + \Omega^H\frac{\partial B}{\partial\mu} = 0(\lambda^2 - c^2) \tag{11.14}$$

$$\frac{\partial B}{\partial\lambda} = 0(1) \qquad \frac{\partial B}{\partial\mu} = 0(1) \tag{11.15}$$

$$W + \Omega^H X = 0(\lambda^2 - c^2) \tag{11.16}$$

$$X = 0(1) \qquad \frac{1}{X} = 0(1) \tag{11.17}$$

as $\lambda \to c$.

Thus our results so far amount to a demonstration of the following key lemma:

To each domain of outer communications $\ll\mathscr{I}\gg$ in a spacetime in which the condition 10 of the previous section satisfied, there corresponds a cannonically defined solution of the system of equations (11.2) to (11.5) on the λ, μ plane in the co-ordinate range $-1<\mu<1, c<\lambda<\infty$ subject to the boundary condition (11.6) to (11.17), and conversely to each solution of this system there corresponds a canonically defined manifold $\ll\mathscr{I}\gg$ which can be extended to form a manifold $\mathscr{M}$ within which $\ll\mathscr{I}\gg$ is the domain of communications, and in which the conditions 10 are satisfied.

In short there is a one-one correspondence between source-free stationary axisymmetric black hole exterior solutions and solutions of the above system. The only known solutions of this system are the pure vacuum family of Kerr (1963), and its electromagnetic generalization given by Newman *et al.* (1965). These solutions are given by

$$\Phi = \frac{Qr - Pa\mu}{r^2 + a^2\mu^2} \tag{11.18}$$

$$B = \frac{P\mu(r^2 + a^2) - Qar(1-\mu^2)}{r^2 + a^2\mu^2} \tag{11.19}$$

$$W = \frac{-a(1-\mu^2)(2Mr - Q^2 - P^2)}{r^2 + a^2\mu^2} \tag{11.20}$$

$$X = (1-\mu^2)\left\{r^2 + a^2 + \frac{a^2(1-\mu^2)}{r^2 + a^2\mu^2}\,[2Mr - Q^2 - P^2]\right\} \tag{11.21}$$

where we have used the standard notation

$$r = \lambda + M \tag{11.22}$$

$$a = \frac{J}{M} \tag{11.23}$$

These solutions are uniquely specified by the values of Q, P, J, M (the gravitational part being invariant under a duality rotation in which Q and P are altered in such a way that the sum of their squares remains constant) which are restricted only by the condition that the boundary valve parameter c, which is given by

$$c^2 = M^2 - a^2 - P^2 - Q^2 \tag{11.24}$$

should remain real and positive. The other parameter involved in the specification of the boundary-conditions, namely the black hole rotation velocity, Ω^H, is given by

$$\Omega^H = \frac{a}{(M+c)^2 + a^2} \tag{11.25}$$

The fundamental conjecture which one would like to verify is that there exists a uniqueness theorem for the system according to which these should be the only solutions. However due to the essential non-linearity of the system it has not yet been possible to attain this objective, except in the static case. It is therefore worthwhile to start by investigating the truth of the weaker conjecture that there exists what has come to be known loosely as a *no-hair theorem* according to which a continuous variation (in a suitably defined sense) of a solution of this system should be uniquely determined by the corresponding continuous variation of the four conserved quantities Q, P, J, M, i.e. any solutions other than those above should also form discrete non-bifurcating families depending on at most these four parameters.

It is evident that if the no-hair conjecture is correct the parameters c, Ω^H and Φ^H which appear in the boundary conditions must be essentially redundant [part of this redundancy being made explicit by the relation (10.56)] merely duplicating the information on the rotation and scale of the black hole already given by M, J and Q. The scale parameter c is necessary to define the manifold on which we are working and therefore cannot easily be eliminated from the problem, so that it is more convenient instead to eliminate the mass parameter M which also controls the overall scale, and which governs only the higher order asymptotic corrections to X. However it turns out that Ω^H and Φ^H can be made to drop out of the problem altogether by recasting the problem in the manner described by Ernst (1967, 1969).

The Ernst method, whose basic purpose is to simplify the field equations, consists of taking advantage of the fact that the equation (11.2) can be interpreted as an integrability condition which is necessary and (under the present global conditions) sufficient for the existence of an electric pseudo-potential E satisfying

$$\left.\begin{aligned}(1-\mu^2)E_{,\mu} &= X\phi_{,\lambda} - WB_{,\lambda}\\ -(\lambda^2-c^2)E_{,\lambda} &= X\phi_{,\mu} - WB_{,\mu}\end{aligned}\right\} \tag{11.26}$$

and hence that (11.4) can be interpreted as an integrability condition which is necessary and sufficient for the existence of a *twist potential* Y satisfying

$$\begin{aligned}(1-\mu^2)Y_{,\mu} &= XW_{,\lambda} - WX_{,\lambda} + 2(1-\mu^2)(BE_{,\mu} - EB_{,\mu})\\ -(\lambda^2-c^2)Y_{,\lambda} &= XW_{,\mu} - WX_{,\mu} + 2(\lambda^2-c^2)(BE_{,\lambda} - EB_{,\lambda})\end{aligned} \tag{11.27}$$

The final stage in the Ernst procedure is to eliminate the variables ϕ and W in favour of the new variable E and Y. It was pointed out by Ernst (1967) that in the pure vacuum case (when E and B are zero) the resulting field equations can be derived from a very simple positive definite Lagrangian. It turns out that in the electromagnetic case the equations can still be derived from a positive definite (but not quite so simple) Lagrangian. The Lagrangian integral to be varied has the form

$$I = \int \mathscr{L} d\lambda\, d\mu \tag{11.28}$$

where the Lagrangian density is

$$\mathscr{L} = \frac{|\nabla X|^2 + |\nabla Y + 2(E\nabla B - B\nabla E)|^2}{2X^2} + 2\frac{|\nabla E|^2 + |\nabla B|^2}{X} \tag{11.29}$$

(the gradient contractions still being expressed in terms of the metric form given by (11.1)) which reduces to the Lagrangian given by Ernst when E and B are set equal to zero. It is to be remarked that this Lagrangian is invariant under a duality rotation in which E and B are replaced by $E\cos\alpha + B\sin\alpha$ and $B\cos\alpha - E\sin\alpha$ respectively, where α is a *constant* duality angle, as also is the resulting system of field equations. These equations consist of two Maxwell equations

$$E_B \equiv \nabla\left\{\frac{\rho\nabla B}{X}\right\} + \frac{\rho}{X^2}\nabla E[\nabla Y + 2(E\nabla B - B\nabla E)] = 0 \tag{11.30}$$

$$E_E \equiv \nabla\left\{\frac{\rho\nabla E}{X}\right\} - \frac{\rho}{X^2}\nabla B[\nabla Y + 2(E\nabla B - B\nabla E)] = 0 \tag{11.31}$$

and two Einstein equations

$$E_Y \equiv \nabla\left\{\frac{\rho}{X^2}[\nabla Y + 2(E\nabla B - B\nabla E)]\right\} = 0 \tag{11.32}$$

$$E_X \equiv \nabla\left\{\frac{\rho\nabla X}{X^2}\right\} + \rho\frac{|\nabla X|^2 + |\nabla Y + 2(E\nabla B - B\nabla E)|^2}{X^3}$$

$$+ 2\rho\frac{|\nabla B|^2 + |\nabla E|^2}{X^2} = 0 \tag{11.33}$$

the two latter being obtained directly by variation with respect to Y and X respectively and the two former being obtained with the aid of (9.9), by variation with respect to B and E respectively.

The asymptotic boundary conditions, necessary and sufficient for asymptotic flatness, (but dropping the higher order restriction the explicity specifying the mass M in the condition on X) are that E, B, Y and $\lambda^{-2}X$, be well behaved function of λ^{-1} and μ in the limit as $\lambda^{-1} \to 0$, and that they satisfy

$$E = -Q\mu + 0(\lambda^{-1}) \tag{11.34}$$

$$B = -P\mu + 0(\lambda^{-1}) \tag{11.35}$$

$$Y = 2J\mu(3 - \mu^2) + 0(\lambda^{-1}) \tag{11.36}$$

$$\lambda^{-2}X = (1 - \mu^2) + 0(\lambda^{-1}) \tag{11.37}$$

as $\lambda^{-1} \to 0$, (with suitable choice of gauge for E and Y). The symmetry axis boundary conditions which are necessary and sufficient for (11.10), (11.11), (11.12), (11.13) to hold are that E, B, Y, X should be well behaved functions of μ and λ and that they satisfy the conditions

$$\frac{\partial E}{\partial \mu} = 0(1) \qquad \frac{\partial E}{\partial \lambda} = 0(1 - \mu^2) \tag{11.39}$$

$$\frac{\partial B}{\partial \mu} = 0(1) \qquad \frac{\partial B}{\partial \lambda} = 0(1 - \mu^2)$$

$$\frac{\partial Y}{\partial \mu} + 2\left(E\frac{\partial B}{\partial \mu} - B\frac{\partial E}{\partial \mu}\right) = 0(1 - \mu^2) \qquad \frac{\partial Y}{\partial \lambda} = 0(1 - \mu^2) \tag{11.40}$$

$$X = 0(1 - \mu^2) \qquad \frac{1}{X}\frac{\partial X}{\partial \mu} = 1 + 0(1 - \mu^2) \tag{11.41}$$

as $\mu \to \pm 1$. The horizon boundary conditions are extremely simple in the present formulation, and are merely that E, B, Y, X be well behaved functions of λ and μ as $\lambda \to c$, with X non-vanishing, (the parameter Ω^H disappearing from the specification completely) i.e.

$$\frac{\partial E}{\partial \lambda} = 0(1) \qquad \frac{\partial E}{\partial \mu} = 0(1) \tag{11.42}$$

$$\frac{\partial B}{\partial \lambda} = 0(1) \qquad \frac{\partial B}{\partial \mu} = 0(1) \tag{11.43}$$

$$\frac{\partial Y}{\partial \lambda} = 0(1) \qquad \frac{\partial Y}{\partial \mu} = 0(1) \tag{11.44}$$

$$X = 0(1) \qquad X^{-1} = 0(1) \tag{11.45}$$

as $\lambda \to c$.

This Ernst formulation of the problem is completely equivalent to the previous formulation, i.e. *there is a one-one correspondence between solutions of (11.30) to (11.33) subject to (11.34) to (11.45), and solution of (11.2) to (11.5) subject to (11.6) to (11.17)* [and hence also a one one correspondence with domains of communications in manifolds satisfying the basic stationary axisymmetric source free black hole conditions (10)].

The values of the Ernst potentials E and Y in the Kerr-Newman solutions are

$$E = \frac{Q\mu(r^2 + a^2) - Par(1 - \mu^2)}{r^2 + a^2\mu^2} \tag{11.46}$$

$$Y = 2aM\mu(3 - \mu^2) - \frac{2a\mu(1 - \mu^2)}{r^2 + a^2\mu^2}[(Q^2 + P^2)r - Ma^2(1 - \mu^2)] \tag{11.47}$$

It is evident that the Ernst formulation of the problem is as effective in simplifying the boundary conditions as it is in simplifying the field equations. This simplification would not have occurred if we had worked with the complementary form of the equations using V instead of X. (On the other hand the actual solution is more complicated this way round.) The condition that the Lagrangian density is positive definite and well behaved also depends on using X, which is non-vanishing. The complementary form of the Lagrangian density, obtained by using $-V$ in place of X contains terms of opposite signs outside the ergosurface, and is singular on the ergosurface. Unfortunately even in the present case the positive definiteness of the Lagrangian cannot easily be used for drawing global conclusions, since the boundary conditions on X as $\mu \to \pm 1$ ensure that the integral I is divergent when taken over the whole domain.

The only case in which proper black hole uniqueness theorems (as opposed to no-hair theorems) are available at present are those in which the angular momentum is zero. In conjunction with theorem 4.1, the theorems of Israel (1969), (1968) and their refinements (cf. Muller zum Hagen, Robinson, Siefert (1972) give an almost complete demonstration that a black hole solution of given mass and charge is unique, (Schwarzschild or Riesmer-Nordstrom), provided that *staticity* is assumed. We shall complete this section by proving the following result.

Theorem 11 If the condition (10) is satisfied and if the angular momentum is zero, then the metric is necessarily static (in the sense that Y and hence also W

are zero) *and if furthermore the magnetic monopole moment is zero, then the electromagnetic field is also static* (in the sense that the magnetic potential B is zero).

Proof: We shall start by considering the special case when the magnetic monopole moment P is taken to be zero. The proof depends on the identity

$$\rho \frac{|\nabla Y + 2(E\nabla B - B\nabla E)|^2}{X^2} + 4\rho \frac{|\nabla B|^2}{X}$$

$$\equiv \nabla\left\{\frac{\rho(Y + 2EB)[\nabla Y + 2(E\nabla B - B\nabla E)]}{X^2} + 4\rho \frac{B\nabla B}{X}\right\}$$

$$-(Y + 2EB)E_Y - 4BE_B \tag{11.48}$$

When the field equations $E_Y = 0, E_B = 0$ are satisfied, the right hand side reduces to a divergence, and hence the integral of the left hand side over the entire domain $\overline{\mathscr{B}}$ can be expressed in terms of a boundary integral. Since the quantities Y, E, B, X^{-1} are all well behaved on the horizon, the presence of the factor ρ in the divergence ensures that the horizon gives no contribution to the boundary integral. The situation on the axis $\mu = \pm 1$ is more critical, since X^{-1} is singular there; however the boundary conditions ensure that Y and B are always constant on the axis, and hence in the particular case when J and P are zero Y and B are respectively zero also on the axis. Hence we see using (9.22) that the axis gives no contribution to the boundary integral when J and P are both zero, and it is also clear that under these conditions the same applies to the asymptotic boundary integral. It follows that each of the two non-negative terms on the left hand side must be zero everywhere. In conjunction with the boundary conditions, the vanishing of the second term implies that B is zero, and hence the vanishing of the first term, in conjunctions with the boundary conditions, implies that Y is zero also.

To see that Y must be zero whenever the angular momentum is zero, even when a magnetic monopole P (and hence also a non-zero magnetic contribution B to the field) is present, we have only to notice that it is always possible to reduce the magnetic monopole moment to zero by a duality rotation which of course leaves Y invariant. This completes the proof.

In conjunction with this result, Israel's theorems provide an almost complete proof of uniqueness for black holes with *zero angular momentum*, subject to the axisymmetry assumption.

[Approaching the same conclusions from a slightly different angle, the closely related generalised Hawking–Lichnerowicz Theorem 6.2, in conjunction with Israels' theorems, provides an almost complete proof of uniqueness for black holes with *zero angular velocity*, subject to the assumption that the hole boundary ergosurface is the outer ergosurface.]

12 The Pure Vacuum No-Hair Theorem

In order to make progress when non-zero angular momentum is present, we shall restrict our attention in this section to the purely gravitational case where there is no electromagnetic field present, i.e. when E and B are zero.

In this case the field equations reduce to

$$G_X(X, Y) \equiv \nabla\left\{\rho \frac{\nabla X}{X^2}\right\} + \rho \frac{|\nabla X|^2 + |\nabla Y|^2}{X^3} = 0 \tag{12.1}$$

$$G_Y(X, Y) \equiv \nabla\left\{\rho \frac{\nabla Y}{X^2}\right\} = 0 \tag{12.2}$$

and the boundary conditions are simply that as $\lambda^{-1} \to 0$, Y and $\lambda^{-2}X$ must be well behaved functions of λ^{-1} and μ with

$$\lambda^{-2}X = (1 - \mu^2)[1 + 0(\lambda^{-1})] \tag{12.3}$$

$$Y = 3J\mu(1 - \mu^2) + 0(\lambda^{-1}) \tag{12.4}$$

that as $\mu \to \pm 1$, X and Y must be well behaved functions of μ and λ with

$$X = 0(1 - \mu^2) \qquad X^{-1}\frac{\partial X}{\partial \mu} = 1 + 0(1 - \mu^2) \tag{12.5}$$

$$\frac{\partial Y}{\partial X} = 0(1 - \mu^2) \qquad \frac{\partial Y}{\partial \mu} = 0(1 - \mu^2) \tag{12.6}$$

and that as $\lambda \to c$, X and Y must be well behaved function of λ, μ with no other restrictions than

$$X = 0(1) \qquad X^{-1} = 0(1) \tag{12.7}$$

$$\frac{\partial Y}{\partial \lambda} = 0(1) \qquad \frac{\partial Y}{\partial \mu} = 0(1) \tag{12.8}$$

Simple as it is, this system remains essentially non-linear, and it is therefore not easy to obtain a solution of the full uniqueness problem. However it turns out that we can, without too much difficulty, prove the truth of the fundamental no-hair conjecture by showing that *continuous variations of these solutions are uniquely determined by the corresponding continuous variations of the scale parameter c and the angular momentum parameter J*. We use the term *continuous* in this statement to denote a restriction on solution families under discussion sufficient to ensure that if $\{X_1, Y_1\}(J, c)$ and $\{X_2, Y_2\}(J, c)$ are two *distinct* continuous families of solutions, which are functions over a set of values of the parameter pairs (J, c), and which tend (in some appropriate topology on the space of functions) to a common limit $\{X_0, Y_0\}$ as (J, c) tends to (J_0, c_0) then there exists at least one subset of values of (J, c) such that over this subset the difference $\{X_1 - X_2, Y_1 - Y_2\}(J, c)$ *has a well defined limit direction*, represented by a non-zero function-space tangent vector $\{\dot{X}, \dot{Y}\}$, as (J, c) tends to (J_0, c_0), (in the sense that there exists some parameter β say which is a function of (J, c) on the subset, tending to zero as (J, c) tends to (J_0, c_0), and such that $\beta^{-1}\{X_1 - X_2, Y_1 - Y_2\}$ tends to $\{\dot{X}, \dot{Y}\}$ as $\{X_1 - X_2, Y_1 - Y_2\}$ tends to zero).

In view of a comment made by Wald we emphasize that this continuity requirement has nothing to do with analyticity or even piece-wise analyticity; in fact in a *finite* dimensional vector space the continuity property postulated above would be an *automatic* consequence of the ordinary continuity condition that any neighbourhood of the point $\{X_0, Y_0\}$ contains members of the families $\{X_1, Y_1\}$ and $\{X_2, Y_2\}$ other than $\{X_0, Y_0\}$ (in consequence of the fact that in a finite dimensional vector space, as opposed to an infinite dimensional Banach space, the unit sphere is compact). We shall not attempt to investigate here the precise restrictivity of this continuity requirement in terms of explicitly defined Banach space structure of function space. For anyone who may be interested in following up such fine mathematical points, we refer to a discussion of such questions in the context of a rather simple kind of non-linear partial differential equation system by Berger (1969).

The continuity property we have postulated implies that if there are two distinct families $\{X_1, Y_1\}$, $\{X_2, Y_2\}$ of solutions bifurcating from some given solution $\{X_0, Y_0\}$ then there will exist some subset of values, parametrized as functions over some corresponding subset of values of a parameter β such that

$$X_2(J, c) = X_1(J, c) + \beta\dot{X} + 0(\beta) \tag{12.9}$$

$$Y_2(J, c) = Y_1(J, c) + \beta\dot{Y} + 0(\beta) \tag{12.10}$$

as $\beta \to 0$ over this subset for some (not everywhere vanishing) functions $\dot{X}$, $\dot{Y}$. [We emphasize again that this condition does not imply any assumption that $\{X_1, Y_1\}$ and $\{X_2, Y_2\}$ or the associated parameter values (J, c) are analytic or even differentiable functions of β, so that the present treatment automatically covers "higher order bifurcation" in which the derivatives of $\{X_2 - X_1, Y_2 - Y_1\}$ with respect to (J, c)—if indeed such derivatives exist—are zero at (J_0, c_0).]

On substitution of the above expression into the operators defining the field equations we obtain

$$G_X(X_2, Y_2) = G_X(X_1, Y_1) + \beta\dot{G}_X(X_1, Y_1; \dot{X}, \dot{Y}) + 0(\beta) \tag{12.11}$$

$$G_Y(X_2, Y_2) = G_Y(X_1, Y_1) + \beta\dot{G}_Y(X_1, Y_1; \dot{X}, \dot{Y}) + 0(\beta) \tag{12.12}$$

as $\beta \to 0$ where the linearized perturbation operators $\dot{G}_X(X, Y; \dot{X}, \dot{Y})$ and

$\dot{G}_Y(X, Y; \dot{X}, \dot{Y})$ are given by

$$\dot{G}_X \equiv \nabla\left\{\rho\,\frac{\nabla\dot{X}}{X^2} - 2\rho\,\frac{\dot{X}\nabla X}{X^3}\right\} + 2\rho\left\{\frac{\nabla X\nabla\dot{X} + \nabla Y\nabla\dot{Y}}{X^3}\right\}$$

$$- 3\rho\left\{\frac{|\nabla X|^2 + |\nabla Y|^2}{X^4}\right\}\dot{X} \tag{12.13}$$

$$\dot{G}_Y \equiv \nabla\left\{\rho\,\frac{\nabla\dot{Y}}{X^2} - 2\rho\,\frac{\dot{Y}\nabla X}{X^3}\right\} \tag{12.14}$$

Using the fact that $\{X_2, Y_2\}$ and $\{X_1, Y_1\}$ are solutions of the exact field equations, $G_X(X_2, Y_2) = G_Y(Y_2, Y_2) = 0$ and $G_X(X_1, Y_1) = G_Y(X_1, Y_1) = 0$, and dividing by β, we obtain

$$\dot{G}_X(X_1, Y_1; \dot{X}, \dot{Y}) + 0(1) = 0 \tag{12.15}$$

$$\dot{G}_Y(X_1, Y_1; \dot{X}, \dot{Y}) + 0(1) = 0 \tag{12.16}$$

Hence we finally deduce that we must have

$$\dot{G}_X(X, Y; \dot{X}, \dot{Y}) = 0 \tag{12.17}$$

$$\dot{G}_Y(X, Y; \dot{X}, \dot{Y}) = 0 \tag{12.18}$$

when X, Y take the values X_0, Y_0 of the solutions at which the hypothetical bifurcation takes place.

Similarly by substituting (12.9), (12.10) into the boundary conditions (12.3), (12.4), (12.5), (12.6), (12.7), (12.8), dividing by β, and taking the limit, we obtain the corresponding linearized boundary conditions on $\dot{X}$, $\dot{Y}$ in the form of conditions that as $\lambda \to \infty = \dot{Y}$ and $\lambda^{-2}\dot{X}$ be well behaved functions of μ and λ^{-1} with

$$\lambda^{-2}\dot{X} = 0(\lambda^{-1}) \tag{12.19}$$

$$\dot{Y} = 0(\lambda^{-1}) \tag{12.20}$$

that as $\mu \to \pm 1$, $\dot{X}$ and $\dot{Y}$ must be well behaved functions of λ, μ with

$$\dot{X} = 0(1 - \mu^2) \tag{12.21}$$

$$\frac{\partial\dot{Y}}{\partial\lambda} = 0(1 - \mu^2) \qquad \frac{\partial\dot{Y}}{\partial\mu} = 0(1 - \mu^2) \tag{12.22}$$

and that as $\lambda \to c$, $\dot{X}$, $\dot{Y}$ must be well behaved functions of λ, μ with no other restrictions than

$$\dot{X} = 0(1) \tag{12.23}$$

$$\frac{\partial\dot{Y}}{\partial\lambda} = 0(1) \qquad \frac{\partial\dot{Y}}{\partial\mu} = 0(1) \tag{12.24}$$

We prove that the families $\{X_1, Y_1\}$ and $\{X_2, Y_2\}$ must be identical, i.e. that there can be no bifurcations by showing that if $\dot{X}$, $\dot{Y}$ satisfy the equations (12.17), (12.18) and the boundary conditions (12.19), (12.20), (12.21), (12.22), (12.23), (12.24) they must always be zero everywhere and so cannot determine a well defined direction in function space. We point out that the converse deduction could *not* be made i.e. although the *non-existence* of a non-zero linearized bifurcation solution *is* sufficient to rule out the existence of a bifurcating family of exact solutions, the *existence* of a non-zero linearized bifurcation solution would *not necessarily* imply the existence of a bifurcating family of solutions of the full non-linear system; it would only be in this latter case that higher order analysis might be relevant.

The proof depends on the identity

$$\rho\left|\nabla\left(\frac{\dot{X}}{X}\right)+\frac{\dot{Y}\nabla Y}{X^2}\right|^2+\rho\left|\nabla\left(\frac{\dot{Y}}{X}\right)-\frac{\dot{X}\nabla Y}{X^2}\right|^2+\rho\left|\frac{\dot{X}\nabla Y-\dot{Y}\nabla X}{X^2}\right|^2$$
$$\equiv\nabla\left\{\rho\left[\frac{\dot{X}}{X}\nabla\left(\frac{\dot{X}}{X}\right)+\frac{\dot{Y}}{X}\nabla\left(\frac{\dot{Y}}{X}\right)\right]\right\} \tag{12.23}$$
$$+\frac{2\dot{X}^2+\dot{Y}^2}{X}G_X+\frac{\dot{X}\dot{Y}}{X}G_Y-\dot{X}\dot{G}_X-\dot{Y}\dot{G}_Y$$

which is analogous to that of the previous section in that the right hand reduces to a divergence when the field equations $G_X = G_Y = 0$ and the linearized field equations $\dot{G}_X = \dot{G}_Y = 0$ are satisfied, so that the integral of the left hand side over the entire domain $\bar{\mathscr{B}}$ can be expressed in terms of a boundary integral, which can be seen to vanish in consequence of the boundary conditions (12.19), (12.20), (12.21), (12.22), (12.23), (12.24). Since each term on the left hand side is non-negative, it follows that they must all three be zero everywhere. Thus we obtain three linear first order differential equations for $\dot{X}$ and $\dot{Y}$, from which, by taking linear combinations we can obtain $\nabla\dot{Y} = \dot{Y}X^{-1}\nabla X$ which implies (by the boundary condition that $\dot{Y}$ is zero on the axis, together with the standard uniqueness theorem for the solution of a homogeneous gradient equation of this type) that $\dot{Y}$ is zero everywhere. The remaining first order differential equations then imply directly that $\dot{X}$ is zero also, which completes the proof of the following result:

THEOREM 12 (No-Hair Theorem) The mathematically possible domains of communications $\ll\mathscr{I}\gg$ of space-time manifolds $\mathscr{M}$ satisfying the conditions 10, fall in to discrete continuous families depending on at most the two parameters J and c.

COROLLARY There is at most one such family for which the angular momentum parameter J can take the value zero.

This family consists of course of the Kerr vacuum solution with $a^2 < M^2$ for which J can take any value without restriction for arbitrary positive valves of C, the mass parameter M being given as a function of C by $M^2 = \frac{1}{2}\{C^2 + (C^2 + 4J^2)^{1/2}\}$. The corollary follows immediately from the staticity Theorem 11 and the appropriate form of the Israel theorem establishing uniqueness in the static case.

We can give a simple explicit proof of the relevant restriction of Israel's theorem (i.e. the restriction to the axisymmetric pure vacuum case) by noticing that in the static case when Y is zero the system $G_X = G_Y = 0$ reduces simply to

$$\nabla\{\rho\nabla(\ln X)\} = 0 \tag{12.24}$$

which is *linear* in $\ln X$, and hence will also be satisfied if X is replaced by the quotient X_1/X_2 of any two different solutions X_1 and X_2. Hence using the further identity

$$\rho\,|\nabla(\ln X)|^2 \equiv \nabla\{\rho \ln X\,\nabla(\ln X)\} - (\ln X)\,\nabla\{\rho\nabla(\ln X)\} \tag{12.25}$$

we can deduce that the quotient of any two solutions of (12.24) satisfies

$$\rho\left|\nabla\left[\ln\left(\frac{X_1}{X_2}\right)\right]\right|^2 = \nabla\left\{\rho\ln\left(\frac{X_1}{X_2}\right)\nabla\left[\ln\left(\frac{X_1}{X_2}\right)\right]\right\} \tag{12.26}$$

By integrating over the whole domain $\overline{\mathscr{B}}$, we can again use the relevant boundary conditions (12.3), (12.5), (12.7) to deduce that the non-negative term on the left hand side is zero everywhere, which proves that $X_1 = X_2$ i.e. that the solution is unique, and must therefore be the relevant Schwarzschild solution for which $X = (\lambda + C)^2(1 - \mu^2)$.

We remark that although this proof is much more restricted than Israel's original (1967) demonstration in that it assumes axisymmetry at the outset, it has the advantage of being independent of Israel's assumption that the gradient of V is nowhere vanishing, instead making use of the condition that X is nowhere vanishing (except on the axis).

(In a *static* as opposed to merely stationary domain, it is clear that X must be non-zero except on the axis, whether or not the global causality requirement is satisfied in $\mathscr{M}$.)

It has in fact recently been shown by Muller-zum-Hager, Robinson and Siefert that Israel's postulate that the gradient of V be non-zero can in fact be dispensed with in the general (non-axisymmetric) vacuum Israel theorem, albeit at the expense of considerable technical complexity in the demonstration. It would be desirable to extend both the general demonstration of Muller-zum-Hagen, Robinson and Siefert, and the very much simpler restricted demonstration given here, to cover the electromagnetic case. Unfortunately, the demonstration given in this section depends essentially on the *linearization* of the system which can be achieved when both the angular momentum and the electromagnetic field are zero, but which fails when even a static electric field is present.

13 Unsolved Problems

Having succeeded in proving the truth of the no-hair conjecture in the pure vacuum case, we now consider the question of the electromagnetic generalization, according to which any two continuous families

$$\{X_1, Y_1, E_1, B_1\}(c, J, Q, P) \quad \text{and} \quad \{X_2, Y_2, E_2, B_2\}(c, J, Q, P)$$

of solutions $\{X, Y, E, B\}$ of the system (11.30) to (11.45), parametrized by the four quantities c, J, Q, P should coincide if they have any solution, $\{X_0, Y_0, E_0, B_0\}$ say, in common. As before we may assume that if they do not coincide there exist non-zero perturbation functions $\dot{X}, \dot{Y}, \dot{E}, \dot{B}$ such that for a subset of values of some parameter there exists a corresponding subset of values of c, J, Q, P for which the equations (12.9) and (12.10) and the further equations

$$E_2 = E_1 + \beta\dot{E} + 0(\beta) \tag{13.1}$$

$$B_2 = B_1 + \beta\dot{B} + 0(\beta) \tag{13.2}$$

as $\beta \to 0$ are satisfied, where $\{X_1, Y_1, E_1, B_1\}$ and $\{X_2, Y_2, E_2, B_2\}$ both tend to $\{X_0, Y_0, E_0, B_0\}$ as $\beta \to 0$. By the same reasoning as used before the perturbation functions $\dot{X}, \dot{Y}, \dot{E}, \dot{B}$ must satisfy the linearized equations

$$E_X(X, Y, E, B; \dot{X}, \dot{Y}, \dot{E}, \dot{B}) \tag{13.3}$$

$$E_Y(X, Y, E, B; \dot{X}, \dot{Y}, \dot{E}, \dot{B}) \tag{13.4}$$

$$E_E(X, Y, E, B; \dot{X}, \dot{Y}, \dot{E}, \dot{B}) \tag{13.5}$$

$$E_B(X, Y, E, B; \dot{X}, \dot{Y}, \dot{E}, \dot{B}) \tag{13.6}$$

when $\{X, Y, E, B\}$ takes the value $\{X_0, Y_0, E_0, B_0\}$, where

$$\begin{aligned}\dot{E}_X \equiv\ & \nabla\left\{\rho\frac{\nabla\dot{X}}{X^2} - 2\rho\frac{\dot{X}\nabla X}{X^2} - 3\frac{\dot{X}}{X^3}[|\nabla X|^2 + |\nabla Y + 2(E\nabla B - B\nabla E)|^2]\right\} \\ & + \frac{2}{X^2}\nabla X\nabla\dot{X} + \frac{2}{X^2}[\nabla Y + 2(E\nabla B - B\nabla E)] \\ & \times [\nabla\dot{Y} + 2(E\nabla\dot{B} - B\nabla\dot{E} + \dot{E}\nabla B - \dot{B}\nabla E)] \\ & - \frac{2\dot{X}}{X^2}[|\nabla E|^2 + |\nabla B|^2] + \frac{2}{X}[\nabla E\nabla\dot{E} + \nabla B\nabla\dot{B}]\end{aligned} \tag{13.7}$$

$$\begin{aligned}\dot{E}_Y \equiv\ & \nabla\left\{\frac{\rho}{X^2}[\nabla\dot{Y} + 2(E\nabla\dot{B} - B\nabla\dot{E} + \dot{E}\nabla B - \dot{B}\nabla E)]\right. \\ & \left. - 2\rho\frac{\dot{X}}{X}[\nabla Y + 2(E\nabla B - B\nabla E)]\right\}\end{aligned} \tag{13.8}$$

$$\dot{E}_E \equiv \nabla\left\{\rho\frac{\nabla\dot{E}}{X} - \rho\frac{\dot{X}\nabla E}{X^2}\right\} - \rho\left(\frac{\nabla\dot{B}}{X^2} - 2\frac{\dot{X}\nabla B}{X^3}\right)[\nabla Y + 2(E\nabla B - B\nabla E)]$$

$$+\frac{\rho}{X^2}\nabla B[\nabla\dot{Y} + 2(E\nabla\dot{B} - B\nabla\dot{E} + \dot{E}\nabla B - \dot{B}\nabla E)] \qquad (13.9)$$

$$\dot{E}_B \equiv \nabla\left\{\rho\frac{\nabla\dot{B}}{X} - \rho\frac{\dot{X}\nabla B}{X^2}\right\} + \rho\left(\frac{\nabla\dot{E}}{X^2} - 2\frac{\dot{X}\nabla E}{X^3}\right)[\nabla Y + 2(E\nabla B - B\nabla E)]$$

$$-\frac{\rho}{X^2}\nabla E[\nabla\dot{Y} + 2(E\nabla\dot{B} - B\nabla\dot{E} + \dot{E}\nabla B - \dot{B}\nabla E)] \qquad (13.10)$$

The corresponding homogeneous boundary conditions, derived from (11.34) to (11.45) are that $\lambda^{-2}\dot{X}$, $\dot{Y}, \dot{E}, \dot{B}$ be well behaved functions of μ, λ^{-1} satisfying

$$\lambda^{-2}\dot{X} = 0(\lambda^{-1}) \qquad (13.11)$$

$$\dot{Y} = 0(\lambda^{-1}) \qquad (13.12)$$

$$\dot{E} = 0(\lambda^{-1}) \qquad (13.13)$$

$$\dot{B} = 0(\lambda^{-1}) \qquad (13.14)$$

as $\lambda^{-1} \to 0$, that $\dot{X}$, $\dot{Y}, \dot{E}, \dot{B}$ be well behaved function of μ, λ satisfying

$$\dot{X} = 0(1-\mu^2) \qquad (13.15)$$

$$\left.\begin{aligned} &\frac{\partial\dot{Y}}{\partial\lambda} = 0(1-\mu^2) \\ &\frac{\partial\dot{Y}}{\partial\mu} + 2\left(\dot{E}\frac{\partial B}{\partial\mu} - \dot{B}\frac{\partial E}{\partial\mu} + \dot{E}\frac{\partial B}{\partial\mu} - \dot{B}\frac{\partial E}{\partial\mu}\right) = 0(1-\mu^2)\end{aligned}\right\} \qquad (13.16)$$

$$\frac{\partial\dot{E}}{\partial\lambda} = 0(1-\mu^2) \qquad \frac{\partial\dot{E}}{\partial\mu} = 0(1) \qquad (13.17)$$

$$\frac{\partial\dot{B}}{\partial\lambda} = 0(1-\mu^2) \qquad \frac{\partial\dot{B}}{\partial\mu} = 0(1) \qquad (13.18)$$

as $\mu \to \pm 1$, and that $\dot{X}$, $\dot{Y}, \dot{E}, \dot{B}$ be well behaved functions of λ, μ subject only to

$$\dot{X} = 0(1) \qquad (13.19)$$

$$\frac{\partial\dot{Y}}{\partial\lambda} = 0(1) \qquad \frac{\partial\dot{Y}}{\partial\mu} = 0(1) \qquad (13.20)$$

$$\frac{\partial\dot{E}}{\partial\lambda} = 0(1) \qquad \frac{\partial\dot{E}}{\partial\mu} = 0(1) \qquad (13.21)$$

$$\frac{\partial\dot{B}}{\partial\lambda} = 0(1) \qquad \frac{\partial\dot{B}}{\partial\mu} = 0(1) \qquad (13.22)$$

as $\lambda \to C$.

I have not found any way of determining whether or not this system can have non-zero solutions $\dot{X}, \dot{Y}, \dot{E}, \dot{B}$ for general values of X, Y, E, B subject to (11.30) to (11.45). If non-zero solutions do exist for some eigenvalues of c, J, Q, P it would suggest†, but would not prove that the solution family bifurcates at the corresponding solutions. There is however a special case which is much more tractable, namely the case where the electromagnetic field in the unperturbed solution is zero, i.e. when $E = B = 0$, since in this case the general form of the Einstein-Maxwell equations ensures (independently of the special symmetry conditions assumed in the present problem) that the linearized equations for the gravitational perturbation decouple from those for the electromagnetic perturbation, the latter reducing simply to a set of pure Maxwell equations in the curved space background. Advantage has been taken of this by Wald (1971), and also independently by Ipser (1971), who have shown that in the particular case of a pure (non-electromagnetic) Kerr solution background the perturbation solution of the Maxwell equations are indeed zero when the corresponding perturbed values of the charge Q and monopole moment P are zero, from which it follows that even if bifurcations from the Kerr-Newman family do exist, they cannot start from the pure vacuum members.

We shall conclude this course by presenting a generalization of this Wald-Ipser theorem showing that the conclusion holds for electromagnetic perturbations about *any* pure vacuum black hole solution family satisfying conditions (10) even if it is not the Kerr family. That is to say *it is true not only for the Kerr-Newman family but for any other family of electromagnetic black hole solutions of the system* (11.30) to (11.45) (if there are any others) *that no bifurcation can take place starting from the pure vacuum members.*

To prove this we use the fact that when E and B are zero, the full linearized equations $\dot{E}_E = 0, \dot{E}_B = 0$ reduce simply to the pure Maxwell equations

$$M_E \equiv \nabla\left\{\rho\frac{\nabla\dot{E}}{X}\right\} - \rho\frac{\nabla Y \nabla\dot{B}}{X^2} = 0 \tag{13.24}$$

$$M_B \equiv \nabla\left\{\rho\frac{\nabla\dot{B}}{X}\right\} + \rho\frac{\nabla Y \nabla\dot{B}}{X^2} = 0 \tag{13.25}$$

(the other two equations $\dot{E}_X = 0, \dot{E}_Y = 0$ reducing to the equations $\dot{G}_X = 0$, $\dot{G}_Y = 0$ which we have already studied).

To establish that the only solution of (13.24), (13.25) subject to (13.17) to (13.22) and (12.1) to (12.8) is $\dot{E} = \dot{B} = 0$, we use the identity

$$\frac{1}{X^3}\{|X\nabla\dot{E} - B\nabla\dot{Y}|^2 + |X\nabla\dot{B} - \dot{E}\nabla Y|^2\} + \frac{1}{X}\left\{\left|\nabla\left(\frac{\dot{E}}{X}\right)\right|^2 + \left|\nabla\left(\frac{\dot{B}}{X}\right)\right|^2\right\}$$

$$\equiv \nabla\left\{\rho\nabla\left(\frac{\dot{E}^2 + \dot{B}^2}{X}\right)\right\} + (\dot{E}^2 + \dot{B}^2)G_X - 2(\dot{E}M_E + \dot{B}M_B) \tag{13.26}$$

† The existence of a perturbation eigenvalue might be a symptom of the setting in of a dynamic instability even where no *exactly stationary* bifurcation exists.

When the Maxwell equations $M_E = 0, M_B = 0$ are satisfied, and when the unperturbed pure vacuum field equation $G_X = 0$ is satisfied, the right hand side reduces to a divergence whose integral over the whole domain is a boundary integral which can easily be seen to vanish by the boundary conditions. It follows that each of the non-negative terms on the left hand side is zero. Hence, with further use of the boundary conditions, it is clear that $\dot{E}$ and $\dot{B}$ themselves must be zero everywhere, as required.

By an examination (based on the work of Vishveshwara (1968)) of stationary perturbations of the Schwarzschild and Reissner Nordstrom solutions it has also been shown by Wald (1972) that there can be no bifurcation *from the spherical members* of the Kerr–Newman family of electromagnetic black hole solutions. (This work is a generalization of an earlier perturbation analysis, now superseded by the Vaccuum No Hair Theorem 12, which was carried out by Hartle and Thorne (1968).)

Although it represents an amusing mathematical challenge, the problem of generalizing these partial results to a complete electromagnetic no-hair theorem—or finding a counter-example—is less important (and probably less difficult) than the problem of generalizing the vacuum no-hair theorem to an absolute uniqueness theorem for vacuum black holes—or finding a counter example, i.e. a family of non-Kerr pure vacuum black holes with the pathological property that the angular momentum cannot be varied to zero. Neither of these problems is as pressing as that of determining the *stability* of the black hole equilibrium states which we have been discussing. (In the unlikely event that non-Kerr stationary pure vacuum black holes are discovered, the physical implications would be less startling if it were to turn out that they were all unstable.)

Bibliography

Bardeen, J. (1970) *Ap. J.* **162,** 71.

Bardeen, J. (1971) *Ap. J.* **167,** 425.

Bardeen, J., Carter, B. and Hawking, S. W. (1972) *Black Hole Mass Formulae.* Preprint Inst. of Astronomy, Cambridge.

Beckenstein, J. (1972) Phd. Thesis, Princeton.

Berger, M. (1969) *Bifurcation Theory and Nonlinear Eigenvalue Problems,* 113. Ed. B. Keller and S. Autman, (Benjamin).

Boyer, R. H. and Price, T. G. (1965) *Proc. Camb. Phil. Soc.* **62,** 531.

Boyer, R. H. and Lindquist, R. W. (1967) *J. Math. Phys.* **8,** 265.

Boyer, R. H. (1968) *Proc. Roy. Soc. A.*

Carter, B. (1968) *Phys. Rev.* **174,** 1559.

Carter, B. (1969) *J. Math. Phys.* **10,** 70.

Carter, B. (1970) *Commun. Math. Phys.* **17,** 233.

Carter, B. (1971a) *General Relativity and Gravitation* **1,** 349.

Carter, B. (1971b) *Phys. Rev. Lett.* **26,** 331.

Carter, B. (1972a) *Domains of Stationary Communication in Spacetime.* Preprint Inst. of Astronomy, Cambridge.

Carter, B. (1972b) *Nature.*

Carter, B. (1972c) *The Electrical Equilibrium of a Black Hole.* Preprint Inst. of Astronomy, Cambridge.

Carter, B. (1972d) *Variation Principle for a Rotating Solid Star.* Preprint Inst. of Astronomy, Cambridge.
Chandrasekhar, S. (1931) *Ap. J.* **74,** 81.
Chandrasekhar, S. (1935) *Mon. Not. R.A.S.* **95,** 207.
Chandrasekhar, S. (1969) *Ellipsoidal Figures of Equilibrium*, Yale.
Chandrasekhar, S. (1970) *Ap. J.* **161,** 561.
Chandrasekhar, S. and Friedman, J. (1972) *Ap. J.* **175,** 379.
Christodoulou D. (1970) *Phys. Rev. Lett.* **25,** 1596.
Christodoulou, D. and Ruffini, R. (1971) *Phys. Rev.* **4,** 2552.
Doroskevitch, A. G., Zel'dovich, Ya. B. and Novikov, L. D. (1966), *Sov. Phys. J.E.T.P.* **22,** 12
Ernst, F. J. (1968a) *Phys. Rev.* **167,** 175.
Ernst, F. J. (1968b) *Phys. Rev.* **167,** 1175.
Geroch, R. P. (1970) *J. Math. Phys.* **11,** 437, 2580.
Hajicek, P. (1971) *General Theory of Ergospheres.* Preprint, Bern.
Hartle, J. B. and Sharp, D. H. (1967) *Ap. J.* **147,** 317.
Hartle, J. B. and Hawking, S. W. (1972) *Commun. Math. Phys.* **27,** 283.
Hartle, J. B. and Thorne, K. S. (1968), *Ap. J.* **153,** 807.
Hawking, S. W. (1967) *Proc. Roy. Soc. A.* **300,** 187. *Commun. Math. Phys.* **27,** 283.
Hawking, S. W. and Penrose, R. (1970) *Proc. Roy. Soc. A.* **314,** 529.
Hawking, S. W. (1971) *Phys. Rev. Lett.* **26,** 1344.
Hawking, S. W. (1972) *Commun. Math. Phys.* **25,** 152.
Ipser, J. R. (1971) *Electromagnetic Perturbations of a Kerr Metric Black Hole. Phys. Rev. Letters,* **27,** 529.
Israel, W. (1967) *Phys. Rev.* **164,** 1776.
Israel, W. (1968) *Commun. Math. Phys.* **8,** 245.
Kerr, R. (1963) *Phys. Rev. Lett.* **11,** 237.
Kundt, W. and Trumper, M. (1966) *Z. Physik* **192,** 419.
Landau, L. D. (1932) *Phys. Zs. Swjetunion* **1,** 285.
Landau, L. D. and Lidshitz (1962) *Classical Theory of Fields,* Addison Wesley.
Lewis, T. (1937) *Proc. Roy. Soc. A* **136,** 176.
Lichnerowicz, A. (1955) *Theories Relativistes de Gravitation et de l'Electromagnetism* (Paris: Mason).
Lynden-Bell, D. (1969) *Nature* **223,** 690.
Muller-Zum-Hagen, H., Robinson, P. and Seifert, H. J. (1972) *Black Holes in Static Vacuum Spacetimes.* Preprint, Kings College, London.
Oppenheimer, J. R., Serber, R. (1938) *Phys. Rev.* **54,** 530.
Oppenheimer, J. R., Volkoff, G. (1939) *Phys. Rev.* **55,** 370.
Ostriker, J. P., Bodenheimer, P. and Lynden Bell, D. (1966) *Phys. Rev. Letters* **17,** 816.
Ostriker, J. P. and Tassoul, J. L. (1969) *Ap. J.* **155,** 987.
Papapetrou, A. (1966) *Ann. Inst. H. Poincaré* **A4,** 83.
Penrose, R. (1965) *Phys. Rev. Lett.* **14,** 57.
Penrose, R. (1967) *Battelle Rencontres* Ed. C. M. De Witt and J. A. Wheeler (Benjamin).
Penrose, R. (1969) *Riv. Nuovo Cimento I,* **1,** 252.
Vishveshwara (1968) *J. Math. Phys.* **9,** 1319.
Vishveshwara (1970), *Phys. Rev.* **D1,** 2870.
Wald, R. (1971), *Static Axially Symmetric Electromagnetic Fields in Kerr Spacetime.* Preprint, Princeton University.
Wald, R. (1972), *J. Math. Phys.* **13,** 490.

Timelike and Null Geodesics in the Kerr Metric

James M. Bardeen

Department of Physics
Yale University

Contents

I Introduction 219
II Orbits in the Equatorial Plane 221
III Photon Orbits 227
A The Equations, 228
B The Apparent Shape of the Black Hole, 229
C The Throat of the Extreme Kerr Metric, 233
D Geometrical Optics, 236
Bibliography 239

I Introduction

One of the best ways to get a feeling for the geometry of a spacetime is to study the timelike and null geodesics. Furthermore, the nature of the geodesic trajectories is of direct astrophysical interest. Models of accretion into black holes, such as the models discussed here by Novikov and Thorne, often assume approximate geodesic motion of the accreting matter. Radiation emitted near the black hole propagates to a distant observer along null geodesics as long as the wavelength of the radiation is sufficiently small compared with the radius of the black hole. In calculating the geodesics I will assume that the Kerr metrics are in fact the generic uncharged black hole metrics.

An important property of the Kerr metrics, as discussed in the lectures of Carter, is the fact that the Hamilton–Jacobi equation for the geodesics separates, so that one has explicitly a complete set of constants of the motion.

I will use this approach to discuss the nature of the photon orbits out of the equatorial plane in the latter part of the chapter, but the constants of the motion generated directly by the two Killing vectors of the Kerr metric are sufficient for trajectories in the equatorial plane.

The first section will deal with trajectories in the equatorial plane, which have been analyzed in some detail by de Felice (1968). It is the circular test particle orbits which are most important in astrophysical calculations, and I will present the simplified formulas for the energy, angular momentum, etc. which have derived recently by Teukolsky (Bardeen, Press, and Teukolsky 1972).

While in the second section I give the equations for general geodesic trajectories out of the plane in a convenient form, only a few specific applications will be discussed. One is the question of which null geodesics are trapped by the black hole and which escape to infinity. Another is the nature of the null geodesics near the horizon in the extreme Kerr metric with $a = m$. I will conclude with a discussion of some of the techniques of geometrical optics which enable one to calculate the appearance of a source of radiation near the horizon as seen by a distant observer.

The behavior of geodesics inside the Kerr horizon is of great importance for obtaining the maximal analytic extension of the Kerr metric (see Carter 1968), but is not relevant for the astrophysical aspects of black holes.

It will be convenient at times to use a standard form for the line element of a stationary, axisymmetric metric:

$$ds^2 = -e^{2\nu}dt^2 + e^{2\psi}(d\phi - \omega dt)^2 + e^{2\lambda}dr^2 + e^{2\mu}d\theta^2 \tag{1}$$

In the usual Boyer–Lindquist coordinates the Kerr Metrics have

$$\begin{aligned} e^{2\nu} &= \Delta A/B, \\ e^{2\psi} &= B\sin^2\theta/A, \\ \omega &= 2amr/B, \\ e^{2\lambda} &= A/\Delta, \qquad e^{2\mu} = A, \end{aligned} \tag{2}$$

where

$$A = r^2 + a^2 \cos^2 \theta,$$
$$B = (r^2 + a^2)^2 - \Delta a^2 \sin^2 \theta, \qquad (3)$$
$$\Delta = r^2 - 2mr + a^2,$$

and m and ma are the mass and angular momentum of the black hole, respectively.

The standard description of the direction of a geodesic at a point in spacetime is through the tangent four-vector

$$p^a = dx^a/d\tau, \qquad (4)$$

where τ is an affine parameter. I follow Carter in normalizing τ so

$$p^a p_a = -\mu^2 \qquad (5)$$

where μ is the rest mass of the test particle. The proper time along a timelike geodesic is $\mu\tau$ and a null geodesic has $\mu = 0$.

The symmetries of the Kerr metric give immediately two constants of the motion along the trajectory, the energy relative to infinity

$$E = -p_t, \qquad (6)$$

and the angular momentum about the symmetry axis

$$\Phi = p_\phi \qquad (7)$$

These are sufficient to determine trajectories in the equatorial plane, $\theta = \pi/2$.

It is often useful to refer to the physical components of the momentum of the test particle as measured by local observers. These depend on the motion of the observer; the natural choice at each point is the observer whose r and θ coordinates are fixed and who has zero angular momentum (Bardeen 1970; Bardeen, Press and Teukolsky 1972). The world lines of these observers are perpendicular to the t = constant spacelike hypersurfaces, but because of the off-diagonal term in the line element the observers rotate with coordinate angular velocities $d\phi/dt = \omega(r, \theta)$ relative to infinity.

Each observer sets up a frame of reference represented by an orthonormal tetrad tied to the r, θ coordinate directions,

$$e_{(t)} = e^{-\nu}\left[\frac{\partial}{\partial t} + \omega \frac{\partial}{\partial \phi}\right],$$
$$e_{(\phi)} = e^{-\psi}\frac{\partial}{\partial \phi},$$
$$e_{(r)} = e^{-\lambda}\frac{\partial}{\partial r}, \qquad (8)$$
$$e_{(\theta)} = e^{-\mu}\frac{\partial}{\partial \theta}.$$

The locally measured energy of the test particle is then

$$p^{(t)} = -e^a_{(t)} p_a = e^{-\nu}(E - \omega\Phi). \tag{9}$$

The physical momentum in the ϕ-direction is

$$p^{(\phi)} = e^{-\psi}\Phi \tag{10}$$

and

$$p^{(r)} = e^{-\lambda} p_r, \qquad p^{(\theta)} = e^{-\mu} p_\theta . \tag{11}$$

Note that a test particle with zero angular momentum moves perpendicular to the ϕ-direction in this frame. For a timelike test particle ($\mu^2 > 0$) we may also define a local velocity three-vector with magnitude v and components $v^{(\phi)}$, $v^{(r)}$, $v^{(\theta)}$ in the usual way.

II Orbits in the Equatorial Plane

The motivation for considering only test particle trajectories lying in the equatorial plane, and in particular circular orbits in the plane, is to a large extent just mathematical simplicity. However, it seems plausible that of all particle orbits with a fixed rest mass μ and a fixed angular momentum Φ and which are *not* trapped by the black hole, the orbit which minimizes the energy is the stable circular orbit with the given value of Φ/μ, as long as Φ/μ is greater than the minimum value for direct circular particle orbits or less than the minimum value for retrograde circular orbits. If Φ/μ is within the range for which there are no circular orbits in the equatorial plane the minimum energy for untrapped orbits is presumably associated with an orbit out of the plane. Also, it seems plausible that a dissipative process, such as gravitational radiation, will cause a test particle to relax to such a minimum energy orbit unless the particle is first trapped by the black hole. Of course, a strong preference for the equatorial plane would require that a/m not be small compared with one. For information on certain types of particle orbits out of the plane, the reader is referred to a paper by Wilkins (1972).

The equation governing radial motion in the equatorial plane is simply equation (5), with the constants of the motion E and Φ inserted. The explicit form for the Kerr metric is

$$\left(\frac{dr}{d\tau}\right)^2 = r^{-2}R(r) = r^{-3}\{[r(r^2+a^2) + 2a^2m]E^2$$
$$- 4amE\Phi - (r-2m)\Phi^2 - \Delta r\mu^2\}. \tag{12}$$

At a given radius r and for a given angular momentum Φ the physically accessible values of E, for which $R(r) \geqslant 0$, lie above a minimum value E_{min}. While

$R(r) > 0$ also for E less than a certain value, these values of E are excluded because they correspond to negative energies in the local observer's frame (see equation (9)). Thus

$$E_{\min} = \frac{\Delta^{1/2}\{r^2\Phi^2 + [r(r^2+a^2)+2a^2m]r\mu^2\}^{1/2} + 2am\Phi}{r(r^2+a^2)+2a^2m}. \tag{13}$$

Considerable insight into the nature of the orbits comes from plotting $E_{\min}$ as a function of r for various values of Φ. The typical behavior is shown in Figure 1. As $r \to \infty$, $E_{\min}/\mu \to 1$. There is a stable, bound circular orbit where $E_{\min}(r)$ has a minimum and an unstable, unbound circular orbit where $E_{\min}(r)$ has a maximum. The range in r for a trajectory with some particular value of E is such that $E \geqslant E_{\min}(r)$ and the turning points are where $E = E_{\min}(r)$. If Φ/μ lies in a certain

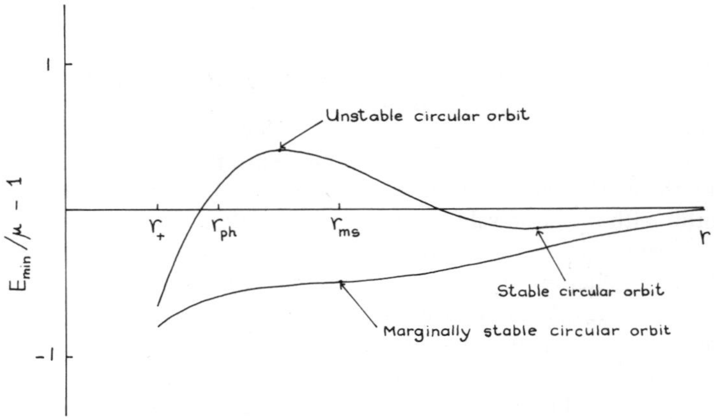

Figure 1. Schematic examples of "effective potential" plots for time-like geodesics in the equatorial plane of the Kerr metric. There is a separate effective potential for each value of the angular momentum per unit mass Φ/μ (see Ruffini and Wheeler 1970).

range about zero the curve $E_{\min}(r)$ decreases monotonically as r decreases and there are no circular orbits.

When $a \neq 0$ and r is sufficiently close to the horizon, $E_{\min}$ is less than zero for retrograde orbits, with $a\Phi < 0$. In this region particles with positive energy in the local observer's frame can have negative energy with respect to infinity; in a sense, their gravitational binding energy exceeds their rest mass energy. These negative energy orbits are only possible in the ergosphere, where $g_{tt} > 0$. Combining equations (9) and (10) gives

$$E = e^{\nu}p^{(t)} + \omega e^{\psi}p^{(\phi)}. \tag{14}$$

Any physical particle must have $|p^{(\phi)}| < p^{(t)}$, so necessary conditions for $E < 0$ are

$$\omega^2 e^{2\psi} - e^{2\nu} = g_{tt} > 0 \tag{15}$$

and $p^{(\phi)} < 0$.

Equation (13) simplifies considerably for photon trajectories (null geodesics), and can be written

$$E_{\min} = \pm\Phi \left\{ \frac{r(r^2 - 2mr + a^2)^{1/2} \pm 2am}{r(r^2 + a^2) + 2a^2 m} \right\}. \tag{16}$$

The upper sign holds if $\Phi > 0$, the lower sign if $\Phi < 0$. It is fairly easy to see from equation (16) that for either direct or retrograde orbits $E_{\min}(r)$ only has a maximum between the horizon at $r_+ = m + (m^2 - a^2)^{1/2}$ and $r = \infty$. Null geodesics in the equatorial plane can have at most one turning point, and the circular photon orbits, one for each sign of Φ, are unstable.

Now I will discuss in some detail circular orbits in the equatorial plane using the simplified formulas derived by S. Teukolsky (see Bardeen, Press, and Teukolsky 1972). Circular orbits are obtained from equation (12) by solving the two simultaneous equations $R = 0$ and $dR/dr = 0$ for E and Φ as functions of r. After considerable algebraic manipulation, one finds

$$E/\mu = \frac{r^{3/2} - 2mr^{1/2} \pm am^{1/2}}{r^{3/4}(r^{3/2} - 3mr^{1/2} \pm 2am^{1/2})^{1/2}}, \tag{17}$$

and

$$\Phi/\mu = \pm \frac{m^{1/2}(r^2 \mp 2am^{1/2}r^{1/2} + a^2)}{r^{3/4}(r^{3/2} - 3mr^{1/2} \pm 2am^{1/2})^{1/2}}. \tag{18}$$

Again, the upper sign is for the direct orbit ($\Phi > 0$) and the lower sign is for the retrograde orbit ($\Phi < 0$). Other physically interesting quantities are the angular velocity of the particle as seen from infinity,

$$\Omega = d\phi/dt = \pm\, m^{1/2}/(r^{3/2} \pm am^{1/2}), \tag{19}$$

and the physical velocity of rotation

$$v^{(\phi)} = \pm \frac{m^{1/2}(r^2 \mp 2am^{1/2}r^{1/2} + a^2)}{\Delta^{1/2}(r^{3/2} \pm am^{1/2})}. \tag{20}$$

At large r ($r \gg m$) both direct and retrograde orbits are bound, with nearly equal binding energies. However, a "spin-orbit coupling" effect which increases the binding energy of the direct orbit and decreases the binding energy of the retrograde orbit relative to the Schwarzschild value becomes important at smaller r.

The maximum binding energy $(1 - E/\mu)$ (and the minimum value of Φ/μ) in each case is reached at the radius where the orbit becomes unstable. This radius of marginal stability, r_{ms}, is

$$\begin{aligned} r_{ms} &= m\{3 + Z_2 \mp [(3 - Z_1)(3 + Z_1 + 2Z_2)]^{1/2}\}, \\ Z_1 &= 1 + (1 - a^2/m^2)^{1/3}[(1 + a/m)^{1/3} + (1 - a/m)^{1/3}], \\ Z_2 &= (3a^2/m^2 + Z_1^2)^{1/2}, \end{aligned} \tag{21}$$

and can be determined directly from equation (12).

When $a = 0$, $r_{ms} = 6m$ for both direct and retrograde orbits, while in the limit $a \to m$, $r_{ms} = m$ for direct orbits and $r_{ms} = 9m$ for retrograde orbits.

At still smaller radii the value of E/μ becomes greater than one and the orbits become unbound, as well as being unstable. The radius of the marginally bound orbits is the minimum perihelion for parabolic orbits,

$$r_{mb} = 2m \mp a + 2m^{1/2}(m \mp a)^{1/2}. \tag{22}$$

All test particles which are non-relativistic at infinity are trapped if they get inside r_{mb}. When $a = 0$, $r_{mb} = 4m$ and when $a = m$, $r_{mb} = m$ for direct orbits and $(3 + 2^{3/2})m = 5.83m$ for retrograde orbits.

In the limit

$$r^{3/2} - 3mr^{1/2} \pm 2am^{1/2} \to 0, \tag{23}$$

or

$$r \to r_{ph} = 2m\left\{1 + \cos\left[\tfrac{2}{3}\cos^{-1}\left(\mp\frac{a}{m}\right)\right]\right\} \tag{24}$$

both E/μ and Φ/μ become infinitely large; there is an unstable circular photon orbit at $r = r_{ph}$. Again, the values are $r_{ph} = 3m$ for $a = 0$, $r_{ph} = m$ for direct $a = m$, $r_{ph} = 4m$ for retrograde $a = m$. Given r_{ph}, the corresponding impact parameter as calculated from equations (21) and (22) is $\Phi/E = \pm \frac{1}{2}r_{ph}^{1/2}m^{1/2}(r_{ph}/m + 3)$. This impact parameter represents where the visual "rim" of the black hole intersects the equatorial plane. Photon trajectories which circle the black hole a large number of times before escaping have only slightly larger impact parameters.

The direct marginally stable orbit, the direct marginally bound orbit, the direct circular photon orbit, and the horizon all seem to coincide at $r = m$ when $a = m$. The apparent conflict with the null character of the horizon is because the coordinate r misrepresents the geometry of the spacetime near $r = m$ when $a = m$. An infinitesimal range of r near $r = m$ can correspond to an infinite range of proper radial distance. From equations (1)–(3) the proper radius R_p is given by

$$R_p = \int r\Delta^{-1/2}\,dr \to r + m\ln(r/m - 1) \tag{25}$$

when $a = m$. Let $a = m(1 - \delta)$; then in the limit $\delta \ll 1$ the horizon is at

$$r_+ \simeq m[1 + (2\delta)^{1/2}],$$

and

$$r_{ph} \simeq m[1 + (8\delta/3)^{1/2}],$$
$$r_{mb} \simeq m[1 + 2\delta^{1/2}], \tag{26}$$
$$r_{ms} \simeq m[1 + (4\delta)^{1/3}].$$

The limiting proper radial distance between r_+ and r_{ph} is $\frac{1}{2}m \ln 3$, that between r_+ and r_{mb} is $m \ln (1 + 2^{1/2})$, and that between r_+ and r_{ms} is $m \ln [2^{7/6}\delta^{-1/6}] \to \infty$. The limiting energies E and velocities $v^{(\phi)}$ as $a \to m$ are

$$E/\mu \to 3^{-1/2}, \qquad v^{(\phi)} \to 1/2 \qquad \text{at} \qquad r = r_{ms},$$
$$E/\mu = 1, \qquad v^{(\phi)} \to 2^{-1/2} \qquad \text{at} \qquad r = r_{mb}. \tag{27}$$

The rotational velocity at $r = r_{ms}$ is the same for $a = m$ as for $a = 0$, even though the orbit is in a much stronger gravitational field when $a = m$.

The results for circular orbits are displayed graphically in Figures 2 and 3. Figure 2 shows how r_+, r_{ph}, r_{mb}, and r_{ms} vary with a/m for both direct and

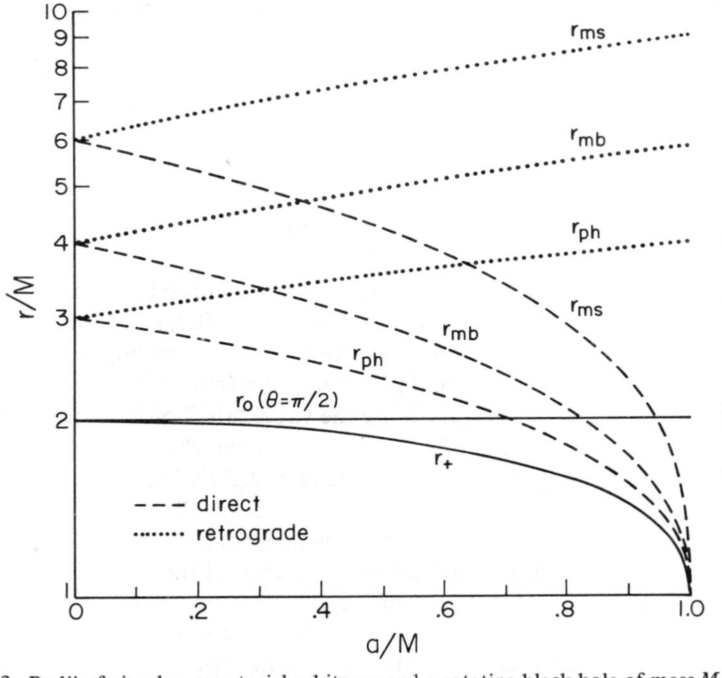

Figure 2. Radii of circular, equatorial orbits around a rotating black hole of mass M, as functions of the hole's specific angular momentum a. Dashed and dotted curves (for direct and retrograde orbits) plot the Boyer–Lindquist coordinate radius of the innermost stable (ms), innermost bound (mb), and photon (ph) orbits. Solid curves indicate the event horizon (r_+) and the equatorial boundary of the ergosphere (r_0). (From Bardeen, Press, and Teukolsky 1972.)

retrograde orbits. Figure 3 illustrates the development of the long cylindrical throat at the horizon as $a \to m$. The equatorial plane is embedded in a Euclidean three-space such that the circumference is the proper circumference of a circle in the equatorial plane, and the length along the tube is the proper radial distance in the equatorial plane. Note how close a must be to m for the throat to have an appreciable length.

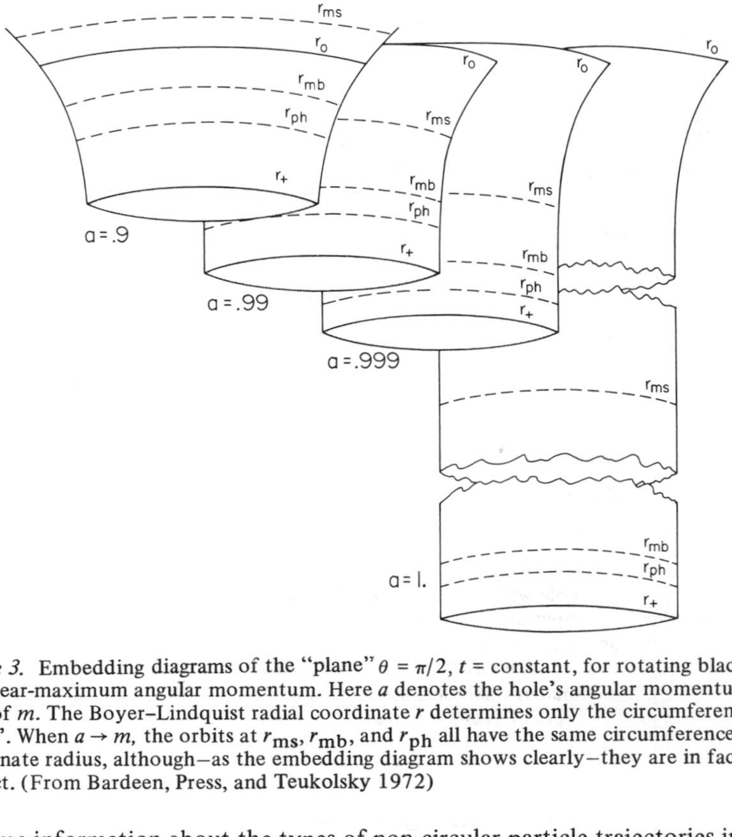

Figure 3. Embedding diagrams of the "plane" $\theta = \pi/2$, t = constant, for rotating black holes with near-maximum angular momentum. Here a denotes the hole's angular momentum in units of m. The Boyer–Lindquist radial coordinate r determines only the circumference of the "tube". When $a \to m$, the orbits at r_{ms}, r_{mb}, and r_{ph} all have the same circumference and coordinate radius, although—as the embedding diagram shows clearly—they are in fact distinct. (From Bardeen, Press, and Teukolsky 1972)

Some information about the types of non-circular particle trajectories in the equatorial plane is provided in Figure 4, for the case $a = 0.95m$. At each radius r the various orbits passing through that point are represented by points in the $v^{(r)}$, $v^{(\phi)}$ plane inside the speed-of-light circle $v^{(r)^2} + v^{(\phi)^2} = 1$. The orbits are classified into bound, stable orbits, denoted (B); orbits which originate at infinity and are captured by the hole, denoted (P); the time-reverse of (P) orbits which escape from the past event horizon to infinity, denoted (E); hyperbolic orbits which are scattered by the hole, denoted (H); bound orbits which plunged into the hole, denoted (C). The orbits on the boundary between regions (H) and (P) are unstable, including the unstable circular orbits where this boundary crosses the line $v^{(r)} = 0$. The shaded regions have $E < 0$; if the black hole captures a particle in such a trajectory the mass of the hole decreases, even though the energy of the particle in the local observer's frame is positive. Such orbits can only exist in the region $r < 2m$ in the equatorial plane where $g_{tt} > 0$, the so-called ergosphere.

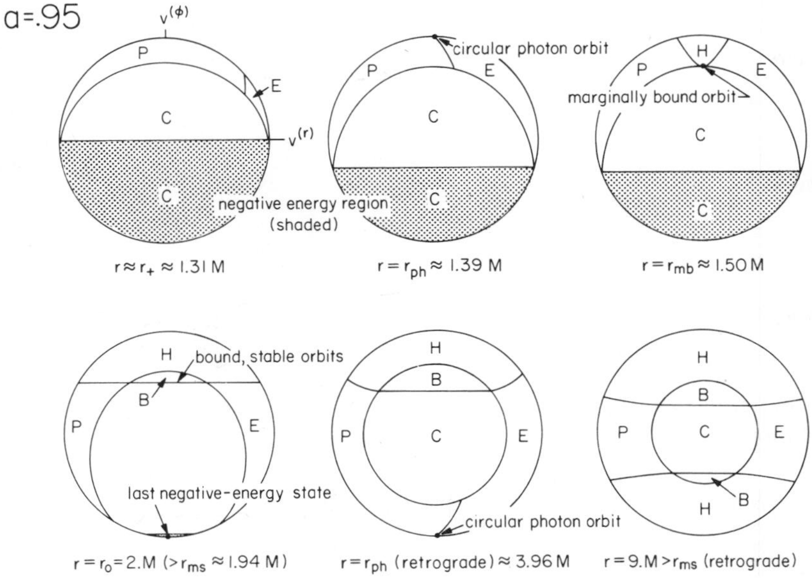

Figure 4. Distribution in velocity space of equatorial orbits passing through various radii r, around a rotating black hole with $a = 0.95$M. Each circle is the "space" of equatorial, ordinary velocities [d(proper distance)/d(proper time)] as measured in the proper reference frame of a locally nonrotating observer. The velocity circles are labeled by the radius r of the observer. The center of each circle is zero velocity; the edge is the speed of light; the $v^{(r)}$ direction corresponds to radial velocities, the $v^{(\phi)}$ direction to tangential velocities. The labelling of the various types of orbits is explained in the text. (From Bardeen, Press, and Teukolsky 1972.)

III Photon Orbits

Since there is no guarantee a distant observer will happen to lie precisely in the equatorial plane of a black hole, any discussion of the appearance of a star or disk near the black hole (or the effect of the black hole geometry on the propagation of gravitational radiation from source) must take into account null geodesic trajectories out of the equatorial plane. The additional constant of the motion obtained by separating the r and θ dependence in the Hamilton–Jacobi equation for the geodesics (Carter 1968) comes into play at this point.

A The Equations

I will not set $\mu = 0$ until after the equations are derived, so the equations will be valid for all types of geodesics. The Hamilton–Jacobi equation has the form

$$-\frac{\partial S}{\partial \tau} = \tfrac{1}{2} g^{ab} \frac{\partial S}{\partial x^a} \frac{\partial S}{\partial x^b}, \tag{28}$$

where

$$\frac{\partial S}{\partial x^a} = p_a \tag{29}$$

and

$$\frac{\partial S}{\partial \tau} = \tfrac{1}{2}\mu^2. \tag{30}$$

From the overt symmetries

$$S = \tfrac{1}{2}\mu^2 \tau - Et + \Phi\phi + S(\theta, r). \tag{31}$$

As Carter has shown, substitution of equation (31) into equation (28) gives the solution

$$S(\theta, r) = \int^r \frac{[R(r)]^{1/2}}{\Delta}\, dr + \int^\theta [\Theta(\theta)]^{1/2}\, d\theta, \tag{32}$$

with

$$\begin{aligned} R(r) = r[r(r^2 + a^2) + 2a^2 m]E^2 - 4amrE\Phi \\ - (r^2 - 2mr)\Phi^2 - \Delta[r^2\mu^2 + Q] \end{aligned} \tag{33}$$

and

$$\Theta(\theta) = Q + a^2(E^2 - \mu^2)\cos^2\theta - \Phi^2 \cot^2\theta. \tag{34}$$

The separation constant Q is zero for trajectories lying in the equatorial plane and is greater than or equal to zero for all trajectories touching or crossing the equatorial plane. Some trajectories confined near the axis of symmetry have $Q < 0$, but

$$K = Q + (aE - \Phi)^2 \tag{35}$$

must always be greater than zero, or zero for trajectories along the axis of symmetry, if $[\Theta(\theta)]^{1/2}$ is to be real.

The momentum components p_r and p_θ are given at each point along the trajectory by

$$p_r = \Delta^{-1}[R(r)]^{1/2}, \qquad p_\theta = [\Theta(\theta)]^{1/2}. \tag{36}$$

The signs of $[R(r)]^{1/2}$ and $[\Theta(\theta)]^{1/2}$ change at turning points in the r and θ motions.

The tangent vector $dx^a/d\tau$ is now completely determined along the trajectory, and the trajectory could be found by a direct integration. However, to avoid integrals which mix θ and r it is better to take advantage of the fact that derivatives of the Hamilton–Jacobi action S with respect to the constants of the motion are themselves constants. The values of $\partial S/\partial E$, etc., represent initial values of the coordinates and will be absorbed as constants of integration. Then $\partial S/\partial E$ gives

$$t = \int^r \frac{[r^2(r^2 + a^2)E + 2amr(aE - \Phi)]}{\Delta[R(r)]^{1/2}} dr + \int^\theta \frac{a^2E^2 \cos^2\theta}{[\Theta(\theta)]^{1/2}} d\theta, \tag{37}$$

$\partial S/\partial \Phi$ gives

$$\phi = \int^r \frac{[r^2\Phi + 2amr(aE - \Phi)]}{\Delta[R(r)]^{1/2}} dr + \int^\theta \frac{\Phi \cot^2\theta}{[\Theta(\theta)]^{1/2}} d\theta, \tag{38}$$

and $\partial S/\partial Q$ gives

$$\int^r \frac{dr}{[R(r)]^{1/2}} = \int^\theta \frac{d\theta}{[\Theta(\theta)]^{1/2}}. \tag{39}$$

These equations suffice for most purposes, but to complete the set we present the equation for the affine parameter, obtained from $\partial S/\partial \mu = 0$,

$$\tau = \int^r \frac{r^2}{[R(r)]^{1/2}} dr + \int^\theta \frac{a^2 \cos^2\theta}{[\Theta(\theta)]^{1/2}} d\theta. \tag{40}$$

It is now straightforward to evaluate these integrals numerically. They can be reduced to standard elliptic integrals.

B The Apparent Shape of the Black Hole

One important question in the propagation of radiation outside the horizon is the range of the parameters E, L, and Q for which null geodesics emitted at some radius r_e are trapped by the hole. For null geodesics only the two ratios

$$\lambda = \Phi/E \tag{41a}$$

and

$$\eta = Q/E^2 \tag{41b}$$

are really independent. Furthermore, λ and η are directly related to the impact parameters which describe the direction of the photon as seen by an observer at a large distance r_o from the black hole along the polar angle θ_o.

Assume that the observer measures directions of photons relative to the center of symmetry of the asymptotic gravitational field. The component of the angular displacement perpendicular to the axis of symmetry, as projected on the plane of the sky of the observer, is $-p^{(\phi)}/p^{(t)}$, where the sign is appropriate to a situation where the near edge of the black hole rotates from left to right as seen by the observer. The component of the apparent angular displacement parallel to the axis of symmetry is $p^{(\theta)}/p^{(t)}$. These angles, for a given trajectory are inversely proportional to r_0, so it is convenient to work in terms of the impact parameters

$$\alpha = -r_o p^{(\phi)}/p^{(t)} = -\lambda/\sin\theta_o \tag{42a}$$

and

$$\beta = r_o p^{(\theta)}/p^{(t)} = p_\theta = [\eta + \cos^2\theta_o - \lambda^2\cot^2\theta_o], \tag{42b}$$

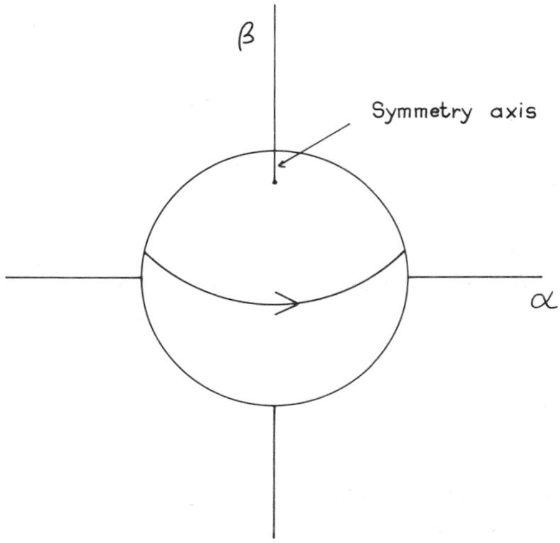

Figure 5. The impact-parameter coordinates α, β in the plane of the sky of an observer slightly above the equatorial plane of the black hole are illustrated here, in relation to the direction of rotation of the black hole. The β-axis is the projection of the axis of symmetry, and the α-axis passes through the projection of the center of symmetry defined by the gravitational field far from the black hole.

which are independent of r_o. Each photon trajectory which reaches the observer is a point in the α–β plane, which represents a small portion of the celestial sphere of the observer. Figure 5 shows the relationship between α, β and the rotation of the black hole.

Now imagine a source of illumination behind the black hole whose angular size is large compared with the angular size of the black hole. As seen by the distant observer the black hole will appear as a "black hole" in the middle of the larger bright source. No photons with impact parameters in a certain range about $\alpha,\beta = 0$ reach the observer. The rim of the "black hole" corresponds to photon trajectories which are marginally trapped by the black hole; they spiral around many times before they reach the observer. It is conceptually interesting, if not astrophysically very important, to calculate the precise apparent shape of the black hole.

In astrophysics one may have a source of radiation near the black hole and want to calculate what portion of the emitted radiation is trapped by the black hole and what portion escapes to infinity. For each direction of emission at the source there is a corresponding pair of values of λ and η which can be calculated by relating the direction cosines of the photon trajectory in the local observer's frame at the position of the source to the covariant momentum components E, Φ and p_θ.

For either of the above questions we want to know the range of the parameters λ and η for which the photon trajectory has a turning point in its radial motion between the horizon at $r = r_+$ and $r = \infty$. Photons which originate behind the black hole relative to a distant observer must have a turning point in radius to get to the observer. Photons emitted by a local source must have a turning point to escape if they are emitted inward, with $p^{(r)} < 0$. On the other hand, if they initially have $p^{(r)} > 0$ a turning point means that they will be trapped by the black hole. No null geodesics have more than one turning point outside the horizon.

The photon trajectories which are on the verge of having a true turning point circle indefinitely at some constant value of r. The constant-r photon trajectories are unstable, since a small perturbation causes the photon either to escape to infinity or to fall into the black hole. To find these trajectories we solve simultaneously $R(r) = 0$ and

$$\frac{d}{dr}\,[r^{-1}R(r)] = 0. \tag{43}$$

The explicit equations are

$$[r(r^2 + a^2) + 2a^2m] - 4am\lambda - (r - 2m)\lambda^2 - r^{-1}(r^2 - 2mr + a^2)\eta = 0 \tag{44a}$$

and

$$3r^2 + a^2 - \lambda^2 - r^{-2}(r^2 - a^2)\eta = 0. \tag{44b}$$

For general a/m it is only feasible to solve for λ and η in terms of the radius r of the orbit, which provides an implicit relation between λ and η. The equations are quadratic in λ. Of the two roots, one,

$$\lambda = (r^2 + a^2)/a, \tag{45}$$

gives

$$\eta = -\,r^4/a^2 \tag{46}$$

and

$$K/E^2 = \eta + (a - \lambda)^2 = 0. \tag{47}$$

The only place $K = 0$ is acceptable is on the axis of symmetry, where λ must be equal to zero. Therefore, this solution must be rejected.

The physical solution is

$$\lambda = \frac{-r^3 + 3mr^2 - a^2(r+m)}{a(r-m)} \tag{48}$$

and

$$\eta = \frac{r^3}{a^2(r-m)^2}\,[4a^2m - r(r-3m)^2]. \tag{49}$$

The constant-r orbits only exist as long as $\eta > 0$, since they all cross the equatorial plane. Therefore, they are confined to a limited range of r, between (see equation (23)) the radius of the direct circular photon orbit in the equatorial plane, r^+_{ph}, and the radius of the retrograde circular photon orbit in the equatorial plane, r^-_{ph}. Let the corresponding extreme values of λ be λ^+ and λ^-, respectively.

Equations (48) and (49) determine implicitly a curve in the η,λ-plane, $\eta = \eta_c(\lambda)$. In the region $\lambda^- < \lambda < \lambda^+$, $0 < \eta < \eta_c(\lambda)$, the photon trajectories do not have turning points and independent of the radius of emission ingoing photons are trapped by the black hole, outgoing photons escape.

Outside this region a turning point does exist. Let $r_c(\lambda)$ be the inversion of equation (48) for $\lambda^- < \lambda < \lambda^+$. For direct orbits with $\lambda > \lambda^+$ let $r_c(\lambda) = r^-_{\text{ph}}$. For all retrograde orbits with $\lambda < \lambda^-$ or with $E < 0$ and $\lambda > 0$ let $r_c(\lambda) = r^-_{\text{ph}}$. If the photon is emitted at a radius $r_{\text{em}} < r_c(\lambda)$ the turning point is outside the radius of emission and both ingoing and outgoing trajectories with the given values of λ and η are trapped by the black hole. Conversely, if the photon is emitted at $r_{\text{em}} > r_c(\lambda)$ both ingoing and outgoing trajectories escape to infinity.

The parameters η and λ are directly related to the direction of emission in the local observer's frame through equations (9)–(11) and equations (34), (36). For instance,

$$p^{(\phi)}/p^{(t)} = e^{\nu-\psi}\lambda/(1-\omega\lambda). \tag{50}$$

It is a straightforward matter to find the range of solid angle over which photons escape at any point outside the horizon. Once $r_{\text{em}} \leqslant r^+_{\text{ph}}$ the only photons which escape have η and λ in the no-turning-point region, $\eta < \eta_c(\lambda)$. In the limit r_{em} is infinitesimally close to the horizon (in proper radius) the no-turning-point region in the λ,η-plane corresponds to an infinitesimal range of solid angle in the local observer's frame, just as is familiar from the Schwarzschild metric. However, we shall see that in the extreme $a = m$ Kerr metric a substantial range of solid angle can escape to infinity when r_{em} is infinitesimally close to the horizon in coordinate radius.

The apparent shape of the black hole is just the boundary of the no-turning-point region, $\eta = \eta_c(\lambda)$, translated into a closed curve in the (α, β) impact parameter plane through equations (42). The apparent shape depends on the polar angle θ_0 of the distant observer and is particularly easy to calculate when the observer is in the equatorial plane of the black hole. Then $\alpha = -\lambda$ and $\beta = \eta^{1/2}$. Furthermore, for the extreme $a = m$ Kerr metric equations (48) and (49) simplify to

$$\lambda = m^{-1}(-r^2 + 2mr + m^2), \tag{51}$$

$$\eta = m^{-2}r^3(4m - r). \tag{52}$$

The non-uniform nature of the limit $a \to m$ allows η to range between 0 and $3m^2$ at $r = m$. Recall that when $a = m$ the single coordinate radius $r = m$ corresponds to an infinite range of proper radius. Equation (51) can be inverted to give

$$r_c(\lambda) = m + (2m^2 - m\lambda)^{1/2}, \tag{53}$$

and

$$\eta_c(\lambda) = m^{-2}[m + (2m^2 - m\lambda)^{1/2}]^3[3m - (2m^2 - m\lambda)^{1/2}]. \tag{54}$$

The apparent shape of the $a = m$ Kerr black hole obtained from equation (54) is plotted in Figure 6. The maximum diameter parallel to the axis of symmetry is almost the same as the diameter of the apparent Schwarzschild black hole, which is $2 \cdot 3^{3/2}m = 10.39m$. However, the effect of the frame dragging induced by the angular momentum of the Kerr black hole is quite apparent.

Similar curves have been calculated by Godfrey (1970). Unfortunately, a mathematical error invalidates most of his results for photon orbits.

C. The Throat of the Extreme Kerr Metric

An interesting, if somewhat unphysical, special case is the behavior of photon orbits emitted at $(r/m - 1) \ll 1$ in the extreme Kerr metric. Far down the cylindrical throat of the black hole the geometry is essentially homogeneous in the radial direction, and the metric reduces to a simple limiting form. We replace r as a radial coordinate by

$$\xi = (r/m - 1). \tag{55}$$

Consider photons emitted at $\xi = \xi_e \ll 1$. All except the photons emitted precisely perpendicular to the ϕ-direction in the local observer's frame have almost the same impact parameter λ (see equation (50)),

$$\lambda \simeq \omega^{-1} \simeq 2m. \tag{56}$$

Almost all trajectories with negative angular momentum have $E < 0$ and are captured by the black hole.

The direction of a photon trajectory in the local observer's frame at the point of emission will be represented by two parameters,

$$\gamma = p^{(\phi)}/p^{(t)} \tag{57}$$

and

$$\delta = p^{(\theta)}/p^{(\phi)} = e^{\psi - \mu} p_\theta / p_\phi \simeq 2\eta^{1/2}/\lambda \tag{58}$$

in the limit $\xi_e \ll 1$. To find the turning points in the photon orbits as a function of γ and δ it is necessary to expand the expression for λ (equation (50)) to order ξ_e^2,

$$\lambda^{-1} = (2m)^{-1}\left[1 + \frac{(1 - .2\gamma)}{2\gamma}\xi_e + \tfrac{1}{4}\xi_e^2 + \ldots\right] \tag{59}$$

as long as $\xi_e/\gamma \ll 1$.

The frequency shift factor between the local observer's frame and infinity for those photons that do escape is

$$g \equiv E/p^{(t)} = e^{\psi}\lambda\gamma \simeq \gamma. \tag{60}$$

The redshifts are not extreme over most of the forward hemisphere. On the other hand, the time dilation factor between infinity and the local observer *is* large,

$$\frac{dt}{ds} = e^{-\nu} \simeq 2/\xi_e. \tag{61}$$

The energy of a typical photon is not greatly attenuated, but the *average* rate photons are received at infinity is reduced by a factor $\xi_e/2$ from the rate at which the photons are emitted as measured in the local observer's frame. The photons emitted continuously by a point source at $\xi_e \ll 1$ are received at infinity in sharp bursts, much like the synchrotron phenomenon for relativistic particles in circular orbits in flat space. However, the gravitational radiation associated with the orbital motion of a particle at $\xi_e \ll 1$ is not affected in this way unless the particle is relativistic in the local observer's frame (Misner 1972; Bardeen, Press, and Teukolsky 1972).

The discriminant $R(r)$ is calculated by taking the limit of equation (33) when $r/m - 1 \ll 1$, with E and Q expressed in terms of γ and δ. In the limit both ξ and ξ_e are small compared with one,

$$R(\xi) \simeq \tfrac{1}{4}m^2\Phi^2[(3 - \delta^2)\xi^2 + 4\gamma^{-1}(1 - 2\gamma)\xi\xi_e + \gamma^{-2}(1 - 2\gamma)^2\xi_e^2]. \tag{62}$$

The zeroth order approximation for λ is sufficient in $\Theta(\theta)$,

$$\Theta(\theta) \simeq \tfrac{1}{4}\Phi^2[\delta^2 + \cos^2\theta - 4\cot^2\theta]. \tag{63}$$

The roots of $R(\xi)$ are at

$$\xi/\xi_e = \frac{2\gamma - 1}{\gamma}\,\frac{2 \pm (1 + \delta^2)^{1/2}}{3 - \delta^2}. \tag{64}$$

A necessary condition for escape of the photon trajectory is that the coefficient of ξ^2 in equation (62) be positive, or $\delta^2 < 3$. All photons emitted inward ($p^{(r)} < 0$)

with $\gamma > 1/2$ must have $\delta^2 < 3$ from the condition

$$(p^{(\phi)^2} + p^{(\theta)^2})/p^{(t)^2} = \gamma^2(1 + \delta^2) < 1. \tag{65}$$

They reach a turning point at $\xi/\xi_e > 0$ and escape. However, inward photons with $\gamma < 1/2$ are trapped. All initially outgoing photons ($p^{(r)} > 0$) with $\delta^2 < 3$ and $\gamma > 0$ escape, since the turning points are at $\xi/\xi_e \leqslant 1$. When $p^{(r)} = 0$, $\gamma = 1/2$ implies $\delta^2 = 3$.

If the photons are actually emitted by a particle in a direct circular orbit at $\xi = \xi_e$, it is physically more relevant to describe the directions of emission and the frequency shift factor in the frame comoving with the particle, rather than the canonical local observer's frame. A Lorentz transformation with a velocity $v^{(\phi)} = 1/2$, appropriate to a stable circular orbit at $\xi_e \ll 1$, gives for the new frequency shift factor

$$g' = \frac{3^{1/2}\gamma}{2 - \gamma} \tag{66}$$

and for the new direction cosine relative to the $+\phi$-direction

$$\gamma' = (2\gamma - 1)/(2 - \gamma). \tag{67}$$

The solid angle of escaping photons is greater than 2π in the comoving frame, and photons with negative energy parameters are restricted to $-1 \leqslant \gamma' < -1/2$.

The variation of θ with radius along a photon trajectory is given by equation (39). As long as $\xi_e \ll \xi \ll 1$ for escaping photons, equation (39) can be written

$$\int_{\pi/2}^{\theta} \frac{d\theta}{[\delta^2 + \cos^2\theta - 4\cot^2\theta]^{1/2}} \simeq (3 - \delta^2)^{-1/2} \int^{\xi} \xi^{-1}\, d\xi. \tag{68}$$

The θ coordinate oscillates between

$$\cos\theta = \pm \{\tfrac{1}{2}[(\delta^2 + 1)(\delta^2 + 9)]^{1/2} - \tfrac{1}{2}(\delta^2 + 3)\}^{1/2}, \tag{69}$$

and in order of magnitude the number of periods of the oscillation is $(2\pi)^{-1}\ln \xi_e^{-1}$. The maximum value of $|\cos\theta|$ obtainable by photons for which $\xi_e/\gamma \ll 1$ is the limit (69) for $\delta^2 = 3$,

$$|\cos\theta| = [(12)^{1/2} - 3]^{1/2} = 0.684,$$

or an angle of 33° from the equatorial plane. This is valid at all *r*.

For an observer in the equatorial plane, almost all the photons he sees coming from a source at $\xi_e \ll 1$ in the extreme Kerr metric have an impact parameter $\alpha = -2m$ and an impact parameter β in the range $-3^{1/2}m < \beta < 3^{1/2}m$. *Any* such source will thus appear as a thin line of length $(12)^{1/2}$ parallel to the axis of symmetry. The brightness distribution along the line will generally be very sharply peaked at the equatorial plane, both because the average frequency shift factor *g*

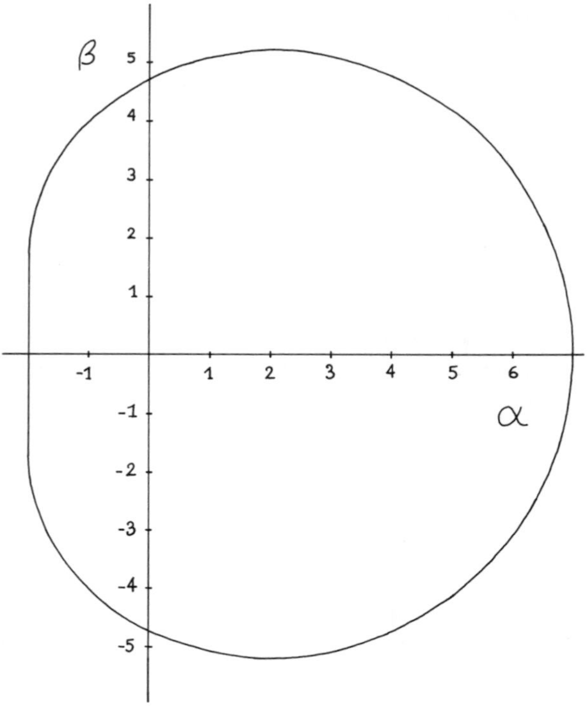

Figure 6. The apparent shape of an extreme ($a = m$) Kerr black hole as seen by a distant observer in the equatorial plane, if the black hole is in front of a source of illumination with an angular size larger than that of the black hole.

is largest there and because of the gravitational focusing effects associated with the bending of the rays toward the equatorial plane. Note that the radiation comes out along the flat portion of the apparent boundary of the extreme black hole as plotted in Figure 6.

D. Geometrical Optics

A detailed calculation of the brightness distribution coming from a source near a Kerr black hole requires more of geometrical optics than the calculation of photon trajectories. I will now review some techniques which are useful in making astrophysical calculations in connection with black holes.

The fundamental principle can be expressed as the conservation of photon density in phase space along each photon trajectory. A phase space element $d^3x\, d^3p$ the product of a proper spatial volume element and a physical momentum-space volume element in a local observer's frame of reference, is a Lorentz invariant, so the particular choice of local observer is arbitrary. The density $N(x^a, p^{(\beta)})$ is defined as a function of position x^a in spacetime and the momentum three-vector of the photon in the local frame, such that $N(x^a, p^{(\beta)})\, d^3x\, d^3p$ is the number of photons in $d^3x\, d^3p$. It is a general relativistic, as well as a Lorentz, scalar.

A more conventional description of the radiation field in astrophysics is by means of the specific intensity $I_\nu(x^a, \nu, n^{(\beta)})$, the energy flux per unit frequency range per sterradian in the beam of radiation with frequency ν and direction along the unit three-vector $n^{(\beta)}$ in some local observer's frame of reference. The specific intensity is the surface brightness per unit frequency range the local observer sees looking backward along the beam of radiation. Within a universal constant of proportionality,

$$I_\nu = N\nu^3 \tag{70}$$

for a given beam of radiation. Therefore, I_ν/ν^3 is the same for any local observer along the beam. Since ν can vary along the beam and with the choice of local observers, I_ν itself is not necessarily constant.

If detailed information about the frequency distribution of the radiation is not required, it may be convenient to work with the intensity I of the radiation defined by

$$I = \int_0^\infty I_\nu\, d\nu. \tag{71}$$

Then I/ν^4 is constant along each beam of radiation as long as there is no scattering, absorption, or emission of radiation out of or into the beam. Such processes require that matter be present.

If the source of the radiation is a concentration of matter which radiates like a blackbody at a temperature T, then I_ν in the rest frame of the matter is just the Planck function $B_\nu(T)$ for every ray leaving the surface.

Now consider a localized source of radiation which is in a circular orbit at a radius r_s in the equatorial plane of a Kerr metric. The source emits radiation isotropically and with luminosity L_s in its own rest frame. The source might be a spot on a disk of accreting matter or it might be a star orbiting the black hole. Let an observer be in a direction represented by the polar angles θ_o, ϕ_o of the Kerr metric, at a radius $r_o \gg m$. The apparent luminosity per unit solid angle at $\theta = \theta_o$, $(dL/d\Omega)_o$, is the energy flux times r_o^2.

The total energy flux is the sum over all the images of the product of the intensity I and the apparent angular size for each individual image seen by the observer. There are an infinite number of such images, since photon trajectories can circle the black hole any number of times before reaching the observer.

If only the time-average of $(dL/d\Omega)_o$ is required, a simple graphical method can be used. For each direction of emission in the local rest frame of the source, let Θ be the angle of the beam with the $+\phi$-direction and let Ψ be the angle with the outward radial direction as projected into the plane perpendicular to the ϕ-direction. The conserved parameters of the photon trajectory are then

$$\lambda = \frac{e^{\psi - \nu}(\cos\Theta + v_s)}{1 + e^{\psi - \nu}(\Omega_s \cos\Theta + \omega v_s)}, \tag{72}$$

$$\eta^{1/2} = r_s e^{-\nu}(1 - \Omega_s\lambda)\sin\Theta\sin\Psi, \tag{73}$$

where Ω_s and v_s are the angular velocity of the source with respect to infinity and the velocity of rotation in the local observer's frame, respectively. Use equation (39) to calculate the asymptotic value of θ at infinity, θ_o, as a function of direction of emission.

The average rate at which photons reach infinity between θ_o and $\theta_o + d\theta_o$ is the rate of emission per unit solid angle in the rest frame of the source, times the solid angle in the rest frame for which the asymptotic polar angle θ_o is in the required range, and corrected for the difference in average time scales between the rest frame and infinity. The time dilation correction factor is just $(u^t)^{-1} = e^{\nu}(1 - v^2)^{1/2}$, where $u^t = dt/ds$ is the time component of the four-velocity of the matter.

The ratio of $(dL/d\Omega)_o$ to the luminosity per unit solid angle in the rest frame is just the ratio of the photon fluxes times the ratio of photon energy at infinity to photon energy in the rest frame. This frequency shift factor is

$$g = g(\Theta) = (1 - v_s^2)^{-1/2}(e^{\nu} + \omega e^{\psi} v_s + \Omega_s e^{\psi}\cos\Theta). \tag{74}$$

Now, plot curves of constant $\cos\theta_o$ on a projection of the sphere of initial photon directions in which Ψ becomes the plane polar angle coordinate and the radius coordinate R is related to Θ by

$$R^2 = 2\int_0^{\Theta} g(\Theta)\sin\Theta\, d\Theta. \tag{75}$$

In the projection planar area is equal to solid angle in the rest frame of the source, weighted by the frequency shift factor g. The planar area between two values of $\cos\theta_o$, times $L_s/(4\pi u^t)$, is the number of ergs per second reaching infinity within that range of directions, which has a solid angle $2\pi\, d(\cos\theta_o)$.

Actually, for each value of $\cos\theta_o$ there are an infinite number of disjoint closed curves in such a diagram, corresponding to photon trajectories which cross the equatorial plane 0, 1, 2, 3, . . ., etc. times between the source and the observer. Normally most of the area is taken up by the zero and perhaps the one-orbit trajectories, but in the limit that the source radius r_s is close to m in the extreme Kerr metric the energy flux at infinity is dominated by photons which have circled the black hole a large number of times before escaping, as we have seen in the last section.

Of course, if the source is part of an optically thick disk, photons whose trajectories intersect the disk will be reabsorbed and will not reach infinity.

Cunningham and Bardeen (1972) have made numerical calculations of this sort for isolated sources orbiting in the extreme Kerr metric. They find that when the source is moderately close to the horizon, say at $r_s = 1.5m$, something like 50 per cent of the energy reaching infinity comes out within 5 degrees of the equatorial plane.

Somewhat different techniques have been used by Campbell and Matzner (1972) to analyze the focusing of radiation from a localized source orbiting in the Schwarzschild metric. The time dependence of $(dL/d\Omega)_o$ and the motion of the images in the plane of the sky is discussed by Cunningham and Bardeen (1972). Some rather bizarre effects occur in the Kerr metric. Images are created and destroyed in pairs, and most of the radiation comes out in one sharp pulse every period when the source is moderately close to the black hole and the observer is close to the equatorial plane.

Bibliography

Bardeen, J. M. (1970) *Astrophys. J.* **162,** 71.
Bardeen, J. M., Press, W. H., and Teukolsky, S. (1972) *Astrophys. J.* **178,** 347.
Campbell, G. A. and Matzner, R. A. (1972) preprint.
Carter, B. (1968) *Phys. Rev.* **174,** 1559.
Cunningham, C. T. and Bardeen, J. M. (1972) *Astrophys. J.* **173,** L 137.
de Felice, F. (1968) *Nuovo Cimento* **57B,** 351.
Godfrey, B. (1970) *Phys. Rev.* **D1,** 2721.
Misner, C. W. (1972) *Phys. Rev. Letters* **28,** 994.
Ruffini, R. and Wheeler, J. A. (1970) in *The Significance of Space Research for Fundamental Physics,* ed. by A. F. Moore and V. Hardy (European Space Research Organization, Paris).
Wilkins, D. C. (1972) *Phys. Rev.* **D5,** 814.

Rapidly Rotating Stars, Disks, and Black Holes

James M. Bardeen

Department of Physics
Yale University

Contents

Introduction 245
I The Equations 246
II Boundary Conditions 249
III The Angular Momentum and Mass 253
IV The Structure of the Matter Configuration 261
V Stability 266
VI Uniformly Rotating Disks 271
Bibliography 289

Introduction

The remainder of my lectures will be devoted to a discussion of rapidly rotating stars and disks in general relativity. By rapidly rotating I mean configurations in which the main support against gravitation is centrifugal force, as opposed to pressure gradient forces. What I hope to impress upon you is that rapidly rotating configurations offer an avenue of approach to black holes which is qualitatively different from the much more intensively studied and much better understood approach of spherically symmetric gravitational collapse, pioneered by Oppenheimer and Snyder (1939).

In the case of spherical or slowly rotating stars the gravitational field can never become extremely strong at the surface as long as the star is in equilibrium. In the extreme relativistic limit the gravitational potential becomes infinite at the center of the star while it is still finite at the surface. Furthermore, for any realistic equation of state the star becomes unstable to gravitational collapse before the gravitational potential is large compared with unity anywhere in the star. The exterior metric is unaffected by spherical collapse, and the process of formation of a Schwarzschild black hole has been followed in great detail (see Harrison *et al.* 1965).

In contrast, equilibrium models of rapidly rotating configurations can exist in which the gravitational potential is arbitrarily large everywhere in the configuration. A true dynamic collapse is almost impossible to calculate in the presence of rotation, but the quasi-stationary contraction of a rotating disk is represented by a sequence of equilibrium models. We will see how the extreme Kerr geometry forms around the disk in the limit that the disk becomes infinitely relativistic.

Rapidly rotating relativistic configurations are also of astrophysical interest. The only way a supermassive star can have a large enough binding energy relative to its rest mass energy to be a possible energy source for a QSO is if it is in rapid rotation (Fowler 1966). When a rapidly rotating supermassive star collapses to form a black hole, a significant fraction of its mass is likely to be left behind in a disk around the black hole, where it can continue to release energy as it loses angular momentum and gradually spirals into the black hole. We will discuss the structure of both completely self-gravitating stars and the structure of disks around black holes. We pay particular attention to the gravitational interaction of the disk and the black hole and its effect on the energy of the black hole.

Disks around black holes are also important in models of accretion into black holes in binary systems. Here the mass in the disk is necessarily very small compared with the diameter of the black hole. The fluid elements in the disk follow approximately geodesic trajectories in the background metric of the black hole.

While for spherical and slowly rotating stars the dominant instability is to overall gravitational collapse, for rapidly rotating stars fragmentation instabilities are perhaps more important, even when the stars are highly relativistic. We will review the various types of fragmentation instabilities in Newtonian theory, and briefly mention some aspects of the stability problem in general relativity.

The discussion of general properties of first the structure and then the stability of rapidly rotating stars will precede the description of the numerical results for the quasi-stationary contraction of a uniformly rotating disk. The basic reference for this material is the paper of Bardeen and Wagoner (1971) on the uniformly rotating disks. Other basic information on the structure of rapidly rotating stars is contained in Bardeen (1970b). Thorne (1971) gives a review of some properties of rotating stars, but oriented primarily toward slow rotation.

I The Equations

As Carter has shown in his lectures, the line element of the gravitational field generated by a stationary, axisymmetric star with purely rotational fluid motions can be put in a form which is explicitly symmetric under simultaneous change of sign of ϕ and t, the coordinates associated with the commuting time and axial Killing vectors, k^a and m^a. By transformation of the remaining two coordinates among themselves the line element can always be reduced to the standard form

$$ds^2 = -e^{2\nu}\,dt^2 + e^{2\psi}(d\phi - \omega dt)^2 + e^{2\mu}(d\rho^2 + dz^2). \tag{1.1}$$

The coordinates ρ and z are cylindrical coordinates at the asymptotically flat infinity and $\rho = 0$ is the axis of symmetry.

In vacuum the Einstein equations allow further transformation of ρ and z which can be used to reduce e^{ψ} to

$$e^{\psi} = \rho e^{-\nu}. \tag{1.2}$$

However, as long as the source of the gravitational field has non-negligible pressure a single coordinate mesh of the type (1.1) requires a more general form,

$$e^{\psi} = \rho B e^{-\nu} \tag{1.3}$$

over the entire spacetime, not just the region where the pressure is non-zero.

The metric functions ν, ω, and ψ in the line element (1.1) can be written as invariant combinations of the Killing vectors,

$$e^{2\psi} = m_a m^a, \tag{1.4a}$$

$$\omega = -(k_a m^a)(m_b m^b)^{-1}, \tag{1.4b}$$

$$e^{2\nu} = -k_a k^a + (k_a m^a)^2(m_b m^b)^{-1}. \tag{1.4c}$$

They also have simple physical interpretations. The invariantly defined zero angular momentum local observers (denoted henceforth as ZAMO's), whose world lines are perpendicular to the t = constant spacelike hypersurfaces, find that e^{ψ} is the proper circumferential radius of a circle around the axis of symmetry. The function ω is the angular velocity of the local ZAMO with respect to infinity, and $e^{-\nu}$ is the time dilation factor between the proper time of the ZAMO and coordinate time (proper time at infinity). The ratio of frequency at infinity to frequency measured by the ZAMO for a zero angular momentum photon is e^{ν}.

The energy-momentum tensor of the star or disk is assumed to be that of a perfect fluid,

$$T^{ab} = (\epsilon + p)u^a u^b + p g^{ab}. \tag{1.5}$$

The energy density ϵ and pressure p are considered functions of two independent thermodynamic variables: the baryon density n and the entropy per baryon s. In terms of the angular velocity of the matter relative to infinity, $\Omega = d\phi/dt$, the velocity of rotation measured by the ZAMO is

$$v = (\Omega - \omega)e^{\psi - \nu} \tag{1.6}$$

and

$$u^t = e^{-\nu}(1 - v^2)^{-1/2}, \qquad u^\phi = \Omega u^t. \tag{1.7}$$

The Einstein equations for a line element of the type (1.1) have been written down many times; they take a particularly simple form if they are projected onto the orthonormal tetrad of the ZAMO. Let $\boldsymbol{\nabla}$ and $\boldsymbol{\nabla}\cdot$ be the gradient and divergence operators in the Euclidean 3-space in which ρ, z, and ϕ are cylindrical coordinates. The $R_{(t)(t)}$ Einstein equation becomes

$$B^{-1}\boldsymbol{\nabla}\cdot(B\boldsymbol{\nabla}\nu) = \tfrac{1}{2}\rho^2 B^2 e^{-4\nu}\,\boldsymbol{\nabla}\omega\cdot\boldsymbol{\nabla}\omega + 4\pi e^{2\mu}\left[(\epsilon + p)\frac{1 + v^2}{1 - v^2} + 2p\right]. \tag{1.8}$$

Equation (1.8) is the Raychaudhuri equation for the world lines of the ZAMO's. The acceleration four-vector A_α of a ZAMO has components

$$A_t = A_\phi = 0, \qquad A_\alpha = \nu_{,\alpha}, \tag{1.9}$$

$\alpha = \rho, z$. The square of the shear of the world lines of the ZAMO's is

$$\sigma^2 = \tfrac{1}{2}\sigma_{ab}\sigma^{ab} = \tfrac{1}{4}e^{2\psi - 2\nu}e^{-2\mu}\,\boldsymbol{\nabla}\omega\cdot\boldsymbol{\nabla}\omega. \tag{1.10}$$

Therefore, equation (1.8) can also be written

$$A^a_{;a} = 2\sigma^2 + R_{(t)(t)}. \tag{1.11}$$

In highly relativistic rapidly rotating stars the shear of the ZAMO world lines can predominate over the matter terms as a source for the relativistic gravitational potential ν.

The source for the function $\omega(\rho, z)$ is the angular momentum of the matter, since an axisymmetric gravitational field carries no angular momentum of its own. This can be seen from the $R_{(t)(\phi)}$ Einstein equation

$$\boldsymbol{\nabla}\cdot(\rho^2 B^3 e^{-4\nu}\boldsymbol{\nabla}\omega) = -16\pi\rho B^2 e^{-2\nu}e^{2\mu}(\epsilon + p)\,v\,(1 - v^2)^{-1}. \tag{1.12}$$

The angular momentum per unit coordinate volume is

$$T^t_a m^a(-g)^{1/2} = \rho^2 B^2 e^{-2\nu} e^{2\mu}(\epsilon + p)\, v\,(1 - v^2)^{-1}. \tag{1.13}$$

Hawking, in his discussion of the initial value equations on (t, ϕ)-symmetric hypersurfaces, writes equation (1.12) in terms of a vector field J_α,

$$J_\alpha = \tfrac{1}{2} e^{2\psi} e^{-\nu} \omega_{,\alpha}. \tag{1.14}$$

The metric function $B(\rho, z)$ is obtained from the $R^{(t)}_{(t)} + R^{(\phi)}_{(\phi)}$ Einstein equation,

$$\nabla\cdot(\rho\,\nabla B) = 16\pi\rho B e^{2\mu} p. \tag{1.15}$$

Since this equation is linear in B, it is relatively easy to solve. In particular, if the pressure is effectively zero over an entire space with Euclidean topology, the solution for B is uniquely B = constant.

The remaining metric function μ only appears multiplying the matter source terms in equations (1.8), (1.12), and (1.15), so it decouples from ν, ω, and B in vacuum. Once ν, ω, and B are known μ can be found by quadratures from some combination of the $R_{(\rho)(\rho)} - R_{(z)(z)}$ and the $R_{(\rho)(z)}$ Einstein equations. Define a metric function $\zeta(\rho, z)$ by

$$\zeta = \mu + \nu. \tag{1.16}$$

Then the $R_{(\rho)(\rho)} - R_{(z)(z)}$ Einstein equation becomes

$$\begin{aligned}\rho^{-1}\zeta_{,\rho} &+ B^{-1}[B_{,\rho}\zeta_{,\rho} - B_{,z}\zeta_{,z}]\\ &= \tfrac{1}{2}\rho^{-2}B^{-1}(\rho^2 B_{,\rho})_{,\rho} - \tfrac{1}{2}B^{-1}B_{,zz} + (\nu_{,\rho})^2 - (\nu_{,z})^2\\ &\quad - \tfrac{1}{4}\rho^2 B^2 e^{-4\nu}[(\omega_{,\rho})^2 - (\omega_{,z})^2],\end{aligned} \tag{1.17a}$$

and the $R_{(\rho)(z)}$ Einstein equation becomes

$$\begin{aligned}\rho^{-1}\zeta_{,z} &+ B^{-1}[B_{,\rho}\zeta_{,z} + B_{,z}\zeta_{,\rho}]\\ &= \rho^{-1}B^{-1}(\rho B_{,z})_{,\rho} + 2\nu_{,\rho}\nu_{,z} - \tfrac{1}{2}\rho^2 B^2 e^{-4\nu}\omega_{,\rho}\omega_{,z}.\end{aligned} \tag{1.17b}$$

Each of these equations is a first-order partial differential equation for ζ (and therefore μ); they are guaranteed to be consistent by the Bianchi identities. Matter source terms are excluded from these equations by our assumption of a perfect fluid stress tensor.

Equations (1.8), (1.12), and (1.15) keep the same form under any explicit coordinate transformation, as long as the operators $\nabla\cdot$ and ∇ are reinterpreted appropriately. For instance, it is often convenient to use spherical polar coordinates r, θ, ϕ defined by

$$\rho = r\sin\theta, \qquad z = r\cos\theta. \tag{1.18}$$

Equations (1.17) change form under such a coordinate transformation, but in a straightforward way. The advantage of cylindrical coordinates at this point is that they simplify equations (1.17), but the spherical coordinates are more convenient in discussing boundary conditions.

Besides the four Einstein equations for the four metric functions, we have two non-trivial equations of hydrostatic equilibrium

$$T^a_{\alpha;a} = 0, \tag{1.19}$$

$\alpha = \rho, z$. These can be written out explicitly, but a variational principle for rotating stars (Bardeen 1970b, Abramowicz 1970) leads to a particularly convenient grouping of terms. Let

$$\Phi = \frac{\epsilon + p}{n}(u^t)^{-1} = \frac{\epsilon + p}{n} e^{\nu}(1 - v^2)^{1/2} \tag{1.20}$$

and

$$j = \frac{\epsilon + p}{n} u_a m^a = \frac{\epsilon + p}{n} e^{\psi}\, v\,(1 - v^2)^{-1/2}. \tag{1.21}$$

The quantity Φ is the energy (including rest mass energy) required to inject a baryon into the star with zero angular momentum, and j is the angular momentum per baryon. The energy required to inject a baryon into a ring with angular velocity Ω and bring it into equilibrium with its surroundings is

$$\Phi + \Omega j = \frac{\epsilon + p}{n}(-u_t) = \frac{\epsilon + p}{n}(-u_a k^a). \tag{1.22}$$

The injection energy for general stationary flows is discussed in the lectures by Thorne.

In general, Φ, Ω and the entropy per baryon s are functions of position in the star and the equations of hydrostatic equilibrium are

$$\nabla\Phi + j\nabla\Omega - T(u^t)^{-1}\nabla s = 0. \tag{1.23}$$

The quantity T is the temperature of a fluid element, defined by

$$T = (\partial\epsilon/\partial s)_n.$$

If the equation of state depends on composition variables as well as the baryon density and entropy, additional terms will appear in equation (1.23), since $T\nabla s = \nabla(\epsilon/n) + p\nabla(1/n)$ only if the composition is uniform.

An immediate result of equation (1.23) is the general relativistic version of rotation on cylindrical shells. If the star is isentropic the surfaces of constant Φ, Ω, and j must coincide. Only in the Newtonian limit are these the coordinate surfaces ρ = constant. The result holds as long as $\nabla\epsilon$ and ∇p are parallel.

II Boundary Conditions

Equations (1.8), (1.11), (1.15), (1.17), and (1.23), plus the equation state $\epsilon = \epsilon(n, s)$, form a complete set. However, to solve them we need boundary conditions. The boundary conditions for equations (1.23) are simply that $p = 0$ at the surface of the matter configuration. In the boundary conditions for the Einstein equations we will allow for the possibility that a black hole is present, in which case there must be boundary conditions on the event horizon as well as at the asymptotically flat infinity and on the axis of symmetry. We will work with the spherical coordinates defined by equation (1.18).

In the limit $r \to \infty$ asymptotic flatness implies that

$$\nu, \omega, \mu \to 0 \tag{2.1a}$$

and

$$B \to 1. \tag{2.1b}$$

The vacuum equations for ν, ω, and B in the weak field limit give

$$\nu = -M/r + \theta(r^{-2}), \tag{2.2}$$

which defines the gravitational mass M,

$$\omega = 2J/r^3 + \theta(r^{-4}), \tag{2.3}$$

which defines the angular momentum J, and

$$B = 1 + \theta(r^{-2}). \tag{2.4}$$

We then conclude from equations (1.17) that

$$\zeta = \theta(r^{-2}). \tag{2.5}$$

On the axis of symmetry, ν, ω, and B must be regular with zero gradients normal to the axis. Equation (1.17a) then guarantees that the same is true for the function ζ (or μ). The constant of integration in ζ is fixed by the requirement of local flatness on the axis, that the linear radius of a circle around the axis of symmetry, $e^{\mu}\delta\rho$, must equal the circumferential radius, $Be^{-\nu}\delta\rho$. On the axis, then,

$$e^{\zeta} = B. \tag{2.6}$$

As long as every z = constant line intersects the axis of symmetry the boundary condition (2.6) applied in equation (1.17a) clearly fixes ζ everywhere. If a black hole is present, rewriting equations (1.17) in spherical coordinates and integrating

$$\zeta = \int_0^{\theta} (\partial\zeta/\partial\theta)\, d\theta + \ln B\,|_{\theta=0} \tag{2.7}$$

determines ζ uniquely at all points outside the event horizon (see below). The functions ν, ω, B, and ζ will be even in $\cos\theta$ for a perfect fluid matter distribution

The remaining boundary conditions are on the functions ν, ω, and B at the inner edge of the range in r. If no black hole is present the usual regularity conditions at the origin of a spherical coordinate system apply at $r = 0$.

When a black hole is present, we rely on the general properties of stationary, axisymmetric event horizons as discussed in the lectures of Hawking and Carter.

Without loss of generality, the coordinate locus of the horizon can be made a sphere of constant radius. The value of the coordinate radius of the horizon,

$$r = h/2 \tag{2.8}$$

is a free parameter of the black hole. In this way we remove the ambiguity in coordinates associated with transformations which leave the form of the line element (1.1) unchanged. Mere regularity of B suffices when a black hole is not present because $r = 0$ is a singular point of the equation for B.

The metric coefficient $e^{2\nu}$ is zero on the horizon. The metric coefficients $e^{2\psi}/\sin^2\theta$ and $r^2e^{2\mu}$ must be regular positive functions of $\cos\theta$ on the horizon, in order that the intersection of the horizon with a spacelike hypersurface have a regular two-geometry. We conclude that

$$r^2B^2 = e^{2\nu+2\psi}/\sin^2\theta = 0 \tag{2.9}$$

on the horizon.

Immediately outside the horizon the space must be vacuum, since any stationary matter configuration would have to resist an infinite gravitational acceleration on the horizon. The general vacuum solution for B consistent with equation (2.9) is

$$B(r,\theta) = \sum_{l=0}^{\infty} b_l r^l [1 - (h^2/4r^2)^{l+1}] T_l^{1/2}(\cos\theta). \tag{2.10}$$

The $T_l^{1/2}(\cos\theta)$ are Gegenbauer polynomials, with $T_0^{1/2}(\cos\theta) = 1$. If there is no matter with significant pressure between the horizon and infinity, the boundary condition at infinity gives $b_0 = 1$ and $b_l = 0$, $l > 0$. However, if a source for B is present outside the horizon the b_l should be adjusted to compensate for it.

Now define a new radius variable λ by

$$r - h^2/4r = (\lambda^2 - h^2)^{1/2} \tag{2.11}$$

or

$$\lambda = r + h^2/4r. \tag{2.11b}$$

One can easily verify that

$$rB = (\lambda^2 - h^2)^{1/2} B_R(\lambda, \cos\theta) \tag{2.12}$$

where $B_R(\lambda, \cos\theta)$ is a regular function of λ and $\cos\theta$ at $\lambda = h$. In particular,

$$B_R(h, \cos\theta) = \sum_{l=0}^{\infty} (l+1) b_l (h/2)^l T_l^{1/2}(\cos\theta). \tag{2.13}$$

Carter (1972) shows that necessary and sufficient conditions for regularity of the event horizon are that $e^{\psi}/\sin\theta$ and ω must be regular functions of λ and $\cos\theta$ in the neighborhood of the horizon, with

$$\left.\frac{1}{r}\frac{\partial\omega}{\partial\theta}\right|_{\lambda=h} = 0. \tag{2.14}$$

Therefore,

$$\nu_R = \nu - \tfrac{1}{2} \ln (\lambda^2 - h^2) \tag{2.15}$$

is a regular function of λ and $\cos \theta$ near the horizon. In terms of ν_R and B_R,

$$e^{\psi} = B_R e^{-\nu_R} \sin \theta . \tag{2.16}$$

Carter restricts himself to non-degenerate horizons, for which $h > 0$, but the regularity conditions in the above form also apply when $h = 0$ and $\lambda = r$. Equations (1.17) guarantee that (re^{μ}) is a regular function of λ and $\cos \theta$ near the horizon if ν, ω, and B satisfy the above conditions.

Hawking (Hawking and Ellis (1972)) and Carter (see his lectures) have shown that certain quantities are necessarily constant over the horizon. One of these is the angular velocity of the black hole,

$$\Omega_H = \omega |_{\lambda=h} . \tag{2.17}$$

Another is a quantity κ, which is denoted by ϵ in Hawking's lectures. It is the rescaled gravitational acceleration of a ZAMO on the horizon. The actual acceleration felt by a ZAMO is

$$(A_a A^a)^{1/2} = e^{-\mu} [(\nu_{,r})^2 + r^{-2} (\nu_{,\theta})^2]^{1/2} . \tag{2.18}$$

The physical acceleration is per unit proper time of the ZAMO; when it is rescaled to per unit coordinate time (proper time at infinity) by a factor e^{ν}, the limit on the horizon is

$$\kappa_H = (e^{\nu})_{,r} e^{-\mu} . \tag{2.19}$$

From equation (2.15),

$$\kappa_H = h e^{\nu_R} / (re^{\mu}) = 2 e^{\nu_R} e^{-\mu} . \tag{2.20}$$

This is a constant, independent of $\cos \theta$.

The black hole has two free parameters of its own, which may be specified in addition to the matter distribution around the black hole. The values of these two parameters may be considered part of the boundary conditions on the horizon. We have already said that it is mathematically convenient to take h as one of these parameters. The other parameter could be either Ω_H or κ_H. In the next section we shall see that the physically most appropriate parameters are the surface area of the black hole and its angular momentum.

To sum up, a complete set of boundary conditions on the horizon, one each for equations (1.8), (1.11), and (1.15), is that at $r = h/2$

(1) $e^{2\nu} = 0$
(2) $\omega = \Omega_H$
(3) $rB = 0$.

The regularity conditions follow naturally out of solving the equations with these boundary conditions.

In terms of the mass m and angular momentum per unit mass a of the Kerr metric, the horizon parameters of a Kerr black hole are $h = (m^2 - a^2)^{1/2}$, $\Omega_H = a/[2m(m + (m^2 - a^2)^{1/2})]$, and $\kappa_H = (m^2 - a^2)^{1/2}/[2m(m + (m^2 - a^2)^{1/2})]$.

III The Angular Momentum and Mass

The angular momentum and gravitational mass of the system as a whole are defined by the asymptotic behavior of the geometry at infinity. We will now obtain explicit integral formulas for these quantities, with particular attention to the case that a black hole is present.

The conventional approach to the angular momentum and gravitational mass of stationary, axisymmetric systems uses the fact that the contraction of the Ricci tensor with a Killing vector is a pure divergence. For the angular momentum, the relevant Killing vector is the axial Killing vector m^a. Consider an integral over a t = constant spacelike hypersurface

$$\int R_a^b m^a \, d\Sigma_b = \int R_\phi^t (-g)^{1/2} \, d^3x = 8\pi \int T_\phi^t (-g)^{1/2} \, d^3x. \tag{3.1}$$

This is equivalent to integrating equation (1.12). The volume integral over the divergence is converted into two surface integrals, one at infinity and one on the horizon, of the form

$$S = \int_0^\pi d\theta \, \sin^3 \theta r^4 B^3 e^{-4\nu} \partial\omega/\partial r \tag{3.2}$$

At infinity

$$\partial\omega/\partial r \simeq -6Jr^{-4}, \tag{3.3}$$

so

$$S_\infty = -8J. \tag{3.4}$$

The volume integral of the right-hand-side of equation (1.12) is

$$-16\pi \int\int r^2 \sin\theta \, dr \, d\theta B e^{2\mu} e^{\psi} (\epsilon + p) v(1 - v^2)^{-1} = -8J_M. \tag{3.5}$$

where J_M is the angular momentum of the matter outside the hole. It is natural to define the angular momentum of the black hole from the surface integral on the horizon, such that

$$J_H = J - J_M = -\tfrac{1}{8} S_H. \tag{3.6}$$

In terms of quantities found to be regular on the horizon in the last section,

$$J_H = -\tfrac{1}{8} \int_0^\pi d\theta \, \sin^3 \theta B_R^3 e^{-4\nu_R} \partial\omega/\partial\lambda \,|_{\lambda=h}. \tag{3.7}$$

The angular momentum of the black hole is completely unambiguous and is conserved in any axisymmetric dynamic process in which no matter crosses the horizon.

For the gravitational mass we have the relation

$$\int R^b_a k^a \, d\Sigma_b = \int R^t_t (-g)^{1/2} \, d^3x = 8\pi \int (T^t_t - \tfrac{1}{2} T^a_a)(-g)^{1/2} \, d^3x. \tag{3.8}$$

This involves a combination of equations (1.8) and (1.12),

$$\begin{aligned} &\int\int r^2 \sin\theta \, dr \, d\theta \, \nabla \cdot [B \nabla \nu - \tfrac{1}{2} r^2 \sin^2\theta B^3 e^{-4\nu} \omega \nabla \omega] \\ &\quad = 4\pi \int\int r^2 \sin\theta \, dr \, d\theta B e^{2\mu} \left[(\epsilon + p) \frac{1 + v^2}{1 - v^2} + 2p \right. \\ &\quad\quad \left. + 2(\epsilon + p) \frac{v \omega e^{\psi - \nu}}{1 - v^2} \right]. \end{aligned} \tag{3.9}$$

Now the surface integrals have the form

$$S = \int_0^\pi \sin\theta \, d\theta \left[r^2 B \frac{\partial \nu}{\partial r} - \tfrac{1}{2} \sin^2\theta r^4 B^3 e^{-4\nu} \omega \partial\omega/\partial r \right] \tag{3.10}$$

At infinity the $\partial\omega/\partial r$ term is negligible compared with the $\partial\nu/\partial r$ term, and from

$$\partial\nu/\partial r \simeq M r^{-2}, \tag{3.11}$$

$$S_\infty = 2M. \tag{3.12}$$

In the surface integral on the horizon

$$r^2 B \partial\nu/\partial r = r B (\lambda^2 - h^2)^{1/2} \partial\nu/\partial\lambda = h B_R \tag{3.13}$$

and $\omega = \Omega_H$. From equation (2.20),

$$\int_0^\pi \sin\theta \, d\theta r^2 B \frac{\partial\nu}{\partial r} = \frac{\kappa_H}{2\pi} \int_0^\pi 2\pi \, d\theta \, e^\psi (r e^\mu) = \frac{1}{2\pi} \kappa_H \mathscr{A}_H \tag{3.14}$$

where $\mathscr{A}_H$ is the surface area of the black hole. Thus

$$M = \frac{1}{4\pi} \kappa_H \mathscr{A}_H + 2\Omega_H J_H + 2\pi \int\int r^2 \sin\theta \, dr \, d\theta B e^{2\mu} [\epsilon + 3p + 2nu^t \Omega j] \tag{3.15}$$

In the vacuum case the mass of the black hole is unambiguously

$$M_H = \frac{1}{4\pi} \kappa_H \mathscr{A}_H + 2\Omega_H J_H. \tag{3.16}$$

Smarr (1972) first showed that the mass of a Kerr black hole could be put in this form. Equation (3.15) was derived by Bardeen, Carter, and Hawking (1972) and generalized to include electromagnetic fields by Carter.

Even when a ring of matter is present around the black hole, the hole's gravitational mass may be defined to be the surface integral on the horizon in the form (3.16). Just as an axisymmetric gravitational field can carry no angular momentum, one may say that a stationary gravitational field has no energy of its own, as opposed to the gravitational potential energy of the matter.

There is an important difference between the angular momentum and gravitational mass. Consider a black hole with mass M_H and angular momentum J_H. If an axisymmetric ring of matter is placed around the hole, the angular momentum of the hole is the same as before, but the mass M_H of the hole given by equation (3.16) changes by an amount first order in the mass of the ring. The net amount of energy required to add the ring is *not* given by the volume integral in equation (3.15). The net amount of energy required to add the ring is astrophysically very important, because subtracted from the rest mass energy of the ring it gives the energy released in an accretion process.

By rearranging terms in the volume integral, equation (3.15) can be rewritten

$$M = \frac{1}{4\pi}\kappa_H \mathscr{A}_H + 2\Omega_H J_H + \iint \Phi\, dN + 2\iint \Omega\, dJ + 2\iint (p/n)(u^t)^{-1}\, dN \tag{3.17}$$

(see Bardeen and Wagoner 1971). The quantities dN and dJ are the number of baryons and the angular momentum in a ring of matter,

$$dN = 2\pi n u^t B e^{2\mu} r^2 \sin\theta\, dr\, d\theta, \tag{3.18}$$

$$dJ = j\, dN, \tag{3.19}$$

and Φ and j are given by equations (1.20) and (1.21). If an infinitesimal ring with negligible pressure is added around a pre-existing configuration (either star or black hole), equation (3.17) would suggest (by neglecting the change in the mass of the black hole or star already present) that the net energy increase would be

$$\delta M = \Phi_{\rm ring}\delta N_{\rm ring} + 2\Omega_{\rm ring}\delta J_{\rm ring}. \tag{3.20}$$

In fact, at least if the original configuration is a star, the actual net energy increase is (Bardeen 1970b)

$$\delta M = \Phi_{\rm ring}\delta N_{\rm ring} + \Omega_{\rm ring}\delta J_{\rm ring} = (\epsilon/n)(-u_t)\delta N_{\rm ring} \tag{3.21}$$

Note that $(\epsilon/n)(-u_t)\delta N_{\rm ring}$ is just the energy parameter for a test particle in a circular geodesic orbit.

The physical meaning of this apparent paradox is transparent in Newtonian theory. Consider a central mass which initially has a total mass-energy M. Add an infinitesimal ring with mass m at radius r. The gravitational potential energy of the ring in the field of the large mass is $-mM/r$, its kinetic energy is $\frac{1}{2}mv^2$, and the new total energy is

$$M_{\rm tot} = M + m - mM/r + \tfrac{1}{2}mv^2.$$

However, this assigns all of the gravitational potential energy of interaction to the small mass. In calculating the gravitational potential energy of a single self-gravitating system one usually splits the potential energy of interaction between pairs of mass elements evenly. Furthermore, the relative contribution to the total energy from potential energy terms, rotational energy terms, and internal energy terms can be redistributed through the Newtonian virial theorem,

$$3\int p\, d^3x + 2E_{\rm rot} + E_{\rm grav} = 0.$$

The conventional Newtonian energy formula assigns a mass energy to the central mass of $M - \frac{1}{2}Mm/r$ and a mass-energy to the ring of $m - \frac{1}{2}Mm/r + \frac{1}{2}mv^2$. The Newtonian limit of equation (3.20) assigns a mass energy to the ring of $m - Mm/r + \frac{3}{2}mv^2$, or $M - \frac{1}{2}Mm/r + \frac{1}{2}mv^2$ plus $E_{\rm grav} + 2E_{\rm rot}$. (The pressure is assumed to be negligible in the ring.)

The decrease in the energy assigned to the black hole which compensates for the excess energy assigned to the ring in equation (3.20) over the amount actually required to add the ring is accomplished by changes in κ_H and Ω_H. As long as the ring is added slowly (reversibly) and axisymmetrically, the area $\mathscr{A}_H$ and the angular momentum J_H of the hole are constant.

To verify that equation (3.21) continues to hold when a black hole is present, and more importantly to find a general expression for the difference in mass between any two neighboring stationary, axisymmetric configurations, including changes in the black hole parameters due to matter or radiation falling in, we generalize the variational principle of Bardeen (1970b) to allow for the presence of a black hole. The action for the variational principle is constructed as an integral over a (t, ϕ)-symmetric spacelike hypersurface,

$$I_1 = \iint (\mathscr{L}_G + \mathscr{L}_M)\, d^3x. \tag{3.22}$$

The Lagrangian density for the matter is

$$\mathscr{L}_M = -\, T^t_t(-g)^{1/2}, \tag{3.23}$$

and the Lagrangian density for the gravitational field is

$$\mathscr{L}_G = -\frac{1}{16\pi} R^*(-g)^{1/2} + W^\alpha{}_{,\alpha}, \tag{3.24}$$

where R^* is the curvature scalar on the (t, ϕ)-symmetric hypersurface with all time derivatives of the metric functions omitted. The quantities W^α are chosen so that (1) $\mathscr{L}_G$ contains at most first derivatives of the metric functions; (2) the integral in I_1 converges automatically if the metric functions have the appropriate asymptotic behavior at infinity, and (3) if the initial value equations are satisfied on the momentarily stationary hypersurface, I_1 can be written as the difference of a surface integral at infinity and a surface integral on the horizon, with the surface integral at infinity equal to the total gravitational mass. The gravitational mass is defined from the asymptotic behavior of the spatial metric, since the metric function ν is not well defined unless the spacetime is globally stationary.

The metric functions in the variational principle are defined by equation (1.1), which is a general form on a momentarily stationary hypersurface, except that the spherical coordinates r, θ are used in place of the cylindrical coordinates ρ, z and the metric coefficient e^{ψ} is represented as

$$e^{\psi} = r \sin \theta e^{\sigma}. \tag{3.25}$$

The explicit expression one obtains for $\mathscr{L}_G$ is

$$\begin{aligned}\mathscr{L}_G = -\frac{1}{8\pi} r^2 \sin \theta \{&(e^{\sigma})_{,r}[(e^{\nu})_{,r} - r^{-1}e^{\nu}] \\ &+ \mu_{,r}[e^{\sigma}(e^{\nu})_{,r} + e^{\nu}(e^{\sigma})_{,r} + r^{-1}e^{\nu+\sigma}] \\ &+ r^{-2}(e^{\sigma})_{,\theta}[(e^{\nu})_{,\theta} - \cot \theta e^{\nu}] \\ &+ r^{-2}\mu_{,\theta}[e^{\sigma}(e^{\nu})_{,\theta} + e^{\nu}(e^{\sigma})_{,\theta} + \cot \theta e^{\nu+\sigma}] \\ &+ \tfrac{1}{4} r^2 \sin^2 \theta e^{3\sigma-\nu}[\omega_{,r}\omega_{,r} + r^{-2}\omega_{,\theta}\omega_{,\theta}]\}.\end{aligned} \tag{3.26}$$

A less restricted form of the line element than (1.1) gives a more complicated expression for $\mathscr{L}_G$ (see Bardeen 1970b). The W^{α} which lead to the expression (3.26) are

$$W^r = -\frac{1}{8\pi} r^2 \sin \theta e^{-\mu}(e^{\nu+\sigma+\mu})_{,r} \tag{3.27a}$$

and

$$W^{\theta} = -\frac{1}{8\pi} \sin \theta e^{-\mu}(e^{\nu+\sigma+\mu})_{,\theta}. \tag{3.27b}$$

The statement of the variational principle is as follows. The action I_1 is stationary under arbitrary, independent variations of the metric functions ν, ω, σ, μ in any bounded region of the spacelike hypersurface (outside the horizon) and of the positions of rings of matter containing specified amounts of baryons, angular momentum, and entropy if and only if ν, ω, σ, μ satisfy four independent Einstein equations in that region and the matter distribution satisfies the equation of hydrostatic equilibrium.

We want to use the variational principle to compare neighboring equilibrium configurations in which corresponding rings of matter contain different amounts of baryons, angular momentum, and entropy. Therefore, we relax the explicit constraints on these quantities in the variational principle by introducing Lagrange multipliers Ψ, for the baryon number dN in the ring, Λ, for the angular momentum dJ in the ring, and $\mathscr{T}$ for the entropy dS in the ring (see equations (3.18) and (3.19)). Note that

$$dS = s\, dN. \tag{3.28}$$

The Lagrange multipliers will in general be different for different rings in the matter configuration. The new form of the action is

$$I_2 = I_1 - \int\int \Psi\, dN - \int\int \Lambda\, dJ - \int\int \mathcal{T} dS. \tag{3.29}$$

The action I_2 is stationary under the same variations of the metric functions and the location of the rings as the action I_1, but it is also stationary under arbitrary variations of the baryon density n, the angular velocity Ω, and the temperature T. When evaluated for a stationary equilibrium configuration the Lagrange multipliers are (Bardeen 1970b)

$$\Psi = [(\epsilon + p)/n - Ts](u^t)^{-1}, \tag{3.30}$$

$$\Lambda = \Omega, \tag{3.31}$$

and

$$\mathcal{T} = T(u^t)^{-1}. \tag{3.32}$$

Now consider two neighboring equilibrium configurations. Denote the difference in baryon number between rings at the same coordinate location in the two configurations and occupying the same coordinate range $dr\, d\theta$ by δdN. Similarly denote the difference in angular momentum by δdJ and the difference in entropy by δdS. The differences in the integrals I_1 and I_2 between the two configurations are ΔI_1 and ΔI_2, respectively. The values of the Lagrange multipliers do in fact change between the two equilibrium configurations, so

$$\begin{aligned}\Delta I_2 = \Delta I_1 &- \int\int \Psi \delta dN - \int\int \Lambda \delta dJ - \int\int \mathcal{T} \delta dS - \int\int \delta\Psi\, dN \\ &- \int\int \delta\Lambda\, dJ - \int\int \delta\mathcal{T} dS.\end{aligned} \tag{3.33}$$

First let us calculate ΔI_1. For each of the equilibrium configurations equations (3.22)-(3.24), plus the initial value equation

$$R^t_t - \tfrac{1}{2}R = R^{*t}_t - \tfrac{1}{2}R^* = 8\pi T^t_t, \tag{3.34}$$

give

$$I_2 = \int\int \left[-\frac{1}{8\pi} R^{*t}_t (-g)^{1/2} + W^\alpha_{,\alpha} \right] d^3x. \tag{3.35}$$

Since $R^{*t}_t(-g)^{1/2}$ is a pure divergence, I_1 can be written as the difference of two surface integrals,

$$I_1 = \tfrac{1}{4} \int_0^\pi r^2 \sin\theta\, d\theta\, [-e^{\nu+\sigma}(\sigma_{,r} + \mu_{,r}) - \tfrac{1}{2}\omega\omega_{,r} r^2 \sin^2\theta\, e^{3\sigma-\nu}] \,\Big|^{r=\infty}_{r=h/2} \tag{3.36}$$

At infinity $\sigma \simeq M/r$, $\mu \simeq M/r$, and the $\omega_{,r}$ term is negligible. On the horizon e^ν is zero, $\omega = \Omega_H$, and re^σ and re^μ are regular. From equation (3.7), then,

$$I_1 = M - \Omega_H J_H \tag{3.37}$$

and

$$\Delta I_1 = \Delta M - J_H \Delta\Omega_H - \Omega_H \Delta J_H. \tag{3.38}$$

Now consider ΔI_2. The variations in n, Ω, and T do not contribute to ΔI_2 in first order, by virtue of the modified variational principle. When the terms in the variation of I_2 containing derivatives of the variations in the metric functions between the two equilibrium configurations are integrated by parts to factor out $\delta\sigma, \delta\mu$, $\delta\omega$, and $\delta(e^\nu)$, the surface integrals on the horizon contribute to ΔI_2. Contrary to the assumption of the variational principle the variations in the metric functions are non-zero there. Over the entire range of coordinates outside the horizons of both spacetimes the remaining volume integral vanishes, because the Einstein equations for a stationary, axisymmetric spacetime are satisfied. There is a residual volume integral contribution to ΔI_2 from the range of coordinate radius which is outside one horizon, but not the other. Finally, there is a contribution to ΔI_2 from the variations in the Lagrange multipliers, which are held constant in the variational principle.

Integrating by parts the radial derivatives of the quantities δe^ν, $\delta\omega$, $\delta\sigma$, and $\delta\mu$, all of which are finite on the horizon as long as $h > 0$, gives a surface integral on the horizon

$$\tfrac{1}{4}\int_0^\pi r^2 \sin\theta\, d\theta\, [e^\sigma(\sigma_{,r} + \mu_{,r})\delta(e^\nu) + e^\sigma(e^\nu)_{,r}(\delta\sigma + \delta\mu) + \tfrac{1}{2} r^2 \sin^2\theta\, e^{3\sigma-\nu}\omega_{,r}\delta\omega]. \tag{3.39}$$

Terms containing a factor of e^ν, which is zero on the horizon, are not included in the expression (3.39). The surface integral at infinity vanishes, due to the asymptotic flatness.

Let $\delta(h/2)$ be the change in the coordinate radius of the horizon. The "Lagrangian" change in e^ν following the horizon is zero, so the "Eulerian" change at the fixed coordinate radius $r = h/2$ is

$$\delta(e^\nu) = -(e^\nu)_{,r}\delta(h/2). \tag{3.40}$$

The condition that the change in proper radius of horizon be small is that $\delta(h/2)$ be small compared with $h/2$, since $\delta(\text{proper radius}) = (re^\mu)\delta r/r$.

The terms in $\mathscr{L}_G$ which are non-zero on the horizon give a residual volume integral

$$\int_{(h+\delta h)/2}^{h/2} dr \int_0^\pi d\theta\, [-\tfrac{1}{4} r^2 \sin\theta (e^\nu)_{,r} e^\sigma(\sigma_{,r} + \mu_{,r})] = \tfrac{1}{4}\int_0^\pi r^2 \sin\theta\, d\theta\, e^\sigma(\sigma_{,r} + \mu_{,r})(e^\nu)_{,r}\delta(h/2). \tag{3.41}$$

This contribution to ΔI_2 cancels the first term in the surface integral contribution (3.39).

The change in area of the horizon is

$$\Delta \mathscr{A}_H = 2\pi \int_0^\pi r^2 \sin\theta \, d\theta e^{\sigma+\mu} \left[\frac{2}{r}\delta r + \delta\sigma + \delta\mu + (\sigma_{,r} + \mu_{,r})\delta r\right]. \tag{3.42}$$

Since re^σ and re^μ are regular functions of the radial coordinate λ, and either $d\lambda/dr = 0$ or $\delta r = 0$ on the horizon,

$$(r^2 e^{\sigma+\mu})_{,r}\delta r = 0 \tag{3.43}$$

on the horizon. The quantity $\kappa_H = (e^\nu)_{,r}e^{-\mu}$ is independent of θ on the horizon, so the second term in the expression (3.39) is $(8\pi)^{-1}\kappa_H \Delta \mathscr{A}_H$.

In the third term of the expression (3.39) $\delta\omega$ is equal to

$$\Delta\Omega_H = \delta\omega + \frac{d\lambda}{dr}\omega_{,\lambda}\delta(h/2), \tag{3.44}$$

again because either $d\lambda/dr = 0$ or $\delta(h/2) = 0$ on the horizon. The third term is just $\Delta\Omega_H$ (which is independent of θ) times $-J_H$.

The final expression for ΔI_2 is

$$\Delta I_2 = \frac{1}{8\pi}\kappa_H \Delta\mathscr{A}_H - J_H \Delta\Omega_H - \iint \delta\Psi \, dN - \iint \delta\Lambda \, dJ - \iint \delta\mathscr{T} dS. \tag{3.45}$$

Now combine equations (3.33), (3.38), and (3.45) to get

$$\Delta M = \frac{1}{8\pi}\kappa_H \Delta\mathscr{A}_H + \Omega_H \Delta J_H + \iint \Psi\delta dN + \iint \Lambda\delta dJ + \iint \mathscr{T}\delta \, dS. \tag{3.46}$$

A more convenient form for some purposes (such as comparing two isentropic stars with the same entropy per baryon) is

$$\Delta M = \frac{1}{8\pi}\kappa_H \Delta\mathscr{A}_H + \Omega_H \Delta J_H + \iint \Phi\delta dN + \iint \Omega\delta dJ + \iint T(u^t)^{-1}\delta s \, dN, \tag{3.47}$$

where we have used equations (3.30)–(3.32) to evaluate the Lagrange multipliers.

Equations (3.46) and (3.47) depend only on the initial and final equilibrium configurations and are independent of any details of a physical process of transition between them. Hartle and Hawking (1972) have previously calculated the contribution to ΔM from $\Delta\mathscr{A}_H$ and ΔJ_H for some specific physical processes.

As it applies to the special case of a Kerr black hole, equation (3.47) has been discussed by Bekenstein (1972). The quantity $\kappa_H/8\pi$ can be interpreted as the "surface tension" of the black hole, or alternatively the area is analogous to entropy and $\kappa_H/8\pi$ is analogous to temperature. Carter shows in his lectures that the thermodynamic analogy can be carried quite far; all four laws of classical thermodynamics correspond to laws of black hole physics.

An alternative derivation of equation (3.46) in more abstract and covariant language, due to Carter and Hawking, can be found in Carter's lectures. Carter has generalized equation (3.46) to include electromagnetic fields and matter with variable composition.

If a number of baryons δdN are *reversibly* injected into a ring of matter and given the same entropy per baryon and angular momentum per baryon as the matter already present, $\Delta\mathscr{A}_H = \Delta J_H = 0$ and the increment in the total energy is

$$\Delta M = (\Phi + \Omega j)\delta dN = \frac{\epsilon + p}{n}(-u_t)\delta dN. \tag{3.48}$$

Equation (3.47) gives directly the net energy required to add a ring of infinitesimal mass around a black hole (or central star) only if the internal energy of the ring is small compared with its gravitational potential energy and rotational energy. When the internal pressure is large enough to play a role in the equilibrium of the ring, the net energy required to add the ring is calculated in the normal way for a body in the fixed background metric of the unperturbed black hole (or central star),

$$\begin{aligned} E_{\text{ring}} &= \iint T^a_b k^b \, d\Sigma_a = \iint (-T^t_t)(-g)^{1/2} \, d^3x \\ &= \iint \Phi \, dN + \iint \Omega \, dJ - \iint \frac{p}{n} (u^t)^{-1} \, dN. \end{aligned} \tag{3.49}$$

Equation (3.49) is the correct formula to use in astrophysical accretion calculations as long as the mass in the disk or ring is small compared with the mass of the central object, so the self-gravitation of the ring can be neglected. Use of equation (3.17) would require an elaborate calculation of metric perturbations to find the changes in κ_H and Ω_H (for a central black hole) produced by the presence of the ring.

IV The Structure of the Matter Configuration

Given a matter configuration, we have seen how to put boundary conditions on the metric functions to obtain a more or less unique solution of the Einstein equations. If a black hole is present, the boundary conditions contain two free parameters characterizing the black hole. The mathematically convenient parameters are the coordinate radius of the horizon $h/2$ and the angular velocity of the hole Ω_H. However, in the last section we have seen that the physically appropriate parameters are the area $\mathscr{A}_H$ and the angular momentum J_H, since these are the parameters that remain constant under the influence of a ring of matter surrounding the hole.

Now we consider what might be appropriate parameters to characterize the matter configuration. At least in a numerical calculation these parameters must be chosen in such a way as to specify a model more or less uniquely. For a uniformly rotating star with a given isentropic equation of state $\epsilon = \epsilon(n)$ there are only two free parameters. The physically appropriate choice, as suggested by the Hartle–Sharp variational principle (Hartle and Sharp 1967), is the total baryon number N and the total angular momentum J. These are good parameters for non-equilibrium axisymmetric configurations as well as equilibrium configurations. However, it is mathematically more straightforward to specify the angular velocity Ω and the zero-angular-momentum injection energy Φ. The integral of the equation of hydrostatic equilibrium is just Φ = constant. Given the metric functions and the values of Ω and Φ, one obtains immediately $(\epsilon + p)/n$ and through the equation of state n as functions of position.

In general differentially rotating stars there are two free functions instead of two free parameters, so things are considerably more complex. Furthermore, certain minimal convective stability criteria should be satisfied. Bardeen (1970b) has argued that the angular momentum per baryon j should not decrease outward from the axis of symmetry on surfaces of constant entropy per baryon s, and that s should not decrease in the same direction that T/u^t decreases (the direction of heat flow) on surfaces of constant j. If either condition is violated mass exchange permitted by conservation of angular momentum and entropy can lower the energy of the star. Of course, some non-axisymmetric convective motions are allowed which do not conserve angular momentum of each fluid element, and on sufficiently long time scales entropy may not be conserved. For either reason the above criteria are only necessary for stability. The Newtonian local convective stability analysis of Goldreich and Schubert (1967) could be adapted to handle some of these questions.

When the star is isentropic, so surfaces of constant j, Ω, and Φ coincide, a physically appropriate specification of the model, as Ostriker has emphasized in a Newtonian context, is to fix the angular momentum per baryon j of each "cylindrical shell" of constant j, Ω, and Φ as a function of the number of baryons $N(j)$ inside the shell. Unfortunately, a rather complicated iterative procedure is necessary to apply this in general relativity, since one doesn't know the surfaces of constant j beforehand.

Mathematically, one wants to simplify as much as possible finding the angular velocity and density at every point of the star given a background gravitational field. The quantity

$$q = j/\Omega\Phi \tag{4.1}$$

is independent of the baryon density n and except in highly relativistic models is rather insensitive to Ω. In the Newtonian limit it is just the square of the cylindrical radius ρ. Therefore, if we specify the functional relation

$$\Omega = F(q) \tag{4.2}$$

and know the metric functions as a function of position in the star, it is a simple matter to solve equation (4.2) for Ω as a function of position in the star. The integral of the equation of hydrostatic equilibrium, equation (1.23), can be put in the form

$$\Phi = \Phi_0 \exp\left[-\int_0^q yF\frac{dF}{dy}dy\right]. \tag{4.3}$$

Given Φ_0, the value of Φ on the axis, as an additional parameter of the star, one has Φ as a function of position in the star and from equation (1.20) and the equation of state $(\epsilon + p)/n$ and n as functions of position in the star. The relation (4.2) gives the rotation law and Φ_0 is a measure of how relativistic the star is.

For a given choice of the relation (4.2) and a given Φ_0 an equilibrium model may not exist. Since $\Phi_0 = m_b e^\nu$ where the axis of symmetry intersects the surface of the star, the range of acceptable values of Φ_0 is $0 < \Phi_0 < m_b$. The quantity m_b is the average rest mass per baryon,

$$m_b = \lim_{p\to 0} [(\epsilon + p)/n]. \tag{4.4}$$

The range of acceptable values of Ω may be quite small for highly relativistic, rapidly rotating stars. The rotational velo ity v is sensitive to small differences in $\Omega - \omega$ when $e^{-\nu}$ is large.

Another possible problem is that q may not increase monotonically away from the axis of symmetry. If it does not, then $F(q)$ in equation (4.2) should be a double-valued function. However, our choice for q minimizes this possibility. The angular momentum j must increase monotonically in a realistic model and from the equation of hydrostatic equilibrium Ω and Φ change in opposite directions. Since

$$\nabla(\Omega\Phi) = (\Phi - \Omega j)\nabla\Omega, \tag{4.5}$$

$\Omega\Phi$ will decrease outward if Ω decreases outward for all except models which are both highly relativistic and rapidly rotating. The choice $q = j$ would avoid the problem completely at the cost of complicating the iteration procedure necessary to find Ω and n as functions of position.

Non-isentropic models introduce the second functional degree of freedom. We will only discuss a limited class of such models, those in which the surfaces of constant s and constant j coincide, so the surfaces of constant j are in neutral convective equilibrium. This class is physically not unreasonable—one expects Coriolis forces to inhibit convection perpendicular to the surfaces of constant j in a rapidly rotating star. Furthermore, it allows an integral of the equation of hydrostatic equilibrium like equation (4.3). Specify the functional relation

$$s = s(j). \tag{4.6}$$

Then equation (1.23) becomes

$$\boldsymbol{\nabla}(\Phi + \Omega j) - \left(\Omega + T(u^t)^{-1} \frac{ds}{dj}\right)\boldsymbol{\nabla} j = 0. \tag{4.7}$$

Surfaces of constant j, $(\Phi + \Omega j)$, and $(\Omega + T(u^t)^{-1}\, ds/dj)$ are now forced to coincide. The rotation law can be given by a relation of the form

$$\Omega + T(u^t)^{-1}\, ds/dj = F(j), \tag{4.8}$$

so that

$$\Phi + \Omega j = \Phi_0 + \int_0^j F(y)\, dy. \tag{4.9}$$

A straightforward interaction process gives Ω, j, Φ, s, T, and n at each point in the star, though it is considerably more complicated than in the isentropic case. Note that $F(j)$ is the angular velocity at the surface of the star, where $T \simeq 0$. In the Newtonian limit, the behavior $ds/dj < 0$, which may be possible in a disk-like configuration, implies that at constant ρ, Ω is larger at the equatorial plane than it is at surface of the disk.

If the matter configuration is a disk or ring around a black hole with mass small compared with the mass of the hole, the background metric is the Kerr metric and the structure of the disk is found by the procedures of this section. However, for a self-gravitating matter configuration the gravitational field is not known in advance. In numerical calculations one may use the self-consistent field method, which has been applied extensively in Newtonian theory by Ostriker and his collaborators (Ostriker and Mark 1968). The technique is to make a guess at the structure of the matter configuration, solve the four Einstein equations in Part I for the metric functions, find the new matter configuration consistent with hydrostatic equilibrium in this gravitational field, and iterate until the process converges. In calculating the new field configuration one can use the old metric functions in the non-linear terms in the Einstein equations, so at each stage one is solving four independent linear partial differential equations.

Actual calculations of this sort for relativistic rotating stars have been carried out by Bonnazola and associates (Bonnazola and Maschio 1970) and by Wilson (1972). Bonnazola has computed uniformly rotating models of neutron stars, and Wilson has computed some rather highly flattened differentially rotating, non-isentropic models. Much more remains to be done.

Some simple analytic results are possible in the limit that the centrifugal forces dominate the pressure forces and the matter configuration takes on a disk shape. When the thickness of the disk is small compared with its diameter the equilibrium perpendicular to the plane decouples from equilibrium in the plane. The energy density ϵ becomes large compared with the pressure p, so $\epsilon \simeq m_b n$ as it is in the Newtonian limit. The difference in the gravitational potentials between a point in the equatorial plane and the adjacent surface of the disk is small. For a self-gravitating disk of mass M and proper equatorial radius R_e the pressure is the order of M^2/R_e^4, just as it is for a spherical star. The energy density ϵ is the order of $(M/R_e^3)(W/R_e)^{-1}$, so the pressure gradient force per unit inertial mass in the plane of the disk,

$$-(\epsilon+p)^{-1}e^{-\mu}\,\nabla p=\mathcal{O}\left(\frac{W}{R_e}\frac{M}{R_e^2}\right), \tag{4.10}$$

is small compared with the gravitational and centrifugal forces. On the other hand, perpendicular to the plane of the disk the centrifugal force is zero and the pressure gradient force balances the gravitational force,

$$-(\epsilon+p)^{-1}e^{-\mu}\,\nabla p=\mathcal{O}\left(\frac{M}{R_e^2}\right). \tag{4.11}$$

Salpeter and Wagoner (1971) have given a more quantitative treatment of the equilibrium perpendicular to the plane of relativistic disks. Let $\bar{z}$ be the proper distance perpendicular to the equatorial plane,

$$\bar{z}=\int e^{\mu}\,dz. \tag{4.12}$$

The proper surface density measured by an observer comoving with the rotating matter is

$$\sigma_p=2\int_0^\infty \epsilon\,d\bar{z}. \tag{4.13}$$

In the comoving frame of the matter the locally Newtonian character of the disk gives

$$\frac{\partial}{\partial\bar{z}}\left[\epsilon^{-1}\frac{\partial p}{\partial\bar{z}}\right]=-4\pi\epsilon. \tag{4.14}$$

Integrate to find

$$\partial p/\partial\bar{z}=-4\pi\epsilon\int_0^{\bar{z}}\epsilon\,d\bar{z} \tag{4.15}$$

and

$$p(\bar{z})-p(0)=-2\pi\left[\int_0^{\bar{z}}\epsilon\,d\bar{z}\right]^2. \tag{4.16}$$

Since $p=0$ at the surface of the disk

$$p(0)=\tfrac{1}{2}\pi\sigma_p^2. \tag{4.17}$$

Equation (4.14) assumes that the self-gravitation of the local part of the disk is a more important source of the potential gradients perpendicular to the plane within the disk than the non-local part of the gravitational field. This condition will usually fail close to the rim of a disk of non-zero thickness.

Equation (4.17) will now be used to estimate how the thickness of the disk depends on the surface density. We start from the rough estimate

$$W \simeq \tfrac{1}{2}\sigma_p/\epsilon(0). \tag{4.18}$$

If the equation of state is isentropic, with a constant adiabatic index γ, there is a unique relation between $p(0)$ and $\epsilon(0)$,

$$p(0) = K[\epsilon(0)]^{\gamma}. \tag{4.19}$$

Equations (4.17) and (4.18) give

$$W \simeq \left[\frac{1}{2\pi}\, p(0)\epsilon(0)^{-2}\right]^{1/2} = \frac{1}{2}\left(\frac{2K}{\pi}\right)^{1/\gamma} \sigma_p^{(\gamma-2)/\gamma}. \tag{4.20}$$

Unless $\gamma \geqslant 2$, W increases as σ_p decreases toward the rim of the disk. For any gaseous disk one expects $4/3 \leqslant \gamma \leqslant 5/3$, so the structure of the disk will tend to become singular at the rim. The half-thickness W increases until the assumption of local self-gravitation breaks down.

The flaring of the disk toward the rim can be avoided in a non-isentropic model in which the entropy decreases outward from the axis of symmetry. Models with $s = s(j)$ and $ds/dj < 0$, as discussed above, have this property without being unstable to convection on the surfaces of constant j.

A basic Newtonian result which remains valid for relativistic disks is that there is a definite boundary perpendicular to the equatorial plane for any equation of state stiffer than an isothermal equation of state.

V Stability

We now consider those instabilities which, in contrast to simple convective instabilities, are potentially capable of radically changing the structure of the equilibrium configuration. The instabilities may be either secular or dynamic instabilities.

In a secular instability equilibrium is never lost. Under the influence of some dissipative mechanism, a small violation of a constraint satisfied by the original equilibrium model grows along a branching sequence of equilibrium models. For instance, the constraint of axisymmetry satisfied by the Maclaurin spheroids is violated when they become secularly unstable under the influence of a non-zero viscosity at the point that the Jacobi sequence of ellipsoids branches off from the Maclaurin sequence. Another possible type of secular instability in rapidly rotating stars is a violation of the constraint of uniform rotation—viscosity may tend to make the rotation non-uniform. The time scale for the growth of a secular instability is set by the dissipative mechanism, and is typically much longer than the hydrodynamic timescale.

Dynamic instabilities involve a genuine loss of equilibrium and generally have much more dramatic and immediate consequences for the star than secular instabilities. The time scale for growth is usually the order of the hydrodynamic timescale. An example is the instability to gravitational collapse of spherical stars when the average adiabatic index falls below a critical value (4/3 in the Newtonian limit, larger for relativistic stars). Other dynamic instabilities which are particularly important for rapidly rotating stars are the fragmentation instabilities. These are of two characteristic types. One is a non-axisymmetric quadrupole mode in which the star becomes elongated into a bar perpendicular to the rotation axis and may fission into a binary system. The other is a more local type of fragmentation in which the star first breaks up into one or more axisymmetric rings.

The effect of rotation is to decrease the instability to overall gravitational collapse (Ledoux and Walraven 1958, Fowler 1966). There are indications that sufficiently rapid rotation can remove this instability entirely for relativistic stars (see Part VI). We will focus our attention on the fragmentation instabilities, since it is these that put a limit on how rapidly rotating and disk-like the star can be.

About the only rapidly rotating stars for which exact results are known are the Newtonian Maclaurin spheroids and the related ellipsoids. The Maclaurin spheroids are axisymmetric, uniform density, uniformly rotating, incompressible fluid configurations. By definition stable against overall collapse, they are the source of most of the lore on fragmentation instabilities. The Newtonian spheroids form a one-parameter sequence. The structure of the model depends on the mass M, the angular momentum J, and the density $(m_b n)$ through the dimensionless parameter

$$y = J^2 (m_b n)^{1/3}/(GM^{10/3}), \tag{5.1}$$

where m_b is the mass per baryon and G is the Newtonian gravitational constant (see Bardeen 1971). A more conventional parameter along the sequence is the eccentricity e, and another useful parameter is the ratio of polar radius to equatorial radius, a_3/a_1. For a given density $m_b n$ and a given mass M the spheroids go from spheres when $J = 0$ to thin disks in the limit $J \to \infty$.

The first critical point along the Maclaurin sequence is at $y = y_1 = 0.0572$, or $a_3/a_1 = 0.5827$. Two sequences of ellipsoids branch off from the Maclaurin sequence here. The Jacobi ellipsoids rotate uniformly, without any internal motions in the rotating frame. The Dedekind ellipsoids have a shape which is fixed in the inertial frame; they do have internal motions with uniform vorticity. The Jacobi ellipsoid and the Dedekind ellipsoid with the same density and with the same values of the semi-principle-axes a_1, a_2, a_3 have the same mass and energy, but the angular momentum of the Dedekind ellipsoid is less than the angular momentum of the Jacobi ellipsoid by a factor $2a_1a_2/(a_1^2 + a_2^2)$. Their common energy is less than the energy of the Maclaurin spheroid with the same density and mass and the angular momentum of the Jacobi ellipsoid.

Under the influence of viscosity a Maclaurin spheroid with y slightly greater than y_1 is secularly unstable, in that the viscosity tends to increase an initial deviation from axisymmetry until the configuration becomes the Jacobi ellipsoid with the same density, mass, and angular momentum (see Chandrasekhar 1969, Press and Teukolsky 1973). The same Maclaurin spheroid is secularly unstable under the influence of gravitational radiation reaction (Chandrasekhar 1970), only now the unstable mode converts the Maclaurin spheroid into the (corresponding) Dedekind ellipsoid. The gravitational radiation emitted carries off angular momentum, consistent with the lower angular momentum of the Dedekind ellipsoid.

In both cases, the final configuration is one in which the dissipative mechanism is neutralized. The Jacobi ellipsoid, but not the Dedekind ellipsoid, has zero shear, and the Dedekind ellipsoid, unlike the Jacobi ellipsoid, does not radiate gravitational waves, since its mass distribution is stationary in the inertial frame.

The mode that becomes secularly unstable to viscosity at $y = y_1$ becomes dynamically unstable at $y = y_2 = 0.1608$, $a_3/a_1 = 0.3033$. The spheroid elongates into a bar perpendicular to the axis of rotation and may eventually fission into a binary system. The time scale for the growth of the dynamical instability is the order of the period of rotation of the spheroid, independent of any dissipative mechanisms that may or may not be present, once $y > y_2$.

The rest of the Maclaurin sequence is only of academic interest. Several additional non-axisymmetric modes become unstable before the first axisymmetric instability. At $y = y_3 = 0.2732$, $a_3/a_1 = 0.1713$ the spheroids become secularly unstable to differential rotation. It is no longer true that a positive viscosity acts to restore uniform rotation. At this same point the post-Newtonian corrections to the structure of the spheroids are singular, signifying a change from sphere-like to disk-like behavior, and a sequence of uniformly rotating rings branches off the Maclaurin sequence (see Bardeen 1971). If the initial perturbation of a spheroid with y slightly greater than y_3 is in the direction of the ring (decreased central condensation) viscosity gradually changes the structure to that of the uniformly rotating ring with the same density, mass, and angular momentum as the spheroid and a lower energy. In the direction of increased central condensation viscosity steadily transfers angular momentum outward and increases the amount of differential rotation.

The first axisymmetric dynamic instability is at $y = y_4 = 0.5779$, or $a_3/a_1 = 0.04($ for the same mode that became secularly unstable at $y = y_3$. Once the mode becomes dynamically unstable, it is again secularly stable in that a deformation purely in this mode leads to a higher energy neighboring equilibrium configuration. The dynamic instability means that the neighboring non-equilibrium configurations in which each ring of matter has the same angular momentum as in the unperturbed configuration have lower energy.

The consequence of this first axisymmetric dynamic instability (ignoring the several unstable non-axisymmetric modes present) is a collapse of the center of the disk coupled with an expansion near the rim, or vice versa. In itself it may not lead to true fragmentation, though in the direction of producing a ring the non-axisymmetric instabilities to fragmentation will be enhanced.

How typical is the behavior of the Maclaurin sequence? Ostriker and Tassoul (1969) and Ostriker and Bodenheimer (1972) have calculated sequences of rapidly rotating polytropes in which the angular momentum of each cylindrical shell of matter is conserved as the "entropy" coefficient in the equation of state decreases and the star contracts. Both secular and dynamic stability to the quadrupole mode associated with points y_1 and y_2 on the Maclaurin sequence is tested using the tensor virial theorem. A Jacobi-type neutral point indicating onset of secular instability always seems to occur when the ratio of kinetic energy of rotation to gravitational potential energy is about the same as at y_1 on the Maclaurin sequence, where K.E./P.E. = 0.1375. The ratio of the same quantities at the onset of dynamic instability at the point corresponding to y_2 is also always close to the Maclaurin value, K.E./P.E. = 0.2738. These results suggest that no configuration in which centrifugal force is the primary support against gravity can be stable.

On the other hand, spiral galaxies seem to have a ratio of K.E./P.E. which is fairly close to the virial theorem limit of 0.5, and an effective a_3/a_1 of perhaps 1/10 or less. An invisible massive halo must be present if they are to satisfy the Ostriker criteria. Perhaps highly centrally condensed disk-like configurations can be stable against fragmentation. The models calculated by Ostriker and collaborators do not have sufficiently extreme central condensations and angular momentum distributions to answer this question conclusively.

Since the stability of rapidly rotating stars is still not well understood in Newtonian theory, it will probably be some time before much progress is made in understanding the stability of highly relativistic rapidly rotating stars. Chandrasekhar and Friedman (1972) have begun to develop a mathematical formalism for dealing with perturbations of rotating stars in general relativity, largely restricted to axisymmetric perturbations. Another approach, based on the use of velocity potentials, is described by Schutz (1972).

Local, small scale fragmentation instabilities in a thin relativistic disk can be studied using a straightforward generalization of the Newtonian analysis of Goldreich and Lynden-Bell (1965). They consider plane wave modes in a rotating medium of finite thickness, but infinite horizontal extent relative to the wavelength of the perturbation. The effective half-thickness W is defined by

$$2W = \left[\int (m_p n)\, dz\right]^2 \Big/ \int (m_p n)^2\, dz. \tag{5.2}$$

The wavelength of maximum instability is $\lambda \simeq 4\pi W$ and a half-thickness

$$W \geqslant \frac{\pi}{2}\, xG\sigma K^{-2} \tag{5.3}$$

is necessary to stabilize all wavelengths. The quantity σ is the surface density, K is the epicyclic frequency, and x is a numerical coefficient which ranges from 0.57 to 1.35 as the polytropic index of the equation of state goes from 0 to ∞. The epicyclic frequency in a rotating disk is the frequency of oscillation of a test particle in a slightly non-circular orbit about an unperturbed circular orbit. The general formula for Newtonian disks rotating with angular velocity Ω at radius ρ is

$$K^2 = 2\Omega\rho^{-1}\frac{d}{d\rho}(\Omega\rho^2). \tag{5.4}$$

The analogue of equation (5.3) in stellar dynamics (Toomre 1964) seems to give a minimum thickness for stability which is roughly the observed thickness of the Galactic disk in the neighborhood of the sun. To compare with the exact results for Maclaurin spheroids, consider a uniformly rotating disk of perfect fluid. The epicyclic frequency is $K^2 = 4\Omega^2$. The surface density is

$$\sigma = 2\pi^{-2}G^{-1}\Omega^2 a(1 - \rho^2/a^2)^{1/2}. \tag{5.5}$$

The half-thickness required for marginal stability according to the local criterion (5.3) is

$$W = \frac{1}{7\pi}a(1 - \rho^2/a^2)^{1/2}, \tag{5.6}$$

which is equal to the actual thickness as a function of ρ for the Maclaurin spheroid with $a_3/a_1 = 1/7\pi = 0.0455$. This is remarkably close to the value $a_3/a_1 = 0.0465$ at the point y_4 where the Maclaurin spheroids first become dynamically unstable to axisymmetric modes, even though the critical wavelength $\lambda \simeq \frac{4}{7}a$ is the order of the radius of the disk.

Since a relativistic disk is locally Newtonian, equation (5.3) still applies if the proper distance perpendicular to the equatorial plane $\bar{z}$ replaces z in equation (5.2) for W, if σ becomes the proper surface density σ_p in the frame comoving with the matter, and if the proper relativistic formula is used for the epicyclic frequency measured in the comoving frame.

Let

$$\bar{\Omega} = (\Omega - \omega)(e^{\psi - \nu})_{,\rho}e^{-\mu} - \tfrac{1}{2}\omega_{,\rho}e^{-\mu}e^{\psi - \nu}(1 + v^2). \tag{5.7}$$

The local angular velocity of a fluid element about its own center of mass is

$$\omega^{(z)} = \bar{\Omega} + \tfrac{1}{2}\Omega_{,\rho}e^{-\mu}e^{\psi - \nu}(1 - v^2)^{-1}. \tag{5.8}$$

The relativistic epicyclic frequency is given by

$$K^2 = 4\bar{\Omega}\omega^{(z)}. \tag{5.9}$$

(see Bardeen 1970a).

The quantity $\omega^{(z)}$ is related to the angular momentum per unit rest mass in the disk j/m_b by

$$\omega^{(z)} = \tfrac{1}{2}e^{-\psi}(1 - v^2)^{1/2}(j/m_b)_{,\rho}e^{-\mu} \,. \tag{5.10}$$

Therefore, a disk in which the angular momentum increases at a slow rate with increasing proper radius is particularly susceptible to fragmentation.

The criterion (5.3) is not likely to be very accurate for relativistic disks or differentially rotating Newtonian disks, but it has the advantage of being easy to apply.

VI Uniformly Rotating Disks

An infinitesimally thin disk is highly unstable to fragmentation, but the mathematical simplifications of these configurations make them attractive for exploratory relativistic calculations. Some of the qualitative features of the highly relativistic disks may hold for more realistic models. Bardeen and Wagoner (1971) have obtained a rather complete picture of the behavior of uniformly rotating disks in the relativistic limit, and we now review some of their results.

Since the pressure is finite over an infinitesimal volume in an infinitesimally thin disk, the solution of equation (1.15) for B is simply $B = 1$. The vacuum equations outside the disk become

$$\nabla^2\nu = \tfrac{1}{2}\rho^2 e^{-4\nu}\, \nabla\omega \cdot \nabla\omega \tag{6.1}$$

and

$$\rho^{-2}\, \nabla\cdot(\rho^2\, \nabla\omega) = 4\, \nabla\nu \cdot \nabla\omega. \tag{6.2}$$

There is no interior solution to contend with; the structure of the disk only enters the field equations through the boundary conditions at $z = 0, 0 \leqslant \rho \leqslant a$, where a is the coordinate radius of the disk. The metric functions ν and ω are continuous across the disk, but the normal derivatives are discontinuous by an amount proportional to a surface density σ defined by

$$\sigma = \int \epsilon e^{2\mu}\, dz. \tag{6.3}$$

Integrate equations (1.8) and (1.12) across the disk to find

$$\left.\frac{\partial\nu}{\partial z}\right|_{0^\pm} = \pm\, 2\pi\sigma(1 + v^2)(1 - v^2)^{-1} \tag{6.4}$$

and

$$\left.\frac{\partial\omega}{\partial z}\right|_{0^\pm} = \mp\, 8\pi\sigma(\Omega - \omega)(1 - v^2)^{-1}. \tag{6.5}$$

A third boundary condition on the disk comes from the equation of hydrostatic equilibrium, which for uniform rotation has the form

$$\Phi = m_b(u^t)^{-1} = m_b e^{\nu}(1 - v^2)^{1/2} = \text{constant}. \tag{6.6}$$

The boundary conditions (6.4)–(6.6), plus the usual asymptotic flatness conditions at infinity, form a complete set. Given values of the angular velocity Ω and the coordinate radius a, they determine a more or less unique solution to equations (6.1) and (6.2). The constant in equation (6.6) is not an independent parameter; the parameter γ defined by

$$\Phi = m_b e^{\nu_c} = m_b(1 - \gamma)$$

is fixed by a requirement that the surface density σ be finite at the rim of the disk.

No knowledge of the metric function μ is required in the solution for ν and ω. Once ν and ω are known, equations (1.17a, b) in some combination can be integrated to find ζ and μ.

The rest mass, angular momentum, and binding energy of the disk can be calculated as integrals over the disk.
The rest mass is

$$M_0 = \int \epsilon u^t(-g)^{1/2}\, d^3x = 2\pi e^{-\nu_c} \int_0^a \sigma\rho\, d\rho, \tag{6.7}$$

and

$$J = \int \epsilon u_\phi u^t(-g)^{1/2}\, d^3x = 2\pi e^{-2\nu_c} \int_0^a (\Omega - \omega) e^{-2\nu} \sigma \rho^3\, d\rho, \tag{6.8}$$

$$E_b/M_0 = 1 - M/M_0 = \gamma - 2\Omega J/M_0. \tag{6.9}$$

Of the two parameters necessary to specify a model (Ω and a or M_0 and J, for instance), one is just a scale parameter. All dimensionless quantities depend on a single dimensionless parameter, which we choose to be γ. The parameter γ is related to the redshift of a photon emitted from the center of the disk, Z_c, by

$$\gamma = Z_c/(1 + Z_c). \tag{6.10}$$

It varies from 0 in the Newtonian limit to 1 in the extreme relativistic limit.

We expand all dimensionless quantities in power series in γ. For instance,

$$\nu = \sum_{l=0}^{\infty} \nu_n(\rho, z)\gamma^n \tag{6.11}$$

and

$$\lambda = \omega/\Omega = \sum_{n=1}^{\infty} \lambda_n(\rho, z)\gamma^n. \tag{6.12}$$

In each order the partial differential equations for ν_n and λ_n are *linear*; the coefficients of γ^n on the right-hand sides of equations (6.1) and (6.2) involve only ν_m and λ_m of lower order.

To solve for the ν_n and λ_n we use oblate spheroidal coordinates ξ, η defined by

$$\rho = a(1+\xi^2)^{1/2}(1-\eta^2)^{1/2}, \tag{6.13}$$

$$z = a\xi\eta. \tag{6.14}$$

The disk is then the coordinate surface $\xi = 0$. Separation of variables in ν_n and λ_n,

$$\nu_n(\xi,\eta) = \sum_{l=0}^{\infty} \nu_{n,l}(\xi)P_{2l}(\eta) \tag{6.15}$$

and

$$\lambda_n(\xi,\eta) = \sum_{l=0}^{\infty} \lambda_{n,l}(\xi)W_{2l}(\eta), \tag{6.16}$$

reduces the partial differential equations to ordinary differential equations. The $P_{2l}(\eta)$ are Legendre polynomials and the $W_{2l}(\eta)$ are defined by

$$W_{2l}(\eta) = (2l+1)^{-1}\, dP_{2l+1}(\eta)/d\eta. \tag{6.17}$$

The $\nu_{n,l}$ and $\lambda_{n,l}$ are zero for $l > n$ in the case of a uniformly rotating disk.

Beyond post-Newtonian order ($n = 2$ in ν, $n = 1$ in λ) the ordinary differential equations for the $\nu_{n,l}(\xi)$ and the $\lambda_{n,l}(\xi)$ were solved numerically on a computer, though an analytic solution is possible through $n = 3$ in ν and $n = 2$ in λ using the Ernst formulation of the field equations. The calculations were carried out to order $n = 11$ in ν and $n = 10$ in λ.

For many quantities the relativistic expansion converges rapidly enough that the truncation at $n = 10$ or 11 is accurate to better than 1 per cent even in the extreme relativistic limit, $\gamma = 1$. The use of Padé approximants seems to increase the accuracy to better than one part in 10^4 in some cases.

Our method of calculation gives us the whole sequence of disks from $\gamma \simeq 0$ to $\gamma = 1$. One physical interpretation of the sequence is that it represents a disk of fixed rest mass M_0 which gradually loses angular momentum, contracts, and becomes more relativistic. The collapse of the disk is slow enough that to a good approximation it proceeds through a sequence of equilibrium models. The relation between J/M_0^2, the dimensionless combination of J and M_0, and the relativistic parameter γ is shown in Fig. 1. The angular momentum decreases monotonically as γ increases and approaches a non-zero limit as $\gamma \to 1$. There is a minimum value of the angular momentum for which equilibrium exists, $J/M_0^2 = 0.393$, barring the existence of a separate sequence of uniformly rotating disks which does not have a Newtonian limit and would not be found by our method of solution.

In the same Figure we see that the ratio of binding energy E_b to M_0 increases monotonically with γ to a value of 0.373 in the limit $\gamma \to 1$. Typical spherical, slowly rotating stars reach a maximum in the binding energy when they are only moderately relativistic. Near this point they become unstable to overall gravitational collapse. It seems likely that the disks are stable against *overall* gravitational collapse for all $\gamma < 1$.

The results shown in Figure 2 support this conjecture. The ratio M^2/J is less than one for all $\gamma < 1$. In any axisymmetric collapse J is constant and M can only decrease. If one believes that gravitational collapse leads to a black hole, rather than a naked singularity, and that all uncharged vacuum black holes are described by the Kerr metrics, the fact that all Kerr black holes have $M^2/J > 1$ means that the entire disk cannot collapse into a black hole. Within the numerical accuracy,

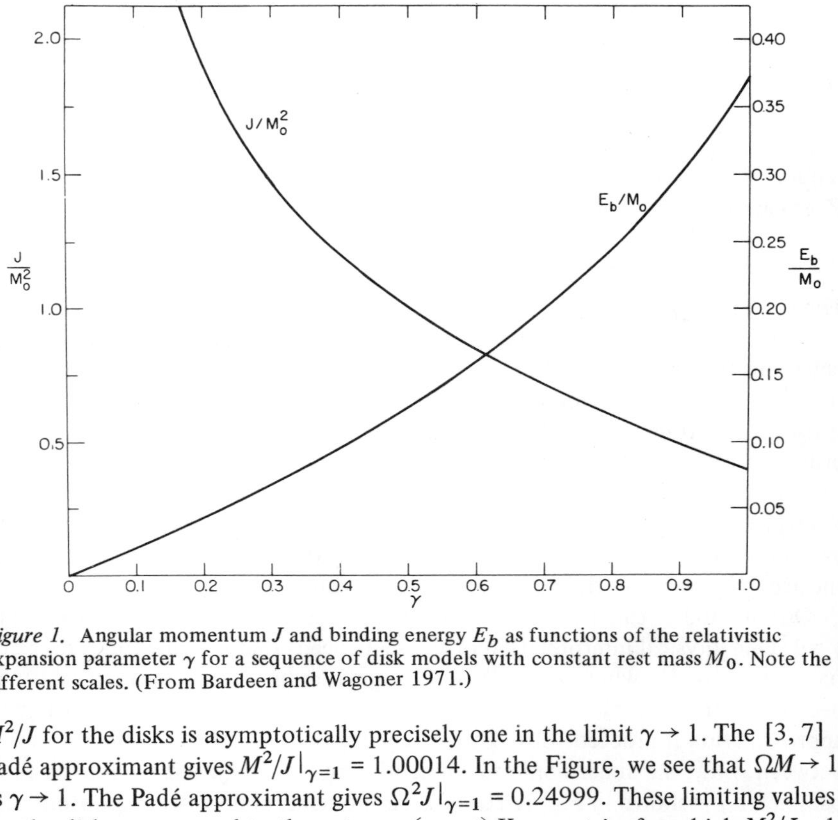

Figure 1. Angular momentum J and binding energy E_b as functions of the relativistic expansion parameter γ for a sequence of disk models with constant rest mass M_0. Note the different scales. (From Bardeen and Wagoner 1971.)

M^2/J for the disks is asymptotically precisely one in the limit $\gamma \to 1$. The [3, 7] Padé approximant gives $M^2/J|_{\gamma=1} = 1.00014$. In the Figure, we see that $\Omega M \to 1/2$ as $\gamma \to 1$. The Padé approximant gives $\Omega^2 J|_{\gamma=1} = 0.24999$. These limiting values for the disks correspond to the extreme ($a = m$) Kerr metric, for which $M^2/J = 1$ and the angular velocity of the horizon relative to infinity is $\Omega_H = 1/2M$.

The scale of proper distances in the disk is represented by $ae^{\mu_c} = ae^{-\nu_c}$. The ratio of the "gravitational radius" M to $ae^{-\nu_c}$ is plotted in Figure 3. Since both M and $M/ae^{-\nu_c}$ have finite, non-zero limits as $\gamma \to 1$, proper distances within the disk are non-zero in the limit even though the coordinate radius a goes to zero. One can also see the advantage of γ over $(\Omega a)^2$ as a relativistic expansion parameter, in that an expansion in powers of $(\Omega a)^2$ would diverge at the point $(\Omega a)^2$ reaches its maximum value.

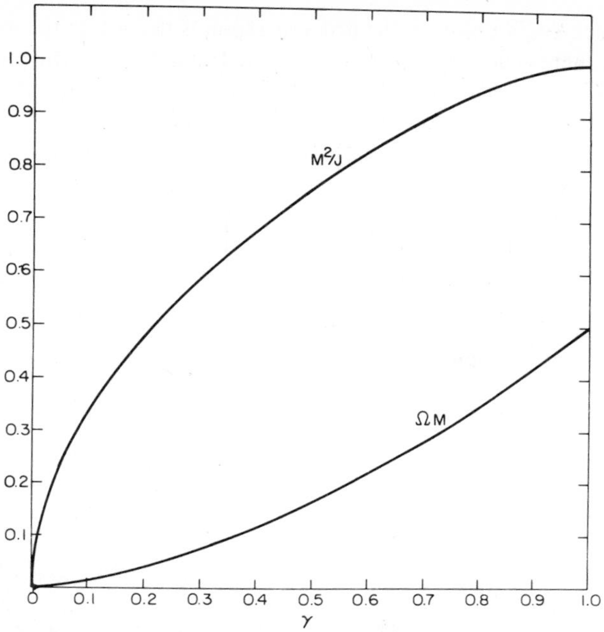

Figure 2. Dimensionless combinations of the parameters measurable at infinity: gravitational mass M, angular momentum J, and angular velocity Ω: versus γ. (From Bardeen and Wagoner 1971.)

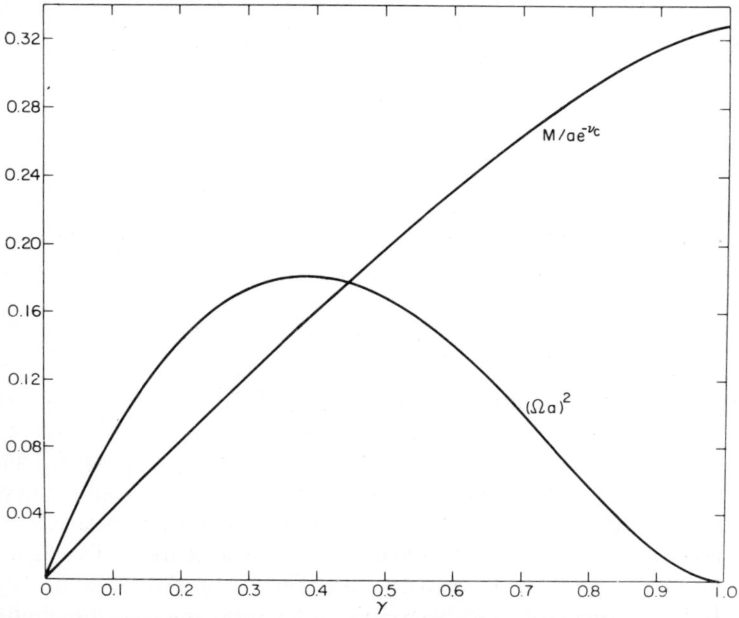

Figure 3. Plot of two dimensionless ratios involving the coordinate radius a of the rim of the disk, illustrating how a vanishes while the scale of proper radius $a \exp(-\nu_c)$ remains finite as $\gamma \rightarrow 1$. (From Bardeen and Wagoner 1971.)

Even though every point on the disk corresponds to $r = 0$ in the extreme relativistic limit, there is no singularity in the spatial geometry. In the oblate spheroidal coordinates ξ, η the spatial line element in the vicinity of the disk is

$$ds^2 = (ae^{-\nu_c})^2\left\{(1 + \xi^2)(1 - \eta^2)e^{2\nu_c - 2\nu}\, d\phi^2 + e^{2\nu_c + 2\mu}(\xi^2 + \eta^2)\left[\frac{d\xi^2}{1 + \xi^2} + \frac{d\eta^2}{1 - \eta^2}\right]\right\}. \tag{6.18}$$

Both $e^{2\nu_c - 2\nu}$ and $e^{2\nu_c + 2\mu}$ are regular at all finite values of ξ in the limit $\gamma \to 1$, except for the usual discontinuities in the normal derivatives across the disk.

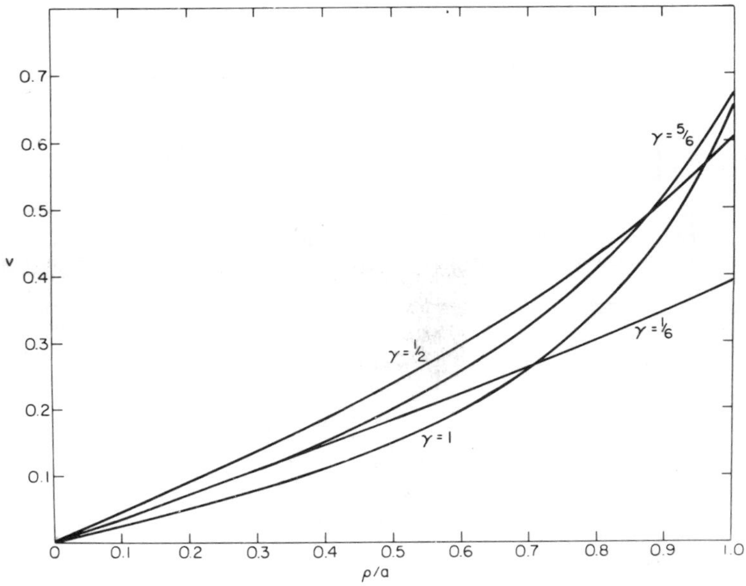

Figure 4. Physical velocity of rotation v as a function of fractional coordinate radius ρ/a for four values of γ. (From Bardeen and Wagoner 1971.)

From the equation of hydrostatic equilibrium, $e^{2\nu_c - 2\nu} = 1 - v^2$ on the disk. The velocity of rotation v is plotted as a function of ρ/a for several values of γ in Figure 4. At fixed ρ/a in the central part of the disk v is largest at $\gamma \simeq 1/2$ and at all ρ/a v is decreasing with increasing γ as $\gamma \to 1$. The velocity of rotation never exceeds 0.7, and the potential differences along the disk remain finite as $\nu \to -\infty$.

In spherical stars the potential difference between the center and surface becomes infinite in the extreme relativistic limit, and the pressure and usually the energy density become infinite at the center. In our disks the mass distribution becomes *less* centrally concentrated as $\gamma \to 1$, as can be seen from the plots of the proper surface density σ_p in Figure 5. The basic reason for this striking qualitative difference between rotating disks and spherical or slowly rotating stars is that in the case of highly relativistic disks the shear of the ZAMO's, rather than the matter terms, is the dominant source of the potential ν. A counter-rotating disk, which is the same as our disk in the Newtonian limit, produces no dragging of inertial frames and has a relativistic limit qualitatively similar to spherical stars (Morgan and Morgan 1969).

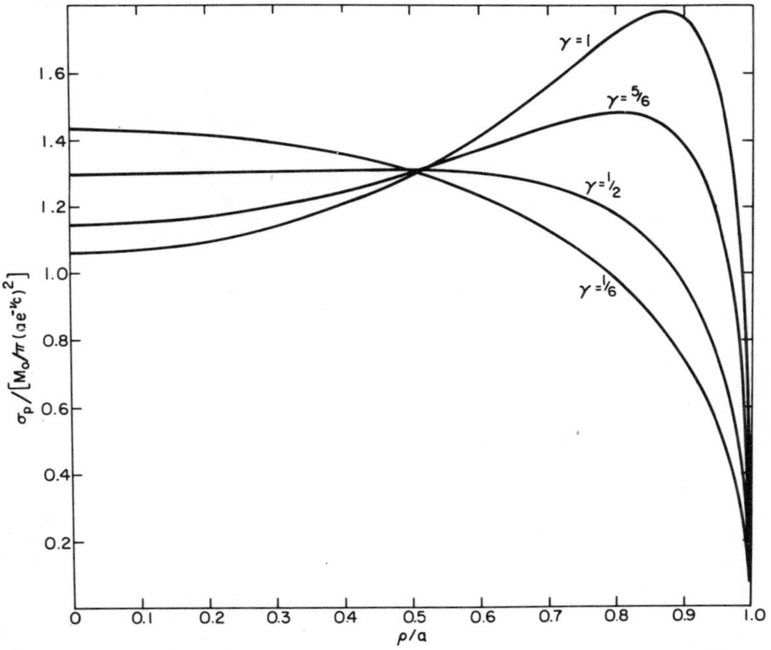

Figure 5. Proper surface density σ_p in the comoving frame, in units of $M_0/\pi[a \exp(-\nu_c)]^2$, versus ρ/a for the same values of γ as Fig. 4. The slope becomes infinite at $\rho = a$ for all values of γ. (From Bardeen and Wagoner 1971.)

At all finite values of ξ the angular velocity of the ZAMO's relative to infinity and the angular velocity of the disk relative to infinity become precisely equal as $\gamma \to 1$. Figure 6 shows how this goes at $\rho/a = 0, 2^{-1/2}$, and 1 on the disk. However, the angular velocity of the disk measured by a ZAMO on the disk,

$$\tilde{\Omega} = (\Omega - \omega)e^{-\nu}, \tag{6.19}$$

is non-zero in the limit. It is the infinite time dilation, $dt/ds = e^{-\nu}$, between the local ZAMO and infinity that produces the infinite frame dragging. The angular velocity $\omega^{(z)}$ of a fluid element about its own center of mass is equal to $\tilde{\Omega}$ at the center of the disk, but increases more rapidly toward the rim (see Figure 7). The local fragmentation instability of a uniformly rotating disk is governed by $\omega^{(z)}$ and σ_p.

A brief look at the spatial geometry outside the disk at finite r/a (finite ξ) in the equatorial plane is given in Figure 8. The quantity $re^{-\nu}$ is the proper circumferential radius of a circle about the axis of symmetry. In the limit $\gamma = 1$ $re^{-\nu}$ reaches a maximum just inside the rim of the disk and seems to approach a constant value as $r/a \to \infty$, $re^{-\nu} = 2M$, identical to the proper circumferential radius at the horizon of the extreme Kerr metric. The accuracy of the numerical

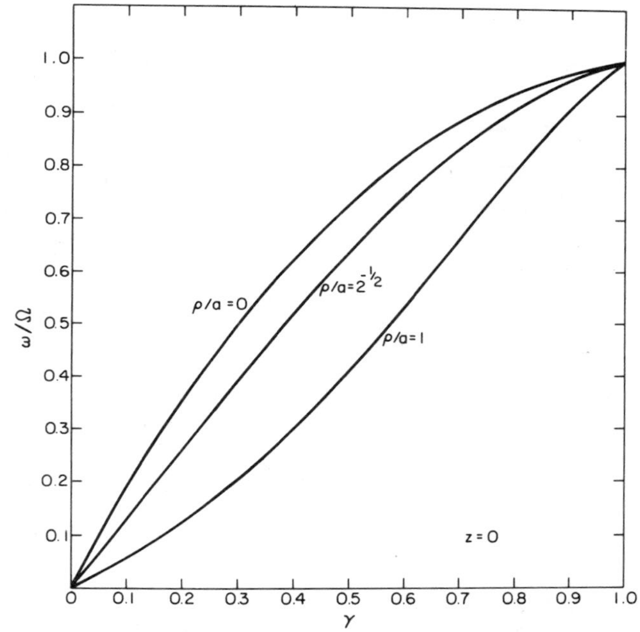

Figure 6. Ratio of the angular velocity ω of the "local inertial frame" to the angular velocity Ω of the matter, both measured at infinity, as a function of γ for three fixed values of ρ/a corresponding to $\eta^2 = 1, \frac{1}{2}, 0$. This ratio approaches unity at all points on the disk as $\gamma \to 1$. (From Bardeen and Wagoner 1971.)

results becomes poor at large values of ξ when $\gamma \simeq 1$, because of slow convergence of the relativistic expansion.

Now we consider how the metric behaves in the limit $\gamma \to 1$ if the coordinate radius r, rather than r/a, is fixed at a non-zero value. In plotting metric quantities the abscissa is rescaled from r/a to

$$r/ae^{-\nu_c} = (r/a)(1 - \gamma). \tag{6.20}$$

We cannot calculate directly the limiting metric, since any finite r corresponds to $\xi = \infty$ in the limit. Our conjecture is that the limiting metric is the $a = m$ Kerr metric, with a mass parameter m equal to the gravitational mass M of the disk in the limit $\gamma = 1$. Our coordinate radius r is related to the Boyer–Lindquist coordinate R by

$$R = r + m, \tag{6.21}$$

and when $\gamma = 1$ at $r > 0$ the polar angle θ is related to the spheroidal coordinate η by

$$\cos\theta = \eta. \tag{6.22}$$

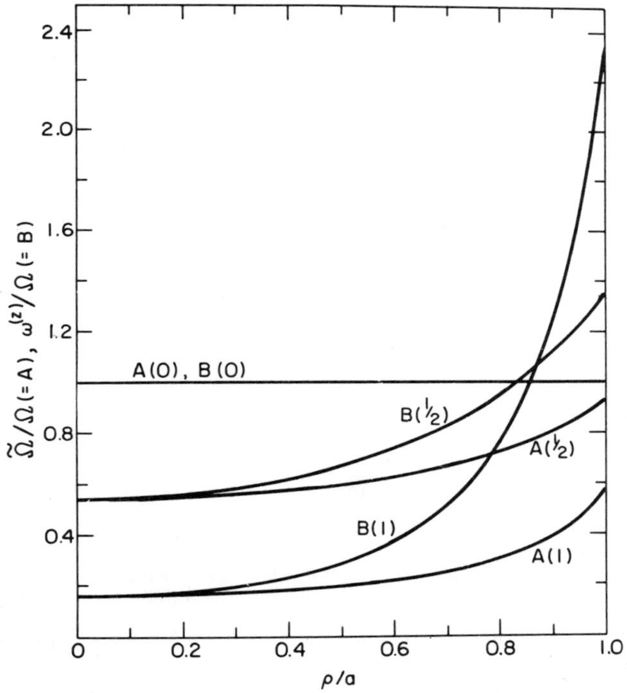

Figure 7. Two local measures of the rotation of the disk. The quantity $\tilde{\Omega}$ is the angular velocity about the axis of symmetry measured by locally non-rotating observers, and the quantity $\omega^{(z)}$ is the angular velocity about its center of mass of a fluid element. Plotted are curves of $\tilde{\Omega}/\Omega \equiv A(\gamma)$ and $\omega^{(z)}/\Omega \equiv B(\gamma)$ for $\gamma = 0, \frac{1}{2}, 1$. (From Bardeen and Wagoner 1971.)

The extreme Kerr metric functions are

$$e^{2\nu} = r^2[(r+m)^2 + m^2\cos^2\theta]\,[((r+m)^2 + m^2)^2 - m^2r^2\sin^2\theta]^{-1}, \tag{6.23}$$

$$\omega = 2m^2(r+m)[((r+m)^2 + m^2)^2 - m^2r^2\sin^2\theta]^{-1}, \tag{6.24}$$

$$B = 1, \tag{6.25}$$

$$e^{2\mu} = [(r+m)^2 + m^2\cos^2\theta]r^{-2}. \tag{6.26}$$

As a first example, consider the proper circumferential radius in the equatorial plane. The curves for $\gamma \leqslant 5/6$ in Figure 9 are identical to those in Figure 8, except for the rescaling of the abscissa. They seem to approach the Kerr metric curve smoothly as $\gamma \to 1$ except at $r = 0$, where the entire $\gamma = 1$ curve in Figure 8 is compressed into a single point. Figure 10 is similar to Figure 9, except that $re^{-\nu}$ is given as a function of r along the axis of symmetry. The relativistic expansion of $e^{\nu_c - \nu}$ converges more rapidly at $\theta = 0$ than at $\theta = \pi/2$, so reasonably accurate results were obtained for $\gamma = 0.95$. The approach to the Kerr metric goes even more smoothly here.

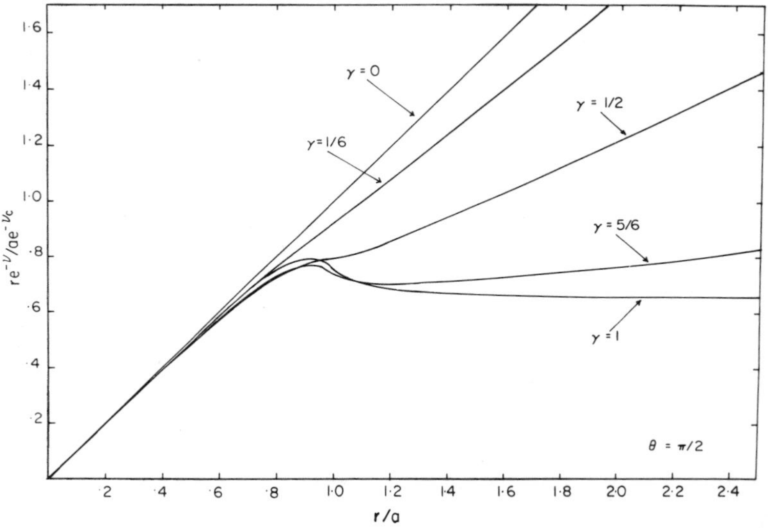

Figure 8. Behavior of $re^{-\nu}$, the proper circumferential radius of a circle around the axis of symmetry in the equatorial plane. In units of $a \exp(-\nu_c)$, this is plotted against r/a for several values of γ. All points on the $\gamma = 1$ curve are at $r/M = 0$. (From Bardeen and Wagoner 1971.)

In the limit $\gamma \to 1$ the proper radial distance $\int e^{\mu}\, dr$ between any finite value of r/a and any non-zero value of $r/ae^{-\nu_c}$ seems to become infinite along all radial directions. It is certainly infinite from $r = 0$ to a finite r/m in the extreme Kerr metric, since

$$\int e^{\mu}\, dr \simeq m(1 + \cos^2 \theta)^{1/2} \ln r \tag{6.27}$$

when $r/m \ll 1$. An infinitely long "cylindrical" region of the t = constant spacelike hypersurfaces, in which the r = constant two-surfaces have constant area, connects the disk region at finite r/a to the Kerr metric region at non-zero r/m. Figure 11 shows $re^{-\nu}/m$ as a function of proper linear radius along both $\theta = 0$ and $\theta = \pi/2$, starting at $r/a = 0$ in the inner, disk region and picking up again in the vicinity of $r/m = 1$ in the Kerr metric region. Also indicated are the transitional values in the cylindrical region.

A particularly interesting aspect of the approach to the Kerr metric is the development of the "ergosphere," the region where $g_{tt} = 0$ and negative energy orbits can exist. The locus of the boundary, where $g_{tt} = 0$, in the $\rho - z$ plane was calculated for two values of γ, $\gamma = 2/3$ and $\gamma = 5/6$, and plotted in Figure 12 along

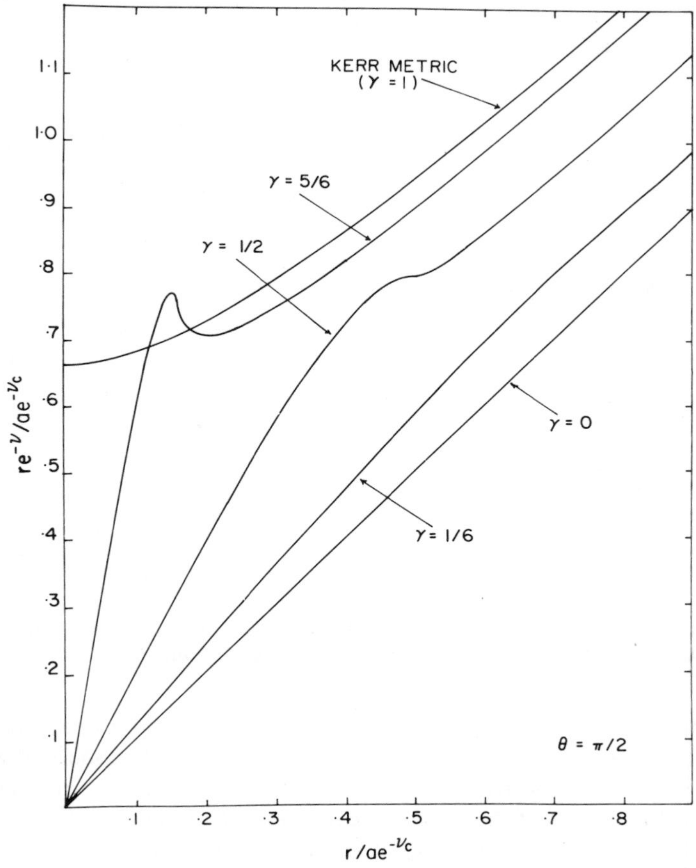

Figure 9. Identical to Fig. 8, except that the abcissa is $r/ae^{-\nu_c} = (1 - \gamma)r/a$ instead of r/a and the $\gamma = 1$ curve is taken from the Kerr metric with $J = M^2$. (Bardeen and Wagoner 1971.)

with the boundary in the Kerr metric. At any value of $\gamma < 1$ the "ergosphere" has a toroidal shape, since $\omega\rho e^{-2\nu}$ cannot be greater than one in a neighborhood of the axis of symmetry as long as $e^{-2\nu}$ is finite. The "ergotoroid" first appears at $\gamma \simeq 0.6$ in the vicinity of the rim of the disk. At $\gamma = 2/3$ the ergotoroid just encloses the rim of the disk at $\rho/ae^{-\nu_c} = 1/3$, but as γ increases the disk shrinks in coordinate radius and the ergotoroid expands to match that of the Kerr metric.

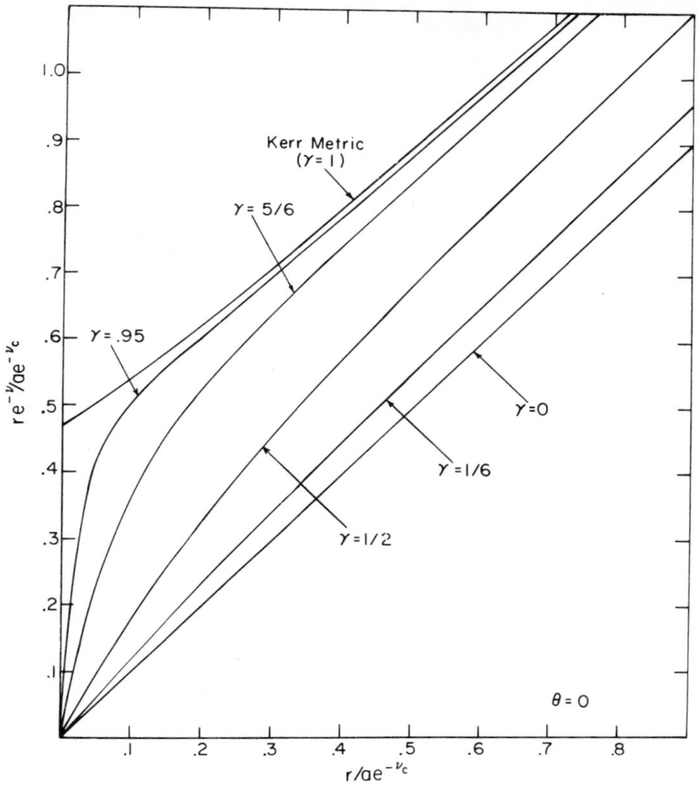

Figure 10. The quantity $re^{-\nu}$ as a function of r along the axis of symmetry, both in units of $ae^{-\nu_c}$. Like Fig. 9, the $\gamma = 1$ curve is the Kerr metric with the same J and M as the $\gamma = 1$ disk. (From Bardeen and Wagoner 1971.)

Perhaps the most precise evidence that the metric at non-zero r approaches the Kerr metric comes from the quadrupole moments Q_ν and Q_λ associated with the potentials ν and λ, respectively. Hartle and Thorne (1968) give an invariant geometrical definition of the quadrupole moments. At large radii the equipotential surfaces have the intrinsic geometry of spheroids of eccentricity e. The quadrupole moments are defined by

$$e^2 = 3|Q_\nu|/Mr^2 \qquad \text{(for } \nu\text{)} \tag{6.28}$$

and

$$e^2 = 5|Q_\lambda|\Omega/2J \qquad \text{(for } \lambda\text{)}. \tag{6.29}$$

The asymptotic behavior of corresponding *flat-space* potentials would then be

$$\nu \sim -Mr^{-1} - Q_\nu r^{-3}P_2(\cos\theta) + \theta(r^{-5}), \tag{6.30}$$

$$\lambda \sim 2J\Omega^{-1}r^{-3} + Q_\lambda r^{-5}W_2(\cos\theta) + \theta(r^{-7}). \tag{6.31}$$

The ellipsoids are prolate if $Q > 0$, oblate if $Q < 0$.

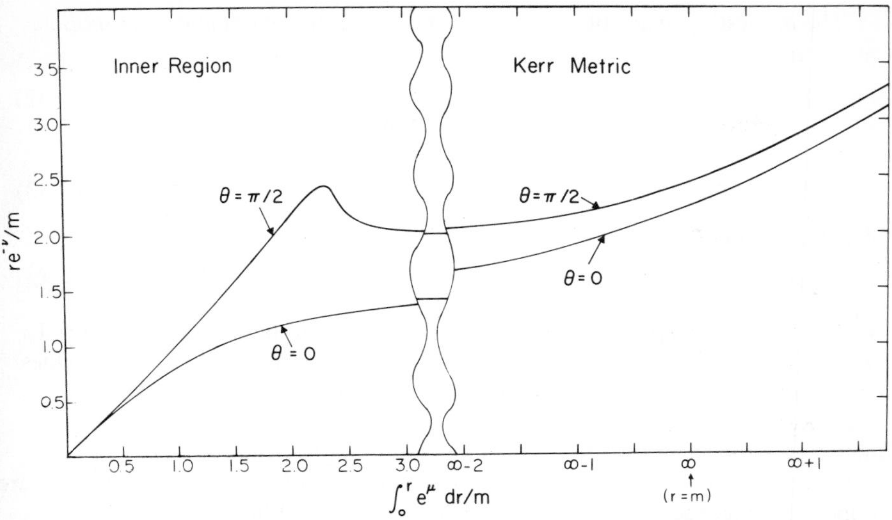

Figure 11. Variation of $re^{-\nu}$ with the proper linear radius from the center of the disk, both along the axis of symmetry and in the equatorial plane, when γ is infinitesimally less than unity. The "inner region" comprises points at finite r/a and finite proper distances from the disk. Points in the "Kerr metric region" are at infinite proper distances from the disk, and non-zero r/m. The point labelled ∞ on the abscissa corresponds to a coordinate radius $r = m$. Also indicated are the transition values of $re^{-\nu}/m$. (From Bardeen and Wagoner 1971.)

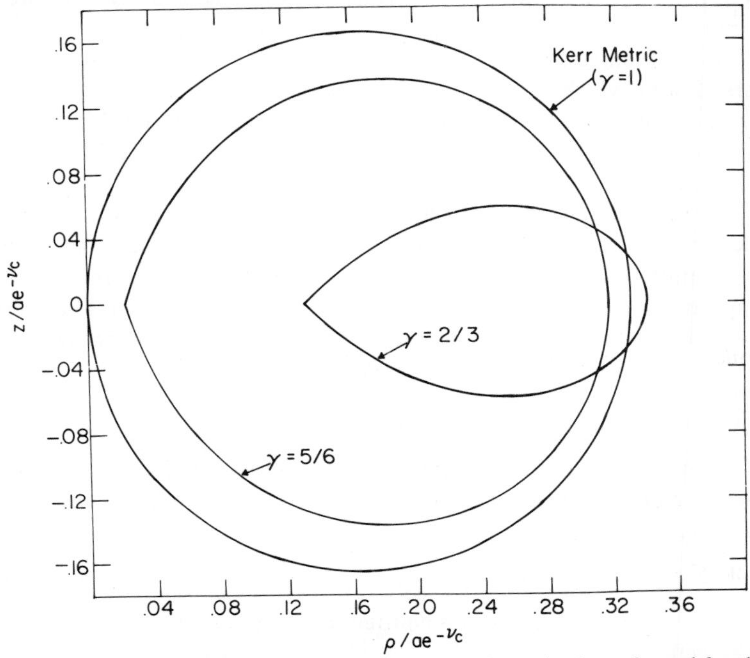

Figure 12. Curves of $g_{00} = 0$ in the (ρ, z)-plane for the indicated values of γ and for the Kerr metric (see text). Within each such curve $g_{00} > 0$ and the t-coordinate axis is a spacelike direction. (From Bardeen and Wagoner 1971.)

The interesting quantities are the ratios of quadrupole moment to monopole moment,

$$K_\nu = Q_\nu / M \tag{6.32}$$

and

$$K_\lambda = \Omega Q_\lambda / 2J. \tag{6.33}$$

The Kerr values are

$$K_\nu^{(\text{Kerr})} = -J^2/M^2 = K_\lambda^{(\text{Kerr})}. \tag{6.34}$$

Figure 13 shows how K_ν and K_λ for the disk approach the Kerr values as $\gamma \to 1$. The disk values for $\gamma < 1$ are always larger in absolute value than the Kerr values corresponding to the same angular momentum and mass, consistent with the arguments of Hernandez (1967) that the quadrupole moments of the Kerr metrics are too small for a perfect fluid distribution to be the source.

The results presented here and in more detail in the Bardeen–Wagoner paper are important evidence that the Kerr metric really is formed in the gravitational collapse of a rapidly rotating star. They are analogous, in a less complete way, to the calculations of Price (1972) which showed that the radiatable multipole moments in nearly spherical gravitational collapse are radiated away to leave a black hole characterized only by its gravitational mass, angular momentum, and charge. In our case the multipole moments are not radiated away; they are swallowed up by the infinitely long cylindrical region that develops as $\gamma \to 1$. The effect of the detailed structure of the disk on the gravitational field dies out as powers of (a/r), which in the limit $\gamma = 1$ means that the gravitational field at non-zero r/m depends only on the value of the angular momentum of the disk (in the absence of charge). The mass has disappeared as a free parameter because of the requirement that the disk be in equilibrium as it contracts.

To get more insight into the relation of the disk and infinity in the extreme relativistic limit we consider photon orbits in the equatorial plane, and in particular photons emitted by the disk. Let ψ be the energy of the photon relative to (asymptotically flat) infinity and let Φ be its angular momentum. The energy ψ' of the photon measured by an observer corotating with the disk is the same at all points in the disk along the photon's trajectory, since

$$\psi' = u^t(\psi - \Omega\Phi) = e^{-\nu_c}(\psi - \Omega\Phi). \tag{6.35}$$

Let α' be the direction cosine of the photon trajectory relative to the forward ϕ-direction measured by a corotating observer. At a given point in the disk ψ/ψ' is a linear function of α',

$$\psi/\psi' = e^{\nu_c} + \Omega \rho e^{-\nu_c}(\alpha' + v). \tag{6.36}$$

Equation (6.36) holds for photons emitted in directions out of the equatorial plane as well.

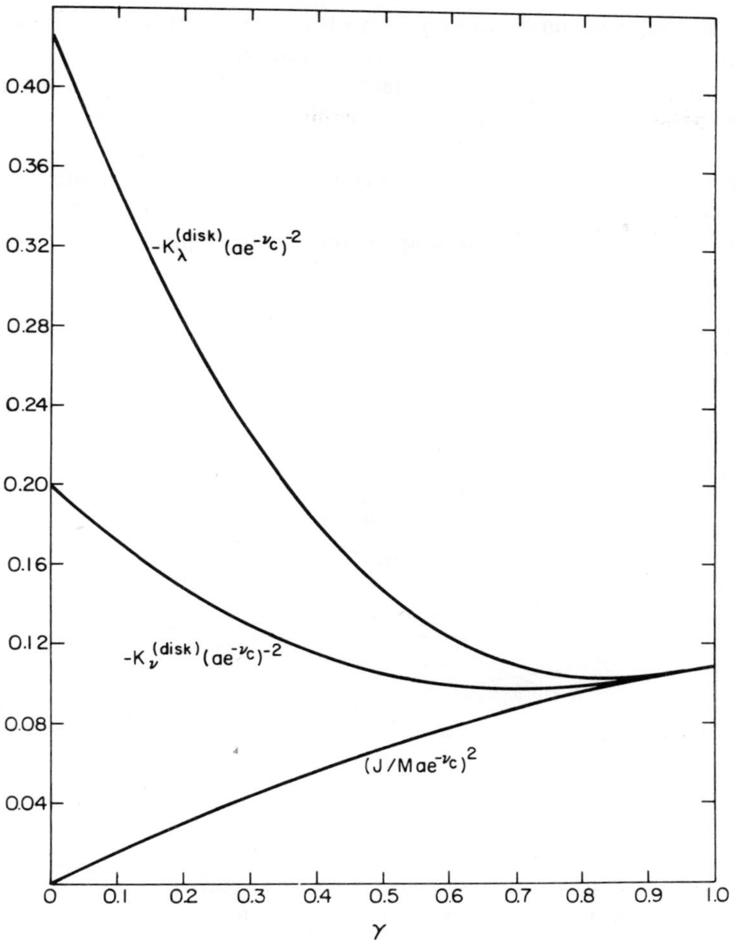

Figure 13. Behavior of the invariantly defined ratios of quadrupole moment to monopole moment for the potential ν, $K_\nu{}^{(disk)}$, and the potential λ, $K_\lambda{}^{(disk)}$. The corresponding ratios for the Kerr metric with the same J and M as the disk are both equal to $-(J/M)^2$. All become equal as $\gamma \to 1$. (From Bardeen and Wagoner 1971.)

In Figure 14 we plot ψ/ψ' for photons emitted in the $+\phi$-direction as a function of ρ/a at the point of emission, and similarly for photons emitted in the $-\phi$-direction. At any $\rho/a < 1$ the value of ψ/ψ' for a particular α' can be found by linear interpolation between the ψ/ψ' at $\alpha' = 1$ and the ψ/ψ' at $\alpha' = -1$. The solid angle in the corotating frame is proportional to α', so one can estimate the average ψ/ψ' of photons emitted isotropically in the corotating frame at any radius. In particular, at the rim of the disk this average is precisely $\psi/\psi' = 1$, independent of γ, and half of the photons emitted have blueshifts. The energy of the photons is not greatly degraded on the average even when the redshift for zero angular momentum photons is infinite. However, the *average* rate the photons are received at infinity is decreased by a factor $(u^t)^{-1} = 1 - \gamma$ relative to the rate they are emitted in the comoving frame. The photons emitted by a given point on the disk are received at infinity in sharp bursts when $\gamma \simeq 1$.

Once $\gamma \gtrsim 1/2$ not all of the photons emitted by the disk reach infinity. The dashed lines in Figure 14 represent, for the respective values of γ, the minimum values of ψ/ψ' for photons that escape in the equatorial plane.

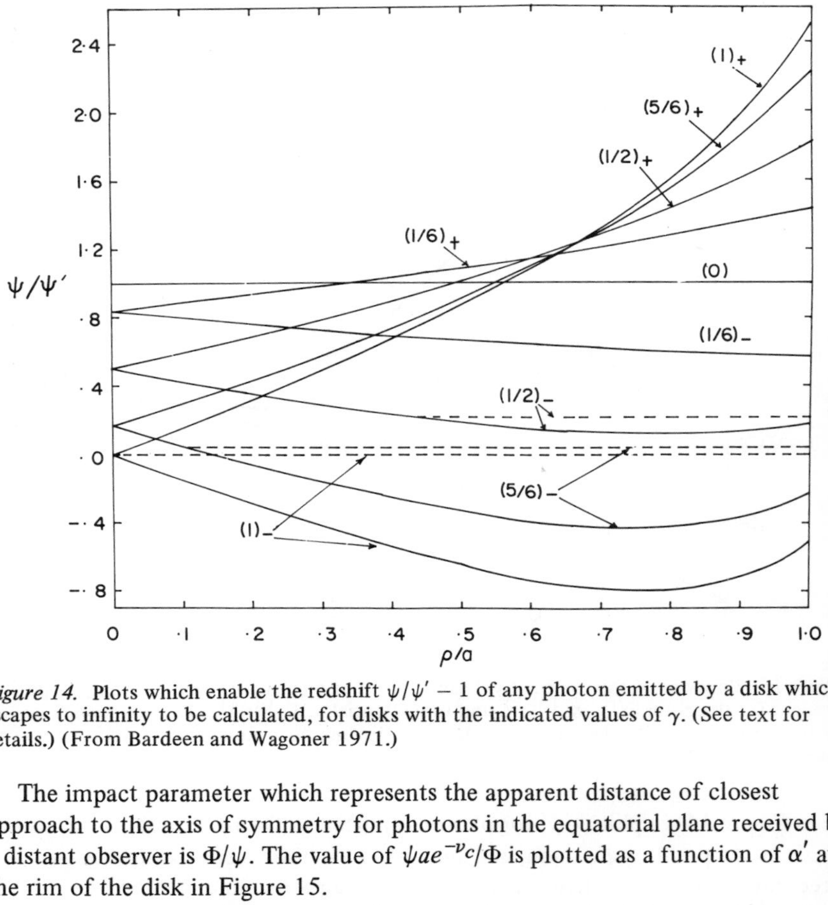

Figure 14. Plots which enable the redshift $\psi/\psi' - 1$ of any photon emitted by a disk which escapes to infinity to be calculated, for disks with the indicated values of γ. (See text for details.) (From Bardeen and Wagoner 1971.)

The impact parameter which represents the apparent distance of closest approach to the axis of symmetry for photons in the equatorial plane received by a distant observer is Φ/ψ. The value of $\psi a e^{-\nu_c}/\Phi$ is plotted as a function of α' at the rim of the disk in Figure 15.

In the limit $\gamma \to 1$ almost all the photons emitted by the disk have $\Phi/\psi = 2M$, whether they are direct $(\Phi > 0)$ or retrograde $(\Phi < 0)$ orbits. This result is qualitatively the same as we found for photons emitted by a particle in a circular orbit at $r/m \ll 1$ in the extreme Kerr metric in the first set of lectures. Therefore, a highly relativistic disk emitting photons isotropically in the corotating frame appears to a distant observer in the equatorial plane as a single spot at an impact parameter of $2M$ on the approaching side of the axis of symmetry.

Figure 16 shows the value of $\psi a e^{-\nu_c}/\Phi$ for which the photon trajectory has a turning point at the given value of ρ/a, both for direct and retrograde orbits. The important point here is the existence of trapped retrograde photon orbits when $\gamma \gtrsim 1/2$. The presence of the disk means that these orbits are stable (there is no black hole to capture them). A stable circular photon orbit is present near the

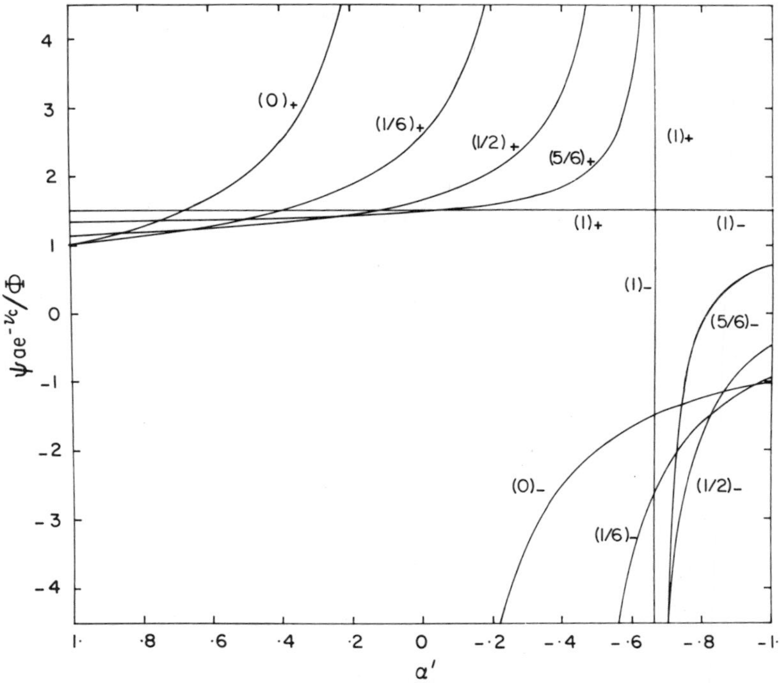

Figure 15. Combination $\psi a \exp(-\nu_c)/\Phi$ of null geodesic parameters versus the direction cosine α' of the photon's trajectory relative to the $+\phi$-direction in the comoving frame of emission, at the rim of the disk. This combination determines the trajectory of photon orbits in the equatorial plane. The direction cosine of the zero-angular-momentum ($\Phi = 0$) photon trajectory is $\alpha' = -\nu$ for each value of γ. (See text for details.) (From Bardeen and Wagoner 1971.)

rim of the disk once $\gamma \gtrsim 1/2$. This is physically more significant than the existence of negative energy photon orbits at larger values of γ. At all values of γ the direct photon trajectories escape to infinity.

Since some photons can always escape, there is no horizon, even in the limit $\gamma \to 1$. However, in the limit the r = constant two-surfaces at $r/a \gg 1$ and $r/m \ll 1$ are marginally trapped surfaces, and the slightest dynamical perturbation will create genuine trapped surfaces and a horizon. Once this happens the Hawking–Penrose theorem guarantees that a singularity will develop. Also, typical photons that are captured by the exterior Kerr metric and have non-zero energy ψ at infinity reach the disk with *infinite* energy ψ' measured by a local corotating observer, since they will not have impact parameters in the infinitesimal range about $\Phi/\psi = 2M$ for which ψ/ψ' is non-zero. Therefore, the limit $\gamma \to 1$ is not physically realizable, even if some stable models of rapidly rotating stars have the same type of formal relativistic limit as the infinitesimally thin disks.

What may be relevant in astrophysical applications are some of the qualitative features of the moderately relativistic models, with γ in the range 0.5 to perhaps 0.9, Wilson (1972) has calculated differentially rotating stars with various degrees

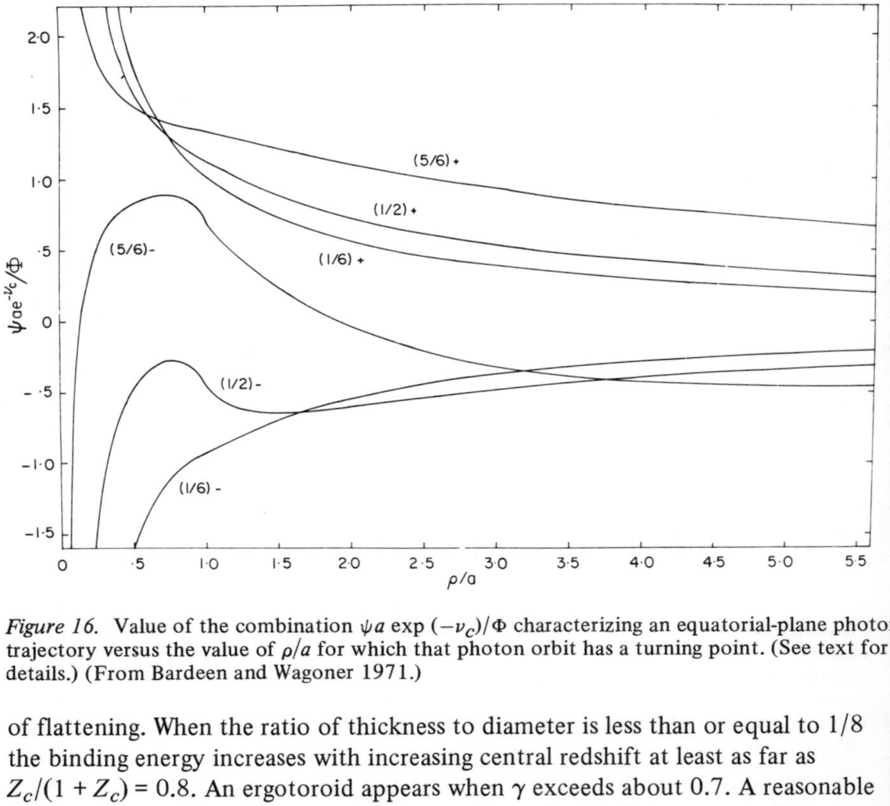

Figure 16. Value of the combination $\psi a \exp(-\nu_c)/\Phi$ characterizing an equatorial-plane photon trajectory versus the value of ρ/a for which that photon orbit has a turning point. (See text for details.) (From Bardeen and Wagoner 1971.)

of flattening. When the ratio of thickness to diameter is less than or equal to 1/8 the binding energy increases with increasing central redshift at least as far as $Z_c/(1 + Z_c) = 0.8$. An ergotoroid appears when γ exceeds about 0.7. A reasonable hope is that models with trapped photon orbits and a binding energy as large as 10 or 15 per cent of the rest mass energy can be stable, at least if the apparent shape of spiral galaxies is a valid indication.

Bibliography

Abramowicz, M. A. (1970) *Astrophys. Letters* **7,** 73.
Bardeen, J. M. (1970a) *Astrophys. J.* **161,** 103.
Bardeen, J. M. (1970b) *Astrophys. J.* **162,** 71.
Bardeen, J. M. (1971) *Astrophys. J.* **167,** 425.
Bardeen, J. M., Carter, B. and Hawking, S. W. (1973) to be published.
Bardeen, J. M. and Wagoner, R. V. (1971) *Astrophys. J.* **167,** 359.
Bekenstein, J. D. (1972), Ph.D. Thesis, Princeton University (unpublished); available from University Microfilms, Inc., Ann Arbor, Michigan.
Bonnazola, S. and Maschio, G. (1971) *The Crab Nebula*, ed. by R. D. Davies and F. G. Smith (D. Reidel Publishing Co., Dordrecht).
Carter, B. (1972) to be published.
Chandrasekhar, S. (1969) *Ellipsoidal Figures of Equilibrium* (Yale University Press, New Haven).
Chandrasekhar, S. (1970) *Astrophys. J.* **161,** 561.
Chandrasekhar, S. and Friedman, J. L. (1972) *Astrophys. J.* **175,** 379; **176,** 745.
Fowler, W. A. (1966) in *High Energy Astrophysics*, edited by L. Gratton (Academic Press, New York).
Goldreich, P. and Lynden-Bell, D. (1965) *M.N.R.A.S.* **130,** 97, 125.
Goldreich, P. and Schubert, G. (1967) *Astrophys. J.* **150,** 571.
Harrison, B. K., Thorne, K. S., Wakano, M., and Wheeler, J. A. (1965) *Gravitation Theory and Gravitational Collapse* (University of Chicago Press, Chicago).
Hartle, J. B. and Hawking, S. W. (1972) *Comm. Math. Phys.* **27**, 283.
Hartle, J. B. and Sharp, D. H. (1967) *Astrophys. J.* **147**, 317.
Hawking, S. W. and Ellis, G. F. R. (1973) *Spacetime in the Large,* to be published.
Ledoux, P. and Walraven, Th. (1958) *Handbuch der Physik* **51,** 458, ed. by S. Flügge (Springer-Verlag, Berlin).
Morgan, T. and Morgan, L. (1969) *Phys. Rev.* **183,** 1097.
Oppenheimer, J. R. and Snyder, H. (1939) *Phys. Rev.* **56,** 455.
Ostriker, J. P. and Bodenheimer, P. (1972) to be published.
Ostriker, J. P. and Mark. J. W.-K. (1968) *Astrophys. J.* **151,** 1075.
Ostriker, J. P. and Tassoul, J. L. (1969) *Astrophys. J.* **155,** 987.
Press, W. H. and Teukolsky, S. (1973) to be published.
Price, R. H. (1972) *Phys. Rev. D* **5,** 2439.
Salpeter, E. E. and Wagoner, R. V. (1971) *Astrophys. J.* **164,** 557.
Schutz, B. F. (1972) *Astrophys. J. Suppl.* **24,** 319.
Smarr, L. (1973) *Phys. Rev. Letters* **30,** 71.
Thorne, K. S. (1971) in *Gravitation and Cosmology, Proceedings of the International School of Physics Enrico Fermi, Course 47*, ed. by R. K. Sachs (Academic Press, New York).
Toomre, A. (1964) *Astrophys. J.* **139,** 1217.

Observations of Galactic X-Ray Sources

Herbert Gursky

American Science and Engineering
Cambridge, Massachusetts

Contents

I Introduction 295
II Observational Techniques 297
- A Proportional Counters, 297
- B Mechanical Collimators, 300
- C Considerations on Statistics, 301
  - 1 Analysis of spectral data, 301
  - 2 X-ray pulsations, 304
  - 3 Association of X-ray sources with known objects, 305
- D Description of the X-ray Satellite UHURU, 306

III Distribution of Sources in the Galaxy 308
- A Longitude Distribution, 310
- B Maximum Luminosity, 311
- C Minimum Luminosity, 311
- D Spectral Content of the Radiation, 314
- E Effect of Anisotropic X-ray Emission, 314
- F Summary, 315

IV Properties of Individual X-ray Sources 315
- A Scorpius X-1, 316
- B Cygnus X-2, 319
- C Cygnus X-1, 320
- D Centaurus X-3, 323
- E Hercules X-1, 325
- F Other X-ray Sources of Interest, 330
- G Summary, 330

Bibliography 335
Appendix 337
- A Galactic X-ray Sources, 337
- B Supernova Remnants, 340
- C Transient Sources, 340
- D Magellanic Cloud Sources, 341

I Introduction

The reason that lectures on observational X-ray astronomy are being included in a course on Black Holes is principally that many astrophysicists are convinced that X-rays may comprise the most significant observable emission from the vicinity of a Black Hole; alternatively, there are many who believe that one of the few ways in which the X-rays can be produced is by mass accretion onto collapsed objects, especially Black Holes. Whichever view one takes, one is compelled to search through the X-ray data for evidence of collapsed objects in general and Black Holes in particular.

By way of introduction it is useful to review some well-established facts concerning the X-ray sources.

1. Other than the supernova remnants, which comprise only ~10 per cent of the observed sources, the sources do not appear to be associated with known kinds of stars. By that we mean simply that if one examines the regions containing X-ray sources, an association can not be established with objects such as old-novae, planetary nebula, Wolf–Rayet stars, etc. In fact based on a few optical identifications the evidence is that an entirely novel kind of stellar system is producing the X-ray emission.
2. The luminosity in X-rays of the sources lies in the range 10^{36}–10^{38} erg/sec, which is so high that the origin of the power is not likely to be nuclear burning. The sun for example radiates no more than 10^{-6} of its power through X-rays and while it has been speculated that certain stars can radiate as much as a few percent of their internally generated power by non-thermal processes, the possibility of a star that can internally generate much more than (10^{36}–10^{38}) erg/sec and radiate a significant fraction in the form of X-rays is not considered seriously.

For these reasons it has been fruitful to examine other sources of energy than nuclear burning and to consider hitherto unrecognized stellar systems. Among others, there are two energy sources which are very attractive. One of these is mass accretion onto a small object; the source of energy is then the gravitational field. Historically, one reason for the interest in this mechanism was that Sco X-1, when it was first identified, was believed to be an old nova for which it was believed that a mass exchange binary system was involved.

Table 1 lists the kinetic energy (KE) accrued by a proton falling onto the surface of various stellar objects along with the rate of mass accretion (dm/dt) required to produce a total energy infall of 10^{36} erg/sec.

There is evidence for mass transfer of order of 10^{-7} $m_\odot$/yr in close binary systems, of mass loss in early-type stars of order 10^{-5} $m_\odot$/yr; and theoretically even larger mass transfer rates are invoked for mass exchange binaries. For these reasons mass accretion, especially onto the degenerate stars, is very attractive.

Another energy source which must be taken seriously is the rotational kinetic energy in rapidly rotating stars; especially, since for one X-ray source, the Crab Nebula pulsar NP0532, this is believed to be the energy source. In the case of this object, accepted to be a neutron star of 1 $m_{\odot}$, stored rotational energy is of order 10^{49} erg. Based on the observed pulsar slowdown rate, the rate of loss of rotational energy is of order 10^{38} erg/sec which just balances the total radiated power of the pulsar and the extended region comprising the Crab Nebula. Of this radiation about 10^{36} erg/sec is in the form of X-rays coming from the pulsar itself.

Extending the introduction a bit farther it is useful to describe very briefly two X-ray sources which appear to be collapsed objects accreting mass. One of these is Cyg X-1. The X-ray emission is highly variable down to a time scale of 0.1 sec thus requiring a very small emitting volume. Optically a 9^m B0I is observed which is a double-line spectroscopic binary with a period of 5.6 days. Based on the ratio of the velocities of the spectral lines, the mass ratio is 1.4. The primary, the B0I, must be of mass of order 20 $m_{\odot}$; the secondary, which is unseen

TABLE 1. Energetics of Mass Accretion

Stellar object	KE	dm/dt (for 10^{36} erg/sec)
Sun	1 keV	10^{-5} $m_{\odot}$/yr
White Dwarf	100 keV	10^{-7} $m_{\odot}$/yr
Neutron Star	100 MeV	10^{-10} $m_{\odot}$/yr
Black Hole	~1 GeV	~10^{-11} $m_{\odot}$/yr

optically, must then be about 15 $m_{\odot}$. The most straightforward interpretation is that the secondary is the source of the X-rays by mass accretion from the B0I. Then since the X-ray source must be a collapsed object, the secondary must be a Black Hole, since the only other known possibilities, neutron stars and white dwarfs, cannot exist with such large masses.

The other X-ray source which is believed to involve a collapsed object is Her X-1. In this object the bulk of the X-ray emission is pulsed with a very stable period of 1.24 sec. The system is an eclipsing binary with a period of 1.7 days. The limiting mass of the X-ray source, based on the mass function $(M_s^3/(M_p + M_s)^2)$ and on the possible dimensions of the Roche Lobes, is 0.2–3 $m_{\odot}$. The most straightforward interpretation of this system is that the X-ray source itself is a magnetic neutron star rotating with a period of 1.24 sec, and the X-rays are produced by the same process as is occurring in NP0532. The energy source now is taken to be mass accretion from the second star.

There are problems with both these objects which will be discussed in detail below, but they indicate the basis for the interest in the X-ray sources.

These lectures are organized into three sections; first, a section on observational techniques in X-ray astronomy; second, a section on the distribution of the X-ray sources in the Galaxy from which can be inferred certain average properties of the sources; and finally, a detailed presentation of the observed characteristics of the X-ray sources.

II Observational Techniques

In this section are described elements of the hardware and data analysis that are most directly related to the X-ray results. We also describe the X-ray satellite "UHURU" which has provided much new data about the sources since its launch in December 1970.

A Proportional Counters

The range of photon energy involved in X-ray astronomy extends from about 0.2 to 200 keV; however, most of what we know about the X-ray sources is based on data gathered below about 20 keV. In this energy range the gas proportional counter has proven to be an ideal detector, and it is employed almost exclusively.

The device consists of an enclosed volume of gas in which there is situated an anode, almost always in the form of a thin (~25 microns) wire operated at positive high voltage (~2500 volts). X-rays enter the gas through a thin window and are absorbed in the gas by the photoelectric effect. The efficiency and bandwidth of the device are determined by the transmission of the window and the absorption in the gas; namely,

$$\epsilon(E) = \exp(-\sigma_w t_w)[1 - \exp(-\sigma_g t_g)] \tag{1}$$

where

$\epsilon(E)$ is the efficiency for photons of energy E,
σ_w, σ_g are the photoelectric absorption coefficients of the window and the gas respectively,
t_w, t_g are the thickness of window and the gas.

At low energies, the window absorption dominates the efficiency. For example, using beryllium of 25μ thickness, the efficiency falls rapidly below about 1.5 keV. Measurements down to 0.2 keV require windows of organic films (e.g. formvar) of about 1μ thickness. At high energy the transmission of the gas dominates the efficiency. Argon, which is a very commonly used gas, can provide high efficiency to about 20 keV; above this energy it is necessary to use a heavy gas such as Xenon. Because of these factors it is not possible to construct a proportional counter that can function efficiently over more than a portion of the X-ray range; e.g., 2–10 keV. Within this restricted range, however, the X-ray detection efficiency is very high, essentially 100 per cent.

If the entire energy of the incident photon is absorbed by the gas, a number of ion pairs, electrons and positive ions, is formed, $E/30$ (E in eV). The electrons drift toward the anode wire under the influence of the high voltage. Within a few wire diameters distance of the anode wire, the electric field is of such an intensity that additional ion pairs are formed and an avalanche results, eventually yielding a multiplication of 10^4 to 10^5 of the number of initial electrons. The electrons are collected on the central wire and are detected externally in appropriate electronics as a pulse of a certain amplitude.

The significance of these characteristics is that the device is automatically a quantum detector; i.e., each absorbed photon yields a detectable signal that can be recorded in external circuits. Also the amplitude of the signal is proportional to the photon energy.

It is on this latter point however that the difficulties begin with proportional counters. For a variety of reasons, one of which is simply that there are statistical fluctuations in the number of ion pairs initially created in the gas, there is a large spread in the amplitude of the detected signals from photons depositing a fixed energy in the gas. This spread is about 25 per cent at 2 keV and varies approximately as $E^{-1/2}$. Furthermore, the energy deposited in the gas, and thus the average number of ion pairs created, is multivalued for photons of fixed energy. The reason for this is that following absorption of a photon of energy E by an atom of gas, one of two things can happen. First, and fortunately, the more probable occurrence is that two electrons are ejected, the sum of whose energy is very close to E. One of these is a photoelectron of energy $E - E_b$ where E_b is the binding energy of the atomic shell in which the photon is absorbed, and the other is an Auger electron of energy E_b. Secondly, instead of the Auger electron, a photon of energy E_b is reradiated by the atom. If this photon is absorbed by the gas it is likely that its total energy, E_b, and hence the total energy, E, of the incident photon appears in the gas. However, there is a significant probability that this second photon will escape, in which case only an energy $E - E_b$ appears in the gas. The net result in Xenon for 5.9 keV incident X-rays is shown in Figure 1. Two peaks appear, one for which all the photon energy appears in the gas; the other at about 4 keV less for which the reradiated photon–here the L-X-ray of Xenon escapes the counter. In principle one must consider all possible atomic shells in which the photon can convert; in practice only the highest energy shell need be considered.

The net effect of the finite width and the multivalue of the amplitude distribution of counter signals is that the determination of the spectrum of incident X-rays from the observed distribution of the amplitudes of the signals is not trivial and generally not unique. It is usual to analyze such data by the X^2 method, whereby a trial spectral function is used to calculate an amplitude distribution. The free parameters in the trial function are varied until X^2 is minimized. The values of the parameters at this point are the most likely values corresponding to that particular function. The uncertainty in the parameters is determined by varying them in the vicinity of the minimum X^2, and the goodness of the fit (i.e., the probability that

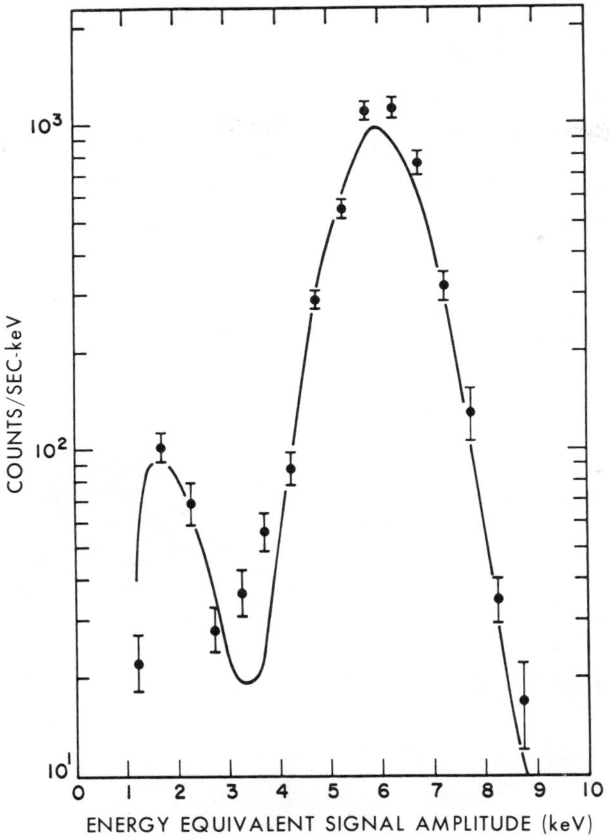

Figure 1. Proportional counter response to 5.9 keV X-rays. The counter gas was Xenon. The peak at 2 keV is caused by the escape of *L* X-rays from the Xenon. The curve is a best fit to the data points based on the known characteristics of the counter and of the gas.

a given spectral function actually fits the data) is determined from the value of X^2 min and the number of degrees of freedom in the data set. The amplitude distribution is calculated from an equation of the form,

$$\frac{dN}{dH} = \int_E \frac{d\phi(E)}{dE} \epsilon(E) [P_1(E, H) + P_2(E, H) + \ldots] \, dE \tag{2}$$

where $d\phi/dE$ is the incident spectrum expressed a number of incident photons of energy E, $\epsilon(E)$ is the detector efficiency and $[P_1 + P_2 + \ldots]$ is the resolution function as shown in Figure 1; that is, the sum of the probabilities that a given photon will yield a given amplitude according to the several processes described. This method of analysis has been described in detail by Gorenstein, Gursky and Garmire.[1]

B Mechanical Collimators

It is necessary to restrict the arrival directions from which X-rays can reach the detectors. For this purpose it has been customary to use a mechanical collimator composed of slats, tubes, wire planes, etc. These are characterized by two dimensions; w, the width of the opening, and l, the length of the slats as is illustrated in Figure 2. Then the total acceptance angle for radiation is $2w/l$ radians. The response function of such a collimator; i.e., the transmission as a function of angular deviation from a central axis, is in the shape of a triangle. The peak transmission defines the central axis of the collimator; outside $\pm w/l$ radians from the central axis, the transmission is zero for a perfect collimator. It is customary to

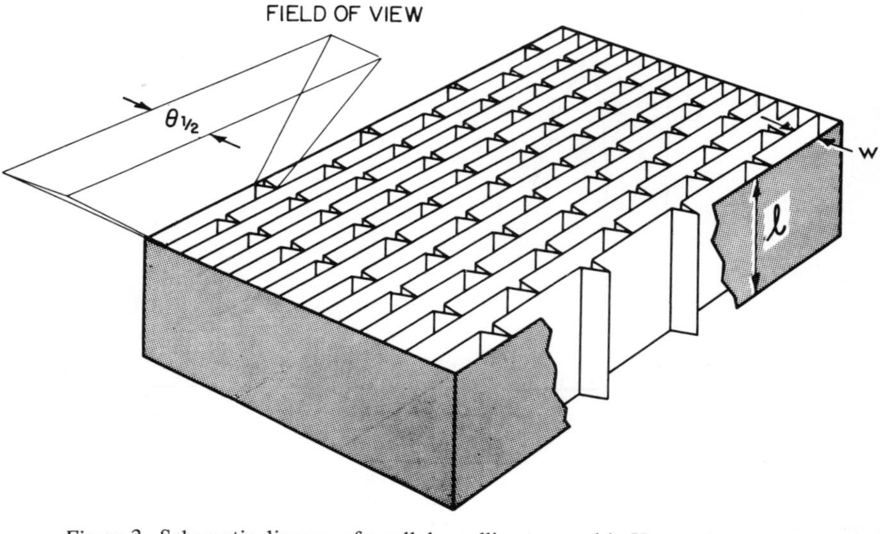

Figure 2. Schematic diagram of a cellular collimator used in X-ray astronomy.

characterize the transmission function by its full width at half the maximum transmission (FWHM) which is just w/l. As shown in Figure 2, a collimator can always be rotated around each of two axes; thus, for completeness the response (FWHM) of the collimator must be specified for both these axes.

If a collimator is rotated past a point X-ray source, a variation in counting rate is observed which is in the shape of the triangular response function. The peak of the counting rate curve is the direction of closest approach to the X-ray source. At this point the X-ray source must lie along a line normal to the direction of collimator rotation; the length of the line is equal to the collimator response in the direction of that line.

The quantity w/l is similar to the resolution of an optical telescope. Adjacent point X-ray sources can be resolved if their separation is greater than w/l and the structure of extended sources can be determined in units of w/l. Also a finite angular size (smaller than w/l) X-ray source can be distinguished from a point source based on the deviation of the observed response of the collimator from the expected response to a point source.

C Considerations on Statistics

As noted, it is usual to detect and record individual X-ray photons; thus, the primary data is a number of recorded events. Since the incident photons arrive at random, the number of events, N, is subject to statistical fluctuations of order $\sqrt{N}$. If N is sufficiently large (at least ~10), $\sqrt{N}$ is the standard deviation. Because of the very low photon fluxes, it is more common than not, that the uncertainties in any given result are dominated by $\sqrt{N}$. In a given experiment there will generally be a continuous background of events which can be expressed N_B, the number of counts recorded per angular resolution element w/l. If any X-ray source is present, it will contribute an additional number of events N_s. Then the uncertainty in the determination of the intensity of the source, I, is just

$$\delta I/I = \sqrt{N_s + N_B}/N_s. \tag{3}$$

Also, the source can be located (i.e., the centroid of the collimator response) with a precision,

$$\delta\theta = (\delta I/I)(w/l) \tag{4}$$

It is fruitless to present all the kinds of analysis that are performed in X-ray astronomy; it is useful however to present three examples of how statistical consideration apply.

1 Analysis of spectral data

As noted above, the amplitude (pulse height) distribution of recorded X-ray photons can be analyzed to uncover the incident photon spectrum using a X^2 technique and a trial function. A typical trial function is

$$E(d\phi/dE) = K[\exp(-E/kT)][\exp(-\sigma_H N_H)] \tag{5}$$

which is the expected distribution law for photons originating by the thermal bremsstrahlung process. There are three free parameters in this function; namely, K, the absolute intensity of the source, kT, the temperature of the source, and N_H, the column density of hydrogen along the line of sight to the source (σ_H is the effective X-ray absorption coefficient per hydrogen atom). The latter term results from the fact that the X-rays are strongly absorbed in the interstellar medium and in material in or near the X-ray source itself. The value of N_H at which the absorption is $1/e$ is plotted in Figure 3. Then $\sigma_H = 1/N_H$.

As noted, in a given experiment, because of the poor energy resolution and finite bandwidth, the amplitude distribution data only has a small number of independent elements not much greater than three. Thus, it is always possible to determine the best value of the three parameters in the above trial function but it is not always possible to determine whether one particular trial function yields a better fit to the data than another trial function.

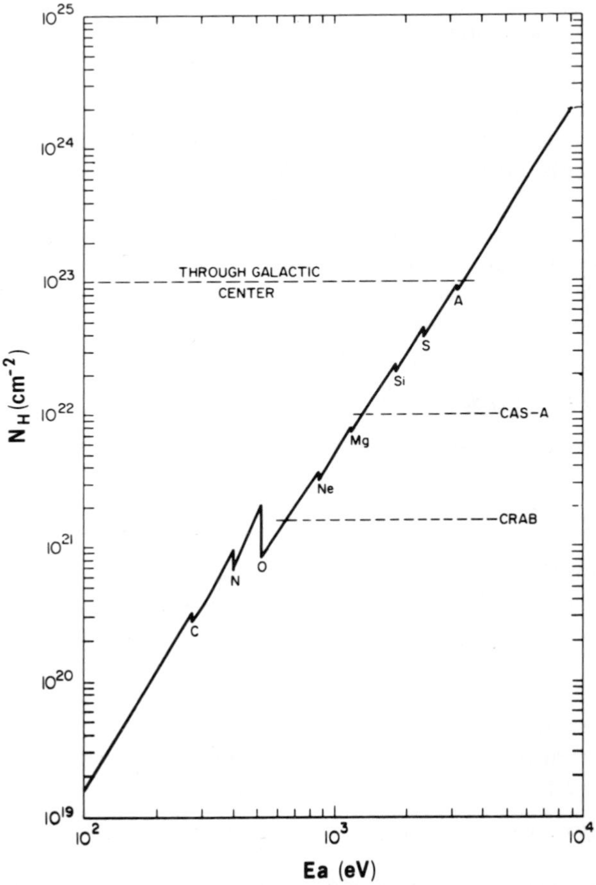

Figure 3. Photoelectric absorption in the interstellar medium. The vertical axis gives the column density in units of hydrogen atoms/cm^2 at which the absorption is $1/e$ at the appropriate photon energy, E_a.

To complicate the problem further, it is known from observations that certain spectra must be composite; i.e., they require a more complex trial function than the one listed above; e.g., a mixture of an exponential plus a power law spectrum. In this case the number of independent free parameters rises to five and it may be impossible to fit the spectrum uniquely, even with data of high statistical precision.

Figure 4 shows an example of an attempt to fit X-ray spectral data from the

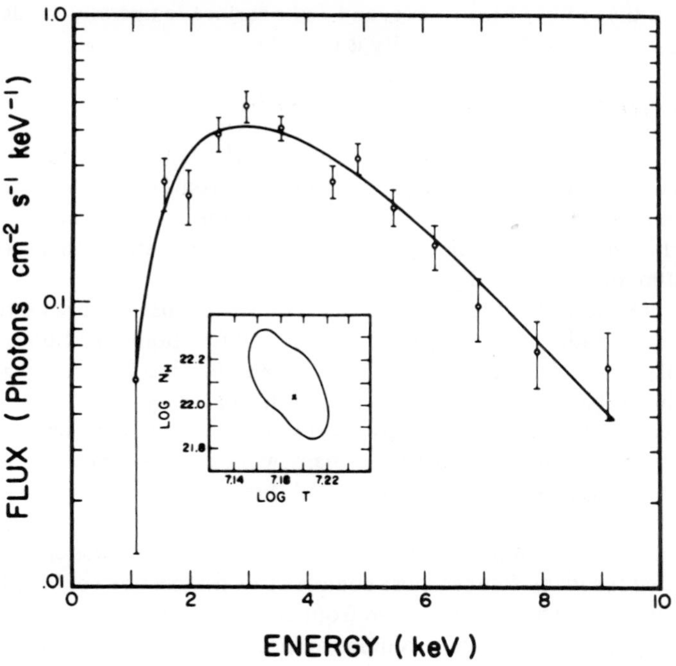

SPECTRAL ANALYSIS OF GX 340+0

Model	T or n*	N_H [cm^{-2}]	Intensity† 1–10 keV [keV cm^{-2} s^{-1}]	Confidence [percent]
Blackbody..........	1.55×10^7 ° K	1.20×10^{22}	8.29 ± 0.42	25.2
Power law...........	-2.25	5.01×10^{22}	9.76 ± 0.50	3.8
Bremsstrahlung.....	7.08×10^7 ° K	3.98×10^{22}	7.85 ± 0.40	8.3

* As used in this paper, n is the exponent describing the photon distribution. The power flux index is $n+1$.

† Errors quoted are statistical only.

Figure 4. X-ray spectrum of GX 340 + 0 (2U 1641–45) as measured by Margon *et al.*[2] Table shows result of fitting various spectral shapes.

X-ray source GX 340-0[2] (2U1641–45). These data are of some interest because they were believed to be evidence that the source was at least the size of a neutron star. The argument was simply that a black-body spectrum was the best fit to the data based on the probability of obtaining a certain X^2 min. Thus, the total X-ray luminosity is given by

$$L_x = \pi R^2 \sigma T^4 \tag{6}$$

where R is the radius of the source, and T the surface temperature is determined from the spectral fit. The luminosity is given by

$$L_x = 4\pi D^2 I \tag{7}$$

where I is the measured X-ray intensity and D is the distance to the source. D can be estimated from the value of N_H obtained again from the spectral fit. On this basis a value of R is obtained of order 10 Km; i.e., appropriate to a neutron star. Thus, the conclusion is that the object is most likely a neutron star with a surface temperature of order 10^7 °K.

There are two difficulties with this analysis. One is that there is no formal criterion for deciding between one of several trial functions based on the probability of obtaining a given X^2. It is common to use a 95% criterion; i.e., as long as the probability of achieving a given value of X^2 exceeds 5 per cent, the fit is considered acceptable. On this basis two trial functions, the black-body and the exponential fit the data. Furthermore, many workers take a more conservative view of such analyses and would say that all three trial functions fit the data adequately.

The second problem is that the simplest spectral function was chosen whereas it was noted for many sources more complex spectra must be applied. In fact this source is one such object; it is known from other experiments to possess a high energy "tail" of X-rays. Thus, a composite spectrum should be used as a trial spectrum, which increased the number of independent parameters to five and reduces the chance of distinguishing one spectral type from another. The purpose of presenting this example is not to dispute the analysis made by Margon *et al.*,[2] which is perfectly correct, but only to point out the kinds of ambiguities that often arise.

2 *X-ray pulsations*

When the X-ray source Cyg X-1 was first observed from the satellite "UHURU", it was found by Oda *et al.*[3] to display large and very rapid bursts of X-rays which appeared to be periodic. In fact these data were used to argue that Cyg X-1 was a black-hole. Subsequently, Terrell[4] showed that purely random bursts of X-rays could yield an apparent periodic component when subject to a fourier analysis.

The difficulty here is that if the X-ray intensity is organized into short bursts occurring at random times, there is a significant probability that a number of the individual bursts will be equally spaced. Upon fourier analysis, these can yield a statistically significant period component. As additional data are analyzed, the periodic component will fade away, and independent sets of data will not reveal the same period. This is approximately what is observed in the case of Cyg X-1. Thus, there are two alternate explanations of the X-ray pulsations from Cyg X-1; one is that there is in fact a small quasi-periodic component to the X-ray emission,[5] or the X-ray pulsations are entirely random.

3 *Association of X-ray sources with known objects*

Finally, we consider a problem which is central to X-ray astronomy at the present time, just as it was to radio astronomy before it was possible to measure positions of radio sources with high accuracy. In X-ray astronomy it is now possible to determine the position of the sources with a precision in the range of arc minutes to a degree. There are only a very few sources for which the position is known with a precision of arc seconds. The result is that for most X-ray sources, the solid angle of uncertainty in the sky in which the X-ray source is located is typically a significant fraction of a square degree. Because of the high density of optical and radio objects in the sky, it is difficult to establish the existence of an X-ray source by positional coincidence alone, especially when there is no *a priori* basis for a particular coincidence. For example, when the binary X-ray source Cen X-3 was first reported, a search through the catalogues of known binary stars revealed an object LR Cen within $1/2^{\circ}$ of Cen X-3 which was a spectroscopic binary with a reported period that agreed with that of Cen X-3 within 1 part per 1000. Many people were convinced that this could not be a chance coincidence, and the small disagreement in period of the two was simply due to an error in the original data on LR Cen (about 40 years old) or was a real change in the orbital period. The latter was considered a real possibility since Cen X-3, based on the X-ray observation was known to be imbedded in a dense cloud of material. However, recent measurements of the period of LR Cen[6] have confirmed the historical data. Improved positional data on Cen X-3 have excluded LR Cen as a possible candidate as well.

There are numerous other examples of this kind of coincidence in the literature. The problem of misidentifications is a serious one and has plagued X-ray astronomy as it did radio astronomy 20 years ago. For this reason, in the discussion of individual objects, positive identifications are claimed only for those objects where there is correlated behavior observed. The problem is whether or not optical candidates are present in a given field. The volume occupied by 1 min^2 solid angle for a distance of 1 Kpc is about 30 pc^3; in the plane for this distance, about 30 stars would be expected to be present. More than this number would be present if the view angle is along a spiral arm.

However, it is possible to make a strong case for a group of identifications based on the positional coincidence between X-ray sources and a unique class of objects. For example eight supernova remnants are coincident with X-ray sources within the positional errors. The likelihood that this is a chance coincidence is negligibly small particularly since for two or three, the identification must be regarded as positive, and the others are among the brightest of the known supernova remnants. If the association were random, the spatial coincidences would occur with the faintest remnants which are the most numerous. Conversely, it can be established that the bulk of the galactic X-ray sources are *not* associated with supernova remnants.

Two other examples of such identifications by associations are noteworthy. One is that there is a class of extended X-ray sources associated with rich clusters of galaxies,[7] and more closely related to these lectures, Gursky[8] has shown that there is a significant association of galactic X-ray sources with bright early type stars.

The only problem with statistical associations of this kind is that it is always possible that one or more members of the association is a chance occurrence.

D Description of the X-ray Satellite UHURU

The first X-ray sources were discovered in 1962[9] and up until 1971, virtually all the observational results in the field were obtained by sounding rockets and balloons. The first satellite devoted exclusively to X-ray astronomy was designed and developed by Giacconi and his colleagues at AS & E for NASA and was launched off the Kenya coast in East Africa into a near-equatorial orbit during December, 1970. The satellite was officially designated "UHURU", the Swahili word for "freedom", in recognition of Kenyan independence day which happened to coincide with the launch date.

The X-ray instrumentation and the operations of the satellite have been described by Giacconi *et al.*[10] and here we recount only certain essential elements.

In any X-ray astronomy payload it is necessary to orient the vehicle to point the X-ray detector axes at particular regions of the sky. The UHURU satellite is spin stabilized and the detectors view normal to the spin axis. There are two sets of detectors each comprising about 10^3 cm^2 of beryllium window proportional counters and mechanical, slat collimators. One set of detectors is collimated to $1/2° \times 5°$; the other to $5° \times 5°$ (FWHM). Thus, as the satellite spins, a 5° wide band in the sky is viewed by the two detectors, one with 1/2° resolution the other with 5° resolution. The scanning geometry is shown in Figure 5. The spin axis can be oriented to a given point in the sky by a magnetic torquing system (a set of coils which set up a magnetic field in the satellite which then reacts against the Earth's magnetic field). To scan a particular object, the spin axis is set to a point 90° from the object. The spin rate is controllable and is normally 10 minutes. Thus, during the course of a day a given band is repeatedly scanned. The entire sky is scanned by moving the spin axis the order of 5° per day. The spin axis is relatively stable (the order of 1°/day deviation from a given orientation). Photoelectric star sensors, coaligned with the detector axes are used to sight known stars during the scan, from which the true orientation of the detectors is determined at each instant of time.

Figure 6 shows a section of the raw data from the satellite. X1 is the accumulated count per 0.096 seconds from the 1/2° detector, during which time interval, the satellite moves about 4 arc minutes. At one point in the X1 data an X-ray source is traversed as evidence by the triangular shaped peak in the accumulated count. This small peak is the primary data of the UHURU satellite. The point

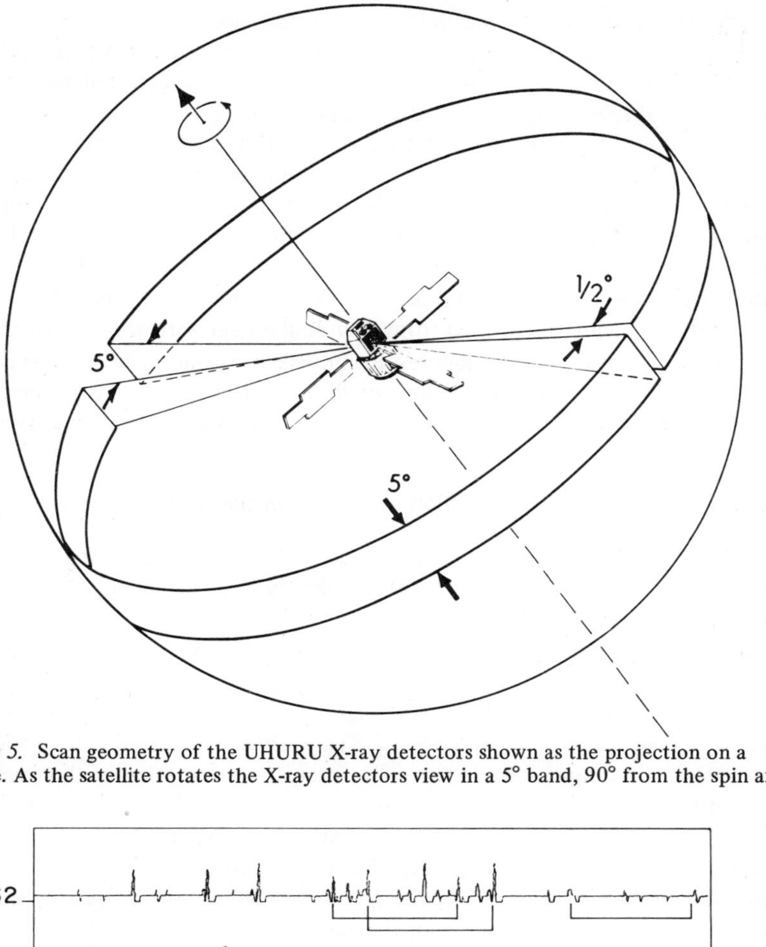

Figure 5. Scan geometry of the UHURU X-ray detectors shown as the projection on a sphere. As the satellite rotates the X-ray detectors view in a 5° band, 90° from the spin axis.

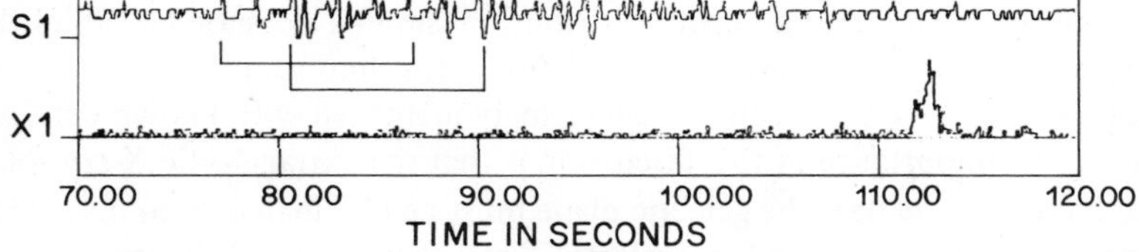

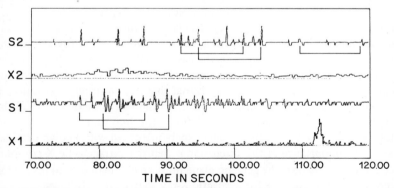

Figure 6. Portion of the raw data from UHURU satellite. S1 and S2 are the outputs of the star sensors; X1 and X2 are from the X-ray detectors from the two sides of the experiment, respectively. Star pulses appear in groups of three with the outer two always at a fixed separation (shown with connecting lines). The X-ray data is in the form of accumulated counts per 0.096 sec in X1 and per 0.384 sec in X2. X-ray sources are traversed from 75–90 sec in X2 and 112-114 sec in X1.

of maximum count rate defines the location of the source to a line of about 5° length in the scan band. The integrated count in the peak yields the intensity. The same X-ray counts are sorted in seven other channels according to amplitude and these yield spectral data. Variations in intensity can be found by looking for variations from sighting to sighting. On a shorter time scale, variations can be found by looking between successive count accumulation elements, taking proper account of the collimator response function. The UHURU satellite can only detect X-rays in the range 2–20 keV and with limited angular resolution. Important experiments continue to be performed in the rest of the X-ray range; particularly, in the range 0.2–2 keV, in which the X-ray sky seems very different. Also X-ray experiments are being performed from sounding rocket with very different instruments—focusing optics, high resolution Bragg crystal spectrometers and polarimeters. While the results to date are limited, these are the instruments of the future. They will at some time dominate the field, just as UHURU results are now.

III Distribution of Sources in the Galaxy

It is useful to look at the X-ray sources as a whole before considering the properties of individual sources. The reason for this is simply that certain characteristics emerge that are not apparent for single sources.

First, it is necessary to establish which X-ray sources we are in fact discussing. The UHURU catalog[11] lists 125 X-ray sources; these are plotted on a sky map in Figure 7. The sources divide themselves naturally into two groups, those that are clustered along the galactic plane ($b < 20°$) and those at high galactic latitude ($b > 20°$). There is a real difference in the intensity distribution of the sources; those on the galactic plane are on the average much more intense than those at high latitude. This is just opposite from what is to be expected if both groups were a single population in the galaxy. The latitude of the sources could be taken as an indication of distance; those at high latitude would be nearby and much brighter than those in the plane, as is the case for common stars. However, just the reverse is seen. In fact, it can be shown that almost all the high latitude sources are extragalactic. The evidence for this is that they are uniformly distributed in b and l, and the log number-intensity distribution shows a slope of 1.5; all of these are appropriate to an extragalactic population. Most importantly about 12 of the weak high latitude sources can be convincingly identified with known external galaxies. The importance of this discussion is that the extragalactic X-ray sources which happen to lie near the galactic plane must be eliminated from discussion. The criterion we adopt here for a galactic X-ray source are as follows.

1. The intensity must be greater than 8 ct/second,
2. The source is not identified with a known external galaxy.

We also eliminate from consideration sources identifiable with supernova remnants of which there are six in the UHURU catalog, transient sources, and the galactic

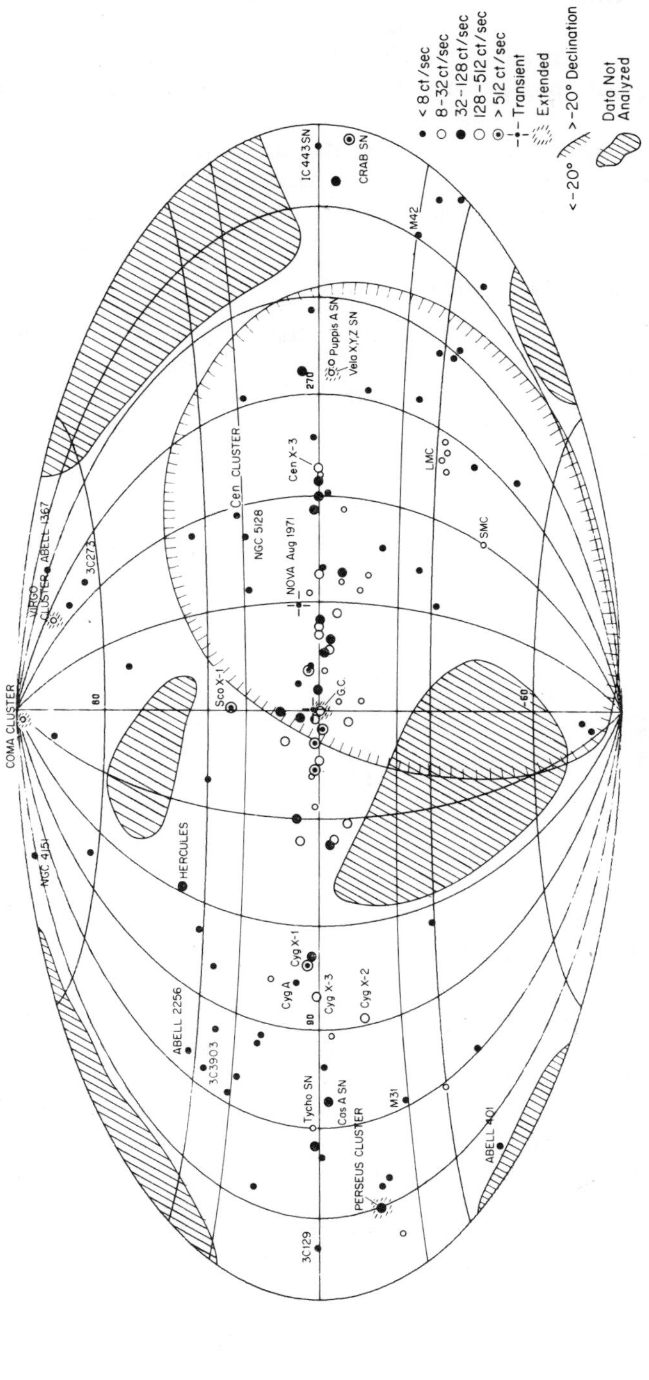

Figure 7. Distribution of X-ray sources from the 2 UHURU catalog, plotted in galactic coordinates.

center. What remains are 55 X-ray sources almost all of which must be within our galaxy. They are close to the galactic plane (the average latitude is about 4°), and their distribution around the plane is highly irregular. These are features associated with a galactic population. Furthermore, it has not been possible to identify any significant number of these with a known class of galactic objects. It is these 55 objects which are of interest here. They are listed in the Appendix. For completeness we list as well the supernova remnants associated with X-ray sources, the transient X-ray sources and the discrete sources observed in the Magellanic Clouds.

A Longitude Distribution

Of the 55 X-ray sources, 32 are concentrated in the Sgr–Sco region between galactic longitudes $19° < l < 322°$. This is to be expected since in these directions we are looking across the bulk of the galaxy. The remaining 23 are distributed highly nonuniformly around the remaining 300 degrees of galactic longitude as shown in Table 2.

TABLE 2. Distribution of Galactic Sources in Longitude

Range of l (°)	Number Observed	Number Expected	Region
322–19	31	–	Sgr–Sco
19–30	None	1	
30–37	5	0.5	Serpens
37–68	1	2.5	
68–92	7	2	Cygnus
92–283	4	15	
283–304	7	1.5	Centaurus
304–322	None	1.5	

The number expected is calculated on the basis of a uniform distribution in l. The observed distribution is far from uniform, but rather shows a concentration in three narrow ranges in l; 30–37°, 68–92°, and 283–304°. Furthermore, these are the very regions where spiral arms are known to occur. The Cygnus and Centaurus regions in particular are believed to define one single spiral arm which contains the sun and is delineated based on 21 cm radio observations and optical spiral arm indicators such as O–B associations and HII regions. The Serpens region appear in radio as a distant but prominent spiral arm. Thus, the conclusion based on the concentration of sources in regions of known spiral arms, is that most of the X-ray sources in the range $20 < l < 320°$ are within the outer spiral arms. It is also likely that many of the sources at low galactic longitude (332–19°) are within the inner spiral arms; however, there is no information on the distribution in these directions and it is possible that many of the low longitude sources are not spiral arm sources.

B Maximum Luminosity

Five discrete sources have been found in the Magellanic Clouds with an intensity of 10–20 ct/sec. (The intensity is listed here as the peak counting rate observed from the X-ray source from UHURU. The conversion to absolute units is approximately 2×10^{-11} erg/cm^{2}per count.) Since the Magellanic Clouds are at a distance of about 60 Kpc, the luminosity of these sources must be on the order 10^{38} erg/sec. We assume that the X-ray sources in the Magellanic Clouds are of the same kind as exist in our own galaxy; thus, 10^{38} erg/sec must be the highest luminosity among the galactic sources.

From consideration of the galactic sources alone, there is independent evidence that 10^{38} erg/sec is the maximum luminosity. The argument is simply as follows. We consider the sources in the Sgr–Sco region between $320 < l < 20°$; we assume that these must be symmetrically placed with respect to galactic center which is at a distance of about 9 Kpc from the Sun. Thus, the average distance of the sources must be 9 Kpc. The spread in distance from the center of the galaxy is then just the maximum l at which the sources are seen; namely 20° on one side and 40° on the other. If these longitudes are taken to define the boundary of a disc centered on the galactic center, the radius of the disc is about 5 Kpc. The sources in this region then range from 5 to 15 Kpc distance. Within a factor 2 then we can find the maximum possible luminosity of these sources by assuming their distance to be 10 Kpc. The brightest sources in this region are seen with an intensity ~700 ct/sec, which at a distance of 10 Kpc, represents a luminosity of 1.4×10^{38} erg/sec. About ten of the other Sgr–Sco sources are seen with such a counting rate that their luminosity would be 10^{38} erg/sec if placed in this inner volume of the galaxy. There is another piece of information regarding these sources that helps place them in the galaxy; namely, the deficiency of low energy photons which is taken to be the result of photoelectric absorption in the interstellar medium. At least seven of the Sgr–Sco sources show significant absorption effects which can be attributed to the interstellar medium.[12] These can be placed at distances of order 10 Kpc and their luminosity is of order 10^{38} erg/sec. They are designated as "Distant" in the source list in the Appendix. There is no way to avoid the $L = 10^{38}$ erg/sec limit. Any significant number of $L = 10^{39}$ erg/sec sources would all be on the far side of the galaxy of order 30 Kpc from us, which is implausible.

It must be kept in mind that it is not possible to tell with certainty the luminosity of a particular one of the seven X-ray sources labelled as distant, since the observed photon deficiency could be intrinsic to the source. Also, there may be other X-ray sources of such high luminosity; for example, Sco X-1, the brightest X-ray source, would be radiating at 10^{38} erg/sec if its distance is 2000 pc.

C Minimum Luminosity

It is apparent that most of the galactic sources can not be radiating at $L = 10^{38}$ erg/sec. They must be substantially less luminous; otherwise, they would be outside the confines of the galaxy. It is not so direct to obtain the minimum luminosity of the X-ray sources, but there are several arguments which indicate that there are few if any sources that radiate at levels below about 10^{36} erg/sec. These arguments all relate to the relative absence of weak, resolved sources or of a diffuse component of radiation along the galactic plane that could be attributed to distant, unresolved sources. Consider for example the Cygnus region between 70 and 90° l in which 7 discrete sources are observed ranging in intensity between 8 to 1000 ct/sec. The UHURU catalog lists no other sources below 8 ct/sec in this region. Thus, on the average there is one per 3° of l; since the fine UHURU collimator is 1/2° FWHM, there is no reason to expect a significant number to be lost by source confusion. These sources are listed in Table 3 with their distance calculated on the assumption that their luminosity is 10^{36} erg/sec.

The only embarrassment is the source 2U(1926 + 43) which would be a height, Z, above the plane of 1600 pc if assigned a luminosity of 10^{36} erg/sec, also a source not listed here 2U(1735 + 43). This source is at $b = 31°$ and a strength of 17 ct/sec.

TABLE 3. Distance of Cygnus Sources

Source	l, b	I (ct/sec)	Distance (pc) R	Distance (pc) Z	Location
(1954 + 31)	68, 1.7	75	2800	90	Cygnus Arm
(1956 + 35)	71, 3	1175	700	35	Cygnus Arm
(1926 + 43)	76, 12	8	8000	1600	?
(2030 + 40)	80, 1	133	1700	30	Cygnus Arm
(2142 + 38)	87, 11	420	1000	200	Cygnus Arm
(2130 ÷ 47)	92 – 3	12	6300	300	Outer Arm

It is listed in the UHURU catalog with a positional uncertainty of 17 deg^2 which is unusually large. Thus, its existence is questionable; also it may be extragalactic.

Ignoring these two objects, four sources which are in the Cygnus Arm are spread along the spiral arm with an average spacing of about 500 pc. If we assume that another spiral arm at about 6 Kpc runs through this region of l (it would be about 11 Kpc from the galactic center which is about the limit of the galaxy), we would be seeing about 2000 pc of length of this arm between 70 and 90° of l, in which four sources might be expected at a count rate of ~12 ct/sec where one or two is observed. Since we do not know about the existence of such a spiral arm or whether to expect the density of sources to be so high in the very outermost spiral arms, this is not a discrepancy. If on the other hand the luminosity of the X-ray sources is dropped to 10^{35} erg/sec the distances drop by a factor of 3, the average separation drops correspondingly and discrepancy between the expected and observed number of weak sources becomes more severe. Furthermore, as will be discussed in the next section, from their observed properties, the distance to 2U(1956 + 35) (Cyg X-1) is estimated to be between 1–2 Kpc and the distance to 2U(2142 + 38) (Cyg X-2) is estimated to be between 500–900 pc which supports the distance estimates above. This argument can be extended to other regions outside the Sgr–Sco region with a similar result. The essential result is that the sources are consistent with an average luminosity of 10^{36} erg/sec and there can not be even an equal number radiating at $L = 10^{35}$ erg/sec.

This argument can also be used to find the total number of sources in the Galaxy. Assuming the radius of the Galaxy to be about 12 Kpc, the region between 20 to 320° l contains more than 1/2 the volume of the Galaxy. In this region are seen 24 sources, thus 24 more are expected in the remaining region of l where 31 are actually reported. Here, however, one must allow for obscuration and confusion effects and for some increase in the source volume density. However, the above picture is consistent with no more than about 100 sources in the galaxy with $L = 10^{36}$ erg/sec and with a smaller number at lower luminosities.

An alternative method of arriving at the minimum luminosity is to look for a diffuse component of radiation along the galactic plane which could be attributed to distant, unresolved sources. The strength of this component is related directly to the average luminosity of the sources and the distance of the farthest resolved source. Arguments of this kind were presented already before the UHURU data by Ryter[13] and by Setti and Woltjer[14] and have been more recently summarized by Salpeter[15].

Assuming a unique class of objects of constant luminosity, L, the basic relation that one obtains is,

$$dj/d\theta = 2Z(\rho L/4\pi) \ln (R_2/R_1) \text{ flux/radian of } l$$

where $2Z$ is the thickness of the galactic disc, ρ is the volume density of X-ray sources, R_2 is the distance to the edge of the galaxy in a particular direction, and R_1 is the distance to the last resolved source.

If j_1 is the intensity of the last resolved source at distance R_1, and N is the number of discrete sources seen within an angular interval θ of longitude, then

$$L = 4\pi R_1^2 j_1 \tag{8}$$

$$\rho = N/R_1^2 \theta Z \tag{9}$$

and the equation for $dj/d\theta$ reduces to

$$dj/d\theta = (2j_1 N/\theta) \ln (R_2/R_1) \tag{10}$$

We can now use this relation to try to estimate the diffuse component in the Sgr–Sco vicinity. Such a component is observed by UHURU in the form of a gradual rise in the background of about 2 ct/sec per 1/2° of l in this vicinity. We ignore the $L = 10^{38}$ erg/sec sources of which we estimated there were 7, since these must all be observed as discrete sources. The total number of sources seen in the Sgr–Sco region is 31 in 60° of l (see Table 2) of which 7 are discounted, leaving a source density, N/θ, of 0.4 per degree. The faintest source in this region is about 12 ct/sec; thus

$$dj/d\theta = 9.6 \ln (R_2/R_1) \text{ ct/sec-deg.}$$

$$= 4.8 \ln (R_2/R_1) \text{ ct/sec-1/2 deg.}$$

If we make $L = 10^{36}$ erg/sec, R_1 appropriate to 12 ct/sec is about 6 Kpc, $R_2/R_1 \approx 3$ and $dj/d\theta$ becomes about 5 ct/sec. Since interstellar absorption will affect this number to some extent, and considering that the statistics are poor, it is possible that this figure is consistent with the 2 ct/sec that is observed. However, if we make $L = 10^{35}$ erg/sec, R_2/R_1 becomes 9, $dj/d\theta$ becomes about 10 ct/sec and the discrepancy with the observed counting rate becomes more serious.

The conclusion is that the Sgr–Sco sources, after removing the $L = 10^{38}$ erg/sec sources are consistent with a class of objects with $L = 10^{36}$ erg/sec. However, the numbers actually favor a slightly higher luminosity, and it is possible that we are seeing a continuous distribution of L between 10^{36}–10^{38} erg/sec. This argument also supports the view that there is a small number of sources in the galaxy.

D Spectral Content of the Radiation

In general, except for absorption effects, the flux density of X-rays is observed to be a monotonically decreasing function of energy. There is no source of X-rays observed above 20 keV that is not also seen with at least equal flux density in the UHURU energy range of 2–10 keV. The converse is not true; most of the galactic X-ray sources observed in the 2–10 keV energy range are seen with much smaller flux densities at the higher energies. If one defines a characteristic energy of the X-ray sources as the energy at which most of the power appears one would state that number of source with characteristic energy between 2–10 keV is much larger than the number that have a characteristic energy greater than 10 keV.

However, this does not continue into the regime less than 2 keV. A survey of the galactic plane in this low photon energy range by the Livermore Radiation Group[16,17] has revealed at most one or two sources in the $<$1 keV energy range not seen by UHURU between 2–10 keV. Interstellar absorption is not a dominant factor since it is possible to see to between 1–2 Kpc at photon energies between 0.5–1 keV.

Thus, it appears that nature prefers making X-ray sources that radiate most of their energy between 2–10 keV, compared to above or below this energy range. It is possible that this is at least partly a result of intrinsic absorption in the sources. There is evidence, as will be discussed in the next section, that a significant number of the sources display intrinsic absorption. Thus, it could be that many of the X-ray sources are accompanied by obscuring gaseous envelopes that prevent the observation of X-rays below 1 keV.

E Effect of Anisotropic X-ray Emission

At least two of the most plausible models of X-ray sources, namely, rotating neutron stars and rotating black holes (discussed in this book by Ruffini and Thorne, respectively) predict that X-ray emission from the source is highly anisotropic. Also, there is observational evidence that in the case of the pulsing X-ray sources, Cen X-3 and Her X-1, the emission is beamed as with the pulsars. If this is the case for many of the X-ray sources, it could affect certain conclusions reached here.

If the X-ray emission from a source is emitted only within a solid angle Ω the probability of seeing the source is $\Omega/4\pi$, and the actual number of sources present must be $4\pi/\Omega$ greater than the number observed. Correspondingly, if this is a property of all the sources, the estimate of the number in the galaxy must be increased accordingly.

The absolute luminosity of the source must be decreased by the same factor $\Omega/4\pi$.

F Summary

Ignoring the effects of anisotropic X-ray emission, the following is the most plausible set of statements that can be made about the galactic X-ray sources;

1. Their luminosity is within the range 10^{36-38} erg/sec.
2. The number in the galaxy is exceedingly small, perhaps no more than 100, of which about 50 are seen.
3. The excitation process by which the X-rays are made favors the production of 2–10 keV photons. Alternately most of the X-ray sources are surrounded by an absorbing envelope of thickness of order 10^{22} atms/cm^2 of hydrogen.
4. The sources in the outer regions of the galaxy are found in the spiral arms.

IV Properties of Individual X-ray Sources

In this section we present the information on the X-ray sources obtained from observations with emphasis on those characteristics that relate most directly to the nature of the object or the stellar system in which it resides. Necessarily the most complete information is available on those sources that are optically identified. We take the view that an optical or radio identification is positive only if there is observed a definite correlation between the X-ray and the optical or radio emission. There are only three such objects, Sco X-1, Cyg X-1 and Her X-1. There are a number of other sources for which optical identifications are suggested based on statistical or similarity arguments. Much of the information here on the X-ray sources have been taken from invited talks by Dr. Harvey Tananbaum at the IAU Meeting on X- and γ-ray Astronomy at Madrid in May, 1972 and at the American Astronomical Society Meeting in East Lansing, Michigan in August, 1972. Also at the Madrid IAU Meeting were presented invited talks by Van den Heuvel and by Kraft on the evolution of binary systems and by Salpeter on compact X-ray sources that form a kind of general picture of the stellar systems that contain the X-ray sources.

A Sco X-1

Sco X-1 was the first X-ray source observed.[9] It was correctly identified in 1966[18] based only on positional data.[19] The identification must be considered positive since correlated optical and X-ray intensity variations have been observed since then. The optical Sco X-1 has been subject to intensive study since its discovery, as has been reviewed by Hiltner.[20] As part of the UHURU observing program, more than 1 week was devoted to correlated X-ray-optical studies of this source.

The X-ray emission in the 1–10 keV energy range can best be fit by an exponential spectrum with $kT \approx 5 \times 10^7$ °K. However, it is seen at energies $>$20 keV to possess a non-thermal tail. The X-ray spectral data themselves can not be taken as positive evidence that the X-ray emission process is thermal bremsstrahlung.

A more plausible case for the thermal mechanism can be made by looking at the entire electromagnetic spectrum. The optical continuum is consistent with an extrapolation of a simple thermal spectrum from X-ray into the visible light range.

An important characteristic has been revealed in the infrared region as observed by Neugebauer *et al.*[21]; namely, that the flux density falls off as expected from free–free absorption in the emitting region, from which the size of the emitting region can be inferred of between 10^4–10^5 Km, depending on the distance to the object. The observed absence of X-ray line emission is consistent with an object of the size; the lines become broadened and weakened by electron scattering.

Optically, Sco X-1 varies between 12–13 m. The intensity varies by a few per-cent on a time scale of minutes (flickering) and flares by a factor of two on a time scale of hours. The spectrum shows a strong continuum plus a number of high excitation emission lines. There is no evidence for a periodic component in the optical intensity or in the Doppler shift of the spectral lines, although the spectral lines do vary in both intensity and position. Also, there is no evidence for the presence of an ordinary star.

In radio, Sco X-1 is the seat of a faint, highly variable radio source radiating from about 0.2 flux units to below present limits of detectability of 0.005 flux units. There are present two weak steady radio sources several arc minutes on either side of Sco X-1. To a precision of arc seconds the line joining these two sources passes through Sco X-1.[22]

Optical photometry and simultaneous X-ray-optical observations performed in 1971 between the UHURU satellite and the Observatory in Cerro Tololo, Chile, a portion of which are shown in Figure 8, have revealed the intensity in Sco X-1 has two well defined states; one in which both the X-ray and optical intensity is relatively steady, and variations occur of the order 20 per cent over times of a day. The second state begins abruptly, both X-ray and optical intensity show continuous variability (flaring?) of a factor of two or more on a time scale of the order of less than 1 hour. The average intensity in both optical and X-ray in this

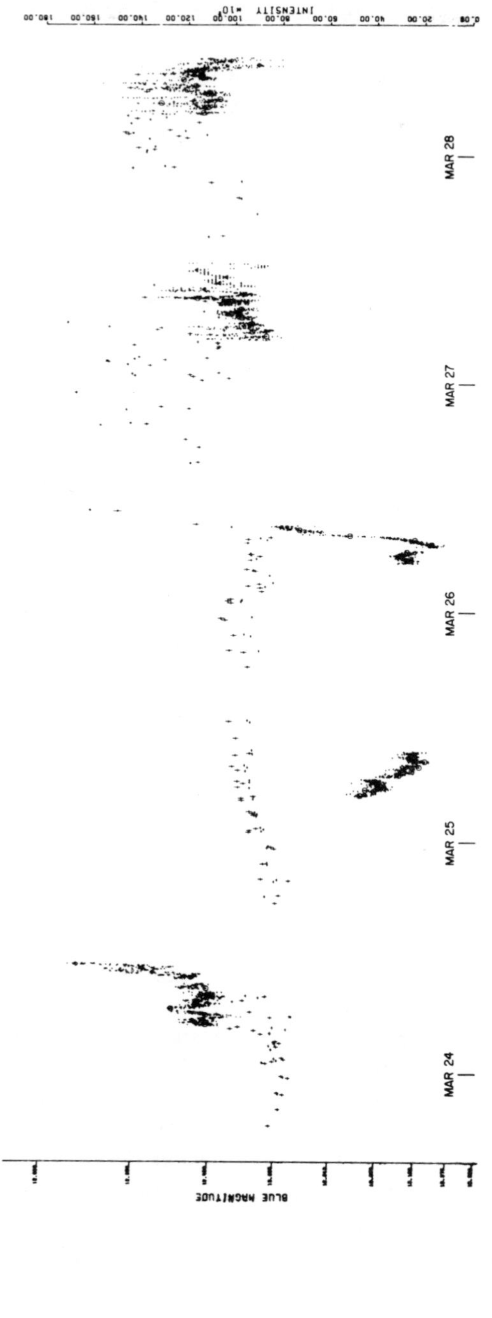

Figure 8. Simultaneous X-ray-optical observations of Sco X-1 taken during 24–28 March 1971. The higher density points are the optical measurements.

second, active state are about a factor of two higher than in the first quiescent state.

There is a correlated spectral change as well. When active, the X-ray spectrum is harder (higher temperature?) and the optical colours become more blue. These are average statements, the point to point values of intensity and spectrum do not show a good correlation. The data in Figure 9 in which the X-ray intensity is plotted against the simultaneously recorded optical intensity (*B*-magnitude), , illustrates the two states very well in that the data points fall only in two sharply

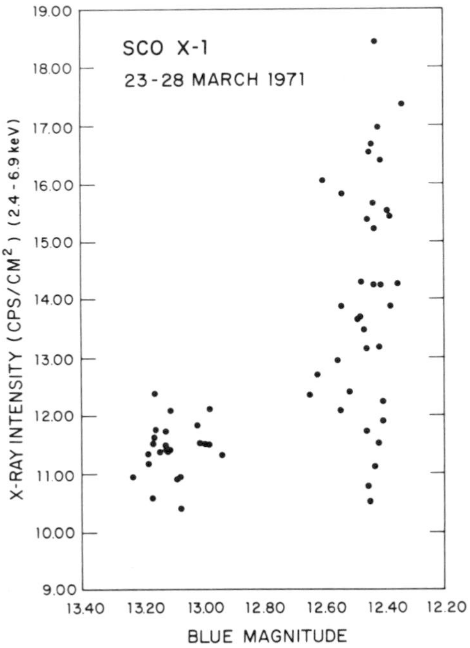

Figure 9. Simultaneous X-ray-optical variability of Sco X-1. Each point represents a single, simultaneous measurement of the X-ray and optical intensity of Sco X-1.

delineated regions; however, within those regions, the points show considerable scatter.

The time of persistence of each state is of the order of days; however, there is no evidence that the occurrence of the states is periodic.

The conventional view of Sco X-1 has been that the X-ray emission is thermal bremsstrahlung and occurs in a volume the size of order 10^3–10^4 Km. The optical emission is produced in an outer region and the radio even further out. If Sco X-1 is radiating in X-rays at 10^{36} erg/sec its distance would be 200 pc. Based on optical distance indicators such as reddening, polarization on CaK absorption, it is at least this distance, but could be much further. Its optical luminosity at 200 pc would be 10^{33} erg/sec, about that of the sun. Any conventional star present in the system must be radiating at a much smaller rate.

Thus, in spite of an enormous amount of observational material, there is only information available on the size of the X-ray emitting region. There is no information on the mass or dimensions of the System. For example, being the strongest source, Sco X-1 has been subject to the most serious attempt to discover emission lines in its X-ray spectrum without success, as most recently exemplified by Kestenbaum, Angel and Novick.[23]

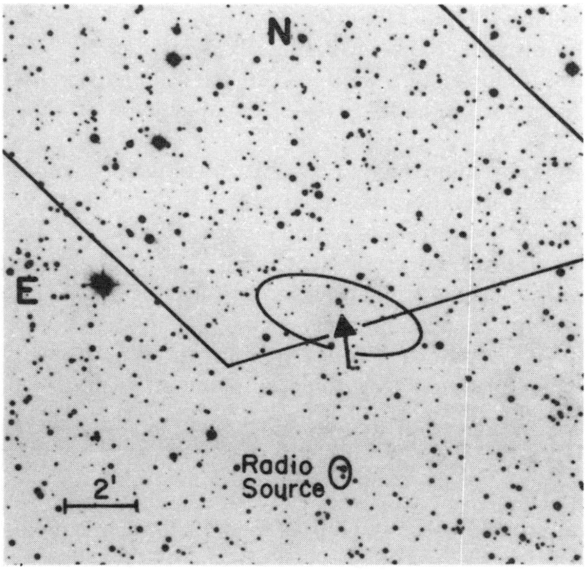

Figure 10. X-ray position of Cygnus X-2. The position from the 2 UHURU catalog is the small ellipse and the arrow points to the blue star believed to be the optical counterpart. The large diamond (partially cutoff) is the 1967 position.

B Cyg X-2

Cyg X-2 is believed to be similar to Sco X-1 except there is not nearly so much information available concerning it. Its intensity is only about 1/40 of Sco X-1. The optical identification was proposed in 1967,[24] based on the discovery of a star with characteristics similar to that of Sco X-1 within the area of uncertainty of the X-ray source. The position of Cyg X-2 has been considerably refined since then[25] as is shown in Figure 10. However, simultaneous X-ray-optical correlations have not been reported; thus the identification cannot be regarded as being positively established.

In both optical and X-ray, variations in intensity of order a factor two are observed on a time scale of a day or less; however, the two states characteristic of

Sco X-1 are not yet reported from Cyg X-2. The X-ray spectrum is best fit by an exponential with $T \approx 4 \times 10^7$ °K, somewhat lower than that of Sco X-1.

The optical Cyg X-2 is actually very dissimilar from Sco X-1 although they share certain common features. The spectral lines show large changes in radial velocity which were thought to be evidence of binary motion; however, Kraft and Demoulin[26] showed that these variations, which are of the order of several hundred Km/sec were not periodic. Furthermore, they found that the bulk of the optical emission could be accounted for by the presence of a G-type subdwarf star based on the absorption-line spectrum. With this information the distance is estimated to be 500–700 pc.

Based on this distance, the X-ray luminosity is of order 10^{36} erg/sec, the optical luminosity is lower by a factor of 100 and is dominated by the G-type star. The optical emission from the X-ray region could be, as in Sco X-1 about 10^{-3} of X-ray emission.

Several other X-ray sources have properties, in terms of the spectrum and variabilities of the X-ray intensity, similar to Sco X-1 and Cyg X-2; however, there has been no other identification of a similar optical object. If, however, the ratio $Lx/L_{opt} = 10^3$ then this is not so surprising. Sco X-1 and Cyg X-2 are not only bright in X-rays but located at high galactic latitude out of regions of high optical obscuration. The visible emission from a source 10^{-2} of Sco X-1 located in the galactic plane would appear of 17 m if its optical emission were 10^{-3} of its X-ray emission. Allowing for several magnitudes of obscuration, such a star would be hopelessly difficult to find.

C Cyg X-1

Cygnus X-1 was probably observed during the first X-ray astronomy experiment in 1962 in which Sco X-1 was observed. It was the first X-ray source observed to be variable.[27] Because it possessed an intensity and spectrum similar to the X-rays from the Crab Nebula, it had been speculated for some time that the object was a supernova remnant that had gone undetected. However, because of the reported variability and the failure to find a radio remnant, this idea was eventually dropped. In 1967, the same experiment which yielded the positional data on Cyg X-2 leading to its identification, yielded equally good positional data for Cyg X-1. However no optical candidate was found at the time that fit the then current idea that the optical emission should be an extension of the observed X-ray continuum. With the launch of UHURU attention focused again on Cyg X-1, and two significant pieces of information were discovered.

1. The X-ray intensity was discovered to vary on a time scale of less than 1 secon The emission seemed to be organized into short pulsations which were thought to be periodic with a period less than 1 second. In fact it was proposed that this object was a Black Hole, on the argument that only a collapsed object could possess such a short periodic component; a neutron

star was excluded on the basis that there was no evidence for a supernova. This latter argument is now known to be naive; in any event, subsequent observations of Cyg X-1 with higher time resolution than was available on UHURU failed to disclose a true periodic component to the radiation.[28,29,30]

2. Based on improved positional data,[31] a variable radio source was discovered by Hjellming and Wade[32] within the Cyg X-1 location which, on the basis of the Sco X-1 radio source, was believed to be a reliable signature of an X-ray source. Only two stars were located within the very small region of uncertainty containing the radio source; one the 9^m B0I star HDE 226868, the other a faint, highly reddened star. This latter object was the subject of much study as the candidate for Cyg X-1 until refinements in the position of the radio source showed that the B0I star was the radio source.

These two developments have generated even more interest in Cyg X-1 and its optical candidate. We can now consider the optical identification positive based on a remarkable X-ray-radio correlation. When the radio source was first discovered, it was thought to be variable because an earlier search failed to reveal significant radio emission from the region. Since its discovery in Spring, 1971 the source has been constant in intensity at about 0.015 flux units. Recently, Tananbaum,[33] after putting together more than a year's worth of UHURU X-ray observations of Cygnus X-1 discovered that the X-ray source underwent a change in its average intensity by a factor of 3 at the same time the radio source came on. As shown in Figure 11, when UHURU was first launched, Cyg X-1 was observed with an intensity of ~1000 ct/sec until March, 1971 when, within 10 days, its intensity dropped to about 300 ct/sec where it has remained for more than a year. The two transitions are coincident within a week, and since no such transition has been seen in radio or X-ray otherwise, these must represent the same phenomena with a high degree of certainty. This then is the basis for the optical identification since the radio source is coincident with the star to a fraction of an arc second.

The general character of the time variations of the X-ray intensity have already been presented. In spite of intensive work by a number of groups, the time variability has failed to reveal a unique characteristic that is indicative of the system in which this radiation originates. Thus, the only statement that can be made is that, since significant variations are seen on a time scale 0.1 sec, the size of the emitting region must be 0.1 light-sec (30,000 Km) or less depending on the propagation velocity. However, it is possible that a number of independent regions comprise the source.

Optically, for some time, the star was not seen to display any characteristics unusual to a B0I. Based on periodic variations in the radial velocity of its spectral lines, the star was determined to be a binary system with a 5.6 day period.[34] The only other peculiarity was the discovery of Hβ emission lines with non-periodic, radial velocities of order 100 Km/sec, indicative of mass loss or transfer. However, neither the binary nature of this system, nor the presence of mass streaming is unusual in such a star. Also, a search through 35 days of continuous observation of Cyg X-1 in X-rays[33] has failed to reveal a 5.6 day variable component in the X-ray emission.

Very recently, however, another spectral line in the star, HeII 4686 in emission has been found to possess a periodic radial velocity change which is out of phase with the radial velocity variations found for the primary star.[35] Since the variations are out of phase the region of emission of the HeII must be moving on the other side of the center of mass from the B0I. If the emission region is assumed to be tied gravitationally to the second object then it is possible to work out completely

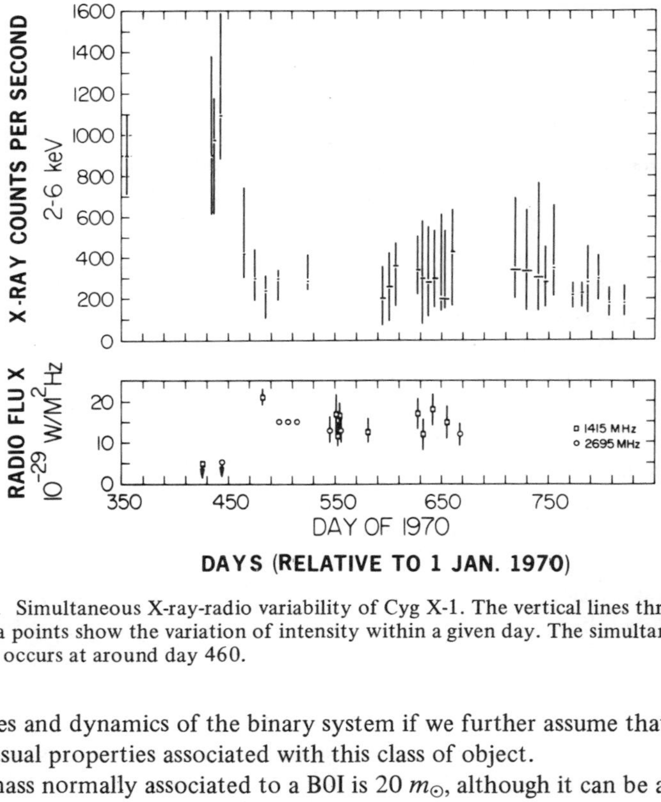

Figure 11. Simultaneous X-ray-radio variability of Cyg X-1. The vertical lines through the X-ray data points show the variation of intensity within a given day. The simultaneous transition occurs at around day 460.

the masses and dynamics of the binary system if we further assume that the B0I has the usual properties associated with this class of object.

The mass normally associated to a B0I is 20 $m_\odot$, although it can be as low as 12 $m_\odot$. Since the velocity of the B0I, based on its radial velocities is 74 Km/sec and the velocity of the unseen component (that of the HeII line) is 100 Km/sec, the ratio of the two masses is about 1.4. Then the mass of the unseen object must be greater than about 8 $m_\odot$ and its most likely value is 14 $m_\odot$. There is no evidence that this is a three body system; in fact, the eccentricity of the orbits, based on the fit of the radial velocity variations to a sine wave, is close to zero.

Now if we accept the view that the X-ray emission is created at the secondary by mass accretion onto a small object (the X-ray variability indicates a size no bigger than a white dwarf), this secondary must be a black hole since its mass is much greater than the limit of stability for either a neutron star or a white dwarf. There are a few problems with this interpretation. Aside from the string of assumptions one must make, the observational material on the B0I is still limited, particularly with respect to the crucial HeII line on which basis the mass of the secondary is determined. It is possible that the HeII is emitted in a region that is moving at an odd velocity with respect to the primary and thus not correctly indicating the velocity of the secondary. However even if we ignore the HeII emission we are left with an uncomfortably high mass secondary. The mass function $(M_s^3/(M_s^2 + M_p^2))$ based on the velocity of the primary alone is $(0.16/\sin^3 i)$ where i is the inclination of the orbital plane of the binary system with respect to the observer ($i = 90°$, the observer's line of sight is in the plane of the orbit). If the mass of primary, M_p, is taken to be only 15 $m_\odot$ the mass of the secondary must be greater than 3.4 $m_\odot$ depending on the inclination of the orbit. This mass is also above the conventional limits for a neutron star or a white dwarf. Thus, there is no way to avoid a massive secondary, provided the identification of the B0I is assumed.

Based on the optical magnitude, Cyg X-1 is placed at a distance of 2000 pc. Also, the inclination of the orbit (accepting the HeII radial velocities) is 26°. The radius of the primary is taken to 16.5 $R_\odot$ typical of a B0I and the separation of the centers of the primary and secondary is $30R_\odot$. Thus, it is not surprising that no X-ray eclipse is observed; based on these dimensions an inclination angle of about 60° would be required before the primary would eclipse the secondary.

It is also not surprising that no optical signature is present that could be evidence for the presence of an X-ray source. If the X-ray emission region of Cyg X-1 were radiating as much in visible light as does Sco X-1 relative to its X-ray intensity, such an amount of light would only be about 10^{-3} of the brightness of the B0I.

At least one other X-ray source, Cir X-1, is observed to exhibit the very short, irregular pulsations characteristic of Cyg X-1; however, it has not been subject to intensive study as yet.

D Centaurus X-3

This was the first X-ray source observed to be pulsing periodically,[36] and the first found to be a binary.[37] Based on the X-ray data alone, it is known that the bulk of the observed X-ray power is in the form of a periodic signal with a period of 4.84239 sec. Furthermore, we know that the source is in a binary system; the X-ray emission is eclipsed every 2.08707 days; but more important, the period of the emission is observed to vary sinusoidally with the same 2-day period. This variation in period has a direct interpretation as a Doppler shift of 415 Km/sec corresponding to the motion of the X-ray source around a central star. Furthermor the center of the eclipse occurs when the Doppler velocity change is zero, consistent with a binary system. The X-ray variations are illustrated in Figure 12.

There are many details of the X-ray emission which clearly relate to the nature of the source of the system in which it is imbedded, but which do not lend

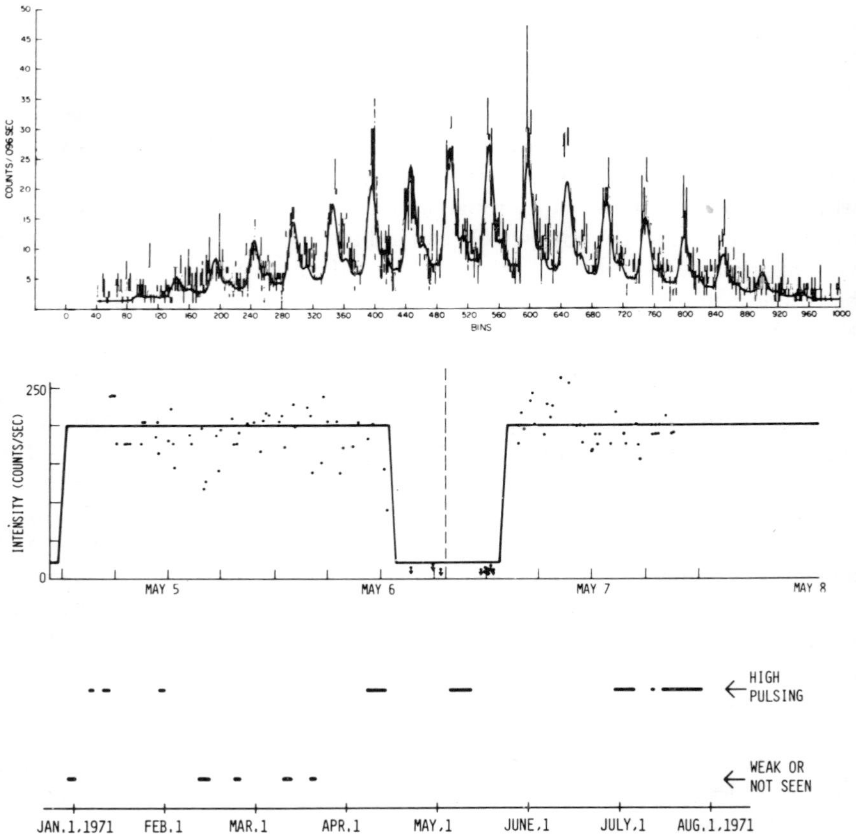

Figure 12. X-ray variability of Cen X-3. The upper curve shows the 4.8 seconds pulsations as they appear in the raw data (each bin is 0.096 sec). The fitted curve is the result of a fourier analysis. The middle curve is the average intensity folded modulo the eclipse period of 2 days. The period and phase of the eclipse were obtained from the Doppler velocity variations of the pulsing. The lower curve is the time variability on a scale of months and shows extended periods (e.g., February–April) when the source is weak or absent when it was predicted to be seen.

themselves to such a clean interpretation. The pulsed portion of the X-ray emission disappears for periods of many days in an apparently erratic fashion. During this time and during the eclipse, continuous X-ray emission is observed whose intensity is about 10 per cent of the peak intensity of the pulsed portion. It can not be determined whether this portion of the radiation is present during the time of the pulsing; however, its presence during the eclipse implies that it is produced independently of the pulsed portion.

The transition into and out of the eclipse require about 0.8 hours; spectral variations (deficiency of low energy photons) during this period indicate that the X-ray flux is being absorbed rather than being cut off by a sharp edge; i.e., the source appears to be "setting" into an atmosphere before going behind the central star. Also, the time of the transition varies by an hour or more, indicating that the "atmosphere" is not steady.

There is no optical candidate available for this object; however, it is located right on the galactic plane in the direction of a spiral arm. Thus, it may be badly obscured optically.

Assuming this to be a binary system, the picture that emerges is of a close binary system of radius 6×10^{11} cm ($\approx 8\, R_\odot$). The separation between the edge of the occulting disk and the X-ray source is only about 10^{11} cm. Based on the Doppler velocity, the mass function is 15 $m_\odot$. The minimum mass of the secondary is 15 $m_\odot$. However, if one accepts the view that this is a mass-exchange binary system in which the unseen object (m_s) fills its Roche Lobe, an independent relation arises between the ratio of the radio and the masses; which in turn allows placing limits on the masses. However, there is some controversy here. Wilson[38] argued that the eclipse duration required a central star that was tidally distorted and derived an upper limit for the mass of the X-ray source of about 0.2 $m_\odot$. Van den Heuvel and Heise,[39] on the other hand, derive an upper limit of 0.7 $m_\odot$ for the object based on the central star filling its Roche Lobe. Ruffini and Leach[40] have derived somewhat higher masses; the main point, insofar as the present discussion is concerned, is that the mass is within the acceptable limits for neutron stars; i.e., one is not compelled to make the X-ray source a black hole. Based on the value of the mass function the other star is of the order of 15 $m_\odot$.

It is generally believed that the X-ray emission from Cen X-3 originates from a rotating, magnetic, neutron star based on the similarity to NP0532. The period of 4.8 seconds of the pulsations is within the range of permissible rotation for a white dwarf; however, the only other X-ray source of this kind we know of, Hercules X-1, has a period of 1.24 seconds which is below the limit of stability for the rotation of normal White Dwarfs.

The shape of the pulse is not consistent with rotation of a source of radiation on the surface which emits isotropically; the pulse is too sharp. The most natural explanation is that it is produced by the same process as is making the X-rays in NP0532. If this is the case, the radiation is beamed, which has important consequences regarding the average luminosity and number of sources of this type.

E Hercules X-1

Hercules X-1 is the only other pulsating X-ray observed as yet.[41] Its essential properties are similar to Cen X-3. The pulse period is 1.24 seconds; the eclipse period is 1.70 days; the actual duration of the eclipse is about 5 hours. The binary nature of the system is confirmed by the observation of the Doppler shift of the 1.24-second pulse rate and the phase of the eclipse.

The X-ray variability is displayed in Figure 13. Like Cen X-3, the source disappears; however, in this case, it is in a low intensity state for a much longer duration than it is on and pulsing. Furthermore, this on/off alternation appears to repeat every 35 days but not with a precise period. During this time the source is on and pulsing for only about 12 days. Contrary to Cen X-3 there has yet to be found any continuous emission at times when the source is not pulsing.

During the 12 or so days that it is "on", the average intensity shows a definite

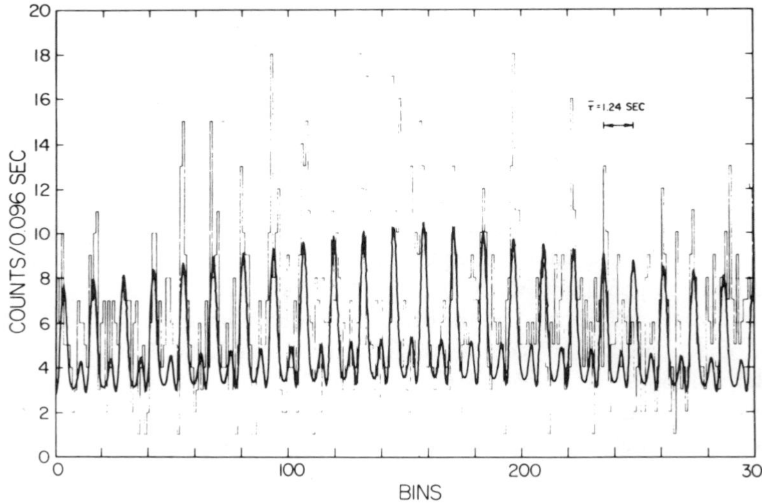

Figure 13. Short time variability of Her X-1 showing the periodic, 1.24 seconds pulsing. The double pulse is characteristic of this source but the relative amplitude of the two varies from time to time.

variation of intensity, not directly connected with the eclipse. As shown in Figure 14, the "on" state comes on abruptly within several hours, it increases slowly for another 3 days, and then declines steadily. It is not likely that this is an absorption effect, since during the entire "on" state, the X-ray spectrum is not observed to vary. If the rise and decline were caused by absorption, a noticeable deficiency of low energy photons would be seen in the spectrum as the intensity declines. Also, prior to entering the eclipse, the intensity is seen to undergo a "dip" in value, a kind of pre-eclipse. This is likely to be an absorption effect, since a spectral change characteristic of absorption is observed. The dips only occur within a restricted range of phase, just prior to eclipse. By contrast, the source makes a sharp transition when it comes out of eclipse with no sign of such an effect.

This source is positively identified with the 13.5 m optical object HZ Hercules,

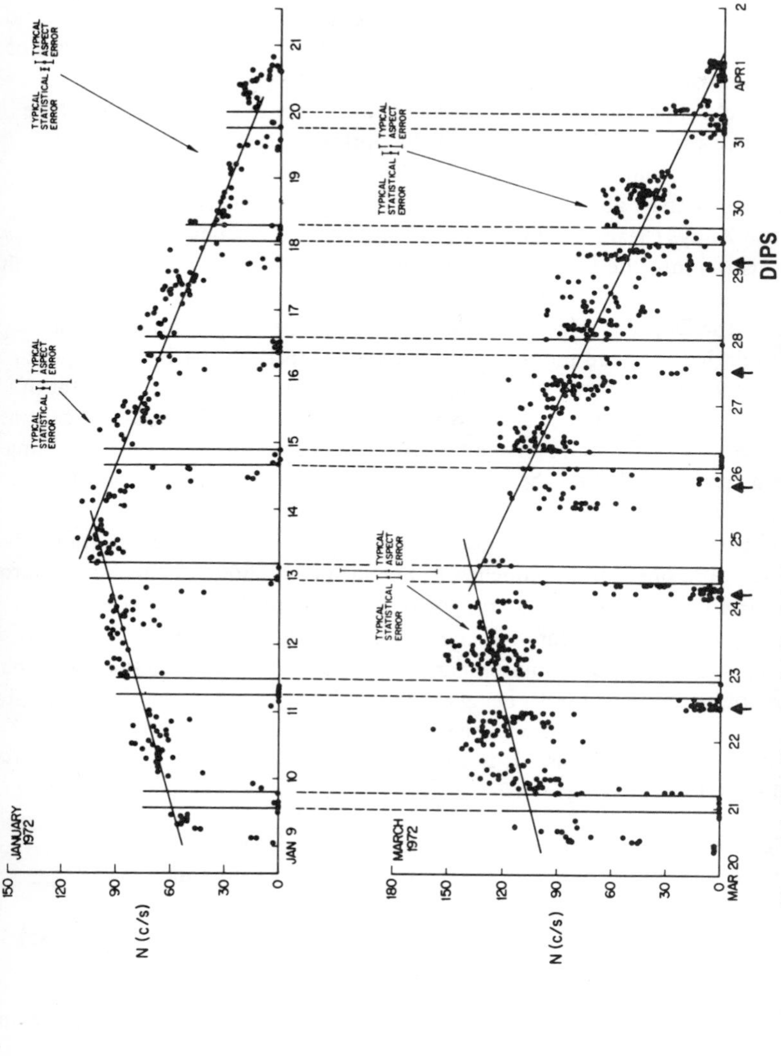

Figure 14. The X-ray variability of Her X-1 during its "on" state. The envelope shows a sudden onset (9 January, 20.5 March) a gradual rise to a peak (14 January, 24 March) and a more gradual decline. The envelope is modulated by the 1.7 day eclipse and by "dips" in intensity just prior to entry into eclipse.

an object catalogued as an irregular variable. Liller first called attention to the presence of this star within the X-ray positional error.[42] The identification must be regarded as positive since it has been observed to eclipse at precisely the same rate as does Her X-1.[43,44] Periodic optical pulsing has been reported as well at very low level (~0.1 per cent of the net power);[45,46] however, many observers have failed to detect such pulses. According to Liller's observations which extend back to 1945 by use of historical material, there is no evidence for dimunition of the optical emission corresponding to the 35-day X-ray cycle.

The optical eclipse, shown in Figure 15 shows a large excess of light in phase with the X-ray emission; i.e., the peak of the optical emission occurs when the X-ray source is in front of the central star. The peak X-ray power exceeds the peak optical power by about a factor of 10; however, considering the short duty cycle of the X-rays (beaming, 35-day cycle) and the probable extension of the optical emission into the UV, it is likely that the average optical emission is much in excess of the average X-ray power.

Based on the Doppler velocity measurements and the eclipse duration it is possible to derive a limited picture of the system comprising Her X-1. The peak velocity is 169 Km/sec, from which $f(M)$ is derived to be 1 $m_\odot$. The diameter of the orbit of the X-ray source is about 8×10^{11} cm and the diameter of the central occulting star is 3×10^{11} cm. As with Cen X-3, the masses of the stars comprising this system can be limited on the assumption that this is a mass transfer binary system and that the diameters are given by considerations of the Roche Lobes. Leach and Ruffini[40] find that the mass of the X-ray object must lie between 0.2–3 $m_\odot$. Here again, one is not compelled to invoke a black hole for the X-ray source.

As discussed already, the most straightforward interpretation for Her X-1 is a rotating, magnetic neutron star–the key point being the very short pulse period. Also, the pulse shape shows some similarity to what is seen in NP053; there being a definite double pulse and both being relatively sharp. Cen X-3 may have a similar double pulse structure; there the pulse shows a shoulder, which could be the superposition of a second pulse. It is useful to summarize the observed characteristics of Her X-1.

1. The pulse shape is double with both pulses being sharp.
2. The average intensity shows an on/off cycle of 35 days which may be periodic. The on state persists for only about 12 days.
3. The average intensity during the 12 day on state (exclusive of the eclipse) shows regular and complex variations.

As discussed in the case of Cen X-3, it is likely that the pulsed portion of the X-ray emission is beamed, which has important implications for any model of this system.

4. The optical object shows a large amount of non-stellar emission which must be originating in or near the X-ray emitting region.

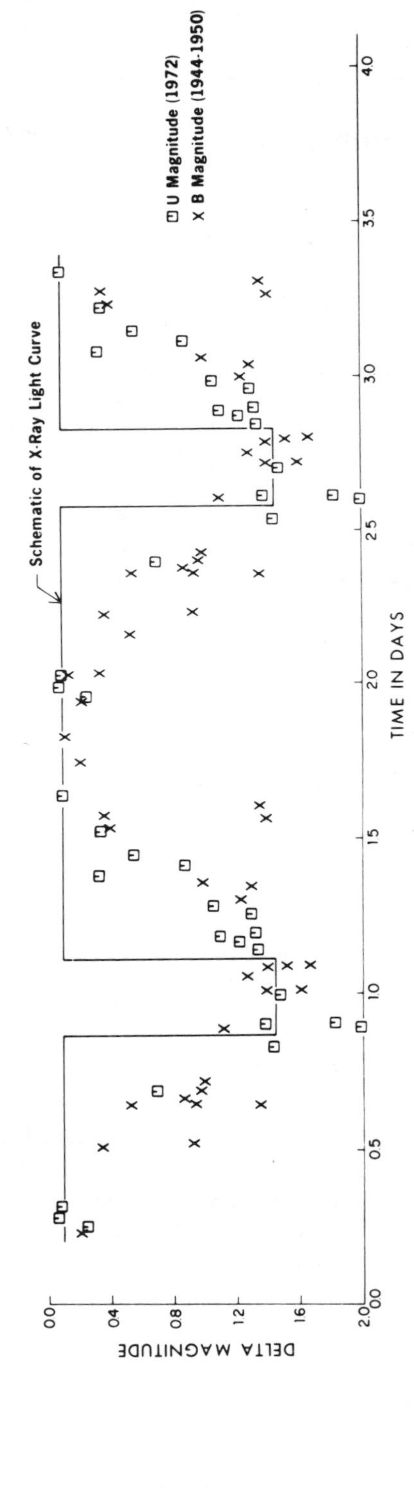

Figure 15. Optical variability of Her X-1 (≡HZ Her) as measured by Liller and Forman. The solid curve is the X-ray variability.

F Other X-ray Sources of Interest

No other single X-ray source has provided the kind of detailed information as is known for the above five objects; however, if we look at the set of sources as a whole (at least the stronger ones) some interesting points emerge.

First, there appears to be a class of X-ray sources associated with bright stars (<10 m). The analysis supporting this statement used 25 galactic X-ray sources with the smallest positional errors. Of these, 10 are associated with stars from the SAO catalogue, whereas fewer than 3 are expected by chance. Furthermore, a preponderant number of these are of an early spectral type. If Cygnus X-1 is not a unique system, such an association must exist.

At least two of these are impressive in their own rights. One is the X-ray source 2U(0900 – 40) for which the optical object may be HD 77581 a 6.9 m B0I. This system has been found by Hiltner and Werner[47] to be a spectroscopic binary with a period of 7 days. The value of the mass function is 0.013 $m_{\odot}$; assuming a mass of 15 $m_{\odot}$ for the B0I, the mass of the secondary (X-ray source?) must be greater than 1.4 $m_{\odot}$, depending on the inclination of the system. A search for an X-ray eclipse is so far partially successful; there is evidence for a seven day eclipse. The X-ray emission is highly variable, but it does not appear to show the kind of pulsations as does Cyg X-1.

There are two observed examples of what appear to be eclipsing X-ray sources. Neither of these objects shows regular pulsing as do Cen X-3 and Her X-1. One of these is 2U(1700 - 37) which has a period of 3.4 days and an eclipse time of 1.3 days. This X-ray source may be associated with the star HD 153919, a $6^m.607$f. The other is the X-ray source in the Small Magellanic Cloud 2U(0115 - 73) which shows a period in X-rays of 4 days and an eclipse time of 0.63 days.

As noted, none of these X-ray sources shows evidence of periodic pulsing; all three however show a spectral shape deficient in low energy X-rays indicative of absorption. In the case of the Small Magellanic Cloud source, this must be intrinsic to the source since the interstellar material between us and the clouds is not sufficient to account for it.

G Summary

In summarizing the observed characteristics of the X-ray sources, one would almost have to say that their outstanding feature is their dissimilarity. Only two X-ray sources, Cen X-3 and Her X-1, reveal themselves as obviously belonging to a distinct class of objects. The others concerning which we have detailed information, Cyg X-1, Sco X-1 and Cyg X-2, really seem to be quite different. This is somewhat at variance with the conclusions reached in section III that the X-ray sources appear to belong to a distinct class of objects with the following overall characteristics.

1. Restricted luminosity range 10^{36}–10^{38} erg/sec.
2. Restricted excitation range. In terms of effective temperature, few if any sources exhibit kT less than about 1 keV.
3. Consistent with a distinct population type; in particular, a strong concentration in the spiral arms.

Also their very small numbers argue for a distinct class of objects. The X-ray sources are very rare. This means that it is a peculiar set of circumstances that bring about an observable X-ray source. To invoke more than one class of objects means invoking more than one set of peculiar circumstances.

Thus, it is useful to study the characteristics of the X-ray sources with a view to seeing if there are present an underlying set of unique characteristics.

Table 4 summarizes certain known properties of the X-ray sources. It should be kept in mind that the identifications are positive for only Sco X-1, Cyg X-1 and Her X-1. The identifications of the other are based on the statistical argument discussed above of the positional association between X-ray sources and bright stars. In addition to 0900–40 and 1700–37, there are eight other galactic X-ray sources associated with stars brighter than 10 m. These must be binary systems. The stars in these objects radiate 10^{37-38} erg/sec from their photospheres in optical power; they cannot also be radiating 10^{36} erg/sec in X-rays. Thus, the X-rays are mostly likely to be produced on a second, unseen object. In fact, Cyg X-1, 2U1700 - 37, and 0900 - 40 are seen to be binary systems.

The conclusion is that there appears to be a class of X-ray sources in binary systems in which there is present a bright, early spectral type (0, B, A) star. The absence of visible light that can be tied directly to the X-ray source may simply be due to the overwhelming power of the normal star. If the visible emission from the X-ray region is no more than about 0.01 of the X-ray power (it is 0.001 in Sco X-1) this would be the case. Based on the analysis by Gursky,[8] as many as half of the galactic X-ray sources that radiate in the range 10^{36} erg/sec may be of this kind.

On the other hand, four X-ray sources are associated with faint stars—Sco X-1, Cyg X-2, Cen X-3, Her X-1. The X-ray characteristics of these objects is very different, Cen X-3 and Her X-1 being periodic and pulsating, Sco X-1 and Cyg X-2 showing no regularity in their X-ray emission. However, here again the binary character of the system is evident; both Her X-1 and Cen X-3 are binaries and in Cyg X-2 a strong case can be made for a binary system because of the presence of a G-type star which cannot be the X-ray source.

Finally, we observe at least one other X-ray source not mentioned above, 2U0115 - 73 to possess X-ray variability which is best interpreted as an eclipse. Aside from adding evidence to the general binary nature of the X-ray sources, this source, being in the Small Magellanic Clouds, radiates at $L = 10^{38}$ erg/sec; it is the only high luminosity source that we can positively identify as being part of a binary system.

TABLE 4. Summary of Characteristics of X-Ray Sources

Object	l, b	I	X-ray variability	X-ray eclipse	Stellar object	Optical binary	Radio	X-ray cutoff >2 keV
Sco X-1	0, 24	17,000	Slow	No	Blue 13m	No evidence	Yes	No
Cyg X-2	87, −11	420	Slow	No	sdG 14m	No evidence		No
Cyg X-1	71, 3	1,175	Fast	No	B0I 9.m	5.6 days	Yes	No
Cen X-3	292, 0	160	4.8 sec	2.1 days	No candidate			Yes
Her X-1	58, 35	100	1.24 sec	1.7 days	HZ-Her 13.5m	1.7 days (eclipse)		Yes
0900–40	263, +4	77	Slow flares	7 days (?)	B0I 7m	7 days		Yes
1700–37	347, 2	102	Slow	3.4 days	07f 6m	?		Yes
1813–14	16, 1	560	Slow	No	No candidate		Yes	Yes
0115–73	300, −43	28	Slow	4 days	No candidate in SMC			Yes

Thus, the evidence is that a binary system is an important element in the existence of the X-ray sources and the present information is consistent with all the X-ray sources being in binary systems with periods of the order of days. It is difficult to make this argument quantitative; of the 55 X-ray sources more than a dozen have been studied in great detail, and of these particular attention has been directed to some that were thought to be eclipsing.

Finally we review the evidence for collapsed objects among these X-ray sources. In the case of Her X-1, the best explanation is that this is a rotating, magnetic, neutron star; however, rotational energy can not be the energy source.

The mass function does not require a massive object. Thus, the picture here is of a mass-transfer binary system, with mass accretion on to a neutron star being the causative agent for the X-ray emission and for the bulk of the optical power.

In the case of Cyg X-1, the mass function requires a secondary of mass greater than 3.4 $m_{\odot}$; if the Doppler velocity of the HeII emission lines is believed, then the mass of the secondary is about 13 $m_{\odot}$, far in excess of what can be a neutron star. Thus, the most straightforward explanation here is of mass-accretion onto a black hole as being the causative agent for the X-ray emission.

Finally, there is one additional observational point that may be of great significance. We discussed low energy cutoffs in connection with the distinction between the distant, high luminosity X-ray sources and nearer, low luminosity sources. The evidence for a cutoff is a deficiency of photons at low energies, compared to a simple extrapolation from the observed spectrum at higher energy. There is evidence for cutoffs far in excess of what can be accounted for absorption in the interstellar medium; these must be caused by material in or near the source. The UHURU instrumentation is sensitive only to cutoffs greater than about 2 keV, corresponding to a column density of Hydrogen of about 3×10^{22} atms/cm^2. Even at a density of 1 cm^{-3}, this corresponds to a distance of 10^4 pc. Thus, it is only possible to account for cutoffs seen by UHURU as interstellar for those sources which can be placed in the vicinity, or beyond the central regions of our galaxy. Otherwise, the cutoffs must be intrinsic to the source.

The remarkable feature of the data is that, as shown in Table 4, every source that displays an X-ray eclipse also shows a cutoff greater than 2 keV. Many of these are at an l, b along which it can be stated firmly that such absorption can not be interstellar. Furthermore, those sources for which no eclipse is observed, Sco X-1, Cyg X-2, and Cyg X-1, are not cutoff as seen by UHURU. Based on data from observations of these sources at lower energies, we know that their cutoffs are below 1 keV, requiring at least a factor of 10 smaller column density of hydrogen compared to the others. Also in these cases, and in the case of 2U(1813 – 14) the observed cutoff can be accounted for by the interstellar medium.

It is possible that we are seeing a view-angle effect here; i.e., if the X-ray emission is coming off a disc, as is expected in the mass accretion model, the spectrum may be expected to show absorption effects if the view direction to the source makes a large angle with respect to spin axis of the disc. The connection to the observation of an eclipse then follows from the fact that the plane of the disc coincides with the orbital plane of the binary system.

Another possibility is that the difference between eclipsing–non-eclipsing and cutoff–non-cutoff sources is simply a matter of separation between the two members of the binary system. As the separation becomes greater the probability of seeing an eclipse decreases as the square of the separation. Furthermore, it is possible that in the larger, widely separated systems the low energy X-ray absorption does not occur, for one of several reasons. First, the absorption may be caused by the atmosphere of the central star and increasing the separation simply removes the X-ray source from the atmosphere. Secondly, the size of the Roche Lobe around the X-ray source increases as the separation gets larger. If the absorbing material is contained in that lobe, its absorbing power (i.e., column density) becomes smaller if the same amount of gas fills the larger lobe; in fact, the column density decreases as the square of the radius of the Roche Lobe for a fixed mass of gas.

In any event there is good reason to expect a very strong dependence of the observed characteristics of the X-ray emission as the geometry of the binary system.

It is apparent that we are just beginning a new field of observational astronomy. It is very encouraging that the X-ray observations are yielding information that appears to be strongly model dependent. Of perhaps more importance is the fact that we seem to be making optical identifications with complex, stellar systems which themselves will yield important information on the distance to the sources, the nature of the stars, the basic geometry of the system and the masses of the components.

Much of the information presented here is as yet unpublished, and I wish to thank the X-ray Astronomy group at American Science and Engineering; in particular, Dr. Riccardo Giacconi, Dr. Harvey Tananbaum, Dr. Steven Murray, and Dr. Ethan Schreier, for permission to use this material. Likewise, I have made use of unpublished, as yet, material of Professor William Liller, Mr. William Forman, and Miss Christine Jones of the Harvard College Observatory, for which I am grateful.

Note Added in Proof

The above article was prepared during August 1972 while the Author was a guest at the Summer School at Les Houches. The Editors have graciously consented to allocate an additional page as a Note to bring the material more up to date.

Further study of the X-ray data of Hercules X-1 has revealed new information on the pulse period, the pulse shape and the 35 day on/off cycle. By analyzing the X-ray emission over four 12-day intervals between January and July 1972, it was found that a significant *decrease* in the pulse period of 1.24 seconds had occurred between each successive interval. The net decrease was 4μ sec over the six month interval. The pulse shape itself is observed to display a complex structure. The MIT group observed Her X-1 in a recent sounding rocket flight and found the main pulse to have a double peaked structure with a pronounced dip in the center. Similar behaviour is seen in the UHURU data which also show that the dip comes and goes on a time scale of minutes. There are also large changes in the interpulse.

The 35 on-off cycle (illustrated in Figure 14) has revealed surprising kinds of regularity. It is found, following the analysis of 11 cycles, that the time between successive onsets, is not fixed, but is either 35.7 or 34.9 days. However, the time between successive peaks in the envelope (which occur 4–5 days following the onset) seem to define a unique and steady period of 34.9 ± 0.01 days. It may be significant that the difference between the two times observed between successive onsets (35.7–34.9) of ~0.8 days is consistent with half the orbital period of 1.7 days.

Also, the "dips" (see Figure 14) seem to have a timing of their own. These features are only seen within a restricted range of orbital phase, just prior to eclipse. The exact phase of the dip may change systematically during the 12-day on interval, moving away from the eclipse by about 10° per 1.7 day orbital period. However, the data is not entirely conclusive.

Additional optical observations of Hz-Hercules has extended and confirmed the results quoted about regarding the light curve; namely, that the visible light varies systematically (almost sinusoidally) between about 13^m and 14.5^m with the peak occurring on the side of the star facing the X-ray source. Spectroscopic observations by Crampton and Hutchings (*Ap. J. Letters*, **178**, L65, 1972) reveal an object that appears to be late A-type at minimum light changing to B-type during the increase to maximum light. Continued study of historical material by Liller has revealed that there are extended time intervals (~ years) when the light variations disappear, leaving behind only the image that we now observe at minimum light.

Finally, I wish to correct an error in attribution. The radio counterpart of Cygnus X-1 was discovered by Braes and Miley (*Nature* **232**, 246, 1971). I apologize to these gentlemen for my error.

Bibliography

1. P. Gorenstein, H. Gursky and G. Garmire, *Ap. J.* **153**, 885, 1968.
2. B. Margon, S. Bowyer, M. Lampton and R. Cruddace, *Ap. J.* (Letters) **169**, L45, 1971.
3. M. Oda, P. Gorenstein, H. Gursky, E. Kellogg, E. Schreier, H. Tananbaum and R. Giacconi, *Ap. J.* (Letters) **166**, L1, 1971.
4. N. James Terrell, Jr., *Ap. J.* (Letters) **174**, L35, 1972.
5. M. Oda, M. Wada, M. Matsuoka, S. Miyamoto, N. Muranaka and Y. Ogawara, *Ap. J.* (Letters) **172**, L13, 1972.
6. J. Kristian, R. J. Burcato and J. S. Westphal, IAU Circ. No. 2395, 1972.
7. H. Gursky, A. Sollinger, E. Kellogg, S. Murray, H. Tananbaum, R. Giacconi and A. Cavaliere, *Ap. J.* (Letters) **173**, L99, 1972.

8. H. Gursky, *Ap. J.* (Letters) **175,** L141, 1972.
9. R. Giacconi, F. R. Paolini, H. Gursky and B. B. Rossi, *Phys. Rev. Lett.* **9,** 439, 1962.
10. R. Giacconi, E. Kellogg, P. Gorenstein, H. Gursky and H. Tananbaum, *Ap. J.* (Letters) **165,** L27, 1971.
11. R. Giacconi, S. Murray, H. Gursky, E. Kellogg, E. Schreier and H. Tanabaum, *Ap. J.* **178,** 281, 1972
12. F. D. Seward, G. A. Burginyon, R. J. Grader, R. W. Hill and T. M. Palmieri, *Ap. J.* **178,** 131, 1972.
13. C. Ryter, *Astron. and Astrophys.* **9,** 288, 1970.
14. G. Setti and L. Woltjer, *Astrophys. and Space Sci.* **9,** 185, 1970.
15. E. E. Salpeter, "Models for Compact X-ray Sources", IAU Symposium on X- and γ-Ray Astronomy, Madrid, 1972.
16. R. W. Hill, G. Burginyon, R. J. Grader, T. M. Palmieri, F. D. Seward and J. P. Stoering, *Ap. J.* **171,** 519, 1972.
17. G. Burginyon, R. Hill, T. Palmieri, J. Scudder, F. Seward, J. Stoering and A. Toor, submitted to *Ap. J.,* June, 1972.
18. A. R. Sandage, P. Osmer, R. Giacconi, P. Gorenstein, H. Gursky, J. Waters, H. Bradt, G. Garmire, B. Sreekantan, M. Oda, K. Osawa and J. Jugaku, *Ap. J.* **146,** 316, 1966.
19. H. Gursky, R. Giacconi, P. Gorenstein, J. R. Waters, M. Oda, H. Bradt, G. Garmire and B. V. Sreekantan, *Ap. J.* **146,** 310, 1966.
20. W. A. Hiltner and D. E. Mook, *Ann. Rev. Astron. and Astrophys.* **8,** 139, 1970.
21. G. Neugebauer, J. Oke, E. Becklin and G. Garmire, *Ap. J.* **155,** 1, 1969.
22. R. M. Hjellming and C. M. Wade, *Ap. J.* (Letters) **164,** L1, 1971.
23. H. Kestenbaum, J. R. P. Angel and R. Novick, *Ap. J.* (Letters) **164,** L87, 1971.
24. R. Giacconi, P. Gorenstein, H. Gursky, P. D. Usher, J. R. Waters, A. Sandage, P. Osmer and J. V. Peach, *Ap. J.* (Letters) **148,** L119, 1967.
25. S. Murray, Private Communication.
26. R. Kraft and M. Demoulin, *Ap. J.* (Letters) **150,** L183, 1967.
27. E. T. Byram, T. A. Chubb and H. Friedman, *Science* **152,** 66, 1966.
28. S. Rappaport, R. Doxsey and W. Zaumen, *Ap. J.* (Letters) **168,** L43, 1971.
29. S. Holt, E. Boldt, D. Schwartz, P. Serlemitsos and R. Bleach, *Ap. J.* (Letters) **166,** L65, 1971.
30. S. Shulman, G. Fritz, J. Meekins and H. Friedman, *Ap. J.* (Letters) **168,** L49, 1971.
31. S. Rappaport, R. Doxsey and W. Zaumen, *Ap. J.* (Letters) **168,** L43, 1971.
32. R. M. Hjellming and C. M. Wade, *Ap. J.* (Letters) **168,** L21, 1971.
33. H. Tananbaum, H. Gursky, E. Kellogg, R. Giacconi and C. Jones, *Ap. J* (Letters) **177,** L5, 1972.
34. C. T. Bolton, *Nature* **235,** 271, 1972.
35. R. J. Brucato and J. Kristian, IAU Circular No. 2421, 1972.
36. R. Giacconi, H. Gursky, E. Kellogg, S. Murray, E. Schreier and H. Tananbaum, *Ap. J.* (Letters) **167,** L67, 1971.
37. E. Schreier, R. Levinson, H. Gursky, E. Kellogg, H. Tananbaum and R. Giacconi, *Ap. J.* (Letters) **172,** L79, 1972.
38. R. Wilson, *Ap. J.* (Letters) **174,** L27, 1972.
39. E. Van den Heuvel and J. Heise, *Nature,* 1972.
40. R. W. Leach and R. Ruffini, "X-ray Sources—A Transient State from Neutron Stars to Black Holes", Preprint, August, 1972.
41. H. Tananbaum, H. Gursky, E. Kellogg, R. Levinson, E. Schreier and R. Giacconi, *Ap. J.* (Letters) **174,** L143, 1972.
42. W. Liller, IAU Circular 2415, 1972.
43. J. N. Bahcall and A. N. Bahcall, *Ap. J.* (Letters) **178,** L1, 1972.
44. W. Forman, C. A. Jones, and W. Liller, *Ap, J.* (Letters) **177,** L103, 1972.
45. D. O. Lamb and J. M. Sorvari, IAU Circular 2422, 1972.
46. A. Davidson, J. P. Henry, J. Middleditch and H. E. Smith, *Ap. J.* (Letters) **177,** L97, 1972.
47. W. A. Hiltner, J. Werner and P. Osmer, *Ap. J.* (Letters) **175,** L19, 1972.

Appendix

A Galactic X-ray Sources

Sources $I > 8$ ct/sec, $b < 20°$ (but including known galactic sources, $b > 20°$)

2U Designation	l, b	I(ct/sec)
0114 + 63	126, 1	70
0352 + 30	163 – 17	20
X-Persei ($6^m.0$, B0) irregular variable		
0613 + 09	200 – 3.5	63
0900 – 40	263, 3.9	77 Vela XR-1
HD 77581 ($6^m.9$, B0 . 5 Ib) spectroscopic binary		
$P = 7$ days,		$f(M) = 0.013$
if $M_p \approx 15\, M_\odot, M_s > 1.5\, M_\odot$		
1022 – 55	283, 1.4	10
1119, –60	292, 0.4	160 Cutoff spectrum, Cen X3
Pulsing, 4.8 sec; binary, 2.087 days, irregular disappearance		
		$f(M) = 15\, M_\odot$
1134 – 61	294, –0.3	8.5
1146 – 61	296, 0	72
1223 – 62	300, 0	32
SAO 251905 ($9^m.9$, AI)		
Highly variable in balloon energy range		
1254 – 69	303, –6	25
1258 – 61	304, +1	47
Highly variable in balloon energy range		
1516 – 56	322, 0	720 $N_H \sim 3 \times 10^{22}$ atm(cm^2), distance, Cir X-1
Irregular X-ray pulsations		
1536 – 52	327, 2	11
1542 – 62	321, –6	35
SAO 253287 ($8^m.7$, B8)		
1556, –60	324, –6	18
SAO 253382 ($8^m.6$, B9)		
1617, –15	359, 24	17,000 Sco X-1
Radio (<0.005 – 0.2) F.U.		
13m, irregular variable, IR cutoff, correlated X-ray-opt. variable		

A Galactic X-ray Sources—*continued*

2U Designation	l, b	I(ct/sec)	
1624 – 49	335, 0	44	
SAO 226781 ($9^m.0$, B8)			
1626 – 67	322, 13	13	
1630 – 47	337, 0	150	
1637, –53	333, 5	256	N_H cutoff, near
SAO 244064 ($9^m.7$, F8)			
1639, –62	326, –11	9.4	
1641, –45	340, 0	400	$N_H > 4 \times 10^{22}$, distant
1658, –46	340, –3	42	
1700 – 37	347, 2	102	
HD 153919 ($6^m.6$, 07f)			
X-ray Binary(?) P ~ 3.4 days			
1701 – 31	353, 5	12	
1702 – 36	349, +3	715	$N_H \approx 10^{22}$ atm/cm^2, dista
1704 – 42	344, –1.5	108	
1705 – 44	343, –2.3	280	No cutoff, near
1705 – 22	0.5, 10.4	42	
1705 + 34	58, 35	100	Cutoff spectrum, Her X-1
Pulsing 1.24 sec; binary, 1.7 days, disappears regularly ~ 35 days.		$f(M) = 0.8\,M_\odot$	
HZ Her (13 m) irregular variable, pulses and eclipses in phase			
1718 – 39	349, –1	16	
1726 – 33	354, 0	73	
SAO 208881 ($6^m.7$, B5)			
1728 – 24	2, 5	60	
1728 – 16	8.5, 9	280	
1735 + 43	69, 31	17	
1744 – 26	2, 1	460	Cutoff?
1757 – 25	5 – 1	1000	$N_H \approx 3 \times 10^{22}$, distant, GX(5 – 1)
1757 – 33	357 – 5	19	
1758 – 20	9 + 1	600	$N_H \approx 1.3 \times 10^{22}$, distant
1811 – 17	13, 0	300	$N_H \approx 3 \times 10^{22}$, distant
1813 – 14	16, 1	560	$N_H \approx 3 \times 10^{22}$, distant, GX(17 + 2)
Radio (<0.005–0.022) FU			

A Galactic X-ray Sources—*continued*

2U Designation	l, b	I(ct/sec)
1813 – 12	18, 2	10
1820 – 30	2 – 8	200
1822, 0	30, 6	40
1822, –37	357, –11	15
1836, +05	36, 5	179
1907, +02	37, –3	42
1908, 00	36, 4	200
1912, –05	31, –7	19
1926, +43	76, +12	8
1954, +31	68, 1.7	75
1956, +35	71, 3	1175 Irregular pulsations, Cyg X-1
Radio (<0.005-0.015) FU, strong and variable at balloon energies, BD + 34° 3815, ($9^m.0$, B0I), double line spectroscopic binary, $M_s/M_p \sim 1.4$ $P = 5.6$ days, $V_p = 74$ Km/s		
2030, +40	80, 1	133 Cyg X-3
Highly cutoff spectrum (~3 keV) SAO 049756 ($9^m.1$)		
2130, +47	92, –3	12
2142, +38	87, –11	420 Cyg X-2
+14m, Irregular variable, absorption line spectrum of G type star, $\Delta\nu \approx 300$ Km/sec		

B Supernova Remnants

Name	Distance	Size	Age	X-ray Power	
Crab Nebula	2.0 Kpc	2.1 pc	900 yr	10^{37} erg/sec	0.5–200 keV
Extended, power law spectrum					
NP 0532	2.0 Kpc	–	900 yr	10^{36} erg/sec	0.5 keV–10^{11} eV
Cas A	3.4 Kpc	4 pc	300 yr	5×10^{36} erg/sec	0.5–5 keV
Thermal spectrum ($kT \sim 2$ keV)					
Tycho	3.5 Kpc	7.1 pc	400 yr	5×10^{36} erg/sec	0.5–5 keV
IC 443	2.0 Kpc	23 pc		5×10^{34} erg/sec	2–10 keV
Seen only by UHURU					
MSH 15-52A	5 Kpc	12 pc		5×10^{35} erg/sec	2–10 keV
Seen only by UHURU					
Puppis A	1.2 Kpc	26 pc	10^{4-5} yr	10^{36} erg/sec	0.2–3 keV
Vela X, Y, Z	0.5 Kpc	35 pc	10^{4-5} yr	10^{36} erg/sec	0.2–3 keV
Extended, resolves into point (?) sources plus extended region					
Cygnus Loop		40 pc	10^{4-5} yr	2×10^{36} erg/sec	0.2–1 keV
Seen only below 1 keV–not seen by UHURU Extended, thermal spectrum ($kT \sim 0.2$ keV)					

C Transient Sources

Name	Position	I(peak) (1–10 keV)
Cen X-2	$13^h\ 8^m$ $-32°\ 15'$	5×10^{-7} ergs/cm^2-sec
Seen for ~80 days (not scanned for 17 months prior to observation)		
Cen X-4	$14^h\ 56^m$ $-32°\ 15'$	5×10^{-7} ergs/cm^2-sec
Seen for ~80 days		
2U 1543 – 47	$15^h\ 43^m\ 50^s$ $-47°\ 33'\ 36''$	3×10^{-8} ergs/cm^2-sec
Seen for 9 months (as of June 1972). Optical emission >18 mag		
2U 1735 – 28	$17^h\ 35^m\ 50^s$ $43°\ 13'\ 12''$	10^{-8} erg/cm^2-sec
Seen only on several occasions during a one week period. May not be a transient source		

D Magellanic Cloud Sources

2U Designation	I(ct/sec)	
0115 – 73	28 Cutoff spectrum	Small cloud
Binary 4 days		
0521 – 72	15	Large cloud
0532 – 66	9.4	Large cloud
0539 – 64	21	Large cloud
0540 – 69	19	Large cloud

Astrophysics of Black Holes

Igor D. Novikov‡

Institute of Applied Mathematics, Academy of Sciences, Moscow

Kip S. Thorne‡

California Institute of Technology, Pasadena

---

† Supported in part by the Acedemy of Sciences of the USSR
§ Supported in part by the U.S. National Science Foundation [GP-27304, GP-28027]

Contents

1 Introductory Remarks 347
2.1 Thermal Bremsstrahlung ("Free-Free Radiation" From a Plasma) . 347
2.2 Free-Bound Radiation 355
2.3 Thermal Cyclotron and Synchrotron Radiation . . . 357
2.4 Electron Scattering of Radiation 360
2.5 Hydrodynamics and Thermodynamics 362
2.6 Radiative Transfer 370
2.7 Shock Waves 379
2.8 Turbulence 383
2.9 Reconnection of Magnetic Field Lines 384
3 The Origin of Stellar Black Holes 386
4 Black Holes in The Interstellar Medium 389
4.1 Accretion of Noninteracting Particles onto a Nonmoving Black Hole 389
4.2 Adiabatic, Hydrodynamic Accretion onto a Nonmoving Black Hole 390
4.3 Thermal Bremsstrahlung from the Accreting Gas 396
4.4 Influence of Magnetic Fields and Synchrotron Radiation . . 397
4.5 Interaction of Outflowing Radiation with the Gas . . . 399
4.6 Validity of the Hydrodynamical Approximation . . . 401
4.7 Gas Flow Near the Horizon 402
4.8 Accretion onto a Moving Hole 404
4.9 Optical Appearance of Hole: Summary 407
5 Black Holes in Binary Star Systems and in the Nuclei of Galaxies . . 408
5.1 Introduction 408
5.2 Accretion in Binary Systems: The General Picture . . . 409
5.3 Accretion in Galactic Nuclei: The General Picture . . . 420
5.4 Properties of the Kerr Metric Relevant to Accreting Disks . . 422
5.5 Relativistic Model for Disk: Underlying Assumptions . . . 424
5.6 Equations of Radial Structure 427
5.7 Equations of Vertical Structure 431
5.8 Approximate Version of Vertical Structure 434
5.9 Explicit Models for Disk 435
5.10 Spectrum of Radiation from Disk 438
5.11 Heating of the Outer Region by X-Rays from the Inner Region . 443
5.12 Fluctuations on the Steady State Model 443
5.13 Supercritical Accretion. 444
5.14 Comparison with Observations 445

6 White Holes and Black Holes of Cosmological Origin 445
6.1 White Holes, Grey Holes, and Black Holes 445
6.2 The Growth of Cosmological Holes by Accretion . . . 446
6.3 Limit on the Number of Baryons in Cosmological Holes . . 447
6.4 Caution 448
References 448

1 Introductory Remarks

To analyze astrophysical aspects of black holes one must bring input from two directions: from the elegant realm of general relativity (lectures of Hawking, Carter, Bardeen, and Ruffini), and from the more mundane realm of astrophysics and plasma physics (§2 of these lectures). We shall here combine these inputs to discuss the origin of stellar black holes (§3); and to analyze observable aspects of black holes in the interstellar medium (§4), in binary star systems and the nuclei of galaxies (§5), and in cosmological contexts (§6).

Readers with strong astrophysical backgrounds will wish to skip over our brief treatment of the fundamentals of astrophysics and plasma physics (§2), and dig immediately into our discussion of black-hole astrophysics (§§3-8). However, for the reader whose previous training has focussed primarily on gravitation theory, the material of §2 is an important prerequisite for understanding the rest of the notes.

The authors are deeply indebted to Mr. Alan Lightman for removing a large number of errors from the manuscript.

2.1 Thermal Bremsstrahlung ("Free-Free Radiation" From a Plasma)

Basic references chapter 6 of Shkarofsky, Johnston, and Bachynski (1966); chapter 15 of Jackson (1962); §4 of Ginzburg (1967).

Notation for fundamental constants

h	Planck's constant
m_e	rest mass of electron
m_p	rest mass of proton
e	charge of electron
c	speed of light
α	fine-structure constant
r_0	classical electron radius
Ry	Rydberg energy
Z	charge of ions in plasma
A	atomic weight of ions
k	Boltzmann's constant
ζ	Euler constant (= 1.78 ....)

Radiation from a single classical collision

Consider a single ion of charge Z, at rest in the laboratory frame. Let a single electron of speed $v \ll c$ (nonrelativistic) scatter off the nucleus (Coulomb scattering), with impact parameter b. Let $I(\omega)$ be the total energy per unit circular frequency, ω, radiated by the electron as it scatters. A simple classical calculation [e.g. Jackson (1962), p. 507] gives

$$I(\omega) \cong \frac{2}{3\pi} \frac{e^2}{c^3} |\Delta \mathbf{v}_\omega|^2, \tag{2.1.1}$$

where $\Delta \mathbf{v}_\omega$ is the change in electron velocity during a time $\tau = 1/\omega$ centered about the electron's point of closest approach to the nucleus.

It is useful to examine two limiting cases: small-angle scattering ($\theta \ll 1$) corresponding to large impact parameter ($b \gg b_{s-l}$); and large-angle scattering ($\theta \sim 2\pi$), corresponding to small impact parameter ($b \ll b_{s-l}$). The impact parameter, b_{s-l}, which separates small-angle from large-angle scattering, is given by

$$Ze^2/b_{s-l} = \tfrac{1}{2} m_e v^2. \tag{2.1.2}$$

For *small-angle scattering* the total scattering angle θ is

$$\theta = \frac{|\Delta \mathbf{v}|}{v} = \frac{Ze^2/b}{\frac{1}{2} m_e v^2} = \frac{b_{s-l}}{b}, \tag{2.1.3}$$

and the time during which the scattering occurs is $\Delta t \approx b/v$. Hence, for $\tau = 1/\omega \gg \Delta t$, the change in velocity during time τ, $|\Delta \mathbf{v}_\omega|$, is the full change $|\Delta \mathbf{v}|$ of (2.1.3); while for $\tau = 1/\omega \ll \Delta t$ the change is essentially zero. Inserting these changes into equation (2.1.1), we obtain

$$\begin{aligned} I(\omega) &= 0 && \text{for} \quad \omega \gg v/b, \\ I(\omega) &= \frac{8}{3\pi} \frac{Z^2 e^2}{c} \left(\frac{c r_0}{v b}\right)^2 && \text{for} \quad \omega \ll v/b. \end{aligned} \tag{2.1.4}$$

For large-angle scattering the electron orbit is very nearly a parabola, with "perihelion" at radius

$$r_p = b^2/b_{s-l} \tag{2.1.5a}$$

and with speed at perihelion

$$v_p = c(2Zr_0/r_p)^{1/2}. \tag{2.1.5b}$$

During time intervals $\tau = 1/\omega \ll r_p/v_p$ a negligible amount of velocity change occurs; hence

$$I(\omega) = 0 \quad \text{for} \quad \omega \gg v_p/r_p = c(2Zr_0 b_{s-l}^3)^{1/2}/b^3. \tag{2.1.6a}$$

During time intervals $\tau = 1/\omega > r_p/v_p$, but $\tau < b_{s-l}/v$, the electron is initially ($t = -\tau/2$) headed directly toward the nucleus with "parabolic" speed

$$v_i = \left(\frac{8}{3} \frac{Zr_0}{\tau} c^2\right)^{1/3} = 2(\tfrac{1}{3} Z r_0 \omega c^2)^{1/3};$$

it swings into a sharp turn around the nucleus; and it emerges at $t = +\tau/2$ with its initial velocity precisely reversed. Consequently, $|\Delta \mathbf{v}_\omega|$ is equal to $2v_i$, and $I(\omega)$ is

$$I(\omega) = \frac{32}{3\pi}\frac{e^2}{c^3}\left(\frac{Zr_0\omega}{3}c^2\right)^{2/3} \quad \text{for} \quad \frac{v}{b_{s-l}} < \omega < \frac{v_p}{r_p} = \frac{c(2Zr_0 b_{s-l}^3)^{1/2}}{b^3} \tag{2.1.6b}$$

Radiation from a nonchromatic beam of electrons

Consider a single ion of charge Z, at rest in the laboratory frame. Bombard the ion with a monochromatic beam of electrons with speed $v \ll 1$ and flux (number per unit time per unit area) S. Examine the radiation of frequency ν (circular frequency $\omega = 2\pi\nu$) emitted by the electrons as they scatter off the nucleus. Let $\mathscr{P}_\nu$ be the total power emitted per unit frequency (ergs sec^{-1} Hz^{-1}); and define the emission cross section per unit frequency, $d\sigma/d\nu$, by

$$\mathscr{P}_\nu = (d\sigma/d\nu)Sh\nu. \tag{2.1.7}$$

The dependence of the emission cross section on photon frequency ν and on electron speed v is most conveniently expressed in terms of the two dimensionless parameters

$$\frac{h\nu}{\frac{1}{2}m_e v^2}, \qquad \frac{\frac{1}{2}m_e v^2}{Z^2 Ry} = \left(\frac{v/c}{\alpha Z}\right)^2. \tag{2.1.8}$$

On a sheet of paper (Figure 2.1.1) plot $(h\nu)/(\frac{1}{2}m_e v^2)$ vertically, and plot $(\frac{1}{2}m_e v^2/Z^2 Ry)$ horizontally. This 2-dimensional plot can be split into several different regions, in each of which the details of the emission process are different.

Forbidden region The individual photons emitted in each scattering cannot have energies greater than the electron kinetic energy—unless the electron gets captured into a bound state around the ion. But capture produces "free-bound radiation" (§2.2) rather than bremsstrahlung. Consequently, the region

$$(h\nu)/(\tfrac{1}{2}m_e v^2) > 1 \tag{2.1.9a}$$

of Figure 2.1.1 is a "forbidden region"; no bremsstrahlung is emitted in this region; $d\sigma/d\nu$ vanishes in this region.

Photon-discreteness region In the region

$$\tfrac{1}{3} \leqslant (h\nu)/(\tfrac{1}{2}m_e v^2) \leqslant 1 \tag{2.1.9b}$$

no more than 3 photons can be emitted by each scattering. This "photon discreteness" has a significant effect on the emission cross-section, cutting it down toward zero as one approaches the edge of the boundary of the forbidden region, $(h\nu)/(\frac{1}{2}m_e v^2) = 1$. No classical calculation can reveal such an effect; the photon-discreteness region must be analyzed in a quantum mechanical manner; see, e.g., Heitler (1954), and see below.

Large-angle region All small-angle scatterings last too long ($\Delta t \sim b/v > b_{s-l}/v$) to produce any radiation of $\tau = 1/\omega < b_{s-l}/v$; cf. eq. (2.1.4). Consequently, in the region

$$\frac{h\nu}{\frac{1}{2}m_e v^2} = \frac{\hbar\omega}{\frac{1}{2}m_e v^2} > \frac{\hbar v/b_{s-l}}{\frac{1}{2}m_e v^2} = \left(\frac{\frac{1}{2}m_e v^2}{Z^2 Ry}\right)^{1/2} \tag{2.1.10}$$

only large-angle scatterings produce radiation. This region is shown in Figure 2.1.1. The power per unit frequency emitted at a fixed frequency ν in this region is

$$\mathscr{P}_\nu = 2\pi \mathscr{P}_\omega = 2\pi \int_0^{b_{\max}} I(\omega) S\, 2\pi b\, db, \tag{2.1.11a}$$

where $I(\omega)$ is given by the large-angle formula (2.1.6), and $b_{\max}$ is the "cutoff" in that formula

$$b_{\max} = (2c^2 Z r_0 b_{s-l}^3/\omega^2)^{1/6}. \tag{2.1.11b}$$

Performing the integration, dividing by $Sh\nu$ to obtain $d\sigma/d\nu$, and rexpressing the result in terms of fundamental atomic constants, one obtains

$$\frac{d\sigma}{d\nu} = \frac{2^{1/3} 16}{3^{5/3}} \frac{\alpha c^2}{v^2} \frac{Z^2 r_0^2}{\nu}.$$

A more exact classical calculation gives a slightly different numerical coefficient:

$$\left(\frac{d\sigma}{d\nu}\right)_{\rm LA} = \frac{16\pi}{3\sqrt{3}} \frac{\alpha c^2}{v^2} \frac{Z^2 r_0^2}{\nu} \tag{2.1.12}$$

(Here "LA" stands for "large-angle region".)

In other regions of Figure 2.1.1, one gets other formulas for $d\sigma/d\nu$ (see below). It is conventional in all regions to lump the deviations from this large-angle cross section into a correction term $G(\nu, v)$ which is called the "Gaunt factor":

$$G(\nu, v) \equiv \frac{d\sigma/d\nu}{(d\sigma/d\nu)_{\rm LA}}. \tag{2.1.13}$$

Thus, in the large-angle region $G = 1$; and in the forbidden region $G = 0$. It turns out that throughout the photon-discreteness portion of the large-angle region, G remains approximately 1; see Figure 2.1.1.

Small-angle, classical region In the region $\tau = 1/\omega \gg b_{s-l}/v$—i.e.

$$\frac{h\nu}{\frac{1}{2}m_e v^2} \ll \left(\frac{\frac{1}{2}m_e v^2}{Z^2 Ry}\right)^{1/2} \tag{2.1.14}$$

–radiation is produced by small-angle scatterings as well as by large-angle scatterings. As a function of impact parameter b, the energy radiated by a single electron is constant in the large-angle region, $b < b_{s-l}$ (eq. 2.1.6b), and decreases as $1/b^2$ in the small-angle region, $b > b_{s-l}$ (eq. 2.1.4). Hence, when one integrates over $2\pi b db$ to get the power radiated, $\mathscr{P}_\nu$, one obtains

$$\frac{(\text{contribution from small angles})}{(\text{contribution from large angles})} = \ln\left(\frac{b_{\max}}{b_{s-l}}\right), \tag{2.1.15}$$

where $b_{\max}$ is the cutoff in the small-angle region (eq. 2.1.4)

$$b_{\max} = v/\omega. \tag{2.1.16}$$

Since the small-angle contribution is logarithmically dominant, one can ignore the contribution from large-angle scatterings and calculate

$$\frac{d\sigma}{d\nu} = \frac{\mathscr{P}_\nu}{Sh\nu} = \frac{1}{Sh\nu} 2\pi \int_{b_{s-l}}^{b_{\max}} I(\omega) S\, 2\pi b db$$

$$= \frac{16}{3} \frac{\alpha c^2}{v^2} \frac{Z^2 r_0^2}{\nu} \ln\left(\frac{b_{\max}}{b_{s-l}}\right). \tag{2.1.17}$$

A more exact classical calculation gives a slightly different argument in the logarithm:

$$\frac{d\sigma}{d\nu} = \frac{16}{3} \frac{\alpha c^2}{v^2} \frac{Z^2 r_0^2}{\nu} \ln\left(\frac{2}{\zeta} \frac{b_{\max}}{b_{s-l}}\right), \tag{2.1.18}$$

where $\zeta = 1.781 \ldots$ is the "Euler constant". Consequently, the Gaunt factor for the small-angle, classical region is

$$G(\nu, v) = \frac{\sqrt{3}}{\pi} \ln\left(\frac{2}{\zeta} \frac{b_{\max}}{b_{s-l}}\right) = \frac{\sqrt{3}}{\pi} \ln\left[\frac{2}{\zeta}\left(\frac{\frac{1}{2}m_e v^2}{h\nu}\right)\left(\frac{\frac{1}{2}m_e v^2}{Z^2 Ry}\right)^{1/2}\right]. \tag{2.1.19}$$

Actually, this small-angle classical result is *not* valid throughout the region (2.1.14). The uncertainty principle requires a modification at electron energies $\frac{1}{2}m_e v^2$ larger than $Z^2 Ry$.

Small-angle, uncertainty-principle region Consider an electron with impact parameter b and speed v. The electron is actually not a classical object; rather, it is a quantum-mechanical wave packet. To scatter with impact parameter b, the wave packet (i) must have transverse dimensions, Δx, less than b

$$\Delta x < b,$$

and (ii) must not spread transversely by more than b during the classical scattering time $\Delta t = b/v$

$$b > \begin{pmatrix}\text{spreading}\\ \text{during } \Delta t\end{pmatrix} = \left(\frac{\Delta p_x}{m_e}\right)\Delta t \gtrsim \left(\frac{\hbar}{m_e \Delta x}\right)\Delta t > \left(\frac{\hbar}{m_e b}\right)\frac{b}{v} = \frac{\hbar}{m_e v}.$$

Thus, the uncertainty principle ("$\gtrsim$" in above equation) prevents the existence of scatterings with b less than the electron de Broglie wavelength

$$b_{dB} \equiv \hbar/m_e v. \tag{2.1.20}$$

If $b_{dB} < b_{s-l}$, this limitation has no effect on small-angle scatterings; and the small-angle, classical Gaunt factor (2.1.19) is valid. If $b_{dB} > b_{s-l}$ the uncertainty principle comes into play, and one must use b_{dB} rather than b_{s-l} as the lower limit on the small-angle integral (2.1.17) for $d\sigma/d\nu$. The result is

$$\frac{d\sigma}{d\nu} = \frac{16}{3}\frac{\alpha c^2}{v^2}\frac{Z^2 r_0^2}{\nu} \ln\left(\frac{b_{\max}}{b_{dB}}\right).$$

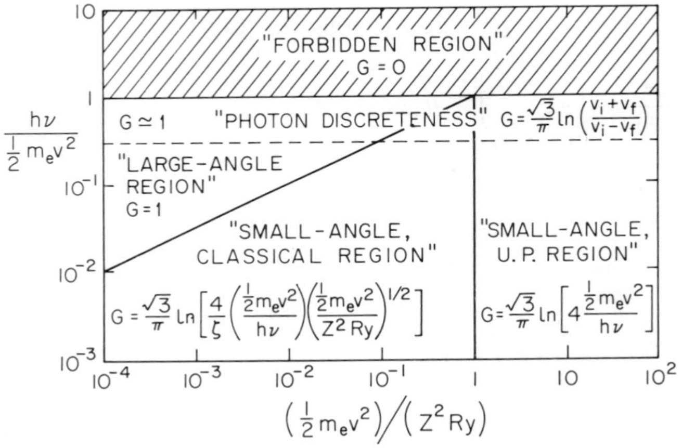

Figure 2.1.1. Regions and Gaunt factors for bremsstrahlung of frequency ν emitted by an electron of kinetic energy $\frac{1}{2}m_e v^2$ when it impinges on an ion of charge Ze.

A more exact, quantum mechanical calculation gives a slightly different argument in the logarithm,

$$\frac{d\sigma}{d\nu} = \frac{16}{3}\frac{\alpha c^2}{v^2}\frac{Z^2 r_0^2}{\nu} \ln\left(\frac{2b_{\max}}{b_{dB}}\right), \tag{2.1.21}$$

corresponding to the Gaunt factor

$$G(\nu, v) = \frac{\sqrt{3}}{\pi} \ln\left(\frac{2b_{\max}}{b_{dB}}\right) = \frac{\sqrt{3}}{\pi} \ln\left(4\,\frac{\frac{1}{2}m_e v^2}{h\nu}\right). \tag{2.1.22}$$

The "small-angle, uncertainty-principle region", in which this expression holds, is separated from the "small-angle, classical region", where (2.1.19) is valid, by the line

$$b_{dB}/b_{s-l} = (\tfrac{1}{2}\, m_e v^2)/(Z^2 Ry) = 1 \tag{2.1.23}$$

See Figure 2.1.1.

From Figure 2.1.1 it is clear that the uncertainty principle cannot have any effect on the large-angle region.

In the "photon-discreteness" portion of the small-angle, uncertainty-principle region, discreteness effects modify the Gaunt factor (2.1.22) into the form

$$G(\nu, v) = \frac{\sqrt{3}}{\pi} \ln \left(\frac{v_i + v_f}{v_i - v_f}\right) = \frac{\sqrt{3}}{\pi} \ln \left[\frac{\frac{1}{2}m_e(v_i + v_f)^2}{h\nu}\right]. \tag{2.1.24a}$$

Here v_i and v_f are the electron speeds before and after the collision:

$$v_i \equiv v, \qquad \tfrac{1}{2}m_e v_f^2 \equiv \tfrac{1}{2}m_e v_i^2 - h\nu. \tag{2.1.24b}$$

[Expression (2.1.24) is the result of a quantum-mechanical Born-approximation calculation, valid throughout the small-angle, uncertainty-principle region.]

Radiation from a plasma

Consider a plasma which may contain a variety of ionic species. Focus attention on radiation from all ions of a particular type, with charge Ze and atomic weight A. Let C_A be the concentration by mass of the species of atom that gives rise to this ion, and let f_i be the fraction of all such atoms ionized to the state of interest. Let f_e be the number of unbound electrons per baryon in the gas. Then

$$\begin{aligned} n_i &= (\text{number of ions per unit volume}) \equiv f_i C_A \rho_0 / A m_p \\ n_e &= (\text{number of electrons per unit volume}) \equiv f_e \rho_0 / m_p. \end{aligned} \tag{2.1.25}$$

Here ρ_0 is the density of rest mass. (Of course, $f_i \leqslant 1; f_e \leqslant 1; C_A \leqslant 1$.) The ions and electrons both have Maxwell velocity distributions; but because $m_e \ll A m_p$, the velocities of the electrons and their accelerations during Coulomb scattering are far greater than those of the ions. Therefore, in calculating bremsstrahlung one can regard the ions as at rest. The number of ions per unit mass of plasma is f_i/m_p, so the total power per unit frequency emitted from one gram of plasma by electron-ion scatterings is

$$\begin{aligned} \epsilon_\nu &= \begin{pmatrix}\text{number of}\\ \text{ions per}\\ \text{unit mass}\end{pmatrix} \int \begin{pmatrix}\text{fractions of all}\\ \text{electrons that have}\\ \text{speeds } v \text{ in range } dv\end{pmatrix} \begin{pmatrix}\text{number}\\ \text{density}\\ \text{of electrons}\end{pmatrix} v \frac{d\sigma}{d\nu} h\nu \\ &= \frac{C_A f_i}{A m_p} \int \left[\frac{\exp(-\frac{1}{2}m_e v^2/kT)}{(2\pi kT/m_e)^{3/2}} 4\pi v^2\, dv\right] \left(\frac{f_e \rho_0}{m_p}\right) v \frac{d\sigma}{d\nu} h\nu. \end{aligned} \tag{2.1.26}$$

Here T is the temperature in the plasma. By inserting into the integrand expressions (2.1.12) and (2.1.13) for $d\sigma/d\nu$ and performing the integration, one obtains

$$\begin{aligned} \epsilon_\nu &= \frac{32\pi c^2}{3} \left(\frac{C_A f_e f_i Z^2}{A}\right) \frac{m_e r_0^3}{m_p^2} \rho_0 \left(\frac{2\pi m_e c^2}{3kT}\right)^{1/2} e^{-h\nu/kT} \bar{G}(\nu, T) \\ &= \left(2.5 \times 10^{10} \frac{\text{erg}}{\text{g sec Hz}}\right) \left(\frac{C_A f_e f_i Z^2}{A}\right) \left(\frac{\rho_0}{\text{g/cm}^3}\right) T_K^{-1/2} e^{-h\nu/kT} \bar{G}(\nu, T). \end{aligned} \tag{2.1.27}$$

Here T_K is temperature measured in °K, and $\bar{G}(\nu, T)$ is the "Maxwell-Boltzmann-averaged Gaunt factor", also called the "mean Gaunt factor":

$$\bar{G}(\nu, T) \equiv \int_0^\infty G\left(\nu, v = \left[\frac{2(ykT + h\nu)}{m_e}\right]^{1/2}\right) e^{-y}\, dy. \tag{2.1.28}$$

Corresponding to the 2-dimensional plot of Gaunt factors (Figure 2.1.1) one can make a 2-dimensional plot of mean Gaunt factors (Figure 2.1.2). It is straightforward to calculate the mean Gaunt factors shown there (up to quantities of order unity) from the Gaunt factors of Figure 2.1.1. Notice that,

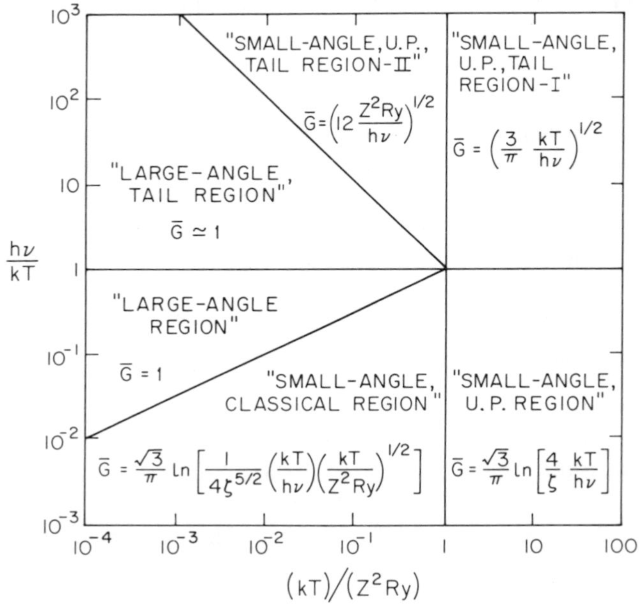

Figure 2.1.2. Mean Gaunt factors for bremsstrahlung of frequency ν emitted by electron-ion collisions in a plasma of temperature T.

aside from quantities of order unity, the mean Gaunt factors in the large-angle and small-angle regions, $h\nu/kT < 1$, are obtained by simply taking the Gaunt factors and making the replacement $\frac{1}{2}m_e v^2 \rightarrow kT$

$$\bar{G}(\nu, kT) \simeq G(\nu, \tfrac{1}{2}m_e v^2 = kT)$$

Notice also that the "forbidden region" of the Gaunt-factor plot is replaced, in the mean-Gaunt-factor plot, by "tail regions". The radiation in these tail regions is produced by the high-energy tail, $\frac{1}{2}m_e v^2 \approx h\nu \gg kT$, of the Maxwell-Boltzmann distribution.

Tables of mean Gaunt factors, calculated with much higher accuracy than Figure 2.1.2, are given by Green (1959) and by Karzas and Latter (1961).

The total energy emitted by electron-ion collisions is obtained by integrating expression (2.1.27) over frequency:

$$\epsilon = \int \epsilon_\nu \, d\nu = \frac{16}{3} \frac{C_A f_e f_i Z^2}{A} \frac{\alpha m_e c r_0^2}{m_p^2} \rho_0 \left(\frac{2\pi kT}{m_e c^2}\right)^{1/2} \bar{\bar{G}}(T) \tag{2.1.29}$$

$$= \left(5.2 \times 10^{20} \frac{\text{erg}}{\text{g sec}}\right) \left(\frac{C_A f_e f_i Z^2}{A}\right) \left(\frac{\rho_0}{\text{g/cm}^3}\right) T_{\text{K}}^{1/2} \bar{\bar{G}}(T).$$

Here $\bar{\bar{G}}(T)$ is the frequency-averaged mean Gaunt factor

$$\bar{\bar{G}}(T) \equiv \int_0^\infty \bar{G}(\nu = xkT/h, T)\, e^{-x}\, dx. \tag{2.1.30}$$

From a value of $\bar{\bar{G}} \approx 1.2$ at $kT/Z^2 Ry \sim 0.01$, it increases gradually to a maximum of 1.42 at $kT/Z^2 Ry \sim 1$, and then decreases gradually toward an asymptotic, high-temperature value of 1.103 at $kT/Z^2 Ry > 100$; see Figure 2 of Green (1959).

In a plasma at nonrelativistic temperatures, the radiation from electron-electron collisions and from ion-ion collisions is negligible compared to electron-ion radiation. This is because (i) when identical particles scatter, the first time derivative of the electric dipole moment is conserved, so only radiation of quadrupole order and higher [with intensity $\sim(v/c)^2$ less than dipole radiation] can be emitted; (ii) the speeds and accelerations of colliding ions are far less than those of electrons because of their much larger masses.

Equations (2.1.27) and (2.1.29), together with Figures 2.1.2 and 2.1.3, are the chief results of astrophysical interest from this section. When dealing with atoms that are only partially ionized, one must sometimes correct these results for screening of the nuclear charge by the bound electron cloud; see, e.g., Ginzburg (1967). For temperatures approaching and exceeding $kT = m_e c^2$, relativistic effects and electron-electron collisions modify the emissivity (2.1.29) to the form

$$\epsilon = \left(5.2 \times 10^{20} \frac{\text{ergs}}{\text{g sec}}\right) \left(\frac{C_A f_e f_i Z^2}{A}\right) \left(\frac{\rho_0}{\text{g/cm}^3}\right) T_{\text{K}}^{1/2} (1 + 4.4 \times 10^{-10} T_{\text{K}}) \bar{\bar{G}}(T) \tag{2.1.31}$$

(Ginzburg 1967).

2.2 Free-Bound Radiation

Basic references §4-17 of Aller (1963); Brussard and van de Hulst (1962).

Basic physics and formulas

When an electron of kinetic energy

$$\tfrac{1}{2} m_e v^2 \lesssim Z^2 Ry$$

impinges on an ion, there is a significant probability that, instead of scattering with the emission of bremsstrahlung, it will get captured into a bound state with an accompanying emission of "free-bound" radiation.

We shall not compute here the details of the radiation. Instead we cite the results of such a computation from the above references.

Consider a plasma at temperature T and density of rest mass ρ_0. Let C_A be the concentration by mass of a particular atom in the plasma; let f_i be the fraction of all such atoms ionized into a given state, with charge Ze; and let f_e be the number of free electrons per baryon [cf. eq. (2.1.25)]. Consider the radiation emitted by the capture of free electrons into a particular bound state about the ion; and let n_q be the principal quantum number of that state, and E_i its ionization energy. Then the emissivity due to such transitions is

$$\epsilon_\nu = 8c^2 \left(\frac{C_A f_e f_i Z^4}{A n_q^5}\right)\left(\frac{\alpha^2 m_e r_0^3}{m_p^2}\right)\rho_0\left(\frac{2\pi m_e c^2}{3kT}\right)^{3/2} e^{-(h\nu - E_i)/kT}\, G_{bf}$$

$$= \left(3.9 \times 10^{15}\ \frac{\text{ergs}}{\text{g sec Hz}}\right)\left(\frac{C_A f_e f_i Z^4}{A n_q^5}\right)\left(\frac{\rho_0}{\text{g cm}^{-3}}\right) T_{\text{K}}^{-3/2} e^{-(h\nu - E_i)/kT} G_{bf}. \tag{2.2.1}$$

Here G_{bf} is a "bound-free Gaunt factor", analogous to the "free-free" Gaunt factors of §2.1. It depends on the kinetic energy

$$\tfrac{1}{2} m_e v^2 = h\nu - E_i \tag{2.2.2}$$

of the electron that is captured, and on the structure of the bound state (which one characterizes by its quantum numbers $n_q, l_q, \ldots$):

$$G_{bf} = G_{bf}(h\nu - E_i; n_q, l_q, \ldots). \tag{2.2.3}$$

Gaunt factors for various bound-free transitions of astrophysical interest are tabulated and graphed by Karzos and Latter (1961). Energy conservation requires that all photons emitted have energies greater than E_i; consequently,

$$G_{bf} = 0 \quad \text{for} \quad h\nu - E_i < 0. \tag{2.2.4}$$

In general, G_{bf} "turns on" discontinuously at $h\nu \doteq E_i$, with an initial value between ~0.5 and ~1.5:

$$G_{bf} \simeq 1 \quad \text{for} \quad 0 \leqslant h\nu - E_i \ll E_i. \tag{2.2.5}$$

For $h\nu - E_i \lesssim Z^2 Ry$, G_{bf} typically remains within an order of magnitude of unity; as $h\nu - E_i$ increases in the region $h\nu - E_i \gtrsim 10\, Z^2 Ry$, G_{bf} falls rapidly. See Karzas and Latter (1961).

The total emissivity, $\epsilon = \int \epsilon_\nu \, d\nu$, due to a given free-bound transition is

$$\epsilon = 8\sqrt{3}\,c^2 \left(\frac{C_A f_e f_i Z^4}{A n_q^5}\right)\left(\frac{\alpha^3 m_e c r_0^2}{m_p^2}\right)\rho_0 \left(\frac{2\pi m_e c^2}{3kT}\right)^{1/2} \overline{G}_{bf}$$

$$= \left(4.2 \times 10^{26}\,\frac{\text{erg}}{\text{g sec}}\right)\left(\frac{C_A f_e f_i Z^4}{A n_q^5}\right)\left(\frac{\rho_0}{\text{g cm}^{-3}}\right) T_{\text{K}}^{-1/2}\overline{G}_{bf}, \tag{2.2.6}$$

where $\overline{G}_{bf}$, the mean Gaunt factor, depends only on temperature and on the bound state, and is approximately equal to one

$$\overline{G}_{bf} = \int_0^\infty G_{bf}(h\nu - E_i = xkT; n_q, l_q, \ldots)\, e^{-x}\, dx \simeq 1. \tag{2.2.7}$$

The total emissivity (2.2.6) due to a given free-bound transition, compared to the total emissivity (2.1.29) of free-free radiation from the same type of ion, is

$$\frac{\epsilon_{fb}}{\epsilon_{ff}} \simeq \frac{Z^2}{n_q^5}\left(\frac{8 \times 10^5}{T}\right). \tag{2.2.8}$$

Ginzburg (1967) states that for astrophysical plasmas the ratio of total free-bound radiation to total free-free radiation (summed over all states and all ions) generally is less than

$$\frac{\epsilon_{fb\ \text{total}}}{\epsilon_{ff\ \text{total}}} < \left(\frac{8 \times 10^5 \text{K}}{T}\right). \tag{2.2.9}$$

2.3 Thermal Cyclotron and Synchrotron Radiation

Basic references Jackson (1962); Ginzberg (1967).

Power radiated by a single electron

Consider an electron of speed v and total mass-energy $\gamma m_e c^2$,

$$\gamma \equiv (1 - v^2)^{-1/2}, \tag{2.3.1}$$

spiralling in a magnetic field of strength B. The Lorentz 4-acceleration of the electron is

$$a^0 = 0, \qquad \mathbf{a} = (e/m_e c)\gamma \mathbf{v} \times \mathbf{B}. \tag{2.3.2}$$

Consequently, the total power radiated—as obtained from the standard Lorentz-invariant equation

$$\frac{dE}{dt} = \frac{2e^2}{3c^3} a^2 \tag{2.3.3}$$

—is

$$\frac{dE}{dt} = \frac{2}{3} \frac{r_0^2}{c} (\gamma v_\perp)^2 B^2. \tag{2.3.4}$$

Here $v_\perp$ is the component of the electron velocity perpendicular to the magnetic field. If the electron is nonrelativistic, $v \ll c$, this radiation is called "cyclotron radiation". If the electron is ultrarelativistic, $\gamma \gg 1$, it is called "synchrotron radiation".

Cyclotron radiation from a nonrelativistic plasma.

Consider a plasma with temperature

$$kT \ll m_e c^2, \text{ i.e. } T \ll 6 \times 10^9 \text{K}, \tag{2.3.5}$$

and with a magnetic field of strength B. The radiation from protons spiralling in the magnetic field is smaller by a factor $(m_e/m_p)^3 \simeq 10^{-10}$ than that from electrons, since

$$\frac{dE}{dt} \propto \frac{v^2}{m^2} \propto \frac{1}{m^3} \tag{2.3.6}$$

(Recall: in thermal equilibrium the mean kinetic energies of protons and electrons are the same.) Hence, we can ignore protons and other ions when calculating cyclotron radiation. Let f_e be the number of unbound electrons per baryon in the plasma. Then the total emissivity (ergs per second per gram of plasma) is

$$\epsilon = \frac{f_e}{m_p} \left\langle \frac{dE}{dt} \right\rangle, \tag{2.3.7}$$

where $\langle \ \rangle$ denotes an average over the Maxwell-Boltzmann velocity distribution of the electrons. Since $\gamma = 1$, since 2 of the 3 spatial directions are orthogonal to B, and since the mean kinetic energy per electron is $\frac{3}{2}kT$, we have

$$\langle (\gamma v_\perp)^2 \rangle = \tfrac{2}{3} \langle v^2 \rangle = 2kT/m_e. \tag{2.3.8}$$

Combining this with equations (2.3.4) and (2.3.7), we obtain

$$\begin{aligned} \epsilon &= \frac{4}{3} \left(\frac{f_e r_0^2 c}{m_p} \right) \left(\frac{kT}{m_e c^2} \right) B^2 \\ &= (0.32 \text{ ergs/g sec}) f_e T_K B_G^2, \end{aligned} \tag{2.3.9}$$

where T_K is temperature in °K and B_G is magnetic field in Gauss.

Each electron emits its radiation monochromatically, with the cyclotron frequency (orbital frequency of electron's spiral motion)

$$\nu_{\text{cyc}} = \frac{eB}{2\pi m_e c} = (2.79 \text{ MHz}) T_K B_G. \tag{2.3.10}$$

Thus, if the magnetic field is uniform, the radiation will be monochromatic with this frequency. But in Nature the magnetic field will always be inhomogeneous, and the spectrum will show a spread proportional to the spread in B^2. (Of course, there are always other sources of spread in the spectrum—e.g., doppler shifts and relativistic generation of harmonics.)

Synchrotron radiation from a relativistic plasma

Consider a plasma with temperature

$$kT \gg m_e c^2, \text{ i.e. } T \gg 6 \times 10^9 \text{K}, \tag{2.3.11}$$

and with a magnetic field of strength B. As in the nonrelativistic case, radiation from protons and ions is totally negligible. The total emissivity due to electrons and positrons (recall: at $kT \gg m_e c^2$ there will be many electron-positron pairs†) is given, as before, by equation (2.3.7); but now the factor f_e must be the number of electrons and positrons per baryon, and the relevant, ultrarelativistic Maxwell-Boltzmann average is

$$\langle (\gamma v_\perp/c)^2 \rangle = \tfrac{2}{3}\langle \gamma^2 \rangle = \tfrac{2}{3}12\left(\frac{kT}{m_e c^2}\right)^2 = 8\left(\frac{kT}{m_e c^2}\right)^2. \tag{2.3.12}$$

[One evaluates $\langle \gamma^2 \rangle$ using the distribution function in phase space

$$\mathcal{N} \equiv \frac{dN}{d^3x\, d^3p} \propto e^{-E/kT} = e^{-\gamma m_e c^2/kT},$$

with

$$d^3p = 4\pi c^{-3} E^2 dE \text{ in ultrarelativistic limit;}$$

in particular

$$\langle \gamma^2 \rangle = \left\langle \left(\frac{E}{m_e c^2}\right)^2 \right\rangle = \frac{1}{(m_e c^2)^2} \frac{\int E^2 e^{-E/kT} E^2\, dE}{\int e^{-E/kT} E^2\, dE} = 12\left(\frac{kT}{m_e c^2}\right)^2 .\Bigg]$$

Combining equations (2.3.12), (2.3.7), and (2.3.4), we obtain for the total emissivity of the plasma

$$\begin{aligned} \epsilon &= \frac{16}{3}\left(\frac{f_e r_0^2 c}{m_p}\right)\left(\frac{kT}{m_e c^2}\right)^2 B^2 \\ &= (2.2 \times 10^{-10} \text{ ergs/g sec}) f_e T_K^2 B_G^2. \end{aligned} \tag{2.3.13}$$

In the ultrarelativistic case the electrons do not emit monochromatically. An electron of energy $\gamma m_e c^2$ beams most of its radiation into a forward cone of half-angle $\alpha = 1/\gamma$ (special-relativistic "headlight effect"). Since the electron is headed toward the observer with speed v when its cone is directed toward him,

† The case of an optically thin plasma is of particular interest in astrophysics. The kinetic theory of the creation and annihilation of positrons in this case is somewhat complex; see, e.g., Bisnovaty-Kogan, Zel'dovich, and Sunyaev (1971).

the cone sweeps past the observer in time

$$\Delta t = (1 - v)\left(\frac{2\alpha}{\omega_{\rm cyc}}\right) = (1 - v^2)\frac{\alpha}{\omega_{\rm cyc}} = \frac{1}{\gamma^3 \omega_{\rm cyc}}. \tag{2.3.14}$$

Here $\omega_{\rm cyc}$ is the angular velocity of the electron in its spiraling orbit

$$\omega_{\rm cyc} = \frac{eB}{\gamma m_e c}. \tag{2.3.15}$$

Thus, the radiation comes in short bursts, of duration $\sim\Delta t = 1/(\gamma^3 \omega_{\rm cyc})$, separated by long intervals of time $2\pi/\omega_{\rm cyc}$. When Fourier analyzed, such radiation must be concentrated near the "critical frequency".

$$\nu_{\rm crit} \equiv \gamma^3 \omega_{\rm cyc} = \gamma^2 eB/m_e c. \tag{2.3.16}$$

A detailed calculation [chapter 14 of Jackson (1962)] reveals a fairly broad spectrum which rises as $\nu^{1/3}$ at low frequencies, which peaks at

$$\nu_{\rm peak} = 0.29\,\nu_{\rm crit}, \tag{2.3.17}$$

and which decays exponentially, $\sim\nu^{1/2} e^{-2\nu/\nu_{\rm crit}}$, at $\nu \gg \nu_{\rm crit}$.

When averaged over the relativistic plasma, such a spectrum will have a broad peak, with maximum located at

$$\nu_{\rm peak,\,plasma} \simeq \begin{pmatrix}\text{value of } \nu_{\rm peak} \text{ for electrons of} \\ \gamma^2 \simeq 12(kT/m_e c^2)^2\end{pmatrix} \tag{2.3.18}$$

$$4\,\frac{eB}{m_e c}\left(\frac{kT}{m_e c^2}\right)^2 \simeq (100\text{ MHz})\,T_{10}^2 B_G.$$

Here T_{10} is temperature in units of 10^{10}K. At $\nu \gg \nu_{\rm peak,\,plasma}$ the spectrum will decay exponentially, and at $\nu \ll \nu_{\rm peak,\,plasma}$ it will rise as a power law.

2.4 Electron Scattering of Radiation

Basic references §14.7 of Jackson (1962); §12.3 of Leighton (1959); Kompaneets (1957); Weyman (1965); Sunyaev and Zel'dovich (1973).

Basic physics and formulas

In discussing electron scattering, we shall confine attention to the nonrelativistic case—i.e. we shall demand that the temperature of the gas and the photon frequencies satisfy

$$kT \ll m_e c^2 (T \ll 6 \times 10^9\text{K});\ h\nu \ll m_e c^2 = 500\text{ keV}. \tag{2.4.1}$$

The differential cross section for a free electron to scatter a photon has the form

$$\frac{d\sigma_{es}}{d\Omega} = \tfrac{1}{2} r_0^2 (1 + \cos^2\theta), \tag{2.4.2}$$

where θ is the angle between incoming photon and outgoing photon. The total cross section ("Thompson cross section"), obtained by integrating over all directions, is

$$\sigma_{es} = (8\pi/3)r_0^2 = 0.665 \times 10^{-24}\ \text{cm}^2. \tag{2.4.3}$$

It is important for astrophysical applications that the scattering cross section is "color blind" (no dependence on frequency). See, e.g., §4. .

It is also important that the differential cross section is symmetric between forward and backward directions. This guarantees, for example, that on the average a scattered photon transmits *all* of its momentum to the electron

$$\langle \Delta \mathbf{p}_e \rangle = \mathbf{p}_\gamma = (h\nu/c)\mathbf{n}_\gamma. \tag{2.4.4a}$$

(Here $\mathbf{n}_\gamma$ is a unit vector in the direction of the initial photon motion). By contrast, only a tiny fraction of the photon energy is transmitted to the electron. In a frame where the electron is initially at rest, it receives a kinetic energy

$$\Delta E_e = (h\nu/m_e c^2)h\nu(1 - \cos\theta). \tag{2.4.4b}$$

(Recall: we have assumed $h\nu/m_e c^2 \ll 1$).

Consider a monochromatic beam of photons, with frequency ν, moving through a thermalized plasma of temperature T. Scattering by protons and ions will be negligible compared to scattering by electrons, since

$$\sigma_{es} \propto r_0^2 = (e^2/m_e c^2)^2 \propto 1/(\text{mass of scatterer})^2. \tag{2.4.5}$$

On the average, how much energy $\langle \Delta E_e \rangle$ is transmitted to the electron in a single scattering? The answer for electrons nearly at rest ($kT \ll h\nu$) is obtained by averaging expression (2.4.4b) over all angles. Because of the forward-backward symmetry of the differential cross section, $\cos\theta$ averages to zero and one obtains

$$\langle \Delta E_e \rangle = (h\nu/m_e c^2)h\nu \quad \text{if} \quad kT \ll h\nu. \tag{2.4.6}$$

One can calculate $\langle \Delta E_e \rangle$ for higher temperatures by invoking conservation of 4-momentum, and averaging over all angles and electron speeds. However, such a calculation is rather long and messy. To obtain the answer more easily, one use the following trick: An examination of the law of 4-momentum conservation convinces one that $\langle \Delta E_e \rangle$ must have the temperature dependence

$$\langle \Delta E_e \rangle \propto (h\nu - \alpha kT),$$

where α is a constant to be calculated. (Thus, if $h\nu > \alpha kT$, the electrons get heated by the photons; if $h\nu < \alpha kT$, they get cooled.) This law will reduce to expression (2.4.6) for $kT \ll h\nu$ if and only if the proportionality constant is $h\nu/m_e c^2$:

$$\langle \Delta E_e \rangle = (h\nu/m_e c^2)(h\nu - \alpha kT). \tag{2.4.7}$$

To calculate the constant α, imagine the following experiment. Place a large number of photons and electrons into a box with perfectly reflective walls.

Require that the photons and electrons interact only by electron scattering. Then photons cannot be created or destroyed; only scattered. Hence, when thermal equilibrium is reached, the photons acquire a Boltzmann energy distribution, rather than a Planck distribution. (Recall that stimulated emissions are responsible for Planckian deviations from the Boltzmann law; §2.5). Since photons have zero rest mass, their Boltzmann distribution

$$\mathcal{N} \equiv \frac{dN}{d^3x\, d^3p} \propto e^{-E/kT} = e^{-h\nu/kT}, \qquad d^3p = 4\pi c^{-3} E^2\, dE$$

corresponds to a number of photons per unit frequency given by

$$\frac{dN}{d\nu} \propto \nu^2 e^{-h\nu/kT}, \tag{2.4.8}$$

and corresponds to

$$\langle h\nu \rangle = 3kT, \qquad \langle (h\nu)^2 \rangle = 12(kT)^2.$$

Thus, for our thought experiment the energy transfer in each collision, expression (2.4.7) averaged over the equilibrium photon distribution, is

$$\langle\langle \Delta E_e \rangle\rangle = (3kT/m_e c^2)(4 - \alpha)kT.$$

But in equilibrium there must be no average energy transfer between photons and electrons; $\langle\langle \Delta E \rangle\rangle$ must be zero. This condition tells us that $\alpha = 4$.

Thus, turning back to the general situation, we conclude that monochromatic photons passing through a plasma of temperature T transfer an average energy per collision

$$\langle \Delta E_e \rangle = (h\nu/m_e c^2)(h\nu - 4kT) \tag{2.4.9}$$

to the electrons. When $4kT \gg h\nu$, the photon energies get boosted by the collisions, and one says that the radiation is being "Comptonized"; see §5.10; see, e.g., Illarionov and Sunyaev (1972).

This concludes, for the moment, our discussion of the interaction between radiation and plasmas. We must point out that we have ignored a number of plasma effects and instabilities which can be important in astrophysical situations. See, e.g., Bekefi (1966), and Kaplan and Setovich (1972).

2.5 Hydrodynamics and Thermodynamics

Basic references §§22.2 and 22.3 of Misner, Thorne, and Wheeler (1973); Ellis (1971); Lichnerowicz (1967) and references cited therein. We adopt the notational conventions of Misner, Thorne, and Wheeler (cited henceforth as MTW), including signature "–+++"; geometrized units ($c = G = 1$); Greek indices ranging from 0 to 3 and Latin from 1 to 3; "hats" on indices that refer to local orthornormal frames, e.g. $u^{\hat{\alpha}}$ and $T^{\hat{0}\hat{j}}$; extra-bold, sans-serif type for 4-vectors and 4-tensors, e.g. $\mathbf{u}$ and $\mathbf{T}$; normal bold-face type for 3-vectors and 3-tensors (local Euclidean geometry).

Parameters describing the "fluid"

"LRF" local rest frame of the baryons; i.e. local orthonormal frame in which there is no net baryon flux in any direction.

$\mathbf{u}$ 4-velocity of LRF; i.e. 4-velocity of the "fluid".

n number density of baryons (number of baryons per unit volume) as measured in LRF.

m_B mean rest mass of a baryon in the fliud; i.e.

$$m_B = \begin{pmatrix}\text{rest mass of}\\ \text{hydrogen atom}\end{pmatrix} \times \begin{pmatrix}\text{fraction of all baryons that are}\\ \text{in the form of hydrogen nuclei}\end{pmatrix} + \frac{1}{4}\begin{pmatrix}\text{rest mass of}\\ \text{helium atom}\end{pmatrix} \times \begin{pmatrix}\text{fraction of all baryons that}\\ \text{are in helium nuclei}\end{pmatrix} + \dots \tag{2.5.1}$$

Note: in this section we restrict ourselves to fluids that are chemically homogeneous and in which no nuclear reactions occur ("standard fluid"). Thus, m_B is constant.

ρ_0 rest-mass density, defined by

$$\rho_0 \equiv m_B n. \tag{2.5.2}$$

V specific volume, i.e. volume per baryon, defined by

$$V \equiv 1/n. \tag{2.5.3}$$

V_0 specific volume, i.e. volume per unit rest mass, defined by

$$V_0 \equiv 1/\rho_0. \tag{2.5.4}$$

ρ total density of mass-energy, as measured in LRF.

Π specific internal energy, defined by

$$\rho = \rho_0(1 + \Pi) \tag{2.5.5}$$

Note: here and throughout this section we set the speed of light equal to one.

p isotropic pressure, as measured in LRF.

T temperature, as measured in LRF.

s entropy per baryon, as measured in LRF.

s_0 entropy per unit mass, as measured in LRF; of course,

$$s_0 = m_B^{-1} s. \tag{2.5.6}$$

μ chemical potential, as measured in LRF; defined by

$$\mu \equiv \left(\frac{\partial \rho}{\partial n}\right)_s = \frac{\rho + p}{n}. \tag{2.5.7}$$

(The second equality follows from the first law of thermodynamics, below.)

$\mathbf{q}$ flux of energy (due to heat conduction, radiation, convection, etc.) as measured in LRF. This flux (ergs $\text{cm}^{-2}\,\text{sec}^{-1}$) is a purely spatial vector as measured in LRF; i.e.

$$\mathbf{q} \cdot \mathbf{u} = 0 \tag{2.5.8}$$

$\mathbf{S}$ entropy density-flux vector. In LRF this vector has time component equal to the entropy density, $s^0 = ns$, and space components equal to the entropy flux, $s^j = q^j/T$. Hence, in frame-independent notation

$$\mathbf{S} \equiv ns\mathbf{u} + \mathbf{q}/T \tag{2.5.9}$$

Note that

$$\nabla \cdot \mathbf{S} = \begin{pmatrix} \text{rate at which entropy is being generated} \\ \text{per unit volume as measured in LRF} \end{pmatrix} \tag{2.5.10}$$

First law of thermodynamics

Follow a fluid element, containing A baryons, along its world tube. The total mass-energy in the fluid element, $\rho A/n$, changes as a result of compresssion (change in volume, A/n) and as a result of influx of heat:

$$d(\rho A/n) = -p d(A/n) + T d(As). \tag{2.5.11}$$

[Here and throughout this section we assume for simplicity no internal generation of entropy in the fluid element—e.g., no irreversible chemical reactions; this enables us to write the influx of heat in terms of the change in entropy, $Td(As)$.]

One often uses A = const. and $\mu = (\rho + p)/n$ to rewrite this first law of thermodynamics in the equivalent forms

$$d\rho = \frac{\rho + p}{n} dn + nT\, ds, \tag{2.5.12}$$

$$d\mu = V dp + T ds. \tag{2.5.12$'$}$$

From the first law one can read off partial derivatives, e.g.

$$(\partial\rho/\partial n)_s = (\rho + p)/n, \qquad (\partial\mu/\partial s)_p = T.$$

Fundamental relation and equations of state

The peculiar thermodynamic properties of the particular fluid being studied are determined by a "fundamental thermodynamic relation"

$$\rho = \rho(n, s) \quad \text{or} \quad \mu = \mu(p, s). \tag{2.5.13}$$

Once this relation has been specified–and once the constant m_B has been specified–one can use the first law of thermodynamics (2.5.12) or (2.5.12′) and definitions (2.5.2)-(2.5.7) to derive explicit expressions for all other thermodynamic variables as functions of n, s or p, s. For example, by combining with the first law one can derive the "equations of state"

$$T(n, s) = \frac{1}{n}\left(\frac{\partial \rho}{\partial s}\right)_n, \qquad p(n, s) = n\left(\frac{\partial \rho}{\partial n}\right)_s - \rho.$$

Adiabatic index and speed of sound

One defines the adiabatic index Γ_1, by

$$\Gamma_1 \equiv \left(\frac{\partial \ln p}{\partial \ln n}\right)_s = -\left(\frac{\partial \ln p}{\partial \ln V}\right)_s = \frac{\rho + p}{p}\left(\frac{\partial p}{\partial \rho}\right)_s, \tag{2.5.14}$$

where the third equality follows from the first law of thermodynamics. It turns out that weak adiabatic perturbations (weak "sound waves") propagate, in the LRF, with ordinary velocity

$$c_S = \left[\left(\frac{\partial p}{\partial \rho}\right)_s\right]^{1/2} = \left(\frac{\Gamma_1 p}{\rho + p}\right)^{1/2}. \tag{2.5.15}$$

Second law of thermodynamics

Equation (2.5.10) allows one to write the second law of thermodynamics in the form

$$\boldsymbol{\nabla} \cdot \mathbf{S} \geqslant 0 \tag{2.5.16}$$

Decomposition of 4-velocity

One decomposes the gradient of the 4-velocity, $\boldsymbol{\nabla}\mathbf{u}$, into its "irreducible tensorial parts"

$$u_{\alpha;\beta} = \omega_{\alpha\beta} + \sigma_{\alpha\beta} + \tfrac{1}{3}\theta h_{\alpha\beta} - a_\alpha u_\beta. \tag{2.5.17}$$

Here $\mathbf{a}$ is the *4-acceleration* of the fluid

$$\mathbf{a} \equiv \boldsymbol{\nabla}_{\mathbf{u}}\mathbf{u}, \text{ i.e. } a_\alpha \equiv u_{\alpha;\beta}u^\beta \tag{2.5.18a}$$

and $-a_\alpha u_\beta$ is that portion of $u_{\alpha;\beta}$ which is *not* orthogonal to $\mathbf{u}$. The remainder of $u_{\alpha;\beta}$ (i.e. $\omega_{\alpha\beta} + \sigma_{\alpha\beta} + \frac{1}{3}\theta h_{\alpha\beta}$) is orthogonal to $\mathbf{u}$; i.e. in the LRF it has only spatial components. This orthogonal part is decomposed into an isotropic expansion, $\frac{1}{3}\theta h_{\alpha\beta}$, where θ is the "*expansion*"

$$\theta \equiv \nabla \cdot \mathbf{u} = u^{\alpha}_{;\alpha} \tag{2.5.18b}$$

and $h_{\alpha\beta}$ is the "*projection tensor*"

$$h_{\alpha\beta} \equiv g_{\alpha\beta} + u_\alpha u_\beta; \tag{2.5.18c}$$

plus a symmetric, trace-free "*shear*"

$$\sigma_{\alpha\beta} \equiv \tfrac{1}{2}(u_{\alpha;\mu}h^{\mu}_{\beta} + u_{\beta;\mu}h^{\mu}_{\alpha}) - \tfrac{1}{3}\theta h_{\alpha\beta}; \tag{2.5.18d}$$

plus an antisymmetric "*rotation*" or "*vorticity*"

$$\omega_{\alpha\beta} \equiv \tfrac{1}{2}(u_{\alpha;\mu}h^{\mu}_{\beta} - u_{\beta;\mu}h^{\mu}_{\alpha}). \tag{2.5.18e}$$

An observer in the LRF sees all fluid elements in his neighborhood to move with low (nonrelativistic) velocities. Let him use standard Newtonian methods to calculate or measure the expansion θ; shear $\sigma_{\hat{j}\hat{k}}$ and rotation $\omega_{\hat{j}\hat{k}}$ of the fluid; and let a relativist calculate θ, $\sigma_{\hat{j}\hat{k}}$, and $\omega_{\hat{j}\hat{k}}$ in the LRF from the above equations. The two calculations will give the same answers. [See, e.g., of Ellis (1971).]

Frozen-in magnetic field

Consider an ionized plasma which contains a "frozen-in" magnetic field. The field is pure magnetic (no electric field) in the LRF. Therefore it can be described by a magnetic-field 4-vector **B** orthogonal to **u**.

$$\mathbf{B} \cdot \mathbf{u} = 0. \tag{2.5.19}$$

[For simplicity we assume that the magnetic permeability of the plasma is the same as that of vacuum; so we do not distinguish between "**B**" and "**M**". For a more general treatment, see Lichnerowicz (1967).] As the fluid moves, carrying with it the frozen-in **B**-field, **B** must change as

$$\frac{DB_\alpha}{d\tau} = u_\alpha a_\beta B^\beta + \omega_{\alpha\beta}B^\beta + (\sigma_{\alpha\beta} - \tfrac{2}{3}\theta h_{\alpha\beta})B^\beta. \tag{2.5.20}$$

The term $u_\alpha a_\beta B^\beta$ is required to keep **B** orthogonal to **u**; by the term $\omega_{\alpha\beta}B^\beta$ the rotation of the fluid rotates the field lines; by the term $(\sigma_{\alpha\beta} - \frac{2}{3}\theta h_{\alpha\beta})B^\beta$, the compression of the fluid orthogonal to **B**, conserving flux, magnifies **B**.

Stress-energy tensor

Consider a fluid with isotropic pressure, with shear and bulk viscosity, with energy flowing between fluid elements, and with a frozen-in magnetic field. The stress-energy tensor for such a fluid is

$$T^{\alpha\beta} = \rho u^\alpha u^\beta + (p - \zeta\theta)h^{\alpha\beta} - 2\eta\sigma^{\alpha\beta} + q^\alpha u^\beta + u^\alpha q^\beta$$
$$+ \frac{1}{8\pi}(\mathbf{B}^2 u^\alpha u^\beta + \mathbf{B}^2 h^{\alpha\beta} - 2B^\alpha B^\beta). \tag{2.5.21}$$

The term $\rho u^\alpha u^\beta$ is the total density of mass-energy (excluding only that of the frozen-in **B**-field) as measured in the LRF. The term $ph^{\alpha\beta}$ is the isotropic pressure that would be measured in the LRF if the gas were not changing volume (if θ were zero). The quantities ζ and η are the *coefficients of bulk viscosity and of dynamic viscosity*, respectively. The term $-\zeta\theta h^{\alpha\beta}$ is the isotropic viscous stress which resists isotropic expansion ($\theta > 0$) or compression ($\theta < 0$) of the fluid. The term $-2\eta\sigma^{\alpha\beta}$ is the viscous shear stress which resists shearing motions. The term $q^\alpha u^\beta + u^\alpha q^\beta$ is the energy flux and momentum flux relative to the LRF. (*Note*: the energy density and stresses associated with the flowing energy **q**, as measured in the LRF, are here neglected by comparison with $\rho u^\alpha u^\beta$ and $ph^{\alpha\beta}$. This neglect is valid with enormous accuracy in most contexts of interest in these lectures. An exception is the radiation pressure which produces "self-regulation" of accretion onto black holes when the inflowing mass is sufficiently large; see §§4.5 and 5.13.) The term

$$T^{\alpha\beta}_{\text{MAG}} = \frac{1}{8\pi}(\mathbf{B}^2 u^\alpha u^\beta + \mathbf{B}^2 h^{\alpha\beta} - 2B^\alpha B^\beta) \tag{2.5.22}$$

is the Maxwell stress-energy associated with the frozen-in **B**-field (in LRF: energy density $\mathbf{B}^2/8\pi$, pressure $\mathbf{B}^2/8\pi$ orthogonal to field lines; tension $-\mathbf{B}^2/8\pi$ along field lines).

Equations of hydrodynamics

The fundamental equations governing the motion of a fluid in a given gravitational field (spacetime geometry) are (i) the *law of baryon conservation*

$$\nabla \cdot (n\mathbf{u}) = 0, \text{ i.e. } dn/d\tau = -\theta n, \tag{2.5.23}$$

or equivalently rest-mass conservation

$$\nabla \cdot (\rho_0 \mathbf{u}) = 0, \text{ i.e. } d\rho_0/d\tau = -\theta\rho_0; \tag{2.5.23$'$}$$

(ii) the *law of local energy conservation*

$$\mathbf{u} \cdot (\nabla \cdot \mathbf{T}) = 0; \tag{2.5.24}$$

(iii) the *Euler equations* (i.e. law of local momentum conservation)

$$\mathbf{h} \cdot (\nabla \cdot \mathbf{T}) = 0; \tag{2.5.25}$$

(iv) the laws of thermodynamics, (2.5.11)–(2.5.16); (v) the law of evolution for the frozen-in **B**-field, (2.5.20); (vi) the laws of energy transport, which govern **q** (see §2.6 eq. 2.6.43).

Law of local energy conservation

When one evaluates the law of local energy conservation, $\mathbf{u} \cdot (\nabla \cdot \mathbf{T}) = 0$, for the stress-energy tensor (2.5.21), one finds that the Maxwell stress-energy gives zero contribution

$$\mathbf{u} \cdot (\nabla \cdot \mathbf{T}_{\text{MAG}}) = 0 \tag{2.5.26}$$

(work done to compress magnetic field is precisely equal to increase in magnetic-field energy); and that the remainder of the stress-energy tensor gives

$$d\rho/d\tau = -(\rho + p)\theta + \zeta\theta^2 + 2\eta\sigma_{\alpha\beta}\sigma^{\alpha\beta} - \nabla \cdot \mathbf{q} - \mathbf{a} \cdot \mathbf{q}. \tag{2.5.27}$$

Here $d/d\tau$ is derivative with respect to proper time along the world lines of the fluid; i.e., in the language of the differential geometer, $d/d\tau \equiv \mathbf{u}$. The term $-(\rho + p)\theta$ is the increase in mass-energy density due to compression. The terms $\zeta\theta^2$ and $2\eta\sigma_{\alpha\beta}\sigma^{\alpha\beta}$ are the increases in mass-energy density due to viscous heating (conversion of relative kinetic energy of adjacent fluid elements into heat). The term $-\nabla \cdot \mathbf{q}$ is the influx of mass-energy from neighboring fluid elements. The term $-\mathbf{a} \cdot \mathbf{q}$ is a special relativistic correction to $\nabla \cdot \mathbf{q}$ associated with the inertia of the energy flux $\mathbf{q}$—or, equivalently, with the "redshift" of $\mathbf{q}$. To understand this term, consider a flux of energy (e.g. photons) that is uniform as viewed in an inertial frame. Examine these photons from the viewpoint of an accelerated fluid ($\mathbf{a} \neq 0$) which does not interact with them. The acceleration gives rise to a redshift (photons become more and more red as time passes; "gravitational redshift") and hence to a nonzero $\nabla \cdot \mathbf{q}$. To compensate for this and keep $d\rho/d\tau = 0$ (no interaction between fluid and photons), one must include the correction factor $-\mathbf{a} \cdot \mathbf{q}$. For a similar reason, a similar relativistic correction appears in the law of heat conduction

$$\mathbf{q} = -\lambda_{\text{th}} \mathbf{h} \cdot (\nabla T + \mathbf{a}T). \tag{2.5.28a}$$

Here λ_{th} is the coefficient of thermal conductivity. This law of heat conduction is merely the "law of energy transport" (2.6.43) rewritten in new notation. By comparing the two laws, one can read off the relation between the coefficient of thermal conductivity and the mean opacity:

$$\lambda_{\text{th}} = \frac{4}{3}\frac{bT^3}{\bar{\kappa}\rho_0}. \tag{2.5.28b}$$

The law of thermal conductivity is valid only in the diffusion approximation—i.e. when the mean-free path of the energy-carrying particles (or turbulent cells) is small compared to other relevant scales of the problem. When one combines the local law of energy conservation (2.5.27) with this law of thermal conductivity (2.5.28), with the first law of thermodynamics (2.5.12), with the law of baryon conservation (2.5.23), and with the definition (2.5.9) of the entropy density-flux vector, one obtains an explicit equation for the rate of generation of entropy due to viscous heating and due to resistance to heat conduction:

$$T\nabla \cdot \mathbf{S} = \zeta\theta^2 + 2\eta\sigma_{\alpha\beta}\sigma^{\alpha\beta} + \frac{1}{T}\lambda_{\text{th}} h^{\alpha\beta}(T_{,\alpha} + Ta_\alpha)(T_{,\beta} + Ta_\beta) \geqslant 0. \tag{2.5.29}$$

[Compare this with equations (2.5.10) and (2.5.16).]

Euler equation for a perfect fluid

Consider a perfect fluid, flowing adiabatically through spacetime ($\zeta = \eta = \mathbf{B} = \mathbf{q} = 0$). In this case the Euler equations $\mathbf{h} \cdot (\nabla \cdot \mathbf{T}) = 0$ reduce to

$$(\rho + p)\mathbf{a} = -\mathbf{h} \cdot \nabla p. \tag{2.5.30}$$

In words:

$$\begin{pmatrix}\text{inertial mass per}\\ \text{unit volume}\end{pmatrix} \times (\text{4-acceleration}) = -\begin{pmatrix}\text{pressure gradient}\\ \text{projected orthogonal to } \mathbf{u}\end{pmatrix}.$$

Bernoulli equation

Consider a perfect fluid undergoing stationary, adiabatic flow in a stationary spacetime. More particularly, set $\zeta = \eta = \mathbf{B} = \mathbf{q} = 0$; assume that spacetime is endowed with a Killing vector field $\boldsymbol{\xi}$,

$$\xi_{\alpha;\beta} + \xi_{\beta;\alpha} = 0, \tag{2.5.31}$$

which need not be timelike; and assume that the flow is adiabatic, $ds/d\tau = 0$, and stationary in the sense that

$$\mathscr{L}_{\boldsymbol{\xi}}\mathbf{u} = 0, \quad \nabla_{\boldsymbol{\xi}}\rho = \nabla_{\boldsymbol{\xi}}p = \cdots = 0. \tag{2.5.32}$$

Here $\mathscr{L}_{\boldsymbol{\xi}}$ is the Lie derivative along $\boldsymbol{\xi}$. In this case the Euler equations (2.5.30), together with the first law of thermodynamics (2.5.12), imply the relativistic Bernoulli equation

$$d(\mu\mathbf{u} \cdot \boldsymbol{\xi})/d\tau = 0; \tag{2.5.33}$$

i.e., $\mu\mathbf{u} \cdot \boldsymbol{\xi}$ is constant along flow lines.

Newtonian limit

The Newtonian limit of relativistic hydrodynamics is obtained when, in a nearly global Lorentz frame, the following approximations hold:

$$g_{00} = -(1 + 2\Phi), \qquad |\Phi| \ll 1;$$
$$p/\rho_0 \ll 1, \qquad \Pi \ll 1, \qquad \mathbf{B}^2/\rho_0 \ll 1, \qquad \mathbf{v}^2 \ll 1. \tag{2.5.34}$$

Here Φ is the Newtonian gravitational potential with sign $\Phi < 0$, and $\mathbf{v}$ is the ordinary velocity of the fluid

$$\mathbf{v} \equiv v^j\mathbf{e}_j = (u^j/u^0)\mathbf{e}_j. \tag{2.5.35}$$

In the Newtonian limit the chemical potential μ reduces to

$$\mu = m_B(1 + w), \tag{2.5.36}$$

where w is the *enthalpy*

$$w = \Pi + p/\rho_0. \tag{2.5.37}$$

The Bernoulli equation (2.5.33) reduces to the familiar form

$$\Phi + \tfrac{1}{2}\mathbf{v}^2 + w = \text{constant along flow lines.} \tag{2.5.38}$$

The Euler equation for a perfect fluid in adiabatic flow, (2.5.30), reduces to

$$\frac{d\mathbf{v}}{d\tau} = -\nabla\Phi - \frac{1}{\rho_0}\nabla p, \tag{2.5.39}$$

where $d/d\tau$, the derivative with respect to proper time along the flow lines, has the Newtonian form

$$d/d\tau = \partial/\partial t + \mathbf{v}\cdot\nabla. \tag{2.5.40}$$

The first law of thermodynamics, (2.5.12) and (2.5.12′), reduces to

$$d\Pi = -p\,dV_0 + T\,ds_0, \tag{2.5.41}$$

$$dw = V_0\,dp + T\,ds_0 \tag{2.5.42}$$

(In Newtonian theory one usually adopts a per-unit-mass viewpoint rather than a per-baryon viewpoint; and thus one uses ρ_0, V_0, s_0 rather than n, V, s.)

2.6 Radiative Transfer

Basic references Appendix 1 of Pacholczyk (1970); Mihalas (1970); Chandrasekhar (1960); Lindquist (1966).

Notation and terminology At an arbitrary event in spacetime pick an arbitrary local Lorentz frame. In that frame pick an arbitrary spatial direction $\mathbf{n}$ ($\mathbf{n}$ is a unit vector; see Figure 2.6.1). Examine the amount of energy dE that is (i) carried by photons across a unit surface area dA orthogonal to $\mathbf{n}$ (the surface $\mathscr{S}_n$ of Figure 2.6.1) during unit time dt, with (ii) the photons having frequencies ν in

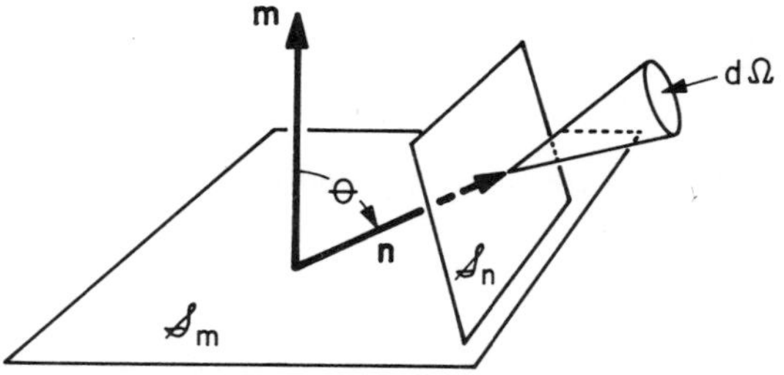

Figure 2.6.1. The surfaces, directions, and solid angle used in the definition of intensity and of flux.

the range $d\nu$, and (iii) the photons being directed into a solid angle $d\Omega$ about $\mathbf{n}$. The ratio

$$I_\nu \equiv \frac{dE}{dt\, dA\, d\nu\, d\Omega} \tag{2.6.1}$$

is called the *specific intensity* or simply the *intensity*. It depends on (a) the event in spacetime; (b) the choice of Lorentz frame; (c) the direction $\mathbf{n}$; (d) the frequency ν. One also defines the *total intensity* I by

$$I \equiv \int I_\nu\, d\nu = \frac{dE}{dt\, dA\, d\Omega}; \tag{2.6.2}$$

the *average specific intensity* or *average intensity* J_ν (averaged over all directions) by

$$J_\nu \equiv \frac{1}{4\pi} \int I_\nu\, d\Omega; \tag{2.6.3}$$

and the *average total intensity* J by

$$J \equiv \frac{1}{4\pi} \int I\, d\Omega = \frac{1}{4\pi} \int I_\nu\, d\nu\, d\Omega. \tag{2.6.4}$$

It is easy to see that the *energy density per unit frequency* in the radiation is

$$\rho_\nu^{(\mathrm{rad})} \equiv \frac{dE}{d^3x\, d\nu} = \frac{4\pi J_\nu}{c}, \tag{2.6.5}$$

and that the *total energy density* in the radiation is

$$\rho^{(\mathrm{rad})} \equiv \frac{dE}{d^3x} = \frac{4\pi J}{c}. \tag{2.6.6}$$

Pick a 2-surface $\mathscr{S}_m$ in the chosen Lorentz frame, with unit normal $\mathbf{m}$ (Fig. 2.5.1). Examine the total energy dE per unit area dA that is carried across the chosen surface during unit time dt, by photons having frequency ν in the range $d\nu$. Make no restrictions on the photon direction or solid angle; but count negatively those photons that cross dA from the front side toward the back. The ratio

$$F_\nu \equiv \frac{dE}{dt\, dA\, d\nu} \tag{2.6.7}$$

is called the *specific flux*, or simply the *flux*. It is easy to see from Figure 2.5.1 that

$$F_\nu = \int I_\nu \cos\theta\, d\Omega. \tag{2.6.8}$$

The specific flux depends on (a) the chosen event in spacetime; (b) the chosen Lorentz frame at that event; (c) the chosen 2-surface $\mathscr{S}_m$ in that Lorentz frame.

The integral of the flux over all frequencies is called the *total flux*

$$F \equiv \int F_\nu d\nu = \int I \cos\theta \, d\Omega. \tag{2.6.9}$$

Notice that the total flux is the "first moment" of the total intensity (the "moment" being taken with respect to the unit normal **m**). One can readily verify that the "second moment" is equal to the radiation pressure that acts across the surface $\mathscr{S}_m$:

$$T^{(\mathrm{rad})}_{mm} \equiv \mathbf{m} \cdot \mathbf{T}^{(\mathrm{rad})} \cdot \mathbf{m} = \int I \cos^2\theta \, d\Omega. \tag{2.6.10}$$

Invariance of I_ν/ν^3

Choose a particular photon that passes through a given event in spacetime. Observe that photon, and all others in the vicinity, from several different Lorentz frames at the given event. The frequency ν of the photon will depend on the choice of Lorentz frame (doppler shift from one frame to another). The specific intensity I_ν in the neighborhood of the photon will also depend on the choice of Lorentz frame. However, the ratio I_ν/ν^3 will be Lorentz-invariant. (Aside from a factor h^{-4}, I_ν/ν^3 is the invariant number density in phase space

$$\mathscr{N} = \frac{dN}{d^3x d^3p} = \frac{1}{h^4}\frac{I_\nu}{\nu^3}; \tag{2.6.11}$$

see, e.g., §22.6 of MTW.)

When photons propagate freely through curved spacetime (no interaction with matter), the ratio I_ν/ν^3 is conserved along the world line of each photon (Liouville's theorem).

Equation of Radiative Transfer

Consider the propagation of photons through a medium (e.g., through gas that is falling into a black hole). Analyze the photon propagation from the viewpoint of observers at rest in the medium (local rest frame; "LRF"). At each event $\mathscr{P}$ denote by **u** the 4-velocity of the medium—and hence also of the LRF. Focus attention on all photons in the (phase-space) neighborhood of a given null geodesic ray. Denote by **p** the 4-momentum of that ray. Then at each event along the given ray **p** is given by

$$\mathbf{p} = h\nu(\mathbf{u} + \mathbf{n}). \tag{2.6.12}$$

Here **n** is a unit vector that (i) is purely spatial as seen in the LRF

$$\mathbf{n} \cdot \mathbf{u} = 0, \tag{2.6.13}$$

and (ii) is interpreted in the LRF as the direction of propagation of the ray (same as 3-vector **n** of Figure 2.6.1). Also, in eq. (2.6.12), ν is the frequency of any photon which propagates along the ray, and $h\nu$ is its energy, as measured in the LRF. Because of the acceleration and shear of the medium, the frequency ν of the chosen ray changes from point to point along the ray. The change $d\nu$, when the ray propagates a proper spatial distance dl as seen in the LRF, is

$$d\nu = \nabla_{\mathbf{u}+\mathbf{n}}(-\mathbf{p}\cdot\mathbf{u}/h)\,dl.$$

(This frame-independent equation is derived easily by geometrical arguments in the LRF.) A straightforward calculation, using the geodesic equation for the ray

$$\nabla_{\mathbf{p}}\mathbf{p} = h\nu\,\nabla_{\mathbf{u}+\mathbf{n}}\mathbf{p} = 0,$$

and using expansion (2.5.17) for $\nabla\mathbf{u}$, reveals

$$d\nu/dl = -\nu(\mathbf{n}\cdot\mathbf{a} + \tfrac{1}{3}\theta + n^\alpha n^\beta \sigma_{\alpha\beta}). \tag{2.6.14}$$

The first term, $-\nu\mathbf{n}\cdot\mathbf{a}$, is the "gravitational redshift" produced by the acceleration of the LRF; the second and third terms, $-\nu(\frac{1}{3}\theta + n^\alpha n^\beta\sigma_{\alpha\beta})$, are the "cosmological redshift" due to the expansion of the medium along the direction of the ray.

If there were no interaction with the medium, then I_ν/ν^3 would be conserved along the ray. Hence, the equation of radiative transfer along the given ray must have the form

$$\frac{dI_\nu}{dl} - \left(\frac{3}{\nu}\frac{d\nu}{dl}\right) I_\nu = \text{(effects of interaction with medium)};$$

i.e.,

$$dI_\nu/dl + (3\mathbf{n}\cdot\mathbf{a} + \theta + 3n^\alpha n^\beta\sigma_{\alpha\beta})I_\nu = \text{(interaction effects)}. \tag{2.6.15}$$

Four types of interaction can occur: spontaneous emission of radiation by the matter; stimulated emission; absorption; and scattering.

We shall assume that on all length scales of interest the medium is isotropic; and we shall denote its emissivity by

$$\epsilon_\nu \equiv \frac{dE}{d\nu\, dt\, dm_0} \equiv \begin{pmatrix}\text{energy emitted spontaneously per unit frequency}\\ \text{during unit time by a unit rest mass, integrated}\\ \text{over all angles, and measured in LRF}\end{pmatrix} \tag{2.6.16}$$

(The emissivities for free-free and free-bound transitions were discussed in §§2.1 and 2.2.) Then spontaneous emission contributes

$$\left(\frac{dI_\nu}{dl}\right)_{\text{spontaneous emission}} = \frac{1}{4\pi}\rho_0\,\epsilon_\nu \tag{2.6.17}$$

to the specific intensity along the given ray. Here ρ_0 is the density of rest mass in the medium.

The rate of absorption and the rate of stimulated emission are both proportional to the intensity of the passing beam. Hence, the effects of absorption and of stimulated emission can be lumped together into a single *absorption coefficient* κ_ν, defined by

$$\left(\frac{dI_\nu}{dl}\right)_{\substack{\text{absorption plus}\\ \text{stimulated emission}}} = -\rho_0 \kappa_\nu I_\nu. \tag{2.6.18}$$

The dimensions of the absorption coefficient are cm^2/g ("absorption cross section per unit mass" at the given frequency). When absorption dominates over stimulated emission, κ_ν is positive; when stimulated emission dominates, κ_ν is negative ("negative absorption").

Scattering is more complicated to treat than emission and absorption. Let

$$\frac{d\kappa_s}{d\Omega'\, d\nu'}(\mathbf{p}, \mathbf{p}')$$

be the differential cross section (as measured in the LRF) for a unit rest mass to convert a photon of momentum $\mathbf{p}$ into a photon of momentum $\mathbf{p}'$; and let

$$\kappa_s(\nu) = \int \frac{d\kappa_s}{d\Omega'\, d\nu'}\, d\Omega'\, d\nu' \tag{2.6.19}$$

be the total scattering cross-section ("scattering opacity") per unit rest mass. For example, in the case of electron scattering by a nonrelativistic plasma ($h\nu \ll m_e c^2$, $kT \ll m_e c^2$; see §4.4), when one neglects the tiny change in photon frequency the cross sections per unit mass are

$$\frac{d\kappa_s}{d\Omega'\, d\nu'}(\mathbf{p}, \mathbf{p}') = \frac{f_e}{m_p}\frac{1}{2} r_0^2 [1 + (\mathbf{n} \cdot \mathbf{n}')^2]\, \delta(\nu' - \nu)$$

$$\kappa_s = (f_e/m_p)(8\pi/3) r_0^2 = (0.40\ \text{cm}^2/\text{g}) f_e \tag{2.6.20}$$

(Recall: $\mathbf{n} = \mathbf{p}/h\nu$, $\mathbf{n}' = \mathbf{p}'/h\nu'$; f_e is the number of free electrons per baryon in the plasma.) Scattering, as described by the appropriate differential cross section, can increase I_ν (scattering into the beam from other directions), or can decrease I_ν (scattering out of the beam to other directions):

$$\left(\frac{dI_\nu}{dl}\right)_{\text{scattering}} = +\rho_0 \int \frac{d\kappa_s}{d\Omega\, d\nu}(\mathbf{p}', \mathbf{p}) I'_\nu\, d\Omega'\, d\nu' - \rho_0 \kappa_s I_\nu. \tag{2.6.21}$$

By combining equations (2.6.15)-(2.6.21), we obtain our final form of the equation of radiative transfer

$$dI_\nu/dl + (3\mathbf{n} \cdot \mathbf{a} + \theta + 3n^\alpha n^\beta \sigma_{\alpha\beta}) = (4\pi)^{-1} \rho_0 \epsilon_\nu - \rho_0 \kappa_\nu I_\nu \tag{2.6.22}$$

$$+ \rho_0 \int \frac{d\kappa_s}{d\Omega\, d\nu}(\mathbf{p}', \mathbf{p}) I'_\nu\, d\Omega'\, d\nu' - \rho_0 \kappa_s I_\nu.$$

Relationship between emissivity and absorption

Into an insulated box place a material medium and a radiation field; and then wait until complete thermodynamic equilibrium is achieved. If T is the equilibrium temperature, then the radiation field as seen in the LRF will be isotropic with the standard black-body intensity, $I_\nu = B_\nu$, where

$$B_\nu \equiv \frac{(2h/c^2)\nu^3}{e^{h\nu/kT} - 1}. \qquad (2.6.23)$$

In equilibrium there must be no net change of I_ν along any ray: scattering into the beam must be completely balanced by scattering out of the beam; and emission must be completely balanced by absorption. The requirement of "detailed balance" for emission and absorption can be met only if the emissivity and the absorption coefficient are related by

$$\frac{dI_\nu}{dl} = \rho_0 \kappa_\nu \left(\frac{\epsilon_\nu}{4\pi\kappa_\nu} - I_\nu\right) = \rho_0 \kappa_\nu \left(\frac{\epsilon_\nu}{4\pi\kappa_\nu} - B_\nu\right) = 0;$$

i.e.,

$$\epsilon_\nu / 4\pi\kappa_\nu = B_\nu. \qquad (2.6.24)$$

Since ϵ_ν and κ_ν depend only on the thermodynamic state of the matter, and have nothing to do with the state of the radiation, relation (2.6.24) must be satisfied not only inside our insulated box, but also in all other cases where the matter by itself is in thermodynamic equilibrium but the radiation might not be.

For matter not in thermodynamic equilibrium one can use a more sophisticated version of the principle of detailed balance ("Einstein A and B coefficients") to derive a more complicated relationship between the emissivity and the absorption coefficient. See, e.g., Chandrasekhar (1960).

As an application of the equilibrium relationship $\epsilon_\nu/4\pi\kappa_\nu = B_\nu$, consider free-free transitions in an ionized gas. Whenever the free electrons (which do the emitting and absorbing) are in thermodynamic equilibrium with each other (Maxwell-Boltzmann velocity distribution), the absorption coefficient—as derived from the emissivity (2.1.27)—must be

$$\kappa_\nu^{\text{ff}} = (1.50 \times 10^{25}\ \text{cm}^2/\text{g}) \left(\frac{f_e f_i Z^2}{A}\right) \left(\frac{\rho_0}{\text{g cm}^{-2}}\right) T_k^{-7/2} \bar{G} \left(\frac{1 - e^{-x}}{x^3}\right), \qquad (2.6.25)$$

$$x \equiv h\nu/kT.$$

See §2.1 for notation.

Notice that for matter in thermodynamic equilibrium the relationship $\epsilon_\nu/4\pi\kappa_\nu = B_\nu$ permits one to rewrite the equation of radiative transfer (2.6.22) in the form

$$\frac{dI_\nu}{dl} + (3\mathbf{n}\cdot\mathbf{a} + \theta + 3n^\alpha n^\beta \sigma_{\alpha\beta}) = \rho_0\kappa_\nu(B_\nu - I_\nu) + \left(\frac{dI_\nu}{dl}\right)_{\text{scattering}}; \tag{2.6.26}$$

or, equivalently,

$$\frac{d(I_\nu/\nu^3)}{dl} = \rho_0\kappa_\nu\left(\frac{B_\nu}{\nu^3} - \frac{I_\nu}{\nu^3}\right) + \left[\frac{d(I_\nu/\nu^3)}{dl}\right]_{\text{scattering}}. \tag{2.6.27}$$

Optical depth

Consider radiation propagating out of a medium into surrounding empty space. Follow a given ray "backward", from "infinity" into the medium. Let the ray have frequency ν_∞ at infinity; then its frequency at location l (l = proper distance measured in LRF of medium) is

$$\nu(l) = \nu_\infty \exp\left[-\int_l^\infty (\mathbf{n}\cdot\mathbf{a} + \tfrac{1}{3}\theta + n^\alpha n^\beta \sigma_{\alpha\beta})\, dl\right]. \tag{2.6.28}$$

[cf. eq. (2.6.14)]. In calculating the change of intensity along the ray, it is often useful to replace the proper-length parameter l by the "optical-depth" parameter

$$\tau_\nu \equiv \int_l^\infty \rho_0\kappa_\nu\, dl. \tag{2.6.29}$$

In the integration κ_ν must be evaluated at the frequency $\nu(l)$. The optical depth τ_ν depends on (i) the world line of the ray in spacetime; (ii) location along that world line; and (iii) the frequency of the ray at "infinity", ν_∞.

One can also introduce an optical depth for scattering radiation out of the beam, τ_s:

$$\tau_s \equiv \int_l^\infty \rho_0\kappa_s\, dl. \tag{2.6.30}$$

In terms of optical depths, the law of radiative transfer (2.6.27) reads

$$-\frac{d(I_\nu/\nu^3)}{d\tau_\nu} = \frac{B_\nu}{\nu^3} - \frac{I_\nu}{\nu^3} - \left[\frac{d(I_\nu/\nu^3)}{d\tau_\nu}\right]_{\text{scattering}}, \tag{2.6.31}$$

where

$$\left[\frac{d(I_\nu/\nu^3)}{d\tau_\nu}\right]_{\text{scattering}} = -\frac{\kappa_s}{\kappa_\nu}\left[I_\nu - \frac{1}{\kappa_s}\int \frac{d\kappa_s}{d\Omega\, d\nu}(\mathbf{p}', \mathbf{p}) I'_\nu\, d\Omega'\, d\nu'\right]. \tag{2.6.32}$$

One says that a medium is *optically thick* to emission and absorption (or to scattering) along a given ray if optical depths $\tau_\nu \gg 1$ (or $\tau_s \gg 1$) are achieved along the ray. One says that the medium is *optically thin* if everywhere along the ray $\tau_\nu \ll 1$ (or $\tau_s \ll 1$).

Consider a medium that (i) has its matter in thermodynamic equilibrium with a spatially uniform temperature, (ii) is optically thick along a chosen ray, and (iii) has emission and absorption dominant over scattering, i.e.,

$$\kappa_\nu \gg \kappa_s \tag{2.6.33}$$

along that ray. Then the radiation emerging to infinity along the chosen ray must have the blackbody form

$$\frac{I_\nu}{\nu^3} = \frac{B_\nu}{\nu^3} = \frac{2h/c^3}{e^{h\nu/kT} - 1}. \tag{2.6.34}$$

[One can prove this easily from the equation of transfer (2.6.31).] Moreover, at a point in the medium where all rays are optically thick, the radiation will have a blackbody intensity at all frequencies and in all directions; so the energy density and pressure in the radiation will be

$$\rho^{(\mathrm{rad})} = 3p^{(\mathrm{rad})} = bT^4, \tag{2.6.35}$$

where b is the universal constant

$$b = \frac{8\pi^5 k^4}{15c^3 h^3} = 7.56 \times 10^{-15} \frac{\mathrm{ergs}}{\mathrm{cm}^3\,\mathrm{K}^4}. \tag{2.6.36}$$

Consider, alternatively, a medium that is optically thin along a chosen ray, and has negligible scattering ($\tau_s \ll 1$) along that ray. Then the equation of transfer (2.6.31) predicts for the radiation emerging to infinity

$$I_\nu/\nu^3 = \int_{\text{entire ray}} (B_\nu/\nu^3)\, d\tau_\nu = (1/4\pi) \int (\epsilon_\nu/\nu^3)\rho_0\, dl \tag{2.6.37}$$

$$\simeq \left(\int \rho_0\, dl\right)(1/4\pi)\,(\epsilon_\nu/\nu^3)_{\text{in region of strongest emission}}. \tag{2.6.38}$$

In summary, the radiation from an optically thin source with negligible scattering has the same spectrum as the spontaneous emissivity of the source, $I_\nu \propto \epsilon_\nu$, and has an intensity proportional to the amount of matter along the line of sight. But radiation from an optically thick source in thermodynamic equilibrium with negligible scattering has the blackbody form independent of the nature of its emissivity. Media with non-negligible scattering will be studied in §5.10.

Radiative transfer in the diffusion approximation

Consider the interior of an optically thick medium $\tau_\nu \gg 1$ which is in local thermodynamic equilibrium. Suppose that the medium has a temperature gradient and/or an acceleration, but that the characteristic length scale

$$l_T \equiv \frac{T}{|\nabla T + \mathbf{a}T|} \tag{2.6.39}$$

over which the temperature changes are long compared to the mean-free path of a photon,

$$l_{fp} \simeq 1/\kappa_\nu \rho_0 \ll l_T. \tag{2.6.40}$$

Then the radiation distribution will consist of a large, isotropic, blackbody component, plus a tiny "correction" due to the temperature gradient

$$I_\nu = B_\nu + I_\nu^{(1)}; \qquad I_\nu^{(1)} \ll B_\nu. \tag{2.6.41}$$

There is no net flux F across any surface associated with the blackbody component B_ν; but the "correction" term $I_\nu^{(1)}$ will lead to a flux in the direction of

$$\mathbf{h} \cdot (\nabla T + \mathbf{a}T). \tag{2.6.42}$$

(Here $\mathbf{h}$ is the projection operator of §2.5.) One can calculate the magnitude of that flux by inserting expression (2.6.41) for I_ν into the equation of transfer (2.6.22), and by then integrating over $\cos\theta \, d\Omega \, d\nu$. (Here θ is the angle between $\mathbf{h} \cdot (\nabla T + \mathbf{a}T)$ and the direction of $d\Omega$.) The result is

$$\mathbf{q} = (1/\bar{\kappa}\rho_0)(\tfrac{4}{3}bT^3)\mathbf{h} \cdot (\nabla T + \mathbf{a}T). \tag{2.6.43}$$

Here $\mathbf{q}$ is the energy flux vector of §2.5, and its magnitude is the flux of energy in the $\mathbf{h} \cdot (\nabla T + \mathbf{a}T)$ direction

$$|\mathbf{q}| = F. \tag{2.6.44}$$

Also, in eq. (2.6.43) $\bar{\kappa}$ is the "Rosseland mean opacity", defined by

$$\frac{1}{\bar{\kappa}} \equiv \frac{\int_0^\infty (\kappa_\nu + \kappa_s)^{-1}(dB_\nu/dT)\, d\nu}{\int_0^\infty (dB_\nu/dT)\, d\nu}. \tag{2.6.45}$$

For free-free transitions by themselves (eq. 2.6.25) the Rosseland mean opacity is

$$\bar{\kappa}_{ff} = (0.645 \times 10^{23}\ \mathrm{cm^2/g}) \left(\frac{C_A f_e f_i Z^2}{A}\right) \bar{\bar{G}} \left(\frac{\rho_0}{\mathrm{g\ cm^{-3}}}\right) T_k^{-7/2}. \tag{2.6.46}$$

See §2.1 for notation. For electron scattering by itself in the nonrelativistic case, the Rosseland mean opacity is

$$\bar{\kappa}_{es} = \kappa_{es} = 0.40\ \mathrm{cm^2/g}. \tag{2.6.47}$$

2.7 Shock Waves

Basic references Zel'dovich and Raizer (1966); Landau and Lifshitz (1959); Taub (1948); Lichnerowicz (1967, 1970, 1971); Thorne (1973a).

Types of shock waves

When gas in supersonic flow encounters an obstacle (e.g., the surface of a star, or the geometric structure of the ergosphere of a black hole), a shock front develops. In the shock front the flow decelerates sharply from supersonic to subsonic, and some of the kinetic energy of the flow gets converted into heat (increase in entropy!).

The structure of the shock depends on the nature of the forces which decelerate the gas particles (atoms, ions, electrons). If those forces are collisions between the particles themselves (the usual case), one has an ordinary shock. But if the particles are decelerated without colliding—e.g., by impact onto the dipole magnetic field of a neutron star, which swings the particles into Larmour orbits—one has a *collisionless shock.* These notes will be confined to ordinary shocks. For the theory of collisionless shocks see, e.g., the end of § 12 of Kaplan (1966).

Shock waves in a partially ionized gas (e.g., the interstellar medium) can have a somewhat different form than ordinary shocks. As gas passes through the shock front, much of its kinetic energy of supersonic flow can be converted into excitation energy of atoms and ions, and can be quickly radiated away as "line" radiation. The result is a bright glow from the shock front itself. See, e.g., Pikel'ner (1961) and § 10 of Kaplan (1966). In this section we shall ignore such energy losses to radiation—i.e., we shall demand that the gas behind the shock have the same total energy per unit rest mass as the gas in front of the shock.

The relativistic Rankine-Hugoniot equations

Pick a particular event $\mathscr{P}$ on a shock front. In the neighborhood of $\mathscr{P}$ introduce a local Lorentz frame ("rest frame of the shock") in which (i) the shock is momentarily at rest; (ii) the shock is the surface $y = z = 0$; and (iii) on both sides of the shock the fluid is moving in the x direction, i.e., perpendicular to the shock front ("normal shock"; see Figure 2.7.1) That such a local Lorentz frame exists in general one can prove quite easily [see, e.g., Taub (1948)]. Denote the "*front*" side of the shock (side *from* which the fluid moves) by a "1", and denote the "*back*" side (side *toward* which the fluid moves) by a "2"; see Figure 2.7.1. Denote the velocity of the fluid, as measured in the rest frame of the shock, by

$$v_1 = (dx/dt)_1 = \text{ordinary velocity on front side,} \tag{2.7.1a}$$

$$v_2 = (dx/dt)_2 = \text{ordinary velocity on back side,} \tag{2.7.1b}$$

$$u_1 = v_1\gamma_1 = v_1/(1 - v_1^2)^{1/2} = \text{"4-velocity" on front side,} \tag{2.7.1c}$$

$$u_2 = v_2\gamma_2 = v_2/(1 - v_2^2)^{1/2} = \text{"4-velocity" on back side.} \tag{2.7.1d}$$

(Note that these "4-velocities" are scalars, not vectors.)

In the rest frame of the shock the law of baryon conservation is equilvalent to continuity of the *baryon flux*:

$$j \equiv n_1 u_1 = n_2 u_2. \tag{2.7.2a}$$

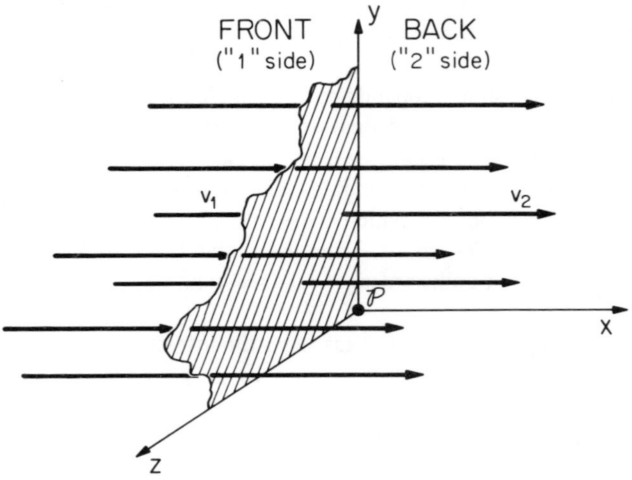

Figure 2.7.1. A shock wave viewed in the local-Lorentz rest frame of the shock front at an event $\mathscr{P}$.

Similarly, energy and momentum conservation are equivalent to continuity of energy flux and momentum flux:

$$\{(\rho + p)\gamma u\}_1 = \{(\rho + p)\gamma u\}_2; \tag{2.7.2b}$$

$$\{(\rho + p)u^2 + p\}_1 = \{(\rho + p)u^2 + p\}_2. \tag{2.7.2c}$$

These junction conditions are more easily understood by writing them in a form analogous to the Rankine-Hugoniot equations of Newtonian theory: First take the law of baryon conservation (2.7.2a) and turn it into equations for the fluid 4-velocities

$$u_1 = jV_1, \qquad u_2 = jV_2. \tag{2.7.3a}$$

(See §2.5 for notation.) Then take the law of momentum conservation (2.7.2c); rewrite it in terms of μ, V, j, and p using the law of baryon conservation (2.7.2a) and the relations $n = 1/V$, $\mu = (\rho + p)V$; and solve for the baryon flux to obtain

$$j^2 = -\frac{p_2 - p_1}{\mu_2 V_2 - \mu_1 V_1}. \tag{2.7.3b}$$

Finally, use the law of baryon conservation (2.7.2a) to rewrite the energy equation (2.7.2b) in the form

$$\mu_1 \gamma_1 = \mu_2 \gamma_2;$$

divide equation (2.7.3b) by $(\mu_1 V_1 + \mu_2 V_2)$, and combine with $j = u_1/V_1 = u_2/V_2$ [eq. (2.7.2a)] to obtain

$$(\mu_2 u_2)^2 - (\mu_1 u_1)^2 = (p_1 - p_2)(\mu_1 V_1 + \mu_2 V_2);$$

then subtract this from the square of the energy equation $(\mu_2\gamma_2)^2 - (\mu_1\gamma_1)^2 = 0$, to obtain

$$\mu_2^2 - \mu_1^2 = (p_2 - p_1)(\mu_1 V_1 + \mu_2 V_2). \tag{2.7.3c}$$

Equations (2.7.3) are Taub's (1948) junction conditions for shock waves. Their Newtonian limits are the standard Rankine-Hugoniot equations:

$$v_1 = j_0 V_{01}, \; v_2 = j_0 V_{02} \tag{2.7.4a}$$

(j_0 = "mass flux" of Newtonian theory);

$$j_0^2 = -\frac{p_2 - p_1}{V_{02} - V_{01}}, \tag{2.7.4b}$$

$$w_2 - w_1 = \tfrac{1}{2}(p_2 - p_1)(V_{01} + V_{02}). \tag{2.7.4c}$$

The form of the Newtonian junction conditions (2.7.4) motivates one to use p and V_0 as one's independent thermodynamic variables when analyzing Newtonian shocks; similarly, the form of the relativistic junction conditions (2.7.3) motivates one to use p and μV as one's independent variables for relativistic shocks.

The Rankine-Hugoniot curve

Consider a family of shocks, each with the same thermodynamic state on the front face (same $\mu_1 V_1, p_1$, etc.), but with different states on the back face (different $\mu_2 V_2, p_2$, etc.). This family of shocks is a one-parameter family. Thus, if one plots all back-face states $(\mu_2 V_2, p_2)$, in the $\mu V - p$ plane, they lie on a single curve—the "Rankine-Hugoniot curve"—passing through the point $(\mu_1 V_1, p_1)$. (In the relativistic case this curve is also called the "Taub adiabat".)

One can also plot, in the $\mu V - p$ plane, the Poisson adiabat (curve of constant entropy) passing through $(\mu V_1, p_1)$. These two curves typically have the relative shapes and locations shown in Figure 4.7.2. In particular, from the Rankine-Hugoniot equations one can derive the following general properties of the Rankine-Hugoniot curve.† [See Thorne (1973a) for derivation.] (i) The Rankine-Hugoniot curve is tangent to the Poisson adiabat, and has the same second derivative at point "1"; i.e., for weak shocks the increase in entropy is third-order in the pressure jump:

$$s_2 - s_1 = \left\{ \frac{1}{12\mu T} \left[\frac{\partial^2(\mu V)}{\partial p^2} \right]_s \right\}_1 (p_2 - p_1)^3 + 0[(p_2 - p_1)^4]. \tag{2.7.5}$$

(ii) As the gas passes from the front of the shock to the back, its entropy, pressure, and chemical potential increase

$$s_2 > s_1, \qquad p_2 > p_1, \qquad \mu_2 > \mu_1; \tag{2.7.6a}$$

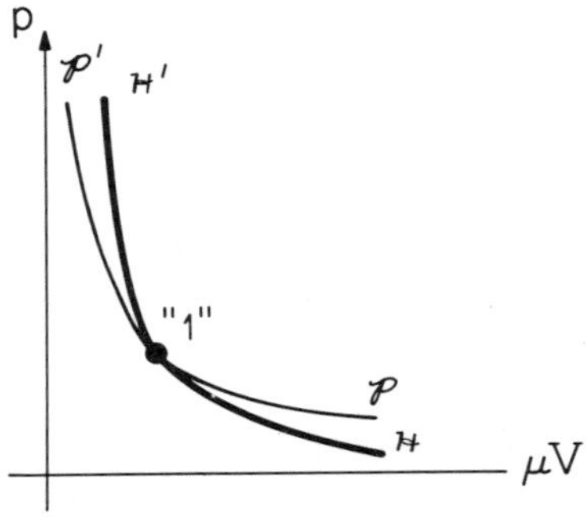

Figure 2.7.2. The Poisson adiabat $\mathscr{P}$-$\mathscr{P}'$ and the Rankine-Hugoniot curve $\mathscr{H}$-$\mathscr{H}'$ plotted in the μV-p plane.

while its specific volume and the product μV decrease

$$V_2 < V_1, \qquad \mu_2 V_2 < \mu_1 V_1. \tag{2.7.6b}$$

(Hence, only the "upper branch" of the Rankine-Hugoniot curve, Figure 2.7.2, is physically relevant.) (iii) The flow on the front side is always supersonic; that on the back side is always subsonic

$$u_1/u_{S1} > 1, \qquad u_2/u_{S2} < 1;$$
$$u_S \equiv c_S/(1 - c_S^2)^{1/2} \tag{2.7.7}$$

† The derivation requires the assumption

$$[\partial^2(\mu V)/\partial p^2]_s \geqslant 0;$$

a condition which is satisfied by all materials of astrophysical or laboratory interest; see Landau and Lifshitz (1959). Some, but not all, of the listed properties remain true without this assumption; see, e.g., Zel'dovich and Rayzer (1966) for nonrelativistic details, and Thorne (1973a) for relativistic details.

(iv) As one moves up the Rankine-Hugoniot curve, away from point "1"–i.e., as one studies a sequence of ever "stronger" shocks–the following quantities increase monotonically:

$$\begin{aligned}&\text{the baryon flux across the shock, } j;\\&\text{the jump in entropy across the shock, } s_2 - s_1;\\&\text{the "relativistic" mach number on the front}\\&\quad\text{side of the shock, } M_1 = u_1/u_{S1}\end{aligned} \tag{2.7.8}$$

Notice that, once the thermodynamic state on the front face of the shock has been specified, the shock has only one free parameter. If one fixes the baryon flux across the shock j, or the speed on the front face u_1, or the pressure on the back face p_2, or any other single parameter, then all other properties of the shock are uniquely determined. To calculate them, one need merely invoke the Rankine-Hugoniot equations (2.7.4), the laws of thermodynamics, and the equation of state of the gas.

Shocks in an ideal gas

Consider, as a special but important case, a nonrelativistic ideal gas with constant adiabatic index $\Gamma = -(\partial \ln p/\partial \ln V_0)$, and with mean rest mass per particle $\overline{m}$, so that

$$pV_0 = kT/\overline{m}. \tag{2.7.9}$$

By a straightforward calculation (§85 of Landau and Lifshitz 1959), one can derive the following relationships between various quantities along the Rankine-Hugoniot curve:

$$\begin{aligned}
\frac{V_{02}}{V_{01}} &= \frac{(\Gamma+1)p_1 + (\Gamma-1)p_2}{(\Gamma-1)p_1 + (\Gamma+1)p_2} \to \frac{\Gamma-1}{\Gamma+1} \text{ for strong shock,}\\
\frac{T_2}{T_1} &= \frac{p_2}{p_1}\frac{(\Gamma+1)p_1 + (\Gamma-1)p_2}{(\Gamma-1)p_1 + (\Gamma+1)p_2} \to \frac{\Gamma-1}{\Gamma+1}\frac{p_2}{p_1} \text{ for strong shock,}\\
j_0{}^2 &= \frac{(\Gamma-1)p_1 + (\Gamma+1)p_2}{2V_{01}} \to \frac{\Gamma+1}{2}\frac{p_2}{V_{01}} \text{ for strong shock,}\\
j_1{}^2 &= \tfrac{1}{2}V_{01}[(\Gamma-1)p_1 + (\Gamma+1)p_2] \to \tfrac{1}{2}(\Gamma+1)V_{01}p_2 \text{ for strong shock,}\\
j_2{}^2 &= \frac{V_{01}}{2}\frac{[(\Gamma+1)p_1 + (\Gamma-1)p_2]^2}{(\Gamma-1)p_1 + (\Gamma+1)p_2} \to \frac{(\Gamma-1)^2}{2(\Gamma+1)}V_{01}p_2 \text{ for strong shock.}
\end{aligned} \tag{2.7.10}$$

Here "strong shock" means "in the limit $p_2/p_1 \gg 1$".

2.8 Turbulence

In the accretion of gas onto a black hole, turbulence probably plays an important role. (See §§4 and 5.) Unfortunately, the theory of turbulence is in a very uncertain state. Little is known with confidence about the astrophysical circumstances under which turbulence should develop, or about the strength of the turbulence in various situations. For overviews of the current state of one's knowledge see, e.g., Chapter 3 of Landau and Lifshitz (1959); also Pikel'ner (1961).

2.9 Reconnection of Magnetic Field Lines

Basic references §5.3 of Cowling (1965); Sonnerup (1970), Yeh (1970), Vainstein and Zel'dovich (1972).

Basic ideas

In flat spacetime consider a plasma that is macroscopically neutral, that has a high electrical conductivity σ, and that is sufficiently dilute for one to ignore its dielectric properties: $\epsilon = \mu = 1$. Let the plasma be endowed with a large-scale magnetic field, and examine the evolution of that field in a Lorentz frame where the plasma has low velocity $|v| \ll c$. Maxwell's equations for the electromagnetic field then read

$$\begin{aligned} &\nabla \cdot \mathbf{E} = \nabla \cdot \mathbf{B} = 0 \\ &\nabla \times \mathbf{E} + (1/c)(\partial \mathbf{B}/\partial t) = 0, \\ &\nabla \times \mathbf{B} - (1/c)(\partial \mathbf{E}/\partial t) = 4\pi \mathbf{J}/c. \end{aligned} \tag{2.9.1}$$

In the local rest frame of the plasma the current is proportional to the electric field, $\mathbf{J} = \sigma \mathbf{E}$. When transformed to the Lorentz frame where the plasma moves with velocity $\mathbf{v}$, this equation says

$$\mathbf{J} = \sigma[\mathbf{E} + (\mathbf{v}/c) \times \mathbf{B}]. \tag{2.9.2}$$

A straightforward calculation from the above equations leads to the following law for the rate of change of magnetic field along the world lines of the plasma:

$$\frac{d\mathbf{B}}{d\tau} = (\boldsymbol{\omega} + \boldsymbol{\sigma} - \tfrac{2}{3}\theta\, \mathbf{1}) \cdot \mathbf{B} - \frac{1}{4\pi\sigma}\left(\frac{\partial^2 \mathbf{B}}{\partial t^2} - c^2 \nabla^2 \mathbf{B}\right). \tag{2.9.3}$$

Here $\mathbf{1}$ is the unit 3-tensor; $\boldsymbol{\omega}$, $\boldsymbol{\sigma}$, and θ are the rotation, shear, and expansion of the plasma [cf. eqs. (2.5.17) and (2.5.18)]; and

$$d/d\tau = \partial/\partial t + \mathbf{v} \cdot \nabla. \tag{2.9.4}$$

Because of the very high electrical conductivity, one can ignore the "wave-equation" part of the evolution law (2.9.3) almost everywhere in the plasma; and the magnetic field evolves in a "frozen-in" manner:

$$d\mathbf{B}/d\tau = (\boldsymbol{\omega} + \boldsymbol{\sigma} - \tfrac{2}{3}\theta\, \mathbf{1}) \cdot \mathbf{B}. \tag{2.9.5}$$

[Cf. Eq. (2.5.20) and associated discussion.] However, in regions of plasma where the field has strong gradients, the "wave-equation" part must come into play.

Consider, as the case of greatest importance, a magnetic field that is chaotic with an ordered structure on some scale l_c (subscript c for "cell size"). At the

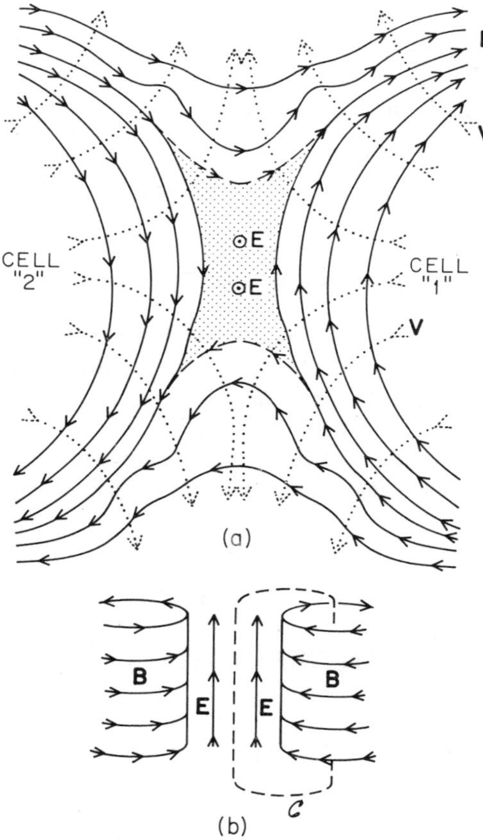

Figure 2.9.1. The region in a plasma where magnetic field lines are reconnecting (schematic picture). In the top view, (a), the reconnection region is stippled; the magnetic field lines are drawn solidly; and the flow lines of the plasma are dotted. The innermost field line is in the process of reconnecting into the dashed form. The perspective view, (b), shows the closed curve $\mathscr{C}$ around which one integrates to derive eq. (2.9.6) for the EMF in the connecting region.

interface between adjacent cells, the magnetic field must reverse sign ("neutral point" or "neutral sheet"), and its gradients will be very large. Hence, at the interface, "wave-equation" behavior can come into play destroying the frozen-in behavior of the field. The result is a reconnection of magnetic field lines, as depicted in Figure 2.9.1. This reconnection gradually converts the two adjacent cells into a single cell, reducing the overall strength of the magnetic field.

In situations of interest (§ §4.4 and 5.2), the decrease of field strength due to reconnection will be counterbalanced by an increase in strength due to compression or shear of the plasma.

The reconnection process creates very strong electric fields. In particular, a simple application of Faraday's law

$$\oint_{\mathscr{C}} \mathbf{E} \cdot dl = -\frac{d}{dt} \int_{\partial \mathscr{C}} \mathbf{B} \cdot d\mathbf{A},$$

with the curve $\mathscr{C}$ as shown in Figure 2.9.1b, shows that the EMF built up in the reconnecting region is

$$\Delta\Phi_e = (\text{rate of reconnection of magnetic flux}) \equiv d\Psi/dt. \tag{2.9.6}$$

In situations of interest to us [§5.12; Lynden-Bell (1969)], these EMF's can be far greater than 10^{10} volts. If the density of the plasma is sufficiently low to permit long mean-free paths for charged particles, and if the reconnecting region is sufficiently straight, then these EMF's can accelerate particles to ultra-relativistic energies; and the particles can then radiate intense synchrotron radiation. In this manner regions of reconnecting field lines may be sources of flares.

3 The Origin of Stellar Black Holes

Basic references: §13.13 of ZN; Peebles (1972).

Basic ideas and issues

How much of the mass of the Galaxy is in the form of black holes? What is the spectrum of black-hole masses? How many black holes are created per year by stellar collapse in our Galaxy? To these questions one has only the vaguest of answers in 1972.

If the spectrum of stellar masses at birth is the same elsewhere as in the solar neighborhood, and always has been the same, then roughly half of the mass of the Galaxy has been cycled through stars of $M > 2M_\odot$, and such stars are being born at a rate of about 0.2 per year. *If* the maximum mass of a neutron star is $2M_\odot$, and *if* stars of $M > 2M_\odot$ lose a negligible amount of mass during their evolutions and deaths, then all of the matter which goes into such stars eventually winds up in black holes. Thus, 5 per cent of the mass of the Galaxy might be in black holes, and new black holes might be forming at a rate of $\sim$0.2 per year.– *Might.* But very probably not, because the if's which go into this result are probably not satisfied.

TABLE 3.1. Summary of Numerical Computations of Supernova Explosions

Initial conditions		Results			
$M_{core}/M_\odot$	? $M_{star}/M_\odot$?	$M_{remnant}/M_\odot$	$E_{kinetic}$ of Envelope, 10^{52} ergs	Main Factor Regulating The Explosion	Authors
2 4	6 12	0.56 2.46	2.3 0.66	Absorption of neutrinos in the envelope (Strong emission of μ-neutrinos)	Arnett (1967)
8 32	24 96	24† 96†	0 0	The core is opaque to e-neutrinos; there is no mass loss (Strong emission of μ-neutrinos)	Arnett (1967)
10	30	9.75	0.025	Oxygen detonation in the envelope during collapse	Ivanova, Imshennik and Nadezhin (1969)
1.4	4	0‡	0.07	Carbon detonation during collapse	Hansen and Wheeler (1969)
				Carbon detonation in the degenerate core at the end of quasistatic evolution	Arnett (1969)
1.5 2 10	3.5 6 30	0.87 0.98 1.8	0.1 0.1 1.6	Absorption of neutrinos in the envelope	Colgate and White (1966)
$\gtrsim 40$	$\gtrsim 100$	0	3	Oxygen detonation in the core at the end of quasi-static evolution	Fraley (1968)

† In these calculations the initial core and final remnant have the same masses: $8M_\odot$ in one case, and $32M_\odot$ in the other. But since we have assumed that $M_{star} = 3M_{core}$, and since the models give no mass loss, we must take $3 \times 8 = 24M_\odot$ and $3 \times 32 = 96M_\odot$ as the masses of the "true" remnants.

‡ In recent calculations by Bruenn (1972) the thermal explosion of a star of $2M_\odot$ produced a remnant. (This does not change the conclusions given in the text about the formation of black holes.)

In particular, both observation and theory suggest that mass loss is very significant during the late stages of stellar evolution and during the death throes of massive stars. Mass loss might be so important that black holes are rare beasts, indeed!

Let us focus attention on mass loss during the death throes, as predicted by the best numerical calculations to date. But in doing so, let us keep in mind the very primitive state of the modern computations—for example, their typical neglect of the effects of rotation.

Modern computations conclude that stars whose masses exceed the Chandrasekhar limit, $M > 1.2M_\odot$, at the endpoint of their evolution should explode as supernovae. A supernova explosion can leave behind two remnants: an expanding gas cloud (the outer part of the original star), and a remnant "star". The remnant "star" will be a neutron star if its mass, M_{remnant}, is less than $\sim 2M_\odot$,† but will be a black hole if its mass is greater than $\sim 2M_\odot$.

We summarize in Table 3.1 the results of various calculations of the supernova process.

In Table 3.1 M_{core} is formally the mass of the entire star used in the numerical calculations. However, all the calculations assumed a homogeneous initial model (typically polytropic or isothermal), whereas the theory of stellar evolution predicts a highly inhomogeneous structure for presupernova stars. In particular, stellar evolution predicts a central core of iron which is more or less homogeneous, surrounded by a huge diffuse envelope of oxygen, helium, hydrogen, and other elements. Thus, we must imagine the initial stars of the supernova calculations to be the central, homogeneous core; and we must regard those calculations as neglecting the envelope. Such neglect is reasonable when one studies supernova dynamics, because the envelope is so large that it has not enough time to do anything while the core is collapsing and reexploding. The envelope will generally be thrown off, with little expenditure of work, by the reexploding parts of the core. The mass of the envelope might be twice that of the core—or perhaps only the same as the core, or perhaps even less. The theory of stellar evolution is far from definitive on this point. Hence the question marks in column 2 of Table 3.1.

It is clear, from the diverse results shown in Table 3.1, that the theory of supernova explosions is far from perfect in 1972. Nevertheless, one can form the following very tentative conclusions. (i) For stars which approach the ends of their quasistatic evolution with masses less than ~ 12 to $30M_\odot$, the supernova explosion may produce a neutron star. (ii) For stars with masses greater than ~ 12 to $30M_\odot$, the explosion may produce a black hole. If this tentative conclusion is correct, then no more than ~ 1 per cent of the mass of the Galaxy should be in the form of black holes today; and new black holes should be created at a rate no greater than ~ 0.01 per year.

---

† Here one should not ignore the large mass defect for massive neutron stars; see Zel'dovich and Novikov (1971).

4 Black Holes in the Interstellar Medium†

4.1 Accretion of Noninteracting Particles onto a Nonmoving Black Hole

Consider a black hole at rest in the interstellar medium; and temporarily treat the interstellar gas as though it were made up of noninteracting particles [collisions neglected; mean free paths large compared to the region over which the hole's gravity makes itself felt, $l_{fp} \gg 2GM/v_\infty^2$; see below]. In the case of a Schwarzschild hole, all particles with angular momentum per unit mass $\tilde{L} < 2r_g c$ eventually get captured by the hole (see Box 25.6 of MTW or Figure 12 of ZN). Here $r_g \equiv 2GM/c^2$ is the gravitational radius of hole and M is the hole's mass. If the particle speeds far from the hole are v_∞, then this condition for capture corresponds to an impact parameter

$$b = \tilde{L}/v_\infty < b_{\text{capture}} \equiv 2r_g(c/v_\infty). \tag{4.1.1}$$

Consequently, the "capture cross section" of the hole is

$$\sigma = \pi(b_{\text{capture}})^2 = 4\pi r_g^2 (c/v_\infty)^2. \tag{4.1.2}$$

For a Kerr hole the capture cross section is of this same order of magnitude. The rest mass per unit time crossing inward through a sphere of $r \gg b_{\text{capture}}$, with particles directed into the capture region (solid angle $\Delta\Omega = \sigma/r^2$), is

$$\dot{M}_0 = \left(\frac{\rho_\infty v_\infty}{4\pi}\right) 4\pi r^2 \Delta\Omega = \rho_\infty v_\infty \sigma = 4\pi r_g^2 \rho_\infty c^2/v_\infty. \tag{4.1.3}$$

This is the rate at which the hole accretes rest mass. Rewritten in typical astronomical units, this accretion rate is

$$\frac{d(M_0/M_\odot)}{d(t/10^{10}\,\text{yrs})} = 10^{-13}\left(\frac{\rho_\infty}{10^{-24}\text{ g cm}^{-3}}\right)\left(\frac{M}{M_\odot}\right)^2\left(\frac{v_\infty}{10\text{ km sec}^{-1}}\right)^{-1}. \tag{4.1.3$'$}$$

[The typical densities and speeds of interstellar gas particles in ionized, "H II" regions are 10^{-24} g cm^{-3} and 10 km sec^{-1}; see e.g. Kaplan and Pikel'ner (1970)]. Even if 100 per cent of the inflowing rest mass were somehow converted into outgoing radiation, the total luminosity would be only

$$L_{\max} = \left(10^{24}\,\frac{\text{erg}}{\text{sec}}\right)\left(\frac{\rho_\infty}{10^{-24}\text{ g cm}^{-3}}\right)\left(\frac{M}{M_\odot}\right)^2\left(\frac{v_\infty}{10\text{ km sec}^{-1}}\right) \tag{4.1.4}$$

–a value much too small to be astronomically interesting. Therefore it is fortunate that the electrons, ions, and magnetic fields of the interstellar gas interact strongly enough to make the accretion process obey fluid-dynamic laws rather than the laws of noninteracting particles (see §4.6, below).

† Material for this section is drawn largely from Chapter 13 of ZN, and from Schwartzman (1971).

4.2 Adiabatic, Hydrodynamic Accretion onto a Nonmoving Black Hole

Switch, then, from a noninteracting description of accretion to a hydrodynamic description. Assume, as above, that the black hole is at rest with respect to the gas and is a Schwarzschild hole, so that the accretion is spherically symmetric. In the hydrodynamic case the accretion rate $\dot{M}_0$ and all other basic characteristics of the flow are governed by the gravitational field at distances much greater than the gravitational radius. (This is because the flow at small radii is supersonic and therefore cannot influence conditions at large radii.) Thus, one can calculate the mass flow and other quantities at large radii accurately using the Newtonian theory of gravitation. Phenomena close to the gravitational radius will be treated later. The motion of the gas will be assumed adiabatic. Deviations from adiabatic flow due to radiative losses and radiative transport between gas elements can be taken into account later by suitably modifying the adiabatic index Γ.

The form of the flow is governed by two fundamental equations: conservation of rest mass ("continuity equation"), which we write in the form†

$$4\pi r^2 \rho u = \dot{M}_0 = \text{constant, independent of } r; \tag{4.2.1}$$

and the Euler equation

$$u\frac{du}{dr} = -\frac{1}{\rho}\frac{dp}{dr} - \frac{GM}{r^2}. \tag{4.2.2}$$

[Cf. eqs. (2.5.23′) and (2.5.39).]
Here $\dot{M}_0$ is the total rate of accretion of rest mass, u is the radial velocity, and M is the mass of the hole. We assume that the gas has constant adiabatic index Γ, so that during the accretion the pressure and density are related by the adiabatic law $p = K\rho^\Gamma$, and the speed of sound is given by $a = (\Gamma p/\rho)^{1/2}$. Let the constants K and Γ be given. Our task is to determine the accretion rate $\dot{M}_0$ and to compute as well the distributions of density $\rho(r)$ and velocity $u(r)$ in the flow.

In place of the Euler equation (4.2.2) we shall use the Bernoulli equation (2.5.38) rewritten in the form

$$\tfrac{1}{2}u^2 + \frac{1}{\Gamma - 1}a^2 - \frac{GM}{r} = \text{constant} = \frac{1}{\Gamma - 1}a_\infty^2. \tag{4.2.3}$$

---

† Notice that our notation differs from that in § 2. Throughout § 4 we denote (for ease of eyesight)

(speed of sound) = a, not c_s
(density of rest mass) = ρ, not ρ_0

No confusion is likely, since nowhere in § 4 shall we deal with 4-accelerations **a** or with total mass-energies $\rho = \rho_0(1 + \pi)$; and because $\Pi \ll 1$ in all regions of the accreting gas. We shall also retain factors of G, c, and k (i.e. use cgs units) throughout § 4.

[Here the enthalpy has been written in the form $w = \int \rho^{-1}\,dp = a^2/(\Gamma - 1)$; cf. eq. (2.5.42).] The constant has been determined by conditions at infinity, where the gas is at rest.

Rewrite the law of mass conservation (4.2.1) with density expressed in terms of sound speed

$$u = \frac{\dot{M}_0}{4\pi\rho_\infty r^2}\left(\frac{a_\infty}{a}\right)^{2/(\Gamma-1)} . \tag{4.2.4.}$$

The only unknown parameter in the coupled equations (4.2.3) and (4.2.4) is the accretion rate $\dot{M}_0$. Once $\dot{M}_0$ is known, one can readily solve for the radial distributions of velocity, sound speed, and density.

The mass flux is determined in the following way. Consider the system of equations (4.2.3) and (4.2.4). In the $u - a$ plane, the Bernoulli equation (4.2.3) for fixed radius r defines an ellipse: value of r corresponds to a different ellipse. See Figure 4.2.1. Similarly, the equation of mass conservation (4.2.4) for each value of r defines a hyperbola of fractional power, $ua^{2/(\Gamma-1)}$ = const. Now, let the accretion rate $\dot{M}_0$ be chosen arbitrarily. For every value of r (4.2.3) and (4.2.4) are 2 equations with 2 unknowns, u and a. Solving these equations–i.e. finding the intersection point of the ellipse with the hyperbola–gives u and a for that particular radius. In other words, in the u, a plane the curve $u(a)$ is determined parametrically by the intersections of corresponding ellipses and hyperbolae. It is clear (see Figure 4.2.1) that for every ellipse-hyperbola pair, there are either two intersection points, or one point of tangency, or no intersection at all. If, for a particular chosen $\dot{M}_0$, there exists any r at which the curves do not intersect, then the two equations are incompatible for the $\dot{M}_0$, and such flow cannot occur. Rejecting this case, we have 2 remaining possibilities, for a given $\dot{M}_0$ (see Figure 4.2.2):

1. The ellipse and hyperbola intersect twice for every value of r (Figure 4.2.2a). In this case we have 2 separate curves, $u(a)$, corresponding to 2 families of intersection points.

2. The ellipse and hyperbola intersect twice for every value of r, except one (call it r_S), at which they meet tangentially (Figure 4.2.2b). In this case the two families of intersection points $u(a)$ cross each other at $r = r_S$. We shall show below that this crossing point necessarily lies on the "bisectrix" of the graph, $u = a$. In both the cases, 1 and 2, the curves $u(a)$ describe possible flows of gas in the gravitational field. The curves begin on the innermost ellipse, $r = \infty$. On the upper curve $u(a)$, at this ellipse we have $a = 0$, but $u \neq 0$. These boundary conditions are not of interest for the accretion problem† because accretion requires $u_\infty = 0, \rho_\infty \neq 0, a_\infty \neq 0$.

The lower curve $u(a)$ begins at the point $u_\infty = 0, a_\infty \neq 0$, which corresponds

† This curve describes the "stellar wind" by which some stars eject matter into interstellar space; see Chapter 13 of ZN.

to our desired boundary condition. Hence we shall restrict attention to the lower curve.

We must still decide which type of lower curve is reasonable—one that remains always below the bisectrix (case 1), or one that crosses the bisectrix (case 2). In case 1 the flow is subsonic at all radii, $u < a$. But this requires a large back-pressure at all radii to retard the inflow—back pressure that can never be provided near the gravitational radius of a black hole. The gas must cross the gravitational

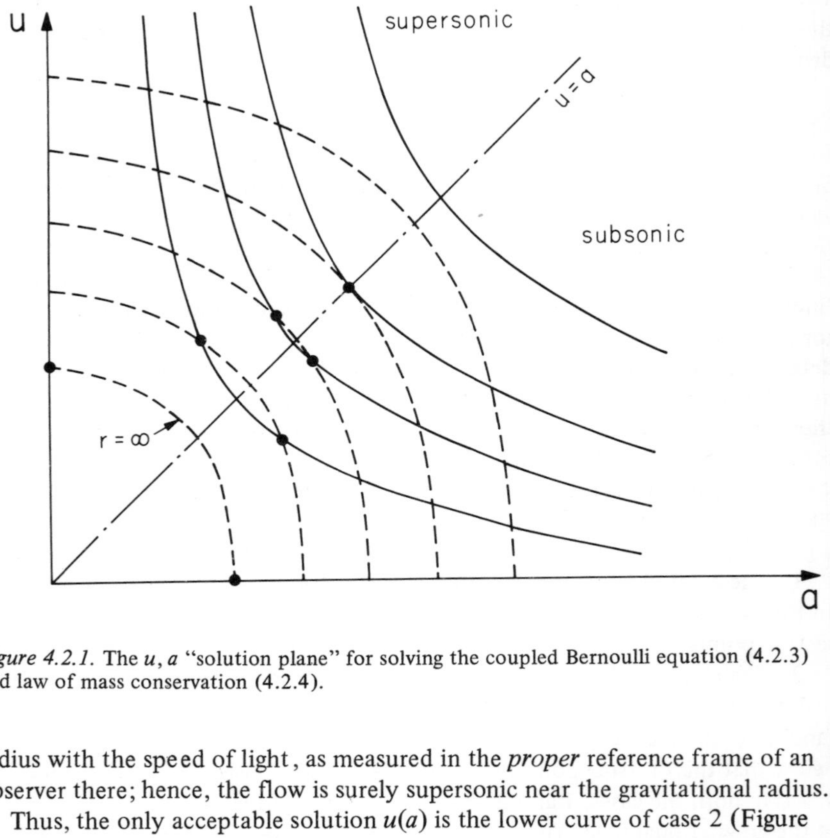

Figure 4.2.1. The u, a "solution plane" for solving the coupled Bernoulli equation (4.2.3) and law of mass conservation (4.2.4).

radius with the speed of light, as measured in the *proper* reference frame of an observer there; hence, the flow is surely supersonic near the gravitational radius.

Thus, the only acceptable solution $u(a)$ is the lower curve of case 2 (Figure 4.2.2b). This curve begins at $u = u_\infty = 0$, $a = a_\infty \neq 0$; and it crosses from the subsonic region into the supersonic region at $r = r_S$. This transition to supersonic flow is crucial to the accretion process. It governs the accretion rate $\dot{M}_0$, in the same manner as the transition to supersonic flow in the throat of a rocket nozzle governs the rate of mass flow through the nozzle. In both cases one and only one mass flow rate is compatible with the required transition to supersonic flow.

Let us calculate the required accretion rate $\dot{M}_0$. We begin by calculating the

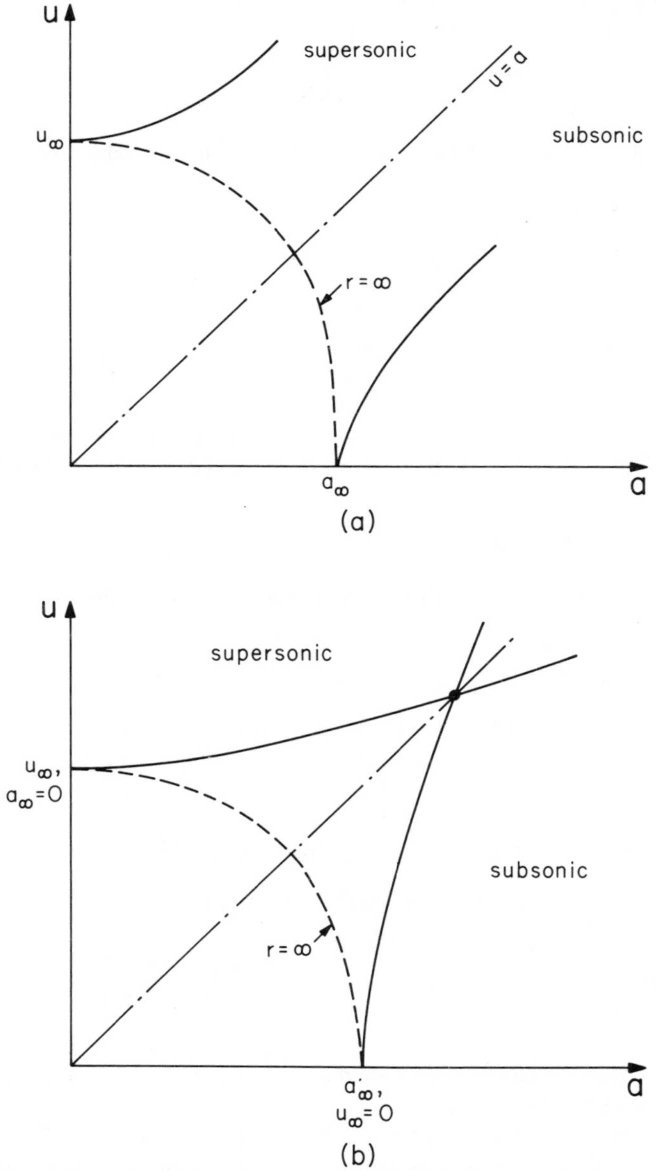

Figure 4.2.2. Possible forms of adiabatic gas flow in a spherical, Newtonian gravitational field. Curves (a) [case 1 in text] correspond to flow that remains subsonic or supersonic everywhere. Curves (b) [case 2 in text] correspond to flow for which there is a sonic point, $u = a$.

flow velocity at the transition radius ("sonic radius") r_S. To do this, we rewrite the Euler equation (4.2.2) in the form

$$u\frac{du}{dr} = -\frac{a^2}{\rho}\frac{d\rho}{dr} - \frac{GM}{r^2} \tag{4.2.5}$$

We differentiate the law of mass conservation (4.2.1) with respect to r and put du/dr from it into (4.2.5). As a result we obtain

$$\frac{d\rho}{dr}\left(\frac{a^2 - u^2}{\rho}\right) = -\frac{GM}{r^2} + \frac{2u^2}{r}, \tag{4.2.6}$$

which shows that at the sonic point, where $a = u$,

$$\frac{2u_S^2}{r_S} = \frac{GM}{r_S^2}, \quad \text{or} \quad u_S^2 = a_S^2 = \tfrac{1}{2}\frac{GM}{r_S}. \tag{4.2.7}$$

By combining this result with the Bernoulli equation (4.2.3) we obtain the speed of sound at the sonic point in terms of the speed of sound at infinity

$$a_S = u_S = a_\infty\left(\frac{2}{5 - 3\Gamma}\right)^{1/2}; \tag{4.2.8}$$

and using equation (4.2.7) we obtain the radius at the sonic point

$$r_S = \left(\frac{5 - 3\Gamma}{4}\right)\frac{GM}{a_\infty^2}. \tag{4.2.9}$$

(Notice that the gravitational potential GM/r_S at the sonic point is of order a_∞^2.) Now it is straightforward to calculate the accretion rate from equation (4.2.4):

$$\begin{aligned}\dot{M}_0 &= 4\pi u_S r_S^2 (a_S/a_\infty)^{2/(\Gamma-1)}\rho_\infty \\ &= 4\Gamma^{3/2}\alpha G^2 M^2 \rho_\infty/a_\infty^3 = \alpha r_g^2 c \rho_\infty (m_p c^2/kT_\infty)^{3/2}.\end{aligned} \tag{4.2.10}$$

Here α is a constant of order unity which depends on Γ:

$$\alpha \equiv \frac{\pi}{4\Gamma^{3/2}}\left(\frac{2}{5 - 3\Gamma}\right)^{\frac{1}{2}(5-3\Gamma)/(\Gamma-1)} \begin{array}{lll} \simeq 1.5 & \text{for} & \Gamma = 1, \\ 1.2 & \text{for} & \Gamma = 1.4, \\ 0.3 & \text{for} & \Gamma = 5/3; \end{array} \tag{4.2.11}$$

and we have used the equation of state $p_\infty = (\rho_\infty/m_p)\,kT_\infty$ corresponding to a mean mass per particle of m_p.

This expression for the accretion rate is the main result of our calculation. Notice that $\dot{M}_0$ depends on Γ only very weakly. At a typical temperature of $T_\infty \sim 10^4$ K the interstellar gas will be partially ionized, so when it is compressed some energy will go into ionization. This means that the value of Γ outside and near the sonic point will be somewhat below 5/3. A value of $\Gamma = 1.4$ might be reasonable for use in equations (4.2.10) and (4.2.11).

It is now straightforward to derive from equations (4.2.3) and (4.2.4) the radial distributions of all quantities. The general picture is as follows. Outside the "radius of influence"

$$r_i \equiv 2GM/a_\infty^2 = r_g(c/a_\infty)^2$$
$$= 10 r_S \text{ for } \quad \Gamma = 1.4. \tag{4.2.12}$$

the pull of the hole hardly makes itself felt, so ρ and a are nearly equal to their values at "infinity", $\rho = \rho_\infty$, $a = a_\infty$. At $r \simeq r_i$ the gas begins to fall with significant velocity, $u \sim a_\infty$; and the density and sound velocity begin to rise. After the gas passes through the sonic point r_S, it is in near free-fall

$$u \simeq (2GM/r)^{1/2} = a_\infty(r_i/r)^{1/2} \quad \text{at} \quad r < r_S; \tag{4.2.13a}$$

so the law of mass conservation requires the density to increase as

$$\rho = \frac{\dot{M}_0}{4\pi r^2 u} = \frac{\alpha\Gamma^{3/2}}{4\pi}\rho_\infty\left(\frac{r_i}{r}\right)^{3/2}$$
$$\simeq 0.2\rho_\infty(r_i/r)^{3/2} \quad \text{if} \quad \Gamma = 1.4 \text{ near sonic point.} \tag{4.2.13b}$$

For adiabatic compression, temperature rises as $T \propto \rho^{\Gamma-1}$; consequently, at $r < r_S$

$$T = \left(\frac{\alpha\Gamma^{3/2}}{4\pi}\right)^{(\Gamma-1)} T_\infty\left(\frac{r_i}{r}\right)^{\frac{3}{2}(\Gamma-1)} \simeq 0.5\, T_\infty\left(\frac{r_i}{r}\right)^{\frac{3}{2}(\Gamma-1)}. \tag{4.2.13c}$$

The above solution, (4.2.10)-(4.2.13), for adiabatic hydrodynamic accretion is due originally to Bondi (1952).

Rewrite the accretion rate (4.2.10) for hydrodynamic flow in a form similar to that (eq. 4.1.3) for noninteracting particles

$$\dot{M}_0 \simeq r_g^2 c\rho_\infty(c/a_\infty)^3. \tag{4.2.14}$$

Direct comparison, and use of the approximate equality between speed of sound a_∞ and proton speeds v_∞, shows that the hydrodynamic accretion rate is larger by a factor $(c/v_\infty)^2 \simeq 10^9$ than the accretion rate for independent particles. The physical reason for this is clear: gas is distinguished from independent particles by the frequent collisions of its atoms; these collisions limit the growth of tangential velocities during infall, but permit radial velocities to grow.

Other, useful forms for the hydrodynamic accretion rate are

$$\frac{d(M_0/M_\odot)}{d(t/10^{10}\,\text{yrs})} \simeq 10^{-5}\left(\frac{M}{M_\odot}\right)^2\left(\frac{\rho_\infty}{10^{-24}\text{ g cm}^{-3}}\right)\left(\frac{a_\infty}{10\text{ km sec}^{-1}}\right)^{-3}, \tag{4.2.15a}$$

$$\dot{M}_0 \simeq \left(1 \times 10^{11}\frac{\text{g}}{\text{sec}}\right)\left(\frac{M}{M_\odot}\right)^2\left(\frac{\rho_\infty}{10^{-24}\text{ g cm}^{-3}}\right)\left(\frac{T_\infty}{10^4\text{K}}\right)^{-3/2}. \tag{4.2.15b}$$

4.3 Thermal Bremsstrahlung from the Accreting Gas

The above idealized case of spherical, adiabatic, hydrodynamic accretion can be complicated by various physical processes. Consider, first, the effects of energy loss due to thermal bremsstrahlung. To calculate the bremsstrahlung most easily, we shall assume that the energy loss has a negligible effect on the thermal energy density of the gas, and then we shall check that this assumption is valid.

When the effects of bremsstrahlung losses are neglected, then the temperature, density, and velocity distributions retain their adiabatic forms (4.2.13). One can readily check that for $T_\infty \sim 10^4$ K and $\Gamma \simeq 1.4$ the gas remains nonrelativistic, $kT < m_e c^2$ (but just barely so) down to $r \simeq r_g$. Consequently, the rate at which one gram of gas at radius r radiates thermal bremsstrahlung is [cf. eq. (2.1.29)]

$$\epsilon_{ff} = (5 \times 10^{20} \text{ ergs/g sec})(\rho/\text{g cm}^{-3}) T_{\text{K}}^{1/2}$$
$$= \left(5 \times 10^{-3} \frac{\text{ergs}}{\text{g sec}}\right)\left(\frac{\rho_\infty}{10^{-24}\,\text{g cm}^{-3}}\right)\left(\frac{T_\infty}{10^4\,\text{K}}\right)^{1/2}\left(\frac{r}{r_i}\right)^{-(3/4)(\Gamma+1)} \tag{4.3.1}$$

For comparison, the rate at which adiabatic compression increases the energy of one gram of gas is [eq. (2.6.41)]

$$\epsilon_{\text{ad. heating}} = \frac{p}{\rho^2}\frac{d\rho}{d\tau} = -\frac{p}{\rho^2}\frac{d\rho}{dr}u = \frac{3}{2}\frac{p}{\rho}\frac{u}{r} \simeq \frac{p_\infty}{\rho_\infty}\frac{a_\infty}{r_i}\left(\frac{r}{r_i}\right)^{-3\Gamma/2}. \tag{4.3.2}$$

Using expression (4.2.12) for r_i, and the thermodynamic relations for a hydrogen gas

$$a_\infty^2 = (\Gamma p_\infty/\rho_\infty) \simeq \Gamma(kT_\infty/m_p), \tag{4.3.3}$$

we can bring this heating rate into the form

$$\epsilon_{\text{ad. heating}} = \left(3 \times 10^4 \frac{\text{ergs}}{\text{g sec}}\right)\left(\frac{T_\infty}{10^4\,\text{K}}\right)^{5/2}\left(\frac{M}{M_\odot}\right)^{-1}\left(\frac{r}{r_i}\right)^{-3\Gamma/2}. \tag{4.3.4}$$

This heating rate greatly exceeds the free-free loss rate (4.3.1) at all radii. Hence, we are justified in our use of the adiabatic forms of $T(r)$, $\rho(r)$, and $a(r)$. The total power radiated as thermal bremsstrahlung is

$$L_{ff} \simeq \int_{2r_g}^{r_i} \epsilon_{ff}\rho 4\pi r^2\, dr$$
$$\sim \left(5 \times 10^{17} \frac{\text{ergs}}{\text{sec}}\right)\left(\frac{M}{M_\odot}\right)^3\left(\frac{\rho_\infty}{10^{-24}\,\text{g cm}^{-3}}\right)^2\left(\frac{T_\infty}{10^4\,\text{K}}\right)^{-3.3} \quad \text{for } \Gamma = 1.4. \tag{4.3.5}$$

Perhaps a more realistic value for Γ would be a little less than 5/3 down to $r \sim 10^3 r_g$ at which point $kT \sim m_e c^2$, and then $\Gamma \simeq 4/3$ below that radius. In this case L_{ff} would be increased by several orders of magnitude [cf. the relativistic corrections in eq. (2.1.31)]. Even with such an increase, and even for $M = 100\,M_\odot$, L_{ff} is so small that it is of little or no observational interest. For this reason, we shall not attempt a more rigorous calculation of it.

4.4 Influence of Magnetic Fields and Synchrotron Radiation

Up to now we have ignored the fact that the interstellar gas possesses a magnetic field. For interstellar temperatures of interest, $T_\infty \sim 10^4$ K, the magnetic field will be frozen into the gas (cf. §2.5). The strength of the typical intergalactic field, $B_\infty \sim 10^{-6}$ G, is such that its energy density and pressure are not far below those of the gas

$$\frac{B_\infty^2}{8\pi} \sim 4 \times 10^{-14}\,\frac{\text{ergs}}{\text{cm}^3}, \qquad \rho_0 \Pi \simeq \frac{3\rho_\infty}{m_p} kT_\infty \sim 2 \times 10^{-12}\,\frac{\text{ergs}}{\text{cm}^3}. \tag{4.4.1}$$

At large radii the field may be small enough that one can ignore its influence on the accretion. In this case the standard hydrodynamical inflow will stretch each fluid element out radially (cross-sectional area of fluid element $\propto r^2$, hence diameter $\propto r$; volume $\propto \rho^{-1} \propto r^{3/2}$, hence radial diameter $\propto$ volume$/r^2 \propto r^{-1/2}$). This radial stretch and tangential compression must quickly convert the initial field in the fluid element into a nearly radial field with strength

$$B \propto (\text{cross sectional area})^{-1} \propto r^{-2}$$

(conservation of flux). Hence, the magnetic energy in a given fluid element will rise as

$$E_{\text{mag}} = \frac{B^2}{8\pi} \cdot (\text{volume of element}) \propto r^{-4} r^{3/2} \propto r^{-5/2}.$$

Not long after the gas crosses the sonic point, this magnetic energy will become so large as to exceed the thermal energy in the fluid element

$$E_{\text{therm}} = 2 \cdot \frac{3}{2}\frac{kT}{m_p} \cdot (\text{mass of element}) \propto r^{-(3/2)(\Gamma - 1)}.$$

At this point magnetic pressures must come into play, and the flow will cease to obey the standard adiabatic hydrodynamic laws of §4.2.

The form of the modified magnetohydrodynamic flow is not understood at all well today. The best guesses and calculations to date are those of Schwartzman (1971). They suggest a turbulent flow, with reconnection of field lines between adjacent "cells" (cf. §2.9), and with a rough equipartition of the gravitational potential energy between magnetic fields, kinetic infall of gas, turbulent energy, and thermal kinetic energy of gas particles:

$$-\begin{pmatrix}\text{gravitational energy}\\ \text{per unit mass}\end{pmatrix} = \frac{GM}{r} \sim 4\,\frac{B^2}{8\pi\rho} \sim 4\tfrac{1}{2}u^2 \sim 4\tfrac{1}{2}v_{\text{turb}}^2 \sim 4\,\frac{3kT}{m_p}. \tag{4.4.2}$$

The mass density corresponding to this equipartition will be determined by the law of mass conservation, $4\pi r^2 \rho u = \dot{M}_0$. Because $\dot{M}_0$ is determined by conditions outside and at the sonic point, where the magnetic field is not yet large enough to strongly influence the flow, we can use equation (4.2.15b) for $\dot{M}_0$ and write

$$\rho \simeq \left(6 \times 10^{-12} \frac{\text{g}}{\text{cm}^3}\right)\left(\frac{\rho_\infty}{10^{-24}\ \text{g cm}^{-3}}\right)\left(\frac{T_\infty}{10^4\,\text{K}}\right)^{-3/2}\left(\frac{r}{r_g}\right)^{-3/2} \tag{4.4.3}$$

Accepting this educated guess as to the nature of the flow, we can ask what types of radiation might be emitted. As before (§4.3) bremsstrahlung is too weak to be interesting. However, synchrotron radiation can be rather strong. Most of the synchrotron radiation will come from the high-temperature, strong-field region near the Schwarzschild radius. There $kT \gg m_e c^2$, so the relativistic formula (2.3.13) for synchrontron radiation is applicable:

$$\epsilon_{\text{synch}} \simeq \left(2.2 \times 10^{-10} \frac{\text{ergs}}{\text{g sec}}\right) T_{\text{K}}^2 B_G^2; \tag{4.4.4}$$

and, because the gas turns out to be optically thin, the total luminosity is

$$L_{\text{synch}} \simeq \int_{2r_g}^{\infty} \epsilon_{\text{synch}} \rho 4\pi r^2\, dr$$
$$\sim \left(10^{29} \frac{\text{ergs}}{\text{sec}}\right)\left(\frac{M}{M_\odot}\right)^3\left(\frac{\rho_\infty}{10^{-24}\ \text{g cm}^{-3}}\right)^2\left(\frac{T_\infty}{10^4\,\text{K}}\right)^{-3} \tag{4.4.5}$$

For comparison, the rest mass-energy being accreted is

$$\dot{M}_0 c^2 \simeq \left(1 \times 10^{32} \frac{\text{ergs}}{\text{sec}}\right)\left(\frac{M}{M_\odot}\right)^2\left(\frac{\rho_\infty}{10^{-24}\ \text{g cm}^{-3}}\right)\left(\frac{T_\infty}{10^4\,\text{K}}\right)^{3/2} \tag{4.4.6}$$

Since most of the radiation comes from $r \sim 2r_g$ where the thermal energy is $\sim$3 to 10 per cent of the rest mass-energy, the synchrotron losses for $M \gtrsim 10\, M_\odot$ are a significant fraction of the thermal energy available. However, for $M < 100\, M_\odot$ the effects of the energy losses on the temperature and thence on the luminosity will not exceed a factor $\sim$10. [See Schwartzman (1971) for detailed calculation of those effects.]

Equations (4.4.5) and (4.4.6) predict that a hole of $M \sim 10 M_\odot$ will convert $\sim$1 per cent of the rest mass of the infalling gas into synchrotron radiation, producing a luminosity of $\sim 10^{32}$ ergs/sec, which is about 3 per cent of the luminosity of the sun. The radiation emitted from $r \simeq 2r_g$ will have a spectrum with a broad maximum located at [cf. eq. (2.3.18)]

$$\nu_{\text{peak}} \simeq (7 \times 10^{14}\ \text{Hz})\left(\frac{\rho_\infty}{10^{-24}\ \text{g cm}^{-3}}\right)^{1/2}\left(\frac{T_\infty}{10^4\,\text{K}}\right)^{-3/4} \tag{4.4.7}$$

Since the radiation from $r \simeq 2r_g$ strongly dominates the synchrotron output, the composite spectrum of all the radiation will also have a broad maximum at

$\nu \sim 7 \times 10^{14}$ Hz ($\lambda \sim 4000$ Å); and it will die out exponentially for $\nu \gg 7 \times 10^{14}$ Hz; cf. §2.3. As Schwartzman (1971) remarks, such a spectrum and luminosity resemble those of "DC white-dwarf stars" (white dwarfs with continuous, line-free spectra). This has led Schwartzman to speculate that "perhaps some of the objects heretofore regarded as type DC white dwarfs are actually black holes".

Instabilities and inhomogeneities in the flow (due, e.g., to reconnection of field lines) might lead to marked fluctuations in the luminosity of the hole. The timescales for such fluctuations might be of the order of the gas travel time from $\sim 10r_g$ to $\sim 2r_g$; i.e.

$$\Delta t_{\text{fluctuations}} \sim (10^{-3} \text{ to } 10^{-4} \text{ sec})(M/M_\odot). \tag{4.4.8}$$

Such rapid fluctuations in an object so faint should be very difficult to detect, even with the 200-inch telescope. For further details on the fluctuations, see the end of §4.7.

4.5 Interaction of Outflowing Radiation with the Gas

As the luminosity L pours out from the region $r \sim 2r_g$, it interacts with the inflowing gas. For the luminosities $L \sim 10^{32}$ ergs/sec and frequencies $\nu \sim 10^{14}$ Hz derived in the last section, one can readily verify that the infalling gas is optically thin to free-free absorption and to synchrotron reabsorption. (See §2.6 for basic concepts and equations used in such a calculation.)

What of electron scattering? Electron scattering can exert 3 types of influence: (i) If the electron concentration is sufficiently high, it can act as a "blanket" to impede the outflow of the radiation. To test for such blanketing one can examine the optical depth of the gas for electron scattering.

$$\begin{aligned} \tau_{es}(r = 2r_g) &= \int_{2r_g}^{\infty} \kappa_{es}\rho\, dr = \int_{2r_g}^{\infty} (0.4 \text{ cm}^2/\text{g})\rho\, dr \\ &\simeq (1 \times 10^{-6})(M/M_\odot)(\rho_\infty/10^{-24} \text{ g cm}^{-3})(T_\infty/10^4 \text{ K})^{-3/2}. \end{aligned} \tag{4.5.1}$$

(A more careful calculation would replace the nonrelativistic absorption coefficient $\kappa_{es} = 0.4$ cm^2/g by the coefficient appropriate to relativistic electrons, since $T \sim 10^{12}$ K at $r \sim r_g$; such a calculation would also reveal extreme optical thinness). (ii) Since the outgoing photons have energies $h\nu \sim 10$ eV far less than the electron kinetic energies, Compton scattering can modify the spectrum, producing high-energy photons [see eq. (2.4.9)]. However, the extreme optical thinness guarantees that only about one photon in a million scatters; so this effect can be ignored. (iii) The outpouring photons exert a pressure on the infalling gas when they scatter, and this effect can retard the inflow. Notice that the magnitude of the first two effects is governed by the density of the inflowing gas. By contrast, the magnitude of the pressure effect is governed by the luminosity of the outpouring radiation. Let us calculate this effect explicitly, at radii sufficiently large that the electrons are nonrelativistic.

Because the electron scattering cross-section $d\sigma/d\Omega = \frac{1}{2}r_0^2(1 + \cos^2\theta)$ is symmetric between forward and backward angles, on the average a scattered photon gives all of its momentum, $h\nu/c$, to the electron. Hence, the time-averaged photon force acting on an electron at radius r is

$$\langle \text{Force} \rangle = \left\langle \frac{d(\text{momentum})}{dt} \right\rangle = \int \frac{d(\text{number of photons})}{d(\text{area})\, dt\, d\nu} h\nu\sigma_{es}\, d\nu \qquad (4.5.2)$$

$$= \sigma_{es}F = \sigma_{es}(L/4\pi r^2),$$

where $\sigma_{es} = (8\pi/3)r_0^2 = 0.657 \times 10^{-24}\ \text{cm}^2$ is the total scattering cross section, F is the radiation flux (ergs cm^{-2} sec^{-1}) and L is the total luminosity of the hole. Notice that this time-averaged force is completely independent of the spectrum of the radiation (so long as most of the photons have $h\nu \ll m_e c^2$ as seen in the electron rest frame, so that nonrelativistic cross sections are applicable). It is also independent of the optical thickness—being equally valid for $\tau \gg 1$, in which case

$$F \ll J = (4\pi)^{-1}c \quad \text{(energy density of radiation)},$$

and for $\tau \ll 1$, in which case $F \simeq J$ (see §2.6 for notation).

The pull of gravity must work against this photon force. The photon force acts almost entirely on the electrons ($\sigma_{es} \propto 1/\text{mass}^2$, so each proton feels a photon force 3×10^6 times weaker than that felt by an electron). By contrast, the acceleration of gravity acts equally on electrons and photons. Hence, the electrons move outward slightly relative to the photons, creating an electric field that transmits the photon force $\sigma_{es}L/4\pi r^2$ from electrons to protons. The net force that then acts on each proton (assuming a completely ionized hydrogen gas) is

$$\langle \text{Force} \rangle_{\text{total}} = \frac{\sigma_{es}L}{4\pi r^2} - \frac{GMm_p}{r^2}. \qquad (4.5.3)$$

Thus, there is a critical luminosity ["Eddington" (1926) limit]

$$L_{\text{crit}} \equiv \frac{4\pi GM\dot{m}_p}{\sigma_{es}} = (1.3 \times 10^{38}\ \text{ergs/sec})\,(M/M_\odot) \qquad (4.5.4)$$

such that for $L \ll L_{\text{crit}}$ gravity dominates at all radii, but for $L \gg L_{\text{crit}}$ photon pressure dominates at all radii. Eddington derived this limit for the case of stellar equilibrium. Zel'dovich and Novikov (1964) developed the analogous theory for the cases of quasars and accretion.

For the case of current interest—a black hole of $M \sim 1$ to $100\, M_\odot$ swallowing interstellar gas—the luminosity is far below the critical value, so the photon pressure can be ignored.

However in other contexts the accretion rate $\dot{M}_0$ may be so great, and the efficiency ζ for converting the inflowing mass-energy into outgrowing luminosity L may be sufficiently high that

$$\frac{L}{L_{\text{crit}}} = \frac{\zeta \dot{M}_0 c^2}{L_{\text{crit}}} = \frac{\zeta}{0.1}\left(\frac{\dot{M}_0}{10^{18}\ \text{g/sec}}\right) \tag{4.5.5}$$

may temporarily exceed unity. When this happens, the photon pressure will impede further mass inflow, thereby reducing the luminosity to $L \lesssim L_{\text{crit}}$, and thereafter maintaining roughly the correct $\dot{M}_0$ to make $L \sim L_{\text{crit}}$. Such accretion is said to be "*self-regulated*". Self-regulated accretion, with modifications due to lack of spherical symmetry, may be important for black holes in close binary systems; see §5.13. However, it is unimportant for the case at hand.

The outflowing radiation from a black hole in interstellar space ($L \sim 10^{32}$ ergs/sec, $\nu \sim 10^{14} - 10^{15}$ Hz) can also interact with the surrounding interstellar gas. One can verify that the radiation is sufficiently intense to keep the gas at a temperature $T_\infty \sim 10^4$ K in the neighborhood of $r = r_i$, and to keep it partially ionized.

4.6 Validity of the Hydrodynamical Approximation

The above model for accretion onto an interstellar black hole relies crucially on the validity of the hydrodynamical approximation at and outside the sonic point, and on equipartition assumptions inside the sonic point. If the gas were to behave like independent particles rather than like a fluid, then the accretion rate would be greatly reduced (see §4.1). Thus, it is crucial to check the validity of the hydrodynamic approximation.

In an ionized hydrogen plasma *without* magnetic fields, the deflection of protons away from straight-line paths is caused primarily by the cumulative effects of many small-angle Coulomb scatterings. The total distance a proton must travel before the "random-walk" deflections add up to an angle of ~90° is [see, e.g. Shkarofsky, Johnston, and Bachynski (1966) or Pikel'ner (1961)]

$$\lambda_p = \frac{9}{\pi}\frac{(kT)^2}{n_p e^4 L_c} \simeq (7 \times 10^{12}\ \text{cm})\left(\frac{\rho}{10^{-24}\ \text{g cm}^{-3}}\right)^{-1}\left(\frac{T}{10^4\ \text{K}}\right)^2. \tag{4.6.1}$$

Here L_c is a "Coulomb-logarithm" (analogue of Gaunt factor in theory of Bremsstrahlung) given by

$$L_c \simeq 23 + \tfrac{3}{2}\ln(T/10^4\ \text{K}) - \tfrac{1}{2}\ln(n_e/\text{cm}^{-3}); \tag{4.6.2}$$

n_e and n_p are the number densities of electrons and (ionized) protons; and we have evaluated λ_p for the case of a fully ionized hydrogen plasma. The plasma will behave like a fluid on scales $l \gg \lambda_p$, but like independent particles on scales

$l \ll \lambda_p$. The scale of interest for determining the accretion rate is the sonic radius

$$r_S = \frac{5-3\Gamma}{4}\frac{GM}{a_\infty^2} \simeq (10^{13}\ \text{cm})\left(\frac{M}{M_\odot}\right)\left(\frac{T_\infty}{10^4\,\text{K}}\right)^{-1}, \tag{4.6.3}$$

at which point $T \sim T_\infty$, $\rho \sim \rho_\infty$, so

$$\frac{\lambda_P(r_S)}{r_S} \sim 0.7\left(\frac{M}{M_\odot}\right)^{-1}\left(\frac{\rho_\infty}{10^{-24}\ \text{g cm}^{-3}}\right)^{-1}\left(\frac{T_\infty}{10^4\,\text{K}}\right)^{3}. \tag{4.6.4}$$

Thus, if one ignores the magnetic field, then the gas will not quite behave like a fluid near the sonic point–and it may fail even more to be fluid-like at small radii; cf. § 13.4 of ZN.

The interstellar magnetic field, of strength $B_\infty \sim 10^{-6}$ G, can "save" the hydrodynamic approximation, It deflects protons away from straight-line motion in a distance of the order of the Larmour radius

$$\lambda_L = \frac{m_P c v}{eB} \simeq (1 \times 10^8\ \text{cm})\left(\frac{B}{10^{-6}\,\text{G}}\right)^{-1}\left(\frac{T}{10^4\,\text{K}}\right)^{1/2}, \tag{4.6.5}$$

and this is true even when the thermal pressure of the plasma exceeds the magnetic pressure so that the field gets dragged about by the gas. The Larmour radius is small compared to all macroscopic scales of interest (e.g. small compared to r_S when one uses the values $B \sim B_\infty$ and $T \sim T_\infty$ appropriate to r_S). Hence, the hydrodynamical approximation is fairly well justified.

4.7 Gas Flow Near the Horizon

Near the horizon, $r \sim r_g$, relativistic effects will modify the Newtonian picture of accretion. It is not difficult to solve for the relativistic flow using the relativistic Bernoulli equation (2.5.33) and the law of mass conservation (2.5.23′). Such a solution is analogous to the Newtonian hydrodynamical solution studied in §4.2. However, such a solution is partially irrelevant because it ignores the role of the magnetic field; and such a solution is not needed because the intensity of the gravitational pull, as viewed in stationary frames near the horizon, is so great as to guarantee supersonic flow with qualitatively the same form as free-particle fall. Thus from a knowledge of geodesic orbits one can infer the main features of the flow.

Consider, first, spherical accretion onto a nonrotating hole (Schwarzschild gravitational field). A fluid element in the accreting gas, like a freely falling particle, reaches the horizon in finite proper time. The moment of crossing the horizon is in no way peculiar, as seen by the fluid element. The density does not reach infinity there; and the temperature does not reach infinity. In fact, because the equation for radial geodesics in the Schwarzschild geometry

$$ds^2 = -(1 - r_g/r)c^2 dt^2 + (1 - r_g/r)^{-1} dr^2 + r^2 d\Omega^2. \tag{4.7.1}$$

has the first integral

$$(dr/d\tau)^2 = 2GM/r \text{ (for fall from near rest at } r \gg r_g), \tag{4.7.2}$$

and because the law of rest-mass conservation $\boldsymbol{\nabla} \cdot (\rho \mathbf{u}) = 0$ has the first integral

$$4\pi \sqrt{-g}\, \rho(dr/d\tau) = \dot{M}_0, \tag{4.7.3}$$

the density of rest mass must vary as

$$\rho = \frac{\dot{M}_0}{4\pi r^2}\left(\frac{r}{2GM}\right)^{1/2} \simeq (6 \times 10^{-12}\, \text{g/cm}^3)\,(r/r_g)^{-3/2}. \tag{4.7.4}$$

This must be as true near and inside the horizon, $r \lesssim r_g$, as it is in the Newtonian realm $r \gg r_g$. Similarly, the temperature will retain its Newtonian form (4.4.2):

$$T \simeq (10^{12}\,\text{K})(r_g/r); \tag{4.7.5}$$

and the synchrotron emissivity will retain its Newtonian value (4.4.4). However, the radiation which reaches infinity will be sharply cut off as the gas element passes through the horizon

$$\begin{pmatrix}\text{fraction of emitted radiation that} \\ \text{reaches } r = \infty \text{ from infalling gas at radius } r\end{pmatrix} \simeq \frac{27}{64}\left(1 - \frac{r_g}{r}\right). \tag{4.7.6}$$

In previous sections we have taken rough account of this cutoff and of the accompanying redshift by retaining the Newtonian luminosity down to $r = 2r_g$, and then arbitrarily ignoring all radiation from $r < 2r_g$.

In the case of a rotating (Kerr) hole, near and inside the ergosphere the dragging of inertial frames will swing the infalling gas into orbital rotation about the hole. As the gas approaches the horizon, its angular velocity as seen from infinity must approach the angular velocity of the horizon,

$$\begin{aligned}\Omega \to \Omega_{\text{horizon}} &= \frac{a}{r_+^2 + a^2} \quad \text{(geometrized units, } c = G = 1) \\ &= \frac{c^3}{2GM} = \left(\frac{10^5}{\text{sec}}\right)\left(\frac{M}{M_\odot}\right)^{-1} \quad \text{for "maximally rotating hole", } a = M.\end{aligned} \tag{4.7.7}$$

Here a is the hole's angular momentum per unit mass. Any extra-luminous hot spot in the gas (e.g. a region where magnetic field lines are reconnecting; cf. §2.9) must orbit the hole with this angular velocity as it falls inward. The brightness of the hole as seen at Earth will be modulated with a period $P \simeq 4\pi/\Omega \gtrsim (10^{-4}\,\text{sec})(M/M_\odot)$ as a result of the Doppler shift of the spot's light and the bending of its rays. (One factor of $2\pi/\Omega$ is the orbital period; the other factor of $2\pi/\Omega$ is the light travel time between the radius at which the spot is located at the beginning of one period and the radius to which it has fallen at the end of the period.) Thus, one might expect quasiperiodic fluctuations in the light from a black hole living alone in the interstellar medium.

4.8 Accretion onto a Moving Hole

Turn attention now from a hole at rest in the interstellar medium, to a hole that moves with a speed much larger than the sound speed, $u_\infty \gg a_\infty$. Examine the resulting accretion in the rest frame of the hole. At radii $r \gg 2GM/u_\infty^2$ the kinetic energy per unit mass, $\frac{1}{2}u_\infty^2$, is far larger than the potential energy, GM/r; so the pull of the hole has no effect on the gas flow. At $r \sim 2GM/u_\infty^2$, the pull of the hole becomes significant. Because the flow is supersonic the gas will respond to that pull in the same manner as would noninteracting particles; its pressure cannot have an influence. (One can see this, for example, in the Euler equation

$$\text{const.} = \tfrac{1}{2}\mathrm{v}^2 - \frac{GM}{r} + \frac{a^2}{\Gamma - 1} \simeq \tfrac{1}{2}\mathrm{v}^2 - \frac{GM}{r}, \tag{4.8.1}$$

where the speed-of-sound (enthalpy) term—which accounts for pressure effects —is negligible compared to the kinetic-energy term.) Thus, the flow lines will be identical to the trajectories of test particles: they will be hyperbolae (Fig. 4.8.1a).

After passing around the hole, the trajectories of test particles intersect (particle collisions; dotted line of Figure 4.8.1a). In the gas-dynamic case such intersection of flow lines is prevented by rapidly mounting pressure. Consequently, a shock front must develop around the black hole (Fig. 4.8.1b). Outside the shock front the flow lines will be hyperbolic test-particle trajectories. Behind the shock front, they will not. The shock will be located roughly where potential energy equals kinetic energy, so its characteristic size l_S as shown in Figure 4.8.1b will be

$$l_S \sim GM/u_\infty^2 \tag{4.8.2}$$

In the shock, the gas will lose most of its velocity perpendicular to the shock front ["strong shock" in terminology of §2.7; "strong" because u_∞/a_∞ = (speed of gas)/(speed of sound) $\gg 1$ on front side]; but it will retain all of its velocity parallel to the front (denote this by $v_\parallel$). For a given gas element, if the remaining kinetic energy behind the front greatly exceeds the potential energy, $\frac{1}{2}v_\parallel^2 \gg GM/r$, then the gas element will escape the pull of the hole. If $\frac{1}{2}v_\parallel^2 \ll GM/r$, then it cannot escape. The dividing line between escape and capture is a dotted line in Figure 4.8.1b; and the corresponding impact parameter is labelled b_{capture}.

We can calculate b_{capture} in order of magnitude by idealizing the shock as confined to the thin dotted region of Figure 4.8.1a, (Hoyle and Littleton 1939). The value of b_{capture} will correspond to an orbit for which

$$\tfrac{1}{2}v_x^2 = GM/x \quad \text{at} \quad y = 0.$$

Straightforward examination of Kepler orbits reveals that

$$b_{\text{capture}} = 2GM/u_\infty^2. \tag{4.8.3}$$

Since the mass flux in the gas at large radii is $\rho_\infty u_\infty$, the accretion rate is

$$\dot{M}_0 = (\pi b_{\text{capture}}^2)\rho_\infty u_\infty = 4\pi G^2 M^2 \rho_\infty / u_\infty^3. \tag{4.8.4}$$

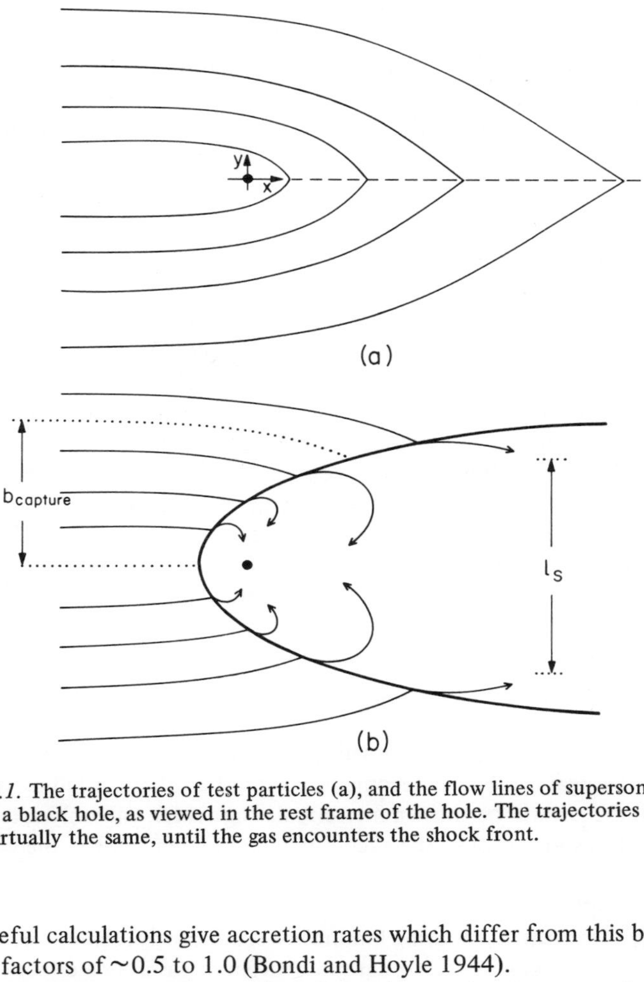

Figure 4.8.1. The trajectories of test particles (a), and the flow lines of supersonic gas (b) relative to a black hole, as viewed in the rest frame of the hole. The trajectories and flow lines are virtually the same, until the gas encounters the shock front.

More careful calculations give accretion rates which differ from this by multiplicative factors of ~0.5 to 1.0 (Bondi and Hoyle 1944).

Notice that the accretion rate (4.8.4) for the supersonic case, and the rate (4.2.10) for a hole at rest are identical except for the replacement of sound speed a_∞ by speed of flow u_∞, and except for a numerical coefficient of order unity. In the intermediate case of $u_\infty \sim a_\infty$ one can use the hybrid formula

(Bondi 1952, Hunt 1971)

$$\dot{M}_0 \simeq \frac{4\pi G^2 M^2 \rho_\infty}{(u_\infty^2 + a_\infty^2)^{3/2}}$$

$$\simeq \left(1 \times 10^{11} \frac{\text{g}}{\text{sec}}\right) \frac{(M/M_\odot)^2(\rho_\infty/10^{-24}\ \text{g cm}^{-3})}{[(u_\infty/10\ \text{km sec}^{-1})^2 + (a_\infty/10\ \text{km sec}^{-1})^2]^{3/2}}. \tag{4.8.5}$$

[Salpeter (1964) gave the first application of these formulas to black holes; all previous work dealt with normal stars.]

It is worth noting that a significant fraction of the kinetic energy released in the shock

$$\frac{dE}{dt} \simeq \tfrac{1}{2} u_\infty^2 \dot{M}_0 \simeq \left(10^{24} \frac{\text{ergs}}{\text{sec}}\right)\left(\frac{M}{M_\odot}\right)^2 \left(\frac{\rho_\infty}{10^{-24}\ \text{g cm}^{-3}}\right)\left(\frac{u_\infty}{10\ \text{km sec}^{-1}}\right)^{-1/2} \tag{4.8.6}$$

may go into excitation of unionized atoms and thence into light as the atoms decay, thus producing a shock front that glows (albeit weakly). See, e.g., Pikel'ner (1961), and § 10 of Kaplan (1966).

The motion of the gas behind the shock is practically unknown, particularly when one takes account of magnetic fields. Many different types of instabilities may occur, both here and in the case of a stationary black hole. For example at radii $r \ll l_S$, where the inflow has become supersonic again, turbulence may create shock waves which convert kinetic energy into heat (Bisnovaty-Kogan and Sunyaev 1971). A variety of plasma instabilities may also arise. And the reconnection of magnetic field lines will surely be important. One can only guess that the overall, time-averaged picture will resemble the equipartition model of Schwartzman (§4.4).

One factor that may be important near the gravitational radius, where most of the luminosity is produced, is angular momentum. Suppose that the accreting gas has nonzero angular momentum about the direction of motion of the hole ($-x$ direction in Fig. 4.8.1). If the angular momentum per unit mass, $\tilde{L}$, exceeds $r_g c$, then centrifugal forces will become important before the infalling gas reaches the horizon; the gas will be thrown into circulating orbits; and only after viscous stresses have transported away the excess angular momentum will the gas fall down the hole. (See §5 for a situation similar to this.) The resulting viscous heating and lengthened time spent outside the hole, as well as the changed flow pattern, may affect the outpouring luminosity significantly. Moreover, when the hole moves from a region with angular momentum of gas in one direction to a region with angular momentum in another, the gas flow may become particularly violent for a while near the horizon; and a strong flare of luminosity may be produced (Salpeter 1964, Schwartzman 1971). Almost nothing is known today about these issues.

Is the specific angular momentum of the accreting gas likely to exceed $r_g c$? Interstellar gas is accreted from a region of diameter

$$2b_{\text{capture}} = \frac{2r_g}{u_\infty^2} = (3 \times 10^{14}\ \text{cm})\left(\frac{M}{M_\odot}\right)\left(\frac{u_\infty}{10\ \text{km sec}^{-1}}\right)^{-2}. \tag{4.8.7}$$

(For $u_\infty < a_\infty$, we must replace u_∞ by a_∞.) The interstellar gas is turbulent with the turbulent velocity v_{turb} on scales l given roughly by

$$v_{\text{turb}} \simeq \left(10^6\ \frac{\text{cm}}{\text{sec}}\right)\left(\frac{l}{3 \times 10^{20}\ \text{cm}}\right)^q \tag{4.8.8}$$

where q is a coefficient believed to lie between 1/2 and 1 (Kaplan and Pikel'ner 1970), and the coefficient 10^6 cm/sec is taken from astronomical observations. Let us take $q = 0.75$ as a best guess. Then the specific angular momentum associated with this turbulence, for a gas element of size l, is

$$\tilde{L}_{\text{turb}} \simeq v_{\text{turb}}\, l \simeq \left(3 \times 10^{26}\ \frac{\text{cm}^2}{\text{sec}}\right)\left(\frac{l}{3 \times 10^{20}\ \text{cm}}\right)^{1.75}. \tag{4.8.9}$$

Consequently, the specific angular momentum of the gas that accretes from a region of size $l = 2b_{\text{capture}}$ should be

$$\frac{\tilde{L}_{\text{turb}}}{r_g c} \simeq 1 \times \left(\frac{M}{M_\odot}\right)^{3/4}\left(\frac{u_\infty}{10\ \text{km sec}^{-1}}\right)^{-7/2}. \tag{4.8.10}$$

Evidently, the angular momentum may be important in some cases and may not be in others.

4.9 Optical Appearance of Hole: Summary

In summary, a black hole alone in the interstellar medium may be expected to emit $\sim 10^{29}$ to 10^{35} ergs/sec of synchrotron radiation, concentrated in the optical part of the spectrum ($\sim 10^{14}$ Hz to 10^{15} Hz) (§4.4). The output may fluctuate on timescales of $\sim 10^{-2}$ to 10^{-4} seconds due to instabilities, and to orbital motion of hot spots about the hole (§§4.4, 4.7, 4.8). There may also be strong flares when the hole moves out of one turbulent cell of interstellar gas into another. The time interval between flares would be about

$$\Delta t_{\text{flares}} \simeq 2b_{\text{capture}}/u_\infty \simeq (10\ \text{years})\left(\frac{M}{M_\odot}\right)\left(\frac{u_\infty}{10\ \text{km sec}^{-1}}\right)^{-3}. \tag{4.9.1}$$

It will probably be difficult observationally to distinguish accreting, isolated black holes from accreting, old neutron stars without magnetic fields. The only distinguishing feature will be emission from the surface of the neutron star.

5 Black Holes in Binary Star Systems and in the Nuclei of Galaxies

5.1 Introduction

Turn attention now from black holes alone in the interstellar medium to (i) black holes in orbit about normal stars ("star-hole binary systems") and to (ii) supermassive holes ($10^7 M_\odot \lesssim M \lesssim 10^{11} M_\odot$) which might reside at the centers of some galaxies. As for isolated holes, so also here, the phenomenon of interest is the accretion of gas and the accompanying emission of radiation. But the general picture of the accretion is quite different here–different in two ways:

First, the accretion rate and resulting luminosity may be much larger than those ($\sim 10^{-15} M_\odot$/yr, $\sim 10^{31}$ ergs/sec) for an isolated hole. In the binary case gas can flow from the atmosphere of the ordinary star onto its companion hole. One knows that variable stars of the β-Lyrae type eject mass continuously from their atmospheres at rates $\sim 10^{-5} M_\odot$/yr. Roughly half of this mass might fall onto the companion (if there is a companion.) For typical observed binary systems that emit X-rays (e.g., Cyg X–1 and Cen X–3) the observations and models suggest that the normal star is dumping gas onto its companion at a rate of $M_0 \sim 10^{-9} M_\odot$/yr, and that this gas radiates a luminosity of $\sim 10^{37}$ ergs/sec. A supermassive hole at the center of a galaxy, by virtue of its high mass and the large gas density there, will accrete much more than a hole of ordinary mass in a normal interstellar region. The accretion rate and luminosity might be $M_0 \sim 10^{-3} M_\odot$/yr and $\sim 10^{43}$ ergs/sec. (See §5.3 below.)

Second, the accreting gas in a binary system and in the center of a galaxy has very high specific angular momentum, $\tilde{L} \gg r_g c$. As a result, the accretion is far from spherical; and the considerations of the last chapter are inapplicable. Instead of falling inward radially or roughly radially, the gas elements go into Keplerian orbits around the hole, forming a gas disk analogous to Saturn's rings. However, the density in the accreting disk is far greater than that in Saturn's rings; and viscosity is important. The viscosity removes angular momentum permitting the gas to spiral gradually into the hole. The viscosity also heats the gas, causing it to radiate. The radiation is largely X-rays in the binary case; and ultraviolet and blue light, in the supermassive case.

Hayakawa and Matsuoko (1964) were the first to propose that X-rays might be produced by accretion of gas in close binary systems. However, they discussed not accretion onto compact companions, but rather accretion into the atmosphere of a normal companion star, with the formation of a hot shock front. Novikov and Zel'dovich (1966), and Shklovsky (1967) were the first to point out that accretion onto neutron stars and black holes in binary systems should produce X-rays. They also inferred from observational data that Sco X-1 might be a neutron star in a state of accretion. [For further details on this early history see Burbidge (1972).] The essential role of the angular momentum of the gas in binary accretion was first emphasized by Prendergast [see Prendergast and Burbidge (1968)]. He built models for disk-type accretion onto white dwarfs in binary systems. Later Shakura (1972), Pringle and Rees (1972), and Shakura and Sunyaev (1972) built models for disk-type accretion onto neutron stars and black holes. All of these binary accretion models were Newtonian; Thorne (1973), and Novikov, Polnarev, and Sunyaev (1973) have calculated the effects of general relativity on the inner regions of the accreting disk.

Lynden-Bell (1969) was the first to argue that galaxies might have supermassive holes at their centers, and to analyze disk-type accretion onto such holes. Subsequently Lynden-Bell and Rees (1971) extended this work.

The analysis of disk-type accretion given in these lectures is based primarily on the calculations of Shakura and Sunyaev (1972) and of Thorne (1973b); but it has been strongly influenced also by the earlier work of Prendergast, of Lynden-Bell (1969), and of Pringle and Rees (1971). Our analysis will make extensive use of the mathematical tools of general relativity. Therefore, throughout it we shall use geometrized units (gravitation constant G, speed of light c, and Boltzmann constant k all equal to unity).

5.2 Accretion in Binary Systems: The General Picture

Consider a close binary system with one component a "normal" star and the other a "compact star" (black hole or neutron star or white dwarf). "Close" means that the separation a between the centers of mass is within a factor 2 or so of the radius R_N of the normal star

$$a \lesssim 2R_N. \tag{5.2.1}$$

For such a system, the interaction between the stars, acting over astronomical time scales, may have produced a circular orbit and may have brought the normal star into co-rotation with its companion ("same face" always turned toward companion). For this reason, and to simplify the discussion, we assume a circular orbit and and co-rotation of the normal star. By Kepler's laws, the angular velocity of the stars about each other is

$$\mathbf{\Omega} = [(M_N + M_c)/a^3]^{1/2}\, \mathbf{e}_z, \tag{5.2.2}$$

where $\mathbf{e}_z$ is a unit vector perpendicular to the orbital plane, and where M_N and M_c are the mass of the normal star and the compact star respectively. For cases of interest (e.g., the binary systems associated with Cyg X-1 and Cen X-3),

$$\begin{aligned} &M_N \sim M_c \sim 1 \text{ to } 20 M_\odot, \\ &a \sim 2R_N \sim 10^{11} \text{ to } 10^{12} \text{ cm} \\ &\text{orbital period} = 2\pi/\Omega \sim 1 \text{ to } 5 \text{ days.} \end{aligned} \tag{5.2.3}$$

Analyze the flow of gas from the normal star to its compact companion using a coordinate system that co-rotates with the binary system (noninertial frame!). In the Newtonian realm (i.e., everywhere except near the surface of the compact star), the Euler equation (2.5.30) for the flowing gas reduces to

$$\frac{d\mathbf{v}}{d\tau} = -\nabla\Phi_{gc} - \mathbf{\Omega} \times \mathbf{v} - \frac{1}{\rho_0}\nabla p + \frac{2}{\rho_0}\nabla\cdot(\eta\sigma). \tag{5.2.4}$$

Here $d/d\tau$ is the time derivative moving with a fluid element

$$d/d\tau = \partial/\partial t + \mathbf{v}\cdot\nabla; \tag{5.2.5}$$

$\mathbf{v}$ is the gas velocity relative to the rotating frame; p and ρ_0 are pressure and rest-mass density; Φ_{gc} is the "gravitational-plus-centrifugal" potential

$$\Phi_{gc} = -\frac{M_N}{|\mathbf{r}-\mathbf{r}_N|} - \frac{M_c}{|\mathbf{r}-\mathbf{r}_c|} - \tfrac{1}{2}(\mathbf{\Omega}\times\mathbf{r})^2; \tag{5.2.6}$$

and we have included a viscous shear stress, with η the coefficient of dynamic viscosity and $\boldsymbol{\sigma}$ the shear. In the Euler equation the term $-\nabla\Phi_{gc}$ gives rise to gravitational and centrifugal accelerations; and the term $-\mathbf{\Omega}\times\mathbf{v}$ gives rise to Coriolis accelerations.

Notice that when pressure and viscous accelerations are unimportant (test-particle motion), the equation of motion (5.2.4) is identical to that for a particle with electric charge q moving in an electric field with potential Φ_{gc}/q, and a uniform magnetic field with strength $\mathbf{B} = \mathbf{\Omega}/q = (\Omega/q)\mathbf{e}_z$. Conservation of energy in this case requires $\Phi_{gc} + \frac{1}{2}\mathbf{v}^2 = \text{const}$. When pressure forces are taken into account, energy conservation [Bernoulli equation (2.5.38)] is modified to read

$$\Phi_{gc} + \tfrac{1}{2}\mathbf{v}^2 + w = \text{constant along flow lines.} \tag{5.2.7}$$

When viscous stresses are taken into account, one cannot write such a simple conservation law.

To deduce the qualitative nature of the gas flow from the "normal" star to its companion, one must have a clear picture of the potential Φ_{gc}. Figure 5.2.1 gives such a picture for the orbital plane. Of particular interest is the equipotential curve marked "Roche lobe". If the surface of the normal star is inside its Roche lobe, then the only way it can dump gas onto its compact companion is by means of a "stellar wind", which blows gas off the star supersonically in all directions with only that gas blown toward the companion being captured. But if the surface of the normal star fills its Roche lobe, then it can dump gas continuously through the "Lagrange point" L_1 (Fig. 5.2.1) onto its companion. In this case the steady-state flow, as governed by the Euler equation (5.2.5) and by the law of mass conservation

$$\nabla\cdot(\rho_0\mathbf{v}) = -\partial\rho_0/\partial t = 0, \tag{5.2.8}$$

will have the qualitative form shown in Fig. 5.2.2.

The gas falls from the surface of the normal star, over the "lip" of the potential (Lagrange point L_1), toward the compact companion. As the gas picks up speed, Coriolis forces swing it to the right in Figure 5.2.2, and then gravitational forces swing it back leftward into roughly circular motion about the compact companion. Gas previously dumped onto the companion is now in an orbiting disk. The incoming gas interacts viscously with the gas of the disk. Some of the incoming gas

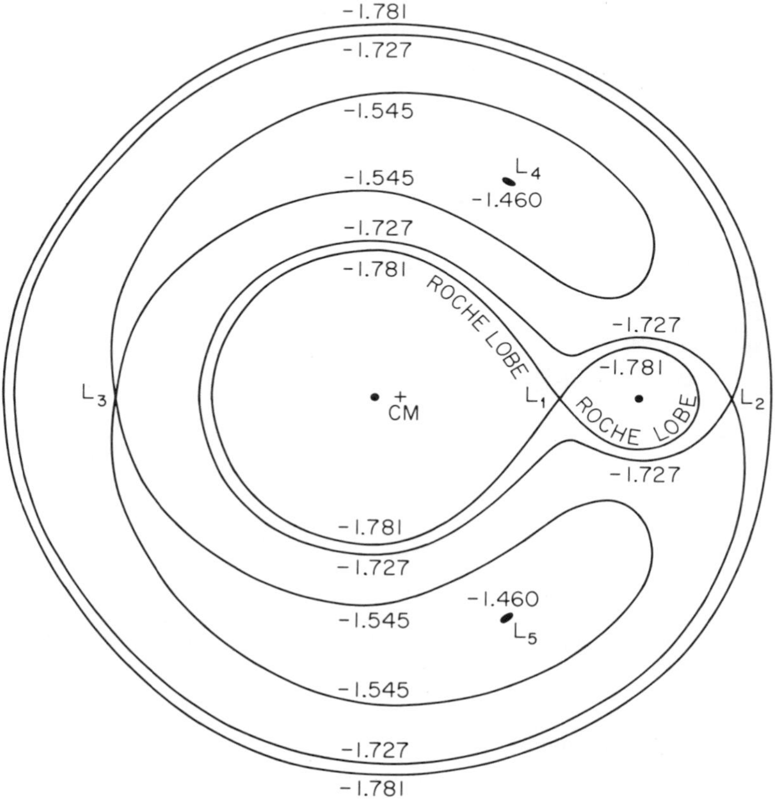

Figure 5.2.1. The equipotentials Φ_{gc} = const. for the Newtonian-plus-centrifugal potential in the orbital plane of a binary star system with a circular orbit. For the case shown here the stars have a mass ratio $M_N{:}M_c = 10{:}1$. The equipotentials are labelled by their values of Φ_{gc} measured in units of $(M_N + M_c)/a$, where a is the separation of the centers of mass of the two stars. The innermost equipotential shown is the "Roche lobe" of each star. Inside each Roche lobe, but outside the stellar surface, the potential $\tilde{\Phi}$ is dominated by the "Coulomb" $(1/r)$ field of the star, so the equipotentials are nearly spheres. The potential Φ_{gc} has local stationary points ($\nabla\Phi_{gc} = 0$), called Lagrange points", at the locations marked L_J.

It is instructive to figure out why, even though L_4 and L_5 are maxima of the potential, they are stable points for test-particle orbits. [Particles placed at L_4 or L_5, when perturbed slightly, go into stable orbits about L_4 or L_5. The key to this stability is the role of Coriolis forces; see eq. (5.2.4).] The "trojan asteroids" orbit the Lagrange points L_4 and L_5 of the Sun-Jupiter system.

gets deposited into the disk. Other incoming gas is fed angular momentum from the disk by means of viscous stresses, and thereby gets ejected out of the disk region and back onto the normal star or through the Lagrange point L_2 into interstellar space.

This qualitative picture of the deposition of gas into the disk and removal of angular momentum from the disk is based on qualitative examination of the

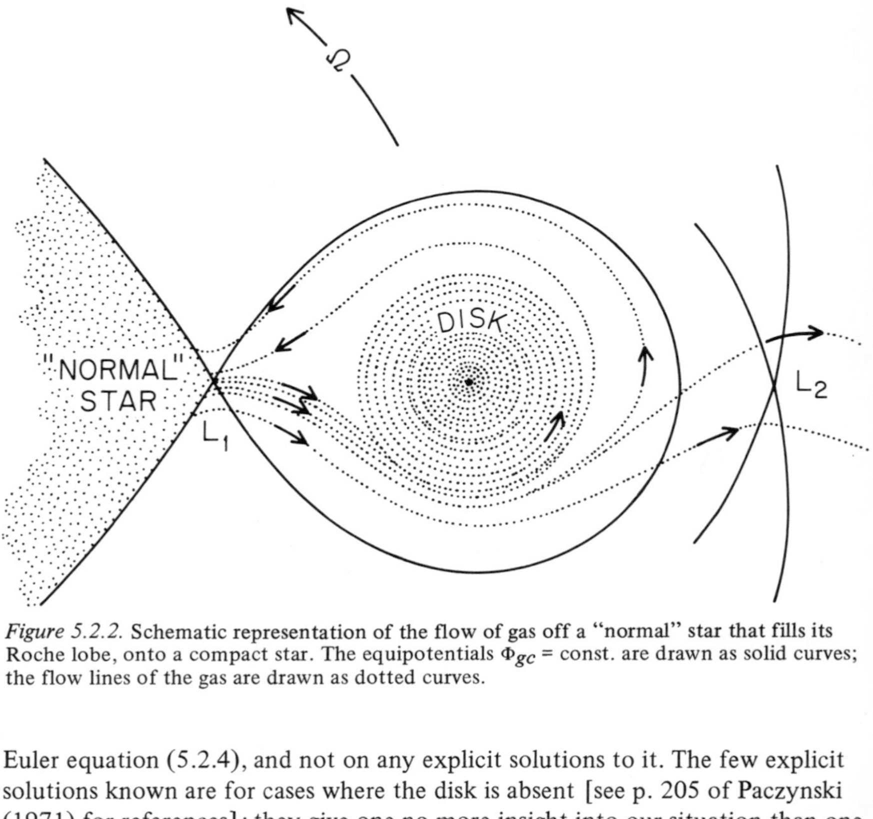

Figure 5.2.2. Schematic representation of the flow of gas off a "normal" star that fills its Roche lobe, onto a compact star. The equipotentials Φ_{gc} = const. are drawn as solid curves; the flow lines of the gas are drawn as dotted curves.

Euler equation (5.2.4), and not on any explicit solutions to it. The few explicit solutions known are for cases where the disk is absent [see p. 205 of Paczynski (1971) for references]; they give one no more insight into our situation than one gets from qualitative considerations.

One can analyze quantitively the steady-state structure of the disk, even though one has no quantitative understanding of the deposition process. The key point is that most of the gravitational energy is released and most of the luminosity is emitted from the inner parts of the disk $|\mathbf{r} - \mathbf{r}_c| \ll a$; whereas the deposition of gas and removal of angular momentum occur in the outer parts. Sections 5.4–5.11 will be devoted to a detailed analysis of the inner disk structure. But before launching into that analysis, we shall examine the structure qualitatively.

Viscous stresses in the orbiting disk gradually remove angular momentum from each gas element, permitting it to gradually spiral inward toward the compact star. The angular momentum removed is transported by the viscous stresses from the inner parts of the disk to the outer parts, and is then picked up and carried away by passing gas (see above). At the same time, the shearing orbital motion of the gas in the inner parts of the disk, acting against the viscous stresses, produces heat ("frictional heating"). Much more heat is generated than the gas can store, so most of the heat gets radiated away from the top and bottom faces of the disk.

The total energy radiated by a unit mass of gas during its passage inward through the disk must equal, approximately,† the gravitational binding energy of the unit mass when it reaches the inner edge of the disk, $\tilde{E}_{\text{bind}}$. For a white dwarf or neutron star the inner edge of the disk is near the star's surface, so

$$\begin{aligned} \tilde{E}_{\text{bind}} &\simeq \tfrac{1}{2}(M_c/R_c) \\ &\simeq 10^{-4} \text{ for white dwarf} \\ &\simeq 0.05 \text{ for neutron star.} \end{aligned} \tag{5.2.9a}$$

For a black hole the inner edge of the disk is at the last stable circular orbit, which has

$$\begin{aligned} \tilde{E}_{\text{bind}} &\simeq 0.057 \text{ for nonrotating hole} \\ \tilde{E}_{\text{bind}} &\simeq 0.42 \text{ for maximally rotating hole.} \end{aligned} \tag{5.2.9b}$$

Thus, in order of magnitude the total luminosity of the disk must be

$$L \sim 10^{-4}\dot{M}_0 \sim (10^{34} \text{ ergs/sec})\left(\frac{\dot{M}_0}{10^{-9}M_\odot/\text{yr}}\right) \text{ for white dwarf,} \tag{5.2.10}$$

$$L \sim 0.1\,\dot{M}_0 \sim (10^{37} \text{ ergs/sec})\left(\frac{\dot{M}_0}{10^{-9}M_\odot/\text{yr}}\right) \text{ for neutron star or hole.}$$

Here $\dot{M}_0$ is the accretion rate. In the case of a neutron star or white dwarf, the gas hits the star (or its magnetic field) with a kinetic energy roughly equal to the binding energy $\tilde{E}_{\text{bind}}$. In a steady state the energy liberated by this collision must all get radiated away, so the star itself will emit a luminosity of the same order as that emitted by the disk. In the case of a black hole, by contrast, the disk is the only source of radiation.

If the total luminosity approaches the "Eddington limit"

$$L_{\text{crit}} \simeq (1 \times 10^{38} \text{ ergs/sec})(M/M_\odot) \tag{5.2.11a}$$

---

† "Approximately" because the compact star itself can feed energy into the disk by means of viscous stresses, and that energy can subsequently be converted to heat and radiated away; see below.

(eq. 4.5.4), then radiation pressure will destroy the disk, and the general nature of the accretion will be quite different from that depicted above. For details see §5.13, below. We shall concentrate, until §5.13, on the "subcritical case" $L \ll L_{\rm crit}$, with a well-defined, thin disk. Note that this subcritical case corresponds to an accretion rate of

$$\begin{aligned}\dot{M}_0 \ll \dot{M}_{0\,\rm crit} &\sim 10^{-5} M_\odot/\mathrm{yr} \text{ for white dwarf} \\ &\sim (10^{-8} M_\odot/\mathrm{yr})(M/M_\odot) \text{ for neutron star or black hole.}\end{aligned} \tag{5.2.11b}$$

For a semiquantitative, Newtonian analysis of the subcritical disk structure, introduce an inertial frame centered on the compact star and ignore the tidal gravitational forces of the normal star. Let the rate at which gas is deposited into the disk, $\dot{M}_0$ ("accretion rate"), be given.

The specific angular momentum of the gas at radius r is

$$\tilde{L} = (M/r^3)^{1/2} r^2 = (Mr)^{1/2}, \tag{5.2.12}$$

where M—previously denoted M_c—is the mass of the compact star. Since the radius of the compact star is far less than that of the outer edge of the disk, a gas element must lose nearly all of its initial angular momentum by the time it reaches the star. Thus, the rate at which angular momentum is removed from the disk by passing gas must be

$$\begin{aligned}\dot{J} &= \dot{M}_0 \times (\tilde{L} \text{ evaluated at outer edge of disk, } r_0) \\ &= \dot{M}_0 (M r_0)^{1/2}.\end{aligned} \tag{5.2.13}$$

In a steady state the accretion rate and the outer edge of the disk will adjust themselves until this relation is satisfied, with angular momentum continually being removed from the outer edge by passing gas.

Return to the inner regions of the disk. Let ρ_0 be the mass density, let $2h$ be the disk thickness, and let $\Sigma = 2h\rho_0$ be the surface density (g/cm^2) at radius r. Let $v^{\hat{r}}$ be the radial velocity of the gas. (Note: $v^{\hat{r}} < 0$.) The orbital velocity $v^{\hat{\phi}}$ and angular velocity Ω are

$$v^{\hat{\phi}} = \Omega r = (M/r)^{1/2}. \tag{5.2.14}$$

The viscosity will never get strong enough to allow large radial velocities; i.e., $|v^{\hat{r}}|$ will always be far smaller than $v^{\hat{\phi}}$. Let $t_{\hat{\phi}\hat{r}}$ be the viscous stress (components relative to an orthonormal frame at radius r). This stress is related to the shear of the circular Keplerian orbits by

$$t_{\hat{\phi}\hat{r}} = -2\eta\sigma_{\hat{\phi}\hat{r}}, \tag{5.2.15}$$

$$\sigma_{\hat{\phi}\hat{r}} = -\tfrac{3}{4}\Omega = -\tfrac{3}{4}(M/r^3)^{1/2}, \tag{5.2.16}$$

where η is the coefficient of dynamic viscosity (see §2.5). The sources of viscosity will be discussed below. Let F be the flux of radiation (ergs/cm^2 sec) off the upper face of the disk. An equal flux F will come off the lower face.

The steady-state structure of the disk is governed by four conservation laws (conservation of mass, angular momentum, energy, and vertical momentum); by the nature of the viscosity–i.e., value of η–and by the law of radiative transfer from the inside of the disk to its surface.

Rest-mass conservation

Mass must flow across a cylinder of radius r at a rate equal to the accretion rate

$$-2\pi r\, \Sigma\, v^{\hat{r}} = \dot{M}_0. \tag{5.2.17}$$

Angular-momentum conservation

The rate at which angular momentum is carried inward across radius r by inflowing gas must equal the rate at which viscous stresses carry angular momentum out of radius r, plus the rate $\dot{J}_c$ at which angular momentum is deposited in the compact star

$$\dot{M}_0(M\hat{r})^{1/2} = 2\pi r \cdot 2h \cdot t_{\hat{\phi}\hat{r}} \cdot r + \dot{J}_c$$

The specific angular momentum $\tilde{L}_c$ deposited into the compact star cannot exceed the Keplerian angular momentum at the inner edge of the disk r_I; hence

$$\dot{J}_c = \beta \dot{M}_0 (M r_I)^{1/2} \quad \text{for some} \quad |\beta| \leqslant 1.$$

Solving for the product of stress and disk thickness, we obtain

$$2ht_{\hat{\phi}\hat{r}} = \frac{\dot{M}_0}{2\pi r^2}\left[(Mr)^{1/2} - \beta (M r_I)^{1/2}\right].$$

$$\simeq \frac{\dot{M}_0 (Mr)^{1/2}}{2\pi r^2} \quad \text{for} \quad r \gg r_I. \tag{5.2.18}$$

Notice that in a steady state the product $2ht_{\hat{\phi}\hat{r}}$ is determined uniquely by the accretion rate and the mass of the hole. If, at some radius, $2ht_{\hat{\phi}\hat{r}}$ is (temporarily) smaller than (5.2.18), then the viscous stresses there will not be sufficient to handle the mass flow $\dot{M}_0$. As a result, mass will accumulate in the deviant region, increasing thereby the disk thickness h and/or the stresses $t_{\hat{\phi}\hat{r}}$ there, and returning the deviant region to a steady-state structure. For a quantitative analysis of this "approach to steady state" see Thorne (1973b).

Energy conservation

Heat is generated in the disk by viscosity at a rate, per unit volume, given by

$$\epsilon = 2\eta \boldsymbol{\sigma}^2 = 4\eta(\sigma_{\hat{\phi}\hat{r}})^2 = -2t_{\hat{\phi}\hat{r}}\sigma_{\hat{\phi}\hat{r}}$$

(see §2.5). Thus, the heat generated per unit area is

$$2h\epsilon = (2ht_{\hat{\phi}\hat{r}})(-2\sigma_{\hat{\phi}\hat{r}}) = \frac{3\dot{M}_0}{4\pi r^2}\frac{M}{r}\left[1 - \beta\left(\frac{r_I}{r}\right)^{1/2}\right];$$

and the heat generated between radii r_1 and r_2 is

$$\int_{r_1}^{r_2} 2h\epsilon 2\pi r\, dr = \tfrac{3}{2}\dot{M}_0 \left\{\frac{M}{r_1}\left[1 - \frac{2}{3}\beta\left(\frac{r_I}{r_1}\right)^{1/2}\right] - \frac{M}{r_2}\left[1 - \frac{2}{3}\beta\left(\frac{r_I}{r_2}\right)^{1/2}\right]\right\}$$

$$\simeq \frac{3}{2}\dot{M}_0\left(\frac{M}{r_1} - \frac{M}{r_2}\right) \quad \text{for} \quad r_2 > r_1 \gg r_I.$$

It is instructive to examine the origin of this heat energy by asking at what net rate energy is being deposited in the region $r_1 < r < r_2$. Gravitational potential energy is released at a rate $\dot{M}_0(M/r_1 - M/r_2)$; but only half of this energy can go into heat. The other half must go into orbital kinetic energy (virial theorem). Thus, the rate at which gravitational energy gets converted into heat is

$$\tfrac{1}{2}\dot{M}_0(M/r_1 - M/r_2).$$

The viscous stresses transport outward not only angular momentum, but also energy. The rate at which energy is transported across radius r is

$$\dot{E} = \Omega \dot{J} = \Omega(2\pi r \cdot 2h \cdot t_{\hat{\phi}\hat{r}} \cdot r) = \dot{M}_0(M/r)[1 - \beta(r_I/r)^{1/2}].$$

Thus, energy is deposited by viscosity between radii r_1 and r_2 at a rate

$$\dot{M}_0\left\{\frac{M}{r_1}\left[1 - \beta\left(\frac{r_I}{r_1}\right)^{1/2}\right] - \frac{M}{r_2}\left[1 - \beta\left(\frac{r_I}{r_2}\right)^{1/2}\right]\right\}.$$

This energy deposition rate plus the deposition rate for gravitational energy is equal to the total heating rate. Notice that at radii $r \gg r_I$, gravity accounts directly for only one-third of the heat. The remaining two-thirds is transported into the heating region by viscous stresses. Much of the early literature on disk accretion, e.g., Lyndon-Bell (1969), failed to take account of energy transport by viscous stresses, and therefore underestimated by a factor 3 the heating at $r \gg r_I$.

If none of the heat were radiated away, the thermal energy of the disk would be 3/2 times its gravitational potential energy; and its temperature would thus be

$$T \sim (M/r)m_p \sim (10^{13}\ \text{K})(r/10^5\ \text{cm})^{-1}(M/M_\odot).$$

Such temperatures are absurdly high. Thermal bremsstrahlung and other radiative processes will never permit such temperatures to arise. Instead, they will quickly remove almost all of the heat that is generated. This means that the total flux $2F$ from the top and bottom faces of the disk must equal the heating rate per unit area, as calculated above (the heat flows out of the disk vertically rather than radially because the disk is so thin):

$$F = \frac{3\dot{M}_0 M}{8\pi r^2 r}\left[1 - \beta\left(\frac{r_I}{r}\right)^{1/2}\right]. \tag{5.2.19}$$

The total power radiated is thus

$$L = \int_{r_I}^{\infty} 2F \cdot 2\pi r\, dr = (\tfrac{3}{2} - \beta)\dot{M}_0 \frac{M}{r_I}. \tag{5.2.20}$$

Of this total power radiated, $\frac{1}{2}\dot{M}_0 M/r_I$ comes from gravity, while $(1 - \beta)\dot{M}_0 M/r_I$ comes from the rotational energy of the star.

Spectrum

A detailed treatment of the spectrum radiated will be given in §5.10. Here we only note that, if the radiation is blackbody, then the surface temperature of the disk at radius r must be

$$T_s = \left(\frac{4F}{b}\right)^{1/4} \simeq (3 \times 10^7\,\mathrm{K})\left(\frac{\dot{M}_0}{10^{-9}M_\odot/\mathrm{yr}}\right)^{1/4}\left(\frac{M}{M_\odot}\right)^{-1/2}\left(\frac{M}{r}\right)^{3/4} \times \left[1 - \beta\left(\frac{r_I}{r}\right)^{1/2}\right]. \tag{5.2.21}$$

Because most of the radiation is emitted from $r/M \sim 10$ for a black hole or neutron star, and from $r/M \sim 10^4$ for a dwarf, and because the black-body spectrum peaks at a photon energy of

$$h\nu_{\max} \simeq (2.44 \times 10^{-4}\,\mathrm{eV})(T_s/°\mathrm{K}), \tag{5.2.22}$$

the spectrum of the total radiation from the disk should peak at

$$h\nu_{\max} \simeq (1\,\mathrm{keV})\left(\frac{\dot{M}_0}{10^{-9}M_\odot/\mathrm{yr}}\right)^{1/4}\left(\frac{M}{M_\odot}\right)^{-1/2} \quad \text{for neutron star or hole}$$

$$h\nu_{\max} \simeq (0.01\,\mathrm{keV})\left(\frac{\dot{M}_0}{10^{-9}M_\odot/\mathrm{yr}}\right)^{1/4} \quad \text{for white dwarf.} \tag{5.2.23}$$

Thus, for reasonable accretion rates the disk can emit strongly in the X-ray region (~1 to 10 keV); but the X-ray spectrum should fall fairly rapidly with increasing energy. It actually turns out that electron-scattering opacity impedes the emission of blackbody radiation, and causes the spectrum to peak at energies higher than that (5.2.23). (See §5.10.) However, the above estimates are still accurate to within a factor ~10.

Vertical pressure balance

The thickness of the disk is governed by a balance between vertical pressure force and the tidal gravitational force of the compact star ("vertical momentum conservation"):

$$\frac{dp}{dz} = \rho_0 \times (\text{"acceleration of gravity"}) = \rho_0 \frac{Mz}{r^3}. \tag{5.2.24}$$

The approximate solution to this equation is

$$h \simeq (p/\rho_0)^{1/2}(r^3/M)^{1/2} \simeq c_s/\Omega. \tag{5.2.25}$$

Here h is the half-thickness of the disk, c_s is the speed of sound in the gas, and $\Omega = (M/r^3)^{1/2}$ is the angular velocity of the gas.

Sources of viscosity

All of the above conclusions are based on conservation laws alone. To proceed further, one must make some assumption about the magnitude of the viscosity. The dominant sources of viscosity are probably chaotic magnetic fields, and turbulence in the gas flow.

For cases of interest the normal star (typically a B0 supergiant) could have a surface magnetic field of $B_s \sim 100$ gauss. Field lines will be dragged, by the flowing gas, off of the normal star and into the disk. The deposited field should be rather chaotic because there is no preferred direction for the field in the gas that flows off of the normal star. Turbulence, if present in the disk, will also make the field chaotic. Once the chaotic field has been deposited in the disk, the shear of the gas flow will magnify it at a rate (eq. (2.5.20))

$$dB_{\hat\varphi}/d\tau = \sigma_{\hat\varphi\hat r} B_{\hat r} \simeq \Omega B_{\hat r}, \qquad dB_{\hat r}/d\tau = 0 \tag{5.2.26}$$

corresponding to an increase of $B_{\hat\varphi}$ by the amount $B_{\hat r}$ with every circuit around the compact star. This growth of field will be counterbalanced, at least in part, by reconnection of field lines at the interfaces between chaotic cells (see §2.9), and perhaps also by a bulging of field lines out of the disk, pinch-off of field lines, and escape of "magnetic bubbles." From a macroscopic viewpoint (scale large compared to chaotic cells) the effects of the field can be described by pressure and viscosity. The magnetic pressure will be of order $B^2/8\pi$, where B is the mean field strength. Because shearing of the chaotic field will tend to string it out along the ϕ-direction, its shear stress $t_{\hat\phi\hat r}$ will be somewhat (perhaps a factor 10?) smaller than its pressure

$$t_{\hat\phi\hat r}^{(\text{mag})} < p^{(\text{mag})} = B^2/8\pi. \tag{5.2.27}$$

The magnetic pressure cannot exceed the thermal pressure

$$p^{(\text{mag})} \lesssim p^{(\text{therm})} \simeq \rho_0 c_S^2; \tag{5.2.28}$$

otherwise the field lines would bulge out of the disk, reconnect, and escape ("bubbles"). Thus, the magnetic viscous stresses will satisfy

$$t_{\hat{\phi}\hat{r}}^{(\text{mag})} \lesssim p \simeq \rho_0 c_S^2. \tag{5.2.29}$$

(Here c_S is the speed of sound, and p is the total pressure.)

The gas flow in the disk may well be turbulent. The coefficient of dynamic viscosity associated with turbulence is

$$\eta \simeq \rho_0 v_{\text{turb}} l_{\text{turb}}, \tag{5.2.30}$$

where v_{turb} is the speed of the turbulent motions relative to the mean rest frame of the gas, and l_{turb} is the characteristic size of the largest turbulent cells [see, e.g., §31 of Landau and Lifshitz (1959)]. If the turbulent speed ever exceeds the sound speed, then shocks develop and quickly convert the turbulent energy into heat. Thus, $v_{\text{turb}} \lesssim c_S$. The turbulent scale is limited by the disk thickness, $l_{\text{turb}} \lesssim h$. Consequently, the shear stress due to turbulence is bounded by

$$t_{\hat{\phi}\hat{r}}^{(\text{turb})} \simeq \eta \sigma_{\hat{\phi}\hat{r}} \lesssim (\rho_0 c_S h)\Omega \simeq \rho_0 c_S^2 \simeq p. \tag{5.2.31}$$

Inequalities (5.2.29) and (5.2.31) on the magnetic and turbulent stresses are identical. They suggest that one dump one's lack of knowledge about the true magnitude of the viscosity into a single parameter α defined by

$$t_{\hat{\phi}\hat{r}} = \alpha p. \tag{5.2.32}$$

Someday, perhaps ten years hence, when one understands the magnetic and turbulent viscosities better, one can insert into the formalism a reliable value of α. In the meantime one only knows that

$$\alpha \lesssim 1. \tag{5.2.33}$$

If $\alpha \simeq 1$, the disk will have a rather mottled structure on scales of order h; if $\alpha \ll 1$ it will be rather smooth on such scales.

Radiative transport

The heat generated by viscosity must be transported vertically to the surface of the disk before it can be radiated. The disk turns out to be optically thick ($\tau_{ff} + \tau_{es} \gg 1$ at $z = 0$). Hence, one calculates the energy transport using the diffusion approximation [eq. (2.6.43) reduced to Newtonian form]:

$$\frac{d}{dz}\left(\tfrac{1}{3} b T^4\right) = \bar{\kappa} \rho_0 q^{\hat{z}}$$

The approximate solution to this equation of transport is

$$b T^4 \simeq \bar{\kappa}\, \Sigma F. \tag{5.2.34}$$

Opacity and equation of state

The dominant source of opacity in the outer parts of the disk will be free-free transitions, and also (of comparable but not much larger magnitude) bound-free transitions and lines. Thus, in the outer regions one must take

$$\bar{\kappa} \simeq \bar{\kappa}_{ff} \simeq 0.64 \times 10^{23} \left(\frac{\rho_0}{\text{g cm}^{-3}}\right)\left(\frac{T}{^\circ\text{K}}\right)^{-7/2} \text{cm}^2/\text{g} \tag{5.2.35a}$$

[eq. (2.6.46)]. In the inner regions, where the temperature is higher, electron scattering is the dominant source of opacity, so

$$\bar{\kappa} \simeq \bar{\kappa}_{es} \simeq 0.40 \text{ cm}^2/\text{g} \tag{5.2.35b}$$

[eq. (2.6.47)]. Throughout most of the disk gas pressure dominates over radiation pressure, so

$$c_S^2 \simeq p/\rho_0 \simeq p^{(\text{gas})}/\rho_0 \simeq T/m_p \simeq T/10^{13}\,\text{K}. \tag{5.2.36a}$$

But in the innermost regions of the disk, where the temperature is particularly high, radiation pressure dominates

$$c_S^2 \simeq p/\rho_0 \simeq p^{(\text{rad})}/\rho_0 \simeq \tfrac{1}{3} bT^4/\rho_0. \tag{5.2.36b}$$

Summary

The steady-state structure of the disk is governed by the law of mass conservation (5.2.17), the law of angular momentum conservation (5.2.18), the law of energy conservation (5.2.19), the law of vertical pressure balance (5.2.25), the magnitude of the viscosity (5.2.32), the law of radiative transport (5.2.34), the magnitude of the opacity (5.2.35), the equation of state (5.2.36), and the relation $\Sigma = 2\rho_0 h$. It is straightforward but tedious to combine these equations algebraically and obtain explicit expressions for the structure of the disk. The resulting explicit expressions are given, along with relativistic corrections, in §5.9 below. The reader can examine them now, or he can ignore them until after studying the relativistic theory of the disk structure (§§5.4-5.8).

If $\alpha \simeq 1$, one expects rather large deviations from the steady-state structure on length scales of the order of the disk thickness. We shall discuss such ‘ mottling" and flares in §5.12.

5.3 Accretion in Galactic Nuclei: The General Picture

When one builds models for quasars and for the violent activity observed in the nuclei of some galaxies, one typically is forced to invoke a strong energy source confined to a small region. Whether one invokes as the energy source a dense cluster of stars, a single supermassive star, a large black hole, or some combination of these, one is led to speculate that it will evolve toward a black-hole—or will explode completely—in a time less than the age of the universe. [See, e.g.,

Lynden-Bell (1969) and Gold, Axford, and Ray (1965) for details.] Thus, it becomes attractive to suppose that when a quasar dies, or when the violence in a galactic nucleus terminates, one remnant left behind is a supermassive black hole. In the case of a quasar the hole might have a mass of $\sim 10^{10}$ to $10^{11} M_\odot$. In the case of a "normal" galaxy such as as our own, it might have a mass of $\sim 10^7$ to $10^8 M_\odot$. (These numbers are based on the total energies involved in the quasar phenomenon and in galactic outbursts.) Moreover because (i) violent activity and quasars are fairly abundant in our universe [see Lynden-Bell (1969) for details], and (ii) because quasars may well reside in the nuclei of galaxies, it is quite possible that a large fraction of all galaxies once supported violent nuclear activity and now possess black holes of masses $\sim 10^7$ to $\sim 10^{11} M_\odot$.

Such a large black hole, residing at the center of a galaxy, will accrete gas from its surroundings. The accreted gas, like the galaxy itself, will typically have specific angular momentum $\tilde{L}$ far larger than that for a circular orbit near the horizon of the hole, $\tilde{L} \gg r_g c$. Consequently, the gas will form an orbiting disk about the hole with a structure similar to that for disk accretion in binary systems.

It is difficult to estimate the accretion rate for a supermassive hole in the center of a galaxy, because one knows so little about the ambient conditions there. The accretion rate would presumably not exceed the rate at which all stars in the galaxy shed mass into the interstellar medium ($\sim 1 M_\odot$/yr for our Galaxy). But it might be much less than this. One can use observational data to place a limit on the accretion rate. If the total power output from the center of a galaxy is L, and if all of that power is supplied by accretion onto a hole with ~ 10 per cent efficiency for converting mass into energy, then $\dot{M}_0 \lesssim 10L$. For the most violent of quasars, $L \simeq 10^{47}$ ergs/sec, so $\dot{M}_0 \lesssim 10 M_\odot$/yr. For the nucleus of our own Galaxy, $L \simeq 10^{42}$ ergs/sec, so $\dot{M}_0 \lesssim 10^{-4} M_\odot$/yr.

The equations of structure for an accreting disk around a supermassive hole in the nucleus of a galaxy are the same as for the binary accretion problem of the last section. In particular, if the disk is optically thick and radiates as a blackbody, then it will emit most of its radiation near the frequency (5.2.23)–which we rewrite as

$$\nu_{\max} \simeq (1 \times 10^{15}\ \mathrm{Hz}) \left(\frac{\dot{M}_0}{10^{-3} M_\odot/\mathrm{yr}}\right)^{1/4} \left(\frac{M}{10^8 M_\odot}\right)^{-1/2} \tag{5.3.1}$$

Thus, the radiation will be concentrated largely in the ultraviolet and optical region of the spectrum. If the disk cannot build up a blackbody spectrum, either because of optical thinness or because of high electron-scattering opacity (see §5.10 for details), then the radiation will be concentrated at somewhat higher frequencies than (5.3.1).

The strong optical and UV radiation ($\sim 10^{42}$ ergs/sec to $\sim 10^{47}$ ergs/sec, depending on the accretion rate) probably cannot escape from the neighborhood of the disk. The accreting gas presumably contains dust, and radiation pressure is likely to expell the dust from the disk. As a result, a thick cloud of dust may build up around the disk. Such a cloud would absorb the optical and ultraviolet radiation and would reemit it in the far infrared. [See Lynden-Bell and Rees (1971) for details.] Hence, as seen from Earth the hole would be a strong source of infrared radiation ($L \sim 10^{42}$ to 10^{47} ergs/sec depending on accretion rate). Just such infrared sources are observed in the nucleus of our Galaxy ($L \sim 3 \times 10^{41}$ ergs/sec), in the nuclei of many other galaxies, and in quasars. Unfortunately, however, an accreting hole is not the only type of object that can generate such radiation. Any strong source of optical or ultraviolet radiation, surrounded by dust, will emit strongly in the infrared.

We shall see in §5.12 that the disk around a supermassive hole may also emit a significant flux of radio waves, which propagate to Earth relatively freely.

5.4 Properties of the Kerr Metric Relevant to Accreting Disks

Before examining the details of the structure of an accreting disk (§5.9), we shall extend our equations of structure from Newtonian theory to general relativity. In this extension, we shall assume that the spacetime geometry outside the hole is that of Kerr, and we shall assume that the disk lies in the equatorial plane of the Kerr metric.

To treat the structure of such a disk, we shall need a number of properties of "direct" circular orbits in the equatorial plane of the Kerr metric (orbits that rotate in the same direction as the black hole). Most of our formulas are taken from Bardeen, Press and Teukolsky (1972), or Bardeen (1973)—or are readily derivable from results quoted there.

For simplicity in splitting formulas into Newtonian limits plus relativistic corrections, we shall introduce the following functions with value unity far from the hole:

$$\mathscr{A} \equiv 1 + a_*^2/r_*^2 + 2a_*^2/r_*^3, \tag{5.4.1a}$$

$$\mathscr{B} \equiv 1 + a_*/r_*^{3/2}, \tag{5.4.1b}$$

$$\mathscr{C} \equiv 1 - 3/r_* + 2a_*/r_*^{3/2}, \tag{5.4.1c}$$

$$\mathscr{D} \equiv 1 - 2/r_* + a_*^2/r_*^2, \tag{5.4.1d}$$

$$\mathscr{E} \equiv 1 + 4a_*^2/r_*^2 - 4a_*^2/r_*^3 + 3a_*^4/r_*^4, \tag{5.4.1e}$$

$$\mathscr{F} \equiv 1 - 2a_*/r_*^{3/2} + a_*^2/r_*^2, \tag{5.4.1f}$$

$$\mathscr{G} \equiv 1 - 2/r_* + a_*/r_*^{3/2}, \tag{5.4.1g}$$

$$\mathscr{J} \equiv \exp\left[-\tfrac{3}{2} \int_{r_*}^{\infty} \mathscr{B}^{-1}\mathscr{C}^{-1}\mathscr{F} r_*^{-2}\, dr_*\right], \tag{5.4.1h}$$

$$\mathscr{L} \equiv \frac{\tilde{L} - \tilde{L}_{ms}}{(\mathrm{Mr})^{1/2}} = \frac{\mathscr{F}}{\mathscr{C}^{1/2}} - \frac{\tilde{L}_{ms}}{(Mr)^{1/2}}, \tag{5.4.1i}$$

$$\mathcal{Q} \equiv \mathcal{L} - \frac{3}{2r_*^{1/2}} \mathcal{I} \int_{r_{ms_*}}^{r_*} \frac{\mathcal{F}\mathcal{L}}{\mathcal{B}\mathcal{C}^- \mathcal{I}} \frac{dr_*}{r_*^{3/2}}. \tag{5.4.1j}$$

Here M and a are the mass and specific angular momentum of the black hole; r is radius; r_* and a_* are dimensionless measures of r and a

$$r_* \equiv r/M, \qquad a_* \equiv a/M;$$

$\tilde{L}_{ms}$ is a constant defined below, and $\tilde{L}$ is a function of r defined below.

The properties of direct circular orbits that we shall need are the following. (i) Form of Kerr metric in and near equatorial plane ($|\theta - \pi/2| \ll 1$):

$$ds^2 = -\frac{r^2\Delta}{A} dt^2 + \frac{A}{r^2}(d\phi - \omega\, dt)^2 + \frac{r^2}{\Delta} dr^2 + dz^2 \tag{5.4.2a}$$

$$\Delta \equiv r^2 - 2Mr + a^2 = r^2\mathcal{D}, A \equiv r^4 + r^2a^2 + 2Mra^2 = r^4\mathcal{A}$$

$$\omega = 2Mar/A = (2Ma/r^3)\,\mathcal{A}^{-1}. \tag{5.4.2b}$$

[We have replaced the usual angular coordinate θ by $z = r\cos\theta \simeq r(\theta - \pi/2)$.] Note that the square root of the determinant of $g_{\alpha\beta}$ is $(-g)^{1/2} = r$. (ii) Angular velocity of orbit:

$$\Omega = \frac{d\phi}{dt} = \frac{M^{1/2}}{r^{3/2} + aM^{1/2}} = \frac{M^{1/2}}{r^{3/2}} \frac{1}{\mathcal{B}}. \tag{5.4.3}$$

(iii) Linear velocity of orbit relative to "locally nonrotating observer":

$$\mathcal{V}_{(\phi)} = \frac{A}{r^2\Delta^{1/2}}(\Omega - \omega) = \frac{M^{1/2}}{r^{1/2}} \frac{\mathcal{F}}{\mathcal{D}^{1/2}\mathcal{B}} \tag{5.4.4a}$$

(iv) "γ-factor" corresponding to this linear velocity:

$$\gamma = (1 - \mathcal{V}_{(\phi)}^2)^{-1/2} = \frac{\mathcal{B}\,\mathcal{D}^{1/2}}{\mathcal{A}^{1/2}\mathcal{C}^{1/2}}. \tag{5.4.4b}$$

(v) Orthonormal frame attached to an orbiting particle with 4-velocity u ("*orbiting frame*"):

$$\mathbf{e}_{\hat{0}} = \gamma(A/r^2\Delta)^{1/2}\left(\frac{\partial}{\partial t} + \Omega\frac{\partial}{\partial\phi}\right) = \frac{\mathcal{B}}{\mathcal{C}^{1/2}}\left(\frac{\partial}{\partial t} + \frac{M^{1/2}}{r^{3/2}}\frac{1}{\mathcal{B}}\frac{\partial}{\partial\phi}\right),$$

$$\mathbf{e}_{\hat{\phi}} = \gamma\left(\frac{r^2}{A}\right)^{1/2}\frac{\partial}{\partial\phi} + \gamma\mathcal{V}_{(\phi)}\left(\frac{A}{r^2\Delta}\right)^{1/2}\left(\frac{\partial}{\partial t} + \omega\frac{\partial}{\partial\phi}\right) \tag{5.4.5a}$$

$$= \frac{\mathcal{B}\mathcal{D}^{1/2}}{\mathcal{A}\mathcal{C}^{1/2}}\frac{1}{r}\frac{\partial}{\partial\phi} + \frac{\mathcal{F}}{\mathcal{C}^{1/2}\mathcal{D}^{1/2}}\left(\frac{M}{r}\right)^{1/2}\left(\frac{\partial}{\partial t} + \omega\frac{\partial}{\partial\phi}\right),$$

$$\mathbf{e}_{\hat{r}} = \frac{\Delta^{1/2}}{r}\frac{\partial}{\partial r} = \mathcal{D}^{1/2}\frac{\partial}{\partial r}, \qquad \mathbf{e}_{\hat{z}} = \frac{\partial}{\partial z}.$$

(vi) Corresponding othonormal basis of one-forms

$$\omega^{\hat{0}} = \frac{\mathscr{G}}{\mathscr{C}^{1/2}}\,\mathbf{d}t - \frac{\mathscr{F}}{\mathscr{C}^{1/2}}\left(\frac{M}{r}\right)^{1/2} r\,\mathbf{d}\phi,$$

$$\omega^{\hat{\phi}} = \frac{\mathscr{B}\mathscr{D}^{1/2}}{\mathscr{C}^{1/2}}\left[r\,\mathbf{d}\phi - \left(\frac{M}{r}\right)^{1/2}\frac{1}{\mathscr{B}}\,\mathbf{d}t\right], \tag{5.4.5b}$$

$$\omega^{\hat{r}} = \mathscr{D}^{-1/2}\,\mathbf{d}r, \qquad \omega^{\hat{z}} = \mathbf{d}z.$$

(vii) Shear of the congruence of circular, equatorial geodesics (congruence with 4-velocity $\mathbf{u} = \mathbf{e}_{\hat{0}}$)

$$\sigma^{(EG)}_{\hat{r}\hat{\phi}} = \sigma^{(EG)}_{\hat{\phi}\hat{r}} = \frac{1}{2}\frac{A}{r^3}\gamma^2\Omega_{,r} = -\frac{3}{4}\frac{M^{1/2}}{r^{3/2}}\frac{\mathscr{D}}{\mathscr{C}}, \tag{5.4.6}$$

all other $\sigma^{(EG)}_{\hat{\alpha}\hat{\beta}}$ vanish.

(viii) Angular momentum per unit mass for circular orbit:

$$\tilde{L} = u_\phi = M^{1/2} r^{1/2} \mathscr{F}/\mathscr{C}^{1/2}. \tag{5.4.7a}$$

(ix) Energy per unit mass for circular orbit:

$$\tilde{E} = |u_0| = \mathscr{G}/\mathscr{C}^{1/2}. \tag{5.4.7b}$$

(x) Minimum radius for stable circular orbits ("marginally stable" orbits) r_{ms}:

$$r_{ms}^2 - 6Mr_{ms} + 8aM^{1/2}r_{ms}^{1/2} - 3a^2 = 0,$$

$$r_{ms} = M\{3 + Z_2 - [(3 - Z_1)(3 + Z_1 + 2Z_2)]^{1/2}\}, \tag{5.4.8a}$$

$$Z_1 \equiv 1 + (1 - a^2/M^2)^{1/3}[(1 + a/M)^{1/3} + (1 - a/M)^{1/3}],$$

$$Z_2 \equiv (3a^2/M^2 + Z_1^2)^{1/2}. \tag{5.4.8b}$$

(xi) Angular momentum per unit mass for last stable circular orbit:

$$\tilde{L}_{ms} = \frac{2M}{3^{1/2}x}(3x - 2a), \qquad x = M^{1/2}r_{ms}^{1/2}. \tag{5.4.9}$$

5.5 Relativistic Model for Disk: Underlying Assumptions

In order to avoid confusion, we shall lay more careful foundations for our relativistic analysis of disk structure than we did for the Newtonian analysis of §5.2.

We shall split the calculation of disk structure into three parts: the radial structure (§5.6), the vertical structure (§§5.7, 5.8) and the propagation of radiation from the disk's surface to the observer.

In calculating the disk structure, we shall make the following assumptions and idealizations:

(i) The central plane of the disk coincides with the equatorial plane of the black hole.

(ii) The companion star in the binary system has negligible gravitational influence on the disk structure (assumption valid in inner part of disk; not at outer edges).

(iii) The disk is thin; i.e., its proper thickness $2h$, as measured in the orbiting frame of the disk's matter (same as proper thickness measured by any other observer who moves in the equatorial plane), satisfies

$$h(r) \ll r. \tag{5.5.1}$$

(iv) The disk is in a quasisteady state. This assumption can be made precise in terms of an averaging process. Pick a particular height $z(|z| \lesssim h \ll r)$ above the central plane of the disk, and pick a particular quantity Ψ (e.g., Ψ = density or temperature of 4-velocity of gas) to be measured at height z. Average Ψ over all ϕ (Lie dragging it along $\partial/\partial\phi$ while averaging, if it is a vector or tensor); average over a proper radial distance of order $2h$; and average over the time interval required for gas to move inward a distance $2h$:

$$\langle \Psi \rangle \equiv \frac{1}{2\pi \Delta r\, \Delta t} \int_{r-\Delta r/2}^{r+\Delta r/2} \int_{t-\Delta t/2}^{t+\Delta t/2} \int_{0}^{2\pi} \Psi \, d\phi \, dt \, dr; \tag{5.52}$$

$$|g_{rr}|^{1/2} \Delta r = 2h, \qquad (\mathscr{C}^{1/2}/\mathscr{B}) \bar{v}^{\hat{r}} \Delta t = 2h;$$

($\bar{v}^{\hat{r}}$ will be defined below). The assumption of a quasisteady state means that $\langle \Psi \rangle$ is time-independent—i.e.,

$$\partial \langle \Psi \rangle / \partial t = 0 \quad \text{if} \quad \langle \Psi \rangle \text{ is a scalar,}$$

$$\mathscr{L}_{\partial/\partial t} \langle \Psi \rangle = 0 \quad \text{if} \quad \langle \Psi \rangle \text{ is a vector or tensor,} \tag{5.5.3}$$

where $\mathscr{L}_{\partial/\partial t}$ is the Lie derivative along the Killing vector $\partial/\partial t$. The disk structure may be very far from steady state on scales $\lesssim h$; for example, it may have turbulence, flux reconnection, and flares on such scales. Such local violence will not invalidate the analysis that follows.

(v) When viewed "macroscopically" (turbulence removed by the above averaging process), the gas of the disk moves (very nearly) in direct, circular, geodesic orbits—i.e., with 4-velocity $\langle \mathbf{u} \rangle \cong \mathbf{e}_{\hat{0}}$. Superimposed on this orbital motion is a very small radial flow, produced by viscous stresses, and an even smaller flow in the vertical direction, as required by the variation of disk thickness with radius:

$$\langle \mathbf{u} \rangle = \frac{\mathbf{e}_{\hat{0}}}{[1 - (v^{\hat{r}})^2 - (v^{\hat{z}})^2]^{1/2}} + v^{\hat{r}} \mathbf{e}_{\hat{r}} + v^{\hat{z}} \mathbf{e}_{\hat{z}}, \tag{5.5.4}$$

$$|v^{\hat{z}}(r, z)| \ll |v^{\hat{r}}(r, z)| \ll \mathscr{V}_{(\phi)} \simeq (M/r)^{1/2}. \tag{5.5.5}$$

By assuming that the orbital motion of the gas is very nearly geodesic, we are automatically requiring that the gravitational pull of the hole dominates over radial pressure gradients and over shear stresses. In order of magnitude the gravitational pull of the hole ("gravitational acceleration") is the gradient of the gravitational binding energy, $(1 - \tilde{E})_{,r}$; and the acceleration due to pressure gradients and shears is $\sim (T^{\hat{j}\hat{k}}_{,r}/\rho_0) \sim (T^{\hat{j}\hat{k}}/\rho_0)_{,r}$. Thus, we are automatically demanding that the stress tensor in the fluid's rest frame ($\simeq$ orbiting orthonormal frame) satisfy

$$\frac{T^{\hat{j}\hat{k}}(r, z)}{\rho_0(r, z)} \ll 1 - \tilde{E}(r) \overset{N}{=} \frac{1}{2}\frac{M}{r}. \tag{5.5.6}$$

(Here $\overset{N}{=}$ means "equals in the Newtonian limit".)

Because the specific internal energy Π is approximately equal to $T^{\hat{j}\hat{k}}/\rho_0$, the assumption of nearly geodesic orbits implies the condition of "negligible specific heat":

$$\Pi(r, z) \ll 1 - \tilde{E}(r) \overset{N}{=} \frac{1}{2}\frac{M}{r}. \tag{5.5.7}$$

In words—the density of internal energy, $\rho_0 \Pi$ (including thermal energy of gas, energy of turbulent motions, and magnetic-field energy) is much smaller than the density of gravitational binding energy $\rho_0(1 - \tilde{E})$.

Notice that conditions (5.5.6) and (5.5.7) imply

$$\Pi \ll 1, \qquad T_{\hat{j}\hat{k}}/\rho_0 \ll 1. \tag{5.58}$$

at all r and z—even near the black hole. This means that, although general relativistic effects (spacetime curature; deviations from Newtonian gravity) are very important near the hole, one can everywhere ignore special relativistic corrections to the local thermodynamic, hydrodynamic, and radiative properties of the gas. Contrast this situation with the interior of a neutron star and the early stages of the Universe, where not only is gravity relativistic but so is the matter ($\Pi \gtrsim 1, p/\rho_0 \gtrsim 1, T_{\hat{j}\hat{k}}/\rho_0 \gtrsim 1$).

Turn now from the underlying assumptions of the model to the reference frames to be used in the calculations. Three reference frames will be needed: (i) The Boyer-Lindquist coordinate frame [(t, r, z, ϕ) inside and near disk; (t, r, θ, ϕ) when $|\theta - \pi/2|$ is not $\ll 1$; see eqs. (5.4.2a)]. (ii) The orbiting orthonormal frame $[\mathbf{e}_{\hat{0}}, \mathbf{e}_{\hat{r}}, \mathbf{e}_{\hat{z}}, \mathbf{e}_{\hat{\phi}})$; eqs. (5.4.5)]. (iii) The mean local rest-frame of the gas

$$\begin{aligned} &\mathbf{e}_{\tilde{0}} \equiv \langle \mathbf{u} \rangle \quad \text{as given by eq. (5.5.4)}, \\ &\mathbf{e}_{\tilde{\phi}} = \mathbf{e}_{\hat{\phi}}, \\ &\mathbf{e}_{\tilde{r}} = (\mathbf{e}_{\hat{r}} + v^{\hat{r}} \mathbf{e}_{\tilde{0}}) / | \mathbf{e}_{\hat{r}} + v^{\hat{r}} \mathbf{e}_{\tilde{0}} |, \\ &\mathbf{e}_{\tilde{z}} = (\mathbf{e}_{\hat{z}} + v^{\hat{z}} \mathbf{e}_{\tilde{0}} - v^{\hat{r}} v^{\hat{z}} \mathbf{e}_{\tilde{r}}) / | \mathbf{e}_{\hat{z}} + v^{\hat{z}} \mathbf{e}_{\tilde{0}} - v^{\hat{r}} v^{\hat{z}} \mathbf{e}_{\tilde{r}} |. \end{aligned} \tag{5.5.9}$$

The mean local rest frame is nearly identical to the orbiting orthonormal frame.

5.6 Equations of Radial Structure

The relativistic equations of radial structure for our disk will be expressed in terms of the following parameters: (i) The steady-state accretion rate $\dot{M}_0$; (ii) The surface density of the disk

$$\Sigma \equiv \int_{-h}^{+h} \langle \rho_0 \rangle \, dz, \tag{5.6.1a}$$

where ρ_0 is the density of rest mass as measured in the rest frame of the gas. (iii) The integrated shear stress

$$W \equiv \int_{-h}^{+h} \langle T_{\hat{\phi}\hat{r}} \rangle \, dz, \tag{5.6.1b}$$

where $T_{\hat{\phi}\hat{r}}$ is the component of the stress-energy tensor on the orbiting orthonormal basis vectors $\mathbf{e}_{\hat{\phi}}$ and $\mathbf{e}_{\hat{r}}$. (iv) The mass-averaged radial velocity of the gas

$$\bar{v}^{\hat{r}} \equiv (1/\Sigma) \int_{-h}^{+h} \langle v^{\hat{r}} \rho_0 \rangle \, dz. \tag{5.6.1c}$$

(v) The flux of radiant energy off the upper face of the disk (equal also to flux off lower face)

$$F \equiv \langle T^{\hat{0}\hat{z}}(z = h) \rangle = \langle -T^{\hat{0}\hat{z}}(z = -h) \rangle. \tag{5.6.1d}$$

(vi) The mass M and specific angular momentum a of the hole. (vii) The radial coordinate r.

The laws governing the radial structure are conservation of rest mass, conservation of angular momentum, and conservation of energy.

Conservation of rest mass

The amount of rest mass that flows inward across of cylinder a radius r during coordinate time Δt, when averaged by the method of equation (5.5.2), must equal $\dot{M}_0 \Delta t$ (conservation of rest mass). The mass transferred can be written as the flux integral

$$\dot{M}_0 \Delta t = \int_{\mathscr{S}} \langle \rho_0 \mathbf{u} \rangle \cdot \mathbf{d}^3\Sigma = (-2\pi r \mathscr{D}^{1/2} \Delta t) \int_{-h}^{h} \langle \rho_0 v^{\hat{r}} \rangle \, dz,$$

where $\mathscr{S}$ is the 3-surface $\{0 \leqslant \varphi \leqslant \pi, -h \leqslant z \leqslant h, 0 \leqslant t \leqslant \Delta t\}$. Rewritten in terms of the mass-averaged radial velocity [eq. (5.6.1c)], this equation becomes

$$\dot{M}_0 = -2\pi r \, \Sigma \, \bar{v}^{\hat{r}} \mathscr{D}^{1/2} = (\text{constant, independent of } r \text{ and } t) \tag{5.6.2}$$

Conservation of angular momentum

The law of conservation of angular momentum can be written in the form

$$\boldsymbol{\nabla} \cdot \mathbf{J} = 0, \qquad \mathbf{J} \equiv \mathbf{T} \cdot (\partial/\partial\phi), \tag{5.6.3}$$

where $\partial/\partial\varphi$ is the Killing vector associated with rotation about the symmetry axis. Without any loss of generality, we can write the stress-energy tensor $\mathbf{T}$ in the form

$$\mathbf{T} \equiv \rho_0(1 + \Pi)\mathbf{u} \otimes \mathbf{u} + \mathbf{t} + \mathbf{u} \otimes \mathbf{q} + \mathbf{q} \otimes \mathbf{u}, \tag{5.6.4a}$$

where Π is specific internal energy, $\mathbf{t}$ is the stress tensor as measured in the local rest frame of the baryons, $\mathbf{q}$ is the energy flux relative to the local rest frame (see §2.5), and

$$\mathbf{u} \cdot \mathbf{t} = \mathbf{u} \cdot \mathbf{q} = 0. \tag{5.6.4b}$$

The corresponding density of angular momentum is

$$J^{\alpha} = \rho_0 u_{\varphi} u^{\alpha} + t^{\alpha}_{\varphi} + u_{\varphi} q^{\alpha} + q_{\varphi} u^{\alpha},$$

where we have dropped the angular momentum $\rho_0 \Pi u_{\varphi} u^{\alpha}$ associated with the internal energy because $\Pi \ll 1$. The law of angular-momentum conservation thus reads

$$\boldsymbol{\nabla} \cdot \mathbf{J} = 0 = \rho_0 \, du_{\varphi}/d\tau + r^{-1}(r t^{\alpha}_{\varphi})_{,\alpha} + u_{\varphi} \boldsymbol{\nabla} \cdot \mathbf{q} + \mathbf{q} \cdot \boldsymbol{\nabla} u_{\varphi} + \boldsymbol{\nabla} \cdot (q_{\varphi} \mathbf{u}).$$

[Here we have used the law of rest-mass conservation, $\boldsymbol{\nabla} \cdot (\rho_0 \mathbf{u}) = 0$.] Average this equation over t, φ, r; and integrate over z. The result is

$$\int_{-h}^{h} \langle \rho_0 \rangle \langle du_{\varphi}/d\tau \rangle \, dz + r^{-1} \left(r \int_{-h}^{h} \langle t^{r}_{\varphi} \rangle \, dz \right)_{,r} + 2 \langle u_{\varphi} \rangle F = 0.$$

[Here we have invoked the thinness of the disk to infer that $\langle \mathbf{q} \rangle$ is in the vertical, $\mathbf{e}_{\hat{z}}$, direction; we have used stationarity, axial symmetry, and reflection symmetry about $z = 0$ to discard several terms; and we have used the relation $\langle q^{\hat{z}}(r, h) \rangle = \langle -q^{\hat{z}}(r, -h) \rangle = F$.] Express the coordinate-frame component of the stress, $\langle t^{r}_{\varphi} \rangle$, in terms of the orbiting orthonormal components $\langle t_{\alpha\beta} \rangle$—which are the same to high accuracy as components in the mean frame of the gas, $\langle t_{\tilde{\alpha}\tilde{\beta}} \rangle$:

$$\langle t^{r}_{\varphi} \rangle = r \mathscr{B} \mathscr{C}^{-1/2} \mathscr{D} \langle t_{\hat{\varphi}\hat{r}} \rangle.$$

Inserting this and the relation

$$\langle du_\varphi/d\tau\rangle = \langle u^r u_{\varphi,r}\rangle = \mathscr{D}^{1/2}\bar{v}^{\hat{r}}\tilde{L}_{,r}$$

into the conservation law, obtain

$$\mathscr{D}^{1/2}\,\Sigma\,\bar{v}^{\hat{r}}\tilde{L}_{,r} + r^{-1}(r^2\mathscr{B}\mathscr{C}^{-1/2}\mathscr{D}W)_{,r} + 2\tilde{L}F = 0. \tag{5.6.5}$$

The first term is the rate of increase of angular momentum in the gas; the second is the rate at which shear stresses carry off angular momentum; the third is the rate at which photons carry off angular momentum. When combined with the law of mass conservation (5.6.2), this law of angular momentum conservation takes on the simpler form

$$(-\dot{M}_0\tilde{L}/2\pi + r^2\mathscr{B}\mathscr{C}^{-1/2}\mathscr{D}W)_{,r} + 2r\tilde{L}F = 0. \tag{5.6.6}$$

Conservation of energy

Turn now to the law of energy conservation

$$\mathbf{u}\cdot(\nabla\cdot\mathbf{T}) = 0. \tag{5.6.7}$$

Rewrite this law in the form

$$\nabla\cdot(\mathbf{u}\cdot\mathbf{T}) - u_{\alpha;\beta}T^{\alpha\beta} = 0. \tag{5.6.8}$$

Use the general stress-energy tensor (5.6.4) to write

$$\mathbf{u}\cdot\mathbf{T} = -\rho_0(1+\Pi)\mathbf{u} - \mathbf{q};$$

and combine this with the law of rest-mass conservation $\nabla\cdot(\rho_0\mathbf{u}) = 0$ to obtain

$$\nabla\cdot(\mathbf{u}\cdot\mathbf{T}) = -\rho_0\,d\Pi/d\tau - \nabla\cdot\mathbf{q}. \tag{5.6.9}$$

Decompose $u_{\alpha;\beta}$ into its irreducible tensorial parts

$$u_{\alpha;\beta} = \omega_{\alpha\beta} + \sigma_{\alpha\beta} + \tfrac{1}{3}\theta h_{\alpha\beta} - a_\alpha u_\beta$$

[eq. (2.5.17)], and contract it into the stress-energy tensor (5.6.4a) to obtain

$$u_{\alpha;\beta}T^{\alpha\beta} = \sigma_{\alpha\beta}t^{\alpha\beta} + \tfrac{1}{3}\theta t^\alpha_\alpha + \mathbf{a}\cdot\mathbf{q}. \tag{5.6.10}$$

Then use relations (5.6.9) and (5.6.10) to rewrite the law of energy conservation (5.6.8) in the form

$$\rho_0\,d\Pi/d\tau + \nabla\cdot\mathbf{q} = -\sigma_{\alpha\beta}t^{\alpha\beta} - \tfrac{1}{3}\theta t^\alpha_\alpha - \mathbf{a}\cdot\mathbf{q}. \tag{5.6.11}$$

The various terms in this equation have simple interpretations. The left-hand side represents the fate of the energy being generated locally in the gas: $\rho_0\,d\Pi/d\tau$ is the energy going into internal forms; $\nabla\cdot\mathbf{q}$ is the energy being transported out of the region of generation. The right-hand side represents the rate at which energy is generated: $\sigma_{\alpha\beta}t^{\alpha\beta}$ is the energy being generated by "frictional" (viscous) heating; $-\tfrac{1}{3}\theta t^\alpha_\alpha$ is the energy being fed in by compression. The remaining term, $\mathbf{a} \cdot \mathbf{q}$, is a special relativistic correction associated with the inertia of the flowing energy $\mathbf{q}$ (see §2.5).

To put the law of energy conservation (5.6.11) into a form relevant to the radial structure of the disk, (i) we drop the internal energy and compressional work terms, $\rho_0\, d\Pi/d\tau$ and $\frac{1}{3}\theta t^{\alpha}_{\alpha}$, because our condition of "negligible specific heat" (5.5.8) guarantees that they are negligible:

$$\int \langle \rho_0\, d\Pi/d\tau \rangle\, d\tau \sim \int \langle \tfrac{1}{3}\theta t^{\alpha}_{\alpha} \rangle\, d\tau \sim \rho_0 \Pi \sim p$$
$$\ll \text{(gravitational energy released)}$$
$$\sim \text{(energy generated by frictional heating);}$$

(ii) we drop the special relativistic correction $\mathbf{a} \cdot \mathbf{q}$ because the gas is in nearly geodesic (unaccelerated) orbits; (iii) we average and integrate vertically, and use the relation $\langle q^{\hat{z}}(r, h) \rangle = \langle -q^{\hat{z}}(r, -h) \rangle = F$; (iv) we replace the averaged shear of the gas $\langle \sigma_{\alpha\beta} \rangle$ by the shear of the equatorial geodesic orbits. The result is

$$2F = -\,\sigma^{(\mathrm{EG})}_{\alpha\beta} \int_{-h}^{h} \langle t^{\alpha\beta} \rangle\, dz = -2\sigma^{(\mathrm{EG})}_{\hat{\phi}\hat{r}}\, W.$$

Using expression (5.4.6) for the shear, we obtain finally

$$F = \tfrac{3}{4}(M/r^3)^{1/2} \mathscr{C}^{-1} \mathscr{D} W. \qquad (5.6.12)$$

Manipulation of conservation laws

The conservation laws (5.6.6) and (5.6.12) for angular momentum and energy, when combined, give a differential equation for the integrated stress

$$(-\dot{M}_0 \tilde{L}/2\pi + r^2 \mathscr{B} \mathscr{C}^{-1/2} \mathscr{D} W)_{,r} + \tfrac{3}{2}(M/r)^{1/2} \tilde{L} \mathscr{C}^{-1} \mathscr{D} W = 0. \qquad (5.6.13)$$

All quantities in this differential equation are explicit functions of r that describe properties of the Kerr-metric (§5.4), except the constant accretion rate $\dot{M}_0$ and the unknown integrated stress $W(r)$. It is straight-forward to integrate this differential equation. The constant of integration is fixed by the physical fact that once the gas reaches the stable circular orbit of minimum radius, $r = r_{ms}$, the gas will "fall out" of the disk and spiral rapidly down the hole. Consequently, the gas density at $r < r_{ms}$ is virtually zero compared to that at $r > r_{ms}$—which means that no viscous stresses can act across the surface $r = r_{ms}$; i.e., W must vanish at $r = r_{ms}$. The solution that satisfies this boundary condition is

$$W = \frac{\dot{M}_0}{2\pi} \left(\frac{M}{r^3} \right)^{1/2} \frac{\mathscr{C}^{1/2} \mathscr{Q}}{\mathscr{B} \mathscr{D}}. \qquad (5.6.14a)$$

The corresponding value of the flux, as obtained from the law of energy conservation (5.6.12), is

$$F = \frac{3\dot{M}_0}{8\pi r^2}\frac{M}{r}\frac{\mathscr{Q}}{\mathscr{B}\mathscr{C}^{1/2}}. \tag{5.6.14b}$$

The final equations of radial structure are these two equations for W and F, plus the law of rest-mass conservation (5.6.2):

$$\dot{M}_0 = -2\pi r\, \Sigma\, \bar{v}^{\hat{r}} \mathscr{D}^{1/2}. \tag{5.6.14c}$$

These are the relativistic, black-hole versions of the Newtonian equations of radial structure (5.2.17), (5.2.18), and (5.2.19).

The equations of radial structure can be viewed as three equations linking the four radial functions Σ (surface density), W (integrated stress), $\bar{v}^{\hat{r}}$ (mass-averaged radial velocity), and F (radiant flux). Note that these equations determine W and F explicitly; but they determine only the product $\Sigma\,\bar{v}^{\hat{r}}$, not the individual functions Σ and $\bar{v}^{\hat{r}}$. To calculate Σ and $\bar{v}^{\hat{r}}$ individually—and to calculate other features of the disk such as thickness $2h$, internal temperature T, etc.—one must build a model for the vertical structure.

5.7 Equations of Vertical Structure

The equations of radial structure (5.6.14) are based on conservation laws, without any reference to the detailed properties of the gas in the disk (no reference to equation of state, or to nature of viscous stresses, or to nature of opacity, or to nature of turbulence and magnetic fields). By contrast, the equations of vertical structure, discussed below, require explicit assumptions about the properties of the gas. Hence, almost all of the uncertainties and complications of the model are lumped into the vertical structure. Ten years hence one will have a much improved theory of the vertical structure, whereas the equations of (averaged, steady-state) radial structure will presumably be unchanged.

Inside the disk the characteristic scale on which the vertical structure changes is h, while the characteristic scale for changes in the radial structure is $r \gg h$. Consequently, with good accuracy an observer inside the disk can regard the local variables (temperature, density, etc.) as functions of height, z, only. To analyze the local vertical structure most conveniently, one performs calculations in the local orbiting orthonormal frame $\mathbf{e}_{\hat{0}}$, $\mathbf{e}_{\hat{r}}$, $\mathbf{e}_{\hat{z}}$, $\mathbf{e}_{\hat{\varphi}}$, which is located in the central plane of the disk ($z = 0$). Aside from rotation about the $\mathbf{e}_{\hat{z}}$ axis—which produces Coriolis and centrifugal forces—this frame is (locally) inertial. The Coriolis and centrifugal forces have no influence on the vertical structure; hence we can ignore them and throughout this section can regard the orbiting frame as inertial.

The laws of physics in the orbiting inertial frame at any given t, r, φ are those of special relativity (equivalence principle). And because $\Pi \sim T_{\hat{j}\hat{k}}/\rho_0 \ll 1$ those special relativistic laws take on their standard Newtonian forms. At least they do so if one uses the correct Kerr-metric value for the tidal gravitational acceleration which compresses the disk into its "pancake" shape:

$$(\text{"acceleration of gravity"}) \equiv g = R^{\hat{z}}{}_{\hat{0}\hat{z}\hat{0}}\, z \tag{5.7.1}$$

(cf. eq. 5.2.24). An explicit expression for $R^{\hat{z}}{}_{\hat{0}\hat{z}\hat{0}}$ can be obtained by a transformation of the "LNRF" components, $R^{(\alpha)}{}_{(\beta)(\gamma)(\delta)}$, of the Riemann tensor as given in Bardeen, Press and Teukolsky (1972):

$$R^{\hat{z}}{}_{\hat{0}\hat{z}\hat{0}} = \frac{M}{r^3}\gamma^2\left[\frac{(r^2+a^2)^2 + 2\Delta a^2}{(r^2+a^2)^2 - \Delta a^2}\right] = \frac{M}{r^3}\frac{\mathscr{B}^2\mathscr{D}\mathscr{E}}{\mathscr{A}^2\mathscr{C}} \quad . \tag{5.7.2}$$

Thus, a Newtonian astrophysicist can build a relativistically correct vertical structure without knowing any relativity at all. He need merely follow standard Newtonian procedures and theory; but he must use expression (5.7.1) for the vertical acceleration of gravity, rather than the Newtonian formula $g = (M/r^3)z$.

We shall describe the vertical structure in terms of the following functions of height—which we tacitly assume have all been averaged over t, r, φ by the methods of equation (5.5.2):

$$\begin{aligned}
&\rho_0(r,z) = \text{density of rest mass;}\\
&p(r,z) = \text{vertical pressure, } T_{\hat{z}\hat{z}}\\
&t_{\hat{\varphi}\hat{r}}(r,z) = \text{shear stress;}\\
&T(r,z) = \text{temperature;}\\
&q^z(r,z) = \text{flux of energy;}\\
&\bar{\kappa}(r,z) = \text{Rosseland mean opacity.}
\end{aligned} \tag{5.7.3}$$

These 6 vertical structure functions are governed by the following 6 "equations of vertical structure":

Vertical pressure balance

$$\begin{aligned}
\frac{dp}{dz} &= \rho_0 \times (\text{"acceleration of gravity"}) = \rho_0 R^{\hat{z}}{}_{\hat{0}\hat{z}\hat{0}}\, z\\
&= \rho_0 \frac{Mz}{r^3}\frac{\mathscr{B}^2\mathscr{D}\mathscr{E}}{\mathscr{A}^2\mathscr{C}} .
\end{aligned} \tag{5.7.4a}$$

Sources of viscosity

$$t_{\hat{\varphi}\hat{r}} = \left\{\begin{matrix}\text{some explicit expression which depends on}\\ \text{the explicit assumptions of the model}\end{matrix}\right\} . \tag{5.7.4b}$$

Energy generation [cf. eq. (5.6.11)]

$$\frac{dq^{\hat{z}}}{dz} = -2\sigma^{(\mathrm{EG})}_{\hat{\varphi}\hat{r}} t_{\hat{\varphi}\hat{r}} = \tfrac{3}{2}(M/r^3)^{1/2} t_{\hat{\varphi}\hat{r}} \mathscr{C}^{-1}\mathscr{D}. \tag{5.7.4c}$$

Energy transport

$$q^{\hat{z}} = \left\{\begin{matrix}\text{some explicit expression which depends on the} \\ \text{nature of the transport assumed by the model}\end{matrix}\right\}. \tag{5.7.4d}$$

The energy transport may be by radiative diffusion in some models, but by turbulent gas motions in others. All models to date have assumed radiative transfer with large optical depth, $\tau_{es} + \tau \gg 1$ (see §2.6). In this case

$$q^{\hat{z}} = -\frac{1}{\bar{\kappa}\rho_0}\frac{d}{dz}(\tfrac{1}{3}bT^4). \tag{5.7.4d$'$}$$

Equation of state for vertical pressure

$$p = \left\{\begin{matrix}\text{some explicit expression which depends on} \\ \text{the explicit assumptions of the model}\end{matrix}\right\}. \tag{5.7.4e}$$

Possible contributors to vertical pressure are thermal gas pressure, radiation pressure, magnetic pressure, and turbulent pressure. Most models to date have ignored magnetic pressure and turbulent pressure, and have therefore taken

$$p = \rho_0(T/\mu_{mm} m_p)T + \tfrac{1}{3}bT^4, \tag{5.7.4e$'$}$$

where μ_{mm} is the mean molecular weight (0.5 for an ionized hydrogen gas).

Equation for opacity

$$\bar{\kappa} = \left\{\begin{matrix}\text{some explicit expression which depends on} \\ \text{the explicit assumptions of the model}\end{matrix}\right\}. \tag{5.7.4f}$$

Of these 6 equations of vertical structure, 3 are differential equations. They must be subjected to 3 boundary conditions. As in the theory of stellar structure, the boundary conditions at the surface, $z = h$, must be a join to an atmosphere in which the diffusion approximation for radiative transfer (eq. 5.7.4d$'$) is abandoned. Alternatively, one can use "zero-order boundary conditions" which ignore the atmosphere—and then "tack" an atmosphere onto the model afterwards. The zero-order version of the boundary conditions is as follows: (i) define the surface of the disk, $z = h$, to be that point at which the density goes to zero

$$\rho_0 = 0 \quad \text{at} \quad z = h; \tag{5.7.5a}$$

then (ii) temperature must also go to zero at the surface

$$T = 0 \quad \text{at} \quad z = h; \tag{5.7.5b}$$

(iii) the flux must vanish on the central plane of the disk

$$q^{\hat{z}} = 0 \quad \text{at} \quad z = 0; \tag{5.7.5c}$$

(iv) the integrated stress must equal the expression (5.6.14a) calculated from the theory of radial structure

$$2\int_0^h t_{\hat{\varphi}\hat{r}}\, dz = W. \tag{5.7.5d}$$

The equations of *vertical* structure (5.7.4) and these boundary conditions automatically guarantee that 3 of the equations of *radial* structure—(5.6.14a, b)—are satisfied. The vertical structure also provides one with a value for

$$\Sigma = 2\int_0^h \rho_0\, dz,$$

which one can insert into the third radial equation (5.6.14c) to obtain the mean radial velocity $\bar{v}^{\hat{r}}$. Then the entire structure, both radial and vertical, is known.

5.8 Approximate Version of Vertical Structure

Because of the great current uncertainties about the roles and forms of turbulence and magnetic fields, there is no justification in 1972 for building sophisticated models of vertical structure. Therefore, let us solve for the vertical structure in the same very approximate manner as we used in the Newtonian treatment of §5.2. In particular, let us replace the vertical functions $\rho_0, p, t_{\hat{\varphi}\hat{r}}, T$, and $\bar{\kappa}$ by their mean values in the disk interior, and let us rewrite the equations of vertical structure (5.7.4) in the following vertically-averaged form. (i) *Vertical pressure* balance:

$$h = (p/\rho_0)^{1/2}(r^3/M)^{1/2}\mathscr{A}\mathscr{B}^{-1}\mathscr{C}^{1/2}\mathscr{D}^{-1/2}\mathscr{E}^{-1/2}. \tag{5.8.1a}$$

(ii) *Source of viscosity:*

$$t_{\hat{\varphi}\hat{r}} = \alpha p \tag{5.8.1b}$$

[see eq. (5.2.32) and preceding discussion]. (iii) *Energy generation*: we merely replace $q^{\hat{z}}$ by a mean value of $\frac{1}{2}F$, where F is the surface flux as calculated from the theory of radial structure. (iv) *Energy transport*: we assume that radiative transport dominates over turbulent energy transport, so that

$$bT^4 = \bar{\kappa}\,\Sigma F. \tag{5.8.1c}$$

(v) *Equation of state*: Turbulent pressure cannot exceed thermal pressure; if it did, the turbulence would be supersonic and would quickly dissipate into heat.

Magnetic pressure cannot exceed thermal pressure; if it did, the magnetic field would break free of the disk. Therefore, we are not far wrong in ignoring turbulent and magnetic contributions to the pressure, and setting

$$p = p^{(\mathrm{rad})} + p^{(\mathrm{gas})};$$

$$p^{(\mathrm{rad})} = \tfrac{1}{3}bT^4, \qquad p^{(\mathrm{gas})} = \rho_0(T/m_p). \tag{5.8.1d}$$

(vi) *Opacity*: Ignoring line opacity and bound-free opacity, which are less than or of the order of free-free opacity at the high temperatures of our disk, we set,

$$\bar{\kappa} = \bar{\kappa}_{ff} + \bar{\kappa}_{es}, \tag{5.8.1e}$$

$$\bar{\kappa}_{ff} = (0.64 \times 10^{23}) \left(\frac{\rho_0}{\mathrm{g/cm^3}}\right)\left(\frac{T}{{}^\circ\mathrm{K}}\right)^{-7/2} \frac{\mathrm{cm^2}}{\mathrm{g}}, \qquad \bar{\kappa}_{es} = 0.40 \frac{\mathrm{cm^2}}{\mathrm{g}}.$$

5.9 Explicit Models for Disk

We shall here combine our approximate equations of vertical structure (5.8.1) with our "exact" equations of radial structure (5.6.14) to obtain explicit disk models. These models are due originally to Shakura and Sunyaev (1972)–except for relativistic corrections, which are due to Thorne (1973b).

In these models we shall express M in units of $3M_\odot$ (a typical black-hole mass); we shall express $\dot{M}_0$ in units of 10^{17} g/sec $\simeq 10^{-9} M_\odot$/yr (a value that will produce a total X-ray luminosity typical of the strength of galactic X-ray sources, $L \sim 10^{37}$ ergs/sec); and we shall express r in units of the radius of the extreme-Kerr horizon:

$$M_* \equiv M/3M_\odot, \qquad \dot{M}_{0*} \equiv \dot{M}_0/10^{17}\ \mathrm{g\ sec^{-1}},$$

$$r_* = r/M = (r/4.4 \times 10^5\ \mathrm{cm})M_*^{-1}. \tag{5.9.1}$$

Thus, for galactic X-ray sources, reasonable values are

$$M_* \sim \dot{M}_{0*} \sim 1; \tag{5.9.2a}$$

for a possible supermassive hole at the center of our Galaxy ($M \sim 3 \times 10^7\ M_\odot$, $\dot{M}_0 \sim 10^{-4} M_\odot$/yr., see §5.3), reasonable values are

$$M_* \sim 10^7, \qquad \dot{M}_{0*} \sim 10^5. \tag{5.9.2b}$$

In addition to the quantities that appear explicitly in the equations of structure, we shall calculate the optical depth at the center of the disk,

$$\tau = \bar{\kappa}\,\Sigma\ ; \tag{5.9.3}$$

a rough limit on the strength of the chaotic magnetic field,

$$B \lesssim (8\pi\alpha p)^{1/2}; \tag{5.9.4}$$

and the characteristic timescale for the gas to move inward from radius r to the inner edge of the disk

$$\Delta t(r) = -r/\bar{v}^{\hat{r}}. \tag{5.9.5}$$

The disk can be divided into 3 regions: an "*outer region*" (large radii) in which gas pressure dominates over radiation pressure, and in which the opacity is predominantly free-free; a "*middle region*" (smaller radii) in which gas pressure dominates over radiation pressure, but opacity is predominantly due to electron scattering; and an "*inner region*" (smallest radii) in which radiation pressure dominates over gas pressure, and opacity is predominantly due to electron scattering. Depending on the size of the mass flux, the inner and middle regions may or may not exist. For this reason, we shall use the fully relativistic equations for the radial structure in all regions, rather than take Newtonian limits from the beginning for the outer and middle regions.

Outer region

$p = p^{(\mathrm{gas})}$, $\bar{\kappa} = \bar{\kappa}_{ff}$. In this region straightforward algebraic manipulations of equations (5.8.1) and (5.6.14) yield the following radial profiles:

$$\begin{aligned}
F &= (0.6 \times 10^{26}\ \mathrm{erg/cm^2\,sec})(M_*^{-2}\dot{M}_{0*})r_*^{-3}\mathscr{B}^{-1}\mathscr{C}^{-1/2}\mathscr{Q},\\
\Sigma &= (2 \times 10^{5}\ \mathrm{g/cm^2})(\alpha^{-4/5}M_*^{-1/2}\dot{M}_{0*}^{7/10})r_*^{-3/4}\mathscr{A}^{1/10}\mathscr{B}^{-4/5}\mathscr{C}^{1/2}\\
&\quad\times \mathscr{D}^{-17/20}\mathscr{E}^{-1/20}\mathscr{Q}^{7/10},\\
h &= (9 \times 10^{2}\ \mathrm{cm})(\alpha^{-1/10}M_*^{3/4}\dot{M}_{0*}^{3/20})r_*^{9/8}\mathscr{A}^{19/20}\mathscr{B}^{-11/10}\mathscr{C}^{1/2}\\
&\quad\times \mathscr{D}^{-23/40}\mathscr{E}^{-19/40}\mathscr{Q}^{3/20},\\
\rho_0 &= (8 \times 10^{1}\ \mathrm{g/cm^3})(\alpha^{-7/10}M_*^{-5/4}\dot{M}_{0*}^{11/20})r_*^{-15/8}\mathscr{A}^{-17/20}\mathscr{B}^{3/10}\\
&\quad\times \mathscr{D}^{-11/40}\mathscr{E}^{17/40}\mathscr{Q}^{11/20},\\
T &= (8 \times 10^{7}\ \mathrm{K})(\alpha^{-1/5}M_*^{-1/2}\dot{M}_{0*}^{3/10})r_*^{-3/4}\mathscr{A}^{-1/10}\mathscr{B}^{-1/5}\mathscr{D}^{-3/20}\\
&\quad\times \mathscr{E}^{1/20}\mathscr{Q}^{3/10},\\
\tau_{ff} &= (2 \times 10^{2})(\alpha^{-4/5}\dot{M}_{0*}^{1/5})\mathscr{A}^{-2/5}\mathscr{B}^{1/5}\mathscr{C}^{1/2}\mathscr{D}^{-3/5}\mathscr{E}^{1/5}\mathscr{Q}^{1/5}, \qquad (5.9.6)\\
B &\lesssim (7 \times 10^{8}\ G)(\alpha^{1/20}M_*^{-7/8}\dot{M}_{0*}^{17/40})r_*^{-21/16}\mathscr{A}^{-19/40}\mathscr{B}^{1/20}\mathscr{D}^{-17/80}\\
&\quad\times \mathscr{E}^{19/80}\mathscr{Q}^{17/40},\\
\left(\frac{p^{(\mathrm{gas})}}{p^{(\mathrm{rad})}}\right) &= 4(\alpha^{-1/10}M_*^{1/4}M_{0*}^{-7/20})r_*^{3/8}\mathscr{A}^{-11/20}\mathscr{B}^{9/10}\mathscr{D}^{7/40}\mathscr{E}^{11/40}\mathscr{Q}^{-7/20},\\
\left(\frac{\tau_{ff}}{\tau_{es}}\right) &= 3 \times 10^{-3}(M_*^{1/2}\dot{M}_{0*}^{-1/2})r_*^{3/4}\mathscr{A}^{-1/2}\mathscr{B}^{2/5}\mathscr{D}^{1/4}\mathscr{E}^{1/4}\mathscr{Q}^{-1/2},
\end{aligned}$$

$$\Delta t(r) = (2\ \text{sec})(\alpha^{-4/5}M_*^{3/2}\dot{M}_{0*}^{-3/10})r_*^{5/4}\mathscr{A}^{1/10}\mathscr{B}^{-4/5}\mathscr{C}^{1/2}\mathscr{D}^{-7/20}$$
$$\times \mathscr{E}^{-1/20}\mathscr{Q}^{7/10}.$$

It is worth noting that the relativistic correction $\mathscr{Q}$ goes to zero smoothly at the inner edge of the disk

$$\mathscr{Q} \to 0, \mathscr{Q}_{,r} \to 0 \text{ as } r \to r_{ms}.$$

The transition to the middle region occurs where $\tau_{ff}/\tau_{es} \sim 1$–i.e., at

$$r_* = r_{om*} \equiv 2 \times 10^3(M_*^{-2/3}\dot{M}_{0*}^{2/3})\mathscr{A}^{2/3}\mathscr{B}^{-8/15}\mathscr{D}^{-1/3}\mathscr{E}^{-1/3}\mathscr{Q}^{2/3}. \quad (5.9.7)$$
$$\simeq 100 \text{ for "supermassive case."}$$

Middle region

$p = p^{(\text{gas})}$, $\bar{\kappa} = \bar{\kappa}_{es}$. In this region the equations of structure (5.8.1) and (5.6.14) yield:

$$F = (0.6 \times 10^{26}\ \text{erg/cm}^2\ \text{sec})(M_*^{-2}\dot{M}_{0*})r_*^{-3}\mathscr{B}^{-1}\mathscr{C}^{-1/2}\mathscr{Q},$$
$$\Sigma = (5 \times 10^4\ \text{g/cm}^2)(\alpha^{-4/5}M_*^{-2/5}\dot{M}_{0*}^{3/5})r_*^{-3/5}\mathscr{B}^{-4/5}\mathscr{C}^{1/2}\mathscr{D}^{-4/5}\mathscr{Q}^{3/5},$$
$$h = (3 \times 10^3\ \text{cm})(\alpha^{-1/10}M_*^{7/10}\dot{M}_{0*}^{1/5})r_*^{21/20}\mathscr{A}\mathscr{B}^{-6/5}\mathscr{C}^{1/2}\mathscr{D}^{-3/5}$$
$$\times \mathscr{E}^{-1/2}\mathscr{Q}^{1/5},$$
$$\rho_0 = (10\ \text{g/cm}^3)(\alpha^{-7/10}M_*^{-11/10}\dot{M}_{0*}^{2/5})r_*^{-33/20}\mathscr{A}^{-1}\mathscr{B}^{3/5}\mathscr{D}^{-1/5}\mathscr{E}^{1/2}\mathscr{Q}^{2/5},$$
$$T = (3 \times 10^8\ \text{K})(\alpha^{-1/5}M_*^{-3/5}\dot{M}_{0*}^{2/5})r_*^{-9/10}\mathscr{B}^{-2/5}\mathscr{D}^{-1/5}\mathscr{Q}^{2/5}, \quad (5.9.8)$$
$$\tau_{es} = (2 \times 10^4)(\alpha^{-4/5}M_*^{-2/5}\dot{M}_{0*}^{3/5})r_*^{-3/5}\mathscr{B}^{-3/5}\mathscr{C}^{1/2}\mathscr{D}^{-4/5}\mathscr{Q}^{3/5},$$
$$B \lesssim (1 \times 10^9\ \text{G})(\alpha^{1/20}M_*^{-17/20}\dot{M}_{0*}^{2/5})r_*^{-51/40}\mathscr{A}^{-1/2}\mathscr{B}^{1/10}\mathscr{D}^{-1/5}$$
$$\times \mathscr{E}^{1/4}\mathscr{Q}^{2/5},$$
$$\left(\frac{p^{(\text{gas})}}{p^{(\text{rad})}}\right) = (0.02)(\alpha^{-1/10}M_*^{7/10}\dot{M}_*^{-4/5})r_*^{21/10}\mathscr{A}^{-1}\mathscr{B}^{9/5}\mathscr{D}^{2/5}\mathscr{E}^{1/2}\mathscr{Q}^{-4/5},$$
$$\left(\frac{\tau_{ff}}{\tau_{es}}\right) = (0.6 \times 10^{-5})(M_*\dot{M}_{0*}^{-1})r_*^{3/2}\mathscr{A}^{-1}\mathscr{B}^{2}\mathscr{D}^{1/2}\mathscr{E}^{1/2}\mathscr{Q}^{-1},$$
$$\tau^* \equiv (\tau_{ff}\tau_{es})^{1/2}$$
$$= 50(\alpha^{-4/5}M_*^{-9/10}\dot{M}_{0*}^{11/10})r_*^{3/20}\mathscr{A}^{-1/2}\mathscr{B}^{2/5}\mathscr{C}^{1/2}\mathscr{D}^{-11/20}\mathscr{E}^{1/4}\mathscr{Q}^{1/10},$$
$$\Delta t(r) = (0.7\ \text{sec})(\alpha^{-4/5}M_*^{8/5}\dot{M}_{0*}^{-2/5})r_*^{7/5}\mathscr{B}^{-4/5}\mathscr{C}^{1/2}\mathscr{D}^{-3/10}\mathscr{Q}^{3/5}.$$

The transition from the middle region to the inner region occurs where $p^{(\text{gas})}/p^{(\text{rad})} \sim 1$–i.e., at

$$r_* = 40(\alpha^{2/21}M_*^{-2/3}\dot{M}_{0*}^{16/20})\mathscr{A}^{20/21}\mathscr{B}^{-36/21}\mathscr{D}^{-8/21}\mathscr{E}^{-10/21}\mathscr{Q}^{16/21}$$
$$\simeq 4 \text{ for supermassive case.} \quad (5.9.9)$$

Thus, in the supermassive case the "middle region" extends all the way—or almost all the way—into the inner edge of the disk.

Inner region

$p = p^{(\mathrm{rad})}$, $\bar{\kappa} = \bar{\kappa}_{es}$. In this region the equations of structure (5.8.1) and (5.6.14) yield:

$$\begin{aligned}
F &= (0.6 \times 10^{26}\ \mathrm{erg/cm^2\ sec})(M_*^{-2}\dot{M}_{0*})r_*^{-3}\mathscr{B}^{-1}\mathscr{C}^{-1/2}\mathscr{Q},\\
\Sigma &= (20\ \mathrm{g/cm^2})(\alpha^{-1}M_*\dot{M}_{0*}^{-1})r_*^{3/2}\mathscr{A}^{-2}\mathscr{B}^{3}\mathscr{C}^{1/2}\mathscr{E}\mathscr{Q}^{-1},\\
h &= (1 \times 10^{5}\ \mathrm{cm})(\dot{M}_{0*})\mathscr{A}^{2}\mathscr{B}^{-3}\mathscr{C}^{1/2}\mathscr{D}^{-1}\mathscr{E}^{-1}\mathscr{Q},\\
\rho_0 &= (1 \times 10^{-4}\ \mathrm{g/cm^3})(\alpha^{-1}M_*\dot{M}_{0*}^{-2})r_*^{3/2}\mathscr{A}^{-4}\mathscr{B}^{6}\mathscr{D}\mathscr{E}^{2}\mathscr{Q}^{-2}, \qquad (5.9.10)\\
T &= (4 \times 10^{7}\ \mathrm{K})(\alpha^{-1/4}M_*^{-1/4})r_*^{-3/8}\mathscr{A}^{-1/2}\mathscr{B}^{1/2}\mathscr{E}^{1/4},\\
\tau_{es} &= 8(\alpha^{-1}M_*\dot{M}_{0*}^{-1})r_*^{3/2}\mathscr{A}^{-2}\mathscr{B}^{3}\mathscr{C}^{1/2}\mathscr{E}\mathscr{Q}^{-1},\\
B &\lesssim (7 \times 10^{7}\ \mathrm{G})(M_*^{-1/2})r_*^{-3/4}\mathscr{A}^{-1}\mathscr{B}\mathscr{E}^{1/2},\\
\left(\frac{p^{(\mathrm{gas})}}{p^{(\mathrm{rad})}}\right) &= (5 \times 10^{-5})(\alpha^{-1/4}M_*^{7/4}\dot{M}_{0*}^{-2})r_*^{21/8}\mathscr{A}^{-5/2}\mathscr{B}^{9/2}\mathscr{D}\mathscr{E}^{5/4}\mathscr{Q}^{-2},\\
\tau^* &\equiv (\tau_{es}\tau_{ff})^{1/2}\\
&= (2 \times 10^{-3})(\alpha^{-17/16}M_*^{31/16}\dot{M}_{0*}^{-2})r_*^{93/32}\mathscr{A}^{-25/8}\mathscr{B}^{41/8}\mathscr{C}^{1/2}\mathscr{D}^{1/2}\mathscr{E}^{25/16}\\
&\quad \times \mathscr{Q}^{-2},\\
\Delta t(r) &= (2 \times 10^{-4}\ \mathrm{sec})(\alpha^{-1}M_*^{3}\dot{M}_{0*}^{-2})r_*^{7/2}\mathscr{A}^{-2}\mathscr{B}^{3}\mathscr{C}^{1/2}\mathscr{D}^{1/2}\mathscr{E}\mathscr{Q}^{-1}.
\end{aligned}$$

Notice that for the 2 "typical" cases of interest (galactic X-ray sources with $M_* \simeq \dot{M}_{0*} \simeq 1, \alpha \lesssim 1$; supermassive hole with $M_* \sim 10^7, \dot{M}_{0*} \sim 10^5$) the disk is everywhere thin in the sense that $h/r \lesssim 0.1$. In any case where the model predicts $h/r \gtrsim 1$, the model is self-inconsistent.

5.10 Spectrum of Radiation from Disk

Outer region

In the outer region electron-scattering opacity is negligible compared to free-free opacity, and the disk is optically thick. Consequently, the disk's surface is able to build up a blackbody spectrum (see §2.6). The "surface temperature" of the disk, which characterizes the blackbody spectrum, is

$$\begin{aligned}
T_s = \left(\frac{4F}{b}\right)^{1/4} &= (3 \times 10^{7}\ \mathrm{K})(M_*^{-1/2}\dot{M}_{0*}^{1/4})r_*^{-3/4}\mathscr{B}^{-1/4}\mathscr{C}^{-1/8}\mathscr{Q}^{1/4}\\
&\simeq (1 \times 10^{5}\ \mathrm{K})(\dot{M}_{0*}^{-1/4})(r_*/r_{om*})^{-3/4}. \qquad (5.10.1)
\end{aligned}$$

Here r_{om*} is the inner edge of the outer region [eq. (5.9.7)]. Thus, the outer region of the disk's surface emits a blackbody spectrum with temperature $\sim 10^4$ K to $\sim 10^5$ K. For typical galactic X-ray sources ($M_* \sim \dot{M}_{0*} \sim 1$) the inner edge of the outer region is located at $r_{0m} \simeq 10^9$ cm—compared to a size $r \sim 3 \times 10^{11}$ cm for the Roche lobe of the disk region, and compared to a size $r \sim 10^6$ cm for the inner edge of the disk if the central object is a black hole or neutron star, and $r \sim 10^9$ cm if it is a white dwarf.

Middle region and inner region

In the middle region and inner region electron-scattering opacity modifies the emitted spectrum so it is no longer blackbody. An unsophisticated way to calculate the modification is this: (i) At some fixed radius consider all photons that emerge from the disk with some fixed frequency ν. After being formed by free-free emission or some other process, these photons must "random-walk" their way through scattering electrons before they can emerge from the disk's surface. Let y_ν be the depth in the disk at which these photons are formed, as measured in g/cm^2;

$$y_\nu = \int_{\text{formation point}}^{\infty} \rho_0 \, dz.$$

The mean-free path of the photons between scatterings, as measured in these same units, is $\lambda = 1/\kappa_{es} = 2.5$ g/cm^2 (independent of photon frequency). The total depth travelled, y_ν, is the product of this mean-free path with the square-root of the number of scatterings $N_{\nu s}$

$$y_\nu = \lambda N_{\nu s}^{1/2} = \kappa_{es}^{-1} N_{\nu s}^{1/2}$$

("standard square-root factor for random walks"). Hence, the total number of scatterings between emission and emergence from disk is

$$N_{\nu s} = (\kappa_{es} y_\nu)^2 = [\tau_{es}(\text{emission point})]^2.$$

The depth of the emission point y_ν is determined by the demand that the total free-free optical depth traversed along a photon's tortured trajectory be unity:

$$\begin{aligned} 1 &= \kappa_\nu^{ff}(N_{\nu s}\lambda) = \kappa_\nu^{ff}(y_\nu N_{\nu s}^{1/2}) \\ &= [\tau_\nu^{ff}(\text{emission point})\,\tau_{es}(\text{emission point})]^{1/2}. \end{aligned}$$

Here τ_ν^{ff}, like τ_{es}, is measured not along the tortured random-walk photon path, but rather along the straight-line path of standard radiative transfer theory (§2.6). Let us summarize: In an atmosphere dominated by electron scattering, those photons of frequency ν that escape from the surface are generated at a depth [Zel'dovich and Shakura (1969), Shakura (1972)]

$$\begin{aligned} \tau_{\nu *}(\text{emission pt.}) &\equiv [\tau_\nu^{ff}(\text{emission pt.})\,\tau_{es}(\text{emission pt.})]^{1/2} \\ &= (\kappa_\nu^{ff}\kappa_{es})^{1/2} y_\nu = 1. \end{aligned} \qquad (5.10.2)$$

[Note: throughout this discussion we have ignored the tiny changes in photon frequency at each scattering; see §2.6 and see below.]

Because κ_ν^{ff} decreases with increasing photon frequency

$$\kappa_\nu^{ff} \propto \frac{1-e^{-x}}{x^3} \simeq x^{-3} \quad \text{for} \quad x \gg 1$$

$$\simeq x^{-2} \quad \text{for} \quad x \ll 1, \tag{5.10.3}$$

$$x \equiv h\nu/T$$

[eq. (2.6.25)], the high frequency part of the spectrum gets formed at greater depths than the low-frequency spectrum.

At the formation point, the specific intensity will be blackbody, $I_\nu = B_\nu$. The specific flux that crosses outward through the surface at depth y_ν (ignoring for the moment the flux that crosses back inward) thus has the blackbody form

$$F_\nu(y_\nu) = \int_0^{\pi/2} B_\nu \cos\theta \, d\Omega = 2\pi B_\nu.$$

But of all the photons in this specific flux, only a fraction $1/(N_{\nu s})^{1/2}$ ever reaches the surface of the disk ("standard $\sqrt{N}$ random-walk factor"). All the rest eventually get scattered back to depths greater than y_ν, and eventually get absorbed there. Hence, the specific flux emerging from the surface of the disk will be

$$F_\nu = 2\pi B_\nu (N_{\nu s})^{-1/2} = 2\pi B_\nu [\tau_{es}(\text{at pt. where } \tau_{\nu *} = 1)]^{-1}. \tag{5.10.4}$$

This is the "modified spectrum" which emerges from the middle and inner parts of the disk.

In the above derivation we tacitly assumed a homogeneous atmosphere. However, a more sophisticated derivation, with an atmosphere in which temperature and density (and hence κ_ν^{ff}) vary with height, yields essentially the same spectrum as (5.10.4). (See Shakura and Sunyaev 1972). In the case of a homogeneous atmosphere, eqs. (5.10.2) and (5.10.3) show that the "spectrum modification factor" has the form

$$[\tau_{es}(\tau_{\nu *} = 1)]^{-1/2} \propto (\kappa_\nu^{ff})^{1/2} \propto x^{-3/2}(1-e^{-x})^{1/2} \tag{5.10.5}$$

so the spectrum has the form

$$F_\nu \propto \frac{x^{3/2} e^{-x/2}}{(e^x - 1)^{1/2}}. \tag{5.10.6}$$

The total flux in this case works out to be

$$F = (1.54 \times 10^{-4} \text{ erg/cm}^2 \text{ sec}) \left(\frac{\rho_0/m_p}{\text{cm}^{-3}}\right)^{1/2} \left(\frac{T_s}{{}^\circ\text{K}}\right)^{9/4} \tag{5.10.7}$$

If the atmosphere is not homogeneous, the spectrum and flux have somewhat different forms than these. (See Shakura and Sunyaev 1972.)

Assuming a homogeneous atmosphere with density roughly equal to that in the central regions of the disk, and combining eq. (5.10.7) for the flux with expressions (5.9.8) and (5.9.9) for the structures of the middle and inner regions, we obtain surface temperatures of

$$T_s = (5 \times 10^7\ \mathrm{K})(\alpha^{28/80} M_*^{-1/5} \dot{M}_{0*}^{16/45}) r_*^{-87/90} \mathscr{A}^{2/9} \mathscr{B}^{-26/45} \mathscr{C}^{-2/9} \times \mathscr{D}^{2/45} \mathscr{E}^{-1/9} \mathscr{Q}^{16/45} \quad \text{for middle region} \tag{5.10.8}$$

$$T_s = (6 \times 10^8\ \mathrm{K})(\alpha^{2/9} M_*^{-10/9} \dot{M}_{0*}^{8/9}) r_*^{-17/9} \mathscr{A}^{8/9} \mathscr{B}^{-16/9} \mathscr{C}^{-2/9} \times \mathscr{D}^{2/9} \mathscr{E}^{4/9} \quad \text{for inner region.}$$

In the inner region these surface temperatures are roughly an order of magnitude higher than would occur if the disk could radiate as a blackbody [cf. eq. (5.2.21)].

The above estimates assume, of course, that the disk is optically thick in the sense that $\tau_* > 1$ at the central plane $z = 0$. However, in the innermost regions of the disk this may not be the case. In fact, our model [eq. (5.9.10)] predicts

$$\tau_*(z=0) \simeq (1 \times 10^{-3}) \alpha^{-17/16} M_*^{31/16} \dot{M}_{0*}^{-2}) r_*^{93/32} \mathscr{A}^{-25/8} \mathscr{B}^{41/8} \times \mathscr{C}^{1/2} \mathscr{D}^{1/2} \mathscr{E}^{25/16} \mathscr{Q}^{-2}.$$

Thus, for galactic X-ray sources with $\alpha \sim 1, M_* \sim \dot{M}_{0*} \sim 1, a = M$ ("maximal Kerr"), the disk will be optically thin over a narrow region between $r \sim 2M$ and $r \sim 10M$. Inside this region it will be thick (because $\mathscr{Q} \to 0$ as $r \to M$); outside it will also be thick.

In such an optically thin region the disk must still radiate the huge flux (5.6.14b) required by energy conservation. It can do so only at temperatures far above the blackbody value for the given flux. All the photons emitted will escape, so the spectrum will have the free-free form

$$F_\nu \propto \epsilon_\nu^{ff} \propto e^{-x} \tag{5.10.9}$$

(see §§2.1 and 2.6). At least this would be the case if Comptonization were negligible. However, at such high temperatures ($T \sim 10^9$ K), Comptonization must be taken into account. The free-free spectrum has many more low-energy photons than high; and when scattered, each low-energy photon gets boosted in energy by a fractional amount

$$\Delta h\nu/h\nu \simeq 4(T/m_e) \simeq (T/2 \times 10^9\ \mathrm{K}). \tag{5.10.10}$$

As a result the low-energy end of the spectrum gets depleted and the high-energy end gets augmented. [See Shakura and Sunyaev (1972) for quantitative estimates of this effect].

The total spectrum as observed from Earth is the integral of F_ν over the entire disk—with corrections in the innermost regions for the capture of some

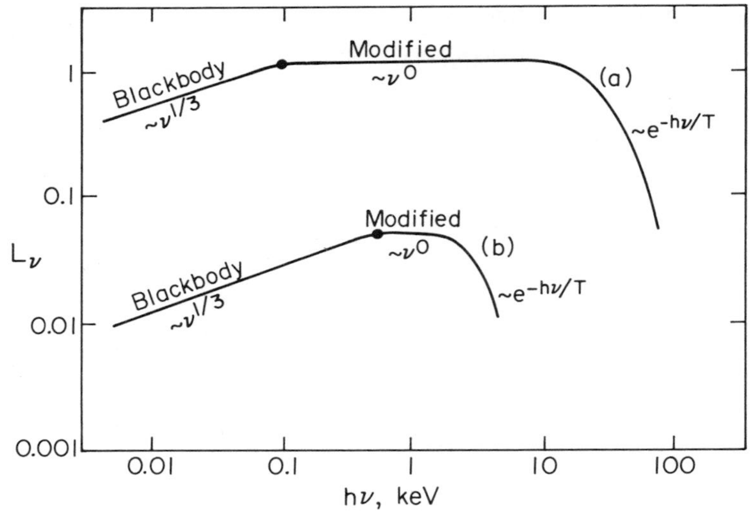

Figure 5.10.1. The total power per unit frequency L_ν emitted by two model disks around black holes, as calculated by Shakura and Sunyaev (1972) without taking account of relativistic corrections or capture of radiation by the hole. In both models the hole was assumed to be nonrotating, so the inner edge of the disk was at $r_* = r/M = 6$. Model (a) corresponds to

$$\alpha \sim 10^{-3}, \quad M = M_\odot, \quad \dot{M}_0 = 10^{-8} M_\odot/\text{yr},$$
$$L \simeq L_{\text{crit}} = 10^{38}\ \text{erg/sec};$$

model (b) corresponds to

$$\alpha \sim 10^{-2} \text{ to } 1 \text{ (spectrum insensitive to } \alpha),$$
$$M = M_\odot, \quad \dot{M}_0 = 10^{-6} M_\odot/\text{yr}, \quad L \simeq 10^{36}\ \text{erg/sec}.$$

The portion of the spectrum marked "blackbody" is generated primarily by the outer, cool region of the disk where electron scattering is unimportant. The portion marked "modified" is generated by the middle and inner regions where electron scattering is the dominant source of opacity. The temperature of the exponential tail is the surface temperature of the innermost region.

of the radiation by the black hole. The foundations for calculating the capture corrections are given in Bardeen's lectures in this volume.

Figure 5.10.1 shows the total spectrum as calculated by Shakura and Sunyaev (1972) for two typical cases, without taking account of relativistic effects or capture by the black hole. In both cases the disk is optically thick everywhere. In the X-ray region (1 to 100 keV) these spectra are similar to those observed for typical galactic X-ray sources; see lectures of Gursky in this volume.

5.11 Heating of the Outer Region by X-Rays from the Inner Region

Thus far we have ignored any radiative energy transfer from one portion of the disk to another. As long as the disk is optically thick, no significant transfer can occur through its interior. However, radiation can skim along the surface of the disk from the hot, inner region toward the cooler outer region: Because the disk is thick with increasing radius

$$h = \text{constant in inner region}$$

$$h \propto r_*^{21/20} \text{ in middle region}$$

$$h \propto r_*^{9/8} \text{ in outer region,}$$

such radiation will get captured in the middle and outer regions and will heat their surface layers. As a result, the outer regions will thicken even more than is predicted by the above model. Rough estimates by Shakura and Sunyaev (1972) suggest that for typical binary accretion models ~1 per cent of the total X-ray luminosity can be captured and reradiated. About 10 per cent of the reradiated flux comes off in spectral lines in the optical region, and about 90 per cent comes off in ultraviolet free-bound and free-free emission.

5.12 Fluctuations on the Steady-State Model

Superimposed on the steady-state disk structure will be local fluctuations due to turbulence, magnetic flux reconnection, and various plasma instabilities. Such fluctuations will probably be large if $\alpha \sim 1$, and unimportant if $\alpha \ll 1$.

Local fluctuations will presumably create local, extra-luminous hot spots. Sunyaev (1972) estimates that such hot spots can last for many orbital periods. Doppler shifts and focusing by the black hole will cause the frequency and intensity of the radiation from a hot spot as seen at Earth to fluctuate with the orbital period. The minimum possible fluctuation period is the orbital period at the last stable circular orbit:

$$P_{\min} = 12\pi\sqrt{6}M = (0.5 \text{ msec})(M/M_\odot) \text{ for nonrotating hole,}$$

$$P_{\min} = 4\pi M = (0.06 \text{ msec})(M/M_\odot) \text{ for maximally rotating hole.}$$

In a binary system one can measure with some confidence the mass of the compact, accreting object by examining the Doppler shifts of spectral lines (lines from the normal star and from the outer regions of the disk, not from the inner disk!) If one could also find quasiperiodic fluctuations in the X-ray output and place a lower limit on their periods, one might be able to distinguish whether the black hole is nonrotating [$P_{\min} \sim (0.5 \text{ msec})(M/M_\odot)$] or rotating [$(0.06 \text{ msec})(M/M_\odot) \lesssim P_{\min} \lesssim (0.5 \text{ msec})(M/M_\odot)$]. This "test for rotation" was devised by the Moscow group of Zel'dovich, Novikov, Polnarev, and Sunyaev (see Sunyaev 1972).

An alternative test for rotation devised by Thorne (1973b) involves the shape of Doppler-broadened spectral features. The "dragging of inertial frames" by the rotation of the hole will produce a strong asymmetry in the Doppler broadening of any spectral features emitted from in or near the ergosphere. Three factors make such asymmetries a less attractive test for rotation then the Moscow test: (i) the spectrum from in and near the ergosphere should be rather featureless because of the high temperatures; (ii) the Doppler broadening is so great that it would smear into the background all features except abnormally strong ones; (iii) Comptonization can also produce asymmetries in spectral features.

Local fluctuations ("hot spots") may well produce nonthermal radiation. A case of particular interest is the synchrotron radiation emitted by charged particles that are accelerated in regions of reconnecting magnetic flux (see §2.9). For supermassive holes at the centers of galaxies, Lynden-Bell (1969) has estimated the spectrum of such synchrotron radiation. He argues that the mean free paths of electrons against Coulomb scattering will be too small to permit acceleration in the reconnection regions; but that protons, having larger inertia, can get accelerated significantly. The result is a large amount of proton synchrotron radiation in the radio band. Efforts are being made to compare the predictions of this model with radio-frequency observations of the galactic center (Ekers and Lynden-Bell 1970).

5.13 Supercritical Accretion

If the accretion rate onto a black hole were larger than

$$\dot{M}_{0\,\mathrm{crit}} \sim (10^{-8} M_{\odot}/\mathrm{yr})(M/M_{\odot}), \tag{5.13.1}$$

the luminosity produced by our model would exceed the "Eddington limit",

$$L_{\mathrm{crit}} \simeq (1 \times 10^{38}\ \mathrm{erg/sec})(M/M_{\odot}). \tag{5.13.2}$$

[See eq. (4.5.4) and §5.1.] Shakura and Sunyaev (1972) suggest the following general picture for disk accretion in this "supercritical" case: In the outer regions of the disk the gas is shielded from radiation pressure by high opacity, so it accretes in the usual manner. However, as the gas nears the black hole, radiation pressure builds up, and ultimately becomes strong enough to eject the gas out of the disk, in the z-direction. Only a small fraction of the accreting matter ever reaches the hole; and the total luminosity is self-regulated at the Eddington critical value (5.13.2).

Rough computations by Shakura and Sunyaev suggest that, if the accretion is strongly supercritical

$$\dot{M}_0 > \dot{M}_{0\,\mathrm{crit}} \times (10^3 \alpha M_{\odot}/M)^{2/3},$$

the outflowing gas becomes opaque and reprocesses the radiation emitted near the hole from high frequencies to lower frequencies. Most of the energy comes off in the ultraviolet and optical regions; and the outflowing matter achieves velocities

$$v \sim (10^5 \text{ cm/sec})\, \alpha(\dot{M}_0/\dot{M}_{0\ \mathrm{crit}}).$$

The opaque region in this case is $\sim 10^{10}$ to 10^{12} cm in radius, and may thus cover the normal star as well as the disk and the hole.

5.14 Comparison with Observations

Currently the time scale for 100 per cent improvements in the observations of X-ray sources is less than a year. Therefore, it would be foolhardy to present in these notes a detailed comparison of the above models with X-ray observations. We shall merely remark that as of August 1972 the comparisons are very promising. In particular, the X-ray sources Cyg X-1 and 2U 0900-40 are excellent binary black-hole candidates.

The comparison of these models with observations of galactic nuclei is much more difficult than comparison with X-ray source observations. Obscuration by dust, and the high density of radiation sources in galactic nuclei create great difficulties. However, the observations are not in conflict with the hypothesis that a supermassive hole of $M \lesssim 10^8 M_\odot$ resides at the center of our Galaxy. (See Lynden-Bell and Rees 1971.)

6 White Holes and Black Holes of Cosmological Origin

6.1 White Holes, Grey Holes, and Black Holes

All black holes studied in previous sections were regarded as remnants of the relativistic collapse of massive stars. However, the formation of stars from rarefied gas is possible only in relatively late stages of the evolution of the Universe ($t \gtrsim 10^7$ years). In earlier stages, according to current cosmological models, radiation pressure impeded the growth of condensations; so individual bodies could not form from *small* perturbations of the primeval gas. In the absence of star formation, there correspondingly should have been no black-hole formation.

However, this conclusion might be wrong. It is quite possible that near the beginning of the cosmological expansion the matter distribution and metric were *highly* inhomogeneous and anisotropic. In this case the condensation of individual bodies would have been possible. Moreover, because in the early stages of the hot universe the mass-energy in radiation and in particle-antiparticle pairs greatly exceeded that in baryons, bodies which condensed then must have been made primarily of photons and pairs. However, because such an ultrarelativistic gas has an adiabatic index $\Gamma \leq 4/3$, any body made from it is unstable against gravitational collapse. Thus, one is led to imagine primordial condensations that, like

Wheeler's (1955) geons, were made primarily from photons and pairs, but unlike geons, quickly collapsed to form black holes.

Another possible type of strong inhomogeneity in the early universe is a "white hole" (Novikov 1964; Ne'eman 1965). Suppose that some isolated portions of the universe failed to begin their expansion at the same moment as the rest of the universe. As measured by external observers, the delay in the initial expansion of a given region (or "core") might be arbitrarily great; and the delays might vary from core to core. When a "lagging core" eventually begins to expand, and emerges through its gravitational radius, external observers should see an explosion with a release of tremendous energy. Such an object is, in some sense, a concrete realization of Ambartsumyan's (1961, 1964) concept of a superdense "D-body". (Thus, one can regard "D-body", "white hole", and "lagging core" as different names for the same concept.)

Finally, if a lagging core, when it begins to expand, has insufficient energy to emerge through its gravitational radius into the external universe, then one can regard it as a "grey hole". Thus, in the idealized spherical case, the matter of a grey hole is "born" in the initial $r = 0$ singularity of the Krustal metric; it moves upward in the Kruskal diagram, staying always to the left of the center line so it is always separated from the external universe by a past horizon or a future horizon; and it finally dives to its death in the terminal $r = 0$ singularity.

Might all three types of cosmological holes actually exist in Nature? What would be their properties? Is there any observational way to rule out their existence? Theorists have worked very little on these issues; and little is known about them.

In these notes we shall dwell on only two points: (i) the growth of cosmologica holes by accretion, and (ii) a limit on how many baryons can be inside cosmologic holes.

6.2 The Growth of Cosmological Holes by Accretion

If cosmological holes have existed since near the beginning of the universe, then during the era when the energy density in radiation exceeded that in matter, accretion of radiation onto the holes must have been important (Zel'dovich and Novikov 1966). For a rough estimate of the effects of such accretion we can use equation (4.2.14) for the accretion of gas onto a stationary hole, taking the velocity of sound to be $a_\infty = c/\sqrt{3} \simeq c$:

$$dM/dt \simeq r_g^2 c \rho^{(\mathrm{rad})}. \tag{6.2.1}$$

In standard cosmological models the radiation energy density varies as

$$\rho^{(\mathrm{rad})} = 1/Gt^2. \tag{6.2.2}$$

Putting this into the growth equation (6.2.1), and integrating from the moment t_0 when a hole of initial mass M_0 first forms, we obtain for the final mass, at

$t \gg t_0$,

$$M = M_0(1 - GM_0/c^2 t_0)^{-1}. \tag{6.2.3}$$

Notice that the final mass diverges for $t_0 \to GM_0/c^3$. However, the method of calculation breaks down in this same limit because, for $t_0 \sim GM_0/c^3$ the characteristic time scale for the accretion is the same as that for the change of $\rho^{(\mathrm{rad})}$, so the accretion violates the "steady-state" hypothesis on which equation (6.2.1) is based. To determine whether the accretion is catastrophically great at early times ($t_0 \lesssim GM_0/c^3$), one must solve the accretion problem in a nonsteady, cosmological context. One would be surprised if the final answer depends sensitively on the particular choice of initial conditions ($t_0 = GM_0/c^3$ or $t_0 = 0.01\ GM_0/c^3$ or $t_0 = 10^{-10}\ GM_0/c^3$). But the problem has not yet been solved.

6.3 Limit on the Number of Baryons in Cosmological Holes

Independently of the issue of accretion, one can place a tight limit on what fraction α of all baryons in the Universe were swallowed into cosmological holes in the early stages.

Suppose that prior to some particular early moment t_*, a fraction α of all baryons had been swallowed into holes. At the moment t_*, the ratio of rest mass-energy in baryons to mass-energy in radiation and pairs was tiny

$$\beta \equiv \left[\frac{\rho_0}{\rho^{(\mathrm{rad})} + \rho^{(\mathrm{pairs})}}\right]_{t_*} \sim 10^{-7}(t_*/1\ \mathrm{sec})^{-1/2} \ll 1. \tag{6.3.1}$$

However, as the universe subsequently expanded, this ratio increased until today it is far greater than unity. Moreover, the total mass-energy in holes divided by the volume of the universe (call this $\rho^{(\mathrm{holes})}$) changed in the same manner as $\rho_0(\propto \mathrm{vol}^{-1})$, if no new holes were formed after t_* and if accretion was negligible. Otherwise it increased relative to ρ_0. This means that

$$\frac{\alpha}{\beta(1-\alpha)} \leqslant \left[\frac{\rho^{(\mathrm{holes})}}{\rho_0}\right]_{t_*} \leqslant \left[\frac{\rho^{(\mathrm{holes})}}{\rho_0}\right]_{\mathrm{today}} < 80 \tag{6.3.2}$$

Here $\alpha/[\beta(1-\alpha)]$ is the value of $[\rho^{(\mathrm{holes})}/\rho_0]_{t_*}$ if the holes were all created from primeval plasma at the moment t_*; if they were created even earlier, when rest mass was even less important, $[\rho^{(\mathrm{holes})}/\rho_0]_{t_*}$ would be even bigger. The number 80 is a generous observational upper limit on the ratio of nonluminous matter to luminous matter in the universe today.

Equation (6.3.2) for $\beta \ll 1/80$ can be rewritten as

$$\alpha < 1/80\beta \sim 10^{-5}(t_*/1\ \mathrm{sec})^{-1/2}. \tag{6.3.3}$$

This is a very tight limit on the amount of rest mass that could have been down black holes at early times t_*.

One can also place a tight limit on the amount of mass-energy that white holes have spewed forth as radiation in recent times (at cosmological redshifts $z \lesssim 10$). That mass-energy cannot exceed the total mass-energy in radiation today,

$$\rho_{\text{from white holes}} < \rho^{(\text{rad})} = 5 \times 10^{-13}\ \text{erg/cm}^3$$
$$= 6 \times 10^{-34}\ \text{g/cm}^3 \qquad (6.3.4)$$

6.4 Caution

We conclude with a word of caution. This cosmological section has dealt with objects about which *both* theory and observation are equivocal. There is no firm theoretical reason for believing that cosmological holes exist, though theory surely permits them.

By contrast, theory virtually demands the existence of black holes formed by stellar collapse. It would be astonishing indeed if every star in the Galaxy somehow managed to avoid the black-hole fate! [See Hoyle, Fowler, Burbidge, and Burbidge (1964).]

The clear discovery of black holes, we expect, is only a few months or years away. (The X-ray source Cyg X-1 is an excellent candidate.) But cosmological holes might never be found and might, in fact, not exist at all. On the other hand, we might eventually discover that they are the energizers of quasars and of explosions in galactic nuclei.

References

Aller, L. H. 1963, *Astrophysics–The Atmospheres of the Sun and Stars* (New York: The Ronals Press Company).
Ambartsumyan, V. A. 1960, *Collected Works*, **2** (Yerevan).
Ambartsumyan, V. A. 1964, *Rapport 13 Conseilde Physique Solvay* (Brussels: Stoops.
Arnett, W. D. 1967, *Canadian J. Phys.* **45**, 1621.
Arnett, W. D. 1969, *Ap. Space. Sci.* **5**, 180.
Bardeen, J. M. 1973. This volume.
Bardeen, J. M., Press, W. H., and Teukolsky, S. A. 1972, *Ap. J.*, in press.
Bekefi, G. 1966, *Radiation Processes in Plasma.*
Bisnovaty-Kogan, G. S., and Sunyaev, R. A. 1972, *Soviet Astron.–A. J.* **49**, 243.
Bisnovaty-Kogan, G. S., Zel'dovich, Ya. B., and Sunyaev, R. A. 1971, *Soviet Astron.-A. J.* **48**, 24.
Bondi, H. 1952, *Mon. Not. Roy. Astron. Soc.* **112**, 195.
Bondi, H. and Hoyle, F. 1944, *Mon. Not. Roy. Astron. Soc.* **104**, 273.
Bruenn, S. 1972, *Ap. J. Supp.* **24**, 283.
Brussaard, P. J. and van de Halst, H. C. 1962, *Rev. Mod. Phys.* **34**, 507.
Burbidge, G. R. 1972, *Comments Ap. Space Phys.*, in press.
Chandrasekhar, S. 1960, *Radiative Transfer* (New York: Dover Publications, Inc.).
Colgate, S. A. and White, R. H. 1966, *Ap. J.* **143**, 626.
Cowling, T. G. 1965, Chapter 8 of *Stellar Structure*, L. H. Aller and D. B. McLaughlin eds. (Chicago: University of Chicago Press).

Eddington, A. S. 1926, *The Internal Constitution of the Stars* (Cambridge: Cambridge University Press).
Ekers, R. and Lynden-Bell, D. 1971, *Astrophys. Lett.* **9**, 189.
Ellis G. F. R. 1971, in *Relativity and Cosmology*, R. Sachs ed. (New York: Academic Press).
Felton, J. E. and Rees, M. J. 1972, *Astron. Ap.* **17**, 226.
Fraley, G. C. 1968, *Ap. Space Sci.* **2**, 96.
Ginzburg, V. L. 1967, in *High Energy Astrophysics, Vol. 1*, C. DeWitt, E. Schatzman, P. Véron eds. (New York: Gordon and Breach), p. 19.
Gold, T., Axford, W. I. and Ray, E. C. 1965, Chapter 8 of *Quasistellar Sources and Gravitational Collapse*, I. Robinson, A. Schild and E. L. Schucking eds. (Chicago: University of Chicago Press).
Green, J. 1959, *Ap. J.* **130**, 693.
Hansen, C. J. and Wheeler, J. C. 1969, *Ap. Space Sci.* **3**, 464.
Hayakawa, S. and Matsuoko, M. 1964, *Prog. Theor. Phys. Suppl.* **30**, 204.
Heitler, W. 1954, *The Quantum Theory of Radiation* (Oxford: Clarendon Press).
Hoyle, F. and Lyttleton, R. A. 1939, *Proc. Camb. Phil. Soc.* **35**, 405.
Hoyle, F., Fowler, W. A., Burbidge, G. R. and Burbidge, E. M. 1964, *Ap. J.* **139**, 909.
Hunt, R. 1971, *Mon. Not. Roy. Astron. Soc.* **154**, 141.
Illarionov, A., and Sunyaev, R. A. 1972, *Soviet Astron.-A. J.* **49**, 58.
Ivanova, L. N., Imshennik, V. S. and Nadezhin, O. K., *Sci. Inf. Astr. Council USSR Acad. Sci.* **13**.
Jackson, J. D. 1962, *Classical Electrodynamics* (New York: John Wiley and Sons, Inc.).
Kaplan, S. A. and Pikel'ner, S. B. 1970, *The Interstellar Medium* (Cambridge: Harvard University Press).
Kaplan, S. A. 1966, *Interstellar Gas Dynamics* (Oxford: Pergamon Press).
Kaplan, S. A. and Setovich, 1972, *Plasmenaya Astrofizika* (Moscow: Izdatel'stvo Nauka).
Karzas, W. J. and Latter, R. 1961, *Ap. J. Suppl.* **6**, 167.
Kompaneets, A. S. 1957, *Soviet Phys.-JETP* **4**, 730.
Landau, L. D. and Lifshitz, E. M. 1959, *Fluid Mechanics* (London: Pergamon Press).
Leighton, R. B. 1959, *Principles of Modern Physics* (New York: McGraw-Hill).
Lichnerowicz, A. 1967, *Relativistic Hydrodynamics and Magnetohydrodynamics* (New York: W. A. Benjamin, Inc.).
Lichnerowicz, A. 1970, *Physica Scripta* **2**, 221.
Lichnerowicz, A. 1971, lectures in *Relativistic Fluid Dynamics*, proceedings of July 1970 summer school at Bressanone, C. Cattaneo ed. (Rome: Edizioni Cremonese).
Lindquist, R. W. 1966, *Ann. Phys. (USA)* **37**, 487.
Lynden-Bell, D. 1969, *Nature* **223**, 690.
Lynden-Bell, D. and Rees, M. J. 1971, *Mon. Not. Roy. Astron. Soc.*, **152**, 461.
Mihalas, D. 1970, *Stellar Atmospheres* (San Francisco: W. H. Freeman and Co.).
Misner, C. W., Thorne, K. S. and Wheeler, J. A. 1973, *Gravitation* (San Francisco: W. H. Freeman and Co.); cited as MTW.
Ne'eman, Y. 1965, *Ap. J.* **141**, 1303.
Novikov, I. D. 1964, *Astron. Zhur.* **41**, 1075.
Novikov, I. D., Polnarev and Sunyaev, R. A., 1973, in preparation.
Novikov, I. D. and Zel'dovich, Ya. B. 1966, *Nuovo Cim. Suppl.* **4**, 810, addendum 2.
Pacholczyk, A. G. 1970, *Radio Astrophysics* (San Francisco: W. H. Freeman and Co.).
Paczynski, B. 1971, *Ann. Rev. Astron. Ap.* **9**, 183.
Peebles, P. J. E. 1972, *Gen. Rel. Grav.* 3, 63.
Pikel'ner, S. B. 1961, *Foundations of Cosmical Electrodynamics* (Moscow: Fizmatgiz) (in Russian).
Prendergast, K. H. and Burbidge, G. R. 1968, *Ap. J. Letters* **151**, L83.
Pringle, J. E. and Rees, M. J. 1972, *Astron. Astrophys.*, in press.
Salpeter, E. 1964, *Ap. J.* **140**, 796.
Schwartzman, V. F. 1971, *Soviet Astronomy–A. J.* **15**, 377.
Shakura, N. I. 1972a, *Soviet Astron.-A. J.* **49**, 642.
Shakura, N. I. 1972b, *Soviet Astron.-A. J.* **49**, 495.

Shakura, N. I. and Sunyaev, R. A. 1972, *Astron. Astrophys.*, in press.
Shkarofsky, I. P., Johnston, T. W. and Bachynski, M. P. 1966, *The Particle Kinetics of Plasmas* (Reading Massachusetts: Addision Wesley).
Shklovsky, I. S. 1967, *Ap. J. Letters* **148,** L1.
Sonnerup, B. U. Ö. 1970, *J. Plasma Phys.* **4,** part 1, 161.
Sunyaev, R. A. 1972, *Astron. Zhur.* in press.
Sunyaev, R.A. and Zel'dovich, Ya. B. 1973, *Ann. Rev. Astron Ap.*, in press.
Taub, A. 1948, *Phys. Rev.* **74**, 328.
Thorne, K. S. 1973a, *Ap. J.*, in press.
Thorne, K. S. 1973b, *Ap. J.*, in preparation.
Vainstein, S. I. and Zel'dovich, Ya. B. 1972, *Uspekhi Fiz. Nauk* **106,** 431.
Weyman, R. 1965, *Phys. Fluids* **8**, 212.
Wheeler, J. A. 1955, *Phys. Rev.* **97**, 511.
Yeh, T. 1970, *J. Plasma Phys.* **4,** part 2, 207.
Zel'dovich, Ya. B. and Novikov, I. D. 1966, *Astron. Zhur.* **43**, 758.
Zel'dovich, Ya. B. and Novikov, I. D. 1971, *Relativistic Astrophysics, Vol. 1* (Chicago: University of Chicago Press); cited as ZN.
Zel'dovich, Ya. B. and Novikov, I. D. 1964, *Doklady Akad. Nauk. SSSR* **158,** 811.
Zel'dovich, Ya. B. and Raizer, Yu. P. 1966, *Physics of Shock Waves and High Temperature Hydrodynamic Phenomena*, Vols. I and II (New York: Academic Press).
Zel'dovich, Ya. B. and Shakura, N. I. 1969, *Soviet Astron.-A. J.* **46,** 225.

On the Energetics of Black Holes†

Remo Ruffini
Joseph Henry Laboratories, Princeton, N.J. 08540

† Work partially supported by N.S.F. Grant GP30799X to Princeton University.

Contents

PART I On the Critical Mass of a Neutron Star

1.1 Introduction 455
1.2 Special Relativity and the Concept of Critical Mass. . . . 457
1.3 Nuclear Interaction and the Value of the Critical Mass . . . 466
1.4 Critical Mass Without any Detailed Knowledge of the Equation of State at Supranuclear Densities 482
1.5 The Critical Mass and the Late Stage of Evolution of Stars . . 483
1.6 X-ray Sources and the Determination of the Critical Mass. . . 488

References 495

Appendix 1.1 Hagedorn Equation of State in Neutron Stars
C. Rhoades and R. Ruffini R1

Appendix 1.2 On the Masses of X-ray Sources
R. Leach and R. Ruffini R7

Appendix 1.3 On the Maximum Mass of a Neutron Star
C. Rhoades and R. Ruffini R19

PART II Electromagnetic Radiation in Static Geometries

2.1 Introduction 497
2.2 Timelike Geodesics in Schwarzschild and Reissner-Nordstrøm Geometries 499
2.3 Electromagnetic Perturbations in a Static Geometry . . . 508
2.4 Electromagnetic Radiation from a Particle Falling Radially into a Schwarzschild or Reissner-Nordstrøm Black Hole 511
2.5 Electromagnetic Radiation from a Particle in Circular Orbits in a Schwarzschild or a Reissner-Nordstrøm Geometry 516
2.6 In what Sense can we Speak of an Object Endowed with a Net Charge in Astrophysical Problems?. 525

References 527

Appendix 2.1 On the Energetics of Reissner-Nordstrøm Geometries
G. Denardo and R. Ruffini R33

Appendix 2.2 Electromagnetic Field of a Particle Moving in a Spherically Symmetric Black-Hole Background
R. Ruffini, J. Tiomno and C. Vishveshwara R45

Appendix 2.3 Fully Relativistic Treatment of the Brehmstrahlung Radiation from a Charge Falling in a Strong Gravitational Field.
R. Ruffini. R51

Appendix 2.4 Lines of Force of a Point Charge near a Schwarzschild Black Hole
R. Hanni and R. Ruffini R57
Appendix 2.5 Ultrarelativistic Electromagnetic Radiation in Static Geometries
R. Ruffini and F. Zerilli R75
Appendix 2.6 On a Magnetized Rotating Sphere
R. Ruffini and A. Treves R89

PART III Gravitational Radiation in Schwarzschild Geometries

Introduction 528
References 538
Appendix 3.1 Gravitational Radiation
R. Ruffini and J. A. Wheeler. R95
Appendix 3.2 Gravitational Radiation in the Presence of a Schwarzschild Black Hole. A Boundary Value Search
M. Davis and R. Ruffini R117
Appendix 3.3 Gravitational Radiation from a Particle Falling Radially into a Schwarzschild Black Hole
M. Davis, R. Ruffini, W. Press and R. H. Price . . R121
Appendix 3.4 Pulses of Gravitational Radiation of a Particle Falling Radially into a Schwarzschild Black Hole
M. Davis, R. Ruffini and J. Tiomno . . . R125
Appendix 3.5 Gravitation Radiation from a Mass Projected into a Schwarzschild Black Hole
R. Ruffini. R129
Appendix 3.6 Can Synchrotron Gravitational Radiation Exist?
M. Davis, R. Ruffini, J. Tiomno and F. Zerilli . . R133
Appendix 3.7 Polarization of Gravitational and Electromagnetic Radiation from Circular Orbits in Schwarzschild Geometry
R. Ruffini and F. Zerilli R137

PART IV Energetics of Black Holes

Introduction 539
References 546
Appendix 4.1 Reversible Transformations of a Charged Black Hole
D. Christodoulou and R. Ruffini R147
Appendix 4.2 On the Electrodynamics of Collapsed Objects
D. Christodoulou and R. Ruffini R151
Appendix 4.3 Metric of Two Spinning Charged Sources in Equilibrium
L. Parker, R. Ruffini and D. Wilkins R161

PART I On the Critical Mass of a Neutron Star

1.1 Introduction

The physics of a neutron star is by far one of the richest we can find in the entire field of physics. See Figure 1. At least five different regimes are encountered as we go from the outermost crust toward the inner core of the star.

(I) Plasma physics characterizes the magnetosphere and the outermost few centimeters of the crust. The presence of huge magnetic fields of $\sim 10^{12}$ gauss

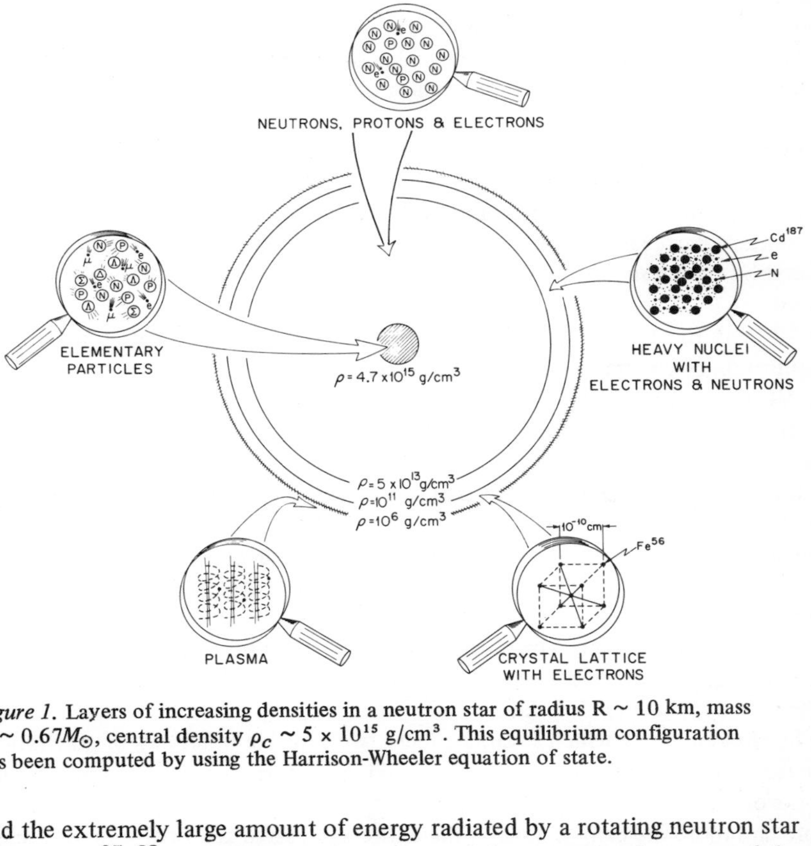

Figure 1. Layers of increasing densities in a neutron star of radius R ~ 10 km, mass $M \sim 0.67 M_{\odot}$, central density $\rho_c \sim 5 \times 10^{15}$ g/cm³. This equilibrium configuration has been computed by using the Harrison-Wheeler equation of state.

and the extremely large amount of energy radiated by a rotating neutron star ($dE/dt \sim 10^{37-38}$ erg/sec) make the analysis of the magnetosphere one of the most challenging opportunities to apply the concepts and knowledge of plasma physics obtained from investigations performed under "normal" conditions in laboratories on the earth.

(II) Solid state physics describes the next layer of material. This region can be as large as tens of kilometers in the less massive neutron stars and as thin as a few hundred meters in the more massive one. Lattices of iron nuclei or possibly the existence of compounds in the presence of a degenerate background of electrons make this region a natural place to apply concepts and techniques acquired in the analysis of solid state properties of matter under normal conditions of pressure and density.

(III) Low energy nuclear physics is the domain of the third layer. Neutron rich nuclei are here in equilibrium with a relativistic electron gas. The nuclei are so heavy that neutron and proton drip occur from them.

(IV) Nuclear physics is needed to describe the main core of the neutron star where $\sim 10^{57}$ neutrons are collected together in an enormous nucleus due to the effects of the gravitational field and the presence of a degenerate gas of protons and electrons allowing the core to be stable. As we will see, this is by far the most important region for the determination of the macroscopic features of the neutron star.

(V) High energy nuclear physics dominates the innermost core. All kinds of resonances and hadrons are generated due to the enormous compression and the ultrarelativistic Fermi energies. The usual knowledge of the spectrum of elementary particles must be used here under much more difficult conditions caused by the presence of collective effects and ultrarelativistic degenerate matter.

All these five regimes are not yet sufficient to describe completely the equilibrium configuration of a neutron star.

(VI) General relativity must also be used.

All the leading fields of physics participate in the description of this unique object in our universe.

Black Holes are markedly different in the regimes of theoretical physics needed for their descriptions: No equation of state, or strong interactions, or degenerate gas of elementary particles here! Purely carved in the empty geometry of spacetime, they are solely described by Einstein's theory of General Relativity.

Yet so similar in their dimensions: a neutron star with a mass near its critical value is expected to have a mass $\sim 1M_\odot$ and a radius of ~ 10 km, a black hole of $M \sim 2M_\odot$ a radius of ~ 5.88 km.

The key quantity which separates these two possible outcomes of gravitational collapse is the critical mass of a neutron star. First introduced by the pioneering works of Landau[1] and Chandrasekhar,[2] the existence of a critical mass has become more and more *the* central issue in the entire analysis of the final state of a star at the endpoint of thermonuclear evolution. The need, therefore, of establishing as clearly as possible the major physical parameters which, quite apart from minor details, determine its numerical value is the reason for this introductory chapter.

More precisely:

(I) how can we infer the existence of a critical mass just from special relativistic effects?
(II) how is the value of the critical mass affected by the detailed knowledge of the interactions between nucleons, and which are the most promising directions to attack this problem?
(III) can we establish, out of first principles and basic physical requirements, an absolute upper limit for the value of the critical mass?
(IV) can we measure experimentally the effective value of the critical mass?

Only after these basic questions are answered will we proceed to a detailed analysis of the "energetics of black holes" and all the associated processes of emission of radiation (both electromagnetic and gravitational) in the three subsequent chapters. Here is the place to mention how much the recent discovery of X-ray sources in binary systems has contributed to a new understanding in this field.[3] We finally have an extremely powerful experimental tool to verify many of the most advanced theoretical predictions in this field.

1.2 Special Relativity and the Concept of Critical Mass

As far as 1927 E. Fermi[4] and L. Thomas[5] presented a simple model for the description of an atom. Their approximation was particularly good for atoms with a large number of electrons. The main assumptions in this model can be easily summarized:

(a) The electrons are described by a degenerate gas of Fermi particles. The total number of particles N in a given volume of phase space is simply given by

$$2\,\frac{4\pi P_F^3}{3h^3}\,V = \frac{P_F^3}{3\pi^2\hbar^3}\,V = N \tag{1}$$

$$P_F = (3\pi^2\hbar^3 n)^{1/3} \tag{2}$$

where P_F is the Fermi momentum of the electron gas and n the number of particles per unit of volume.

(b) All the interaction between electron are neglected apart from their coulomb interaction with the nucleus of charge Z. The electric potential ϕ fulfills the equation

$$\Delta\phi = 4\pi ne \tag{3}$$

with

$$P_F^2/2M - e\phi = \text{const} \tag{4}$$

For a neutral atom we have

$$n = \frac{(2me\phi)^{3/2}}{3\pi^2\hbar^3} \tag{5a}$$

$$\Delta\phi = \frac{8\sqrt{2}(em)^{3/2}}{3\pi\hbar^3}\phi^{3/2} \tag{5b}$$

which is the well known Fermi-Thomas equation. In dimensionless units equation (5) becomes

$$d^2\chi/dx^2 = \chi^{3/2}/x^{1/2} \tag{6}$$

with

$$r = \tfrac{1}{2}\left(\frac{3\pi}{4}\right)^{2/3}\frac{1}{Z^{1/3}}\frac{\hbar^2}{me^2}x \tag{7a}$$

and

$$\phi = \frac{Ze}{r}\chi(x) \tag{7b}$$

and boundary conditions

$$\chi(0) = 1 \tag{8a}$$

and

$$\chi(\infty) = 0 \tag{8b}$$

Once the universal equation (6) is tabulated (see e.g. Landau and Lifshitz[6]) then the electron distribution for *any* atom with $Z \gg 1$ can be found from Eq. (5a) just by a scaling law obtained from Eqs (8a) and (8b). In particular we can see that the majority of the electrons are confined, in an atom of atomic number Z, to a distance from the nucleus $\propto Z^{-1/3}$: the larger the Z the more compact the electron distribution is. The success of this simple model is based mainly on its good agreement with the experimental results: ionization potentials, charge distributions can be predicted and directly compared with experimental data. The agreement is striking. To a more sophisticated many-bodies treatment, an Hartre-Fock approximation, is left the theoretical justification for the validity of the Fermi-Thomas model and the analysis of the detailed interactions taking also into account the effects of the spin of the electrons. The important point for us here is to remark how much can be done with so simple a model and also to stress the clear existence of scaling laws.

Can we develop a similar simple model for neutron stars? Again we can make some drastic simplifying assumptions

(1) describe the neutron gas by a degenerate system of neutrons neglecting all possible interactions between them apart from their mutual gravita-

tional attraction. The density of particles is therefore still related to the Fermi momentum by Eq. (2)

(2) the gravitational interaction can be simply described by the newtonian and non-relativistic equation

$$\Delta g = 4\pi G m n \tag{9a}$$

and

$$\frac{P_F^2}{2m} = mg \tag{9b}$$

and therefore

$$n = \frac{P_F^3}{3\pi^2\hbar^3} = \frac{(2m^2 g)^{3/2}}{3\pi^2\hbar^3} \tag{10a}$$

$$\Delta g = -\frac{8\pi\sqrt{2}Gm}{3\pi^2\hbar^3}(m^2 g)^{3/2} \tag{10b}$$

in dimensionless units Eq. (9a) reduces to

$$\frac{d^2\chi}{dx^2} = -\frac{\chi^{3/2}}{\sqrt{x}} \tag{11}$$

with

$$r = \frac{1}{2}\left(\frac{3\pi}{4}\right)^{2/3}\frac{1}{N^{1/3}}\frac{h^2}{Gm^3}x \tag{12.1}$$

and

$$g = \frac{NmG}{r}\chi \tag{12.2}$$

The integration of Eq. (11) has been done numerically, assuming the normalization condition

$$\int_0^{x_{\max}} \chi^{1/2} x^{3/2}\, dx = 1 \tag{13}$$

and the regularity of the potential g at the origin of the coordinate system. The results of the integration are given in Figure 2; like in the case of an atom an equilibrium configuration exists here for an *arbitrary* number N of particles. No collapse, no critical mass are present! From a solution with a number of particles N_1 it is possible to obtain a new solution with a larger number of particles N_2 just shrinking the dimensions of the equilibrium configuration by a factor $(N_1/N_2)^{1/3}$, following Eq. (12.1), and deepening the gravitational potential by a factor N_2/N_1 as from Eq. (12.2).

Is this treatment really valid? What agreement have we with experimental data? For a neutron star from the observation of pulsar timing we expect to have an object of radius $\sim 10^{12}$ kilometers and mass of the order of 1 solar mass at nuclear or supernuclear density. Assuming for m the mass of the neutron and $N \sim 10^{57}$, the equilibrium configuration obtained on the basis of the Fermi-Thomas model will have a radius $R \sim 6.53$ km, a mass $M \sim 1.7 \times 10^{33}$ gr and a central density $\rho_c \sim 8 \times 10^{15}$ g/cm^3, in quite good agreement with the dimensions to be expected for a neutron star. What happens, however, for a still larger number of particles? From the scaling laws given by Eqs (12.1) and (12.2) we see that the system should shrink and become nearly pointlike! A clear breakdown of our approximation has, therefore, to occur. Special relativistic effects cannot be any longer neglected once the Fermi energy of the neutrons becomes comparable with the rest mass energy of the neutron and general relativistic effects cannot be neglected once the system approaches its own Schwarzschild radius.

The first condition, namely a significant contribution of special relativistic effects, is reached once we have a critical number N of particles such that

$$mg = N^{4/3} m^5 G^2 \hbar^{-2} \left(\frac{\chi}{x}\right)_0 \sim N^{4/3} m^5 G^2 \hbar^{-2} \sim mc^2 \tag{14}$$

or

$$N_{\text{crit}} \sim \left(\frac{m_{\text{Planck}}}{m}\right)^3 \sim 10^{57} \tag{15}$$

Here as usual we assume $m_{\text{Planck}} = (\hbar c/G)^{1/2}$, m being the mass of a neutron. At the same time, for a configuration of equilibrium corresponding to a number of particles $N \sim N_{\text{crit}}$, we have

$$\frac{GmN_{\text{crit}}}{Rc^2} \sim 0.1 \tag{16}$$

We have indicated by R the radius of the system of self-gravitating neutrons. We, therefore, conclude that for $N \sim N_{\text{crit}}$ the system is expected to be very near its own Schwarzschild radius and general relativistic effects should be far from negligible.

To properly evaluate the contributions due to special and general relativity we still describe the neutron star material as an ideal, degenerate, self-gravitating, perfect gas of Fermions, neglecting all the nuclear interactions. Instead of using the Fermi-Thomas model of the atom, however, we will now use the more familiar equations of hydrostatic equilibrium and describe the fluid by an equation of state $p = p(\rho)$.

For a degenerate relativistic system of Fermions we have (see e.g. Landau and Lifshitz[6])

$$P = \left(\frac{m^4 c^5}{32\pi^2 \hbar^3}\right) \left(\tfrac{1}{3} \sinh \xi - \tfrac{8}{3} \sinh \tfrac{1}{2}\, \xi + \xi\right) \tag{17.1}$$

$$\rho = \frac{E}{V} = \frac{m^4 c^5}{32\pi^2 \hbar^3} (\sinh \xi - \xi) \tag{17.2}$$

and

$$n = \frac{N}{V} = (mc/\hbar)^3 (1/3\pi^2) \sinh^2 (\xi/4) \tag{17.3}$$

Here $\xi = 4 \sinh^{-1}(p_F/mc)$. We have indicated with n the density of particles while with ρ we indicate the total energy per unit of volume, including the rest mass and the relativistic contribution due to the kinetic energy. The profound difference introduced by special relativity in this equation of state can be most easily focused by going to the extreme relativistic and non-relativistic limits.

In the non-relativistic limit the dependence of the pressure from the number density of particles is

$$P = (3\pi^2)^{2/3} (\hbar^2/5m)(N/V)^{5/3} \tag{18}$$

while in the extreme relativistic regime

$$P = (3\pi^2)^{1/3} (\hbar c/4)(N/V)^{4/3} \tag{19}$$

To have pointed out, as far as 1931, the deep significance of this "*softening*" of the equation of state due to special relativistic effects for the final configuration of a star at the endpoint of thermonuclear evolution, has been the great merit of L. D. Landau.

As the density increases, Landau[1] pointed out, the dependence of the pressure from density smoothly varies from the expression given by Eq. (18) to the one given by Eq. (19). The system, if endowed with a mass larger than a critical value, reaches a density so high that special relativistic contributions are not negligible anymore. As a direct consequence, the pressure is not strong enough to support the system in an equilibrium configuration and *gravitational collapse* occurs. The phenomenon of gravitational collapse was, therefore, discovered and tied to a pure special relativistic effect. To a deeper analysis due to Oppenheimer and Volkoff[7] and Oppenheimer and Snyder[8] in 1939 was left the task of systematically analyzing the contributions of a general relativistic treatment to the equilibrium configuration of a neutron star and to explore for the first time the physical implications of a continued gravitational contraction of a collapsing star up to its own Schwarzschild radius.

Let us focus here first on the effects due to special relativity. The Newtonian equilibrium configuration for any selected value of the central density is immediately obtained by integrating the equations

$$\frac{dm(r)}{dr} = 4\pi r^2 \rho(r) \tag{20.1}$$

$$\frac{dp}{dr} = -\frac{m(r)}{r^2}\rho(r) \tag{20.2}$$

Here and in the following we use geometrical units $G = c = 1$. Different from a classical Newtonian treatment, our special relativistic treatment has to take into account the contribution to the gravitational field due to the kinetic energy of the particles. Therefore, instead of using as usual $\rho(r) = n(r) \times mc^2$, we use the density of energy of the degenerate neutron gas as given by Eq. (17.2). The mass that comes out from the Newtonian treatment

$$M = \int_0^R 4\pi\rho r^2 \, dr$$

is neither the strict Newtonian mass (including as it does mass energy of compression) nor the strict special relativistic value. It fails to correct for the mass energy equivalent of gravitational binding:

$$E_{gr} = -\int_0^R \frac{m(r)}{r} \, dm(r) \tag{21}$$

It is simple to make the necessary corrections and have

$$M_{\mathrm{ToT}} = M + E_{gr} \tag{22}$$

Results of the integration are given in Figure 2.

The general relativistic equilibrium configuration, for any value of the central density ρ_0 and for the same relativistic equation of state, can be obtained by direct integration of Eq. (17) and Eq. (20.1) and the general relativistic generalization of Eq. (20.2).

$$\frac{dp(r)}{dr} = -\frac{(\rho + p)(m + 4\pi r^2 p)}{r(r - 2m)}. \tag{23}$$

Most important in this treatment is the effect of the "self regeneration" of pressure present in Eq. (23) and the divergence in the strength of the gravitational interaction for $r = 2m$. Both these effects, however, have only a limited effect in our idealized model for the computation of the equilibrium configuration of a cold catalyzed star due to the special relativistic instability discovered by Landau.

Results and comparison between the three different treatments here considered are given in Figure 2 and Table 1. The main conclusions can be summarized as follows:

(1) In a pure Newtonian non relativistic treatment a configuration of equilibrium exists for *any* number of particles. Simple scaling laws allow us to go from one equilibrium configuration relative to a number N_0 of particles to one with a different number of particles (arbitrary!). No critical mass against gravitational collapse can possibly exist.

(2) The introduction of special relativity alone "softens" the equation of state at high densities. The existence of simple scaling laws disappears and the maximum critical mass against gravitational collapse is introduced.

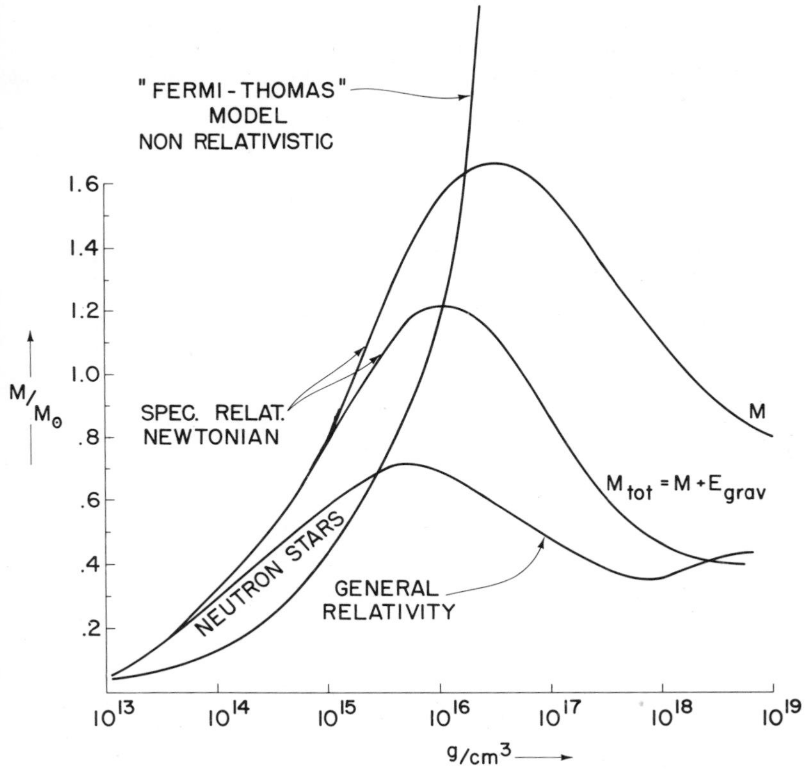

Figure 2. Equilibrium configurations of a neutron star computed for selected values of the central density, using the equation of state of a fully degenerate neutron gas interacting only gravitationally. The upper curve represents the configurations of equilibrium computed out of a non-relativistic "Fermi-Thomas" gravitational model. Next curve down, Newtonian treatment including rest mass and mass energy of compression computed from a special relativistic equation of state. Middle curve, same as previous one corrected for mass-energy of gravitational binding. Lower curve computations from general relativistic computations for the same fully degenerate and relativistic neutron gas.

(3) General relativity gives very significant quantitative contributions but does not substantially modify the general results of the special relativistic treatment. The critical density at which the maximum mass is reached is moved from a density of $\sim 10^{16}$ g/cm^3 down to a density of 6×10^{15} g/cm^3. It also reduces the critical mass for gravitational collapse from $1.2 M_\odot$ down to $0.7 M_\odot$, (see Figure 2).

The entire treatment we have here presented for neutron stars is heavily based on two major assumptions:

(a) The neutron star material can be described by a degenerate gas of self-gravitating neutrons in which all possible strong interactions are neglected. Is

TABLE 1. General relativity versus Newtonian gravitation theory for configuration of hydrostatic equilibrium, in both cases for the equation of state of a degenerate relativistic neutron gas

ρ (g/cm^3)	General Relativity		Newtonian Theory	
	$M/M_\odot$	R(km)	$M/M_\odot$	R(km)
7.781×10^{13}	0.260	41.9	0.279	42.1
2.729×10^{14}	0.423	21.4	0.495	22.3
6.127×10^{14}	0.528	16.7	0.671	17.8
1.082×10^{15}	0.593	14.1	0.816	15.4
1.660×10^{15}	0.642	12.3	0.961	13.8
2.361×10^{15}	0.674	11.0	1.090	12.6
3.214×10^{15}	0.700	10.0	1.232	11.7
4.400×10^{15}	0.715	9.2	1.355	11.0
5.852×10^{15}	0.715	8.6	1.445	10.4
7.653×10^{15}	0.708	8.0	1.516	9.9
9.993×10^{15}	0.694	7.6	1.564	9.4
1.238×10^{16}	0.676	7.0	1.598	8.9
1.543×10^{16}	0.657	6.7	1.627	8.5
1.927×10^{16}	0.637	6.4	1.648	8.2
2.362×10^{16}	0.616	6.2	1.660	7.9
2.895×10^{16}	0.595	5.9	1.666	7.6
3.558×10^{16}	0.574	5.7	1.666	7.3
4.379×10^{16}	0.554	5.6	1.658	7.1
5.358×10^{16}	0.534	5.4	1.644	6.9
6.463×10^{16}	0.515	5.3	1.624	6.6
7.834×10^{16}	0.497	5.2	1.601	6.4
9.518×10^{16}	0.479	5.1	1.574	6.2
1.148×10^{17}	0.463	5.1	1.514	6.1
1.386×10^{17}	0.448	5.0	1.512	5.9

Column 1, central density in g/cm^3; columns 2 and 4, mass-energy as sensed at infinity, in units of the solar mass $M_\odot = 1.987 \times 10^{33}$ g; columns 3 and 5, radius in km.

this a meaningful approximation? How significant are the modifications introduced by a more accurate description of the composition of the neutron star crust (ions, heavy nuclei, lattice structure, etc.) and of the core of a neutron star (protons, neutrons, electrons in equilibrium against beta decay and all kinds of resonances in the innermost core)? How much is the "softness" of the equation of state modified by the strong interactions between nuclei? To these issues we will twin in the next two paragraphs.

(b) Both in the Fermi-Thomas model of the atom and in our treatment of the neutron star we idealize the system of many Fermions by a perfect, degenerate gas solely determined by its pressure and density. To prove the validity of this model in the case of the atom is straightforward—we can describe the system of many electrons in a self-consistent way by the Hartre-Fock approximation. The results of the Fermi-Thomas model are reobtained to a very good approximation. Can this treatment be extended to the case of a system of self-gravitating neutrons? This has indeed been done. Ruffini and Bonazzola[9] have developed a method of self-consistent fields to study the equilibrium configurations of a system of self-gravitating scalar bosons or spin $\frac{1}{2}$ fermions in the ground state. The many-particle system is described by a second quantized free field, which in the boson case satisfies the Klein-Gordon equation in general relativity,

$$\nabla_\alpha \nabla^\alpha \phi = \mu^2 \phi, \text{ here } \mu = \frac{mc}{h},$$

m being the mass of the boson field and $\nabla_\alpha(\nabla^\alpha)$ representing as usual covariant (controvariant) differentiation. In the Fermion case the many-particle system is described by a spin $\frac{1}{2}$ field fulfilling the general relativistic generalization of the Dirac equation, $\gamma_\alpha \nabla^\alpha \psi = \mu\psi$. In turn the coefficients of the metric $g_{\alpha\beta}$ are determined by the Einstein equations with a source term given by the mean value $\langle\phi|\, T_{\mu\nu}\, |\phi\rangle$ of the energy momentum tensor operator constructed from the scalar or the spinor field. The state vector $\langle\phi|$ corresponds to the ground state of the system of many particles. In the bosons case it is found that the concept of an equation of state completely breaks down for *any* value of the density of the system or for *any* number of particles. In the Fermion case, instead, it is shown that the concept of an equation of state makes sense up to a density of 10^{42} g/cm^3! This treatment confirms beyond any possible doubt the Oppenheimer-Volkoff treatment and the approximation of using an equation of state. The two approaches are equivalent to an extremely high approximation. Similar to the Harte Fock approximation in which an electromagnetic spin orbit coupling exist, in the gravitational case a gravitational spin-orbit coupling is evidenced, but its magnitude is generally negligible.

Much has been said recently about competing theories of gravity. It is, therefore, of some interest to analyze here the influence on the equilibrium configuration of a neutron star of the scalar tensor theory of gravity as given by Jordan[10] and Brans and Dicke[11]. The equations of hydrostatic equilibrium that this theory leads to are more complicated than either the Newtonian one or the simple standard Einstein theory. The necessary equations in Schwarzschild coordinates are

$$\frac{dm(r)}{dr} = 4\pi r^2 \left[\frac{\rho}{\phi} + \frac{\omega\phi'^2}{16\pi\phi^2}\frac{r-2m}{r} + \frac{1}{\phi}\frac{(3p(r)-\rho(r))}{3+2\omega}\right] r^2 \tag{24}$$

for the Jordan-Brans-Dicke version of Eq. (20.1).

$$-\frac{dp(r)}{dr}=\frac{p+\rho(p)}{r(r-2m)}\left[m+4\pi r^3\left(\frac{p}{\phi}-\frac{\omega}{4^2}\left(1-\frac{2m(r)}{r}\right)\phi'^2\right)\right] \tag{25}$$

is the Jordan-Brans-Dicke version of Eq. (23), and finally we have the scalar wave equation with source term

$$\phi''+\left[\frac{3}{r}-\frac{r-2m}{r^2}+\frac{4\pi}{\phi}(p-\rho)-\frac{\omega\phi'^2}{2\phi^2}(r-2m)-\frac{8\pi}{\phi}\frac{r(3p-\rho)}{(3+2\omega)}\right]\phi'$$

$$=\frac{8\pi}{(3+2\omega)}(3p-\rho)\frac{r}{(r-2m)} \tag{26}$$

Here we indicate the scalar field with ϕ and with the prime its derivative with respect to r. The quantity ω is a dimensionless constant for which Dicke favours a value in the range $4 \leqslant \omega \leqslant 6$. To recover the general relativistic equations it is sufficient to make $\omega \to \infty$.

The integration of this system of equations with $\omega = 4$ gives results which are qualitatively identical to the general relativistic ones. The only quantitative difference is an increase of approximately two per cent ($M_{\text{crit}} = 0.730M_{\odot}$) in the value of the critical mass for a neutron star and an increase of a similar order of magnitude in the radius of the neutron star!

We can, therefore, conclude: The presence of a critical mass in the equilibrium configuration of neutron stars is due to pure special relativistic effects. Different gravitational theories introduce *no qualitative* changes in the description of the family of stable equilibrium configurations of neutron stars. The *numerical* value of the critical mass can change within a multiplicative factor of 2 depending upon the particular gravitational theory used.

1.3 Nuclear Interaction and the Value of the Critical Mass

In the last paragraph we have emphasized the contribution of different gravitational theories to the existence of a critical mass of gravitational collapse in the computation of the equilibrium configuration of a neutron star. To focus on the main difference introduced by alternative gravitational theories (newtonian, general relativity, Jordan-Brans-Dicke theory) we described the neutron star material by an idealized system of completely degenerate, self-gravitating fermions neglecting all interactions between them. The aim in this paragraph is quite different: assuming that the equilibrium configuration can, indeed, be computed in the framework of general relativity by a perfect fluid approximation to analyze how the introduction of strong interactions between nucleons can affect the value of the critical mass. At the same time we will inquire into the effects of introducing a more realistic composition of the material of a neutron star.

In particular: (a) the influence of a lattice structure of nuclei and possibly of compounds in the crust and (b) the generation of all kinds of resonances and hadrons in the innermost core of a neutron star.

We can divide the entire material of the neutron star into three main regions;

(1) the crust with a density ρ $5.7\ \mathrm{g/cm^3} \lesssim \rho \lesssim 10^{13}\ \mathrm{g/cm^3}$

(2) the core with a density ρ $10^{13}\ \mathrm{g/cm^3} \lesssim \rho \lesssim 5 \times 10^{15}\ \mathrm{g/cm^3}$

(3) the innermost core with a density ρ $5 \times 10^{15}\ \mathrm{g/cm^3} \lesssim \rho$

The crust of a neutron star can, itself, be subdivided into a multiple set of concentric layers of various composition.

The following regimes are encountered as one goes down in depth from the "atmosphere": magnetic fields of the order of 10^{12} gauss are expected to exist on the surface. In the outermost layers of matter, as suggested by Ruderman,[12] matter might exist in the form of very dense (average density of the order of $10^4\ \mathrm{g/cm^3}$) one dimensional "hairs" parallel to the lines of force of the magnetic field. The next shell of material in the density range between $10^4\ \mathrm{g/cm^3} \lesssim \rho \lesssim 10^7\ \mathrm{g/cm^3}$ is expected to behave as a lattice of nuclei embedded in a degenerate gas of relativistic electrons (white dwarf material!), the reason being that the Fermi energy of the electrons is very much higher than the ionization energy of the atoms. Since the material is expected to be at the complete endpoint of nuclear evolution, the nuclei are thought to be mainly iron nuclei. However, as pointed out by F. Dyson,[13] in this material some incompleteness of combustion could occur (as, for example, H burned to Helium, but not burned to iron) and the formation of some compounds could still be possible (see Table 2). The Fe-He compound listed in the last column is purely illustrative. The details of the thermonuclear reactions taking place in the collapse of a core of a white dwarf material are not known with the necessary details to state what nuclear material and in what amount should be expected in the upper layers of a neutron star. The important point is just to give a conceivable example of incompleteness of combustion, implying the presence in any given layer of more than one nuclear species. The simpler example considered by Dyson clearly shows that a particularly stable configuration is given by a lattice with NaCl structure and with Fe-He composition.

At densities between $10^7\ \mathrm{g/cm^3} \lesssim \rho \lesssim 10^{11}\ \mathrm{g/cm^3}$ relativistic electrons transform bound protons into neutrons. Under normal circumstances, in fact, the total packing of a nucleus, under the two conflicting effects of nuclear and electrostatic forces is minimized for a value of $Z = 28$ and $A = 56$. A relativistic electron transmutes a nucleus of charge Z and atomic number A by inverse beta decay

$$e + (Z, A) \rightarrow (Z - 1, A) + \nu$$

The nuclei becomes neutron rich compared to nuclei unpressured by electrons. For these neutron rich nuclei the mass number $A = 56$ no longer represents the point of maximum stability. Stability shifts to higher A values. The details of this shifting process are far from being well understood. For any electron pressure there corresponds a nucleus with a fixed value of Z and A which is in beta equilibrium with the electrons and has the most favorable packing fraction. At still higher densities in the range 10^{11} g/cm$^3 \lesssim \rho \lesssim 5 \times 10^{12}$ g/cm^3 nuclei become so heavy ($A \sim 122$) and so neutron rich ($N/Z \sim 83/39$) that neutron "drip" occurs. An atmosphere of unbound free neutrons is formed. With a further increase in density the Fermi energy of the electrons increases and

TABLE 2. Properties of a superdense star of central density 3.5×10^{13} g/cm^3. (Calculations based on HW equation of state.[14] The calculated mass is $0.18 M_\odot$)

Schwarzschild radial coordinate (km) (approx.)	Density (g/cm^3)	Pressure (g/cm sec^2)	Dominant nucleus in an idealized neutron star at the absolute endpoint of thermonuclear evolution	Sample of conceivable constitution for an actual neutron star (incomplete thermonuclear combustion)
210 (top ~1 cm)	gaseous	gaseous	26 Fe 56	gaseous Fe
210 (next few cm)	7.85	"0"	26 Fe 56	Fe-He compound
170	8.00E 6	8.96E 23	26 Fe 56	Fe-He compound (crust)
50	1.67E 10	1.56E 28	31 Ga 78	Fe-He compound
30	3.18E 11	5.83E 29	39 Y 122	Fe-He compound
20	4.5E 12	6.62E 30	Fermi gas neutrons protons electrons	Superfluid neutrons plus superconducting protons plus degenerate electrons
2	3.49E 13	1.85E 32		

Reproduced from Cosmology from Space Platform—R. Ruffini and J. A. Wheeler—ESRO Book SP52.

the nuclei become even more neutron rich. The number of free electrons increases further. Three different components characterize this range of densities (a) an ultrarelativistic degenerate gas of electrons, (b) a system of heavy nuclei, (c) a degenerate neutron gas. The contribution of the nuclei to the pressure is always negligible while the contribution of neutrons becomes more and more relevant with the increase of density. At densities already of the order of $\sim 5 \times 10^{12}$ g/cm^3 the pressure of the degenerate gas of neutrons, extremely large by comparison to the one of the nuclei, is comparable to the one of the ultrarelativistic degenerate electron gas. To a further small increase in the density corresponds the disappearance of nuclei as such. The material of the star is uniquely formed of electrons, neutrons and protons in equilibrium against beta decay.

The properties of the material of the crust of a neutron star have been analyzed in depth in recent years largely using notions of solid state physics. Major contributions in this analysis have been made by D. Pines[15] and collaborators M. Ruderman[16] as well as F. Dyson[13] and R. Smoluchowsky[17] and C. Rhoades.[18] The major directions of research have been toward the determination of the composition, strength and conductivity of the material contained in the crust. The strength of this material is so small when compared with gravitational forces existing at the surface of the neutron star, that only "mountains" of a few centimeters or less could be supported on the surface. This entire analysis could, indeed, prove to be of extreme importance for the explanation of the tiny observed "spin up" in the period of Pulsars ($\Delta P/P \sim 2 \times 10^{-6}$ in the Vela Pulsar and $\Delta P/P \sim 10^{-9}$ in the case of the Crab Nebula Pulsar) as well as in the understanding of the electrodynamic processes taking place near the surface of a neutron star.

However, from the point of view of the main issue we are addressing ourselves in this chapter, namely the determination of the critical mass of gravitational collapse of the neutron star all these details can be completely neglected. The reason is simply stated: The crust of a neutron star which extends a few tenths of kilometers in the case of a configuration of equilibrium corresponding to a central density $\rho_c \sim 10^{14}$ g/cm^3, becomes extremely thin for configurations of equilibrium with larger values of the central density. For a neutron star with a central density $\rho_c \sim 5 \times 10^{15}$ g/cm^3 the entire configuration of equilibrium has shrunk to a radius of ~10 km, the crust is only a few hundred meters thick and only a few per cent of the total mass of the star is contained at a density $\rho \sim 10^{13}$ g/cm^3. See Figure 3.

We can, therefore, neglect completely for our purpose the contribution due to the crust and we will only focus on the physics of the core and innermost core.

The core of the neutron star can, indeed, be thought of as a big nucleus of $\sim 10^{57}$ particles. From traditional nuclear physics a reliable equation of state can be given up to densities of the order of $\sim 2 \times 10^{14}$ g/cm^3.

As a first approximation the system can be simply described by a mixture of three non interacting degenerate gases of neutrons, protons and electrons in equilibrium against beta decay. This approach followed by Harrison and Wheeler can be summarized in a few formulas

(a) the condition of electrical neutrality implies

$$n^{(e)} = [P_F^{(e)}]^3/3\pi^2\hbar^3 = (m^{(e)}c/\hbar)^3 \sinh^3(t^{(e)}/4)/3\pi^2 = n^{(p)}$$
$$= [P_F^{(p)}]^3/3\pi^2\hbar^3 = (m^{(p)}c/\hbar)^3 \sinh^3(t^{(p)}/4)/3\pi^2 \qquad (27)$$

where the superscript indicates the kind of particle under consideration. (e) electrons, (p) protons.

(b) to have stability against beta decay we need to have that the sum of the Fermi energies of the electron and proton gas be equal to the Fermi energy of

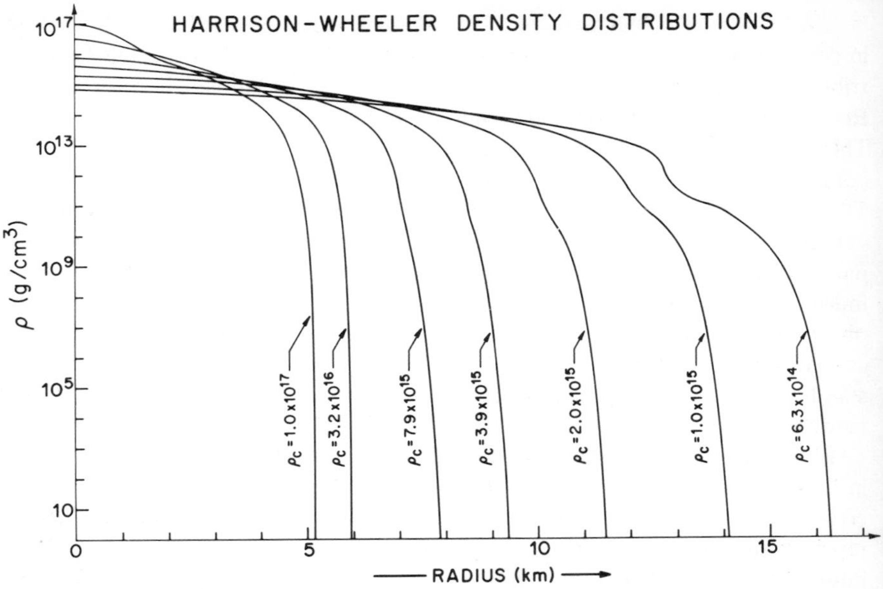

Figure 3. Density distribution of configurations of equilibrium of neutron stars for selected values of the central density. The equation of state used is the one of Harrison and Wheeler,[14] given in Table 4. As the central density increases the configuration of equilibrium shrinks. The radius of the star becomes increasingly smaller and the thickness of the crust ($\rho \lesssim 2 \times 10^{13}$ g/cm^3) drastically decreases. This last effect is even larger for smaller values of the central density.

the neutron gas

$$m^{(e)}c^2 \cosh(t^{(e)}/4) + m^{(p)}c^2 \cosh(t^{(p)}/4) = m^{(n)}c^2 \cosh(t^{(n)}/4) \tag{28}$$

(c) the pressure of the system of the three non interacting gases is simply given

$$\begin{aligned} P = {} & (m^{(e)^4} c^5/32\pi^2\hbar^3)(\tfrac{1}{3}\sinh t^{(e)} - \tfrac{8}{3}\sinh \tfrac{1}{2}t^{(e)} + t^{(e)}) \\ & + (m^{(p)^4} c^5/32\pi^2\hbar^3)(\tfrac{1}{3}\sinh t^{(p)} - \tfrac{8}{3}\sinh \frac{t^{(p)}}{2} + t^{(p)}) \\ & + (m^{(n)^4} c^5/32\pi^2\hbar^3)(\tfrac{1}{3}\sinh t^{(n)} - \tfrac{8}{3}\sinh (t^{(n)}/2) + t^{(n)}) \end{aligned} \tag{29}$$

(d) the density

$$\begin{aligned} \rho = {} & (m^{(e)^4} c^5/32\pi^2\hbar^3)(\sinh t^{(e)} - t^{(e)}) + (m^{(p)^4} c^5/32\pi^2\hbar^3)(\sinh t^{(p)} - t^{(p)}) \\ & + (m^{(n)^4} c^5/32\pi^2\hbar^3)(\sinh t^{(n)} - t^{(n)}) \end{aligned}$$

It is, therefore, straightforward to obtain the tabulation of the pressure versus the density and at the same time the determination of the abundance of protons, neutrons and electrons.

This treatment gives a value of the critical mass practically identical to the one obtained in the primitive work of Oppenheimer and Snyder.[8] The reasons are clearly understood: the contribution of the electron and proton gases, essential in guaranteeing the charge neutrality of the system as well as the stability against beta decay, are completely negligible in the evaluation of the total pressure and density. The major contributions, in fact, comes uniquely from the neutron gas. Recently an alternative picture has been introduced by Sawyer[19] and Scalapino.[20] They give evidence that a mixture of three gas of neutron, protons, and π^- could be energetically more favorable than the one traditionally adopted. However, this modification, while important as a matter of principle, cannot sizably modify the value of the critical mass.

The previous treatment, extremely handy for any estimate of the size and dimensions of a neutron star, in first approximation, neglects completely all the strong interactions between nucleons and fails in four major respects from being a realistic description of the material in the core of a neutron star:

(a) at densities below $\rho \sim 2 \times 10^{14}$ g/cm^3 the attractive effects of nuclear interactions is not taken into proper account. The equation of state will give a value of the pressure always larger than the one to be expected on the ground of a realistic treatment of nuclear interactions. In different words this equation of state can safely be considered the *upper limit* in hardness for all the realistic equations of state in the range of density 10^{13} g/cm$^3 \lesssim \rho \lesssim 2 \times 10^{14}$ g/cm^3. Consequently it will give upper limits to the masses of the equilibrium configurations computed up to nuclear densities.

(b) at densities $\rho \gtrsim 4 \times 10^{14}$ g/cm^3 this equation of state neglects the hard core repulsive forces. The value of the pressure computed will always be an *underestimate* of the pressure to be expected in any realistic analysis. Consequently, the value of the mass for the equilibrium configurations is also underestimated.

(c) No collective interaction between the many particle system is taken into account.

(d) the entire approximation completely breaks down at densities $\rho \gtrsim 5 \times 10^{15}$ g/cm^3 (innermost core). In this range of densities, as predicted and computed by Saakian and Vartanian,[21] the entire spectrum of hadrons is generated. Moreover, the strong interactions between the particles cannot be neglected without totally departing from a realistic analysis.

One of the first attempts to describe the interactions between nucleons in matter at nuclear density was done by Cameron *et al.*[22] as far as 1966. Their description was, indeed, extremely important in emphasizing the dependence of the value of the critical mass on the particular kinds of interactions considered between nuclei. At the same moment, however, it was failing from being a realistic treatment in a very important respect: the speed of sound in the nuclear material at supernuclear densities was found to be larger than the speed of light (causality violation!). The most recent attempt at giving a description of matter in this regime has been done following totally different approaches by V. R. Pandharipande[23] and by R. Hagedorn[24].

In the Pandharipande[23] approach the system of N nucleons ($N \sim 10^{57}$!) is described by a Schrödinger equation of the form

$$-\sum_{mi} \frac{\hbar^2}{2M_n} \nabla_i^2 - \frac{1}{2} \sum_{m,n,i,j} v_{mn}(r^{ij})\psi(r^{mi} \ldots) = E\psi(r^{mi} \ldots) \tag{31}$$

here and in the following the indices m, n, o, etc., denote the different kinds of hadrons considered, while i, j, k, etc., label the single particles or states.

Two main assumptions have already been made at this point of the treatment:

(1) the possibility of using the Schrödinger equation, (a *non relativistic* equation) in the description of the highly relativistic system of hadrons at nuclear and supernuclear densities.

(2) the use of a *two body* potential $v_{mn}(r^{ij})$.

Both of these assmptions could very well reveal to be inadequate for the solution of our problem; the failure of any one of these two will clearly invalidate the entire approach.

Equation (31) is unsolvable for $N > 3$. To solve the problem a further approximation is needed.

(3) we can adopt a cluster expansion[25] of Eq. (31). Neglecting for simplicity the indices m, n, we can then expand

$$E = \sum_{n=1}^{N} E_{on} \tag{32}$$

where

$$E_{o1} = \sum_{i=1}^{N} C_1(i) = \sum_{i=1}^{N} \frac{(\psi_1(r_i), H_1(r_i)\psi_1(r_i))}{(\psi_1(r_i), \psi_1(r_i))}$$

$$E_{o2} = \sum_{1 \leqslant i < j \leqslant n} C_2(i, j) = \sum_{1 \leqslant i < j \leqslant n} \frac{(\psi_2(r_i, r_j), H_2(r_i, r_j)\psi_2(r_i, r_j))}{(\psi_2(r_i, r_j), \psi_2(r_i, r_j))} - C_1(i) - C_1(j) \tag{33}$$

and more generally for an arbitrary n

$$E_{on} = \sum_{\{n\} c N} C_n(\{n\}) = \sum_{\{n\} c N} (\psi_n(\{n\}), H_n(\{n\})\psi_n(\{n\}))/(\psi_n(\{n\}), \psi_n(\{n\})) - \sum_{m=1}^{n-1} \sum_{\{m\} c \{n\}} C_m(\{m\}) \tag{34}$$

Here we have indicated by $\{n\} cN$ the n particle subset of N particles, and

$$H_n = -\frac{h^2}{2M} \sum_{i=1}^{n} \nabla_i^2 + \sum_{1 \leqslant i < j \leqslant n} v_{ij} \tag{35}$$

A further approximation is introduced at this stage.

(4) the expansion (35) is truncated at the two body clusters. This truncation, certainly valid in the low density regime ($\rho \leqslant 2 \times 10^{14}$ g/cm^3), should be very carefully and critically examined in the regime of nuclear and supernuclear densities where also distant neighbors might be strongly correlated and, therefore, have an effect far from negligible on the energy E.

The two body potential will usually depend on the angular momentum l and will also differ for different nucleons. We are confronted with a further approximation.

(5) the determination of the potential $v^l_{mn}(r, j)$. In Figure 4 different possibilities are presented for the $l = 0$ state.

Clearly all the three possible choices are in perfect agreement for inter nucleon distances larger than 1 Fermi where the potential is still attractive, while they strongly differ in their short range behavior. This short range behavior is, indeed, the *only* one which determines the main features of the equation of state at nuclear and supernuclear density. The knowledge of the detailed form of the potential is still poorer for larger l values! Clearly a potential of the Hamada Johnston type will necessarily introduce a supraluminous speed of sound at high densities as a consequence of the infinitely repulsive core.

Pandharipande chooses for the v^i_{mn} expressions of the Reid[26] form

$$v^l_{mn} = \Sigma_j\, a^l_{mnj} \frac{e^{-j\mu r}}{\mu r} \qquad j = 1, 2, 4, 6, 7 \tag{36}$$

TABLE 3. a^l_{nni} from Reid potential in MeV

j	1	2	4	6	7
$l = 0$	−10.463	0	−1650.6	0	+6484.2
$l =$ odd	+ 3.448	0	− 933.48	+4152.1	0
$l =$ even $\neq 0$	−10.463	−12.332	−1112.6	0	+6484.2

with the a^l_{mnj} given in Table 3.

What about the interactions between the many different hadrons in the core or in the innermost core? A further approximation is needed.

(6) universal repulsion and intermediate range attraction are chosen for all pairs. We therefore, have

$$a^l_{mni} = a^l_{nni} \tag{37}$$

for all m and n and for $i = 6, 7$ (repulsive terms) and $i = 2, 4$ (intermediate range attraction). This last assumption has been relaxed in the recent work of V. Canuto[27] where more realistic interactions of hyperonic matter are made on the basis of recent experimental data analysis.

Important for us here is to emphasize the large number of major approximations adopted in this entire treatment. The breakdown of any one of them would make the entire model largely uninteresting and unreliable. Moreover, even assuming the validity of the entire set of successive approximations the knowledge of the two body interactions even in lower order is extremely poor, see Figure 4. Before inferring the major consequences of the equation of state obtained following this approach, let us look at a totally different one: the Hagedorn thermodynamical approach. The usual description based on the many body Schrödinger equation explores the domain of densities $\rho \gtrsim 10^{13}$ g/cm^3. The Hagedorn work applies to densities $\rho \gtrsim 5 \times 10^{15}$ g/cm^3. Emphasis on the interactions in the previous approach, total emphasis on the spectrum of the particle and resonances

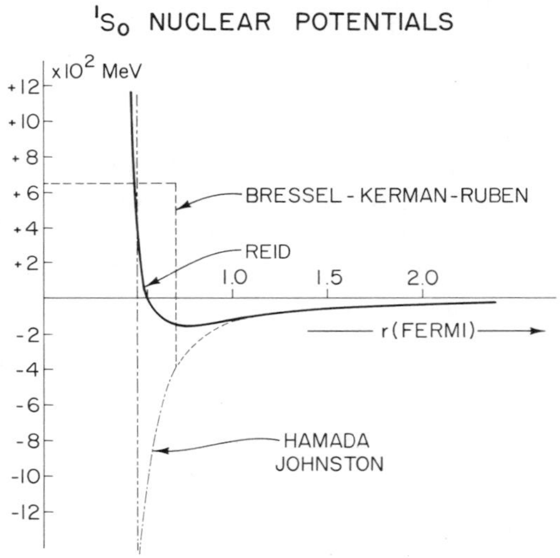

Figure 4. Nuclear potentials for two bodies interaction $l = 0$ state. For the determination of the equation of state at supranuclear densities the knowledge of the potential at distances smaller than 1.0 Fermi is essential. The indeterminacy in the knowledge of the $l = 0$ two body potential is well summarized in the present figure.

here with a complete neglect of the interactions! A model which originates not from nuclear physics but from elementary particle physics. This statistical treatment was first introduced by Fermi[28] in the analysis of high energy collisions. No better explanation of the physical reasons at the basis of this approach than the words of Fermi himself:

"When two nucleons collide with very great energy in their center of mass system this energy will be suddenly released in a small volume surrounding the two nucleons. We may think pictorially of the event as of a collision in which the nucleons with their surrounding retinue of pions hit against each other so that all the portion of space occupied by the nucleons and by their surrounding pion field will be suddenly loaded with a very great amount of energy. Since the interactions of the pion field are strong we may expect that rapidly this energy will be distributed among the various degrees of freedom present in this volume according to statistical laws. One can then compute statistically the probability that in this tiny volume a certain number of pions will be created with a given energy distribution. It is then assumed that the concentration of energy will rapidly dissolve and that the particles into which the energy has been converted will fly out in all directions.

It is realized that this description of the phenomenon is probably as extreme, although in the opposite direction, as is the perturbation theory approach. On the other hand, it might be helpful to explore a theory that deviates from the unknown truth in the opposite direction from that of the conventional theory. It may then be possible to bracket the correct state of fact in between the two theories. One might also make a case that a theory of the kind here proposed may perhaps be a fairly good approximation to actual events at very high energy, since then the number of possible states of the given energy is large and the probability of establishing a state to its average statistical strength will be increased by the very many ways to arrive at the state in question.

The statement that we expect some sort of statistical equilibrium should be qualified as follows. First of all there are conservation laws of charge and of momentum that evidently must be fulfilled. One might expect further that only those states that are easily reachable from the initial state may actually attain statistical equilibrium. So, for example, radiative phenomena in which photons could be created will certainly not have time to develop. The only type of transitions that are believed to be fast enough are the transitions of the Yukawa theory. A succession of such transitions starting with two colliding nucleons may lead only to the formations of a number of charged or neutral pions and also presumably of nucleon-anti-nucleon pairs. The discussion shall be limited, therefore, to these particles only. Notice the additional conservation law for the difference of the numbers of the nucleons and the anti-nucleons.

The proposed theory has some resemblance to a point of view that has been adopted by Heisenberg† who describes a very high energy collision of two nucleons by assuming that the pion "fluid" surrounding the nucleons is set in some sort of turbulent motion by the impact energy. He uses qualitative ideas of turbulence in order to estimate the distribution of energy of this turbulent motion among eddies of different sizes. Turbulence represents the beginning of an approach to thermal equilibrium of a fluid. It describes the spreading of the energy of motion to the many states of larger and larger wave number. One might say, therefore, in a qualitative way that the present proposal consists in pushing the Heisenberg point of view to its extreme consequences of actually reaching statistical equilibrium".

The approach of Hagedorn is based on the assumption that, indeed, a thermodynamic description of material compressed to supranuclear density is possible.

In statistical mechanics[29] if we have an ensemble A of systems M with energy $\{E_1, E_2, \ldots, E_k \ldots\}$ we can characterize an actual state of the ensemble by saying how many systems (n_1) have an energy E_1, how many (n_2) energy E_2, etc., or by a vector

$$\mathrm{n} = (n_1, n_2, \ldots, n_k, \ldots) \tag{38}$$

Here n_j and E_j are subject to the conservation law

$$N = \sum_i n_i \quad \text{and} \quad E^{(N)} = \sum_i n_i E_i \tag{39}$$

The average energy of a system is

$$\bar{E} = \left(\sum_i E_i e^{-BE_i}\right) / \sum_i e^{-BE_i} = -\frac{d}{dB} \ln \left(\sum_i e^{-BE_i}\right) \tag{40}$$

† W. Heisenberg. *Nature* **164**, 65, 1949; *Zs. fur Phys.* **126**, 569, 1949.

and the average occupation number

$$\frac{\bar{n}_j}{N} = w^j = \frac{e^{-BE_j}}{\sum_i e^{-BE_i}} = -\frac{1}{B}\frac{\partial}{\partial E_j} \ln\left(\sum_j e^{-BE_j}\right) \tag{41}$$

If we now consider a thermodynamic treatment we can then ascribe to each system M a volume V, a temperature T, pressure P, energy E, entropy S etc. We then have

$$P = T\frac{\partial}{\partial V}[\ln Z(V, T)] \tag{42.1}$$

$$\bar{E} = T^2 \frac{\partial}{\partial T}[\ln Z(V, T)] \tag{42.2}$$

$$S(E, V) = \frac{\partial}{\partial T}[T \ln Z(V, T)] \tag{42.3}$$

$$w_j = -T\frac{\partial}{\partial E_j}[\ln Z(V, T)] \tag{42.4}$$

where $Z(V, T) = \Sigma_i\, e^{-E_i B}$, with $B = 1/T$, is the partition function of the system. We can define the level density $\sigma(E, V)$, $\sigma(E, V)\, dE$ being the number of levels lying between E and $E + dE$. We then have

$$Z(V, T) = \int_0^\infty \sigma(E, V)\, e^{-E/T}\, dE \tag{42.5}$$

or $Z(V, T)$ is the Laplace transform of $\sigma(E, V)$. Therefore, from the knowledge either of $Z(V, T)$ or $\sigma(E, V)$ we can determine the thermodynamic properties of the system.

To determine the function $\sigma(E, V)$ in the specific case of matter at supranuclear density much has to be known on the spectral distribution and on the interactions of all elementary particles and resonances.

Here, however, following the work of Bethe and Uhlenbeck[30] and of Belenskij,[31] Hagedorn introduces a drastic simplification: eliminate all the interactions between the particles and express them by the change of level density caused by additional terms due to the interaction phase shift.

This fundamental idea can be exemplified as follows: Let us have N non interacting particles of mass $m_1, m_2, \ldots, m_N$ with a level density

$$\sigma_N(E, V, m_1, \ldots, m_N). \tag{42.6}$$

Now let a force act between particle 1 and 2 such that the two particles are bound together in a new state m_{12}—we can describe the system by using the new level density

$$\sigma_{N-1}(E, V, m_{12}, m_3, \ldots, m_N) \tag{42.7}$$

In other words, instead of considering the detailed features of the interaction we can simply consider again a system of *non* interacting particles with a change in the level density.

In a similar way, if we consider in a quantum mechanical system the transition probability of a system in an initial status $|i\rangle$ and energy E to a final state $|f\rangle$ with a N particles is given by

$$P(N, E) = \int |\langle f|S|i\rangle|^2 \, \delta\left(E - \sum_{i=1}^{N} E_i\right) \delta^3\left(\sum_{i=1}^{N} \vec{p}_i\right) \prod_{i=1}^{N} d^3\vec{p}_i$$

$$= \int |\langle f|S|i\rangle|^2 \, dr_N(E, m_1, m_2, \ldots, m_N). \tag{43}$$

Suppose now that particles 1 and 2 have a resonant state $(1, 2)^*$ with an invariant mass m^*, then instead of expression (43) we can write

$$P(N, E) = \int |\langle f|S'|i\rangle|^2 \, dR_N(E, m_1, m_2, \ldots, m_N)$$

$$+ \int |\langle f|S'|i\rangle|^2 \, dR_{N-1}(E, m^*, m_3, \ldots, m_N) \tag{44}$$

where in S' the interaction leading to the resonance has been eliminated.

The problem of taking into proper account strong interactions is, therefore, solved if we could know the *detailed* spectrum of all particles and resonances, or in other words the function $\rho(m)\,dm$ = number of distinct particles and resonances in the mass interval (m, dm). In fact, the Beth-Uhlenbeck-Belenskij formalism allows Hagedorn to eliminate all the interactions from $|S|^2$ and to express them as a change in the level density. Formally the gas is ideal (non interacting particles), in reality strong interactions are fully contained in the hadronic spectrum and in the freedom to create and annihilate the particles described by the function $\rho(m)$. The problem in this approach is, therefore, reduced to obtain a detailed knowledge of the function $\rho(m)$. The knowledge of this function from experimental data is, however, far from complete.

To overcome this difficulty, Hagedorn imposes the "bootstrap" condition that for very high energies

$$\lim_{m \to \infty} \frac{\rho(m)}{\sigma(m, V_0)} = 1 \tag{45}$$

m being the mass of the very massive or energetic "particle" or "fireball". "Somewhere where the mass becomes smaller and smaller, selection rules begin to make a thermodynamic description less and less applicable, and fireballs gradually obtain, according to their quantum numbers individual personalities. They become what we call resonances. On the other hand, we know that with increasing mass, resonances begin to decay statistically into more and more decay channels, some of which contain resonances themselves. In other words: resonances with greater and greater masses gradually become what we call fireballs. Thus fireballs are resonances and vice versa—we call all of them fireballs".[34]

From this bootstrap condition a mass spectrum is obtained.

$$\rho(m) \underset{m\to\infty}{\Rightarrow} cm^{-\alpha} e^{m/T_0} \tag{46}$$

Hagedorn first suggested a value for $\alpha = \frac{5}{2}$ and $T_0 = m_\pi = 160$ Mev. In the meantime much more has been analyzed on the nature of the spectra of elementary particles. An exponential mass spectrum has been confirmed by Fubini and Veneziano,[32] Fubini, Gordon and Veneziano[33] on the basis of a different approach to a bootstrap theory. The most reliable value for the index α after the analysis of Frautschi and Hamer[34] and Frautschi[35] appears to be $a = 3$.

If we take the initial value proposed by Hagedorn we can obtain two different equations of state for matter at supranuclear densities:

in the limit $T \to T_0$

$$p = p_0 + \rho_0 + \ln(\rho/\rho_0)$$

$$\rho_0 = 1.253 \times 10^{14} \text{ g/cm}^3 \quad \text{and} \quad p_0 = 0.314 \times 10^{14} \text{ g/cm}^3 \tag{47}$$

in the limit $T \to 0$

$$p = \rho/\ln(\rho/\rho_0). \tag{48}$$

Our main interest here is to analyze the effects of these equations of state on the equilibrium configuration of a neutron star. A tabulation of the equations of state of Harrison-Wheeler, Pandharipande, and Hagedorn is given in Table 4. The interest in the equation (47) is clearly explained: it is the most extreme dependence of pressure from density not violating Le Chatelier's principle envisaged up to now for ultradense matter, a weaker dependence on ρ than any γ-law could possibly predict. From this the analysis of Rhoades and Ruffini presented in Appendix 1.1, to explore the effects of this most extreme dependence of pressure from density.

Can we understand in a simple way the physical reasons for the results presented there? The Hagedorn equation of state, or for that matter any equation of state which takes into account production of hyperons and resonances at high densities, is very much *softer* (lower value of γ) than the equation of state obtained neglecting this effect. As a direct consequence all the configurations of equilibrium with a central density $\rho_c \gtrsim 5 \times 10^{15}$ g/cm^3 have a mass nearly constant and insensitive to any change in the value of the central density. The physical reasons for this is simply illustrated in Figure 5. The density in the central innermost core increases but the equation of state is so soft that this change of density and pressure is not transmitted to the main body of the star (compare and contrast Figure 5 with Figure 3). Only a small part of the innermost core is affected by the change in density, the remaining part of the star is left unchanged.

This drastic difference of behavior cannot, however, affect in any way the stable equilibrium configuration of a neutron star. Only the *unstable* equilibrium configurations are largely modified (see Appendix 1.1). The only detectable difference in the stable equilibrium configuration is a small decrease (of the order of a few per cent) in the value of the critical mass.

TABLE 4. Equations of State for Neutron Star Matter

Harrison–Wheeler		Pandharipande-Reid potential		Hagedorn $T = 0$	
P/c^2 g/cm^3	ρ g/cm^3	P/c^2 g/cm^3	ρ g/cm^3	P/c^2 g/cm^3	ρ g/cm^3
2.54×10^8	1.00×10^{11}	0.52×10^{12}	8.46×10^{13}	–	–
4.11	1.58	1.30	1.53×10^{14}	–	–
6.18	2.51	2.55	2.22	–	–
8.90	3.98	4.26	2.91	–	–
1.29×10^9	6.32	6.96	3.61	–	–
2.03	1.00×10^{12}	1.05×10^{13}	4.30	–	–
2.10	1.58	2.71	6.27	–	–
3.42	2.51	4.97	7.95	–	–
6.06	3.98	6.65	8.90	–	–
1.23×10^{10}	6.32	8.69	9.85	–	–
2.54	1.00×10^{13}	1.10×10^{14}	1.08×10^{15}	–	–
8.21	2.00	2.43	1.50	–	–
3.55	5.00	4.49	1.96	9.46×10^{14}	3.00×10^{15}
1.11×10^{12}	1.00×10^{14}	5.80	2.21	1.36×10^{15}	5.00
3.50	2.00	8.18	2.61	1.74	7.00
1.47×10^{13}	5.00	1.10×10^{15}	3.03	2.11	9.00
4.39	1.00×10^{15}	2.44	4.69	2.28	1.00×10^{16}
1.31×10^{14}	2.00	3.99	6.36	5.47	3.00
5.30	5.00	5.42	7.80	8.36	5.00
1.31×10^{15}	1.00×10^{16}	7.14	9.42	1.12×10^{16}	7.00
3.27	2.00	9.05	1.12×10^{16}	1.37	9.00
9.91	5.00	1.56×10^{16}	1.70	1.50	1.00×10^{17}
2.23×10^{16}	1.00×10^{17}	2.68	2.64	3.86	3.00
4.85	2.00	3.34	3.20	6.02	5.00
1.36×10^{17}	5.00	4.10	3.83	8.10	7.00
2.81	1.00×10^{18}	4.94	4.51	1.01×10^{17}	9.00
5.78	2.00	5.90	5.29	1.11	1.00×10^{18}

In Figure 6 compared and contrasted are the computations of the mass of the stable equilibrium configurations for three different equation of state: Harrison-Wheeler, Pandharipande, Hagedorn ($T = 0$). It is in particular clear that, as predicted in Appendix 1.1, the modifications in our results introduced by the change from the Hagedorn equation of state with $T = T_0$ to the one $T = 0$ are *not* substantial (compare Figure 6 with Figure 2 in Appendix 1.1). What can we conclude on the value of the critical mass? From the analysis of the different equilibrium configurations for selected values of the central density, we have seen that in the determination of the critical mass we can always neglect densities

$$\rho \lesssim 10^{13}\ \text{g/cm}^3.$$

We have also seen that due to the "softening" of the equation of state in the innermost core we can neglect the regions with

$$\rho \gtrsim 5 \times 10^{15}\ \text{g/cm}^3.$$

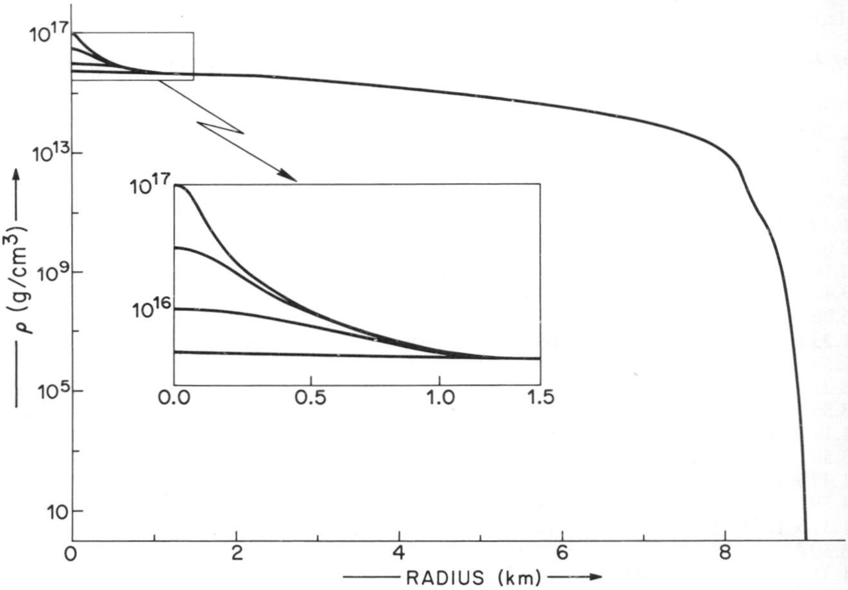

Figure 5. Density distribution of configurations of equilibrium of neutron stars for slected values of the central density at supranuclear density. For density $\rho \gtrsim 5 \times 10^{15}$ g/cm^3 the Hagedorn equation of state has been used. To an increase of the central density only the innermost core of the neutron star is affected; no sizable changes in the mass or the radius of the equilibrium configuration occurs.

(This proof will be made even stronger in next paragraph by the introduction of the "domain of dependence".)

The *only* important region, from the point of view of the determination of the numerical value of the critical mass is the one with a density range

$$10^{13}\ \text{g/cm}^3 \lesssim \rho \lesssim 5 \times 10^{15}\ \text{g/cm}^3.$$

From the different assumptions adopted in this range of densities the value of the critical mass can have values (with equations of state not violating causality

$$0.69 \lesssim M_{\text{crit}}/M_\odot \lesssim 1.5.$$

Again the range of variation is of the order of a multiplicative factor 2 (see previous paragraph).

To describe this region two radically different approaches are available. One, the perturbation treatment of Pandharipande, with emphasis on the interactions

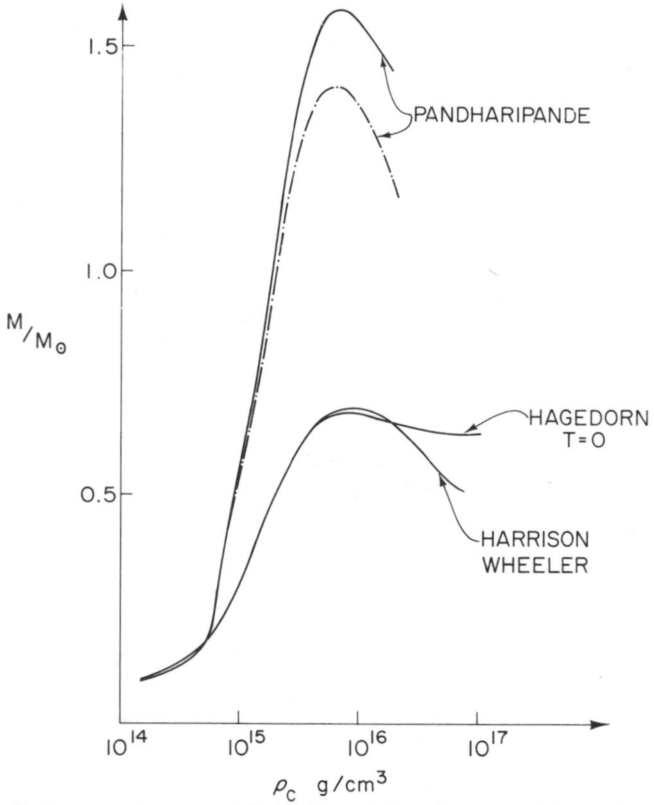

Figure 6. Masses of the equilibrium configuration of neutron stars plotted as a function of the central density for different equations of state (see also Table 4). The Pandharipande equation of state takes into account the strong interactions between nucleons. The Harrison-Wheeler equation of state neglects *all* the nuclear interactions and is substantially a "free particle" approximation. The Hagedorn equation of state applies only asymptotically for $\rho \gtrsim 5 \times 10^{15}$ g/cm^3. No matter what the different assumptions in the equation of state are the values of the critical mass oscillate in a finite range $0.69 \lesssim m_{crit} \lesssim 1.45$.

between the particles, presents us with serious difficulties in the proper treatment of collective and relativistic effects as well as with the treatment of simple two body interactions at supranuclear densities. The second, the treatment due to Hagedorn in which, however, we have difficulty in extrapolating the results from extreme relativistic supranuclear density regime to normal nuclear matter. It might very well be that these two treatments "bracket" the true status of thinks.

To extrapolate the Hagedorn approach to smaller and smaller densities and carefully examine the region in which the thermodynamic approach fails to apply or, vice versa, extrapolate with a proper relativistic treatment the Pandharipande approach to higher densities, appears today, to be the most important work yet to be done in this fundamental field of research.

In the next paragraph we are going to introduce a completely different approach to the determination of an absolute *upper* limit to the value of the critical mass without focusing on *any* detail of the equation of state at supra-nuclear densities.

1.4 Critical Mass Without any Detailed Knowledge of the Equation of State at Supranuclear Densities

The existing indeterminacy in the equation of state to be used at nuclear and supranuclear densities in neutron star matter is well represented in Figure 7 where the pressure is plotted versus the density for some of the most frequently examined possible equations of state. With this large unknown in the description of matter, particularly at supranuclear densities, we can nevertheless conclude that neutron stars can only exist in a limited range of densities,

$$10^{13} \lesssim \rho \lesssim 5 \times 10^{15} \ \mathrm{g/cm^3}$$

and their maximum mass can vary

$$0.69 \lesssim M_{\mathrm{crit}}/M_{\odot} \lesssim 1.5.$$

If this uncertainty in mass can be found by trails of different equations of state, how can we be sure that we could not arrive to a critical mass higher by a whole order of magnitude by still further exploration of physically admissible equation of state? Or, for that matter, why not an infinite critical mass and no instability at all? Such a conclusion would throw doubt on the possible existence of completely collapsed objects or black holes.

Instead of pursuing a more detailed analysis of the physics of matter at supranuclear densities we answer to the previous question by following a radicall different approach:

We establish an *absolute upper* limit to the value of the critical mass of a neutron star using *only* the following assumptions:

(a) general relativistic equations for hydrostatic equilibrium.
(b) an equation of state with a speed of sound always smaller there or equal to the speed of light (conservation of causality) and such that $dp/d\rho > 0$ (Le Chatelier's principle).

We can then directly conclude that *no matter the details of an equation of state at nuclear or supranuclear densities a neutron star can never have a mass larger than* $3.2M_{\odot}$. Moreover, if we know the equation of state *up to* a given density ρ_{0}

associated with that density then there exists a "*domain of dependence*" inside which have to be contained all values of the critical mass and of the critical central densities for all possible configurations of equilibrium of the neutron star.

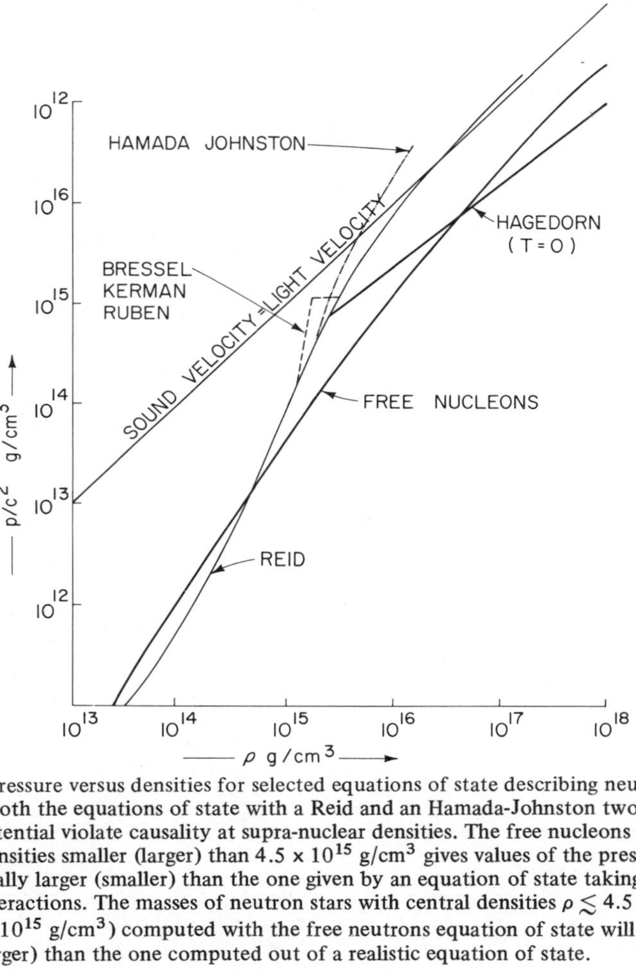

Figure 7. Pressure versus densities for selected equations of state describing neutron stars material. Both the equations of state with a Reid and an Hamada-Johnston two bodies nuclear potential violate causality at supra-nuclear densities. The free nucleons approximation for densities smaller (larger) than 4.5×10^{15} g/cm^3 gives values of the pressure systematically larger (smaller) than the one given by an equation of state taking into account nuclear interactions. The masses of neutron stars with central densities $\rho \lesssim 4.5 \times 10^{15}$ g/cm^3 ($\rho \gtrsim 4.5 \times 10^{15}$ g/cm^3) computed with the free neutrons equation of state will always be smaller (larger) than the one computed out of a realistic equation of state.

Details of both these results obtained by Rhoades and Ruffini are given in Appendix 1.2.

1.5 The Critical Mass and the Late Stage of Evolution of Stars

After all the thermonuclear sources of energy are exhausted, a star undergoes gravitational collapse. The main features of this phenomenon have been outlined and investigated by the work of Colgate, May and White[37] at Livermore. The final outcome is clear: a supernova event occurs, a condensed collapsed core is formed and a shell of material moving at relativistic velocities is expelled. That a realistic supernova event could depart substantially from the general picture

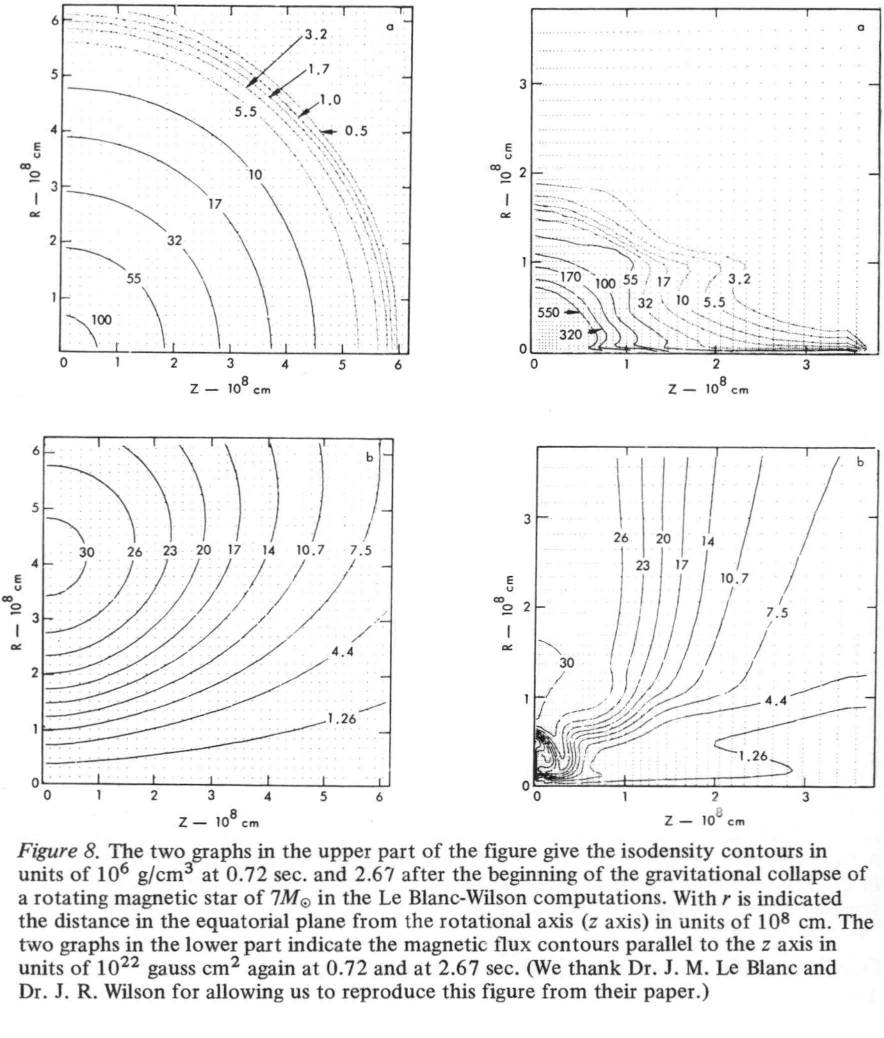

Figure 8. The two graphs in the upper part of the figure give the isodensity contours in units of 10^6 g/cm^3 at 0.72 sec. and 2.67 after the beginning of the gravitational collapse of a rotating magnetic star of $7M_\odot$ in the Le Blanc-Wilson computations. With r is indicated the distance in the equatorial plane from the rotational axis (z axis) in units of 10^8 cm. The two graphs in the lower part indicate the magnetic flux contours parallel to the z axis in units of 10^{22} gauss cm^2 again at 0.72 and at 2.67 sec. (We thank Dr. J. M. Le Blanc and Dr. J. R. Wilson for allowing us to reproduce this figure from their paper.)

presented in Ref. (37) has been clearly evidenced by the recent results of LeBlanc and Wilson[38] and of Paczinsky.[39]

Le Blanc and Wilson[38] have considered the evolution of a rotating magnetic star of $7M_\odot$ after all the material has been burned to the endpoint of thermo-

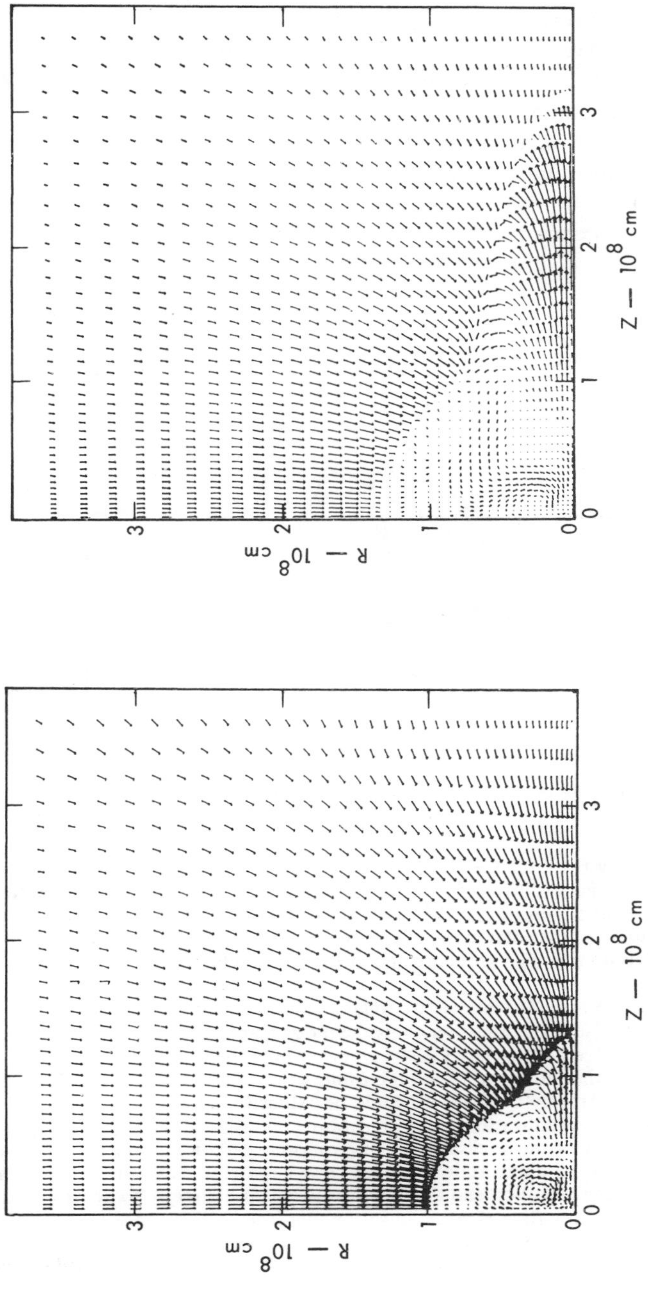

Figure 9. Velocity vector in the jets developed in the collapse of a $7M_{\odot}$ rotating magnetic star. *R* gives the distance in the equatorial plane from the rotation axis (*z* direction) in units of 10^8 cm. The graph in the left part of the figure refers to a configuration 2.61 sec. after $t = 0$ (start of gravitational collapse). The maximum velocity magnitude is 2.43×10^9 cm/sec. The graph in the right part of the figure refers to a configuration 2.64 sec. after $t = 0$ with a maximum velocity of 4.38×10^9 cm/sec. (We thank Dr. J. M. Le Blanc and Dr. J. R. Wilson for allowing us to reproduce this figure from their paper.)

nuclear evolution. The star is assumed to be initially in uniform rotation with a value of the angular velocity $\omega \sim 0.7$ rad/sec and an angular momentum of $\sim 4.6 \times 10^{50}$ g cm^2/sec. The magnetic field parallel to the z axis is assumed to be of the order of 0.025% the gravitational energy of the star. The computations are started from an equilibrium configuration. After a few seconds enough energy has been lost by neutrino emission that the star starts to collapse. As the collapse proceeds, the temperature of the interior region rises above the iron

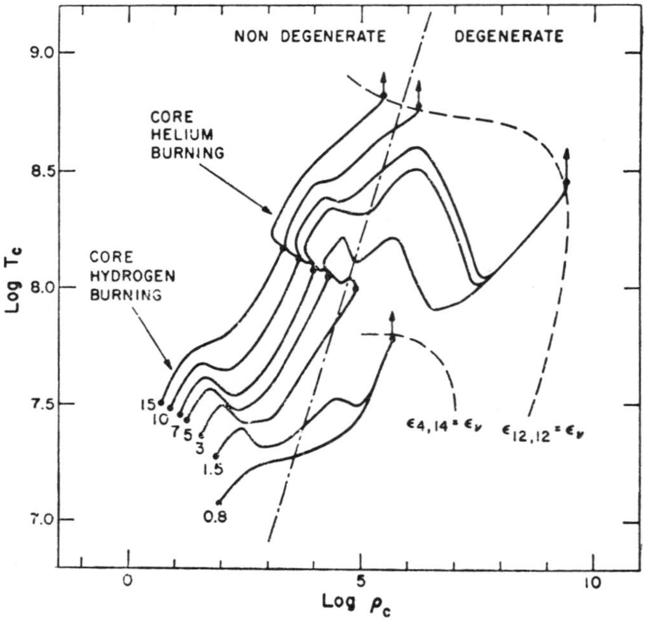

Figure 10. Evolutionary track for the clusters of stellar models proceeding toward the endpoint of thermonuclear evolution. The temperature at the center of the star is given as a function of the central density. At the beginning of each track the value of the mass of the star is given in units of the solar mass. The large dots indicate the position of the centers of the models on the main sequence and at the times helium and carbon ignition occur.

The cores of the 3, 5 and $7M_\odot$ models are smaller and do not contract too rapidly after the helium exhaustion. Neutrino emission cools them down and carbon ignition takes place when the density at the center reaches 3×10^9 g/cm^3. Along the broken lines neutrino energy losses balance either nitrogen + helium burning or carbon burning. The ignition of carbon as suggested by Arnett [41] (1968–69) could be explosive. However, in these computations the effects of crystallization, electron capture, and general relativity were not taken into account. Some of the nuclear reactions governing carbon burning rate are also uncertain. The effects of these uncertainties could be so large as to make the system unable to cause a thermonuclear explosion and the core would collapse directly to neutron star density. Certainly the most striking feature in this model is the convergence of the evolutionary tracks for $3M_\odot \lesssim M \lesssim 7M_\odot$ into a common track after the exhaustion of helium in the core has taken place. See also Table 5. (Figure 10 and caption based on the paper in Ref. (39).)

decomposition temperature and the collapse rapidly accelerates. During the collapse, no radial motions develop due to the increasing centrifugal forces and the angular velocity approaches a vortex configuration. The shear in the magnetic field generates large magnetic fields along the axis of rotation. The multiplication of the strength of the magnetic field by this process is a hundred times larger than that which we would expect from simple compression effects (flux conservation). The collapse is stopped at a density of 10^{11} g/cm^3; the combined effects of rotation and magnetic field produce a jet in the axial direction with a velocity of the order of a tenth of the speed of light. These jets carries a mass of the order of 2.1×10^{31} gr, a kinetic energy of 1.6×10^{50} ergs, and a considerable amount of magnetic energy (3.5×10^{49} erg with fields $\sim 10^{11}$ gauss) see Figure 8 and Figure 9.

TABLE 5. (from B. Paczinsky[39]) Mass of the core for selected values of the stellar mass at these evolutionary phases (1) helium ignition (2) helium exhaustion and (3) carbon ignition (see also Fig. 10).

$M_{star}/M_{\odot}$	$M_{core}/M_{\odot}$		
	(1)	(2)	(3)
0.8	0.39	–	–
1.5	0.40	–	–
3.0	0.45	0.51	1.39
5.0	0.56	0.95	1.39
7.0	0.83	1.45	1.39
10.0	1.35	2.32	2.32
15.0	2.54	3.89	3.91

Paczinsky[39] has recently analyzed the details of the evolution of population I stars with masses 0.8, 1.5, 3, 4, 10 and $15M_{\odot}$. In all these computations the evolutionary tracks of the center of stellar models were computed and some of the results are here reproduced in Figure 10. The most striking aspect of these computations is the formation of a standard size degenerate core of material (white dwarf material) after the exhaustion of helium in the center of the star. The size of this core for initial configurations with $3 \lesssim M/M_{\odot} \lesssim 7$, is always the same $M_{core} \sim 1.39M_{\odot}$. The remaining mass of the star, distributed in a large envelope, should be expelled when the core becomes unstable and undergoes gravitational collapse. If these computations are confirmed they could be used as a natural initial condition with which to analyze the outcome of gravitational collapse. In that case, the final outcome will strongly depend upon the value of the critical mass. If

$$M_{core} > M_{crit}$$

the system will necessarily collapse to a black hole unless a substantial part of the imploding material is expelled during the dynamical phases of the collapse. In this last case the imploding core will settle down in a stable neutron star configuration. If, instead

$$M_{\text{core}} < M_{\text{crit}}$$

only neutron stars could be formed by direct collapse. The formation of black holes would then be possible only if the implosion occurs with an amount of kinetic energy large enough to overcome the barrier against gravitational collapse, or by a multistep process by successive accretion of material on the already formed neutron star (see e.g. Leach and Ruffini[40]). Needless to say, the Paczinsky conclusions are applicable only up to a mass $M \sim 15M_\odot$ and little is yet known about the possibility of the same phenomenon occurring for larger mass stars.

1.6 X-ray Sources and the Determination of the Critical Mass

The existence of an absolute upper limit to the possible mass of a neutron star is the most powerful tool we have today to discriminate between different families of collapsed objects. No hope exists of determining the masses of neutron stars from pulsars. In the more than sixty pulsars known up to this moment, none is known to be in a binary system. The presence of a pulsar in a binary system can, in fact, be easily inferred by the modulation in the intrinsic period due to the doppler effect of the orbital motion. Any other mass determination can only be indirectly done by the analysis of "glitches", sudden changes in the period of slowing down; however, the interpretation of any one of these phenomena is extremely model dependent and no definitive answer is possible.

Quite on the contrary, no hope exists of finding out the presence of an isolated black hole in space. Ruffini and Wheeler[42] pointed out in 1971, "Of all objects one can conceive to be travelling through empty space, few offer poorer prospects of detection than a solitary black hole of a solar mass. No light comes directly from it. Can it be seen by its lens action or other effect on a more distant star? Not by any simple means! It is difficult enough to see Venus 12,000 km in diameter swimming across the disk of the sun without looking for a 15 km object moving across a stellar light source almost infinitely far away." Furthermore, "the possibility to capitalize on such a double star system is more favorable when the black hole is so near to a normal star that it draws in matter from its companion. Such a flow from one star to another is well known in close binary systems, but no unusual radiation emerges. However, when one of the components is a neutron star or a black hole a strong emission in the X-ray region is expected."

The recent discovery of X-ray sources has radically changed this state of affairs. From the satellite UHURU, a completely new family of galactic X-ray sources has been discovered. They all emit an amount of energy of $\sim 10^{37}$–10^{38} erg/sec and they all appear to be members of binary systems[3]. The enormous amount of energy emitted in the X-rays (10^3–$10^4 L_\odot$) and the fact that they are members of close binary systems, implies that the energy emitted comes from accretion onto a collapsed object. The rotational energy of the collapsed object is far too small for explaining the energetics of these objects (see Appendix 1.3).

For the first time we will be able not only to measure the mass of a collapsed object with unprecedented accuracy, but also to conclusively discriminate between neutron stars and black holes.

What can we say about the formation of such a binary system with a collapsed object as one of its members? Initially we have a binary system with two normal stars as components, then one of the two stars undergoes gravitational collapse. On this basis we can already advance some general considerations.

If we indicate by M_1 and M_2 the masses of the two stars and by a their initial separation, the angular velocity of the system around the center of mass is given by

$$\omega^2 = G\frac{M_1 + M_2}{a^3} \tag{49}$$

G being the constant of gravity. Here and in the following we assume for simplicity that the orbital motion of the two masses is exactly circular. During the process of gravitational collapse a large amount of the mass of one of the stars is expelled away and if the time of collapse is very much shorter than the orbital period, the system will be endowed with a new value of the total energy

$$E^{\text{ToT}} = \frac{GM_rM_2}{a}\,[(M_1 + M_2)/2(M_r + M_2) - 1] \tag{50}$$

here

$$M_r = M_1 - M_1^{\text{out}}$$

represents the mass of the remnant and M_1^{out} the expelled mass during the collapse of M_1. There exists a critical value for the mass of the remnant given by the expression

$$M_1 - M_2 = 2M_r{}^{\text{crit}} \tag{51}$$

If $M_r < M_r^{\text{crit}}$ the system will become unbound and the collapsed object will leave behind a runaway star; if instead $M_r > M_r^{\text{crit}}$, then the collapsed object will still be a member of a binary system. We can also conclude that unless M_1 is larger than M_2 the collapsed object *will never become unbound.* If $M_r > M_{\text{crit}}$ the two stars M_r and M_2 will still be a binary system with a separation b satisfying the equation

$$M_1^{\text{out}} = M_1 - M_r = (M_1 + M_2)(1 - 1/(2 - a/b)) \tag{52}$$

For values of $M_r > M_{crit}$ the value of the ratio b/a is usually fairly small: $b/a \lesssim 10$ (see Figure 11). Only for a value $M_r \sim M_{crit}$ can the star reach distances far away from the companion *without* becoming unbound. We should, therefore, expect

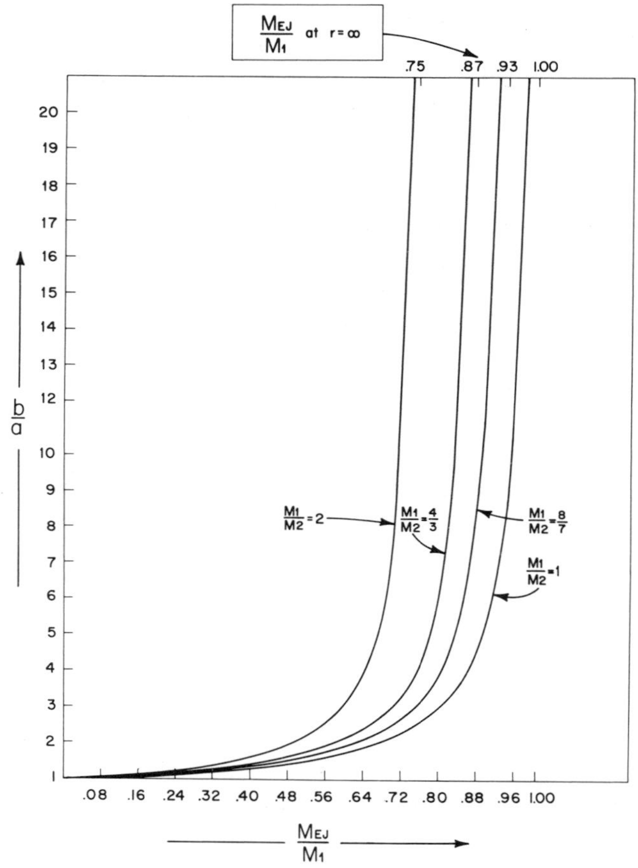

Figure 11. Ratio of the distance between the centers of the stars in a binary system of initial masses M_1 and M_2 with the mass M_1 undergoing gravitational collapse. M_{ej} is the amount of mass ejected by the star M_1. By a (b) we have indicated the distance between the centers of the two stars before (after) gravitational collapse has occurred. Values of $10 < b/a < \infty$ are unlikely to occur. We should therefore expect that by mass ejection the system becomes unbound or will form a new binary system with a period P similar to the initial period P_0: $P_0 \lesssim P \lesssim 10P_0$.

that binary systems with periods P_1 largely different from their original period P_0 ($P_1 \gtrsim 10P_0$) are extremely unlikely.

In Figure 12 the mass of the remnant to be expected from a binary system with masses M_1 and M_2 has been plotted for selected values of the ratio b/a. It is interesting to notice that to become unbound, a system must lose a very large fraction of its mass and the ensuing collapsed object is an ideal candidate to form a Pulsar associated with a runaway star. If, however, the mass M_r is still larger than the maximum critical mass of a neutron star, the collapsed object will coalesce to a black hole isolated in space, with an exceedingly small chance of detectability!

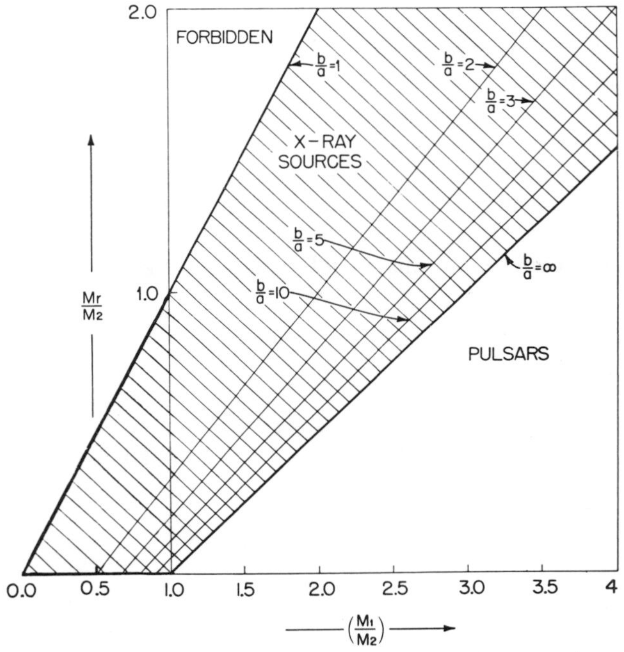

Figure 12. Mass of the remnant M_r plotted as a function of the ratio M_1/M_2 in a binary system with masses M_1 and M_2. The mass M_1 is lowered to the value M_r by ejection of material. We indicate by b/a the ratio of the distances of the centers of the two stars after and before the occurrence of gravitational collapse. Only if the mass of the collapsing star is larger than the mass of the companion ($M_1 > M_2$) can the system become unbound by a large enough ejection of matter. In this case the collapsed object will either form a pulsar or an isolated black hole in space. The two different outcomes depend upon the value of the mass of the collapsed object. If an isolated black hole is indeed formed by this mechanism no realistic detection possibilities exist. (Figure from R. Leach and R. Ruffini[46].)

It is worthwhile to emphasize here that it is, indeed, possible in a binary system that the star which appears to be *less* massive undergoes gravitational collapse well before the one which appears to be the more massive component. This does not contradict our usual notion that the more massive star evolves *faster* than the less massive one. From the original work of Morton[43] it has become more and more clear that in binary systems the transfer of material between the two member stars can go so far as to invert the original mass ratio. In other words, in a binary system with initial masses $M_1^0 > M_2^0$ the mass M_1^0 undergoes thermonuclear evolution very much faster than the other mass. During this evolution, see e.g. as Paczinsky[39] a large amount of the star expands and forms a diffuse envelope, the remaining mass condensing in an inner core. This outer envelope flows over the Roche lobe of the binary system, matter is accreted by the less evolved secondary and the system ends with a mass

$$M_1 < M_2$$

The mass M_1 then, already more evolved, can at the endpoint of thermonuclear evolution undergo gravitational collapse before the less evolved star M_2.

Some of the recently discovered binary X-ray sources are pulsating with a sharply defined period of ~ 1 second. From the Doppler variation of this intrinsic period and from the Doppler shift in the emission and absorption lines of the normal star, we can deduce the velocity of the components of the binary system. By direct application of Kepler law some information on the parameters of the system can then be immediately inferred. Knowing the period of the binary T we have

$$\frac{G(m_x + M)}{a^3} = \left(\frac{2\pi}{T}\right)^2 \tag{53}$$

m_x being the mass of the X-ray source, a the separation between the centers of the two stars and M the mass of the "normal" star. If we know the velocity of the X-ray source

$$v_x = r\omega \sin i \tag{54.1}$$

$$r = \frac{M}{m_x + M} a \tag{54.2}$$

i being the inclination of the orbit with respect to the plane of observation (at $i = 90°$ we are looking *in* the equatorial plane of the system) we can determine

$$\frac{M^3}{(m + M)^2} \sin^3 i = \left(\frac{T}{2\pi}\right) \frac{v_x^3}{G} \tag{55}$$

On the other hand, if we know the projected velocities of both stars and the orbital period

$$(v \sin i)_x \tag{56.1}$$

and

$$(v \sin i)_2 \quad \text{and} \quad T \tag{56.2}$$

we can derive:

$$m_x \sin^3 i = [(v \sin i)_x + (v \sin i)_2]^2 \frac{(v \sin i)_2}{\omega G} \tag{57}$$

$$M \sin^3 i = [(v \sin i)_x + (v \sin i)_2]^2 \frac{(v \sin i)_2}{\omega G} \tag{58}$$

Some of these formulae have been used in determining the numerical values given in Appendix 1.3. To make a model for these X-ray sources we follow here a very simple and straightforward approach: we assume that the binary system can be described by the Roche potential, the X-ray source being *a pointlike* source in the secondary lobe and the normal star completely filling up the primary lobe.

In assuming the Roche model we adopt the following four main simplifying assumptions[44]:

(I) The concentration of mass is so large that the gravitational attraction is considered identical to the one generated by a point mass.

(II) The two components are assumed to revolve around the center of mass of the system in circular orbits.

(III) Their axis of rotation is perpendicular to the orbital plane.

(IV) The period of axial rotation of the normal star is identical to the one of orbital revolution. (This assumption can at least, in principle, be experimentally verified.)

We choose (see Figure 13) a system of coordinate centered at the center of mass of the normal star, with the x axis directed toward the center of the X-ray source and the y axis in the plane of the orbit. The z axis is parallel to the axis of rotation of the system. The potential U, therefore, has the form

$$U = \frac{Gm_x}{r} + \frac{GM}{R} + \tfrac{1}{2}\omega^2(x^2 + y^2) - \frac{m_x a \cdot x}{m_x + M}\omega^2 + \tfrac{1}{2}\omega^2 \left(\frac{am_x}{m_x + M}\right)^2 \tag{59}$$

Here a is the separation of the centers of the two stars and ω the orbital angular velocity. It is useful, instead of using Eq. (59) to use the dimensionless quantity

$$\Omega = \frac{Ua}{GM} - \tfrac{1}{2}(1+q)^{-19} = \frac{1}{R} + \frac{q}{r} + \tfrac{1}{2}(1+q)(\hat{x}^2 + \hat{y}^2) - q\hat{x} \tag{60}$$

here $q = m_x/M; R^2 = x^2 + y^2 + z^2, \hat{x} = x/a, \hat{y} = y/a, \hat{z} = z/a$ and $r^2 = (\hat{x} - 1)^2 + \hat{y}^2 + \hat{z}^2$. Some of the critical equipotentials are shown in Figure 13. A complete table of equipotentials and characteristic dimensions of the lobes has been given by Plavec[44] and Kopal[45]. Important for us here, is to compute the occultation angle of the X-ray source by the eclipse of the "normal" star. We are assuming that to give a sizable amount of accretion on the X-ray source the normal star has at least to fill up its own Roche lobe. We are going, therefore, to assume that the surface of the "normal" star follows the equipotential surface given by $\Omega = \Omega_0$ (see Table 1 in Appendix 1.3). Since the X-ray source is expected to be either a neutron star or a black hole, we can safely assume it to be pointlike. The occultation angle *cannot*, therefore, be computed as is usually done by means of the osculating cone at the Lagrangian point L_0 (see Figure 13). The usual discussions of occultation angle cannot be accepted here. We have instead computed the tangent to the equipotential surface $\Omega = \Omega_0$, called by Kuiper the "innermost contact surface", crossing through the point (1, 0, 0) corresponding to the location of the X-ray source. We, therefore, have at the intersection point of the tangent with the surface $\Omega = \Omega_0$

$$\frac{dy}{dx}_{P_{\rm int}} = \left[\frac{-x/R^3 - (x-1)/r^3 + (1+q)x - q}{y/R^3 + qy/r^3 - (1+q)y}\right]_{P_{\rm int}} \tag{61}$$

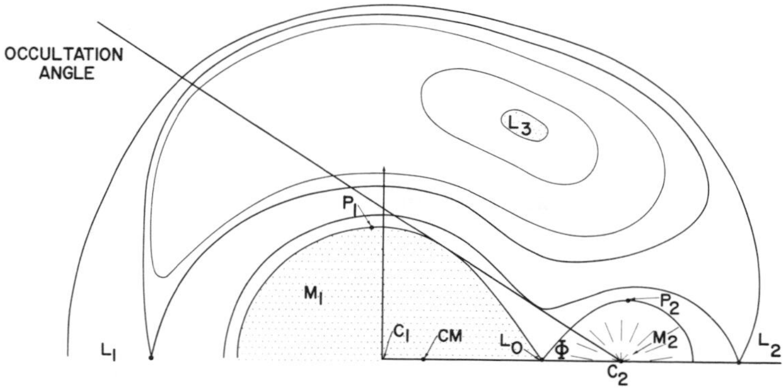

Figure 13. Equipotentials of the Roche model in the orbital plane for an X-ray source with masses $M_2/M_1 \sim 0.2$. With L_0, L_1, L_2, L_3 we have indicated the Langrangian points of the system (no net force under the combined effects of gravitation and centrifugal forces). The normal star is assumed to fill its "innermost contact surface" and inject material in the lobe of the companion collapsed object through the Lagrangian point L_0. C_1 and C_2 are the center of mass of the star mass M_1 and M_2 respectively. *CM* is the center of mass of the system. Tabulations of the critical values of the potential and of the points P_1 and P_2 are given by Kopal[45] and Plavec and Kratochvil[44]. The occultation angle ϕ is computed by assuming the X-ray sources to be pointlike; a table with values of the occultation angles for selected values of the ratio M_2/M_1 and of the inclination angle is given in Appendix 1.3 These computations give a *lower* limit on the mass of the X-ray source. See text.

with the constraint

$$y_{P_{\rm int}} = [(dy/dx)(1-x)]_{P_{\rm int}} \tag{62}$$

For the more general case in which the plane of the orbit is inclined at an angle i with respect to the line of sight (i being the angle between the axis of rotation of the system and the line of sight) we will still use the previous procedure with the constraint

$$z = (x-1)\cos i$$

We have solved the system iteratively. (63)

The occultation angles are reported in Table 1 of Appendix 1.2 for selected values of the inclination. Here are the main conclusions:

On the basis of this very simple model we can advance the following general conclusion: *Pulsating X-ray sources appear to be systematically less massive than non pulsating ones. The first should be identified with neutron stars, the second with black holes.* There will be, of course, the possibility of finding some neutron stars with $M \sim M_{\text{crit}}$ and non pulsating due to the alignment of the magnetic field with the rotation axis or to a particularly weak magnetic field. These borderline cases are by far the most interesting to be investigated. See Appendix 1.2.

We are, therefore, finally able for the first time to infer the mass of a collapsed object! Moreover, the result that the critical mass of a neutron star must *necessarily* be less, then $3.2M_{\odot}$ give us an extremely powerful tool to discriminate between neutron stars and black holes.

This is only the starting point of an entire field of research. From now on let us focus on a detailed treatment:

(a) on the orbital structure of particles in the field of a collapsed object (Schwarzschild-Reißner-Nordstrøm, Part II, Kerr, generalized Kerr, Part IV).
(b) on the radiation emitted by particle falling into a collapsed object through electromagnetic (Part II) and gravitational (Part III) Brehmstrahlung.
(c) on the fully relativistic treatment of electromagnetic (Part II) and gravitational (Part III) radiation emitted from particles in circular orbits in static geometries.
(d) on the detectability and polarization features of gravitational radiation from the previous sources (Part III).
(e) on the electromagnetic properties of magnetic black holes (Part IV).
(f) and finally on the "Energetics of Magnetic Black Holes" (Part IV).

References

1. Landau, L. (1932) *Phys. Zeit. of Soviet Union* **1**, 285.
2. Chandrasekhar, S. (1931) *M. N. of Royal Astro. Society* **91**, 456.
3. See e.g. Gursky, H., *Les Houches Lectures,* in this volume.
4. Fermi, E. (1927) *Rend. Lincei* **6**, 602.
5. Thomas, L. (1927) *Proc. Cambridge Phil. Soc.* **23**, 542.
6. Landau, L. and Lifshitz, L. (1970) *Mecanique Quantique*—Editions de la paix, Moscow.
7. Oppenheimer, J. R. and Volkoff, G. M. (1939) *Phys. Rev.* **55**, 374.
8. Oppenheimer, J. R. and Synder, H. (1939) *Phys. Rev.* **56**, 455.
9. Ruffini, R. and Bonazzola, S. (1969) *Phys. Rev.* **187**, 1767.
10. Jordan, P. (1955) Schwerkraft und Weltall-Braunschweig.
11. Brans, C. and Dicke, R. H. (1960) *Phys. Rev.* **124**, 925.
12. Ruderman, M. (1971) *Phys. Rev. Lett.* **27**, 1306, and ref. mentioned there.
13. Dyson, F. *Annals of Physics*—to be published.

14. Harrison, B. K., Thorne, K. S. Wakano, M. and Wheeler, J. A. (1965) *Gravitation Theory and Gravitational Collapse*, The Univ. of Chicago Press.
15. Baym, G. and Pines, D. (1971) *Annals of Phys.* **66**, 816.
Baym, G. *et al.* (1969) *Nature* **224**, 872.
16. Ruderman, M. (1970). *Nature* **225**, 838.
17. Smoluchowsky, R. (1970) *Phys. Rev. Lett.* **24**, 923.
18. Rhoades, C. (1971) Ph.D. Thesis–Princeton University–unpublished.
19. Sawyer, R. Preprint–Submitted to *Ap. J.*
20. Scalapino, D. J. preprint–Submitted to *Phys. Rev. Lett.*
21. Ambartsumian V. A. and Saakian, G. S. (1961) *Soviet Ast.* **6**, 601.
22. Tsuruta S. and Cameron, A. G. W. (1966) *Can. J. Phys.* **44**, 1895.
23. Pandharipande, V. R. (1971) *Nuclear Phys.* **A178**, 123.
24. Hagedorn, R. (1965) *Nuovo Cimento Sup.* **3**, 147.
Hagedorn, R. (1968) *Nuovo Cimento* **56A**, 1027.
Hagedorn, R. (1970) *Astron. and Astrophys.* **5**, 184.
Hagedorn, R. CERN lecture notes–71–72.
25. vanKampen N. G. (1961) *Physica* **27**, 783.
Nosanov, L. H. (1966) *Phys. Rev.* **146**, 120.
26. Reid, R. V. (1968) *Ann. of Physics* **50**, 411.
27. Canuto, V.–to be published.
28. Fermi, E. (1950) *Prog. Theor. Phys.* **5**, 570.
29. Schrödinger, E. (1946) *Statistical Thermodynamics,* Cambridge Univ. Press.
30. Beth, E., Uhlenbeck, G. E. (1937) *Physica* **4**, 915.
31. Belenskij, S. Z. (1956) *Nuclear Phys.* **2**, 259.
32. Fubini, S. and Veneriano, G. (1969) *Nuovo Cimento* **64A**, 811.
33. Fubini, S., Gordon, D. and Veneriano, G. (1969) *Phys. Rev. Lett.* **29B**, 679.
34. Hamer, C. J. and S. C. Frautschi, Submitted to *Phys. Rev.*
35. Frautschi, S. C. (1971) *Phys. Rev.* **D3**, 2831.
36. Personal communication of Hagedorn, see e.g. Note added in proof in Appendix 1.1.
37. Colgate, S. A. and White R. H. (1966) *Ap. J.* **143**, 626.
May, M. M. and White, R. H. (1966) *Phys. Rev.* **141**, 1232.
38. Le Blanc, J. M. and Wilson, J. R. (1971) *Ap. J.*
39. Paczinsky, (1970) *Acta Astronomica,* **20**, 2.
40. Leach, R. and Ruffini, R. (1972) "X-Ray Sources a Transient State from Neutron Stars to Black Holes", preprint.
41. Arnett, W. D. (1968) *Nature* **219**, 1344.
Arnett, W. D. (1969) *Astrophy. and Space Science* **5**, 180.
42. Ruffini, R. and Wheeler, J. A. (1971) *Cosmology from Space Platform,* ESRO Book SP52.
43. Morton, D. C. (1960) *Ap. J.* **132**, 146.
Kippenham, R., Thomas, H. C. and Weigert, A. (1966) *Zeit. Ap.* **64**, 395.
Kippenham, R., Kohl, K. and Weigert, A. (1967) *Zeit. As.* **66**, 58.
Giannone, P., Giannuzzi, M. A. (1972) *Astron and Astrophys.* **19**, 289.
44. Plavec, M. and Kratochvil, P. *BAC* (1963) **15**, 165.
45. Kopal, Z. (1969) *Astrophys. Space Science* **5**, 360.
Kopal, Z. (1959) *Close Binary Systems*, Chapman Hall and John Wiley, London, New York.
46. Leach, R. and Ruffini, R.–to be published.

A 1.1

THE ASTROPHYSICAL JOURNAL, 163:L83–L87, 1971 January 15

HAGEDORN EQUATION OF STATE IN NEUTRON STARS*

CLIFFORD E. RHOADES, JR.
Joseph Henry Laboratories, Princeton University, Princeton, New Jersey 08540
AND
REMO RUFFINI
Joseph Henry Laboratories, Princeton University, Princeton, New Jersey 08540, and Institute for Advanced Study, Princeton, New Jersey 08540
Received 1970 November 27

ABSTRACT

At a density of $\sim 5 \times 10^{12}$ g cm^{-3} neutron-star matter can be described (Harrison and Wheeler) by a system of three noninteracting degenerate gases—namely, electrons, protons, and neutrons. At a density of $\sim 10^{14}$ g cm^{-3}, strong interactions between particles can give a significant contribution (Cameron, Cohen, Langer, and Rosen). At densities of $\gtrsim 10^{15}$ g cm^{-3}, production of new particles has to be taken into account. Hagedorn has recently given an analytic equation of state for this regime. We have explored the effect of this new equation of state for supranuclear densities on the equilibrium configurations of neutron stars.

That the configurations of equilibrium of a neutron star do not depend by a factor larger than 2 on the particular gravitational theory used has been recently shown by Ruffini and Wheeler (1970).

More relevant to neutron-star models is the effect of the particular equation of state used to describe neutron-star matter at nuclear and supranuclear densities. Unfortunately, the physics of the region near nuclear densities and above is not well understood because of the absence of experimental data on bulk nuclear matter and because of the theoretical difficulties associated with the strong interactions between hadrons, and the creation of new particles.

At densities greater than about 4.5×10^{12} g cm^{-3}, there are several different approaches which have been taken to describe the equation of state.

Harrison and Wheeler (Harrison *et al.* 1965; Hartle and Thorne 1968) consider high-density matter to be composed of a mixture of three noninteracting Fermi gases—namely electrons, protons, and neutrons—all in β-equilibrium. At densities greater than 10^{14} g cm^{-3}, the Fermi momenta of even nucleons increase to relativistic values, and the creation of new particles and the strong interactions have to be taken into account. For the sake of definiteness and simplicity, Harrision and Wheeler neglect altogether both of these effects and consider matter to be a mixture of the three Fermi gases even to the highest densities. Thus,

$$p = \tfrac{1}{3}\rho\,, \quad \rho \to \infty\,.$$

Cohen *et al.* (1969) have treated the effect of particle-particle interactions on the equation of state by considering a two-particle interaction between nucleons as given by the velocity-dependent potential of Levinger and Simmons (1961). Their equation of state differs markedly from that of Harrison and Wheeler in the region of density greater than 10^{14} g cm^{-3}, and eventually reaches a region where

$$p = \rho\,, \quad \rho \to \infty\,.$$

* Work supported in part by NSF grant GP7669 and Air Force Office of Scientific Research grant AF49(638)-1545 to Princeton University.

Hagedorn (1965, 1967, 1968*a*, *b*, 1969) has recently suggested an analytic formula for an equation of state which takes into account particle creation and which is valid asymptotically for densities greater than about 10 times nuclear density. This equation has the form

$$p = p_0 + \rho_0 \log_e \rho/\rho_0 ,$$

where

$$\rho_0 = 1.253 \times 10^{14} \text{ g cm}^{-3}$$

$$p_0 = 0.314 \times 10^{14} \text{ g cm}^{-3} .$$

This formula was obtained from the laws of statistical physics by taking the increase in the number of species of particles with mass between m and $m + dm$ to have the form

$$dN = \frac{a}{(m_0^2 + m^2)^c} \exp (m/T_0) dm .$$

If dN increased any faster as $m \to \infty$ than the above formula, the partition function would not converge. Also, the partition function converges only if the temperature is less than T_0. Thus, T_0 is the effective highest temperature for any system. Hagedorn obtains $T_0 = 160$ MeV and $c = \frac{5}{4}$ from interpreting the transverse momentum distribution of secondaries in very high-energy collisions in terms of the statistical model. The constants $a = 2.63 \times 10^4$ MeV$^{3/2}$ and $m_0 = 500$ MeV are determined by fitting to the 1432 particles and resonances known in 1967 January.

Figure 1 is a logarithmic plot of pressure in units of grams per cm³ versus density in grams per cm³ for the Harrison-Wheeler (HW), Cameron-Cohen-Langer-Rosen (CCLR), and Hagedorn (HAG) equations of state. In addition, the line $p = \rho$, which represents, as suggested by Zel'dovich (1961), the hardest equations of state consistent with the speed of sound less than or equal to the speed of light, is also plotted in this figure.

The effect of the strong interactions is easily seen from this graph in the differences between the HW and CCLR equations of state. The extreme softness of the Hagedorn equation of state is also evident from Figure 1.

With constants originally chosen by Hagedorn, the Hagedorn equation of state crosses the HW and CCLR equations of state in the region of 10 times nuclear density.

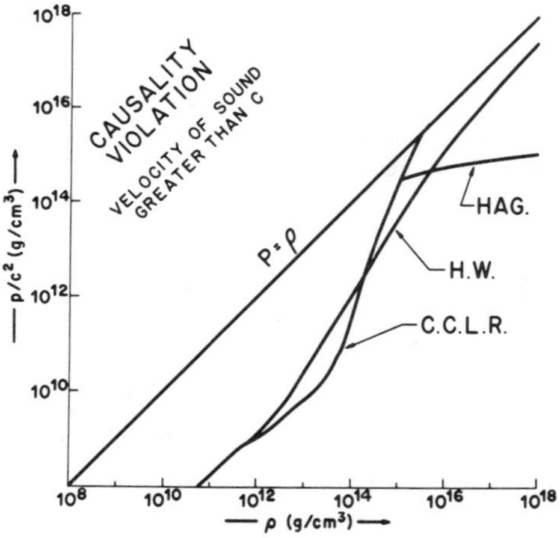

FIG. 1.—Pressure versus energy density

[R2]

Since, as pointed out by Hagedorn, his equation of state is expected to be valid above about 10 times nuclear density, we have matched the Hagedorn equation to that of Harrison and Wheeler and to that of Cameron, Cohen, Langer, and Rosen at about 10 times nuclear density. A detailed treatment of this matching will be given elsewhere (Rhoades 1971).

Computations of the mass, radius, and period or *e*-folding time of neutron-star equilibrium configurations were performed for the following four different sets of equations of state in order to make appropriate comparisons: (*a*) HW; (*b*) HW for $\rho \leq 4.73 \times 10^{15}$ g cm^{-3}, Hagedorn for $\rho > 4.73 \times 10^{15}$ g cm^{-3}; (*c*) CCLR; and (*d*) CCLR for $\rho \leq 1.13 \times 10^{15}$ g cm^{-3}, Hagedorn for $\rho > 1.13 \times 10^{15}$ g cm^{-3}.

The results of our integrations of the Einstein equations of general relativity in the Schwarzschild coordinate (Oppenheimer and Volkoff 1939) are shown in Figure 2. The numerical integrations are calculated by using the standard Runge-Kutta method starting at the origin for a given value of the central density and continuing outward until

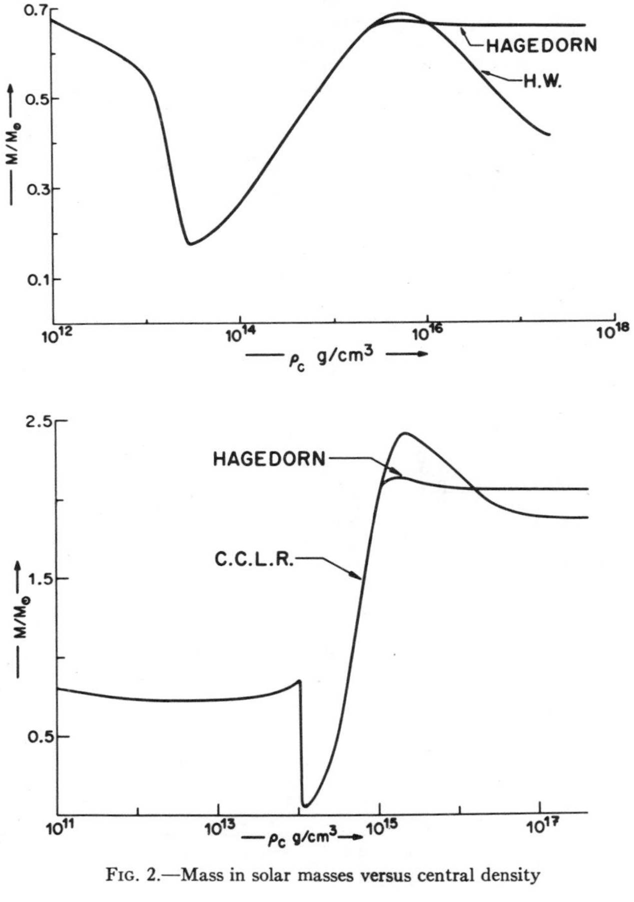

FIG. 2.—Mass in solar masses versus central density

the pressure drops to zero. The total mass is then determined from the asymptotic form of the g_{00} component of the Schwarzschild metric. The top half of Figure 2 shows the total mass in solar masses versus the logarithm of the central density in grams per cm^3 for the HW equation of state and for the Hagedorn equation of state matched to the HW equation of state near the region of 4.73×10^{15} g cm^{-3}. The bottom half of Figure 2 shows a similar plot for the CCLR equation of state and for the Hagedorn equation of state matched to the CCLR equation of state near the region of 1.13×10^{15} g cm^{-3}.

It is evident from this figure that the change in the maximum equilibrium configuration caused by the introduction of the Hagedorn equation of state is very small. The value of the critical mass changes by about 3 percent in the HW case and by about 11 percent in the CCLR case.

The major changes are introduced in the region beyond the maximum mass equilibrium configurations in which a clear asymptotic behavior is introduced and neither a large decrease in mass nor additional turning points in the (mass, central density)-curve are present. This latter effect is in contradistinction to the Harrison-Wheeler case as discussed in chapter 5, especially theorem 9, of Harrison *et al.* (1965)! The physical reasons behind this behavior will be discussed elsewhere (Rhoades 1971).

The question of the stability of the equilibrium configurations has also been analyzed by calculations of the pulsation period for small radial oscillations. The pulsation period

TABLE 1

PERIODS OR *e*-FOLDING TIMES OF THE FUNDAMENTAL NORMAL RADIAL MODES OF OSCILLATION FOR VARIOUS VALUES OF CENTRAL DENSITY

ρ_c (g cm^{-3})	Mass ($\mathfrak{M}_\odot$)	Radius (km)	Period or *e*-folding time (msec)	Mass ($\mathfrak{M}_\odot$)	Radius (km)	Period or *e*-folding time (msec)
	Harrison-Wheeler			Hagedorn		
1.00*E*15	0.555	14.1	0.894	0.555	14.1	0.894
1.26*E*15	0.582	13.1	0.843	0.582	13.1	0.845
1.58*E*15	0.608	12.2	0.804	0.608	12.2	0.813
2.00*E*15	0.631	11.4	0.780	0.631	11.4	0.806
2.51*E*15	0.651	10.6	0.780	0.649	10.7	0.867
3.16*E*15	0.666	10.0	0.827	0.661	10.1	1.53
3.98*E*15	0.677	9.3	0.998	0.670	9.5	*U* 0.746
5.01*E*15	0.685	8.7	2.63	0.670	8.9	*U* 0.401
6.31*E*15	0.685	8.3	*U* 1.01	0.669	8.9	*U* 0.367
7.94*E*15	0.680	7.8	*U* 0.633	0.668	8.8	*U* 0.311
1.00*E*15	0.670	7.4	*U* 0.469	0.666	8.8	*U* 0.267
	Cameron-Cohen-Langer-Rosen			Hagedorn		
3.98*E*14	0.788	12.4	0.419	0.788	12.4	0.419
5.01*E*14	1.144	12.9	0.389	1.144	12.9	0.389
6.31*E*14	1.437	13.0	0.416	1.437	13.1	0.416
7.94*E*14	1.724	13.0	0.443	1.724	13.0	0.443
1.00*E*15	1.995	12.7	0.468	1.995	12.7	0.471
1.26*E*15	2.203	12.3	0.526	2.116	12.5	0.606
1.58*E*15	2.333	11.8	0.650	2.130	12.4	0.818
2.00*E*15	2.394	11.3	1.07	2.129	12.4	*U* 1.94
2.51*E*15	2.398	10.8	*U* 1.22	2.121	12.3	*U* 0.769
3.16*E*15	2.362	10.3	*U* 0.658	2.109	12.3	*U* 0.602
3.98*E*15	2.330	10.0	*U* 0.559	2.098	12.3	*U* 0.527

NOTE.—Values of the total mass and radius of the equilibrium configurations are also tabulated. Note that for central density, computer notation is used; i.e., 1.00*E*15 stands for 1.00×10^{15}. Unstable modes are distinguished by a *U* beside the *e*-folding time. Stable modes have no distinguishing mark beside the period.

or *e*-folding time in the case of unstable stellar models can be obtained by perturbing the star and solving an eigenvalue equation for the perturbations. The pulsation equation has previously been derived directly from Einstein's equations by Chandrasekhar (1964). The numerical method for obtaining the pulsation periods will be given elsewhere (Rhoades 1971).

The fundamental pulsation period in milliseconds is given in Table 1 for the various equations of state for several values of the central density. The radius and mass of the equilibrium configurations are also given. Unstable modes are distinguished by a *U* beside the *e*-folding time, whereas stable modes have no distinguishing mark beside the period. The adiabatic indexes used in these calculations were those obtained directly from the equations of state.

In conclusion, no substantial qualitative or quantitative changes are produced in the stable neutron-star configurations by the introduction of the Hagedorn equation of state—or, for that matter, by any soft equation of state where pressure depends on density in a logarithmic way at densities greater than 10 times nuclear density.

The only changes are in the values of the critical mass, which are slightly lowered. However, a completely new asymptotic behavior is introduced in the high-density region greater than 10^{15} g cm^{-3}: the (mass, central density)-curve does not decrease or oscillate at higher densities as in previous models, but reaches a maximum and maintains a nearly constant value. All the configurations beyond this maximum have been found to be unstable against their lowest mode of radial pulsation. A similar behavior has been found to apply not only to the Hagedorn equation of state but to any equation of state with an adiabatic index decreasing as $1/\log \rho$ or faster as $\rho \rightarrow \infty$.

Note added in proof.—The equation of state we have used at supranuclear densities corresponds to the most extreme case of softness, and it will be valid for $T \sim T_0$ (Hagedorn limiting temperature). In the limit $T \rightarrow 0$ the same equation of state will give $P \propto \rho/\log(\rho/\rho_0)$, as kindly communicated to us by Hagedorn. In the light of our computations we can conclude that even this less extreme equation of state will not introduce any significant qualitative or quantitative change in the value of the critical mass, and the typical asymptotic behavior we have found will still be present.

For their continuous interest and suggestions, we are grateful to Professors Rolf Hagedorn and John A. Wheeler. We are also indebted to Dr. Jeffrey Cohen for enlightening discussions of the CCLR equation of state.

REFERENCES

Chandrasekhar, S. 1964, *Phys. Rev. Letters*, **12**, 114 and 437.
Cohen, J. M., Langer, W. D., Rosen, L. C., and Cameron, A. G. W. 1970, *Ap. and Space Sci.*, **6**, 228.
Hagedorn, R. 1965, *Nuovo Cimento Suppl.*, **3**, 147.
———. 1967, *Nuovo Cimento*, **52A**, 1336.
———. 1968*a*, *Nuovo Cimento Suppl.*, **6**, 311.
———. 1968*b*, *Nuovo Cimento*, **56A**, 1027.
———. 1969, CERN Rept. No. TH 1027.
Harrison, B. K., Thorne, K. S., Wakano, M., and Wheeler, J. A. 1965, *Gravitation Theory and Gravitational Collapse* (Chicago: University of Chicago Press).
Hartle, J. B., and Thorne, K. S. 1968, *Ap. J.*, **153**, 807.
Levinger, J. S., and Simmons, L. H. 1961, *Phys. Rev.*, **124**, 916.
Oppenheimer, J. R., and Volkoff, G. M. 1939, *Phys. Rev.*, **55**, 374.
Rhoades, C. E., Jr. 1971, in preparation.
Ruffini, R., and Wheeler, J. A. 1970, in *The Significance of Space Research for Fundamental Physics* (Paris: European Space Research Organization), chapter on "Relativistic Cosmology and Space Platforms."
Zel'dovich, Ya. B. 1961, *Zh. Eksper. i Teoret. Fiz.* (USSR), **41**, 1609 (English transl. in *Soviet Phys.—JETP*, **14**, 1143).

A 1.2

ON THE MASSES OF X-RAY SOURCES

Robert W. Leach

Remo Ruffini

Joseph Henry Laboratories, Princeton, N. J. 08540

ABSTRACT

An analysis of x-ray sources based on the Roche model is here presented. On this basis we can conclude that pulsating sources appear to have systematically smaller masses than nonpulsating sources. We suggest identifying the first objects as neutron stars and the second as totally collapsed objects or black holes. Detailed predictions are presented. Discriminating features between neutron stars and black holes are also given.

* Work supported in part by National Science Foundation Grant GP30799X to Princeton University..

The recent great interest in the physics of x-ray sources is due to a fortunate coincidence of very significant results in the experimental and theoretical fields. From an experimental point of view it has become clear that for the first time we will be able to determine the mass of a collapsed object to within a few percent. This achievement is made possible by the concurrence of two very important events: (a) many of the most powerful x-ray sources ($dE/dt \sim 10^{36-38}$ erg/sec in the x-rays region) are indeed collapsed objects. Evidence for this is clearly given by arguments based on the energetics and on the short time scale which modulates the x-ray emission; (b) nearly all these very energetic x-ray sources show clear evidence of being members of binary systems. Therefore from the orbital parameters and the inclination the mass of the collapsed object can be inferred, or limits can be placed on it.

From the theoretical point of view it has become more evident than ever (see e.g. Ruffini 1972) that mass is indeed the most powerful tool we have today for differentiating between different families of collapsed objects. This for good reasons! We divide all collapsed objects into three main families: (1) neutron stars (2) black holes and (3) others. To be more explicit, we call a black hole any object collapsing asymptotically in time toward a horizon which is assumed to meet all the usual requirements of regularity. It appears more plausible than before that this family is uniquely populated by the Kerr-Newman solution and its subcases. In "others" we will consider such examples as naked singularities and non-stationary black holes. Our point here is that, at least for the moment, no experimental evidence forces us to accept the existence of this third class of objects, which will from now on be discarded from our consideration.

How can we differentiate between neutron stars and black holes? No differentiation is possible, based

(i) on detailed electrodynamic processes: Christodoulou and Ruffini (1972) and Treves and Ruffini (1972) have explicitly shown striking similarities between the electrodynamic processes in all collapsed objects;

(ii) on spatial dimensions: a neutron star of $1M_\odot$ has a radius $R \sim 10$km; a black hole of $2M_\odot$ is expected to have a radius of $R \sim 5.88$km!

Positively, however, neutron stars differ from black holes in two main respects:

(i) The critical mass of a neutron star can only be as large as $0.7M_\odot$ if all the strong interactions between nucleons are neglected (Oppenheimer and Volkoff 1939).

or $1.5M_\odot$ if nuclear interactions are taken into account (V. Pandharipande 1971 and references mentioned there).

It has anyway to be always <u>less</u> than $3.2M_\odot$ no matter what equation of state is used which does not violate causality, the Chatelier's principle or the general relativistic equations of equilibrium (C. Rhoades and R. Ruffini 1972).

(ii) While regular pulsation processes have been conceived in black hole physics (D. Wilkins 1972) no way has yet been conceived to keep that regularity over a long time period. Regular pulsations on a long time scale should therefore be expected to be associated with neutron stars while black holes could have only bursts with a sharply defined period ($P \sim MG/c^3$) lasting $T \sim 10^2 - 10^3 MG/c^3$ (see e.g. Ruffini 1972)

Assuming a model based on the Roche model we can explain some of the major features of x-ray sources. Moreover, the pulsating sources appear to have masses consistent with neutron stars, while non-pulsating sources seem to have systematically larger masses ($m > m_{crit}$) typical of black holes.

In Table I we present the computations of the ratio of the mass $q = m_x/M$ (m_x is the mass of the x-ray source, M is the mass of the "normal" occulting star) to be expected on the basis of the occultation angle ϕ for selected values of the inclination ($i = 90°$ line of sight lying in the equatorial plane of the system). The Ω_o is the critical value for the Roche potential corresponding to a contact configuration, where

$$(1) \qquad \Omega_o = \frac{1}{r} + \frac{q}{r'} + \frac{(q+1)}{2}\,[(x-q/(q+1))^2 + y^2] - q^2/[2(1+q)]$$

The separation between M and m_x is taken as unity; r' and r indicate the distances of an arbitrary point from m_x and M respectively. The occultation angle has been computed assuming the source of the x-rays to be pointlike and the normal companion star just filling up its Roche lobe. Using these data we are able to determine the minimum mass to be expected for the x-ray source. In other words, a necessary condition to have an x-ray source is that material flows from the primary through the inner lagrangian point into the secondary. The energy in the x-rays has in fact to be originated by accretion (Ruffini and Wheeler 1971 (a) and (b) a references mentioned there). The rotational energy is by far insufficient to explain the x-ray power. In the case of Centaurus $P \sim 4.8$sec $I \sim 2 \times 10^{46}$gr×cm^2 $\frac{dP}{dt} < 10^{-10}$ we have $\left(\frac{dE}{dt}\right)_{rot} \lesssim 10^{35}$ erg/sec compared to the $\frac{dE}{dt} \sim 10^{37}$erg/sec actually detected! In this respect all these x-ray sources strongly depart from the usual pulsating radio sources (pulsars!).

TABLE I

Occultation angles for selected values of the inclination of the orbit and selected value of the ratio $q = m_x/M$.

$q = \frac{m}{M}$	Ω_O	$\phi_{i=90°}$	$\phi_{i=80°}$	$\phi_{i=70°}$	$\phi_{i=60°}$	$\phi_{i=50°}$	$\phi_{i=40°}$	$\phi_{i=30°}$	$\phi_{i=20°}$
1.0	3.7500	22.00	19.81	10.46	-	-			
0.8	3.41697	23.30	21.29	13.25	-	-			
0.6	3.06344	25.03	23.22	16.40	-	-	no occultation		
0.4	2.67810	27.56	26.00	20.41	-	-	possible		
0.3	2.46622	29.42	28.00	23.09	9.58	-			
0.2	2.23273	32.09	30.85	26.69	17.15	-	-	-	-
0.15	2.10309	34.00	32.88	29.16	21.16	-	-	-	-
0.1	1.95910	36.72	35.73	32.52	26.04	10.77	-	-	-
0.05	1.78886	41.32	40.52	37.97	33.15	24.50	-	-	-
0.02	1.65702	47.16	46.54	44.60	41.13	35.58	26.70	-	-
0.01	1.59911	51.29	50.77	49.17	46.37	42.10	35.92	27.01	12.59
0.005	1.56256	55.13	54.69	53.35	51.04	47.63	42.98	36.97	29.83
0.001	1.52148	62.87	62.55	61.63	60.06	57.87	55.07	51.83	48.54

The expected valuesof mass based on the Roche model in the case of Centaurus and Hercules are given in Table II and Table III. Given as well are the radii and mass of the "visible" star and the separation of the components and the value of their velocities. It is evident from these tables that in both cases the x-ray sources should be expected to be simply and clearly neutron stars! White dwarfs, although they have masses in this range, do not have gravitational fields strong enough to produce x-rays. We therefore disagree with recent suggestions (Wilson 1972) that the x-ray source could not be a collapsed object or that the basic Roche model as a minimum criterion on the mass should be relaxed (Hall et al., 1972 , in particular in these computations the value of the occulting radius of the star was not taken into proper account). In the case of Centaurus the "visible" star should be expected to be a BO star with a luminosity $\lg L/L_{\odot} = 4.0$ $\lg \bar{\rho} = -1.3$ ($\bar{\rho}$ in g/cc). The fact that it has not yet been seen could be due to absorption by galactic material.

So much for the most typical x-ray sources to be expected to be neutron stars. Diametrically opposite is the case of Cygnus XI. Using the analysis of Brucato and Kristian and Bolton gives values for the velocities of the primary star and of a region emitting excited helium lines associated with the secondary as well as the orbital period of the system. The region emitting helium lines lies somewhere between the inner Lagrangian point of the binary system and the collapsed secondary, so the velocity of the secondary is known within an adjustable parameter (for details see Brucato and Kristian 1972). A lower limit on the mass comes from assuming the emitting region coinciding with the inner lagrangian point. We obtain $m_x > 8M_{\odot}$

TABLE II

haracteristic parameters computed for selected values of the inclination for entaurus X3. We have assumed that the period of occultation is $T_{occ} = 0.558$ days, he period of the binary system $T = 2.0871$ days and $v_{mx} \sin i = 415.1$ km/sec.

$q = \frac{m}{M}$	$\frac{m}{M_\odot}$	$\frac{M}{M_\odot}$	$\frac{a}{R_\odot}$	$\frac{R}{R_\odot}$	$v_m \frac{km}{sec}$	$v_M \frac{km}{sec}$
0.0172	0.275	16.0	17.4	12.7	415.1	7.14
0.015	0.250	16.6	17.7	13.3	421.5	6.32
0.0115	0.220	19.1	18.4	14.1	441.7	5.08
0.008	0.194	24.2	19.9	15.5	479.3	3.83
0.0047	0.163	35	22.5	17.8	541.9	2.55
0.003	0.176	59	26.7	22.4	645.8	1.94
0.0015	0.186	120	34.3	29.1	830.2	1.25

TABLE III

aracteristic parameters computed for selected values of the inclination for rcules X1. We have assumed $T_{occ} = 0.24$ days, the period of the binary $T = 1.70$ days d $v_{mx} \sin i = 169$ km/sec.

$q = \frac{m}{M}$	$\frac{m}{M_\odot}$	$\frac{M}{M_\odot}$	$\frac{a}{R_\odot}$	$\frac{R}{R_\odot}$	$v_m \frac{km}{sec}$	$v_M \frac{km}{sec}$
0.57	1.20	2.1	8.9	3.8	169.0	96.3
0.43	0.78	1.8	8.2	3.7	171.6	73.8
0.24	0.38	1.6	7.5	3.7	179.8	43.2
0.105	0.17	1.6	7.2	4.1	195.2	20.5
0.048	0.10	2.1	7.8	4.8	220.6	10.6
0.022	0.074	3.3	9.0	6.1	262.9	5.78
0.011	0.076	7.0	11.5	8.2	338.0	3.72

Our further point here is that there are good reasons to believe that some of the other non-pulsating binary x-ray sources are endowed with a high mass collapsed object. In the case of 2U(0115-73) the occultation angle ϕ is small (28°); as a direct consequence from Table I we see that the ratio $q = \frac{m_x}{M}$ is equal to 0.4 for an inclination of $i = 90°$ and 0.1 at 60°. This suggests that indeed this system is a very good candidate for a substantial collapsed mass; this will be verified as additional parameters become available. Also in the case of Cygnus X-3 we can make a similar inference if we consider the two sources to be members of a binary system of period $P \sim 4.8$ hours. Either the system is formed by stars far apart and in that case they have to be fairly massive or the stars are near each other but then the lifetime of the system should be fairly short.

One of the most interesting objects to explore the correlation between non-pulsating sources and mass over the critical value for gravitational collapse is 2U(0900-40) observed by Hiltner <u>et al.</u> Assuming for the mass of the primary 15 $M_\odot$, from the velocity of the primary and the orbital period we can obtain a lower limit on the collapsed object of $1.4M_\odot$.

Only one object seems not to fit this model based on the Roche model: 2U(1700-37) whose value of the occultation angle ($\phi = 69°$) is much too large to give reasonable results.

DISCUSSION

The most important objects to be explored are clearly pulsating objects with mass (a) $m > 0.7M_\odot$ or (b) $m > 1.5M_\odot$. The existence of such x-ray sources would strongly suggest in the case (a) that in the computations of neutron star equilibrium configuration the effects of strong interactions between nucleons should be taken into serious account and that the "free nucleons" model is completely inadequate. In case (b) we should conclude that a perturbation expansion for the treatment of nuclear interactions of the kind usually adopted in the analysis of neutron star physics is not satisfactory and a new approach is necessary. In this light, particularly appealing is the treatment first supported by Fermi and further developed by Hagedorn (see e.g., Ruffini 1972).

Vice versa, also of the greatest interest is the determination of the minimum mass of non-pulsating sources. The absence of pulsation could be due either to (a) a neutron star with a magnetic field endowed with symmetry around the rotation axis or to a very weak magnetic field ($B \ll 10^{12}$ Gauss) or (b) a black hole with mass $m < m_{crit}$ and formed by a star imploding with enough kinetic energy to overcome the barrier against gravitational collapse (see e.g., Ruffini and Wheeler 1971a and b) or a normal black hole with mass near the critical value m_{crit}.

A theoretical analysis to predict more detailed features to discriminate between these four possibilities is currently under way and some preliminary results have been presented elsewhere (R. Ruffini 1972). We can already say, however, that the most interesting object from this point of view to perform an experimental check is 2U(0900-40); in this case the mass of the collapsed object could indeed be very near the upper value for the critical mass $\sim 1.5 M_\odot$.

Some further predictions can be obtained from the present model. If the "normal" star could emit part of its envelope as observed in CV

Serpentis by Kuhi and Schweizer (we are indebted to S. Ames for pointing out to us this interesting example) the star could have low periods in which material will not be dumped on the collapsed object, the star being contained inside its Roche lobe up to a value of $\Omega<\Omega_o$. The x-ray emission as well as all the possible optical phenomena associated with accretion will then simply stop and only the "normal" star will be observed with possibly a pulsar-like object around. On the contrary, a too large rate of accretion from the normal star into the collapsed object could create a piling up of material in the lobe and associated again will be a disappearance of the x-ray source. This material could be absorbed periodically by collapse into the x-ray source.

Let us finally emphasize the importance of these binary systems both for the analysis of neutron stars and black holes. Knowing the mass of the neutron star from a detailed analysis of the sudden changes in period(glitches) of the x-ray sources, much can be inferred on the internal structure of a neutron star and the properties of its crust. In the case of a non-pulsating source an analysis of bursts with a characteristic period of the order $P \sim GM/c^3$sec and lasting $\sim 10^2$-$10^3 GM/c^3$sec could be of the greatest importance in determining the angular velocity of the collapsed object. An angular velocity can indeed be associated with a black hole $w = a/(r_+^2 + a^2)$ (a being the angular momentum or unit mass, r_+ the the radius of the horizon of a black hole, in geometrical units) as shown by Christodoulou and Ruffini (1972) and detectable effects should indeed exist (Wilkins 1971).

References

Bolton, C. T., 1972, preprint.

Brucato, R., Kristian, J., 1972, preprint.

Christodoulou, D., and R. Ruffini, 1972, "Electrodynamics of Collapsed Objects", preprint.

Gursky, H., 1972, "X-Ray Sources", Les Houches, B. DeWitt, editor, Gordon and Breach, in press.

Hall, D., Weedman, D., 1972, Ap. J. Letter, 176, L19.

Hiltner, W., Osmer, P., Werner, J., 1972, Ap. J. Lett., 175, L19.

Huhi, L. V., Schweizer, F., 1970, Ap. J. Lett., 160, L185.

Oppenheimer, J. R., Volkoff, G. M., 1939 Phys. Rev., 54, 374.

Pandharipande, V. R., 1971, Nuclear Physics, A178, 123.

Rhoades, C., R. Ruffini, "On the Maximum Mass of Neutron Stars", Submitted to Phys. Rev. Lett.

Ruffini, R., 1972, "On the Energetics of Black Holes", Les Houches B. DeWitt, ed., Gordon and Breach , in press.

Ruffini, R., Treves, A., 1973, "On a Magnetized Rotating Sphere", Astrop. Lett., in press.

Ruffini, R., J. A. Wheeler, 1971 (a) Physics Today, January.

Ruffini, R., J. A. Wheeler, 1971 (b), Cosmology from Space Platform, ESRO Book SP52, Paris.

Wilkins, D., 1972, Phys. Rev. 5, 814.

Wilson, R., 1972, Ap. J. Lett., 174, L27.

A 1.3

On the Maximum Mass of a Neutron Star[†]

Clifford E. Rhoades, Jr.[††]

Remo Ruffini

Joseph Henry Laboratories

Princeton University

Princeton, N. J. 08540

ABSTRACT

On the sole basis of the Einstein theory of relativity, the principle of causality and Le Chatelier's principle, it is here established that the maximum mass of the equilibrium configuration of a neutron star cannot be larger than 3.2 $M_\odot$. The extremal principle given here applies as well when the equation of state of matter is unknown in a limited range of densities. The concept of "domain of dependence" associated with a given point of the equation of state is introduced. The absolute maximum mass of a neutron star provides a decisive method of observationally distinguishing neutron stars from black holes.

[†]Work partially supported by NSF grant GP-30799X to Princeton University.

[††]Present address: AFWL(DYS), Kirtland AFB, NM 87117

A neutron star with mass above a certain critical limit will inevitably be unstable against complete gravitational collapse to a black hole. Estimates of this critical mass range from .32 to 1.5 $M_\odot$. This is the best present understanding of the situation.

However, if one has found all that much uncertainty in the critical mass by trails so far of different equations of state, how does one know that one cannot arrive at a critical mass higher by a whole order of magnitude by still further exploration of physically conceivable equations of state? And then why not an infinite critical mass and no instability at all? Such a conclusion would throw doubt on the very existence of the phenomenon of gravitational collapse and on the existence of completely collapsed objects (Black Holes).

From a theoretical point of view, it has become more evident over the last few years, that the greatest uncertainty in evaluating the critical mass of a neutron star for gravitational collapse comes from the equation of state at nuclear densities and above. The understanding of the physical properties of neutron star matter at a densities smaller than 10^{13}g/cm^3 is essential to describe the solid state properties of the crust(1) which extends in the less massive neutron stars to a size of a hundred kilometers. While the knowledge of this outer region could in time be of the greatest importance in understanding the sudden changes in the period of the pulsars(2), it is of no relevance whatsoever in the determination of the maximum mass of a neutron star(3). The reason is clear. By increasing the central density, the mass of the star increases and the radius of the star decreases. The neutron star becomes more and more compact and approaches a radius of about 10 km, while its crust becomes only a few tens of meters thick or even less, depending on the models[4].

Unfortunately, it is at nuclear densities and above that the most difficulties in calculating a physical equation of state occur, because of the lack of knowledge of the strong interactions between nucleons[5] just above nuclear density, while at still higher densities the physics is further complicated by the creation of resonances[6]. The situation is even more difficult that appears at first glance. It is, in fact, possible that in the near future much more will be known both on the interaction between nucleons and on the creation of resonances. Very probably, however this will not help to give a definite answer to the problem of an equation of state for neutron star matter at nuclear and supranuclear densities. It is conceivable in fact that the behavior of nucleons, caused by the degeneracy of the system, in the case of a neutron star will radically depart from that of nucleons in normal laboratory physics. Many body interactions (in many respects, a neutron star can be considered "compressed" nucleus of 10^{57} particles) and threshold effects in the creation of resonances[7] because of the unavailability of phase space could change in essential ways some features of an equation of state obtained by a direct extrapolation of the results deduced from laboratory experiments.

From an experimental point of view, it has become evident recently that one of the most powerful tools in determining the difference between a neutron star and a black hole relies on the possible difference in mass of the two objects[8]. Other tests based on the electrodynamic properties of these objects seem, today, largely indecisive because of the large variety of phenomena which take place in the ergosphere of a magnetic black hole[9]. Moreover, the recent discoveries of x-ray sources in binary systems seems to give the first concrete possibility of determining the mass of a collapsed object with unprecedented accuracy[10].

From the above, there is a clear need for establishing on solid ground the maximum mass of a neutron star.

Instead of trying to analyze nuclear forces and elementary particle physics under conditions of high compression with greater accuracy, we here approach the problem from a different point of view.

We disregard everything one knows about these effects and take that most extreme of all conceivable equations of state that produces the maximum critical mass compatible solely with these four conditions:

(1) Standard relation between mass and density in a static general relativity configuration of spherical symmetry.

(2) Standard general relativity equation of hydrostatic equilibrium.

(3) Le Chatelier's principle.

(4) The principle of causality.

While no suggestion is made that the resultant equation of state accurately represents the actual physical behavior of matter, it does illustrate a point of principle by yielding a maximum value for the critical mass.

It is not altogether new to approach the equation of state from the side of hydrostatic theory rather than from the side of the structure of matter. Gerlach[11] has shown that from a set of measurements on a sequence of star configurations (here as always for the "standard case" of cold matter catalyzed to the endpoint of thermonuclear evolution) of stellar radius as a function of mass one can in principle work back to deduce the equation of state, without any call on nuclear or elementary particle theory.

We present here, first, an extremization technique for the case in which we assume that the equation is known everywhere apart from a finite range of pressure and densities

(1a) $\rho_o \leq \rho \leq \rho_1$ (1b) $p(\rho_o) \leq p \leq p(\rho_1)$

in order to avoid "supraluminous" equation of state (velocity of sound greater than the speed of light; violation of causality, c = 1 in our units) we demand

(2) $dp/d\rho \leq 1.$

Also, we require that pressure be a monotonically non decreasing function of the density (Le Chatelier's principle; no spontaneous collapse of matter locally; speed of sound real)

(3) $\frac{dp}{d\rho} \geq 0$

This last condition could, indeed, be relaxed, however the proof is more straightforward assuming the validity of both conditions (2) and (3). We wish to choose the equation of state between ρ_o and ρ_1, such as to extremize the mass of the neutron star. We integrate the standard general relativistic equations

(4.1) $\frac{dm(r)}{dr} = 4\pi\rho r^2 = H(\rho,r)$

(4.2) $\frac{dp}{dr} = - \frac{(\rho+p)}{r(r-2m)} [4\pi r^3 p(r) + m(r)] = G(\rho,p,m,r)$

That is the total mass of the star is given by

(5) $M = \int_0^R 4\pi\rho(r)r^2 dr$, R being the radius of the star.

This problem may be formulated in terms of the standard calculus of variations with inequality constraints (12). We give here an alternate formulation in terms of control theory (13) which yields both necessary and sufficient conditions. We take ρ as the independent variable. Letting the primes denote derivative with respect to ρ we have

(6.1) $u = p'$ (6.2) $u/G = r'$ (6.3) $Hu/G = m'$

where $u(\rho)$ is given by the known equation of state in the range $\rho_o \geq \rho$ and $\rho \geq \rho_1$ and has to have

(6.4) $0 \leq u \leq 1$ in the range $\rho_o \leq \rho \leq \rho_1$

Equations (6) replace equations (1), (2), (3), (4), (5). Here u becomes the so-called control variable. Since we seek the maximum of m we introduce the Hamiltonian

(7) $(\rho,p,m,r,y_1,y_2,y_3,u) = \{y_1 + y_2 H/G + y_3/G\}\, u$

The variables y are Lagrange multipliers which satisfy the equations

(8.1) $y_1' = - [y_2 \frac{\partial}{\partial\rho} (H/G) + y_3 \frac{\partial}{\partial\rho} (1/G)]\, u$

(8.2) $y_2' = - [y_2 \frac{\partial}{\partial m} (H/G) + y_3 \frac{\partial}{\partial m} (1/G)]\, u$

(8.3) $y_3' = - [y_2 \frac{\partial}{\partial r} (H/G) + y_3 \frac{\partial}{\partial r} (1/G)]\, u$

We see immediately without further effort that the desired extremum lies on the boundary of the allowed functions $p = p(\rho)$. That is

(9) $u = 0$ or $u = 1$.

This is shown in fig. 1. To see which one of the two paths on the boundary of the "allowed rhomboid" maximize or minimize the mass we proceed to a direct integration of the equations. The upper path maximizes, the lower path minimizes the mass . We can also show that the extremum cannot lie on any intermediate path fulfilling conditions (9) and contained in the allowed rhomboid (path c in fig. 1). This proof is straightforward: we subdivide our rhomboid in smaller ones and apply in turn our extremizing technique to each one of them (path (d) in Fig. 1).

To apply this general treatment to a concrete example and to obtain the desired maximum mass for a neutron star, we have to assume the knowledge of an equation of state up to a fiducial density ρ_o. In Table I we give the results of the numerical integration performed by assuming the Harrison-Wheeler equation of state at values of the density smaller than the matching densities. In Fig. 2 an explicit example of matching is given for $\rho_o = 2 \times 10^{14} g/cm^3$. How to use these results to establish a firm upper limit to the value of the critical mass? Below $\rho_o = 4.6 \times 10^{14}$ the equation of state of Harrison and Wheeler[(4)] is already one which maximizes the mass of the equilibrium configuration. For the reasons previously given, in fact, the equation of state at densities $\rho < 10^{13} g/cm^3$ does not directly affect the value of the maximum mass. In the density range $10^{13} \leq \rho \leq$ the Harrison-Wheeler ne-

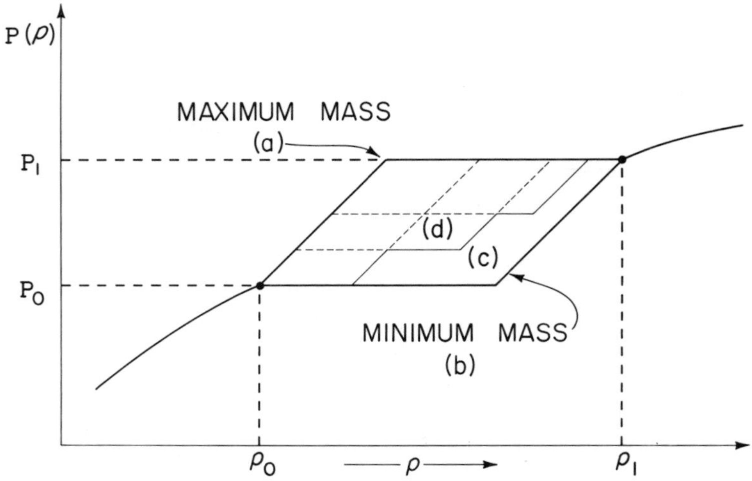

FIGURE 1 "Allowed rhomboid" in the p_1 ρ plane. The equation of state is known for $\rho < \rho_o$ and $\rho > \rho_1$. In the range $\rho_o \leq \rho \leq \rho_1$ all equations of state compatible with the principle of causality and Le Chatelier's principle have to be contained inside the rhomboid. Path a (b) maximizes (minimizes) the mass of the neutron star.

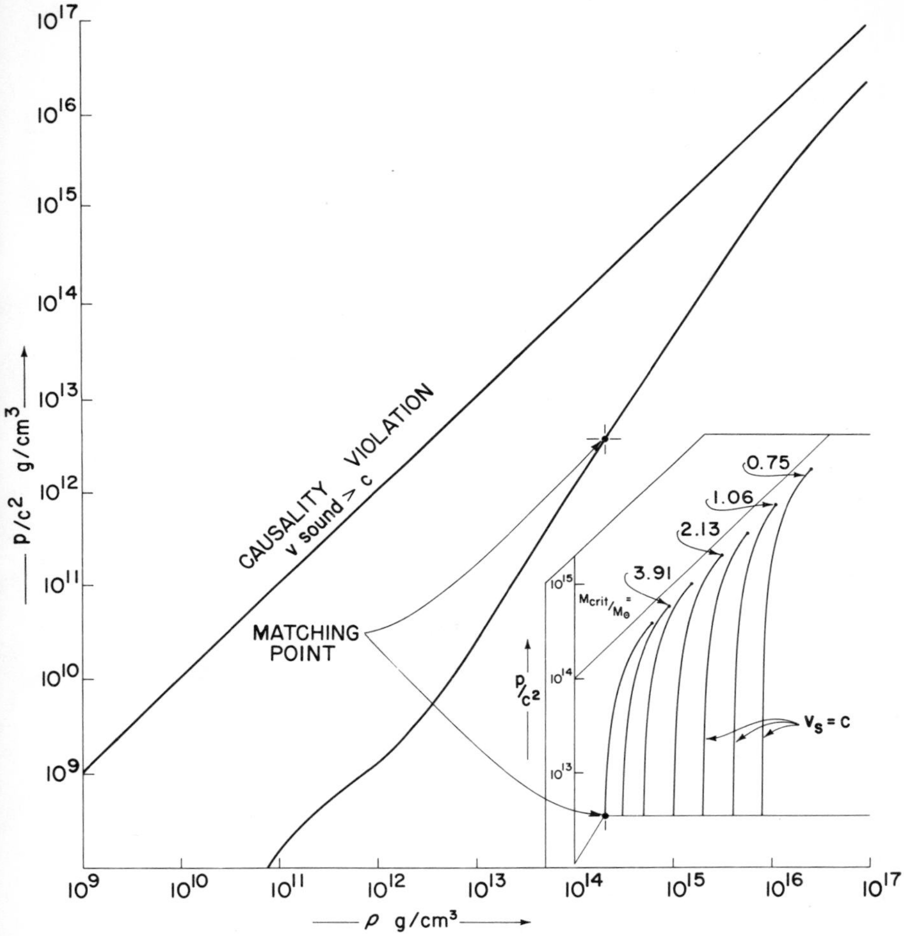

FIGURE 2 Matching of the "allowed rhomboid" to the Harrison-Wheeler equation of state for a value of the density $\rho_o = 2 \times 10^{14} g/cm^3$ (lower right side of the Figure). The different paths followed correspond to an equation of state with $\frac{dp}{d\rho} = 1$ path a in Fig. 1 or to a combination of a path with $\frac{dp}{d\rho} = 0$ and then $\frac{dp}{d\rho} = 1$ (path b in Fig. 1). The apparent difference between Fig. 1 and Fig. 2 is due to the difference in scale: linear in Fig. 1 logarithmic here. In these computations ρ_1 is supposed to be very large ($\rho_1 > 10^{17} g/cm^3$) and the integration are carried out up to the value of critical central density at which the value of the critical mass is reached ($\rho_{cent}^{crit} < 10^{17} g/cm^3$).

glects the nuclear interactions, mainly attractive in this range of densities, and gives a value of larger than the one to be expected on the basis of any realistic treatment of nuclear matter[14]. We can, then, immediately conclude that no matter the details of an equation of state at nuclear or supranuclear density a neutron star can never have a mass as large as $3.2M_{\odot}$. Let us introduce now the concept of "domain of dependence associated with a given point p_o, ρ_o of the equation of state".

Let us assume that we know the equation of state up to a density ρ_o, can we infer anything on the possible range of the value of the critical mass? By integrating along paths with $dp/d\rho = 1$ and $dp/d\rho = 0$ see lower right side of Fig. II, we can define a domain of dependence in the plane of the value of the critical mass M_{crit} versus the value of the central density ρ_{cent}. We can, therefore, establish that: if we know the equation of state up to a value of the density ρ_o we can always define a domain of dependence (M^{crit} vs. ρ_c) such that all the possible values of the critical mass and corresponding central densities are contained in that domain. The domain of dependence corresponding to a density $\rho'_o > \rho_o$ is always contained inside the one corresponding to the lower density ρ_o. From this also follows immediately that the knowledge of an equation of state at supranuclear density ($\rho \gtrsim 10^{15} g/cm^3$) cannot influence substantially the value of the critical mass. The domain of dependence for $\rho_o \simeq 10^{15} g/cm^3$ being extremely narrow on the M_{crit} axis! If we want to increase the value of the critical mass the only way is to act at nuclear densities, there the domain of dependence is still very large (see Fig.3) No matter what we do, however, the mass can never be larger than $M \simeq 3.2\ M_{\odot}$!

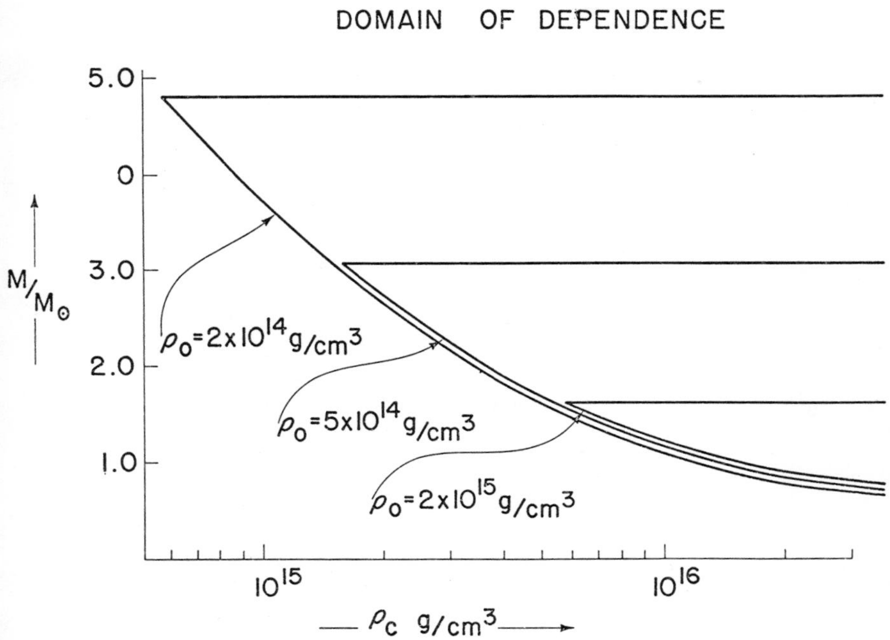

FIGURE 3 Domains of dependence for selected values of the density ρ_o. We have assumed for $\rho < \rho_o$ the equation of state of Harrison and Wheeler, for sake of example. The domain of dependence corresponding to higher values of the density is always contained inside the ones corresponding to lower value of the densities.

TABLE I

ρ matching	$M_{crit}^{max}/M_{\odot}$	ρ_{cent} g/cm^3
2×10^{14} g/cm^3	4.80	5.62 10^{14}
3×10^{14}	3.93	8.91
4×10^{14}	3.41	1.12 10^{15}
5×10^{14}	3.06	1.41
6×10^{14}	2.80	1.78
7×10^{14}	2.59	2.00
8×10^{14}	2.43	2.24

Value of the critical mass and the corresponding critical central density for selected values of the matching density. The Harrison-Wheeler equation of state has been used for densities smaller than the matching density.

References

1. G. Baym, C. J. Pethick, and D. Pines, Nature 224, 674, (1969).
   C. E. Rhoades, Jr., Ph.D. Thesis - Princeton University- 1971- unpublished - Chapters VI, VII, VIII.
   M. Ruderman, Nature 223, 547, (1969) and Nature 218, 1128, (1968).
   R. Smoluchowski, Phys. Rev. Lett. 24, 923, and Phys. Rev. Lett. 24, 1191.
2. P. E. Boynton, E. J. Groth, R. B. Partridge, D. T. Wilkins, Ap. J. Lett. 157, L197, 1969.
3. R. Ruffini, Les Houches Lectures (1972), B. DeWitt, Ed. Gordon and Breach - in Press.
4. See. e.g. B. K. Harrison, K. S. Thorne, M. Waakano, J. A. Wheeler, Gravitation Theory and Gravitational Collapse, The Univ. of Chicago Press, 1965.
5. See e.g. V. R. Pandharipande, Nucl. Physics, A178, 123, 1971, see. also Ref. 3.
6. V. A. Ambartsumian and G. S. Saakian, Astron. Zh. 37, 193 (1960) Englishtrausla: Soviet Astron. 4, 187, (1960).
   R. Hagedorn, Astron. and Astrophysics, 5, 184, (1970). See also Ref. 3.
7. R. Sawyer, to be published in Ap. J. (1972).
8. R. Leach and R. Ruffini, Submitted for publ. in Ap. J. Lett. (1972).
9. D. Christodoulou and R. Ruffini, "On the Electrodynamics of Collapsed Objects", (1972) - Preprint - unpublished.
10. H. Gursky - Les Houches Lectures (1972), B. DeWitt, ed. Gordon and Breach - in Press.
11. U. H. Gerlach, Phys. Rev. 172, 1325, (1968).

12. C. E. Rhoades, Jr., Ph.D. Thesis - Princeton University - May 1971 - unpublished - Chapter V.

F. A. Valentine, "The Problem of Lagrange with Differential Inequalities as Added Conditions" in "Contributions to the Calculus of Variations", 1933-37 - Chicago University Press

13. L. C. Young, "Calculus of Variations and Optimal Control Theory" W. B. Saunders, Co. Philadelphia, 1969.

14. See e.g. Ref. 5 or for an explicit comparison of the different equations of state for neutron star material, see Ref. 3.

PART II Electromagnetic Radiation in Static Geometries

2.1 Introduction

The aim of this part and of the following one is to examine concrete processes of energy extraction from material accreting into black holes. We will mainly focus on a perturbation analysis in a given fixed background metric to analyze processes of emission of electromagnetic radiation; in the following part the emphasis will be on gravitational radiation. For a long time speculations have been advanced as to whether a particle falling in a gravitational field should or should not radiate. Here the spectra, the intensity, the multipole expansion of the radiation field are presented for particles (a) falling radially into static geometries, (b) projected with finite kinetic energy into static geometries, endowed or not endowed with a net charge, (c) spiralling in stable or unstable circular orbits. Possible beaming effects and polarization features are also analyzed.

Let us examine the major questions to which we are addressing ourselves in this part.

In the case of radial infall:

(a) How much energy is radiated during the infall of a particle in the given static geometry?
(b) How much can the amount of energy radiated be affected by the presence of very strong electromagnetic fields (here simulated by the presence of a net charge on the black hole)?
(c) At what frequency is the peak of the radiation emitted?
(d) Is there any hope of emitting radiation at frequencies $\nu \gg c^3/GM$?
(e) Can we predict the detailed features of a burst due to electromagnetic or gravitational brehmstrahlung in the gravitational field of a black hole, and can we predict its detailed polarization features as well?

In the case of circular orbits in flat space one of the most important special relativistic effects characterizing orbits of charged particles moving with speed near the speed of light is the emission of synchroton radiation (see Figure 1). From this arises the need of analyzing:

(a) the emission of electromagnetic radiation from highly relativistic orbits under joint special and general relativistic effects
(b) how can the effect of a curved background modify the beaming due to special relativistic effects
(c) the polarization features of the radiation from highly relativistic circular orbits, both in the case of gravitational and electromagnetic radiation, and giving detailed and observable experimental predictions

(d) "transitions" of relativistic orbits with processes of emission taking place either near or far away from collapsed objects and then recovering the usual special relativistic treatments
(e) the last stable circular orbits (maximum binding energy states) allowed in a given background metric.

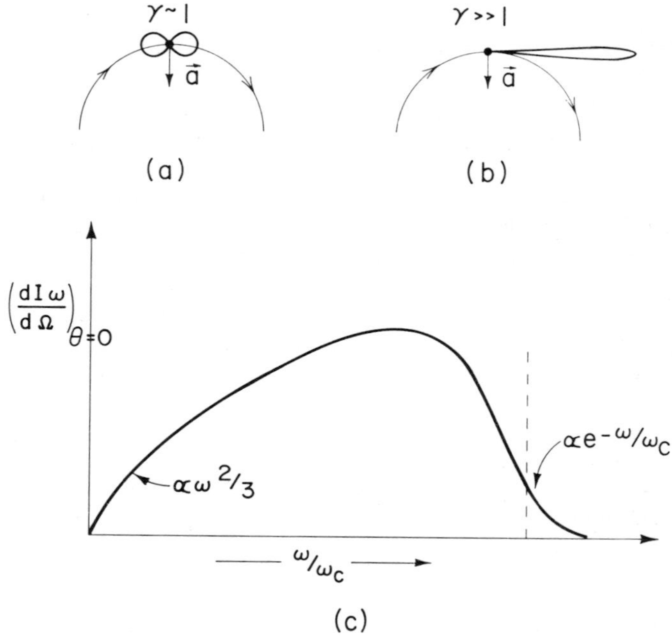

Figure 1. Schematic Representation of Synchroton Radiation.

We will finally address ourselves to a very important problem from an astrophysical point of view. How many parameters should we expect to characterize a collapsed object actually found in nature? In Part IV we will emphasize the importance of the possible presence of a magnetic field and rotation for the energetics of black holes. In the present chapter we are going to approach this problem from a preliminary and totally classic point of view.

A particle moving in a circular orbit of radius R in an extreme relativistic regime ($\gamma = E/mc^2 \gg 1$) emits a sharply focused beam of radiation which sweeps across a far away observer in a time $\Delta t \sim R/c\gamma^3$. This duration is determined by the special relativistic transformation law for the time coordinate t' of an observer moving with the particle and the time coordinate t of an observer at rest at infinity ($dt/dt' \sim 1/\gamma^2$). A pulse of duration Δt contains frequency components up to a critical frequency $\omega_c \sim 1/\Delta t \sim c\gamma^3/R$. In the extreme relativistic regime then $\omega_c \gg \omega_0$. Here ω_0 is the angular velocity of the particle in a circular orbit.

In the orbital plane of the particle the intensity of the radiation is given in the low frequency limit ($\omega \ll \omega_c$) by

$$\frac{dI(\omega)}{d\Omega} \sim e^2(\Gamma(\tfrac{2}{3}))^2\,(\omega R/c)^{2/3}/\pi^2 c$$

and in the high frequency limit is characterized by a cutoff at $\omega \sim \omega_c$

$$\frac{dI(\omega)}{d\Omega} \simeq (3e^2\gamma^2\,\omega\, e^{-2\omega/\omega_c})/2\pi c\omega_c.$$

In the low frequency limit ($\omega \ll \omega_c$) the radiation is confined in angles $\theta < \theta_c \sim (\omega_c/\omega)^{1/3}/\gamma$ around the orbital plane; in the high frequency limit ($\omega \gg \omega_c$) the intensity of the radiation is simply given as a function of θ by

$$\left(\frac{dI(\omega)}{d\Omega}\right)_\theta = \left(\frac{dI(\omega)}{d\Omega}\right)_0 e^{-3\omega\gamma^2\theta^2/\omega_c}$$

where we have indicated by $(dI(\omega)/d\Omega)_0$ the intensity of the radiation in the orbital plane. Therefore, there exists a cutoff angle $\theta_c \sim (\omega_c/\omega)^{1/2}/\gamma$. The radiation for large values of γ is strongly focused in the orbital plane and a considerable amount of the radiation is emitted at frequencies $\omega \gg \omega_0$. This straightforward picture of synchroton radiation obtained from simple special relativistic considerations is essentially modified by the presence of gravitational fields. The radiation is deflected, stored and defocused by the presence of strong gravitational fields (see Appendix 2.5 and Part III) and the beaming is then much less effective.

2.2 Timelike Geodesics in Schwarzschild and Reissner-Nordstrøm Geometries

The metric for a Reissner-Nordstrøm geometry in Schwarzschild like coordinates is given by

$$ds^2 = -e^\nu\, dt^2 + r^2(\sin^2\theta\, d\phi^2 + d\theta^2) + e^{-\nu}dr^2 \tag{1}$$

with $e^\nu = 1 - 2M/r + Q^2/r^2$. Here and in the following we use geometrical units with $G = c = 1$, $M_{\text{conv}} = GM/c^2$ and $Q_{\text{conv}} = G^{1/2}Q/c^2$, Greek indices $\alpha,\beta = 0.3$, latin indices $i = 1,3$. The motion of a test particle of charge q and mass μ moving in the background metric (1) is described by the Lagrangian

$$\mathscr{L} = \tfrac{1}{2} g_{\mu\nu}\dot{x}^\mu\dot{x}^\nu + q/\mu A_\mu \dot{x}^\mu \tag{2}$$

where the dot denotes differentiation with respect to proper time τ and $x^\alpha(\tau)$ is the particle coordinate. The equation of motion of the particle is, then, derived from the Lagrange equation

$$\frac{d^2x^\alpha}{ds^2} + \Gamma^\alpha_{\beta\gamma}\frac{dx^\beta}{ds}\frac{dx^\gamma}{ds} = q/\mu\, F^\alpha_\delta \frac{dx^\delta}{ds} \tag{3}$$

with

$$F_{\mu\nu} = A_{\mu,\nu} - A_{\nu,\mu} \tag{4}$$

and $\Gamma^{\alpha}_{\beta\gamma}$ the Christoffel symbols obtained from the metric (1). In the Reissner-Nordstrøm metric we have

$$A_i = 0 \quad \text{and} \quad A_0 = -Q/r. \tag{5}$$

Since the Lagrangian is independent of the time coordinate t and the angular coordinate ϕ, we have the following conserved quantities

$$P_0 = \frac{\partial \mathscr{L}}{\partial \dot{t}} = -\,(e^{\nu}\dot{t} + Qq/r) \tag{6}$$

$$P_\phi = \frac{\partial \mathscr{L}}{\partial \dot{\phi}} = r^2 \sin^2 \theta \dot{\phi} \tag{7}$$

and the Hamiltonian H

$$H = -e^{\nu}\dot{t}^2 + e^{-\nu}\dot{r}^2 + r^2\dot{\theta}^2 + r^2 \sin^2 \theta\dot{\phi}^2 = -1 \tag{8}$$

for timelike geodesics. We can, as usual, define an effective potential for the motion of the particle in the equatorial plane of the given black hole.

$$E_{\pm} = Qq/r \pm \mu(e^{\nu}(1 + p_{\varphi}^2/\mu^2 r^2))^{1/2} \tag{9}$$

Some of the details and implications of this effective potential in the Schwarzschild geometry are given in Figure 2. The corresponding curves for the Reissner-Nordstrøm geometry are given in Figure 3. The effective potential with the positive sign (positive root states, E_+) in Eq. (9), corresponds to solutions with

$$\lim_{r\to\infty} E_+ = +\mu$$

the ones with the minus sign (negative root states, E_-)

$$\lim_{r\to\infty} E_- = -\mu.$$

The most important difference between solutions with $qQ \geqslant 0$ and the ones with $qQ < 0$, is that in this last case there can exist negative energy states of positive root solutions and as a direct consequence, energy can be extracted from a Reissner-Nordstrøm black hole. The region in which negative energy states of positive root solutions can exist is given by

$$m + (m^2 - Q^2)^{1/2} \leqslant r \leqslant m + (m^2 - Q^2(1 - q^2/\mu^2))^{1/2} \tag{10}$$

and, indeed, its extension depends upon the value of the charge of the test particle (details in Appendix 2.1) (generalized ergosphere). From Eq. (9) we also notice the following important relation

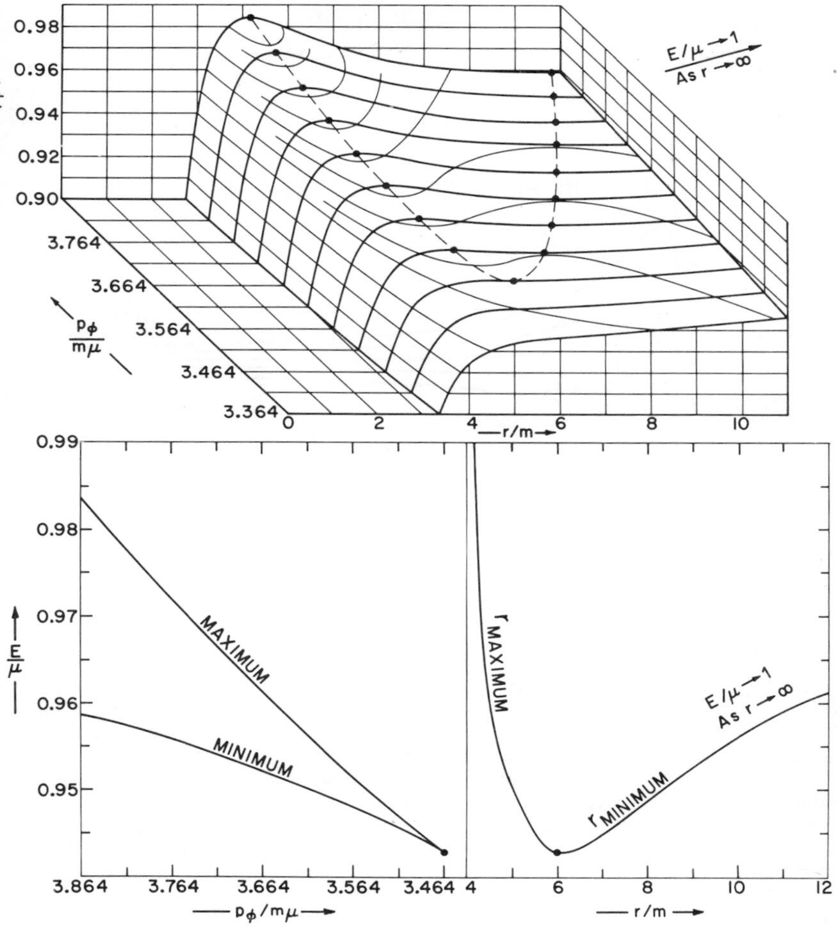

Figure 2. "Effective potential" experienced by test particle moving under influence of Schwarzschild geometry. Shown for values of angular momentum P_ϕ and distance r close to critical region of transition from stable circular orbits (minimum in effective potential) to types of motion in which the capture of the test particle is immediate (no minimum in effective potential). The transition ("last circular orbit") occurs at $p_\phi = (12)^{1/2}\, m\mu$, $r = 6m$ where μ is the mass of the test particle, m is the mass of the center of attraction in geometrical units ($m = 1.47$ km for one solar mass; more generally, $m = (0.742 \times 10^{-28}$ cm/g$)m_{\text{conv}}$) and r is the "Schwarzschild distance" (proper circumference$/2\pi$). The energy at this point is $E = (8/9)^{1/2}\, \mu = 0.943\, \mu$, corresponding to a binding of 5.7 per cent of the rest mass. The effective potential is defined here as that value of E which annuls the expression

$$E^2 = (\mu^2 + p_\phi^2/r^2)(1 - 2m/r).$$

(Figure from Ruffini and Wheeler[1].)

$$E_+(p_\phi, q, r) = -E_-(p_\phi, -q, r) \tag{11}$$

(antiparticles are simply holes in a E_- state!)

and also

$$E^+(p_\phi, q, r) \geqslant E^-(p_\phi, q, r). \tag{12}$$

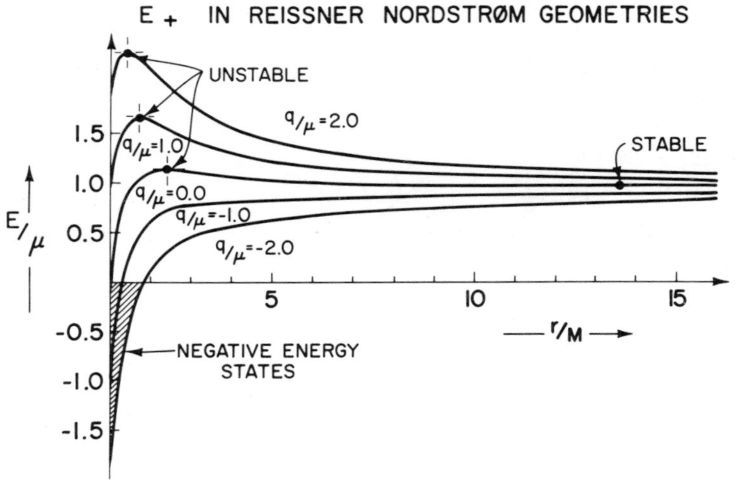

Figure 3. Effective potential $E = Qq/r + \mu(e^\nu(1 - p_\phi^2/r^2))^{1/2}$ in an extreme Reissner-Nordstrøm geometry ($Q/M = 1$) for a particle of mass μ and angular momentum $p_\phi/m\mu = 4.0$. The different curves refer to selected values of the charge of the particle q/μ. The effective potential in Reissner-Nordstrøm geometry differs markedly from the one in Schwarzschild geometry for values of $qQ < 0$, due to the existence of negative energy states. As in the Schwarzschild geometry, however, there still exist stable and unstable circular orbits. Details of the energy and radii of the circular orbits, corresponding to the diagram on the lower right side of Figure 2, in the case of a Reissner-Nordstrøm geometry are presented in Figure 5.

The equal sign is valid only for particles at the horizon

$$r_+ = m + (m^2 - Q^2)^{1/2}. \tag{13}$$

This implies that in the Reissner-Nordstrøm geometry, there exist reversible and irreversible transformations, the reversible ones existing if and only if the particles cross the horizon with zero kinetic energy.

With a set of reversible transformations, a black hole can, in fact, undergo the following two transitions

$M, Q \to M^1, Q^1$ by capture of a charged particle (14.1)

$M^1, Q^1 \to M, Q$ by capture of a particle of opposite sign of the previous one. (14.2)

To the capture of the two particles there is no change in the mass or the charge of the black hole. We can then, using reversible transformations

$$E = Qq/r_+, \tag{15}$$

define a succession of infinitesimal transformations

$$dm = Q\,dQ/(m + (m^2 - Q^2)^{1/2}) \tag{16}$$

and integrating Eq. (16) we obtain the Christodoulou-Ruffini formula governing the energetics of Reissner-Nordstrøm black holes

$$m = m_{ir} + Q^2/4m_{ir}. \tag{17}$$

Here, we have indicated, as usual, by m_{ir} the irriducible mass of the black hole. The fact that, indeed, the irriducible mass is irriducible follows immediately from the following relations

$$4m_{ir}\,\delta m_{ir} = (Er_+^2 - r_+ qQ)/(m^2 - Q^2)^{1/2} \tag{18}$$

This quantity is always positive (irreversible transformations) or null (reversible transformations).

The existence of negative energy states for positive roots solutions implies that processes of energy extraction from charged black holes are possible. A particle P_0 can impinge from infinite distance into the generalized ergosphere, here decay into two particles P_1 and P_2. The particle P_1, with a charge opposite to the one of the black hole, is suitably projected inside the horizon, the other particle P_2 comes out with more energy then the initial particle P_0! The *most important* feature in this process is that energy extraction can be done approaching arbitrarily close to reversible transformations. If an extreme Reissner-Nordstrøm solution, ($Q/m = 1$) is transformed into a Schwarzschild solution ($Q/m = 0$) by reversible transformations up to 50% of the total energy of the black hole can be extracted! More details on the energy extraction processes and on the definition of ergosurface and generalized ergosphere in Reissner-Nordstrøm geometries are given in Appendix 2.1.

Let us now proceed to give some explicit formulae governing a particle moving in circular orbits or infalling radially into a metric given by Eq. (1). For circular orbits in the equatorial plane Eq. (8) gives

$$e^{\nu}\dot{t}^2 = 1 + r_o^2\omega_0^2\dot{t}^2 \tag{19}$$

here by r_o we indicate the radius of the orbit and by $\omega_0 = d\phi/dt$, the angular velocity measured by a far away observer. From the Lagrange equations for the radial coordinate r we obtain

$$e^{\nu}\nu^1\dot{t}^2 = 2\omega_0^2 r_o\dot{t}^2 + 2Qq\dot{t}/r_o^2 \tag{20}$$

From (19), (20) and (7) we finally obtain (see App. 2.5)

$$\dot{t} = [Qq/2r_o + (1 - 3M/r_o + 2Q^2/r_o^2 + Q^2q^2/4r_o^2)^{1/2}]^{-1} \quad (21.1)$$

$$\omega_0^2 = M/r_o^3 - Q^2/r_o^4 - Qq[Qq/2r_o + (1 - 3M/r_o + 2Q^2/r_o^2 + (Qq/2r_o)^2)^{1/2}]/r_o^3 \quad (21.2)$$

$$E/\mu = -p_0 = (1 - 2M/r_o + Q^2/r_o^2)/(Qq/2r_o + (1 - 3M/r_o + 2Q^2/r_o^2 + (Qq/2r_o)^2)^{1/2} + Qq/r_o \quad (21.3)$$

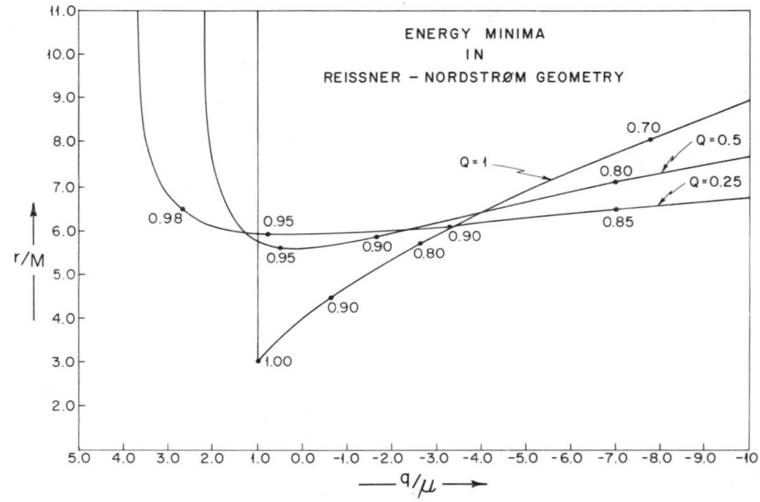

Figure 4. Energy of *last* stable circular orbits in Reissner-Nordstrøm geometries for selected values of the ratio (Q/μ = 1, 0.5, 0.25). The numerical values near the dots on the curves indicate the total energy (binding + rest + kinetic energy) of the particle in circular orbit. By increasing the charge of the test particle the last stable circular orbits moves towards larger values of the radial coordinates and the binding energy of the orbits is also augmented. Compare and contrast these values of the maximum binding energy and of the radii of the circular orbits for selected values of Q/μ and q/μ with the ones given in Figure 2 for the Schwarzschild geometry. (Details in ref. 2.)

In the limit $Q = 0$ Eqs (21) give the usual Schwarzschild results: orbit with maximum binding at $r = 6m$ and $E/\mu = 2\sqrt{2}/3$, last unstable circular orbit at $r = 3m$ and $E/\mu = +\infty$, see Figure 2. In the Reissner-Nordstrøm case stable circular orbits exist if and only if (see Figure 4)

$$Qq/\mu M < 1$$

The radius of the most bounded circular orbit increases with $|Qq|$ and the binding energy *also* increases. In the limit $|Qq| \to +\infty$ the binding energy goes to 100 per cent. We have in fact (see ref. 2), in the limit $|Qq| \to \infty$ from (21.3)

$$E/\mu \simeq -2[(r_o M - Q^2)/(r_o^2 - 3Mr_o + 2Q^2)]Qq/r_o$$
$$- (r_o^2 - 2Mr_o + Q^2)/Qqr_o + 0(r_o/Qq)^3 \tag{22}$$

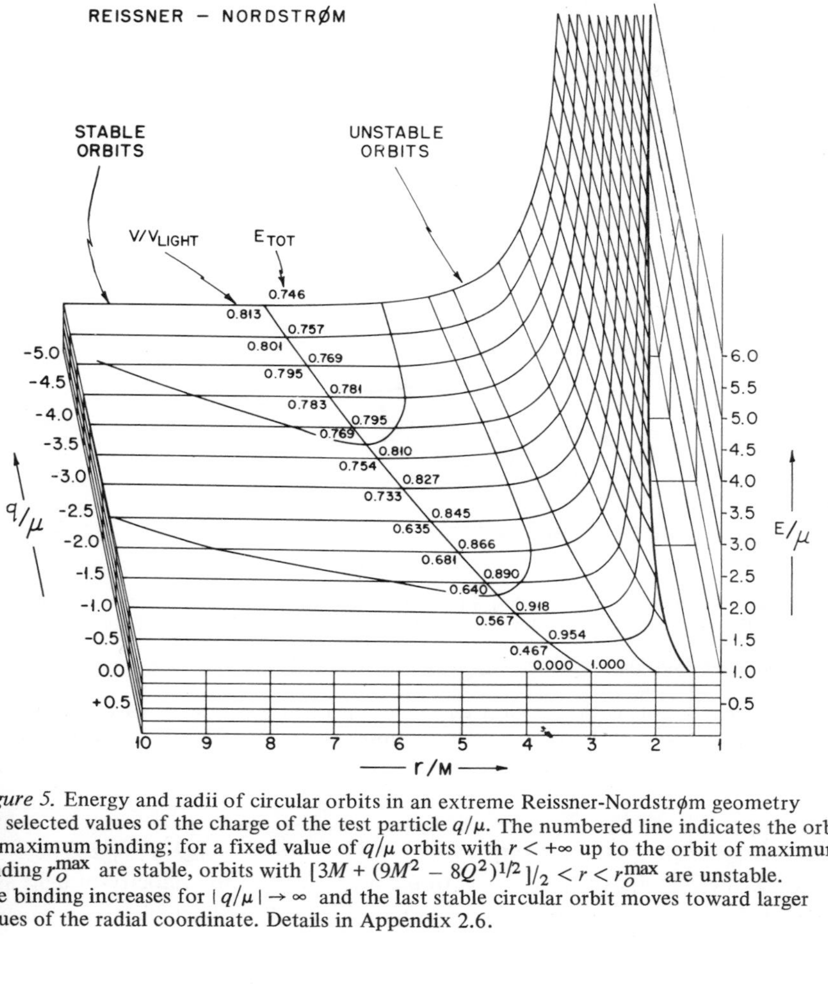

Figure 5. Energy and radii of circular orbits in an extreme Reissner-Nordstrøm geometry for selected values of the charge of the test particle q/μ. The numbered line indicates the orbits of maximum binding; for a fixed value of q/μ orbits with $r < +\infty$ up to the orbit of maximum binding $r_o^{\max}$ are stable, orbits with $[3M + (9M^2 - 8Q^2)^{1/2}]/_2 < r < r_o^{\max}$ are unstable. The binding increases for $|q/\mu| \to \infty$ and the last stable circular orbit moves toward larger values of the radial coordinate. Details in Appendix 2.6.

and for the orbit of maximum binding ($r_o^{\max}$)

$$r_o^{\max} \simeq (\sqrt{2}\,|Qq/\mu|)^{2/3} \tag{23}$$

with an energy

$$E_o^{\max} \simeq 3M(2/|Qq/\mu|)^{1/3}/2 + Q(1/Qq)) \tag{24}$$

An electron orbiting a black hole of one solar mass and charge $Q = 10^{-3}M$ has an $r_o^{max} \simeq 0.39$ light years and $E_{min} \simeq 1.98 \times 10^{-7}\mu$! Some more details on the orbits are given in Appendix 2.5. As in the Schwarzschild case, in the Reissner-Nordstrøm case there exist stable and unstable circular orbits. The stable orbits are the one with radius

$$r > r_o^{max}.$$

the unstable ones

$$[3M + (9M^2 - 8Q^2)^{1/2}]/2 < r < r_o^{max} \tag{25}$$

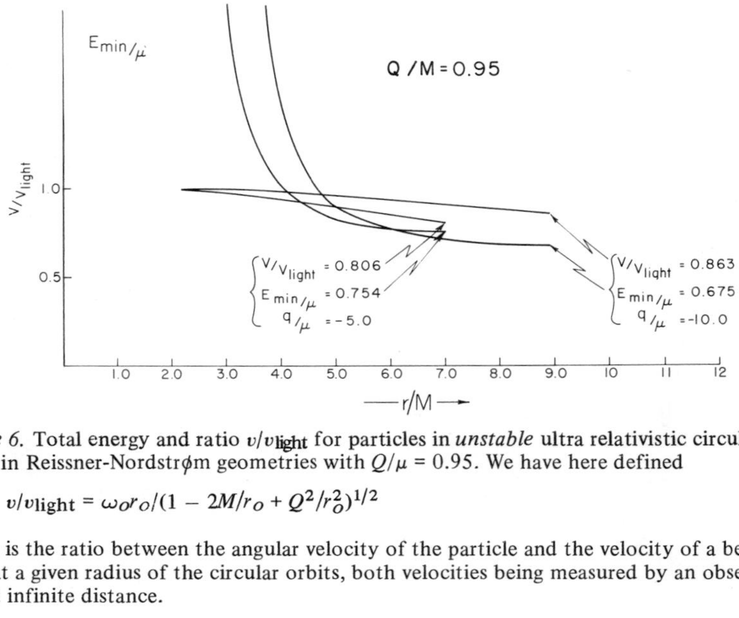

Figure 6. Total energy and ratio v/v_{light} for particles in *unstable* ultra relativistic circular orbits in Reissner-Nordstrøm geometries with $Q/\mu = 0.95$. We have here defined

$$v/v_{light} = \omega_o r_o/(1 - 2M/r_o + Q^2/r_o^2)^{1/2}$$

which is the ratio between the angular velocity of the particle and the velocity of a beam of light at a given radius of the circular orbits, both velocities being measured by an observer at rest at infinite distance.

The energy of the particle diverges for $r_o \to [3m + (9m^2 - 8Q^2)^{1/2}]/2$. The ratios of the velocity of the particle to the local value of the velocity of light as measured by a far away observer are given in Figure 6 for unstable circular orbits and selected values of q/μ and $Q/M = 0.95$.

Let us now examine the motion of a particle in radial infall in the equatorial plane of the given Reissner-Nordstrøm black hole. Equation (8) then gives

$$-e^{\nu}\dot{t}^2 + e^{-\nu}\dot{r}^2 = -1 \tag{26}$$

and from the remaining equations

$$\dot{r} = [(\gamma - Qq/\mu r)^2 - (1 - 2M/r + Q^2/r^2)]^{1/2} \tag{27.1}$$

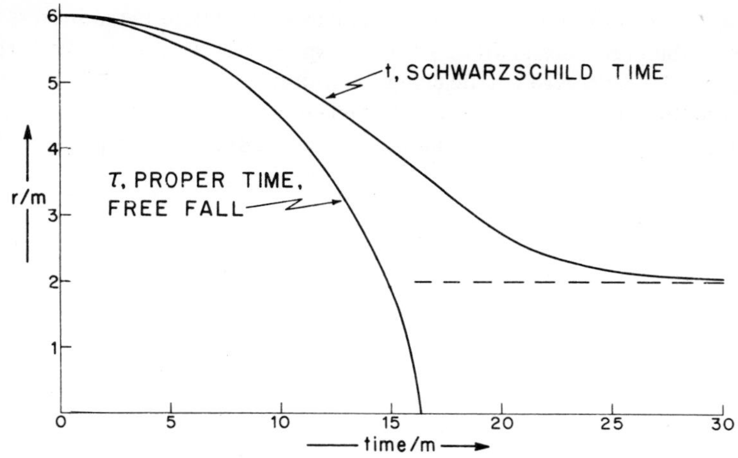

Figure 7. Fall towards a Schwarzschild black hole as seen by comoving (proper time r) and far away observer (Schwarzschild time t). The particle has been assumed to start its implosion initially at rest at $6m$. (Figure from Ruffini and Wheeler[1]).

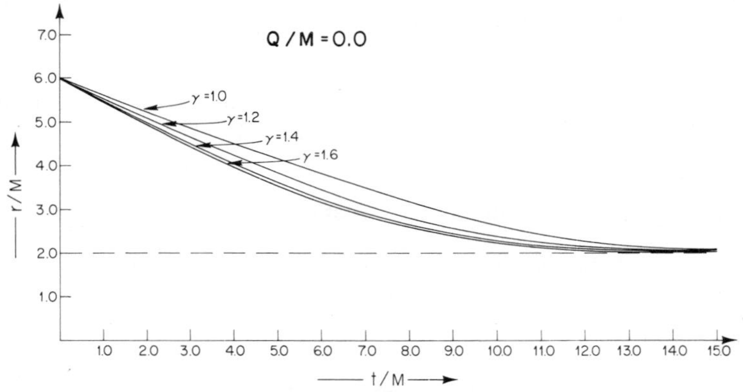

Figure 8. Implosions time for a particle falling radially in the field of a Schwarzschild black hole ($Q/\mu = 0$). The r/M coordinate is the Schwarzschild coordinate of the particle and t/M the Schwarzschild time, both measured in units of the black hole mass. The different curves refer to motion starting at rest ($\gamma = 1.0$) or with a finite value of the kinetic energy at $r = +\infty$ ($\gamma > 1.0$). In each case the time coordinate has been chosen to have the particle at $r = 6M$ for $t = 0$.

$$\dot{t} = (\gamma - Qq/\mu r)/(1 - 2M/r + Q^2/r^2) \tag{27.2}$$

and finally

$$\frac{dT}{dr} = (\gamma - Qq/\mu r)/\{(1 - 2M/r + Q^2/r^2)[(\gamma - Qq/\mu r) - (1 - 2M/r + Q^2/r^2)]^{1/2}\} \tag{28}$$

$T(r)$ being the Schwarzschild coordinate time associated with the motion of the particle. In the previous equations $\gamma = 1$ corresponds to the motion of a particle starting its infall with zero kinetic energy at infinity; $\gamma < 1$, to a particle imploding from finite distance initially at rest and $\gamma > 1$, to a particle imploding with a finite amount of kinetic energy from infinity. Equation (28) can be integrated and an

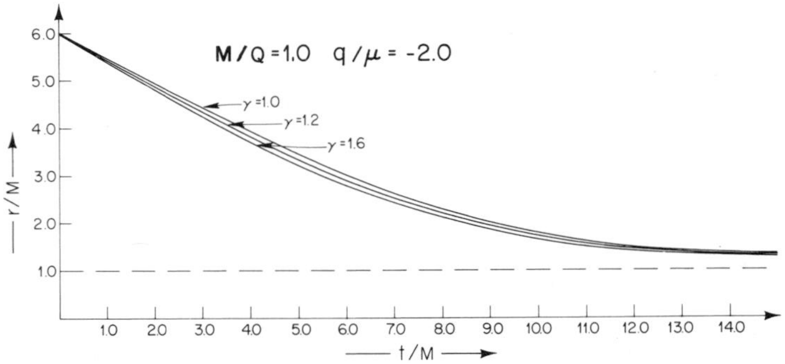

Figure 9. Implosion of a charged particle ($q/\mu = -2.0$) in the field of an extreme Reissner-Nordstrøm geometry ($Q/\mu = 1.0$). $\gamma = 1.0 (\gamma > 1)$ corresponds to a particle starting its motion with zero (finite) kinetic energy at infinity.

explicit formula given.[2] Some details of the $T(r)$ relations are given in Figure 7 and Figure 8. More details are to be found in Ref. (2).

2.3 Electromagnetic Perturbations in a Static Geometry†

If we have in a spherically symmetric background two scalar functions f and g and these functions are solutions of an equation of the form

$$0(f) = g \tag{29}$$

0 being a linear differential operator invariant under rotations, we can always expand the functions f and g into spherical harmonics $Y_l^m(\theta, \phi)$, since the $Y_l^m(\theta, \phi)$ form a complete set of eigenfunctions in θ and ϕ. We then have

$$f(r, \theta, \phi) = \sum_{lm} F_{lm}(r) Y_{lm}(\theta, \phi) \tag{30.1}$$

$$g(r, \theta, \phi) = \sum_{lm} G_{lm}(r) Y_{lm}(\theta, \phi) \tag{30.2}$$

In the case of vector functions in three dimensions defined on a spherically symmetric background metric, we can expand the general vector functions in

† The complete tensorial perturbation of the Reissner-Nordstrøm metric is currently under way and will appear in Ref. 2).

three dimensional vector spherical harmonics[3]

$$\vec{Y}_{jLM} = \sum_{mq} Y_{lm}(\theta, \phi)(l, m, 1, q \mid l, 1, j, M)\vec{e}_q \tag{31}$$

where $(l, m, 1, q \mid l, 1, j, M)$ are the Clebsh-Gordan coefficients and $\vec{Y}_{jLM}$ the eigenfunctions of the total angular momentum $\vec{j} = \vec{L} + \vec{S}$, $\vec{L}$ being the orbital angular momentum and $\vec{S}$ the spin angular momentum.

We then have

$$j^2 \vec{Y}_{jlM} = j(j+1)\vec{Y}_{jlM} \tag{32.1}$$

$$j_z \vec{Y}_{jlM} = M\vec{Y}_{jlM} \tag{32.2}$$

and the normalization condition

$$\int_0^{2\pi}\int_0^{\pi} Y^*_{jlM}(\theta, \phi) Y_{j'l'M'}(\theta, \phi) \sin\theta \, d\theta \, d\phi = \delta_{jj'} \, \delta_{ll'} \, \delta_{MM'}$$

The vectors $\vec{e}q$, with $q = \pm 1, 0$, are the eigenfunctions of the spin angular momentum $\vec{S}$

$$S^2 \vec{e}q = 2\vec{e}q \tag{33.1}$$

$$S_z \vec{e}q = q\vec{e}q \tag{33.2}$$

with

$$e_{+1} = -\frac{2}{\sqrt{2}}(e_x + ie_y) \tag{34.1}$$

$$e_0 = e_z \tag{34.2}$$

$$e_{-1} = \frac{1}{\sqrt{2}}(e_x - ie_y) \tag{34.3}$$

where $\vec{e}_x$, $\vec{e}_y$ and $\vec{e}_z$ are the usual unit vectors in cartesian coordinates.

There are only three different types of vector spherical harmonics with a given j and M

$$\vec{Y}_{jjM} \quad \text{with parity } (-1)^j$$

and

$$\vec{Y}_{jj\pm 1 M} \quad \text{with parity } (-1)^{j+1}$$

Instead of using the expressions (31) for the vector harmonics, it is very helpful in our problem to introduce some linear combinations of the $\vec{Y}_{j,j,M}$ and $\vec{Y}_{j,j\pm 1,M}$. We have, in fact

$$\vec{L} Y_{jM}(\theta, \phi) = [j(j+1)]^{1/2} \vec{Y}_{j,j,M} \tag{35.1}$$

$$\frac{\vec{r}}{r} Y_{jM}(\theta,\phi) = -[(j+1)/(2j+1)]^{1/2}\vec{Y}_{j,j+1,M} + [j/(2j+1)]^{1/2}\vec{Y}_{j,j-1,M} \tag{35.2}$$

$$r\,\vec{\nabla} Y_{jM}(\theta,\phi) = [(j+1)/(2j+1)]^{1/2} j\,\vec{Y}_{j,j+1,M} + [j/(2j+1)]^{1/2}(j+1)\vec{Y}_{j,j-1,M} \tag{35.3}$$

$$\vec{L} = \left(0,\ ir\frac{1}{\sin\theta}\frac{\partial}{\partial\phi},\ -ir\sin\theta\,\frac{\partial}{\partial\theta}\right)$$

and

$$\vec{\nabla} = \left(0,\ \frac{\partial}{\partial\theta},\ \frac{\partial}{\partial\phi}\right)$$

we then immediately obtain for the four dimensional vector spherical harmonics the expression given in Appendix 2.2.

We can now write the equations for the perturbations produced by a test particle moving in the background metric given by Eq. (1) and analyze the electromagnetic radiation emitted. We then obtain for the expression for the flux of radiation through a given surface

$$\left(\frac{dE}{dt}\right)_S = \int T_0^r r^2 \sin\theta\, d\theta\, d\phi \tag{36}$$

In the limit of $r \to \infty$ we have $g_{rr} \sim -g_{00} \sim 1$ (see Appendix 2.2), then

$$4\pi T_0^r r^2 \sin\theta = \sum_{lm}\sum_{l'm'} [(-b_{,0}^{lm} b_{,r}^{l'm'} + a_{,r}^{lm} a_{,0}^{l'm'}) \tag{37}$$
$$(Y_{,\theta}^{l'm'} Y_{,\phi}^{lm}/\sin\theta + Y_{,\theta}^{lm} Y_{,\phi}^{l'm'}\sin\theta) + (a_{,r}^{lm} b_{,r}^{l'm'} - b_{,0}^{lm} a_{,0}^{l'm'}) Y_{,\theta}^{lm} Y_{,\phi}^{l'm'}]$$

where the functions $a_{lm}(r,t)$ and $b_{lm}(r,t)$ fulfill the set of equations

$$a_{,r^*,r^*}^{lm} - a_{,t,t}^{lm} - l(l+1)g^{rr}a^{lm}/r^2 = \alpha^{lm} g^{rr} \tag{38.1}$$

$$b_{,r^*,r^*}^{lm} - b_{,t,t}^{lm} - l(l+1)g^{rr}b^{lm}/r^2 = g^{rr}[(r^2\psi^{lm})_{,r} - n_{,0}^{lm} r^2](l(l+1)) \tag{38.2}$$

We have adopted as usual the r^* coordinate with

$$\frac{dr}{dr^*} = g^{rr}$$

in Eq. (38.1) and (38.2) α^{lm}, ψ^{lm} and n^{lm} are given by

$$\alpha^{lm}(r,t) = q\delta(r - R(t))\left[-\frac{d\phi}{dt}\sin\Theta\, Y_{,\theta}^{lm\,*} + \frac{1}{\sin\Theta}\frac{d\Theta}{dt} Y_{,\phi}^{lm\,*}\right]/l(l+1) \tag{39.1}$$

$$\psi^{lm}(r, t) = q\delta(r - R(t))Y^{lm*}(\Theta, \Phi)/r^2 g^{00} \tag{39.2}$$

and

$$n^{lm}(r, t) = q \frac{dR}{dt} \delta(r - R(t))Y^{lm*}(\Theta, \Phi)/r^2 g^{rr} \tag{39.3}$$

We indicate by $R(t)$, $\Theta(t)$ and $\Phi(t)$ the coordinates of a test particle. We are now able to start a detailed analysis of the radiation emitted by particles moving in the background metric (1). We are going to consider particles in circular orbits and in radial infall.

2.4 Electromagnetic Radiation from a Particle Falling Radially into a Schwarzschild or Reissner-Nordstrøm Black Hole

Since we are dealing with particles moving in a purely radial motion, from Eq. (39.1) it follows that $\alpha_{lm} = 0$ and only the terms of parity $(-1)^{l+1}$ give a contribution. We then have to solve only Eq. (38.2). To analyze the spectral distribution of the radiation we use the Fourier transform equation (38.2), and we obtain the following equation

$$\frac{d^2 b^{lm}}{dr^{*2}} + \left[\omega^2 - g^{rr}\frac{l(l+1)}{r^2}\right] b^{lm}$$
$$= g^{rr}\left[(r^2 \psi^{lm}(\omega, r))_{,r} + i\omega n^{lm}(\omega, r)\right]/(l(l+1)). \tag{40}$$

Since we are dealing with radial motion, only the $m = 0$ terms give contributions. The explicit expressions of $\psi^{lo}(\omega, r)$ and $n^{lo}(\omega, r)$ are given in Appendix 2.3. Equation (40) has been solved numerically using the Green's function technique imposing as boundary conditions purely ingoing waves at the black hole surface ($r^* \to -\infty$) and purely outgoing waves at infinity ($r^* \to +\infty$). We have, in fact,

$$b^{lo}(\omega, r^*) = \int_{-\infty}^{+\infty} S^{lo}(\omega, r^*)G(\omega, r^*, r^{*\prime})\, dr' \tag{41.1}$$

with

$$G(\omega, r^*, r^{*\prime}) = \begin{cases} u^l(\omega, r^*)v^l(\omega, r^*)/W^l(\omega, u, v) & r^* > r^{*\prime} \\ v^l(\omega, r^*)u^l(\omega, r^*)/W^l(\omega, u, v) & r^* < r^{*\prime} \end{cases} \tag{41.2}$$

where $W^l(\omega)$ is the Wronskian of the two linearly independent solutions of Eq. (40) without source term ($S^{lm}(\omega, r^*) = 0$); $u^l(\omega, r^*)$ are the free waves purely outgoing at infinity, and $v^l(\omega, r^*)$ the free waves purely ingoing at $r^* \to -\infty$ (black hole surface). Details of the numerical integration are presented

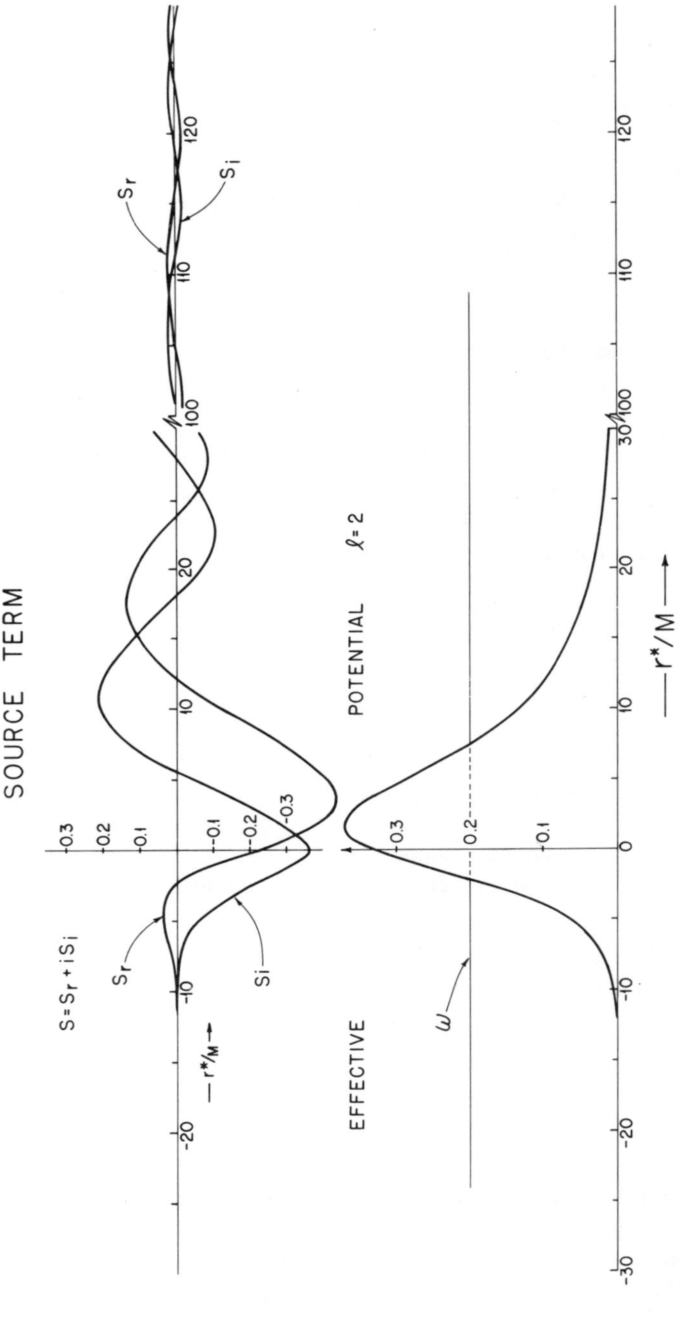

Figure 10.

in Figures 10, 11 and 12. Results are given in Appendix 2.3. It is important for us here to summarize the main results:

(I) If a particle falls in a Schwarzschild metric starting at rest from infinity, the majority of the radiation is emitted in the dipole mode $l = 1$. The peak of the spectrum occurs at a period $P \approx GM/c^3$.

(II) When the particle is either projected inside the black hole (finite kinetic energy at infinite distance) or the particle falls in the field of an oppositely charged Reissner-Nordstrøm black hole, then higher multipole modes are excited and radiation can, indeed, be emitted with period $P \ll GM/c^3$. The total amount of radiation also greatly increases.

What happens when a black hole captures a charged particle?

It would seem *a priori* that some difficulties in this capture process are connected with the fact that as seen by a far away observer the capture of a test particle by a black hole takes an infinite amount of time and really the particle *never* reaches the horizon of the black hole. The problem is most easily treated

---

Figure 10. To evaluate the radiation emitted by a charged particle falling into a black hole we have solved Eq. (40), imposing as boundary conditions purely outgoing waves at infinity

$$\lim_{r^*\to\infty} (b^{lo})^{\text{out}} = B^{lo}(\omega)\, e^{i\omega r^*}$$

and pure ingoing waves at the black hole surface

$$\lim_{r^*\to-\infty} (b^{lo})^{\text{ing}} = I^{lo}(\omega)\, e^{-i\omega r^*}$$

The total amount of radiation per unit frequency and averaged over all angles going inside the black hole is then given by the expression

$$\left(\frac{dE}{d\omega}\right)_{\text{in}} = \frac{\omega^2}{2\pi} \sum_l |I^{lo}|^2,$$

while the radiation outgoing at infinity

$$\left(\frac{dE}{d\omega}\right)_{\text{out}} = \frac{\omega^2}{2\pi} \sum_l |B^{lo}|^2$$

Using the Green's function technique, we obtain from Eq. (40)

$$(b^{lo})^{\text{out}} = B^{lo}(\omega)\, e^{i\omega r^*} = u^l(r^*, \omega) \int_{-\infty}^{+\infty} v^l(r^{*1}, \omega)\, S^{lo}(r^{*1}, \omega)\, dr^{*1}/W(\omega, u, v)$$

$$(b^{lo})^{\text{in}} = I^{lo}(\omega)\, e^{-i\omega r^*} = v^l(r^*, \omega) \int_{-\infty}^{+\infty} u^l(r^{*1}, \omega)\, S^{lo}(r^{*1}, \omega)\, dr^{*1}/W(\omega, u, v).$$

Here we have indicated by $v^l(r^*, \omega)$ $[u^l(r^*, \omega)]$ the solutions of the homogeneous $(S^l(\omega, r^*) = 0)$ Eq. (40) purely ingoing at $r^* \to -\infty$ [outgoing at $r^* \to +\infty$]. $W(\omega, u, v)$ is the Wronskian obtained from the two linearly independent solutions, $u^l(r^*, \omega)$ and $v^l(r^*, \omega)$. The effective potential $V^l(r^*) = g^{rr}l(l+1)/r^2$ in Eq. (40) is here plotted as a function of the coordinate r^* for a value of $l = 2$. Also plotted are the real and imaginary components of the source term of Eq. (40) for $l = 2$.

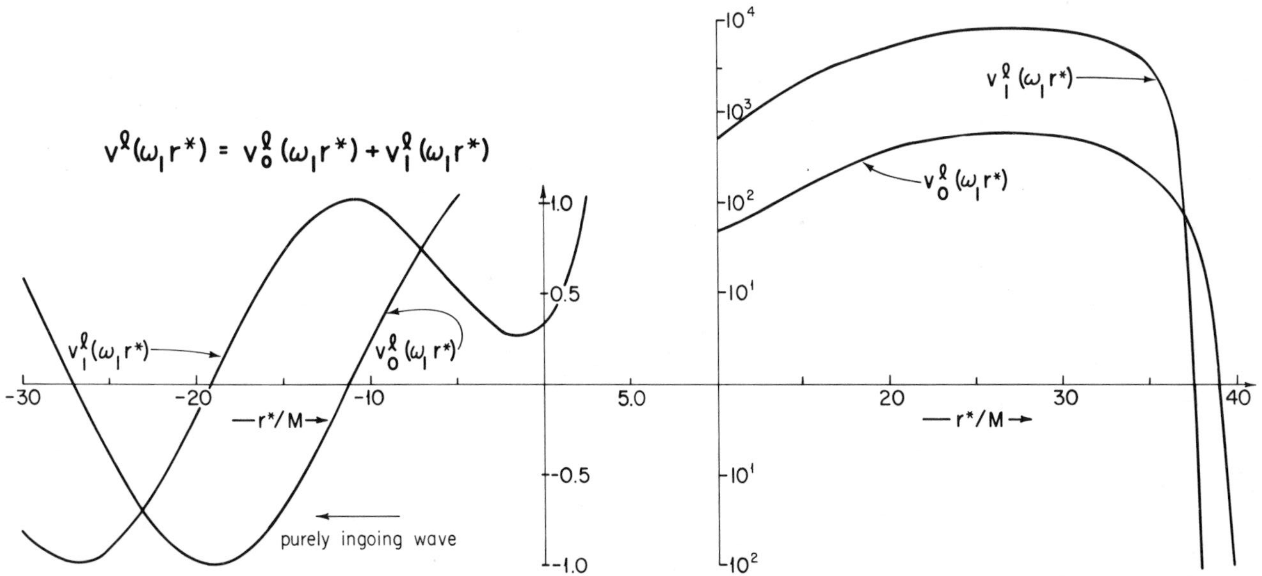

Figure 11. The numerical solution of the homogeneous Eq. (40) purely ingoing at the black hole surface $v^l(\omega, r^*)$ is here plotted as a function of r^* for a selected value of $\omega = 0.2/M$. The function $v^l(\omega, r^*)$ for $r^* \to +\infty$ becomes a mixture of incoming and outgoing waves. The drastic effect of the effective potential on the wave is clearly summarized by a direct comparison of the numerical values of the real $v_0^l(\omega, r^*)$ and imaginary $v_1^l(\omega, r^*)$ part of the function $v^l(\omega, r^*)$ for values of $r^* < 10M$ and $r^* > 10M$. Notice in particular that in the left side of the figure $(r^* < 10M)v^l(\omega, r^*)$ is given on a linear scale while on the right side $(r^* > 10M)$ on a logarithmic scale! The wave $v^l(\omega, r^*)$ is scattered by the effective potential (see Figure 10) and loses its purely ingoing character for $r^* > 0$.

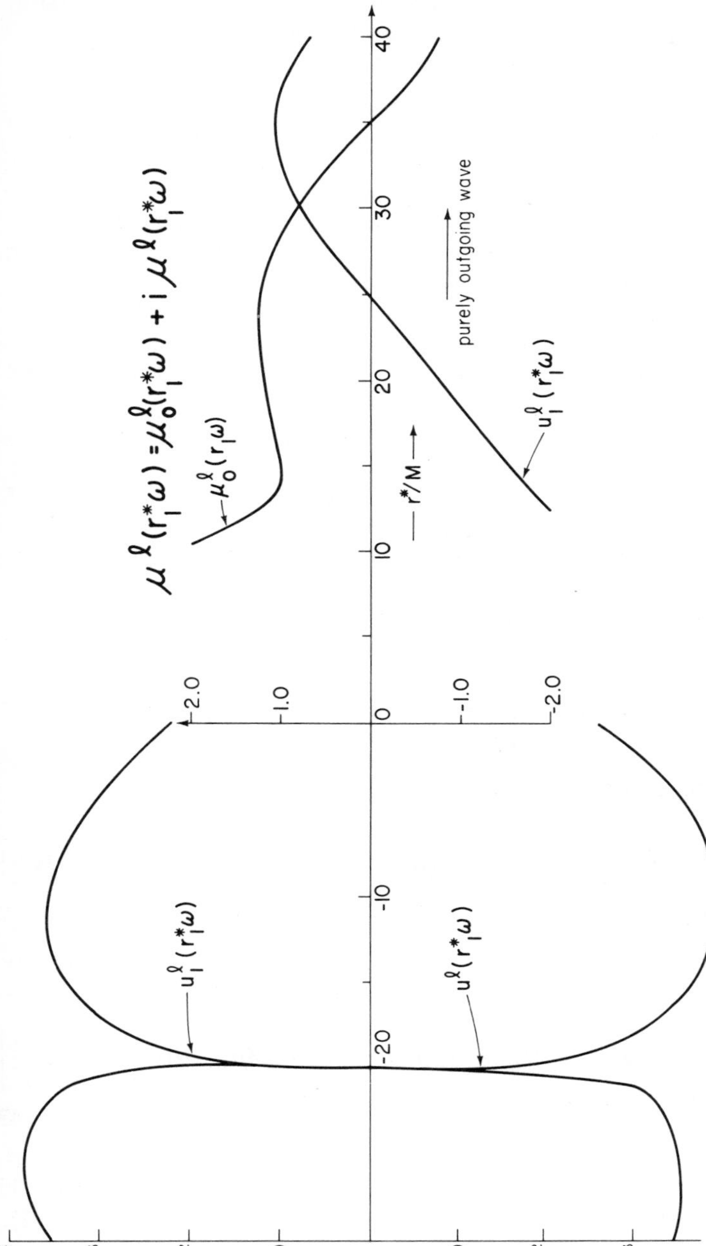

Figure 12. The numerical solution of the homogeneous equations (40) purely outgoing at $r^* \to +\infty$, $u^l(r^*, \omega)$ is here plotted as a function of r^* for a selected value of $\omega = 0.2/M$. The function $u^l(r^*, \omega)$ for $r^* \to -\infty$ becomes a mixture of incoming and outgoing waves, of intensity much larger than the purely outgoing wave for $r^* \to +\infty$. In the left part of the figure ($r^* < 0$) the function $u^l(r^*, \omega)$ is given in logarithmic scale; on the right side ($r^* > 0$ purely outgoing wave) on a linear scale. Compare and contrast this figure with Figures 11 and 10.

by considering a succession of equilibrium configurations (particle momentarily at rest at selected values of the distance from the horizon). We can then

(a) neglect all the complications attached to the emission of radiation (Appendix 2.3).
(b) focus only on the main features of the capture of the particle by the black hole.

The fact that we consider the particle momentarily at rest is, indeed, a very good approximation in the late stages when the particle is very near the horizon. There the particle is practically "frozen" in time! See Figure 7. The details of the treatment are given in Appendix 2.4. There were analyze the case of a charged particle captured by a Schwarzschild black hole and we study the consequent transition of a black hole from a Schwarzschild to a Reissner-Nordstrøm geometry.

The most important feature for us to emphasize here is the fact that the transition *does not* occur abruptly: charges are "induced" on the surface of the black hole and the transition occurs *gradually* and with continuity. The lines of force do, indeed, cross the horizon and at any given moment the far away observer will see the charge localized at the center of the black hole (monopole contribution)! The contribution of higher multipoles ($l \geqslant 1$) always goes to zero as the charge approaches the horizon. A detailed analysis is presented in Appendix 2.4.

2.5 Electromagnetic Radiation from a Particle in Circular Orbits in a Schwarzschild or a Reissner-Nordstrøm Geometry

The non-vanishing terms in the source of Eq. (38.1) and Eq. (38.2) are in the present case

$$\psi^{lm}(r,t) = 4\pi q\delta(r-r_0)Y^{lm*}(\theta,\phi)/r^2 g^{00} \tag{42.1}$$

$$\alpha^{lm}(r,t) = 4\pi q\omega_o \sin\theta\ Y^{lm*}_{,\theta}\ \delta(r-r_o)/l(l+1) \tag{42.2}$$

where we have indicated the orbits radius by r_o, and by ω_o the angular velocity (measured by a far away observer) of the particle

$$\omega_0^2 = M/r_o^3 - Q^2/r_o^4 - Qq/\mu r_o^3\left[Qq/2\mu r_o + \left(1 - 3M/r_o + 2Q^2/r_o^2 + \frac{Q^2q^2}{4\mu^2 r_o^2}\right)\right] \tag{43}$$

Since we are interested in the spectral distribution of the radiation and the particle has a periodic motion we expand Eqs (38.1), (38.1), (42.1) and (42.2) in a Fourier series. We assume that the orbits are in the equatorial plane ($\theta = \pi/2$). We then have

$$\alpha^{lm}(n\omega_0, r) = -q\omega_0 N P_l^{m+1}(\pi/2)\,\delta(m-n)\,\delta(r-r_o)/l(l+1) \tag{44}$$

with

$$N = (-1)^{(m+1)}[l(l+1) - m(m+1)]^{1/2}[(2l+1)(l-m-1)!/(4\pi(l+m+1)!)]^{1/2}$$

and Eq. (38.1) becomes

$$a^{lm}(n\omega_0, r^*)_{,r^*,r^*} + (n^2\omega_0^2 - l(l+1)g^{rr}/r^2)a^{lm}(n\omega_0, r^*) =$$
$$-q\omega_0 N P_l^{m+1}(\pi/2)\,\delta(m-n)\,\delta(r^* - r_o^*) \tag{45}$$

with as usual

$$\frac{dr}{dr^*} = g^{rr} \tag{46}$$

Analogously we have for the Fourier transform of Eq. (38.2)

$$b^{lm}(n\omega_0, r^*)_{,r^*,r^*} + (n^2\omega_0^2 - l(l+1)g^{rr}/r^2)b^{lm}(n\omega_0, r^*) =$$
$$qP^{lm}(\pi/2)N_0\,\delta(r^* - r_o^*)_{,r^*}\,\delta(m-n)/l(l+1) \tag{47}$$

with

$$N_0 = (-1)^m[(2l+1)(l-m)!/4\pi(l+m)!]^{1/2}.$$

We have solved Eqs (45) and (47) by the Green's Function technique $G(\omega, r^*, r^{*1})$ where

$$G(\omega, r^*, r^{*1})_{,r^*,r^*} + [n^2\omega_0^2 - l(l+1)g^{rr}/r^2]G(\omega, r^*, r^{*1}) = \delta(r^* - r^{*1}). \tag{48}$$

We then have

$$G(\omega, r^*, r^{*1}) = \frac{u^l(\omega, r^{*1})v^l(\omega, r^*)}{W(\omega, u, v)} \quad \text{for} \quad r^* < r^{*1} \tag{49.1}$$

$$G(\omega, r^*, r^{*1}) = \frac{u^l(\omega, r^*)v^l(\omega, r^{*1})}{W(\omega, u, v)} \quad \text{for} \quad r^* > r^{*1}. \tag{49.2}$$

$u^l(\omega, r^*)$ and $v^l(\omega, r^*)$ are the solutions of the homogeneous Eqs (45) and (47). $u^l(\omega, r^*)$ is the solution purely outgoing at infinity ($r^* \to +\infty$) and $v^l(\omega, r^*)$ the one purely ingoing at the surface of the black hole $r^* \to -\infty$. $W(\omega, u, v)$ is the Wronskian obtained from the two linearly independent functions $u^l(\omega, r^*)$ and $u^l(\omega, r^*)$.

We obtain from Eqs (45) and (47)

$$a^{lm}(n\omega_0, r^*) = \int_{-\infty}^{+\infty} S^{lm}(n\omega_0, r^{*1})G(\omega, r^*, r^{*1})\,dr^{*1} \tag{50}$$

where with $S^{lm}(n\omega_0, r^{*1})$ we have indicated the source term of Eq. (45); taking into account the explicit form of the source term we obtain from Eq. (50)

$$a^{lm}(n\omega_0, r^*) = -\frac{q\omega_0}{l(l+1)} NP_l^{m+1}(\pi/2)\,\delta(m-n)u^l(\omega, r^*)v^l(\omega, r^*)/W(\omega, u, v) \tag{51}$$

and analogously for $b^{lm}(n\omega_0, r^*)$ we obtain from Eq. (47)

$$b^{lm}(n\omega_0, r^*) = qP_l^m(\pi/2)N_0\,\delta(m-n)u^l(\omega, r^*)(dv^l(\omega, r^*)/dr^*)_{r^{*1}=r_o^*}/ \\ l(l+1)W(\omega, u, v) \tag{52}$$

For the energy radiated integrated over all angles we simply have

$$P(n\omega_0) = (n\omega_0)^2 \sum_{l,m=0}^{l} (a_{lm}^2 + b_{lm}^2)\,\delta(m-n)/2\pi. \tag{53}$$

Since we are interested in ultrarelativistic effects (very high beaming of the radiation), we have examined orbits with $r_o \to r_c = (3M + (9M^2 - 8Q^2)^{1/2})/2$ or $r_o^* \to 0$.

For circular orbits very near $r = r_c$ we have developed an analytical treatment; in the case of a general circular orbit we proceed instead by direct numerical integration.

Since

$$P_l^m(0) = \begin{cases} (-1)^P \dfrac{(2P+2m)!}{2^l p!(p+m)!} & \text{if} \quad l-m=2P \\ 0 & \text{if} \quad l-m=2P+1 \end{cases} \tag{54}$$

where P is an arbitrary integer, we have that the only contribution to Eq. (53) comes from terms with

$$l = m + 2P + 1 \tag{55.1}$$

for $a^{lm}(n\omega_0, r^*)$ (parity $(-1)^P$), and terms with

$$l = m + 2P \tag{55.2}$$

for $b^{lm}(n\omega_0, r^*)$ (parity $(-1)^{l+1}$).

A Analytic Solutions (Orbits with $r_c \simeq (3M + (9M^2 - 8Q^2)^{1/2})/2$ General Case

Both in the case of parity $(-1)^l$ and of parity $(-1)^{l+1}$ the equation determining the radial dependence of the electromagnetic perturbations in the given background metric (1) is of the form

$$\frac{d^2u}{dr^{*2}} + [n^2\omega_0^2 - e^{\nu}l(l+1)/r^2]\,u = A(r_o^*)\,\delta(r^* - r_o^*) \\ + \frac{d}{dr^*}[D(r_o^*)\,\delta(r^* - r_o^*)]. \tag{56}$$

We can integrate equation (46) and choose the constant in such a way as to have identically $r^* = 0$ for $r = r_c$ for arbitrary values of Q and M. In Eq. (56) the effective potential

$$V(r^*) = e^{\nu} l(l+1)/r^2 \tag{57}$$

is a smooth function of r^* and reaches a maximum value for $r = r_c$ or $r^* = 0$. We have in fact

$$\frac{dV}{dr} = -2l(l+1)(r^2 - 3mr + 2Q^2)/r^5 \tag{58}$$

and this quantity goes to zero at $r = r_c$. In §2.2 we have seen that a series of *unstable* circular orbits exists with energy approaching $+\infty$ as the radius $r_o \to r_c$; r_c is also the radius at which a photon can be marginally trapped in a circular orbit (unstable!). To obtain the explicit analytic form of the Green's functions for Eq. (56) we use the method of Ford, Hill, Wakano and Wheeler[4] already used for scalar radiation by Misner *et al.*[5] We expand the potential $V(r^*)$ around $r^* = 0$ (see Figure 13 and Figure 14). Zerilli has given for Eq. (56) and orbits of radius $r_0 \sim r_c$ the following approximate form

$$\frac{d^2u}{d\xi^2} + \left\{kn^2\left(\frac{\omega_0^2}{\omega_\infty^2} - \frac{l(l+1)}{n^2}\right)/(l(l+1))^{1/2} + \xi^2\right\} u = \text{source term.} \tag{59}$$

Here

$$\xi = \omega_\infty (l(l+1))^{1/4} r^*/\sqrt{K} \tag{60.1}$$

$$K = r_c/(3Mr_c - 4Q^2) \tag{60.2}$$

$$\omega_\infty = (Mr_c - Q^2)^{1/2}/r_c^2. \tag{60.3}$$

ω_∞ is the angular velocity of the photon in a circular orbit at r_c, ω_o the angular velocity of the particle given by Eq. (43). Introducing the quantity

$$\epsilon = -kn^2\left(\frac{\omega^2}{\omega_\infty^2} - l(l+1)/n^2\right)/(l(l+1))^{1/2} \tag{61}$$

we can integrate analytically Eq. (59) in the homogeneous case (source term = 0) The equation

$$\frac{d^2u}{d\xi^2} + (\xi^2 - \epsilon)u = 0 \tag{62}$$

is in fact the equation of the well known parabolic cylinder functions. Equation (62) has two linearly independent solutions

$$R(\xi) = D_{-\frac{1}{2}-\frac{i}{2}\epsilon}[(1-i)\xi] \tag{62}$$

$$L(\xi) = R(-\xi). \tag{63}$$

$R(\xi)$ is a purely outgoing wave for large positive values of ξ while $L(\xi)$ is a purely ingoing wave for large negative values of ξ.

The explicit expression of $R(\xi)$ is given by the integral

$$R(\xi) = \frac{1}{\Gamma\left(\frac{1}{2} + i\frac{\epsilon}{2}\right)} \int_0^\infty \exp\left(i\frac{\xi^2}{2} - (1-i)\xi t - t^2/2 - \left(\frac{1}{2} - i\frac{\epsilon}{2}\right)\ln t\right) dt. \tag{64}$$

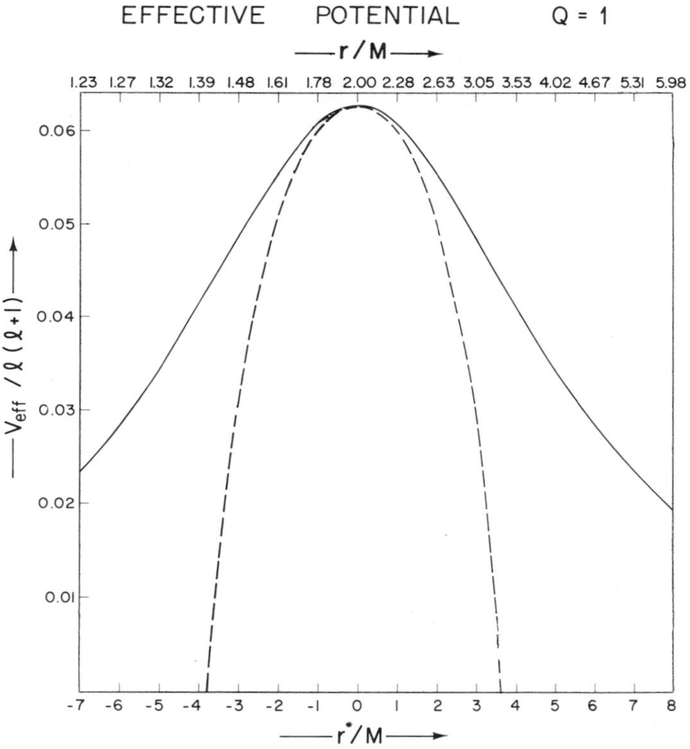

Figure 13. Effective potential of the electromagnetic perturbations in a Reissner-Nordstrøm background (solid line, see Eq. (56) in the text). The dashed line corresponds to the effective potential in the Ford, Hill, Wakano, Wheeler approximation. The solution $R(\xi)$ is matched to a free wave solution at a value of $r^* \sim 3M$ (see text.)

For large positive ξ we have

$$R(\xi) \approx \frac{1}{2^{1/4}\sqrt{\xi}} \exp\left[i\frac{\xi^2}{2} - i\frac{\epsilon}{4}\ln 2\,\xi^2 + \frac{i\pi}{8} - \frac{\pi\epsilon}{8}\right]. \tag{65}$$

For large negative ξ the solution $R(\xi)$ consists of an incident plus a reflected wave.

Using Eqs (60.1) and (60.2) we can now give the explicit form of the Green's functions. Let us first evaluate the Wronskian $W(\omega, u, v)$.

We can for simplicity assume $r^* = 0$, since the value of the Wronskian is independent of the coordinate value at which it is evaluated. We then have

$$W = (1 - i)(2\pi)/\left[2^{i\epsilon/2}\Gamma\left(\frac{3}{4} + i\frac{\epsilon}{4}\right)\Gamma\left(\frac{1}{4} + i\frac{\epsilon}{4}\right)\right]. \tag{66}$$

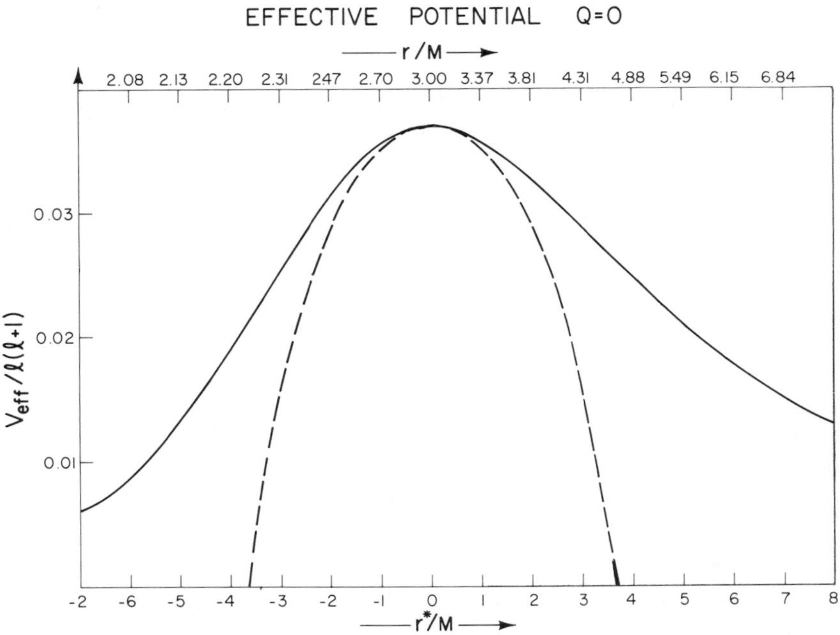

Figure 14. Effective potential of the electromagnetic perturbations in a Schwarzschild background (solid line). The dashed line wave corresponds to the Ford, Hill, Wakano, Wheeler approximation (see text).

to explicitly evaluated Eq. (49.2). We need to know only the asymptotic form of the function $R(\xi)$ for $\xi \gg 0$. We can clearly not extend the analytic solution (64) to large ξ values.

We then (see Figures 13 and 14) match the function $R(\xi)$ to an outgoing wave of the form $A(\omega)\, e^{i\omega r^*}$ or $A(\epsilon)\, e^{i(\xi)}$. Zerilli has given[2] the following explicit form for the Green's functions

$$G(\xi, \xi_0) = -\frac{\Gamma(\frac{3}{4} + i\,\epsilon/4)\Gamma(\frac{1}{4} + i\,\epsilon/4)\, e^{-\pi\epsilon/8} L(\xi_0)\, e^{i\phi(\xi)}}{2\pi(1 - i)2^{-1/2 i\epsilon + 1/4} k^{1/4}(l(l+1))^{1/8}} \tag{67}$$

or in the coordinate r^*

$$G(r^*, r_o^*) = -i\,\frac{\Gamma(\frac{1}{4} + i\epsilon/4)\, e^{-\pi\epsilon/8} k^{1/4}}{4\sqrt{\pi}(l(l+1))^{3/8}\omega_\infty}\, e^{i\tilde{\phi}(r^*)} \tag{68.1}$$

and

$$\frac{\partial G(r^*, r_o^*)}{\partial r_o^*} = -\frac{(1+i)}{\sqrt{8\pi}}\,\frac{\Gamma(\frac{3}{4} + i\epsilon/4)\, e^{-\pi\epsilon/8}}{(l(l+1))^{1/8} k^{1/4}}\, e^{i\tilde{\phi}(r^*)}. \tag{68.2}$$

For values of the radial coordinate r

$$r \sim r_c + M\delta. \tag{69}$$

Here M is as usual the mass of the black hole. We can expand Eq. (61) by keeping in ω_0 only the lowest order term in δ. Thus the electric parity equations become

$$\epsilon^{(e)} = \begin{cases} K\{1 + 4p + Q^2 q^2 |m| / (\mu^2 (Mr_c - Q^2))\} & \text{for} \quad Qq > 0 \\ K\{1 + 4p + Mr_c \delta |m| / k^2 (Mr_c - Q^2)\} & \text{for} \quad Qq = 0 \\ K\{1 + 4p + M^2 (1/k^2 + (9M^2 - 8Q^2)\mu^2 / q^2 Q^2)\,\delta^2 |m| / (Mr_c - Q^2)\} \end{cases} \tag{70}$$

and for the magnetic parity $(-1)^{l+1}$ equations

$$\epsilon^{(m)} = \epsilon^{(e)} + 2K. \tag{71}$$

Details of the spectrum in the general case of a charged particle moving in the field of an extreme Reissner-Nordstrøm solution are given in Appendix 2.5. The major results presented there are simply summarized.

(I) The electric terms with parity $(-1)^l$ give the major contribution to the radiation field and magnetic terms with parity $(-1)^{l+1}$ can always be neglected.

(II) For particles in ultrarelativistic orbits with $r \sim r_c$ a clear enhancement of higher multipoles exists in the radiation field and a sizable part of the radiation is emitted at frequencies $\nu \gg 1/M$.

(III) The spectrum of the radiation drastically differs from the one of synchrotron radiation in flat space, since the enhancement of the highest multipoles is here much less effective.

Let us now consider the treatment here developed in the particular case of a charged particle moving in ultrarelativistic circular orbits in the Schwarzschild geometry.

B Analytic Solutions (Orbits with $r \simeq 3M$) Schwarzschild Case

In the case of the Schwarzschild metric the Eqs (68.1) and (68.2) can be simplified and they acquire the form

$$G(r^*, r_o^*) = \frac{i}{4}\,\omega_0^{-1/2}\, l^{-1/4}\, \frac{e^{-\pi\epsilon/8}}{\pi^{1/2}\sqrt{m\omega_0}}\, e^{-i(\pi/8 + \epsilon/4 \ln \epsilon/4)}\,\Gamma\!\left(\frac{1}{4} + \frac{i}{4}\epsilon\right) e^{i\omega r^*} \tag{72.1}$$

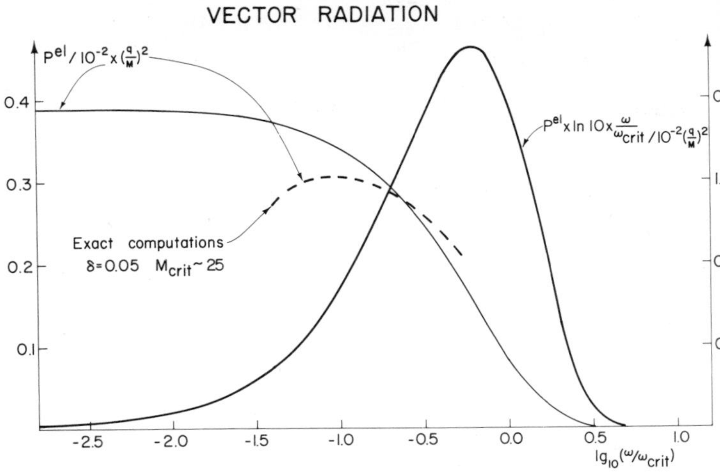

Figure 15. Electromagnetic radiation (P^{el}) emitted by a particle of charge q in an ultra-relativistic circular orbit, with angular velocity $\omega_0 = (M/r_0^3)^{1/2}$ in a Schwarzschild geometry. The orbit of the particle is assumed to be at a radius $r_0 = (3 + \delta)M$ with $\delta \ll 1$. The spectrum is clearly more flat than the one to be expected from a direct extrapolation of classical synchroton radiation (compare and contrast with Figure 1). The frequency of the radiation is here expressed in units of ω_{crit}, we have $\omega/\omega_{crit} = \pi|m|\delta/4$. For direct comparison we have also given the spectrum of the radiation emitted by a particle in circular orbit at $r = 3.05M$ as computed by a direct numerical analysis. On the right side of the figure we have plotted the electromagnetic power in units $10^{-2}(q/M)^2/(\ln 10\ \omega/\omega_{crit})$. Since the diagram here presented is given as a function of the frequency in logarithmic units the area subtended by this curve gives directly the total amount of energy radiated. The asymptotic form of the spectrum for $\omega \gtrsim \omega_{crit}$ is obtained from Eq. (75) by an asymptotic expansion of the gamma functions. This expansion yields

$$P \propto q^2\, e^{-\pi/2}(_\epsilon(e))^{1/2}\, e^{-2\omega/\omega_{crit}/54\pi^{1/2}}$$

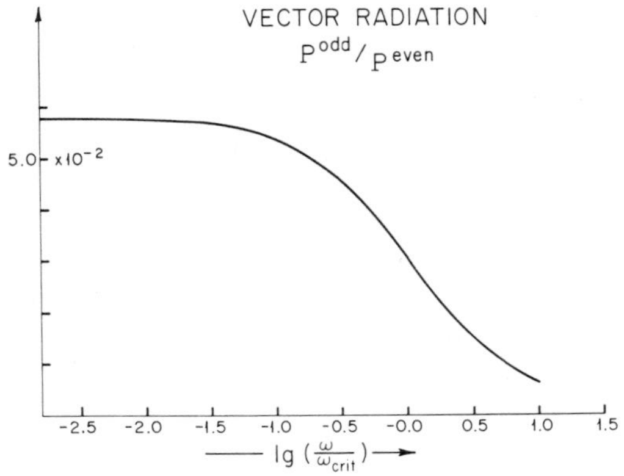

Figure 16. Ratio between electromagnetic energy radiated in odd terms (parity $(-1)^{l+1}$) and even terms (parity $(-1)^l$) (see Eq. (75) in text). The contribution due to the terms of parity $(-1)^{l+1}$ is largely negligible.

$$\frac{\partial G(r^*, r_o^*)}{\partial r_o^*} = \frac{1+i}{\sqrt{8\pi}}\,\omega_0^{1/2}\, l^{1/4}\, \frac{e^{-\pi\epsilon/8}}{\sqrt{m\omega_0}}\, e^{-i(\pi/8+\epsilon/4\ln\epsilon/4)}\,\Gamma\left(\frac{3}{4}+\frac{i}{4}\epsilon\right) e^{i\omega r^*} \tag{72.2}$$

with

$$\epsilon_{ev} = 1 + 4P + |m|\delta \tag{73.1}$$

$$2P = l - |m| \quad \text{for varity } (-1)^l \tag{73.2}$$

and

$$\epsilon_{od} = 3 + 4P + |m|\delta \tag{74.1}$$

$$2P = l - |m| - 1 \quad \text{for varity } (-1)^{l+1}. \tag{74.2}$$

Then for the total energy radiated at infinity as given by Eq. (53), the following analytic formula[6]

$$\begin{aligned} P(m\omega_0) &= \lim_{r^*\to\infty} \sum_{l\geq|m|}^{\infty} (a_{lm}^2 + b_{lm}^2)(m\omega_0)^2/2\pi \\ &\simeq \frac{q^2}{54\pi^{3/2}}\left\{\exp\left(-\frac{\pi}{4}\epsilon_{ev}\right)\left|\Gamma\left(\frac{3}{4}+\frac{i}{4}\epsilon_{ev}\right)\right|^2\right. \\ &\quad \left. + \tfrac{1}{2}\exp\left(-\frac{\pi}{4}\epsilon_{od}\right)\left|\Gamma\left(\frac{1}{4}+\frac{i}{4}\epsilon_{od}\right)\right|^2\right\} \end{aligned} \tag{75}$$

In the above approximate expression only the leading terms in the sum are taken into account ($l = |m|$ for parity $(-1)^l$ and $l = |m| + 1$ for parity $(-1)^{l+1}$) the contribution of the "off diagonal term" being exponentially damped. In Figures 14 and 15 we give the main results of this treatment.

C Numerical Solution—General Case—Charge in Circular Orbit in a Reissner-Nordstrøm Background

If we want to analyze the radiation emitted by a particle in an arbitrary circular orbit in the Reissner-Nordstrøm metric (not necessarily at $r \sim r_c$) the analytic treatment presented in the previous two sections is no longer applicable. We must then proceed to a direct numerical integration of Eqs (51) and (52) in the limit of large r^* values. Therefore it is necessary to evaluate the function $v^l(r_o^*, \omega)$ at the radius r_o^* of the circular orbit of the particle and the function $u^l(\omega, r^*)$ for large values of r^*. We have carried out the numerical integration up to a distance r^* at which both the source term and the effective potential give negligible contributions (see Figure 10):

$$S^l(\omega, r^*) \sim 0 \tag{76.1}$$

$$V^l(r^*) \ll \omega^2. \tag{76.2}$$

The evaluation of the Wronskian and of the functions $v^l(r^*, \omega)$ and $u^l(r^*, \omega)$ is totally analogous to the case already examined in §2.4. This numerical treatment allows us to examine the radiation emitted by particles with circular orbits arbitrarily *far* from the critical one at $r \sim r_c$. These integrations are therefore of the greatest relevance in the understanding of the physical nature of the beaming process of radiation under highly relativistic regimes. In particular they are essential in discriminating between the two competing and equally relevant phenomena of focusing (special relativistic effect) and defocusing (general relativistic effect) of the radiation field (see next section. Detailed results are given in Appendix 2.5 where also a direct comparison is made between orbits with the same ratio of orbital velocity with respect to the local value of the speed of light *but* at different distances from the black hole surface.

D Summary and Conclusions

In the preceding sections we have reached the following important conclusions:

(I) Both in the Reissner-Nordstrøm and in the Schwarzschild metrics particles in unstable ultrarelativistic circular orbits emit radiation at frequencies $\omega \gg 1/M$. Higher multipole radiation ($l \gg 2$) is always excited, and as a consequence the radiation far from being emitted isotropically is beamed.

(II) Contrary to the classical synchroton radiation, the spectrum of the radiation emitted by charged particles in these ultrarelativistic orbits is much more flattened. The reason for this defocusing effect has to be found in the fact that the orbits of the particle are very near r_c, which is also the radius of (unstable) circular orbits for photons. The photons beamed by the usual special relativistic effects are injected in an "effective storage ring" generated by the background geometry and, largely defocused, they leak out from this resonant state. The proof that this is indeed the reason for this defocusing effect is given in the computations presented in Appendix 2.5.

2.6 In what Sense can we Speak of an Object Endowed with a Net Charge in Astrophysical Problems?

It has been known for a long time that objects of astrophysical interest ($M \sim 1M_\odot$) can never be endowed with a net charge of large value. The reason is simply stated: the presence of a net charge on a star will produce selective accretion of opposite charged material. The system will discharge extremely fast, under any reasonable assumption for the density of interstellar material.

These very straightforward considerations can be largely modified if the object under examination is not only endowed with a net charge but also with rotation and strong magnetic fields. The accreting material could then be trapped in the magnetosphere of the rotating object (for simplicity we can assume a dipole magnetic field aligned with the rotation axis), completely screen the net charge and consequently stop any process of selective accretion. Strictly speaking the system as a whole is now neutral, though the "bare" object and the magnetosphere are indeed endowed with large and opposite charges. The possible existence of such an object strongly depends upon two major assumptions:

(a) that the lifetime of charged particles in the magnetosphere is long enough,
(b) that the accretion through the "polar cup" has to be negligible. In this region, in fact, particles will *not* be trapped any longer by the lines of force of the magnetic field and they are free to impinge on the object and discharge it.

Quite apart from these considerations, is there any physical reason why the "bare" object itself should have a net charge? This is the motivation for the analysis presented in Appendix 2.6. We have here analyzed a purely classical model of a rotating magnetized sphere with a magnetic field aligned with the rotation axis. The main conclusions are the following: *The configuration which minimizes the total electromagnetic energy of the system is endowed with a non-zero net charge.* How can we understand the physical reasons for this result? (a) Consider a rotating sphere initially uncharged with a magnetic field aligned with the axis of rotation and of strength B_0 at the pole. (b) Add a net charge to the system. (c) The Coulomb energy (electrostatic) of the system then *increases*. (d) Concurrently for a proper combination of the sign of the charge and the direction of the angular momentum, a new magnetic field is generated, which *decreases* the total magnetic energy of the system. There exists a minimum of the total electromagnetic energy of the system under these two competing effects for a *non-zero* value of the net charge. This effect is quite small in the case of the Earth. However, it is of the greatest importance in the case of highly collapsed objects. The system which minimizes the total electromagnetic energy is endowed with a fixed value of the gyromagnetic ratio. The entire treatment presented in Appendix 2.6 is non-relativistic. However, the relativistic generalization of this physical result appears to be of the greatest interest. In particular, our result seems to suggest that the existence of a fixed gyromagnetic ratio in the case of a magnetic black hole (generalized Kerr-Newmann solution see Part 4) and the fact that a fixed net charge has always to be associated in the presence of a given magnetic field and of a given spin in black hole physics, can be a direct consequence of a minimum energy condition. This extremizing condition should be *automatically* implied by the fulfillment of the complete set of Einstein-Maxwell equations in vacuum.

In the case of a magnetic black hole, the lifetime of particles in the magnetosphere is indeed extremely large: as in the case of a Reissner-Nordstrøm black hole, orbits with 100% binding energy exist (see Part 4). Moreover, particles accreting into a magnetic black hole are mainly concentrated in the equatorial plane: the effective potential in the θ direction becomes, for suitable values of the parameter, infinitely repulsive in the limits $\theta = 0$ and $\theta = \pi$. It is extremely difficult for a particle to accrete in the polar direction.

They are good reasons (energetics—see Part IV) to hope that indeed a magnetic black hole can exist in nature, and computations now in progress suggest that the screening effects are very efficient due to general relativistic effects.[7]

References

1. Ruffini, R. and Wheeler, J. A. *Cosmology from Space θ at form*—ESRO book SP52.
2. Ruffini, R. and Zerilli, F.—paper in preparation.
3. Edmonds, A. R. (1960) *Angular Momentum in Quantum Mechanics* Princeton University Press.
4. Ford, K. W., Hill, D. L., Wakano, M. and Wheeler, J. A. (1959) *Ann. Phys.* **7**, 239.
5. Misner, C. V., Breuer, R. A., Brill, D. R., Chozanowski, P. L. Hughes, H. G. III and Pereira, C. M. (1972) *Phys. Rev. Lett.* **28**, 998.
6. This formula was obtained independently by Breuer, R. A., Ruffini, R., Tionno, H. and Vishveshvara, C. V.
7. Johnston M. and Ruffini, R.—To be published.

A 2.1

ON THE ENERGETICS OF REISSNER NORDSTRØM GEOMETRIES*

Gallieno Denardo

Istituto di Fisica Teorica dell'Universita, Trieste

Institute for Advanced Study, Princeton, N. J. 08540†

Remo Ruffini

Joseph Henry Physical Laboratories, Princeton, N. J. 08540

ABSTRACT

We point out the existence of a generalized ergosphere in the Reissner Nordstrøm geometry and we give an explicit formula to determine its range. These results are compared and contrasted with the ones obtained in the case of the Kerr solution. An explicit process of energy extraction from a Reissner Nordstrøm black hole is given.

* Work partially supported by N. S. F. Grant GP30799X to Princeton University.

† Part of this work was done while a visitor at the Institute for Advanced Study.

[R33]

The recent discovery of binary x-ray sources[1] and their identification as collapsed objects[2] has given raise to interest both in the problem of particles acceleration[3] and energy extraction[4] from black holes. That, indeed, energy can be extracted from black holes has been explicitly pointed out by Floyd and Penrose[5] and Ruffini and Wheeler[6], in the case of a rotating black hole. Christodoulou and Ruffini[7] have pointed out that not only rotational energy, but also electromagnetic energy can be extracted and they established the general formula governing the energetics of magnetic black holes:

$$(1) \quad M^2 = E^2 = (m_{ir} + Q^2/4m_{ir})^2 + L^2/4m^2_{ir}$$

with $a^2 = L^2/M^2 \leq Q^2 + M^2$, m_{ir} being the irreducible mass L the angular momentum, Q the charge, M the mass of the collapsed object. Here and in the following we use geometrized units with $G = c = 1$. This formula clearly shows that as much as 50% of the total mass energy of a black hole can be stored in electromagnetic energy and at least in principle, extracted by reversible transformations.

The region in which this extraction can occur is, here, analyzed in the case of a Reissner-Nordstrøm geometry and by analogy with the Kerr-Newman[7] analysis we call this region the "generalized ergosphere". The aim of this letter is mainly to focus on the major features of this process of energy extraction in the case in which the collapsed object is endowed with a net charge. No reference is, here, made to the possible direct application of this analysis to realistic astrophysical process. However, Ruffini and

Treves[8] have recently pointed out that the most likely solution for a collapsed object endowed with rotation and a magnetic field should be expected to possess a net charge.

The lagrangian for a particle μ and charge q moving in a given background metric is

$$(2) \quad \mathcal{L} = \frac{1}{2} \mu g_{\alpha\beta} \dot{x}^{\alpha} \dot{x}^{\beta} + q A_{\alpha} \dot{x}^{\alpha}$$

here A_{α} is the electromagnetic four potential and the dots represent differentiation with respect to the proper time τ. Greek indices run from 0 to 3, latin indices from 1–3. In our specific case we have

$$(3) \quad ds^2 = e^{-\nu} dt^2 + e^{\nu} dr^2 + r^2(d\theta^2 + \sin^2\theta d\phi^2)$$

with

$$(4.1) \quad e^{-\nu} = (1 - 2M/r + Q^2/r^2)$$

and

$$(4.2) \quad A_i = 0 \quad i = 1,3 \qquad (4.3) \quad A_0 = - Q/r$$

Since we are dealing with a problem endowed with spherical symmetry we can limit ourselves to analyze the orbits in the equatorial plane. We have for the equation of motion of the test particle (<u>not</u> geodesic!)

$$(5)\quad \frac{d^2x^\gamma}{ds^2} + \Gamma^\gamma_{\alpha\beta}\frac{dx^\alpha}{ds}\frac{dx^\beta}{ds} = (q/\mu)\, F^\gamma_\alpha \frac{dx^\alpha}{ds}$$

The Lagrangian being explicitly independent from the time coordinate t and from the azimuthal coordinate ϕ, we have the following two conserved quantities:

$$(6.1)\quad p_o = -E = \frac{\partial \mathcal{L}}{\partial \dot{t}} = -\mu\,(1 - 2\mu/r + Q^2/r^2)\dot{t} - q\,Q/r$$

and

$$(6.2)\quad p_\phi = \frac{\partial \mathcal{L}}{\partial \dot{\phi}} = \mu\, r^2 \sin^2\theta\, \dot{\phi}$$

E being the energy of the particle as measured by the observer at rest at infinity. Since our background metric is static we have a timelike Killing vector field $\xi^\mu(t) = (1,0,0,0)$ and (6.1) is clearly equivalent to $-E = p_\mu\, \xi^\mu(t)$. The fact that E is, indeed, a conserved quantity is an immediate consequence of the equation

$$(7)\quad p^\mu \nabla_\mu (p_\alpha \xi^\alpha) = p^\mu\, p_{\alpha;\mu} \xi^\alpha + p^\mu p_\alpha \xi^\alpha{}_{;\mu} = 0$$

That the first term in the sum is identically zero follows from Eq.(5) and the second term from the equations defining the Killing vector field $\xi_{\mu;\nu} + \xi_{\nu;\mu} = 0$. We have also for the Hamiltonian

$$(8)\quad H = \frac{\mu}{2}\,[e^{-\nu}\dot{r}^2 + r^2\dot{\theta}^2 + r^2\sin^2\theta\dot{\phi}^2 - \dot{t}^2 e^{-\nu}] = -\frac{\mu}{2}$$

then from Eq.(6.1) and (6.2) we have

(9) $e^{-\nu}\dot{r}^2 + r^2\dot{\theta}^2 + p_\phi^2/(\mu^2 r^2 \sin^2\theta) - (p_o/_\mu + Qq/\mu r)^2 e^{-\nu} = -1$

By assuming $\theta = \frac{\pi}{2}$ and $\dot{r} = 0$ we obtain the equation governing the "effective potential" of the orbits in the equatorial plane of the Reissner-Nordstrøm geometry, see Fig. 1. We have

$$(10) \quad E_{eff} = - p_o = Qq/r \pm \mu((1 + p^2\phi/\mu^2 r^2)e^\nu)^{\frac{1}{2}}$$

Here the positive sign corresponds to the positive roots state and the negative sign to the negative roots state. The separation between the positive and negative roots goes to 2μ for $r \to +\infty$. At the horizon $r_+=M+(M^2-Q^2)^{\frac{1}{2}}$ this separation goes to zero and we have

$$(11) \quad \lim_{r \to r_+} E^+ = E^- = Qq/r_+$$

Most important for us here is the existence of the negative energy states for the positive energy solutions which, indeed, allow us to introduce the concept of "effective ergosphere". We also have the following important relation

$$(12) \quad E^+(r,Q,q) = - E^-(r,Q, - q)$$

which interchanges the negative and positive roots state through the transformation

$$(13) \quad t \to - t \qquad\qquad q \to - q$$

Fig. 1

Positive and negative root states for a charge $\varepsilon = q/\mu = -1.0$ in **the field** of an extreme Reissner-Nordstrøm black hole (Q/M = 1). The **negative** energy states of positive roots solution are, here, dashed. The **circular** orbits for $\varepsilon = +1$ obtained by the inversion of the E^+ and E^- roots (see text) are further analyzed in Ref. (10). As usual **minima** in the effective potential indicate stable circular orbits and **maxima** unstable.

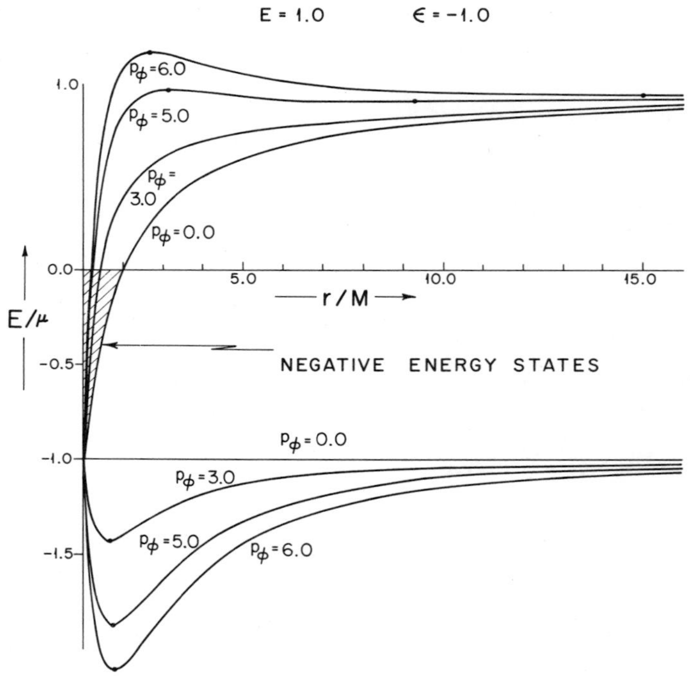

The negative energy states of positive roots are markedly different in the Kerr and in the Reissner-Nordstrøm geometries. In the Kerr case the energy states for $r \to r_+ = M + (M^2 - a^2)^{1/2}$ are given by

$$(14) \quad \lim_{r \to r_+} E^+ = \frac{a p_\phi}{r_+^2 + a^2}$$

Therefore, only counterstating orbits ($p_\phi < 0$) can reach negative energy states and their value is simply proportional to the angular momentum p_ϕ.

In the case here under consideration for $r \to r_+ = M + (M^2 - Q^2)^{1/2}$ the negative states are totally independent from the value of the angular momentum.- a big difference also exists in the extension of the ergosphere. In the Kerr case, the ergosphere does not depend upon any detail of the test particle and extends from the horizon up to the infinite red-shift surface

$$(15) \quad M + (M^2 - a^2)^{1/2} \leq r \leq M + (M^2 - a^2\cos^2\theta)^{1/2}$$

moreover, negative energy states can be reached in the entire range of coordinate given by Eq. 14 only in the limit $p_\phi \to -\infty$. No extraction of energy is possible in the Kerr case for $p_\phi = 0$ (see e.g. Fig. 18 in Ref. 6). In the Reissner-Nordstrøm case the effective ergosphere extends in the range

$$(16) \quad M + (M^2 - Q^2)^{1/2} \leq r \leq M + (M^2 - Q^2(1 - q^2/\mu^2))^{1/2}$$

The effective ergosphere does, indeed, depend upon the value of the test charge Moreover, see Fig. 2, energy extraction is now possible in the entire range of coordinates given by Eq.(15) if and only if $p_\phi = 0$. No extraction is possible in the limit $|p_\phi| \to +\infty$. The entire extraction process is

Fig. 2

Effective potential for a charge $\varepsilon = q/\mu = -2.0$ in the field of an extreme Reissner-Nordstrøm black hole ($Q/m = 1$). The effective potential is unchanged by the transformation $p_\phi \rightarrow - p_\phi$. The ergosurface, surface of $E \leq 0$ in the r, p_ϕ plane, extends up to $r = M + (M^2 + Q^2(1 - q^2/\mu^2))^{\frac{1}{2}}$ in the limiting case of $p_\phi = 0$.

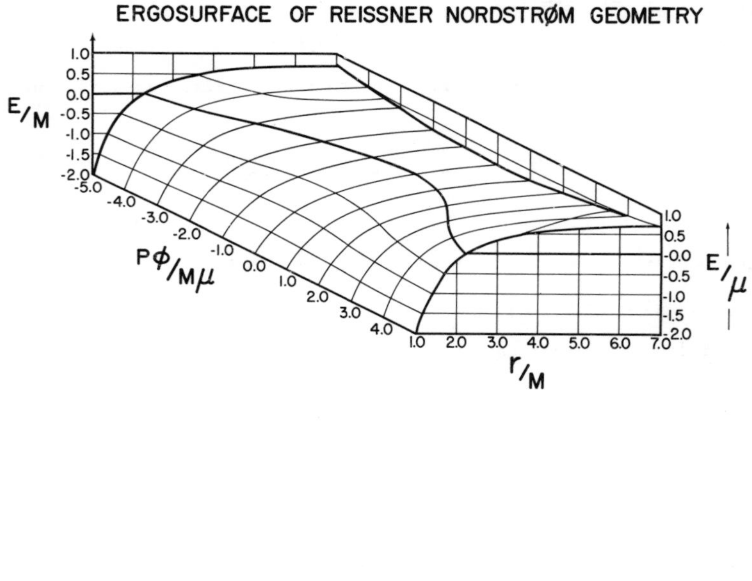

completely symmetric with respect to the interchange $p_\phi \to -p_\phi$.

From the Eq. (1) we have for infinitesimal transformations

(16.1) $dM = Q\,dQ/(M + (M^2 - Q^2)^{1/2})$

The integration of Eq. 16 gives immediately the Christodoulou-Ruffini formula controlling the energetics of black holes in the limit $a \to 0$.

(16.2) $M = m_{ir} + Q^2/4m_{ir}$

Finally, let us give an example of a process of energy extraction. We can consider the process envisaged in Fig. 3. A particle P_o comes in from infinity and infringes the ergosphere of an extreme Reissner-Nordstrøm black hole ($Q = M$). The parameters of the particle P_o are

(17) $\mu_o = 2.197 \quad E_o = 4.5 \quad p_{\phi o} = 2.0 \quad q_o = 6.586$

and the radius of the ergosphere

(18) $r/M = 1 + q/\mu = 3.997$

When the particle P_o attains its turning point at $r = 2M$ decays in two particles:

one P_1 goes inside the black hole

(19) $\mu_1 = 1.0 \quad E_1 = -0.5 \quad p_{\phi 1} = 0 \quad \frac{r}{M} = 2.0 \quad q_1 = -2.0$

the other P_2

Fig. 3.

A process of energy extraction from an extreme Reissner-Nordstrøm black hole. A particle P_o comes in and splits into two debris the particle P_2 coming out has more energy of the initial particle P_o. The process is most favorable for $p_{\phi 1} = 0$, see Fig. 2.

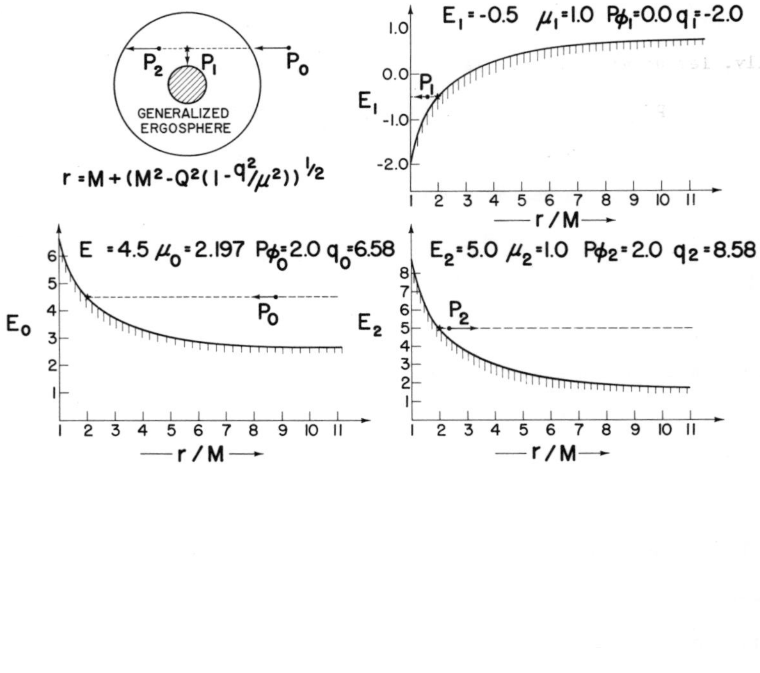

(20) $\mu_2 = 1.0$ $E_2 = 5.0$ $p_{\phi 2} = 2.0$ $\frac{r}{M} = 2.0$ $q_2 = 8.586$

goes back to infinity. The particle P_2 has more energy of the initial particle P_o. The energy gain is due to the loss of electromagnetic energy (Coulomb energy) from the black hole.

It is important, here, to remark that, exactly like in the Kerr case, since the separation between the positive and negative roots states goes to zero at the horizon we can have processes of energy extraction that can approach as much as we like reversibility. Let us finally remark that the energy extraction process in the Reissner-Nordstrøm geometry appears to be much more promising than the one in the Kerr case for two different and equally important reasons: (a) the energy gain per event appear to be very much larger and (b) the reduction in rest mass necessary to make any energy gain process possible is, here, much less stringent than in the Kerr case[9]. Details of this work are going to appear elsewhere[10].

It is a pleasure to thank Professors D. Christodoulou and T. Regge for discussions. One of us (G.D.) is also thankful to Dr. C. Kaysen for the hospitality at the Institute for Advanced Study, where part of this work was done.

References

(1) For a review, see H. Gursky - Les Houches lecture - 1972, Gordon and Breach, in press.

(2) R. Leach and R. Ruffini - submitted for publication in Ap. J. Letters, Oct. 1972.

(3) R. Ruffini, Phys. Lett. 41B, 334, 1972.

(4) M. Rees, R. Ruffini, J.A. Wheeler - Black Holes, Gravitational Waves and Cosmology, Gordon and Breach 1972, in press.

(5) R. Floyd, R. Penrose, Nature 229, PS177, 1972.

(6) R. Ruffini and J. A. Wheeler, "Cosmology from Space Platform", ESRO Book SP52, 1971.

(7) D. Christodoulou, R. Ruffini, Phys. Rev. 4, 3552, 1971.

(8) R. Ruffini, A. Treves, Astrophysical Letters, in press.

(9) B. Mashhoon - has analyzed the details of the reduction in rest mass in the energy gain processes in Kerr geometries, to be submitted for publication.

(10) G. Denardo, R. Ruffini, F. Zerilli - in preparation.

A 2.2

Electromagnetic Field of a Particle Moving in a Spherically Symmetric Black-Hole Background (*).

R. Ruffini and J. Tiomno

Joseph Henry Physical Laboratories - Princeton, N. J.

C. V. Vishveshwara

New York University - New York, N. Y.

(ricevuto il 22 Novembre 1971)

Very much has been speculated recently on the possible existence of collapsed objects, usually called « black holes », and on their possible observation. Christodoulou and Ruffini [1] have shown that « black holes », which had been considered for a long time to be merely sinks for radiation and other forms of energy, could in fact yield up to 50% of their total energy under favourable circumstances. For these reasons it was justified to think of black holes as the strongest storehouses of energy in the Universe [2]. The search for these objects should therefore be directed toward a careful examination of strongly energetic events. From the theoretical point of view a detailed analysis of accretion processes has to be undertaken. Some aspects of this analysis (in the « one-particle approximation »!) have been examined for the emission of gravitational radiation by Zerilli [3], by Davis and Ruffini [4], by Davis, Ruffini, Press and Price [5], and by Davis, Ruffini and Tiomno [6]. Some related aspects concerning the scattering of gravitational waves from black holes have been given by Vishveshwara [7]. In this paper the theoretical basis for the analysis of the electromagnetic radiation emitted by a charge moving in the gravitational field of spherically symmetric black holes is presented.

(*) Work partially supported by NSF GP-30799X and NSF GU-3186.

(1) D. Christodoulou and R. Ruffini: *Phys. Rev.*, in press.
(2) D. Christodoulou and R. Ruffini: *Back Holes, the largest-energy storehouse in the Universe*, in *Gravity Foundation, 1971*, in press.
(3) F. Zerilli: *Phys. Rev. D.*, **2**, 2141 (1970).
(4) M. Davis and R. Ruffini: *Lett. Nuovo Cimento*, in press.
(5) M. Davis, R. Ruffini, W. Press and R. Price: *Phys. Rev. Lett.*, **21**, 1466 (1971).
(6) M. Davis, R. Ruffini and J. Tiomno: submitted for publication.
(7) C. V. Vishveshwara: *Nature*, **227**, 936 (1970).

ISRAEL [8] has shown that the most general spherically symmetric collapsed objects with closed simply connected horizon is the one given by the Reissner Nordström metric, which in Schwarzschild-like co-ordinate assumes the form

$$(1) \qquad \mathrm{d}s^2 = g_{\mu\nu}\mathrm{d}x^\mu \mathrm{d}x^\nu = -\left(1-\frac{2M}{r}+\frac{Q^2}{r^2}\right)\mathrm{d}t^2 + \left(1-\frac{2M}{r}+\frac{Q^2}{r^2}\right)^{-1}\mathrm{d}r^2 + r^2(\mathrm{d}\theta^2 + \sin^2\theta\,\mathrm{d}\varphi^2),$$

here $\mu, \nu = 0, 3$ and $G = c = 1$, M is the mass and Q the charge of the given background metric.

We consider here a black hole with $Q \leqslant M$. We analyze the electromagnetic field generated by a particle of mass m and charge q moving in the above metric (1) under the assumptions that the charge $q \ll M$ and the mass $m \ll M$. The electromagnetic vector potential associated with the moving particle can be expanded in terms of four-dimensional vector spherical harmonics obtained from scalar spherical harmonics by means of the following operations:

$$(2a) \qquad \frac{\boldsymbol{r}}{r} Y^{lm}(\theta, \varphi)\,,$$

$$(2b) \qquad \boldsymbol{\nabla} Y^{lm}(\theta, \varphi)\,,$$

$$(2c) \qquad \boldsymbol{L} Y^{lm}(\theta, \varphi)\,,$$

$$(2d) \qquad e_t Y^{lm}(\theta, \varphi) = (Y^{lm}, 0, 0, 0)\,.$$

Here as usual $\boldsymbol{L}$ is the angular-momentum operator and $\boldsymbol{\nabla}$ the gradient. The parity of eq. (2.a), (2.b), (2.d) is $(-1)^l$ (electric) and the parity of eq. (2.c) is $(-1)^{l+1}$ (magnetic). We obtain then

$$(3) \qquad A_\mu(r, \theta, \varphi, t) = \sum_{lm}\left(\begin{bmatrix} 0 \\ 0 \\ \dfrac{a^{lm}(r,t)}{\sin\theta}\dfrac{\partial Y^{lm}}{\partial\varphi} \\ -a^{lm}(r,t)\sin\theta\dfrac{\partial Y^{lm}}{\partial\theta} \end{bmatrix} + \begin{bmatrix} f^{lm}(r,t)\,Y^{lm} \\ h^{lm}(r,t)\,Y^{lm} \\ k^{lm}(r,t)\dfrac{\partial Y^{lm}}{\partial\theta} \\ k^{lm}(r,t)\dfrac{\partial Y^{lm}}{\partial\varphi} \end{bmatrix}\right).$$

The covariant Maxwell equations to be fulfilled in the given background are given by

$$(4) \qquad F^{\mu\nu}{}_{;\nu} = 4\pi J^\mu \qquad \text{or} \qquad (\sqrt{-g}\,F^{\mu\nu})_{,\nu} = \sqrt{-g}\,4\pi J^\mu\,.$$

In the above equation; (,) indicates covariant (ordinary) derivative and $g = \det g_{\alpha\beta}$. Further we have

$$(5) \qquad F_{\mu\nu} = A_{\nu,\mu} - A_{\mu,\nu}\,.$$

---

(8) W. ISRAEL: *Phys. Rev.*, **164**, 1776 (1967).

The four-current j^μ can itself be expanded in terms of vector harmonics

$$4\pi J_\mu = \sum_{l,m}\left(\begin{bmatrix} 0 \\ 0 \\ \dfrac{\alpha^{lm}(r,t)}{\sin\theta}\dfrac{\partial Y^{lm}}{\partial\varphi} \\ -\alpha^{lm}(r,t)\sin\theta\dfrac{\partial Y^{lm}}{\partial\theta}\end{bmatrix} + \begin{bmatrix}\Psi^{lm}(r,t)\,Y^{lm} \\ \eta^{lm}(r,t)\,Y^{lm} \\ \chi^{lm}(r,t)\dfrac{\partial Y^{lm}}{\partial\theta} \\ \chi^{lm}(r,t)\dfrac{\partial Y^{lm}}{\partial\varphi}\end{bmatrix}\right). \tag{6}$$

The nonvanishing components of the electromagnetic-field tensor $F^{\mu\nu}$ of parity $(-1)^{l+1}$ are given by

$$\begin{cases} F^{0\theta}_{lm} = g^{00}a^{lm}_{,0}\,Y^{lm}_{,\varphi}/(r^2\sin\theta)\,;\; F^{0\varphi}_{lm} = -g^{00}a^{lm}_{,0}\,Y^{lm}_{,\theta}/(r^2\sin\theta)\,, \\ F^{r\theta}_{lm} = g^{rr}a^{lm}_{,r}\,Y^{lm}_{,\varphi}/(r^2\sin\theta)\,;\; F^{r\varphi}_{lm} = -g^{rr}a^{lm}_{,r}\,Y^{lm}_{,\theta}/(r^2\sin\theta)\,, \\ F^{\theta\varphi}_{lm} = l(l+1)\,a^{lm}/(r^4\sin\theta)\,. \end{cases} \tag{7}$$

Equation (4) gives rise to a single differential equation for the radial function

$$(g^{rr}a^{lm}_{,r})_{,r} - g_{rr}\frac{\partial^2 a^{lm}}{\partial t^2} - \frac{l(l+1)}{r^2}a^{lm} = \alpha^{lm}\,. \tag{8}$$

The other equations obtained from eqs. (4) are either identically satisfied or equivalent to (8). Similarly, the nonvanishing components of $F^{\mu\nu}$ with parity $(-1)^l$ are

$$\begin{cases} F^{0r}_{lm} = (f^{lm}_{,r} - h^{lm}_{,0})\,Y^{lm}\,; \qquad F^{0\theta}_{lm} = g^{00}(k^{lm}_{,0} - f^{lm})\,Y^{lm}_{,\theta}/r^2\,, \\ F^{0\varphi}_{lm} = g^{00}(k^{lm}_{,0} - f^{lm})\,Y^{lm}_{,\varphi}/(r^2\sin\theta)\,, \\ F^{r\theta}_{lm} = g^{rr}(k^{lm}_{,r} - h^{lm})\,Y^{lm}_{,\theta}/r^2\,; \qquad F^{r\varphi}_{lm} = g^{rr}(k^{lm}_{,r} - h^{lm})\,Y^{lm}_{,\varphi}/(r^2\sin\theta)\,, \end{cases} \tag{9}$$

and from eqs. (4) we obtain now the following set of differential equations for the radial parts:

$$g_{00}[r^2(f^{lm}_{,r} - h^{lm}_{,0})]_{,r} - l(l+1)\,(k^{lm}_{,0} - f^{lm}) = \Psi^{lm}r^2\,, \tag{10a}$$

$$g_{rr}(h^{lm}_{,0} - f^{lm}_{,r})_{,0} - \frac{l(l+1)}{r^2}(k^{lm}_{,r} - h^{lm}) = \eta^{lm}\,, \tag{10b}$$

$$[(h^{lm} - k^{lm}_{,r})g^{rr}]_{,r} - (k^{lm}_{,0} - f^{lm})_{,0}\,g^{00} = \chi^{lm}\,. \tag{10c}$$

The remaining equations obtained from (4) are either identically satisfied or equivalent to eqs. (10). The equation of conservation of current to be identically fulfilled is

$$J^\mu_{;\mu} = 0$$

This gives the subsidiary equation

$$\frac{1}{r^2}(r^2g^{rr}\eta^{lm})_{,r} + \Psi^{lm}_{,0}\,g^{00} = \frac{l(l+1)}{r^2}\chi^{lm}\,. \tag{11}$$

[R47]

We introduce a new function $b^{lm}(r, t)$ defined by the following equation:

$$h^{lm}_{,0} - f^{lm}_{,r} = \frac{l(l+1)}{r^2}\, b^{lm}\,, \tag{12}$$

and substituting in the expressions (10a) and (10b) we get

$$g^{rr} b^{lm}_{,0} = k^{lm}_{,0} - f^{lm} + \frac{r^2 \Psi^{lm}}{l(l+1)}\,, \tag{12a}$$

$$g^{00} b^{lm}_{,0} = h^{lm} - k^{lm}_{,r} - \frac{r^2 \eta^{lm}}{l(l+1)}\,. \tag{12b}$$

With the help of eqs. (12) we find that the eq. (10c) is identically satisfied in view of (11). Further, integrability condition of these equations is summarized in the compact and simple differential equation

$$(g^{rr} b^{lm}_{,r})_{,r} + g^{00} b^{lm}_{,00} - \frac{l(l+1)}{r^2}\, b^{lm} = \frac{1}{l(l+1)}\,[(r^2 \Psi^{lm})_{,r} - \eta^{lm}_{,0} r^2]\,. \tag{13}$$

Equations (13) and (8) take the same form in vacuum, in which case they were first given by WHEELER [9]. Equation (13) solves completely the problem of the determination of the electric multipole expansion as $F^{\mu\nu}_{lm}$ given by (9) is expressed in terms of b^{lm}, $b^{lm}_{,0}$, $b^{lm}_{,r}$, Ψ^{lm} and η^{lm}. Both eqs. (13) and (8) have to be solved for any given J^μ by assuming as boundary conditions purely ingoing waves at the surface of the black hole.

The four-current J^μ in the case of a point charge moving in the given background. is given by

$$J^\mu = \frac{q}{\sqrt{-g}} \frac{dz^\mu}{dt}\, \delta(\boldsymbol{x} - \boldsymbol{z}(t))\,, \tag{14}$$

$z^\mu \equiv (t, \boldsymbol{z}) \equiv (t, R(t), \Theta(t), \varphi(t))$ describes the trajectory of the particle as given by the equation

$$\frac{d}{ds}\left(\frac{dz^\mu}{dt}\frac{dt}{ds}\right) = F^\mu{}_\alpha \frac{dz^\alpha}{dt}\frac{dt}{ds} + \Gamma^\mu_{\alpha\beta}\frac{dz^\alpha}{dt}\frac{dz^\beta}{dt}\left(\frac{dt}{ds}\right)^2.$$

Both $F^\mu{}_\alpha$ and $\Gamma^\mu_{\alpha\beta}$ are given by the background geometry. From (4) and (6) we obtain

$$\Psi^{lm}(r, t) = \frac{q}{r^2 g^{00}}\, \delta(r - R)\, Y^{lm*}(\Theta, \Phi)\,, \tag{15a}$$

$$\eta^{lm}(r, t) = \frac{q}{r^2 g^{rr}} \frac{dR}{dt}\, \delta(r - R)\, Y^{lm*}(\Theta, \Phi)\,, \tag{15b}$$

$$\alpha^{lm}(r, t) = \frac{q}{l(l+1)}\left[-\frac{d\Phi}{dt} \sin\Theta\, Y^{lm*}{}_{,\Theta} + \frac{1}{\sin\Theta}\frac{d\Theta}{dt}\, Y^{lm*}{}_{,\Phi}\right] \delta(r - R)\,. \tag{15c}$$

These are the relevant source terms to be used when solving eqs. (8) and (13).

---

[9] J. A. WHEELER: *Geometrodynamics* (New York, 1962), p. 203. Similar results for the sourceless case were also obtained by L. FAVELLA - *Tesi di Laurea*, 1957, Torino University. We thank Prof. T. REGGE to have pointed out this interesting work.

Finally we have to give the expression for the energy radiated. The tensor momentum energy for the electromagnetic field is

$$T^{\mu}{}_{\nu} = (F^{\mu\varrho} F_{\varrho\nu} + \tfrac{1}{4}\delta^{\mu}{}_{\nu} F^{\varrho\sigma} F_{\varrho\sigma})/4\pi\,.$$

The flux of energy through a surface S is given by the expression

$$\left(\frac{\mathrm{d}E}{\mathrm{d}t}\right)_S = \int_S T^r_0 r^2 \sin\theta \,\mathrm{d}\theta\,\mathrm{d}\varphi\,. \tag{16a}$$

Here

$$T^r_0 r^2 \sin\theta = -g^{rr} \sum_{lm}\sum_{l'm'} [a^{lm}{}_{,r} a^{l'm'}{}_{,0} + (k^{lm}_{,r} - h^{lm})(k^{l'm'}{}_{,0} - f^{l'm'})]\,(Y^{lm}_{,\theta} Y^{l'm'}_{,\theta} \sin\theta + \tag{16b}$$

$$+\, Y^{lm}_{,\varphi} Y^{l'm'}_{,\varphi}/\sin\theta) + g^{rr} \sum_{lm}\sum_{l'm'} [a^{lm}_{,r}(f^{l'm'} - k^{l'm'}_{,0}) + (k^{lm}_{,r} - h^{lm})\, a_{l'm',0}](Y^{lm}_{,\varphi} Y^{l'm'}{}_{,\theta} - Y^{lm}{}_{,\theta} Y^{l'm'}{}_{,\varphi})\,.$$

The spectral distribution of the radiation is immediately obtained by substituting the Fourier transform of eqs. (3) into (16) and taking the usual time average. Detailed computations of the energy spectrum and amount of radiation emitted by a charged particle falling into a black hole have been done by two of us (R.R. and J.T.) and will be published elsewhere.

* * *

It is a pleasure to acknowledge discussions with F. Zerilli.

A 2.3

FULLY RELATIVISTIC TREATMENT OF THE BREHMSTRAHLUNG RADIATION FROM A CHARGE FALLING IN A STRONG GRAVITATIONAL FIELD *

R. RUFFINI
Joseph Henry Laboratories, Princeton University, Princeton, New Jersey 08540, USA.

Received 24 July 1972

The details of the spectral distribution of the radiation emitted by a particle freely falling in a gravitational field are given in the case of the most general static asymptotically flat solution with regular horizon. No contradiction exists with the principle of equivalence. A clear enhancement of high multipoles is shown to exist.

The radiation emitted by a charge in uniform acceleration has been treated with contrasting results. Born [1] and Pauli [2] purported that such charge should not radiate at all, while Bondi and Gold [3] reached opposite conclusions. Directly related by the principle of equivalence to the previous problem is the analysis of the brehmstrahlung radiation emitted by a particle freely falling in a gravitational field. Also, this problem has been debated with contrasting opinions focusing on different interpretations of the equivalence principle. A lucid analysis of the recent situation in this field and a detailed bibliography has been given by Ginzburg [4]. It is surprising that the only approach to the solution of this problem has been attempted in a weak-field, slow-motion approximation of general relativity, see e.g. De Witt–De Witt [5], and De Witt and Brehme [6].

In this paper we present the main results of a complete relativistic treatment valid in *strong* gravitational fields. This method of analysis suggested by Regge and Wheeler [7] has been already successfully applied by Zerilli [8], Ruffini et al. [9] to the problem of gravitational radiation. The amount of energy radiated, as well as the spectral distribution are, here, given in the case of a charge falling from infinite distance (with zero or positive initial velocity) in the most general static gravitational field of a collapsed object [10]. We will also show that from a point of principle the existence of the radiation does not invalidate any existing theoretical framework. The need for a fully relativistic treatment in very strong gravitational fields has been dictated by the recent discovery of collapsed objects, where this radiation might be observed.

We have to solve the coupled system of Maxwell–Einstein equations under the simplifying assumption that the particle and the radiation emitted are a small perturbation on a given static asymptotically flat geometry. The most general static metric with regular horizon [10] is the Reissner–Nordstrom geometry

$$ds^2 = g_{\mu\nu}\,dx^\mu dx^\nu = -\left(1-\frac{2M}{r}+\frac{Q^2}{r^2}\right)dt^2 + \left(1-\frac{2M}{r}+\frac{Q^2}{r^2}\right)^{-1} dr^2 + r^2(d\theta^2+\sin^2\theta\, d\phi^2)\,. \tag{1}$$

Here and in the following we use geometrical units $G = c = 1$. Greek indices goes from 0 through 3. M is the mass and Q the charge of the background geometry. The covariant Maxwell equations to be fulfilled in the given metric (1) are

$$F^{\mu\nu}{}_{;\nu} = 4\pi j^\mu\,, \tag{2.1}$$

or

$$(F^{\mu\nu}\sqrt{-g})_{,\nu} = 4\pi j^\mu\sqrt{-g}\,, \tag{2.2}$$

where $g = \det g_{\mu\nu}$,

$$F_{\mu\nu} = A_{\nu,\mu} - A_{\mu,\nu} \tag{3.1}$$

* Work partially supported by National Science Foundation Grant GP–30799X.

[R51]

and

$$j^\mu = \frac{q}{\sqrt{-g}} \frac{\mathrm{d}z^\mu}{\mathrm{d}t} \delta(\boldsymbol{x} - \boldsymbol{z}(t)). \tag{3.2}$$

q is the charge of the particle and

$z^\mu \equiv (t, \boldsymbol{z}(t)) \equiv (t, R(t), \theta(t), \Phi(t))$

describes the trajectory of the particle as given by the geodetic equation in the fixed background geometry:

$$\frac{\mathrm{d}}{\mathrm{d}s}\left(\frac{\mathrm{d}z^\mu}{\mathrm{d}t}\frac{\mathrm{d}t}{\mathrm{d}s}\right) = F^\mu_{\ \alpha} \frac{\mathrm{d}z^\alpha}{\mathrm{d}t}\frac{\mathrm{d}t}{\mathrm{d}s} + \Gamma^\mu_{\alpha\beta} \frac{\mathrm{d}z^\alpha}{\mathrm{d}t}\frac{\mathrm{d}z^\beta}{\mathrm{d}t}\left(\frac{\mathrm{d}t}{\mathrm{d}s}\right)^2 . \tag{4}$$

To solve the problem of the radiation, we exploit the symmetry properties of the background space and of the operators acting on our vector field [11]. Both the current j^μ and the vector potential A^μ can then be expanded in vector spherical harmonics which form in our problem a complete and orthonormal set of eigenfunctions. We have

$$A_\mu(r, \theta, \phi, t) = \sum_{lm} \begin{bmatrix} 0 \\ 0 \\ \dfrac{a^{lm}(r,t)}{\sin\theta} \dfrac{\partial y^{lm}}{\partial \phi} \\ -a^{lm}(r,t) \sin\theta \dfrac{\partial y^{lm}}{\partial \theta} \end{bmatrix} + \begin{bmatrix} f^{lm}(r,t) y^{lm} \\ h^{lm}(r,t) y^{lm} \\ k^{lm}(r,t) \dfrac{\partial y^{lm}}{\partial \theta} \\ k^{lm}(r,t) \dfrac{\partial y^{lm}}{\partial \phi} \end{bmatrix} \tag{5.1}$$

and

$$4\pi j_\mu = \sum_{lm} \begin{bmatrix} 0 \\ 0 \\ \dfrac{\alpha^{lm}(r,t)}{\sin\theta} \dfrac{\partial y^{lm}}{\partial \phi} \\ -\alpha^{lm}(r,t) \sin\theta \dfrac{\partial y^{lm}}{\partial \theta} \end{bmatrix} + \begin{bmatrix} \psi^{lm}(r,t) y^{lm} \\ \eta^{lm}(r,t) y^{lm} \\ \chi^{lm}(r,t) \dfrac{\partial y^{lm}}{\partial \theta} \\ \chi^{lm}(r,t) \dfrac{\partial y^{lm}}{\partial \phi} \end{bmatrix} \tag{5.2}$$

The first column being of parity $(-1)^l$ and the second $(-1)^{l+1}$. Eqs. (2.1) and (2.2) reduce then to the following set of equations [12]

$$(g^{rr} a^{lm}{}_{,r})_{,r} - g_{rr} \frac{\partial^2 a^{lm}}{\partial t^2} - \frac{l(l+1)}{r^2} a^{lm} = \alpha^{lm} \tag{6}$$

for the quantities of parity $(-1)^l$ and

$$(g^{rr} b^{lm}{}_{,r})_{,r} + g^{00} b^{lm}_{,00} - \frac{l(l+1)}{r^2} b^{lm} = \frac{1}{l(l+1)} [(r^2 \psi^{lm})_{,r} - \eta^{lm}_{,0} r^2] , \tag{7.1}$$

$$b^{lm} = r^2 (h^{lm}_{,0} - f^{lm}_{,r})/l(l+1) , \tag{7.2}$$

$$k^{lm}_{,0} - f^{lm} = g^{rr} b^{lm}_{,0} - r^2 \psi^{lm}/l(l+1) \;, \tag{7.3}$$

$$h^{lm} - k^{lm}_{,r} = g^{00} b^{lm}_{,0} + r^2 \eta^{lm}/l(l+1) \;, \tag{7.4}$$

for those of parity $(-1)^{l+1}$.

In the case of a particle falling radially, eq. (6) is identically zero and the only ones to be solved are eqs. (7). From eqs. (3.2) and (5.2), it immediately follows that

$$\psi^{lm}(r,t) = 4\pi q \delta(r - R(t)) y^{lm*}(\theta,\phi)/r^2 g^{00} \;, \tag{8.1}$$

$$\eta^{lm}(r,t) = 4\pi q \frac{\mathrm{d}R}{\mathrm{d}t} \delta(r - R(t)) y^{lm*}(\theta,\phi)/r^2 g^{rr} \;. \tag{8.2}$$

The energy radiated through a surface S is computed from the T^r_0 component of the momentum energy tensor of the electromagnetic field [12], in our case,

$$\left(\frac{\mathrm{d}E}{\mathrm{d}t}\right)_S = \int_S T^r_0 r^2 \sin\theta \,\mathrm{d}\theta\,\mathrm{d}\phi = \sum_{l,m} b^{lm}(r,t)_{,0}\, b^{lm}(r,t)_{,r}\, [l(l+1)]/4\pi \;. \tag{9}$$

Since we are interested in computing the spectral distribution of the radiation, we obtain, by making the Fourier transform of eqs. (7.1), (8.1), (8.2) and (9), the final set of equations ($m = 0$, radial fall!)

$$\frac{\mathrm{d}^2 b^{l0}}{\mathrm{d}r^{*2}} + \left[\omega^2 - \left(1 - \frac{2M}{r} + \frac{Q^2}{r^2}\right)\frac{l(l+1)}{r^2}\right] b^{l0} = \left(1 - \frac{2M}{r} + \frac{Q^2}{r^2}\right)[(r^2\psi^{l0}(\omega,r))_{,r} + \mathrm{i}\omega\eta^{l0}(\omega,r)]/(l+1) \;, \tag{10.1}$$

$$\psi^{l0}(\omega,r) = (2l+\tfrac{1}{2}) q (\mathrm{d}R/\mathrm{d}t) \mathrm{e}^{\mathrm{i}\omega T(r)}/r^2 g^{00} \;, \tag{10.2}$$

$$\eta^{l0}(\omega,r) = (2l+\tfrac{1}{2}) q \mathrm{e}^{-\mathrm{i}\omega T(r)}/r^2 g^{rr} \;, \tag{10.3}$$

$$(\mathrm{d}E/\mathrm{d}t)_S = (\omega^2/\pi) \sum_l |b^{l0}(\omega,r)|^2 \, l(l+1) \;. \tag{10.4}$$

The coordinate r^* is related to the coordinate r by the equation

$$\mathrm{d}r/\mathrm{d}r^* = 1 - (2M/r) + Q^2/r^2 \tag{11}$$

and $T(r)$ is given by the solution of the geodetic equation (4). In the case we are considering, we have simply

$$\frac{\mathrm{d}T}{\mathrm{d}r} = \frac{\gamma - qQ/\mu r}{(1-2M/r+Q^2/r^2)\,[(\gamma - qQ/\mu r)^2 - (1-2M/r+Q^2/r^2)]^{1/2}} \tag{12}$$

where μ is the mass of the particle.

Remembering now that

$$\gamma = \left(1 - \frac{2M}{r} + \frac{Q^2}{r^2}\right)\frac{\mathrm{d}t}{\mathrm{d}s} + \frac{\epsilon e}{\mu r} = \mathrm{const} = E/\mu \tag{13}$$

is a constant of the motion, we can see immediately that $\gamma = 1$ corresponds to the case of an infall starting at rest from infinity, while $\gamma > 1$ corresponds to an infall with initially a finite value of the kinetic energy.

The problem reduces now to the simultaneous solution of the system of eqs. (10), (11), (12), with the initial value (13). The boundary conditions we have chosen are, clearly, purely outgoing waves at infinity

$$\lim_{r^*\to\infty} b^{l0}(\omega, r) \to a(\omega)\, e^{i\omega r^*} , \tag{14.1}$$

purely ingoing waves at the horizon ($r^* = -\infty, r = M + \sqrt{M^2 - Q^2}$)

$$\lim_{r^*\to -\infty} b^{l0}(\omega, r) \to b^l(\omega) e^{-i\omega r^*} . \tag{15}$$

The solution of the problem has been found by the use of Green function techniques:

$$b^{l0}(\omega, r) = \int_{-\infty}^{+\infty} S^{l0}(\omega, r)\, G(\omega, r, r')\mathrm{d}r' , \tag{16}$$

where $G(\omega, r, r')$ are the Green functions obtained from the homogeneous equation (10.1). If we indicate by $\mu^l(\omega, r)$ the free waves purely ingoing at $r^* \to -\infty$ and by $v^l(\omega, r)$ the free waves purely outgoing at $r^* \to +\infty$ we have

$$G(\omega, r, r') = \begin{cases} u^l(\omega, r')\, v^l(\omega, r)/W^l(\omega) , & r > r' \\ v^l(\omega, r')u^l(\omega, r)/W^l(\omega) , & r > r' , \end{cases} \tag{17}$$

where $W^l(\omega)$ is the Wronskian of the two linearly independent solutions. This entire problem has been solved by the use of numerical techniques, the details of which will be given elsewhere.

We turn now to the main results:

(a) Charge q falling into a collapsed object with $Q = 0$ (Schwarzschild solution) with initially $\gamma = 1$. The major amount of radiation (~ 90%) is radiated in the dipole term. The contribution due to the quadrupole and highest multipoles is negligible. The total amount of radiation emitted is $\Delta E \sim 2 \times 10^{-2}\, q(q/M)$. The peak of the spectrum is at a frequency $\omega \sim 0.2/M$. Some of the results are plotted in fig. 1.

(b) Charge q falling into a collapsed object with $Q = 0$, with initially $\gamma > 1$. Two distinct effects are present: increase in the amount of energy radiated and enhancement of highest multipoles. For a $\gamma = 1.4$, the total energy emitted is $\Delta E \sim 6 \times 10^{-2}\, q(q/M)$, the energy emitted in the quadrupole term is now approximately $\frac{1}{3}$ of the dipole radiation. As a direct consequence of the enhancement of the multipoles, the peak of the radiation shifts toward larger values of the frequency. Some of the results are given in fig. 1.

(c) Charge $q < 0$ falling into a collapsed object with $Q > 0$ (Reissner–Nordstrom solution). In contrast to the uncharged case, we notice a considerable increase in the amount of energy radiated and the contribution of the higher multipoles ($l > 1$) is now predominant. For $Q = 0.95$ (nearly extreme R.N.) and $q/\mu = 50.0$, we have $\Delta E \sim q(q/M)$. The peak of the spectrum of the radiation is largely displaced toward higher frequencies (see fig. 2).

A detailed analysis of the results is going to appear elsewhere; however, from the results here presented, we can reach as of now the following conclusions:

(I) A particle falling into a pure gravitational field radiates and its radiation is mainly of dipole type if the particle is initially at rest at infinity.

(II) Higher multipole radiation ($l > 1$) can be enhanced either if the falling particle has an initial non zero energy at infinity or if the collapsed object is charged, or clearly, if both these circumstances occur simultaneously.

(III) For realistic detection of the radiation this second case appears more promising. Let us recall here that

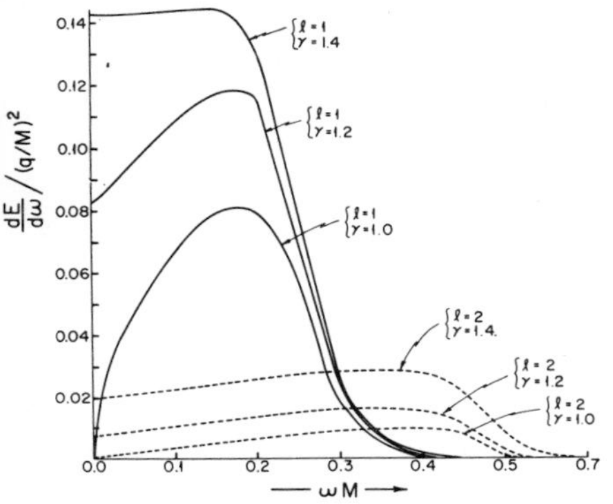

Fig. 1. The spectrum of the electromagnetic energy radiated by a particle falling into a Schwarzschild black hole is here given for selected values of γ (energy at infinity, $\gamma = 1$ the particle start the implosion from rest) and for the dipole and quadrupole term of the radiation. M is here the mass of the collapsed object and q the charge of the particle.

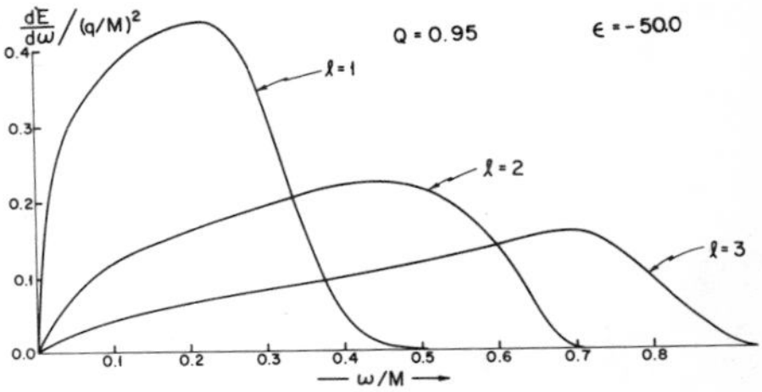

Fig. 2. Energy spectrum for a charge ($\epsilon = q/\mu = -50.0$, μ mass of the particle) falling into an almost extreme Reissner Nordstrφm geometry $Q = 0.95\ M$, (Q is the charge of the collapsed object measured in geometrical units) for selected values of the multipoles ($l = 1$ dipole, $l = 2$ quadrupole, $l = 3$ octupole).

for an electron $q/\mu \sim 2.0 \times 10^{21}$, therefore also a collapsed object with $Q \ll M$ could give rise to the observation of the phenomena we have here presented (enhancement of multipoles, increase in the amount of energy radiated); moreover, extremely strong electromagnetic fields are expected to exist at the surface of a collapsed object [13,14].

(IV) Finally no disagreement can possibly exist between the existence of the radiation and the principle of equivalence. The paradox is easily explained: An observer comooving with the particle in a region small "enough" to allow the application of the principle of equivalence will *not* see any radiation, but only the coulomb part of the field. To see the radiation the observer should look far away, but then the equivalence principle does not apply, its validity only in a local (small enough!) region.

It is a pleasure to acknowledge discussions on some aspects of this problem with T. Gold.

References

[1] M. Born, Ann. d. Phys. 30 (1909) 1.
[2] W. Pauli, Enc. der math. Wissensch. V.(1921) 2.
[3] H. Bondi, T. Gold, Proc. Roy. Soc. 229 (1955) 416.
[4] V.L. Ginzburg, Soviet Physics – Uspekhi 12 (1970) 571, and references quoted there.
[5] C.M. De Witt and B.S. De Witt, Physics 1 (1964) 3.
[6] B.S. De Witt and R.W. Brehme, Ann. Phys. 9 (1960) 220.
[7] T. Regge and J.A. Wheeler, Phys. Rev. 108 (1957) 1063.
[8] F. Zerilli, Phys. Rev. D2 (1970) 2141.
[9] M. Davis and R. Ruffini, Nuovo Cimento Lett. 2 (1971) 1165;
M. Davis, R. Ruffini, W. Press and R. Price, Phys. Rev. Lett. 27 (1971) 1466;
M. Davis, R. Ruffini and J. Tiomno, Phys. Rev. D, May 15 issue (1972);
M. Davis, R. Ruffini, J. Tiomno and F. Zerilli, Phys. Rev. Lett. 28 (1972) 1352.
[10] W. Israel, Phys. Rev. 164 (1967) 1776.
[11] See, e.g., C. Chevalley, Theory of Lie Groups, (Princeton Univ. Press), (1946) 203.
[12] R. Ruffini, J. Tiomno and C.V. Vishveshwara, Nuovo Cimento Lett. 3 (1972) 211.
[13] R. Ruffini and A. Treves, Astr. Letters, in press.
[14] D. Christodoulou and R. Ruffini (to be published).

A 2.4

LINES OF FORCE OF A POINT CHARGE NEAR A SCHWARZSCHILD BLACK HOLE*†

Richard Squier Hanni

and

Remo Ruffini

Joseph Henry Laboratories
Princeton University
Princeton, New Jersey 08540

Abstract

The electric field generated by a charge particle momentarily at rest near a Schwarzschild black hole is analyzed using Maxwell's equations for curved space. By examining the multipole expansion fo the field about the center of the hole, we show that the transition to a Reißner-Nordstrøm hole is continuous. After generalizing the definition of the lines of force to our curved background, we compute them numerically and graph them with the charge at r = 4M, 3M, and 2.2M.

*Partially supported by National Science Foundation Grant #GP-30799X.

†Work partially based on the junior paper of R. S. Hanni submitted at Princeton University on December 11, 1970.

Introduction

The formalism for gravitational perturbations away from a Schwarzschild background has been developed by Regge and Wheeler[1]. It was extended by Zerilli[2], who has shown that perturbations corresponding to a change in the mass, the angular momentum, and the charge of a Schwarzschild black hole are well behaved. The decay of the non-well behaved perturbations has been investigated by Price[3]. He has shown that any multipole $\ell \geq s$, where s is the spin of the field being examined, gets radiated away in the late stage of gravitational collapse and will die as $t^{-(2\ell + 2)}$ for large t.

Instead of analyzing how higher order multipoles are radiated away, we focus on how the allowed transition from a Schwarzschild to a Reißner-Nordstrøm hole takes place through the capture of a charge particle in a given Schwarzschild background. In this paper we neglect the electromagnetic radiation emitted during the fall of the particle and consider a succession of configurations in which the particle is momentarily at rest at decreasing distances from the Schwarzschild horizon ($r = 2m$ in geometrical units $G = c = 1$). The problem of examining the radiation emitted is, indeed, of great interest and has been presented elsewhere.

By considering the charge to be momentarily at rest, we were able to develop its electric field in a multipole expansion centered at the black hole. For any finite separation of the charge from the black hole, the far away observer will detect only the monopole term, the field corresponding to a Reißner-Nordstrøm solution. In the region near the charge, however, the contribution of higher multipoles is important and the lines of force are no longer radial.

As the charge approaches the horizon, the strength of all multipoles, except the monopole term, tends to zero. The lines of force assume more and more their Reißner-Nordstrøm pattern, only a very small region around the particle being significantly affected by the higher multipoles. The strengths of the multipole terms are derived as a function of the distance of the charge from the horizon. We generalize the concept of lines of force to curved space, and show that they are equivalent to the lines of constant flux. Finally, the concept of "induced charge" is introduced, and the smooth transition from the Schwarzschild to a Reißner-Nordstøm field is analyzed and a detailed graphical representation presented.

§1.2 Electrostatic Field in a Schwarzschild Background

The Schwarzschild metric can be expressed as:

$$(1)\ ds^2 = -\left(1-\frac{2M}{r}\right) dt^2 + r^2(d\theta^2 + \sin^2\theta\, d\phi^2) + \left(1-\frac{2M}{r}\right)^{-1} dr^2 \quad .$$

In this background the only non-vanishing components of the electromagnetic field

$$(2.1)\ F = F_{\mu\nu}\, dx^\mu \wedge dx^\nu \qquad \text{with} \qquad (2.2)\ F_{\mu\nu} = A_{\nu,\mu} - A_{\mu,\nu}$$

of a particle momentarily at rest are:

$$(3.1)\ F_{rt} = A_{t,r} = -F_{tr} \qquad \text{and} \qquad (3.2)\ F_{\theta t} = A_{t,\theta} = -F_{t\theta} \quad .$$

Using the relation,

$$(4)\ {}^*F_{\mu\nu} = \frac{\sqrt{-g}}{2}\ \varepsilon_{\mu\nu\delta\gamma}\ F^{\delta\gamma} \quad ,$$

where $\varepsilon_{\mu\nu\delta\gamma}$, we indicate the Levi Civita symbol and $g = \det|g_{\mu\nu}|$, we find that the only non-vanishing components of the dual electromagnetic tensor are

(5.1) $*F_{\theta\phi} = r^2 \sin\theta \, A_{t,r}$ and (5.2) $*F_{\phi r} = \sin\theta \, (1 - \frac{2M}{r})^{-1} \, A_{t,\theta}$.

The only non-vanishing component of the four current is

(6) $j^t = \frac{q}{2\pi r^2} \quad \delta(r - a) \quad \delta(\cos\theta - 1)$

where a is the value of the radial coordinate where the charge is located.

In covariant form Maxwell's equations are:

(7a) $F^{\alpha\beta}{}_{;\beta} = j^\alpha$ (7b) $*F^{\alpha\beta}{}_{;\beta} = 0$.

(Greek indices here and in the following go from 1 to 4). With (5) and (6) they yield a second order differential equation:

(8) $$\frac{\partial(r^2 A_{t,r})}{\partial r} / r^2 + \frac{\partial(\sin\theta \, A_{t,\theta})}{\partial\theta} / r^2 \sin\theta \, (1 - \frac{2M}{r}) = j^t$$

Using the exial symmetry of the problem and its regularity on the axis of symmet: we can expand the solution in terms of Legendre polynomials

(9) $$A_t = \sum_{\ell=0}^{\infty} f_\ell(r) P_\ell(\cos\theta).$$

The functions $f_\ell(r)$ satisfy, then, the second order differential equation

(10) $$\frac{d}{dr} [r^2 \frac{d}{dr} (f_\ell(r))] - \ell(\ell + 1) f_\ell(r)/(1 - \frac{2M}{r}) P^\ell(\cos\theta) = q \, \delta(r-a) \, \delta(\cos\theta-$$

Substituting $u_\ell(r) = rf_\ell(r)$ and $z = r/2M$, Equation (10) takes the hypergeometric form:

$$(11)\quad \frac{d^2u_\ell}{dz^2} - \frac{\ell(\ell+1)}{z(z-1)}\, u_\ell = 0.$$

One of the two independent solutions of this equation is a polynomial of degree $\ell + 1$ with coefficients given by the recursion relation:

$$(12)\quad n\,(n+1)\,\alpha^{\ell}_{n+1} = [n(n-1) - \ell(\ell+1)]\,\alpha^{\ell}_{n}.$$

The first few are:

$$(13.1)\quad u_0 = z \qquad (13.2)\quad u_1 = z(1-z) \qquad (13.3)\quad u_2 = z(z-1)\,(2z-1).$$

To obtain the other independent solution v_ℓ we use the following relation, which follows form (11).

$$(14)\quad v_\ell\, u''_\ell - u_\ell v''_\ell = (v_\ell u'_\ell - u_\ell v'_\ell)' = 0$$

or

$$(15)\quad v_\ell u'_\ell - u_\ell v'_\ell = c$$

Integrating,

$$(16)\quad v_\ell = c\, u_\ell \int \frac{dz}{u_\ell^2}\quad .$$

From the explicit expressions for u_ℓ we obtain for v_ℓ :

$$(17.1)\quad v_0 = 1 \qquad (17.2)\quad v_1 = z(1-z)[2\ln(z/(z-1)) - z^{-1} - (z-1)^{-1}]$$

$$(17.3)\quad v_2 = z(z-1)\,(2z-1)^{-1} - (z-1)^{-1} + 6\ln(z/(z-1)) - z^{-1}]$$

and so on for larger ℓ. The most general solution for the potential can be cast in the form

$$(18) \quad A_t = \sum_\ell \frac{\alpha_\ell u_\ell(r) + \beta_\ell v_\ell(r)}{r} P_\ell(\cos\theta).$$

All of the u_ℓ's except the one corresponding to $\ell = 0$ vanish at the horizon. In the region inside the charge all the β_ℓ's must vanish, for the potential to be regular at the event horizon and vanish far away from the hole. Moreover, since $P_0(\cos\theta) = 1$, we can conclude that the horizon is an equipotential surface (A_t = const.).

For the potential to be regular as $r \to \infty$ all the α_ℓ's must vanish in the region outside the charge. As $r \to \infty$, the term v_0 is constant while all the other terms v_ℓ decreases as $r^{-\ell}$. The monopole term dominates at infinity. Gauss's law gives us the magnitude of the spherically symmetric electric field and thus the weighting coefficient for the monopole term, $\beta_0 = q$. The field far away approaches that of a point charge located at the center of the black hole.

§1.3 Matching Conditions and Strength of the Multipoles

In order to calculate the electric field we must evaluate the weighting coefficients: $\alpha_1, \beta_1, \alpha_2, \beta_2 \ldots$ Substituting the expansion (18) of A_t into Poisson's equation (10) we relate the discontinuity in the slope of each radial function to the amount of charge concentrated at $r = a$, $\theta = 0$, we have,

$$(19) \quad \frac{d^2 u_\ell}{dz^2} - \frac{\ell(\ell+1)}{z(z-1)} u_\ell \, P_\ell(\cos\theta) = \frac{q}{2\pi z} \quad \delta(z - \tfrac{a}{2M}) \; \delta(\cos\theta - 1)$$

Integrating over a thin shell containing the point $z = \frac{a}{2M}$ we obtain

$$(20)\quad \left[\frac{du_\ell}{dz}\right]_{z = \frac{a}{2M}} = -\ (2\ell + 1)q/(a/2M)$$

The continuity of the radial functions and the boundary conditions at the horizon and at infinity determine the weighting coefficients α_ℓ and β_ℓ uniquely. Because of the complexity of higher order terms we resort to numerical integrations to evaluate α_ℓ, β_ℓ, u_ℓ and v_ℓ.

Since the potential is continuous at the charge, we have

$$(21)\quad \alpha^\ell u_\ell\ (a) = \beta^\ell v_\ell(a) \int_o^a \frac{dz}{u_\ell^2(z)} \quad .$$

The discontinuity in the field at the charge gives

$$\alpha^\ell u'_\ell(a) - \beta^\ell v'_\ell\ (a) = \frac{2\ell + 1}{q}$$

or for $\ell = 0$ $\beta^0 = q$ independent from the position of the charge and $\alpha^0 = q/a$

$$(22)\quad \text{for } \ell = 1 \quad \beta^1 = 3q(a - 1) \quad \alpha^1 = \frac{3qv_1(a)}{a}$$

$$(23)\quad \text{for } \ell = 2 \quad \beta^2 = 5q(2a - 1)\ (a - 1) \quad \alpha^2 = \frac{5qv_2(a)}{a}$$

Figure 1 shows the behavior of the radial functions for selected values of ℓ and selected values of the distance of the particle a = 3M, a = 4M and a = 6M. While the monopole coefficient remains constant, all the other multipoles vanish as the particle approaches the horizon. The decay of the higher multipoles (ℓ>0) and the constancy of the monopole term determine the smooth transition from a Schwarzschild to a Reißner-Nordstrøm geometry.

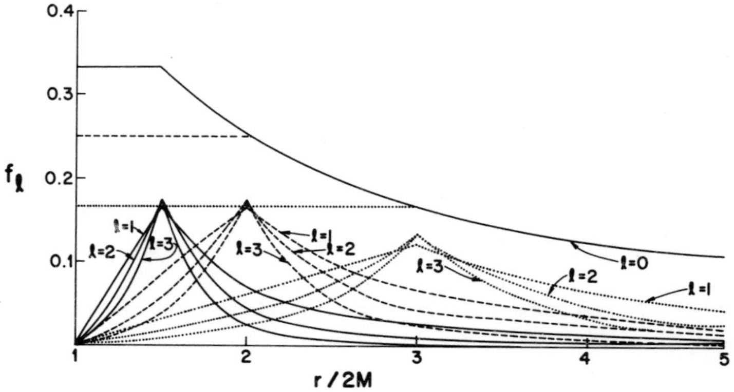

Fig. 1. Radial functions $f_\ell \doteq (\alpha_\ell u_\ell (r) + \beta_\ell v_\ell (r))/r$ plotted as a function of the radial Schwarzschild coordinate r for selected values of ℓ and of the distance of the particle from the Schwarzschild surface ($\frac{a}{2M}$ = 2,1.5, 1.1). The strength of the multipoles $\beta_\ell = f_\ell r^{\ell + 1}$ is constant for = 0 and increases with the distance of the particle from the black hole for ℓ > 0 .

§1.4 Definition of the Lines of Force

The generalization of Gauss's theorem to curved space is (4) given by the equation

(24) $\int_{S} *F = 4\pi q$,

where q is the total charge withing the surface c_2 and

(25) $*F = *F_{\mu\nu}\, dx^{\mu} \Lambda\, dx^{\nu}$.

Only the components (5.1) and (5.2) are non-zero, so

(26) $$*F = [r^2 \sin\theta \frac{\partial A_t}{\partial r}\, d\theta - (1 - \frac{2M}{r})^{-1} \frac{\partial A_t}{\partial \theta}\, dr] \Lambda d\phi .$$

If there exists a function Φ satisfying

(27) $\int_S *F = \int_S d\Phi$,

then

(27.1) $$\frac{\partial \Phi}{\partial \theta} = r^2 \sin\theta \frac{\partial A_t}{\partial r} \quad \text{and}$$

(27.2) $$\frac{\partial \Phi}{\partial r} = -(1 - \frac{2M}{r})^{-1} \frac{\partial A_t}{\partial \theta} \sin\theta .$$

These conditions are met by the following expression for Φ :

(28) $$\Phi = \sum_{\ell} - \frac{r^2}{\ell(\ell + 1)} \frac{d}{dr} (\frac{f_\ell}{r}) \sin\theta \frac{dp_\ell}{d\theta} .$$

We define a line of force as the locus of points with a given value of . At any given point, the slope of the line of force is:

(29) $$\frac{dr}{d\theta} = - \frac{\partial \Phi}{\partial \Phi} / \frac{\partial \Phi}{\partial r} = (1 - \frac{2M}{r}) r^2 \frac{\partial A_t}{\partial r} / \frac{\partial A_t}{\partial \theta} .$$

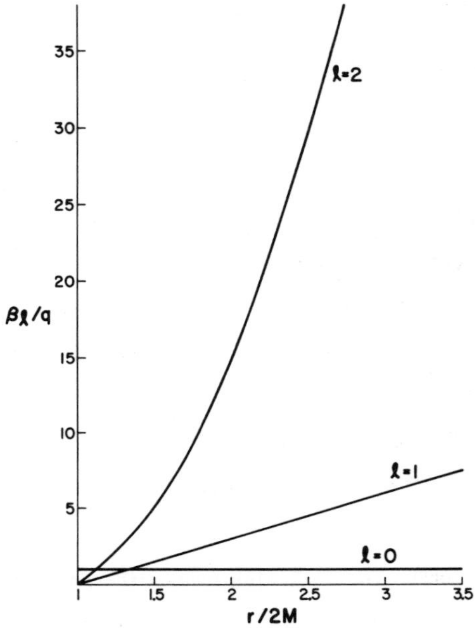

Fig. 2. Strength of the multipoles in units of the charge of the test particle q, as a function of the distance of the charge from the event horizon (r = 2M0. As $\frac{a}{2M} \to 1$, the monopole is constant, while the higher multipoles vanish.

This is equivalent to the operational definition of lines of force, introduced by Christodoulou and Ruffini[5]. They define a line of force as the line tangent to the direction of the electric force measured by an inertial observer momentarily at rest; this definition is also valid for stationary metrics. We then have

(30) $\varepsilon_{ijk} F^{j}{}_{\alpha} \ u^{\alpha} \ dx^{k} = 0$ or

(30.1) $F^{r}{}_{t} u^{t} d\theta - F^{\theta}{}_{t} u^{t} dr = 0,$

which also gives expression (29) for the slope of the lines of force if we assume an inertial observer with four velocity $u^{\alpha} = (0,0,0,1)$.

Figures 3,4, and 5 show the lines of force in Schwarzschild coordinates, as the charged particle approaches the event horizon. When the charge is at r - 2.2m (Fig. 5), the field far away ($r \gtrsim 10M$) from the hole is nearly radial about the center of the hole. The monopole term clearly dominates. The contribution of the higher multipoles is dominant in the region near the charge.

The lines of force intersect the event horizon. We interpret this as a charge induced on the surface of the hole, which is proportional to the electric field normal to the surface. The total flux through the Schwarzschild surface, and thus the net induced charge is zero.

Let us assume that the point charge is positive. At angles less than a critical angle θ_{crit}, the induced charge is negative and the lines of force go toward the even horizon. At the critical angle there is no induced charge and the line of force is tangent to the Schwarzschild surface . At angles greater than the critical angle, the induced charge is positive and the lines of force are directed away from the horizon. Figure 6 shows the ratio

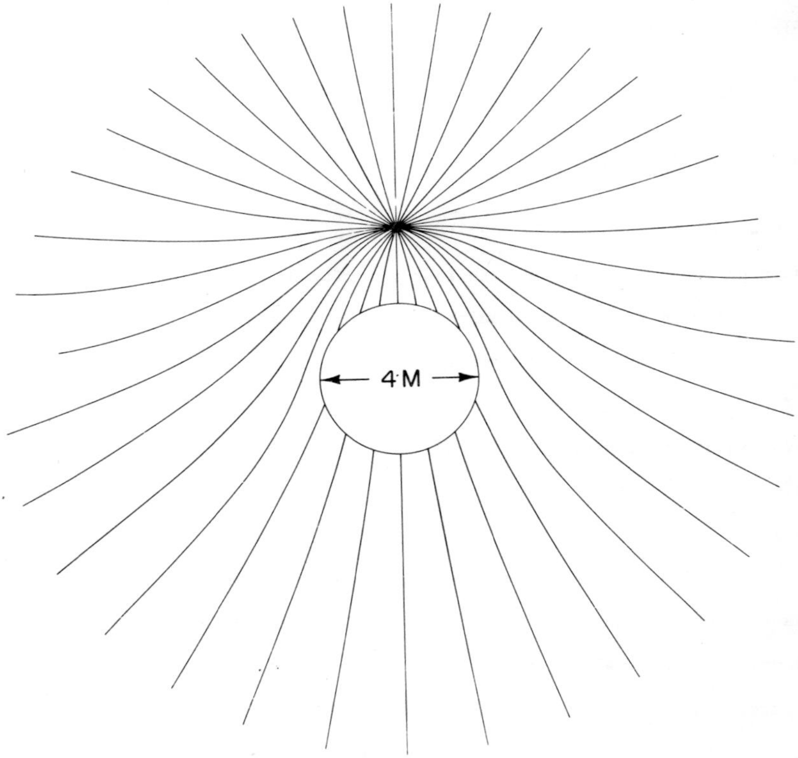

Fig. 3. Lines of force with the test charge momentarily at rest at r = 4M.

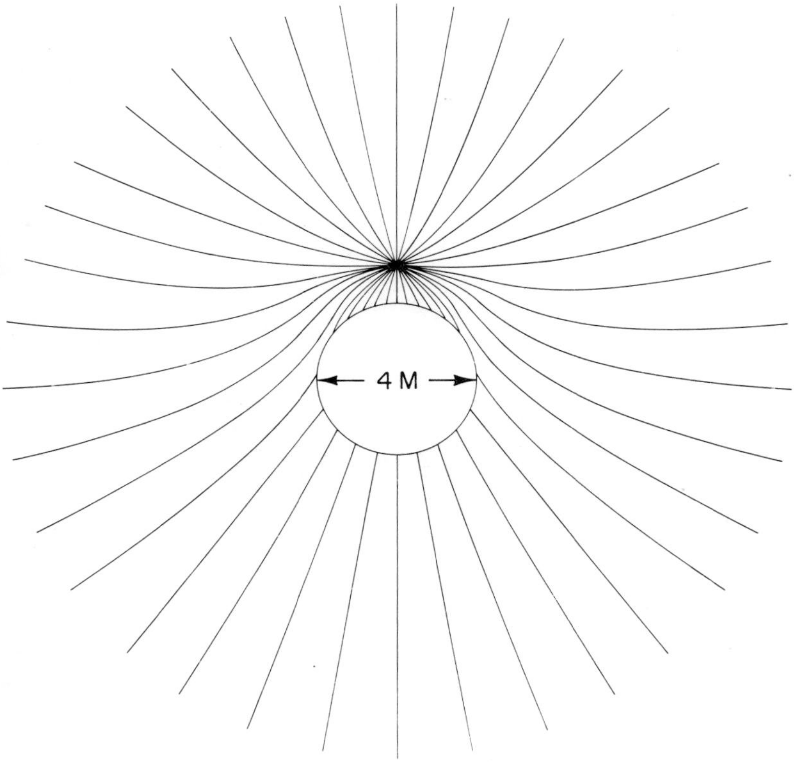

Fig. 4. Lines of force with the test charge momentarily at rest a r = 3M.

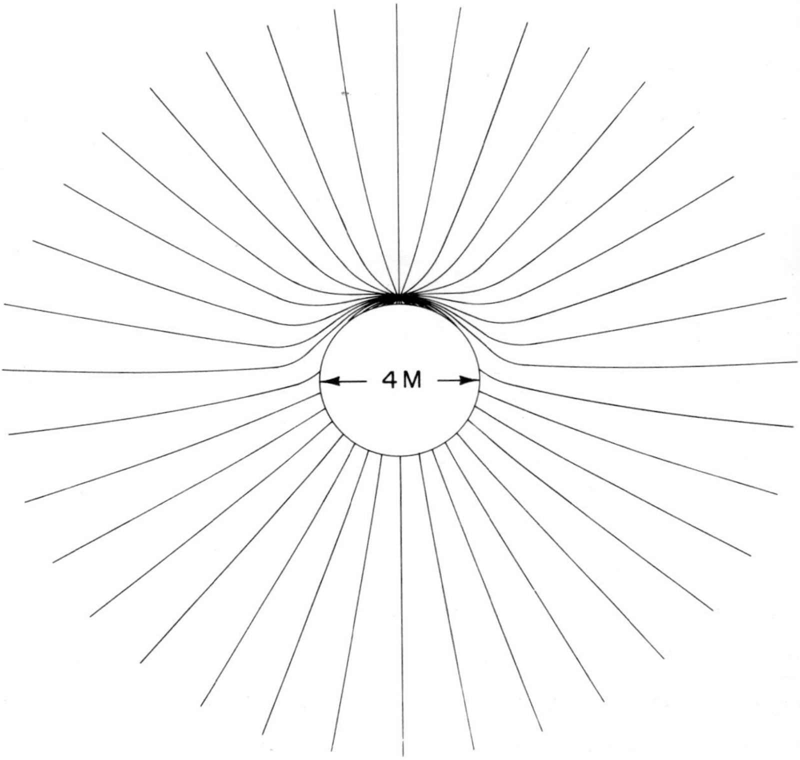

Fig. 5. Lines of force with the test charge momentarily at rest at r = 2.2M.

of the positive charge induced to that of the point charge and the critical angle, as functions of the radial coordinate of the charge.

The behavior of this induced charge shows clearly the smooth transition from a Schwarzschild to a Reißner-Nordstrøm hole. As the charge approaches the horizon, the magnitude of the induced charge approaches that of the point charge and the negative induced charge is crowded into a decreasingly small area around the pole. In contrast, the positive charge disperses itself more and more evenly over the rest of the surface of the sphere. The asymptotic limit of this evolution is a dipole with no strength at the pole, and a positive charge, equal in magnitude to that of the point charge, distributed evenly over the Schwarzschild surface. This charge distribution generates the monopole field of the Reißner-Nordstrøm solution.

It is essential to the interpretation of Figure 6, that the radial coordinate of a freely falling non-radiating particle with no angular momentum about the center of the hole decreases with time as:

$$r = 2m\ (1 + 4\ e^{-8/3} e^{-t/2m})$$

If the particle radiates away some of its energy, then the approach will be even slower. As seen by a far away observer, the charge approaches, but never reaches, the horizon ($r = 2m$). The ratio of the induced charge to the point charge approaches, but never reaches, one. The critical angle approaches, but never reaches zero. The coordinates velocity:

$$\frac{dr}{dt} = -\ 4e^{-8/3} e^{-t/2m}\ ,$$

decreases exponentially with time. Thus the system we have investigated is a good approximation of a charged particle in radial free fall, when the charge is sufficiently close to the hole.

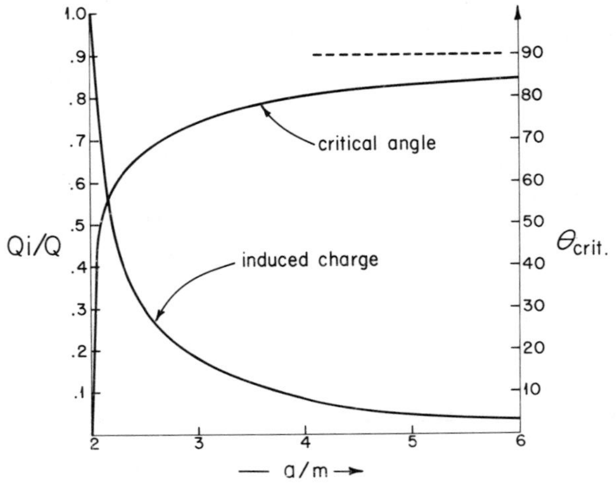

Fig. 6. The critical angle, θ_{crit}, where the line of force is tangent to the Schwarzschild surface and the induced surface charge vanishes, as a function of the radial coordinate of the test particle. Also given is the ratio of the charge induced on the black hole (see text) to the charge of the test particle.

Acknowledgment

It is a pleasure to acknowledge discussion on many aspects of this problem with John A. Wheeler.

References

1. T. Regge and J. A. Wheeler, Phys. Rev. 108, 1063, 1957.

2. F. Zerilli, Phys. Rev., D2, 2141, 1970.

   F. Zerilli, Journ. Math. Phys. 11, 2203, 1970.

   F. Zerilli, Phys. Rev. Lett. 24, 737, 1970.

3. R. H. Price, Phys. Rev. D10, 2419, 1972 and 10, 2439, 1972.

4. J. A. Wheeler, Geometrodynamics, Academic Press, 1962.

5. D. Christodoulou and R. Ruffini, "On the Electrodynamics of Collapsed Objects", unpublished.

A 2.5

Ultrarelativistic Electromagnetic Radiation in Static Geometries*

Remo Ruffini

Joseph Henry Physical Laboratories, Princeton, New Jersey 08540

and

Frank Zerilli†

University of North Carolina, Chapel Hill, North Carolina 27514

Radiation from ultrarelativistic circular orbits are, here, examined in the Reissner-Nordstrøm background, both by analytic and numerical techniques. Stable orbits with 100% binding and $v/v_{light} \sim 1$ exist. New insight and a physical explanation for the spectra recently considered by Davis, Ruffini, Tiomno and Zerilli are presented.

*Work partially supported by National Science Foundation Grant GP-30799X.

†Present address: Department of Astronomy, University of Washington, Seattle, Washington 98105

It has become more evident than ever that the proof for the existence of completely collapsed objects relies heavily on the observation of a continuous flux of radiation emitted by a stream of charged particles accreting toward the horizon[1]. Gravitational radiation, though extremely important, is expected to be emitted only in relatively rare and very short bursts of energy[2,3,4,5]. Quite apart from considering the problem of the energy emitted by a plasma in the ergosphere[6] of a magnetic black hole[7], which still presents superb difficulties both from a mathematical and a physical point of view,[8][9] we are here interested in the radiation emitted by a particle in circular orbit in the most general static geometry with regular horizon[10]. This analysis is also dictated by the need of a deeper understanding of the recent results of Davis, Ruffini, Tiomno, and Zerilli,[11] which clearly shows the existence of different behavior in the radiation field generated by particles of different spin in unbound ultrarelativistic circular orbits. Their results have been confirmed to indeed apply also in the limit of the highest multipoles and a very compact formula has been found to express the main features of the radiation fields [12]. We have, here, shown that the Reissner-Nordstrøm field greatly differs from the Schwarzschild or Kerr one, in the sense that orbit with 100% binding exist, they are stable, and reachable from particles of the characteristic size of an electron. For reasons we will explain we consider these orbits ultrarelativistic. We have also given, using the Ford-Hill-Wakano-Wheeler[13][14] approach the explicit form of the Green's functions for the ultrarelativistic unbound orbits in the Reissner-Nordstrøm geometry. Finally, we have used these two results to give a new insight in the nature of the spectral distribution discovered in Ref. (11).

(a) Orbit in Reissner-Nordstrøm geometry. The metric has the form

(1) $ds^2=g_{\mu\nu}dx^\mu dx^\nu=-(1-2M/r+Q^2/r^2)\ dt^2+r^2(d\theta^2+\sin^2\theta d\phi^2)+dr^2/(1-2M/r+Q^2/r^2)$

Greek indices goes from 0 to 3, G=c=1, M and Q are the mass and the charge characterizing the background geometry. For the energy and angular velocity of a test particle in circular orbit in the given geometry we obtain

(2.1) $\omega_o^2=(d\phi/dt)^2=M/R^3-Q^2/R^4-\ \epsilon Q/R^3[\epsilon Q/2R+(1-3M/R+2Q^2/R^2+\epsilon^2Q^2/4R^2)^{\frac{1}{2}}]$

(2.2) $E/\mu=p_o=(1-2M/R+Q^2/R^2)/[\epsilon Q/2R+(1-3M/R+2Q^2/R^2+\epsilon^2Q^2/4R^2)^{\frac{1}{2}}]+\epsilon Q/R$

Here $\epsilon=q/\mu$ is the charge per unit mass of the test particle. The main results are shown in Fig. 1 for the geometry with Q/M=1. Not shown is an entire family of unstable circular orbits occuring for $\epsilon > 1$ and at radii $r \leq [3M+(9M^2-8Q^2)^{\frac{1}{2}}]/2$. Details elsewhere. It is important to remark here that for every given value of $\epsilon \leq 1$ there exists an orbit of maximum binding which corresponds to the last stable circular orbit; orbits with larger value of the radius are stable, while the ones with smaller radii. are unstable. The binding energy of a particle in the last stable orbits increases very rapidly with decreasing values of ϵ, its velocity tends toward the local value of the velocity of light. (See Fig. 1). In the entire family of unstable circular orbits, for fixed value of ϵ, the ratio between the velocity of the particle and the local value of the velocity of light increases monotonically for decreasing values of the radius and reaches the value of one at $r = r_p = 3M+(9M^2-8Q^2)^{\frac{1}{2}})/2$. Finally for large values of ϵ we can give asymptotic formulae for the radius and energy of the last stable circular orbit:

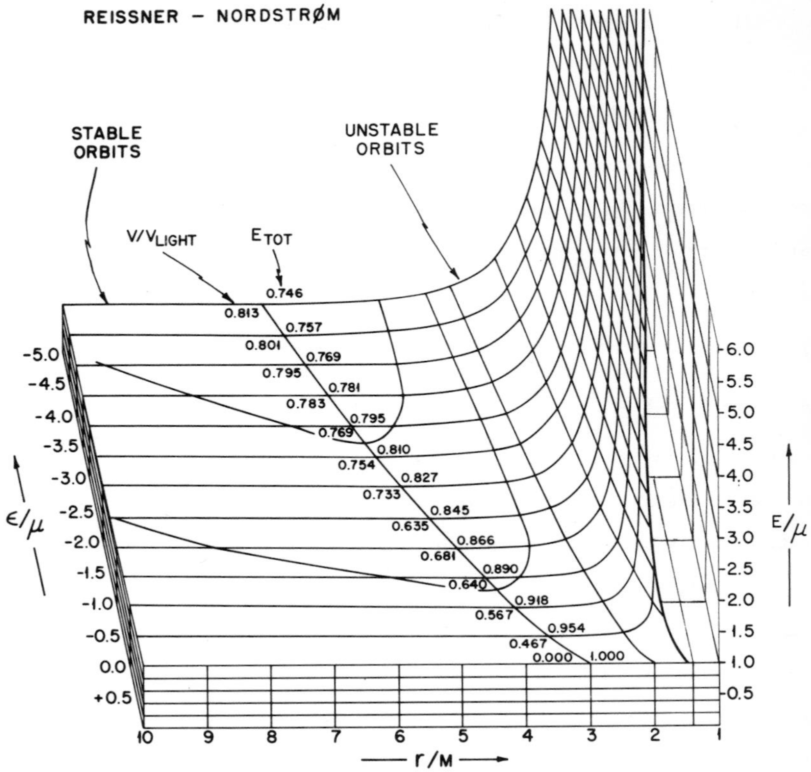

Fig. 1 Energy of circular orbits in an extreme Reissner-Nordstrøm field. E/μ is the energy, r/M the radial coordinate and $\varepsilon = q/\mu$ the charge of the particle; v/v_{light} is the velocity in the circular orbit measured with respects to the local value of the speed of light. The total energy is equal to the covariant p_o component of the momentum. The orbits with radii larger than the one corresponding to the last circular orbits are stable, the remaining ones are unstable.

(3.1) $R_{min} \sim \sqrt{2}\ |\varepsilon Q|)^{2/3}$

(3.2) $E_{min}/\ \mu \sim \frac{3}{2}\ (2/\ |\varepsilon Q|)^{1/3}$.

These formulae apply in the limit $R_{min} >> M$ and $|\varepsilon Q| >> M$. Consider e.g. an electron ($\varepsilon = q/\mu = 2.04 \times 10^{21}$), a black hole of 1 solar mass and a net charge $Q/M \sim 10^{-3} \simeq 5 \times 10^{26}$ e.s.u., then $E_{min}/\mu \simeq 1.89 \times 10^{-7}$ and $R_{min} \simeq 0.39$ light years. For the evaluation of the net charge that should be expected on the surface of a collapsed object see Ref. (8).

(b) radiation from ultrarelativistic unbounded orbits.

Exploiting the symmetry properties of the equations and of the background geometry we can expand the covariant Maxwell equations in vector harmonics[(15)]. The two equations mastering the solution are for parity $(-1)^{\ell+1}$

(4.1) $$b^{\ell m}{}_{,r^*,r^*} + (n^2\omega_o^2 - g_{oo}\ \frac{\ell(\ell+1)}{2})\, b^{\ell m} = (-1)^m\ \frac{4\pi q\omega_o}{\ell(\ell+1)}\ \delta(m-n) P^{m\ell}(o) A^{\ell}_{m}\ \delta(r^* - r_o^*)_{,r^*}$$

and for parity $(-1)^{\ell}$

(4.2) $$a^{\ell m}{}_{,r^*,r^*} + (n^2\omega_o^2 - g_{oo}\ \frac{\ell(\ell+1)}{r^2})\, a^{\ell m} = (-1)^m\ \frac{4\pi q\omega_o}{\ell(\ell+1)}\ \delta(m-n) P^{m+1}_{\ell}(o)\ .$$

where $[\ell(\ell+1)-m(m+1)]^{\frac{1}{2}} A^{\ell}_{m}\ \delta(r^* - r_o^*)$

$$A_m{}^{\ell} = [(2\ell+1)(\ell-m)!/(4\pi(\ell+m)!)]^{\frac{1}{2}}$$

The energy radiated is given by the formula

(5) $$\frac{dE}{d\omega} = \sum_{\ell m} \frac{\omega^2}{2} \ell(\ell+1)\ \{|a_{\ell m}|^2 + |b_{\ell m}|^2\}$$

The solution of the problem is accomplished by finding the Green's functions of Eq. (4.1) and (4.2). We have solved this problem exactly by numerical techniques and we have given analytic expression for the Green's functions[(14)] for orbits near the top of the effective potential. To obtain the analytic form of the approximate Green's function we can expand the effective potential for $r \sim r_b$ into a power series of r^*. Introducing then a new coordinate

$$(6)\quad \xi = \omega_\infty (\ell(\ell+1))^{1/4} r^* / \sqrt{k}$$

with

$$k = r_p (3Mr_p - 4Q^2)^{-1/2} \text{ and } \omega_\infty^2 = (Mr_p - Q^2)/r_p^4$$

the homogeneous part of the equations (4.1) and (4.2) reduce to the well known equations generating the parabolic cylinder functions. The three cases $q > 0$, $q = 0$, $q < 0$ have to be treated separately. For the electric parity equations $(-1)^\ell$ we can introduce the following quantity:

$$(7)\quad \eta^{(e)} \begin{cases} k[1 + 4p + |m|\varepsilon^2 Q^2/(Mr_p - Q^2)] & \text{for } qQ>0 \text{ and } \delta \ll q^2Q^2/\mu^2M^2 \\ k[1 + 4p + M\delta r_p\ |m|/k^2\ (Mr_p - Q^2)] & \text{for } qQ = 0 \\ k[1 + 4p + M^2\delta^2\ |m|\ (\frac{1}{k^2} + \frac{9M^2-8Q^2}{\varepsilon^2 Q^2})/(Mr_p - Q^2)] & \text{for } qQ < 0 \text{ and } \delta \ll q^2Q^2/\mu^2M^2 \end{cases}$$

with $\ell - m = 2p$.

for magnetic parity terms $(-1)^{\ell+1}$ we have

$$\eta^{(m)} = \eta^{(\ell)} + 2k \text{ and } \ell - m = 2p + 1$$

We are now able to write down the two linearly independent solutions of the homogeneous equations (4.1) and (4.2) for $r \sim r_p$:

$$(8)\quad R(\xi)=L(-\xi) = \frac{1}{\Gamma(\frac{1}{2} + i\,\frac{n}{2})} \int_0^\infty \exp[i\,\xi^2/2-(1-i)\,\xi t-t^2/2-(^1/2-i\,^\varepsilon/2)\ell nt]dt$$

where $R(\xi)$ is purely outgoing wave for $r^* \to +\infty$, and $L(\xi)$ a purely ingoing wave for $r^* \to -\infty$. Using Eqs. (6)(7) and (8) we can finally give the explicit form of the Green's functions and their derivatives

$$(9.1)\quad G(r^*,r_o^*) \simeq -\frac{i\,\Gamma(\frac{1}{4}+\frac{i\eta}{4})\,k^{\frac{1}{4}}e^{-\pi\eta/8}e^{i\phi(r^*)}}{[4\sqrt{\pi}\,(\ell(\ell+1))^{3/8}\omega_\infty]}$$

and

$$(9.2)\quad \frac{\partial G(r,r_o^*)}{\partial r_o^*} \simeq -\frac{1+i}{\sqrt{8\pi}}\;\frac{\Gamma(\frac{3}{4}+i\frac{\eta}{4})e^{-\pi\frac{\eta}{8}}}{[\ell(\ell+1)]^{1/8}k^{\frac{1}{4}}}\;e^{i\phi}(r^*)$$

where the phase $\phi(r^*)$ is determined by matching $R(\xi)$ to a purely outgoing wave for large values of r^*. The final expression for the power radiated is

$$(10.1)\quad P_{\ell m}^{(e)} \simeq \frac{\omega_o^2 q^2}{\sqrt{k}\,\pi^{3/2}}\;\frac{(2p)!}{2^{2p}(p!)}\;e^{-\frac{\pi}{4}\eta^{(e)}}\;|\Gamma(\tfrac{3}{4}+\tfrac{i}{4}\eta^{(e)})|^2$$

for the magnetic terms of parity $(-1)^{\ell+1}$

$$(10.2)\quad P_{\ell m}^{(m)} \simeq \frac{\sqrt{k}}{2}\;\frac{\omega_o{}^2}{2\,\omega_\infty}\;\frac{\omega_o^2 q^2}{\pi^{3/2}}\;\frac{(2p+1)!}{2^{2p}(p!)^2}\;e^{-\frac{\pi}{4}\eta^{(m)}}\;|\Gamma(\tfrac{1}{4}+\tfrac{i}{4}\,{}^{(m)}|^2$$

We can then immediately reach from the form of Eq. (10.1) and (10.2) the following general conclusions: the dominant term in the electric multipoles is the one with $\ell = m$ in the magnetic multipoles the one with $\ell - 1 = m$. In both cases the following term is down by a factor $e^{-2\pi k}$ with $k \sim 1$. The ratio of magnetic to electric power is

$$(11)\quad P_{\ell m}^{(m)}/P_{\ell m}^{(e)} = \frac{k}{2}\,(2p+1)\begin{Bmatrix}1-\varepsilon^2Q^2/(Mr_p-Q^2)\\ 1\end{Bmatrix}e^{-\pi k/2}\left|\frac{\Gamma\frac{1}{4}+\frac{1}{4}i\eta)}{\Gamma(\frac{3}{4}+\frac{1}{4}i\eta)}\right|^2\begin{matrix}\text{for } qQ > 0\\ qQ \le 0\end{matrix}$$

which is of the order of 10^{-2} or smaller. (See Fig. 2). For a test charge $q = \varepsilon\mu > 0$ we see that the m = 1 give the dominant contribution, each succeeding term being damped by a factor $\exp(-\pi k\varepsilon^2 Q^2|m|/\ 2(Mr_p - Q^2))$. This behaviour in the spectrum is strictly connected with the fact that the radiation is emitted at values of $r < r_p$. Details will be given elsewhere.

Finally, let us here remark one of the main differences between this case and the one treated in Ref. (11). If $qQ < o$ the critical value of m at which the cutoff in the spectral distribution occurs is given by

$$(12)\quad m_{crit} = 4(Mr_p - Q^2)/[\pi k M^2 (1/k^2 + (9M^2 - 8Q^2)/\varepsilon^2 Q^2)]\delta^2$$

which radically differs from the one in Schwarzschild geometries. Details of the spectral distribution are given in (Fig. 2).

(c) Orbits with a constant value of v/v_{light}. To have a better understanding of the results presented in Ref. (11) we consider a sequence of configurations of equilibrium with a constant value of the ratio between the orbital velocity and the local value of the speed of light.

$$(13)\quad v/v_{light} = \omega_o/\omega_{light} = \omega_o R/(1 - 2M/R + Q^2/R^2)^{\frac{1}{2}}$$

here R is the radius of the circular orbit and ω_o is given by Eq. (2.1). These configurations correspond to different values of the test charge: the largest is the magnitude of the charge, the further away from $r = r_p$ they occur and the nearer to the family of last stable circular orbits. The main results of this analysis are summarized in Fig. 3, where two orbits with $v/v_{light} \sim 0.915$ and $\gamma = 2.48$ are considered. In one case the test charge has been chosen to have $q/\mu = -0.1$ and the orbit has a radius $r = 2.93$ M (very near the top of the potential barrier where our asymptotic

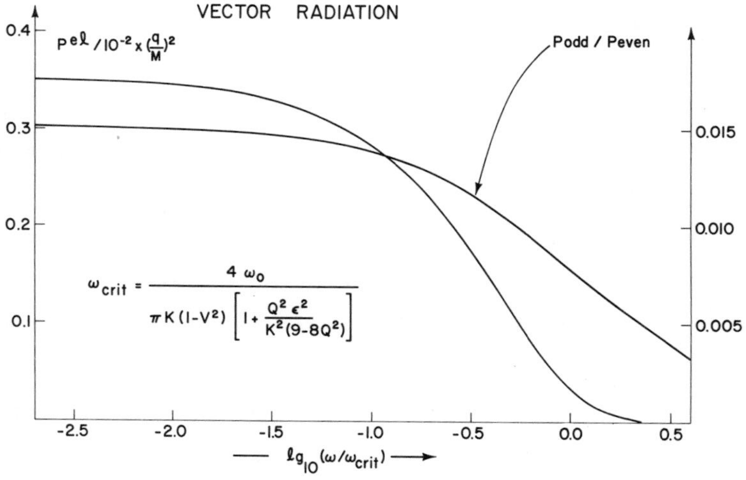

Fig. 2 Spectrum of electromagnetic energy radiated from a particle in ultrarelativistic orbits with radius $r \sim r_p$. The ratio between the odd $(-1)^{\ell+1}$ and even $(-1)^{\ell}$ parity contribution to the power radiated is also given as a function of the frequency.

treatment applies). In the other case, the test charge has been chosen to have $q/\mu = -10.0$ the radius of the orbit occurs now at $r = 6.795M$ where a full treatment of Eqs. (4.1) and (4.2) is needed, the expansion with the parabolic cylinder functions being inadequate in this range of values of r. We have in both cases used numerical techniques of integrations, details of which will be given elsewhere. The results are self-explanatory just by looking at Fig. 3!

It is worthwhile, however, to summarize at least the main conclusions:

(I) With the increase of the absolute value of the test charge the orbits move outward and the spectrum of the radiation emitted, rapidly recovers the main features of synchrotron radiation in flat space.

(II) When the particle is near the top of the potential barrier $r \sim r_p$, the <u>highest</u> multipoles (shorter wavelengths!!) are maximally affected and drastically damped(16). The dipole term is only slightly damped.

(III) The radiation going inside the black hole, is of the order of fifty per cent, in the lowest multipoles limit in the first case considered ($q/\mu = -0.1$) and drops to one part in 10^3 in the case of $q/\mu = -10$.

These results, clearly, point out that the reason of the anomalous behaviour of the spectrum of the radiation for orbits near r_p, cannot be explained in terms of the relativistic velocity of the particle (special relativistic effect). It is, instead, mainly due to the fact that the radiation originates in a region very near the photon circular orbit ($r = r_p$). Photons beamed by the special relativistic effects are channelled in a "storage ring" at $r \sim r_p$ from which they leak out with varying time scales. The higher the multipole, the sharper and more effective the storage process is. This phenomenon, together with the large amounts of energy which goes inside the black hole, are, in our opinion, the

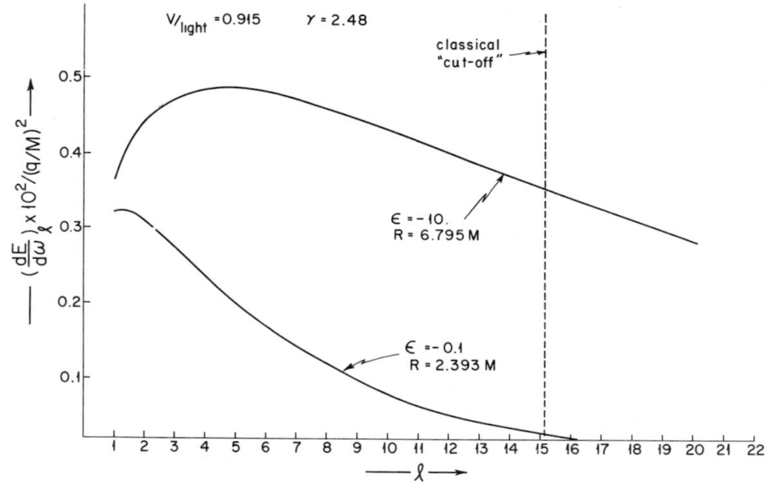

Fig. 3 Compared and contrasted are the spectral distributions of the radiation emitted by a charge in circular orbit near the top of the potential barrier $R \sim r_p$ (Q = -0.1) and one at r = 6.795M (Q = -10). Both orbits have the same value of the velocity measured with respect to the local value of the speed of light. The effects of the background geometry are in the case of Q = -10 is greatly reduced and the spectrum approaches the classical shape of synchrotron radiation.

physical reasons for the defocusing effects pointed out in Ref. 11. These effects clearly diminish when the radius of the orbit of the particle increases. Similar arguments apply to the spin 2 case, as well, the defocusing effect being there very much more pronounced, due to the fact, that the fundamental mode emission of the radiation is quadrupolar.

REFERENCES

(1) R. Ruffini and J. A. Wheeler, Physics Today - January 1971, p. 39.

(2) M. Davis and R. Ruffini, Nuovo Cimento Lett. 2, 1165, 1971.

(3) M. Davis, R. Ruffini, W. Press, R. Price, Phys. Rev. Lett. 27, 1466, 1971.

(4) M. Davis, R. Ruffini, J. Tiomno - Phys. Rev. D12, 2932, 1972.

(5) R. Ruffini, Submitted to Phys. Rev.

(6) For the definition of ergosphere the e.g. R. Ruffini and J. A. Wheeler in "Cosmology from Space Platform" H. Moore and V. Hardy ESRO SP-52-Paris-1971.

(7) The ergosphere has been further generalized to the Kerr-Newman geometry by D. Christodoulou and R. Ruffini, Phys. Rev. 4,3552, 1971.

(8) R. Ruffini and A. Treves - Astrophysical Letters, in press.

(9) R. Leach and R. Ruffini - Ap. J. Lett., Submitted for publication

(10) W. Israel, Phys. Rev. 164 1776, 1967.

(11) M. Davis, R. Ruffini, J. Tiomno, F. Zerilli, Phys. Rev. Lett. 28, 1352, 1972.

(12) R. Breuer, R. Ruffini, J. Tiomno, C. V. Vishveshwara - Submitted to Phys. Rev.

(13) K. W. Ford, D. L. Hill, M. Wakano, J. A. Wheeler, Am. Phys. 7, 239 (1959).

(14) Analogous treatment for the Schwarzschild geometry has been given by C. W. Misner, R. A. Breuer, D. R. Brill, P. L. Chrzanowski, H. G. Hughes,III and C. M. Pereira, Phys. Rev. Lett. 28, 998, 1972.

(15) R. Ruffini, J. Tiomno, C. V. Vishveshwara, Nuovo Cimento Lett. 3, 211, 1972.

(16) This result manifestly contradict recent claim to have explained the "Nature of Gravitational Synchrotron Radiation" and the spectrum first founded in Ref. 11 by the inapplicability of the law of geometrical optics, in this regime. See e.g. D. M. Chitre and R. H. Price, Phys. Rev. Lett. 29, 185, 1972. If this was, indeed, the case the lowest multipoles should have been maximally affected and not viceversa!

A 2.6

Astrophysical Letters, 1973, Vol. 13, pp. 109–113

On a Magnetized Rotating Sphere

REMO RUFFINI *Joseph Henry Physical Laboratories, Princeton, N.J. 08540, USA*

ALDO TREVES *Massachusetts Institute of Technology, Cambridge, Massachusetts 02138, USA*

The issue of the most stable electromagnetic structure associated with a rotating sphere with a magnetic axis, aligned to the rotation axis, is here discussed. With a classical model it is shown that the configuration that minimizes the total electromagnetic energy of the system is endowed with a non-zero net charge. Some implications for the physics of collapsed objects are stressed.

The physics of a rotating, magnetized charged sphere has long attracted the attention of physicists; see e.g. the lucid paper of Schiff (1939) and the paper by Blackett (1952) and references mentioned there. Today the interest in this problem has been enhanced by at least two important results. (a) From the experimental point of view, the discovery of pulsars has given concrete evidence for the existence of rotating, magnetized, collapsed objects. (b) From the theoretical point of view, the existing exact solutions of Einstein's equations have shown that the most general stationary solution, which is asymptotically flat with a regular horizon for a fully collapsed object, has to be rotating and endowed with a net charge. The so-called Kerr–Newman solution, with its properties of the horizon, angular velocity, and structure of the electro-magnetic field, has recently been examined by Christodoulou and Ruffini (1971).

With these two main results an entire set of puzzling facts is associated. Not the least of the unanswered questions is the fact that the gyromagnetic ratio associated with the Kerr–Newman solution is equal to that of the Dirac electron (see e.g. Carter (1968)), while the mass of a black hole could never be made comparable to a microscopic system.

The aim in this work is to make a simple and classical model to focus attention on the fact that purely on energetic grounds a magnetized rotating object should, indeed, be expected to possess a non-zero *net* charge. Our treatment is made in the limit of small rotation, $\omega a/c \ll 1$, where ω is the angular velocity, a is the radius of the sphere, and c is the speed of light; interaction with the surrounding medium and non-radiative solutions are neglected.

We add to the solution of Deutsch (1955) a new solution endowed with rotation and charge. The Maxwell equations being linear we obtain a new solution. We minimize the total electromagnetic energy of the system with respect to the net charge Q. The connection between the minimization of electromagnetic energy, presented here, and the generalization to the case of a generel relativistic treatment are under investigation and the results will be presented elsewhere. The electromagnetic field of a magnetized sphere with axis of rotation aligned to the magnetic axis is, inside the sphere,

$$B_r = +B_0 \cos\theta$$
$$B_\theta = -B_0 \sin\theta$$
$$B_\phi = 0, \tag{1a}$$
$$E_r = -(r\omega B_0/c)\sin^2\theta$$
$$E_\theta = -(r\omega B_0/c)\sin\theta\cos\theta$$
$$E_\phi = 0, \tag{1b}$$

and outside the sphere,

$$B_r = B_0(a^3/r^3)\cos\theta$$
$$B_\theta = (B_0/2)(a^3/r^3)\sin\theta$$
$$B_\phi = 0, \tag{2a}$$
$$E_r = -\tfrac{1}{4}(\omega a/c)B_0(a^4/r^4)(3\cos 2\theta + 1)$$
$$E_\theta = -\tfrac{1}{2}(\omega a/c)B_0(a^4/r^4)\sin 2\theta$$
$$E_\phi = 0, \tag{2b}$$

B_0 being the value of the magnetic field at the pole. The internal electric field has been obtained under conditions of large conductivity from the knowledge of the magnetic field. The charge density inside the sphere is

$$\rho = (\varepsilon/4\pi)\,\mathrm{div}\,E = -\varepsilon B_0\,\omega/2\pi c, \tag{3}$$

while the surface charge and current density are obtained immediately from the matching conditions to be

$$\sigma(\theta) = \omega a B_0(\sin^2\theta - \tfrac{1}{4}(3\cos 2\theta + 1)) \tag{4}$$

and

$$J_\phi(\theta) = (3c/8\pi)B_0 \sin\theta\,\delta(r-a)/\mu. \tag{5}$$

Let us consider, now, a charge uniformly distributed on the surface of the sphere rotating with angular velocity ω. We assume that the charge is in corrotation, and therefore generates a current in the inertial system with origin at the center of the sphere given by

$$J_r = 0, \quad J_\theta = 0, \quad J_\phi = Q\omega \sin\theta/4\pi a. \tag{6}$$

Associated with this current we have outside the sphere a dipole-like magnetic field

$$B_r^{\text{out}} = \frac{2\mu}{3}\left(\frac{\omega a}{c}\right) Qa\cos\theta/r^3;$$

$$B_\theta^{\text{out}} = \frac{\mu}{3}\left(\frac{\omega a}{c}\right) Qa\sin\theta/r^3;$$

$$B_\phi^{\text{out}} = 0. \tag{7}$$

For the solution inside the sphere we require the field to be independent of ϕ, stationary and symmetric with respect to the equatorial plane. The most general solution is therefore of the form

$$B_r = R_1(r)F_1(\cos\theta); \quad B_\theta = R_2(r)F_2(\cos\theta); \quad B_\phi = 0 \tag{8}$$

We can assume as a particular solution $F_1(\cos\theta) = \cos\theta$; the condition that the magnetic field is divergence-free immediately implies that

$$B_r = R_1(r)\cos\theta, \quad B_\theta = -\frac{1}{2r}\frac{\mathrm{d}}{\mathrm{d}r}(r^2 R_1)\sin\theta,$$

where $R_1(r)$ is a completely arbitrary function. We can therefore choose for $R_1(r) = \text{constant} = A$, obtaining the simple form

$$B_r^{\text{ins}} = A\cos\theta; \quad B_\theta^{\text{ins}} = -A\sin\theta; \quad B_\phi^{\text{ins}} = 0. \tag{9}$$

The condition that the field **B** is divergence-free also implies the continuity of the radial component of the field across the boundary of the sphere, which together with equation (9) gives

$$B_r^{\text{ins}} = \frac{2\mu}{3}\frac{\omega Q}{ca}\cos\theta;$$

$$B^{\text{ins}} = -\frac{2\mu}{3}\frac{\omega Q}{ca}\sin\theta;$$

$$B_\phi^{\text{ins}} = 0. \tag{10}$$

Assuming negligible ohmic dissipation inside the sphere, we have

$$E_r^{\text{ins}} = -\frac{2}{3}\mu\frac{\omega^2 Q}{c^2 a} r\sin^2\theta;$$

$$E_\theta^{\text{ins}} = -\frac{2}{3}\mu\frac{\omega^2 Q}{c^2 a} r\sin\theta\cos\theta;$$

$$E_\phi^{\text{ins}} = 0. \tag{11}$$

The solution outside the sphere is simply given, in our case of azimuthal symmetry ($m = 0$), by

$$E_r = \sum_l \frac{4\pi(l+1)}{(2l+1)} \mathscr{R}_{l0}\, y_{l0}(\theta,\phi)/r^{l+2},$$

$$E_\theta = \sum_l \frac{4\pi}{2l+1} \mathscr{R}_{l0} \frac{\partial}{\partial\theta}[y_{l0}(\theta,\phi)]/r^{l+2},$$

$$E_\phi = 0. \tag{12}$$

Imposing the matching conditions we have

$$E_r^{\text{out}} = \frac{Q}{r^2} - \frac{\mu}{3}\left(\frac{\omega a}{c}\right)^2 \frac{Qa^2}{r^4}(3\cos^2\theta - 1)$$

$$= \frac{Q}{r_2} - \frac{\mu}{6}\left(\frac{\omega a}{c}\right)^2 \frac{Qa^2}{r^4}(3\cos 2\theta + 1),$$

$$E_\theta^{\text{out}} = -\frac{2\mu}{3}\left(\frac{\omega a}{c}\right)^2 \frac{Qa^2}{r^4}\cos\theta\sin\theta,$$

$$E_\phi^{\text{out}} = 0. \tag{13}$$

From the discontinuities in some of the components of the **B** and **E** fields, we can now complete the currents and the charge distributions at the surface of the sphere. We have

$$\sigma = E_r^{\text{out}} - E_r^{\text{ins}} = \frac{Q}{r^2} + \frac{2\mu}{3}\frac{Q}{a^2} \times \left(\frac{\omega a}{c}\right)^2 \sin^2\theta - \frac{\mu}{3}\frac{Q}{a^2}\left(\frac{\omega a}{c}\right)^2 (3\cos^2\theta - 1) \tag{14}$$

and for the current on the surface

$$J_\phi = \frac{c}{4\pi}(H_\theta^{\text{out}} - H_\theta^{\text{ins}}) = Q\omega\sin\theta/4\pi a. \tag{15}$$

We notice that the density of charge on the surface differs from the uniform charge distribution in terms $(\omega a/c)^2$; the effects of these additional terms are negligible in our approximation. This correction term could be made arbitrarily small—arbitrary large power in $(\omega a/c)$—by assuming an initial charge distribution endowed with a sufficiently high number of multipoles. The current components J^ϕ do, indeed, agree with the one we had assumed in the beginning of our treatment, providing once again the self-consistency of our approach.

We can now proceed to write the total electromagnetic field of the charged sphere by adding the Deutsch solution to our new solution, the equations of Maxwell being linear. We conclude:

$$E_r^{\text{out}} = \frac{Q}{r^2} - \frac{\mu}{3}\left(\frac{\omega a}{c}\right)^2 \frac{a^2 Q}{r^4}$$

$$\times (3\cos^2\theta - 1) - \frac{1}{4}\frac{\omega a}{c} B_0 \frac{a^4}{r^4}(3\cos 2\theta + 1)$$

$$= \frac{Q}{r^2} - \frac{1}{4}(3\cos 2\theta + 1)\left(\frac{\omega a}{c}\right)\frac{a^4}{r^4} B_{\text{tot}},$$

$$E_\theta^{\text{out}} = -\frac{2}{3}\mu\left(\frac{\omega a}{c}\right)^2 \frac{Qa^2}{r^4}\cos\theta\sin\theta - \frac{1}{2}\frac{\omega a}{c}$$

$$\times B_0 \frac{a^4}{r^4}\sin 2\theta = -\left(\frac{\omega a}{c}\right)\frac{a^4}{r^4} B_{\text{tot}}\sin\theta\cos\theta,$$

$$E_\phi^{\text{out}} = 0; \tag{16}$$

$$E_r^{\text{ins}} = -r\sin^2\theta\left[\frac{2}{3}\mu\left(\frac{\omega a}{c}\right)^2 \frac{Q}{a^3} + \frac{\omega B_0}{c}\right]$$

$$= -r\sin^2\theta\frac{\omega}{c}B_{\text{tot}},$$

$$E_\theta^{\text{ins}} = -r\sin\theta\cos\theta\left[\frac{2}{3}\mu\left(\frac{\omega a}{c}\right)^2 \frac{Q}{a^3} + \frac{\omega B_0}{c}\right]$$

$$= -r\sin\theta\cos\theta\frac{\omega}{c}B_{\text{tot}},$$

$$E_\phi^{\text{ins}} = 0. \tag{17}$$

Here we have introduced the abbreviation

$$B_{\text{tot}} = B_0 + \frac{2}{3}\mu\left(\frac{a\omega}{c}\right)\frac{Q}{a^2}$$

and for the magnetic field

$$B_r^{\text{out}} = \left[B_0 + \frac{2}{3}\mu\left(\frac{a\omega}{c}\right)\frac{Q}{a^2}\right]\frac{a^3}{r^3}\cos\theta = B_{\text{tot}}\frac{a^3}{r^3}\cos\theta,$$

$$B_\theta^{\text{out}} = \frac{1}{2}\left[B_0 + \frac{2}{3}\mu\left(\frac{a\omega}{c}\right)\frac{Q}{a^2}\right]\frac{a^3}{r^3}\sin\theta = \frac{1}{2}B_{\text{tot}}\frac{a^3}{r^3}\sin\theta,$$

$$B_\phi^{\text{out}} = 0; \tag{18}$$

$$B_r^{\text{ins}} = \left[B_0 + \frac{2}{3}\mu\left(\frac{a\omega}{c}\right)\frac{Q}{a^2}\right]\cos\theta = B_{\text{tot}}\cos\theta,$$

$$B_\theta^{\text{ins}} = -\left[B_0 + \frac{2}{3}\mu\left(\frac{a\omega}{c}\right)\frac{Q}{a^2}\right]\sin\theta = -B_{\text{tot}}\sin\theta,$$

$$B_\phi^{\text{ins}} = 0. \tag{19}$$

We can now proceed to the minimization of the total electromagnetic energy of the system. We have

$$W_{\text{tot}} = \frac{1}{8\pi}\int \mathbf{D}\cdot\mathbf{E}\,dv + \frac{1}{8\pi}\int \mathbf{H}\cdot\mathbf{B}\,dV$$

$$= \frac{1}{8\pi}\int \varepsilon E^2\,dv + \frac{1}{8\pi}\int \frac{B^2}{\mu}\,dV. \tag{20}$$

This gives inside the sphere

$$W_{\text{tot}}^{\text{ins}} = \frac{a^3}{6}\left(\frac{1}{\mu} + \frac{2\varepsilon}{5}\left(\frac{a\omega}{c}\right)^2\right)B_{\text{tot}}^2,$$

and outside

$$W_{\text{tot}}^{\text{out}} = Q^2/2a + a^3 B_{\text{tot}}^2/12 + (\omega a/c)^2 a^3 B_{\text{tot}}^2/30 \tag{21}$$

We can extremize the total energy with respect to the charge Q; we obtain for a given configuration a non-zero value of Q given by

$$Q \simeq -[a^2\mu(1+2/\mu)a\omega B_0/(gc)]/[1 + 2\mu^2(1+2/\mu)$$

$$\times (a\omega/c)^2/27] \tag{22}$$

and for the *observable* magnetic field B_{tot} as a function of the 'bare' magnetic field,

$$B_{\text{tot}} = B_0/[1 + 2\mu^2(1+2/\mu)(a\omega/c)^2/27]. \tag{23}$$

From the two previous relations it also follows immediately that

$$Q = -\frac{a^3}{9}\mu(1+2/\mu)(\omega/c)B_{\text{tot}}. \tag{24}$$

Introducing the angular momentum $L = (2/5)Ma^2\omega$ (M is the mass of the sphere) and the magnetic moment of the total magnetic field

$$\mathcal{M} = \tfrac{1}{2}B_{\text{tot}}a^3,$$

we then obtain for equation (24) the form

$$Q = \frac{5\mu}{9c}\left(1+\frac{2}{\mu}\right)\frac{L}{Ma^2}\mathscr{M} \tag{25}$$

and clearly

$$a = \left(\frac{5\mu}{9c}\left(1+\frac{2}{\mu}\right)\frac{L}{M}\frac{\mathscr{M}}{Q}\right)^{1/2};$$

for the gyromagnetic ratio of the sphere we have

$$g = -\frac{45}{4}\frac{c}{\omega^2 a^2(\mu+2)}\frac{Q}{M}. \tag{26}$$

In Figure 1 the total electromagnetic energy is plotted as a function of the net charge in properly chosen units. It is of some interest to evaluate here some numerical examples. To this end we

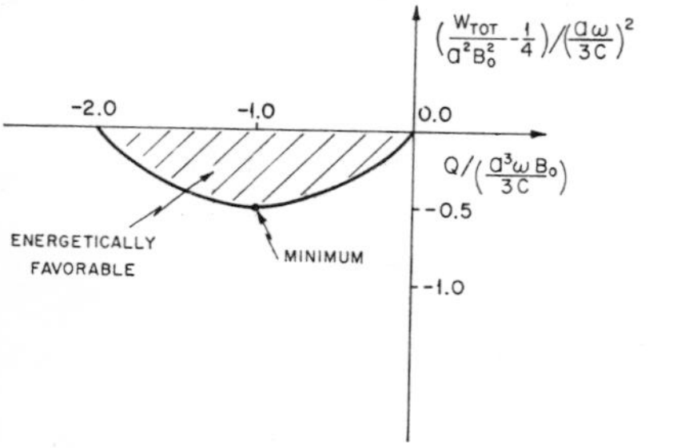

FIG. 1. The total energy minus the 'bare' energy of the magnetic field is given as a function of the net charge. The minimum energy $W_{\text{tot}} = a^2B_0^2[\frac{1}{4}-\frac{1}{2} \times (a\omega/3c)^2]$ is reached for a value $Q = -a^3\omega B_0/3c$ of the net charge.

assume a neutron star of radius $a = 10^6$ cm rotating at an angular velocity $\omega \sim 200$ rad/sec.

In Table 1 the main results are summarized. It becomes immediately clear that (a) the net charge is not extremely large for an object with the size and mass of a neutron star (even in the most extreme circumstances we have 1 electron/10^{25} nucleons!), (b) the net charge is, however, sizable enough to give very strong electric fields on the surface of a neutron star (up to 1.5×10^{16} Volts/m in the example under consideration!), (c) the total electromagnetic energy and initial magnetic field are slightly modified by the introduction of the electric charge.

In the case of the earth, $\omega \sim 7.3 \times 10^{-5}$ rad/sec, $a \sim 6.4 \times 10^8$ cm, $B_0 \sim 1$ Gauss; we then obtain $Q \simeq (\mu+2)\,1.4 \times 10^{20}$ electron charges and $E_{\text{surface}} \simeq (\mu+2) \times 5 \times 10^{-3}$ Volt/m.

Our computation *does not* take into account any interaction between the magnetized sphere and the surrounding medium and applies to the case of a 'bare' magnetized sphere in vacuo. Also clear is that processes of accretion should be taken into account in any realistic astrophysical application. This last problem, however, is much more complicated and quite outside the scope of the present work. We would like, however, to point out that in the realistic case the magnetized sphere could be surrounded by a magnetosphere with an *opposite* net charge, avoiding selective accretion of material from the outside space and preventing the discharge of the net charge of the surface. Moreover, the fact that this net surface charge indeed minimizes

TABLE 1

Value of net charge to be expected on the surface of a magnetized rotating sphere for selected values of the magnetic field.

B_0 (Gauss)	$B_{\text{ind}} = B_0 - B_{\text{tot}}$ (Gauss)	Q Electron charges	E (Volt/m)	W_{tot} (erg)	ΔW (erg)
10^7	10^2	-4.6×10^{25}	1.5×10^9	2.5×10^{31}	2.5×10^{26}
10^9	10^4	-4.6×10^{27}	1.5×10^{11}	2.5×10^{35}	2.5×10^{30}
10^{12}	10^7	-4.6×10^{30}	1.5×10^{14}	2.5×10^{41}	2.5×10^{36}
10^{14}	10^9	-4.6×10^{32}	1.5×10^{16}	2.5×10^{45}	2.5×10^{41}

B_0 is the 'bare' magnetic field of the neutron star, B_{tot} is the observed magnetic field, Q the net charge minimizing the energy, E the surface electric field of the star at an angle $\theta = \pi/4$. W_{tot} is the total electromagnetic energy, while $\Delta W = B_0{}^3 a^3/4 - W_{\text{tot}}$ is the correction to the total electromagnetic energy due to the charge Q. These numbers refer to the case of a neutron star of radius $a = 10^6$ cm, angular velocity, $\omega = 2 \times 10^2$ rad/sec, and with $\mu = 1$.

the total electromagnetic energy also implies that accretion of charged particles of opposite signs will be avoided by the existence of *self-consistent* electromagnetic fields.

The physical interpretation of our result can be summarized very simply: The introduction of a net charge has two different effects; (a) it *increases* the Coulomb (electrostatic energy) of the system, and (b) for a proper combination of the sign of the charge and direction of spin, a magnetic field is generated which *decreases* the total magnetic energy of the system. There exists a minimum of the energy under the two competing effects for a non-zero net charge.

The application of this physical idea to a relativistic system is under examination.

We thank Francis Everitt for helpful discussions. This work was partially supported by National Science Foundation Grant GP-30799X. A. T. is a ESRO NASA Postdoctoral Fellow; his permanent address is the Instituto di Fisica, Universita di Milano, Italy.

REFERENCES

Blackett, P. M. S., 1952, *Phil Trans. Roy. Soc. London*, **245**, 309.

Carter, B., 1968, *Phys. Rev.*, **174**, 1559.

Christodoulou, D., and Ruffini, R., 1971, Internat. Conf. on General Relativity, Copenhagen, July 1971.

Deutsch, A. J., 1955, *Ann. Astrophys.*, **1**, 1.

Schiff, L. I., 1939, *Proc. Nat. Acad. Sci.*, **25**, 391.

Received in original form 6 July 1972
Received in final form 2 October 1972

PART III Gravitational Radiation in Schwarzschild Geometries

Introduction

The aim of this part is mainly to present some recent results of a completely relativistic analysis of the gravitational radiation emitted by particles being captured or spiralling around Schwarzschild black holes. A more general treatment of sources of gravitational radiation of astrophysical interest can be found in Ref. (1). There a detailed analysis of gravitational radiation detectors (cross section, directional properties, etc.) as well as a critical analysis of the energetics involved in gravitational radiation experiments is presented. Also in Ref. (1) are studied some preliminary spectral analyses of bursts of gravitational radiation by a direct generalization of results and techniques usually adopted in electrodynamics. These results are generalized to a complete relativistic treatment in the present part.

The theoretical background for developing a completely relativistic approach to small perturbations in highly relativistic regimes near Schwarzschild black hole was given as far as 1957 by Regge and Wheeler.[2] This basic work was then largely developed in a set of papers by Zerilli[3] who has been able to reduce the spectral analysis of any tensorial perturbation in a Schwarzschild background to only two independent equations of the form:

$$\frac{d^2 f^l(\omega, r^*)}{dr^{*2}} + (\omega^2 - V_l^{\text{eff}}(r^*))f^l(\omega, r^*) = S^l(\omega, r^*)$$

The details of this treatment for a particle falling radially into a Schwarzschild black hole are given by Zerilli in Ref. (3). The analysis for circular orbits is here presented in Appendix 3.5. The perturbation analysis for a tensor field follows a pattern very similar to the one we have presented in the previous chapter for vectorial perturbations. First by using the symmetry properties of the background geometry the angular part of the tensorial perturbations are developed in a complete set of eigenfunctions obtainable, as in the electromagnetic case, from the spherical harmonics $Y^{lm}(\theta, \phi)$ by repeated application of the $\vec{L}$, $\vec{\nabla}$, $\vec{r}/r$ operators. The two sets of perturbations are given by:

$$h_{\mu r} = \begin{vmatrix} 0 & 0 & -h_0(T,r)(\partial/\sin\theta\,\partial\phi)Y_L^M & h_0(T,r)(\sin\theta\,\partial/\partial\theta)Y_L^M \\ 0 & 0 & -h_1(T,r)(\partial/\sin\theta\,\partial\phi)Y_L^M & h_1(T,r)(\sin\theta\,\partial/\partial\theta)Y_L^M \\ \mathrm{Sym} & \mathrm{Sym} & h_2(T,r)(\partial^2/\sin\theta\,\partial\theta\,\partial\phi - \cos\theta\,\partial/\sin^2\theta\,\partial\phi)Y_L^M & \mathrm{Sym} \\ \mathrm{Sym} & \mathrm{Sym} & \frac{1}{2}h_2(T,r)(\partial^2/\sin\theta\,\partial\phi\,\partial\phi + \cos\theta\,\partial/\partial\theta - \sin\theta\,\partial^2/\partial\theta\,\partial\theta)Y_L^M & -h_2(T,r)(\sin\theta\,\partial^2/\partial\theta\,\partial\phi - \cos\theta\,\partial/\partial\phi)Y_L^M \end{vmatrix} \tag{1}$$

for parity $(1)^{l+1}$ and by

$$h_{\mu r} = \begin{vmatrix} (1-2m^*/r)H_0(T,r)Y_L^M & H_1(T,r)Y_L^M & h_0(T,r)(\partial/\partial\theta)Y_L^M & h_0(T,r)(\partial/\partial\phi)Y_L^M \\ H_1(T,r)Y_L^M & (1-2m^*/r)^{-1}H_2(T,r)Y_L^M & h_1(T,r)(\partial/\partial\theta)Y_L^M & h_1(T,r)(\partial/\partial\phi)Y_L^M \\ \mathrm{Sym} & \mathrm{Sym} & r^2[K(T,r) + G(T,r)(\partial^2/\partial\theta^2)]Y_L^M & \mathrm{Sym} \\ \mathrm{Sym} & \mathrm{Sym} & r^2G(T,r)(\partial^2/\partial\theta\,\partial\phi - \cos\theta\,\partial/\sin\theta\,\partial\phi)Y_L^M & r^2[K(T,r)\sin^2\theta + G(T,r)(\partial^2/\partial\phi\,\partial\phi + \sin\theta\cos\theta\,\partial/\partial\theta)]Y_L^M \end{vmatrix} \tag{2}$$

for parity $(-1)^l$

The equations for the perturbations (1) and (2) are obtained by solving the Einstein equations for the metric

$$g_{\mu\nu} = (g_{\mu\nu})_{\text{schwarzschild}} + h_{\mu\nu} \tag{3}$$

with a source term of the form

$$T^{\mu\nu} = m_0 \int_{-\infty}^{+\infty} \delta^{(4)}(x - z(\tau)) \frac{dz^\mu}{d\tau} \frac{dz^\nu}{d\tau} \, d\tau. \tag{4}$$

Here m_0 is the mass, $z^\mu(\tau) = (T(\tau), R(\tau), \Theta(\tau), \Phi(\tau))$ the coordinates and τ the proper time along the world line of a test particle in the given background metric.

The equation governing the radial dependence of the Fourier transformed perturbations of parity $(-1)^l$ is of the form:

$$\frac{d^2 a^{lm}(\omega, r^*)}{dr^{*2}} + (\omega^2 - V_l^{(m)}(r^*))a^{lm}(\omega, r^*) = S^{lm}_{(m)}(r^*) \tag{5}$$

with

$$V_l^{(m)} = (1 - 2M/r)(l(l+1)/r^2 - 6M/r^3) \tag{6}$$

$S^{lm}_{(m)}(r^*)$ is the source term which depends upon the detail of the motion of the test particle and, as usual

$$\frac{dr}{dr^*} = g^{rr}. \tag{7}$$

Equation (5) is the analogous of Eq. (38.1) in Part 2 for electromagnetic perturbations.

For perturbations of parity $(-1)^{l+1}$ we have

$$\frac{d^2 b^{lm}(\omega, r^*)}{dr^{*2}} + (\omega^2 - V_l^{(e)}(r^*))\, b^{lm}\,(\omega, r^*) = S^{lm}_{(e)}(r^*) \tag{8}$$

with (see Figure 1)

$$V_l^{(e)}(r^*) = (1 - 2M/r)(2\lambda^2(\lambda + 1)r^3 + 6\lambda^2 M r^2 + 18M^2\lambda r + 18M^3)/r^3(\lambda r + 3M)^2 \tag{9}$$

and $\lambda = (l - 1)(l + 2)/2$ and $S^{lm}_{(e)}$ the source term determined by the test particle motion. This equation is analogous to Eq. (38.2) in Part 2.

The gravitational radiation far away from the source is computed by a direct evaluation of the energy-momentum pseudo-tensor and is simply given by

$$\frac{dE}{d\omega} = \sum_{lm} |a_{lm}|^2 + |b_{lm}|^2 \tag{10}$$

more details can be found in Ref. (3) or in the Appendices 3.1–3.7.

In the problem of a particle falling radially into a Schwarzschild black hole the entire set of perturbations is totally described only by perturbations of $(-1)^{l+1}$ parity and fulfilling Eq. (8), much like in the electromagnetic case presented in part II. The detailed analysis of this problem is presented in the Appendices 3.2, 3.3, 3.4 and 3.5. In Appendix 3.2 the detailed features and tunnelling properties of the effective potential barrier are analyzed by direct integration of the equations for selected values of the frequency. The essential role of the source term in guaranteeing the possibility of both purely outgoing

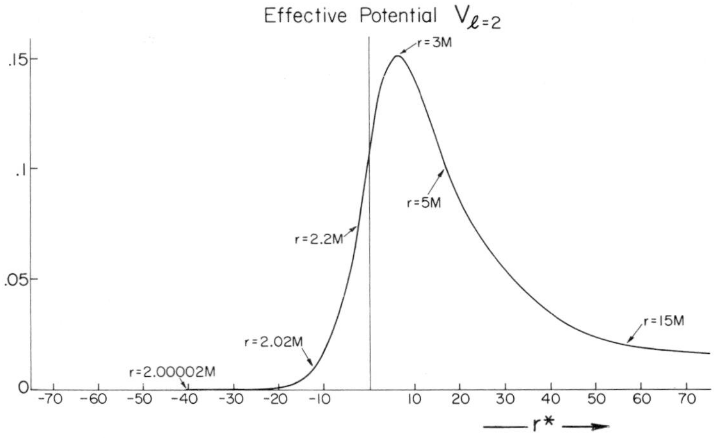

Figure 1. Effective potential for perturbations of parity $(-1)^{l+1}$ as a function of the coordinate r^* for $l = 2$. For a direct comparison we also indicate the corresponding value of the r coordinate.

waves at infinity and purely ingoing waves at the black hole surface is there analyzed in terms of a double eigenvalue search, both in the amplitude and in the phase of the waves.

In Appendix 3.3 the problem is approached from a totally different point of view. As in the electromagnetic case the Green's functions technique is here used. The effective potential (Figure 1) is extremely similar to the one shown in Figure 10 of Part II. Here, also, for each selected value of ω we solve for the two linearly independent solutions $u^l(\omega, r^*)$ (purely outgoing at infinity) and $v^l(\omega, r^*)$ (purely ingoing at the black hole surface). We then build up the Green's functions and by usual techniques we find the solution of equation (8) by integrating the Green function over the source term. The results of this treatment are summarized in Appendix 3.3:

(1) Approximately four times more energy is emitted in this complete relativistic treatment as compared with the treatment presented in Appendix 3.1.
(2) The qualitative features of the spectrum presented in Appendix 3.1 are

indeed confirmed (spectrum goes to zero at low frequencies and peaks at $\nu \propto 1/M$).

(3) Also confirmed are the major features of the energetics: a particle falling into a black hole emits an amount of energy $\Delta E \propto m^2/M$. The larger the black hole the smaller is the amount of energy radiated!

(4) Only quadruple radiation is matter; contributions due to higher multiples are totally negligible.

In Appendix 3.4, this work is taken one step forward. The major aim is to give the *shape* of the burst of gravitational radiation by Fourier transforming the results obtained in Appendix 3.3. The problem is solved and both the shape of the burst of gravitational energy (see Figure 2) and the tide-producing components of the Riemann tensor are obtained. This is perhaps the most important result of this entire analysis. Since the Fourier transform is made in retarded time we can also examine when the different parts of the burst have been emitted. Indeed, we can then differentiate between three different regimes in the pulse:

(1) A precursor, emitted when the particle is still far away from the black hole ($r > 5M$); the energy emitted in this range is negligible.

(2) A sharp burst lasting a few M units emitted when the particle goes from $r \sim 5M$ up to $r \sim 2M$, the majority of the radiation is being emitted in this stage.

(3) A "ringing tail" due to the capture of the particle by the black hole.

The angular distribution of the radiation is also reconstructed: the beaming differs from the one of pure quadruple radiation. The radiation is focused forward, but only of $7\frac{1}{2}$ degrees. Finally the detailed features of the polarization of the radiation are analyzed.

How much radiation goes inside the black hole? This problem is analyzed in Appendix 3.4. Markedly different from the case of radiation propagating outward, in the case of the ingoing radiation, the higher multipoles contribution is not negligible any longer: all the multipoles contribute approximately with the same strength. Moreover, the total radiation going inside the black hole is now *not* any longer dependent upon the ratio m/M and is enormously larger than the radiation going outward. We have, in fact, $\Delta E \lesssim \pi m/8$. The inequality is due to a "natural" cutoff introduced in Appendix 3.4.

In at least two respects the results presented in the previous appendices strongly limit the efficiency of emission of gravitational radiation:

(a) the frequency is always of the order $\nu \sim 1/M$

(b) The amount of energy is extremely low and always reduced by a factor m/M.

If a very large black hole ($\sim 10^8 M_\odot$) should exist then from the point of view of these results its emission of gravitational energy would be largely uninteresting.

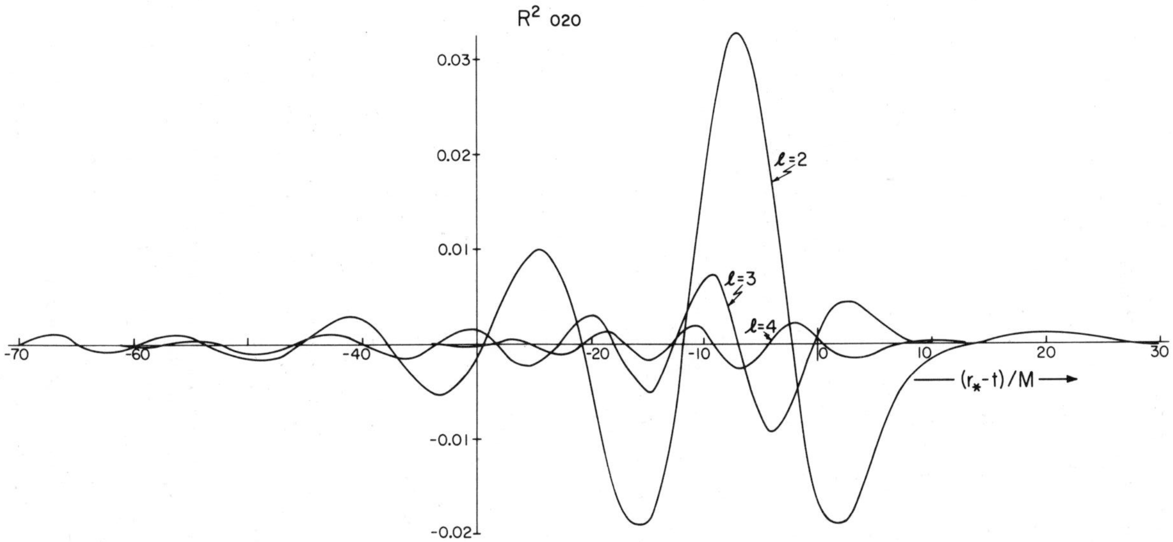

Figure 2. Shape of a burst of gravitational radiation plotted as a function of the retarded time $(r^* - t)/M$. Here the tide producing components R^2_{020} (the significant components for gravitational radiation detectors, see Ref. (1)) are plotted for selected values of l. The three regions of the burst are clearly shown. The first for $(r^* - t)/M \gtrsim 5$ is the precursor, the second (sharp burst) for $-20 \lesssim (r^* - t)/M \lesssim 5$ and the third for $(r^* - t)/M \lesssim -20$ is the "ringing tail".

In Appendix 3.5 an alternative is presented. Though we do not believe the example there presented is of direct astrophysical application, it is interesting from a point of view of principle since it shows a way of overcoming both difficulties (a) and (b).

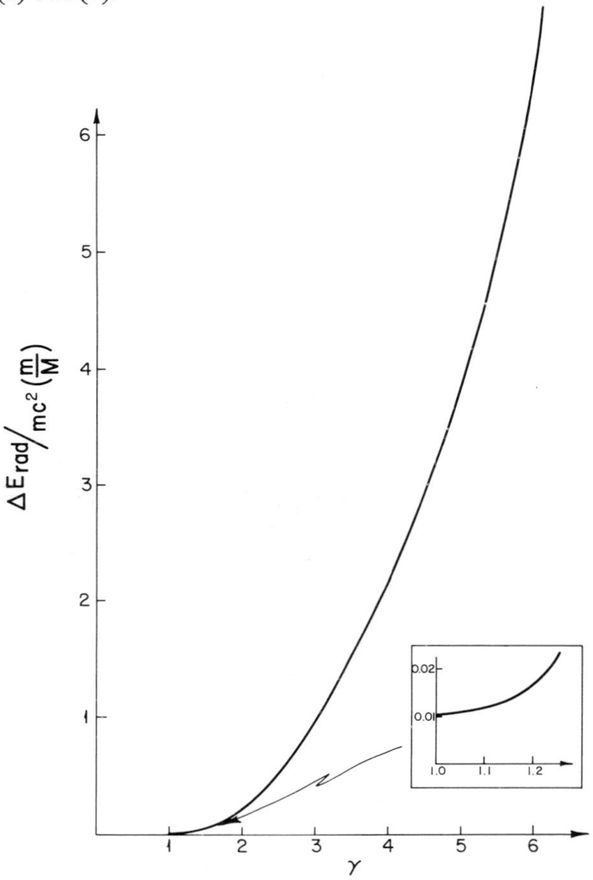

Figure 3. Total energy radiated as a function of the γ of the incoming particle. $\gamma = 1$ corresponds to the case in which a particle implodes starting at rest at infinite distance, $\gamma > 1$ to one in which the particle is initially endowed with a finite value of the kinetic energy. At first glance it would seem that the energy radiated goes to zero at $\gamma = 1$!

The message in Appendix 3.5 is very simply stated: do not let the particle drop starting at rest at infinity, but project the particle with an initial finite value of the kinetic energy. The radiation outward then increases tremendously (see Figure 3) and the spectrum is shifted toward higher frequencies.

From particles falling radially into black holes let us focus our attention now on particles moving in circular orbits. In the last chapter we have seen how the presence of strong gravitational fields could have defocusing effects on the electromagnetic radiation field beamed by pure special relativistic effects. The situation is made even more extreme in the gravitational case. Better than any further word, Figure 4 clearly exemplifies the situation in the case of three different fields (scalar, vector, tensor). In the formal solution of the problem presented in Appendix 3.6, exactly as in the electromagnetic case, both electric (parity $(-1)^l$) and magnetic (parity $(-1)^{l+1}$) terms give a contribution. Again the magnetic terms give a negligible contribution and they can be systematically neglected. A numerical evaluation of the Green's functions has been done for the analysis of circular orbits of arbitrary radius. The main formulae necessary

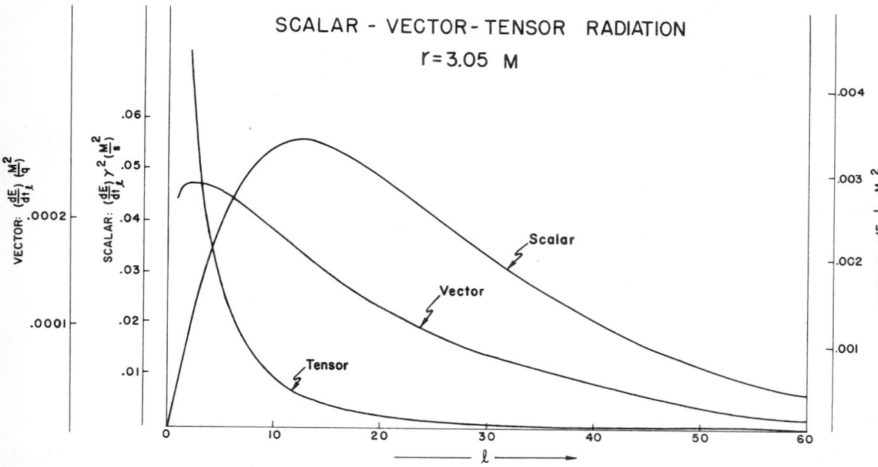

Figure 4. Compared and contrasted are the spectral distribution of scalar, vector and tensor radiation emitted by a particle in an ultrarelativistic circular orbit at $r = 3.05M$. The contribution of the highest multipoles already weakened in the case of vector radiation is further reduced in the case of tensor radiation. In this last case the quadrupole radiation is always predominant with respect to any other multipole. Also compare and contrast these results with Figure 1 of Part II.

for the solution of the problem are summarized in Table I of Appendix 3.6. Once again the evaluation of the radiation emitted at infinity reduces to the evaluation of the function $u^l(\omega, r^*)$ at large r^* values and the function $v^l(\omega, r^*)$ at the radius r_0^* corresponding to the coordinate of the circular orbit of the particle. The problem is therefore solved in complete analogy to the treatment presented in section c of § 2.5.

For orbits very near $3M$ we can again find explicit analytic formulae for the Green's functions. Their form in the present case is identical to the one given by Eqs (72.1) and (72.2) or Eqs (68.1) and (68.2) in part II. Once again we can give analytic expressions for the power radiated from ultrarelativistic (unstable) orbits. We have[4] for the terms of parity $(-1)^l$

$$P_{\text{even}}(m\omega_0) = \mu^2 \frac{\exp\left(-\frac{\pi}{4}\epsilon_{\text{even}}\right)}{54\pi^{3/2} m\delta} \left| \left(1 + \frac{m\delta}{2}\right) \Gamma\left(\frac{1}{4} + \frac{i}{4}\epsilon_{\text{even}}\right) + \frac{\sqrt{2}(1-i)}{\sqrt{3m}} \Gamma\left(\frac{3}{4} + \frac{i}{4}\epsilon_{\text{even}}\right) \right|^2 \tag{11}$$

and

$$P_{\text{odd}}(m\omega_0) = \mu^2 \frac{\exp\left(-\frac{\pi}{4}\epsilon_{\text{odd}}\right)}{54\pi^{3/2} m\delta} \left| (1+i)\,\Gamma\left(\frac{3}{4} + \frac{i}{4}\epsilon_{\text{odd}}\right) + \frac{i}{\sqrt{6m}} \Gamma\left(\frac{1}{4} + \frac{i}{4}\epsilon_{\text{odd}}\right) \right|^2 \tag{12}$$

for the terms of parity $(-1)^{l+1}$

with

$$\epsilon_{\text{even}} = 1 + 4p + |m|\delta, \qquad 2p = l - |m|$$

and

$$\epsilon_{\text{odd}} = 3 + 4p + |m|\delta, \qquad 2p = l - |m| - 1.$$

Some of the results of this numerical analysis are given in Figure 6 and Figure 5.

The main conclusion can be summarized as follows: the focusing and beaming effects are in the gravitational case even less effective than in the electromagnetic case. Moreover, once we turn from the ultrarelativistic unstable circular orbits to the stable circular orbits (the only ones physically meaningful) the beaming effects totally disappear. The reason is simply stated: the orbital velocity of the particle in a circular orbit is always much smaller than the local value of the speed of light; the necessary condition to have beaming $v_{\text{orb}}/v_{\text{light}} \sim 1$ is not fulfilled.

In all these processes of emission of gravitational radiation we have noticed that the spectral distribution of the radiation drastically differs for different sources. The aim is now to capitalize on these differences as a tool with which to identify with experiments possible sources of gravitational radiation. This brings us automatically to the subject treated in Appendix 3.7. There the polarization of electromagnetic and gravitational radiation from particles in circular orbits is analyzed. In both cases the radiation is linearly polarized in the plane of the orbit, circularly polarized at the pole and elliptically polarized in between. The main issue, however, in Appendix 3.7 is *not* to study the polarization of the radiation "in se" but to analyze how a detector can indeed be sensitive to different polarizations and use this new data to discriminate between different sources of gravitational radiation (one more data with which to characterize the

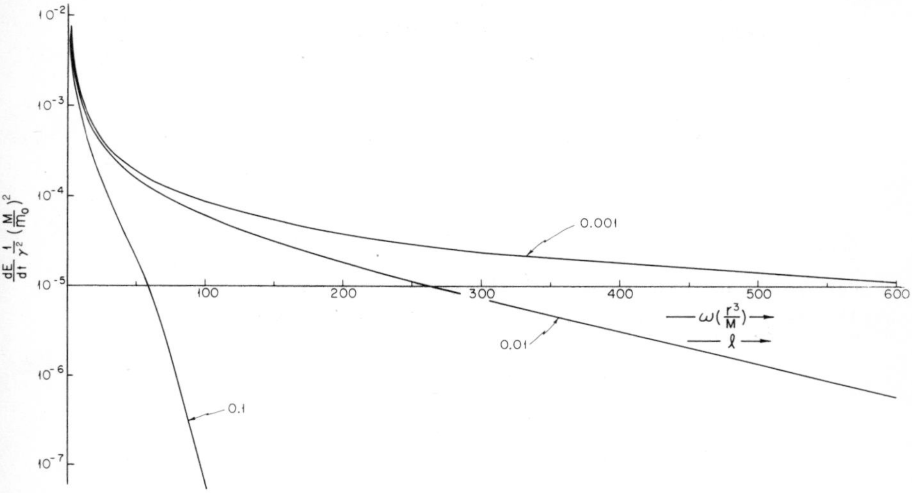

Figure 5. Spectrum of gravitational radiation emitted by a particle in an ultrarelativistic circular orbit at $r = (3 + \delta)M$ for selected values of the parameter $\delta = 0.1, 0.01, 0.001$. Clearly the contribution of the quadrupole term is always larger than the contribution of any other multipole.

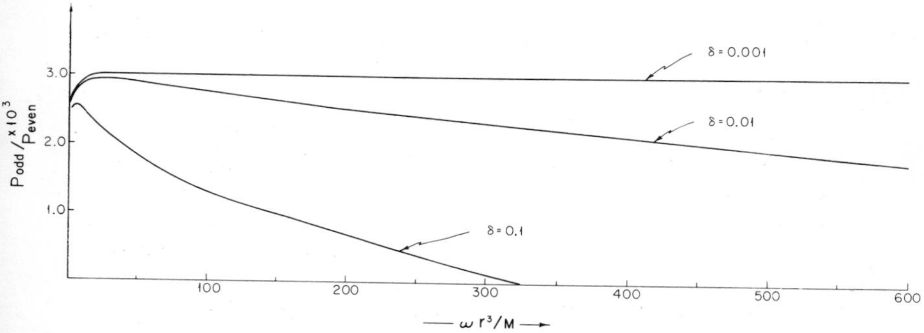

Figure 6. For the spectra given in Figure 6 we here present the ratio between electric terms (parity $(-1)^l P_{\text{even}}$) and magnetic terms (parity $(-1)^{l+1} P_{\text{odd}}$). Magnetic terms give always negligible contributions quite independently of the value of δ.

"signature" of the radiation). The results are self explanatory by a direct look at Figure 2 of Appendix 3.7. The response of a detector to randomly polarized radiation and linearly polarized radiation not only differs markedly in intensity but also drastically differs in its overall features!

We conclude this chapter with the consciousness that detectors already in their operative stage could indeed determine very important features in the gravitational radiation field and possibly discriminate between different sources.

References

1. Rees, M., Ruffini, R. and Wheeler, J. A. (1973) *Black Holes, Gravitational waves and cosmology*, Gordon and Breach. In press.
2. Regge, T. and Wheeler, J. A. (1957) *Phys. Rev.* **108,** 1063.
3. Zerilli, F. (1970) *J. Math. Phys.* **11,** 2213.
   Zerilli, F. (1970) *Phys. Rev. Letters*–**24,** 737.
   Zerilli, F. (1970) *Phys. Rev.* **2,** 2141.
4. Formulae 11 and 12 were obtained independently by Breuer, R. A., Ruffini, R., Tiomno, J. and Vishveshvara, C. V. The expressions obtained were identical apart from numerical mistakes.

A 3.1

REMO RUFFINI(*) and JOHN ARCHIBALD WHEELER(**)

GRAVITATIONAL RADIATION(***)

Radiation Implied by the Principle of Causality.

An accelerated charge must radiate. The principle of causality leaves no escape. A review of the reasoning on this point will help to indicate why gravitational radiation is equally unavoidable when the mass quadrupole moment of a system suddenly changes.

The $1/r^2$ field of a point charge carries not the slightest hint of any ability to transport energy with the speed of light. However, consider a particle accelerated as in Figure 1. The particle moves first with constant velocity β (negative) to the left. In a short interval of time $\Delta\tau$ the particle reverses its velocity and moves with constant velocity β (positive) to the right. Draw a sphere centered on the turning point with radius $r = \tau$. Inside that sphere the lines of force have the normal Coulomb pattern of a charge moving with constant velocity β to the right. Outside that sphere the lines of force have also the normal Coulomb pattern but they are now centered on the place where the particle would have been if it had continued moving to the left in a straight line with uniform velocity. The line of force have to make their way from the one field pattern to the other in the short interval $\Delta\tau$ in which the velocity changes its direction. The change from one pattern to the other is confined to a shell of thickness $\delta r = \Delta\tau$. In this shell the lines of force that would otherwise have a length $\Delta\tau$ are stretched out to a length $r\Delta\beta_\perp$. Here $\Delta\beta_\perp$ is the component of the change in velocity of the particle perpendicular to the line of sight. This "stretching length" is the greater the more distant the observer. Thus the normal field e/r^2 is augmented by the "stretching factor" $r\Delta\beta_\perp/\Delta\tau$ to the value

$$-(e/r^2)\,(r\Delta\beta_\perp/\Delta\tau) = -ea_\perp/r \qquad (1)$$

Here $a_\perp$ is the acceleration perpendicular to the line of sight. One sees in (1) all the characteristic features of electromagnetic radiation: (1) the (1/r)-

(*) Institute for Advanced Study, Princeton, New Jersey.
(**) Joseph Henry Laboratories, Princeton University, Princeton, New Jersey,
(***) Work supported in part by NSF Grant GP-436, and in part by U. S. Air Force Grant AF49-638-1545, to Princeton University.

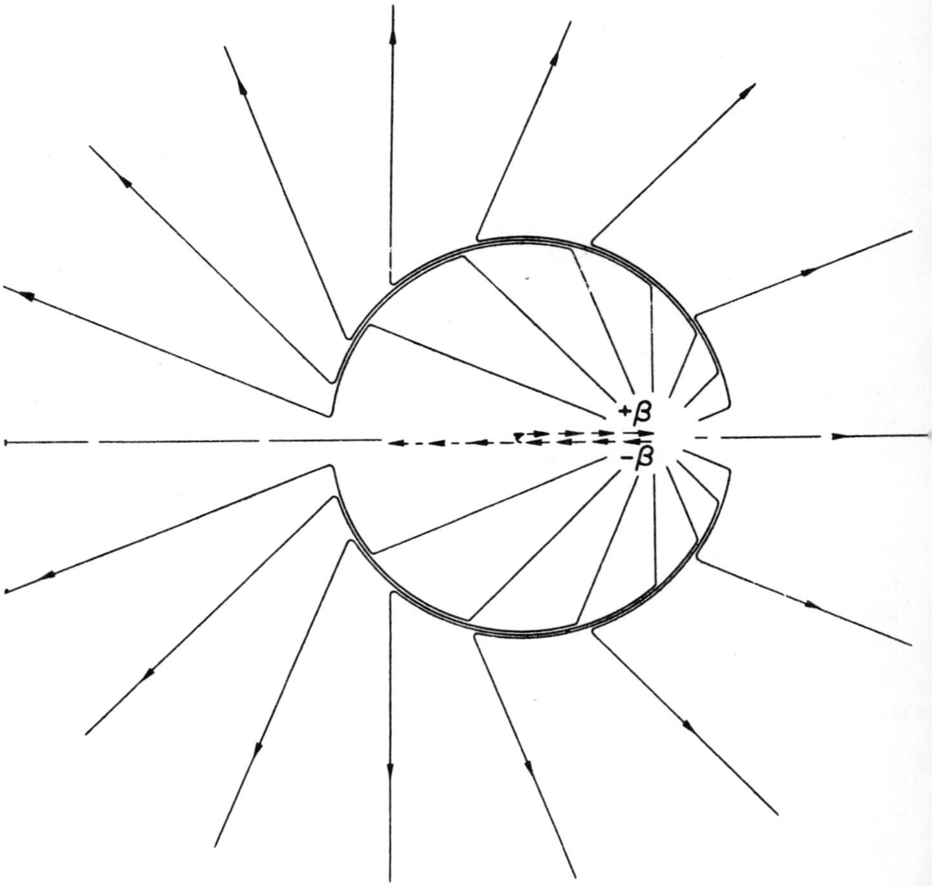

Figure 1 – J.J. Thomson's way to understand why the strength of an electromagnetic wave falls only as the inverse first power of distance r and why the amplitude of the wave varies (in the case of low velocities) as $\sin\theta$ (maximum in the plane perpendicular to the line of acceleration). The charge was moving to the left at uniform velocity. Far away from it the lines of force continue to move as if this uniform velocity were going to continue forever (Coulomb field of point charge in slow motion). However, closer up the field is that of a point charge moving the right with uniform velocity ($1/r^2$ dependence of strength upon distance). The change from the one field pattern to another is confined to a shell of thickness $\Delta\tau$ located at a distance r from the point of acceleration (amplification of field by "stretching factor" r $\sin\theta$ $\Delta\beta_{\perp}/\Delta\tau$;(see text)). We thank C. Teitelboim for the construction of this diagram.

dependence of amplitude upon distance (2) the dependence upon the charge and acceleration of the source (3) the dependence upon angle of observation and (4) the polarization of the radiation.

When we have several charges rather than one single charge the radiation emitted is given by the charge of each particle multiplied by its own acceleration.

When we turn to gravitational radiation and consider the distant field of two masses revolving around each other under the influence of their mutual gravitational attraction alone the situation is very different. The principle of action and reaction tells us that mass dipole moment of the system relative to the center of gravity is identically zero:

$$m_1 a_1 + m_2 a_2 = 0 \tag{2}$$

The radiative effects of the two masses compensate in this approximation. Thus we have no dipole gravitational radiation. However, if we allow for the fact that there is a time delay, that the detector receives a force with a different time delay from one mass than from the other, then no longer do the two effects compensate exactly There is a phase difference $\Delta\varphi$ between the two which is governed by the angular velocity of revolution, ω, and the separation L, of the two objects:

$$\Delta\varphi \sim \omega L$$

The gravitational field of the pair of masses at a distance is of the order

$$\sim \frac{Gma}{c^2 r} \Delta\varphi \sim \frac{Gm\omega^2 L}{c^2 r} \Delta\varphi \sim \frac{Gm\omega^3 L^2}{c^2 r}$$

$$\sim \frac{G}{c^2 r} \cdot \begin{Bmatrix} \text{third time rate of change of mass} \\ \text{quadrupole moment} \end{Bmatrix} \tag{3}$$

The corresponding rate of radiation of energy is

$$-\frac{dE}{d\tau} \sim \frac{G}{c^5} \cdot \begin{Bmatrix} \text{third time rate of change of mass} \\ \text{quadrupole moment} \end{Bmatrix} \tag{4}$$

A more detailed argument revises the conclusions about polarization and angular distribution of the radiation given by this simple argument (actual amplitude $\sim \sin^2\theta$, intensity $\sim \sin^4\theta$ as compared to amplitude $\sim \sin\theta$, intensity $\sim \sin^2\theta$ for electromagnetic radiation) but gives the same order of magnitude for the total emission.

The factor governing the emission is the third time rate of change $\dddot{Q}^{mn}$ of the quadrupole moments

$$\ddot{Q}^{mn} = \int \rho \,(3x^m x^n - \delta^{mn} r^2)\, d\,(\text{volume}) \tag{5}$$

The total rate of emission is

$$-\frac{dE}{d\tau} = \frac{1}{45}\dddot{Q}^{mn}\dddot{Q}^{mn} \tag{6}$$

Here we have converted from cgs units to geometrical units (mass, time, energy in cm; mass of sun $M_\odot = 1.47 \times 10^5$ cm; frequency in cm^{-1}; power in cm of mass-energy per cm of light travel time; conversion to conventional units via the factor $c^5/G = 3.6 \times 10^{59}$ erg/sec).

In the special case of two identical masses revolving about their common center of gravity under the influence of their mutual attraction the standard analysis gives for an "effective emission time" the formula

$$\frac{1}{\tau} = \frac{-dr/dt}{r} = \frac{2}{3}\frac{d\omega/dt}{\omega} = \frac{(-dE/dt)}{-E}$$

or

$$\tau = \frac{5}{128}\frac{r^4}{m^3} \tag{7}$$

For $m = 1\ M_\odot = 1.47$ km and a separation $r = 1000$ km we have an emitted power of 3.2×10^{46} ergs/sec and a time constant τ of 11.4 hr as contrasted to a period of 0.39 sec. For $m = 1\ M_\odot$ and a separation of 10,000 km we have an emitted power of 3.2×10^{41} ergs/sec, $\tau = 13$ year and a period of 12.2 sec.

Spectrum of the Radiation.

In the case of electromagnetism the radiation is distributed in time according to the second derivative of the dipole moment,

$$-\Delta E = \frac{2}{3}\int \ddot{p}^{\,k}(t)\, \ddot{p}^{\,k}(t)\, dt$$

and to the third derivative of the quadrupole moment in the case of gravitation

$$-\Delta E = \frac{1}{45}\int \dddot{Q}^{pq}(t)\, \dddot{Q}^{pq}(t)\, dt \tag{8}$$

(geometric units!). Go from a description in time to a description in frequency via Fourier analysis:

$$\ddot{p}^{\,k}(\omega) = \frac{1}{\sqrt{2\pi}} \int e^{i\omega t} \ddot{p}^{\,k}(t)\, dt$$

$$\dddot{Q}^{\,pq}(\omega) = \frac{1}{\sqrt{2\pi}} \int e^{i\omega t} \dddot{Q}^{\,pq}(t)\, dt \tag{9}$$

When a particle of charge e and mass m flies past a heavy center of attraction and experiences a sudden deflection, it gives off a spectrum of electromagnetic radiation, the low frequency part of which has nothing to do with the details of the deflection mechanism. Thus the loss of energy in the collision is

$$-\Delta E = \frac{4}{3} \int_0^\infty \ddot{p}^{\,k}(\omega)\, \ddot{p}^{\,k*}(\omega)\, d\omega\ , \tag{10}$$

with a low frequency component governed by

$$\underset{\omega \to 0}{\ddot{p}^{\,k}(\omega)} = \frac{1}{\sqrt{2\pi}} \int \ddot{p}^{\,k}(t)\, dt = \frac{\Delta \dot{p}^k}{\sqrt{2\pi}}$$

or

$$-\frac{d\Delta E}{d\nu} = -2\pi \frac{d\Delta E}{d\omega} = \frac{4}{3} (\Delta\, \dot{p})^2 \ \text{(low frequencies)} \tag{11}$$

When the particle passes with a non-relativistic speed β and impact parameter b near a center of attraction of charge Q and mass M it picks up a transverse impulse $2eQ/b\beta$. In this event one has

$$\lim_{\omega \to 0} -\frac{d\Delta E}{d\nu} = \frac{16}{3} \frac{Q^2 e^4}{m^2 c^5 b^2 \beta^2} \tag{12}$$

The distribution in frequency of the emitted radiation (Figure 2) is flat up to a limiting cutoff value of the order

$$\omega_{\text{cutoff}} = 2\pi\nu_{\text{cutoff}} \sim \frac{\left\{\begin{matrix}\text{impact parameter b of} \\ \text{the incoming particle}\end{matrix}\right\}}{(\text{velocity } \beta \text{ of particle})} \tag{13}$$

These well known results for "splash" electromagnetic radiation[1] allow themselves to be extended to splash gravitational radiation (Figure 2, from Ruffini and Wheeler[2]). Here the quantity that governs the intensity of the low frequency part of the spectrum is the change in the second time derivative of the quadrupole moment; thus,

$$\lim_{\omega \to 0} -\frac{d\Delta E}{d\nu} = \frac{1}{45} \Delta\ddot{Q}^{pq}\, \Delta\ddot{Q}^{pq} = \frac{64}{5} \frac{m^2 M^2}{b^2} \tag{14}$$

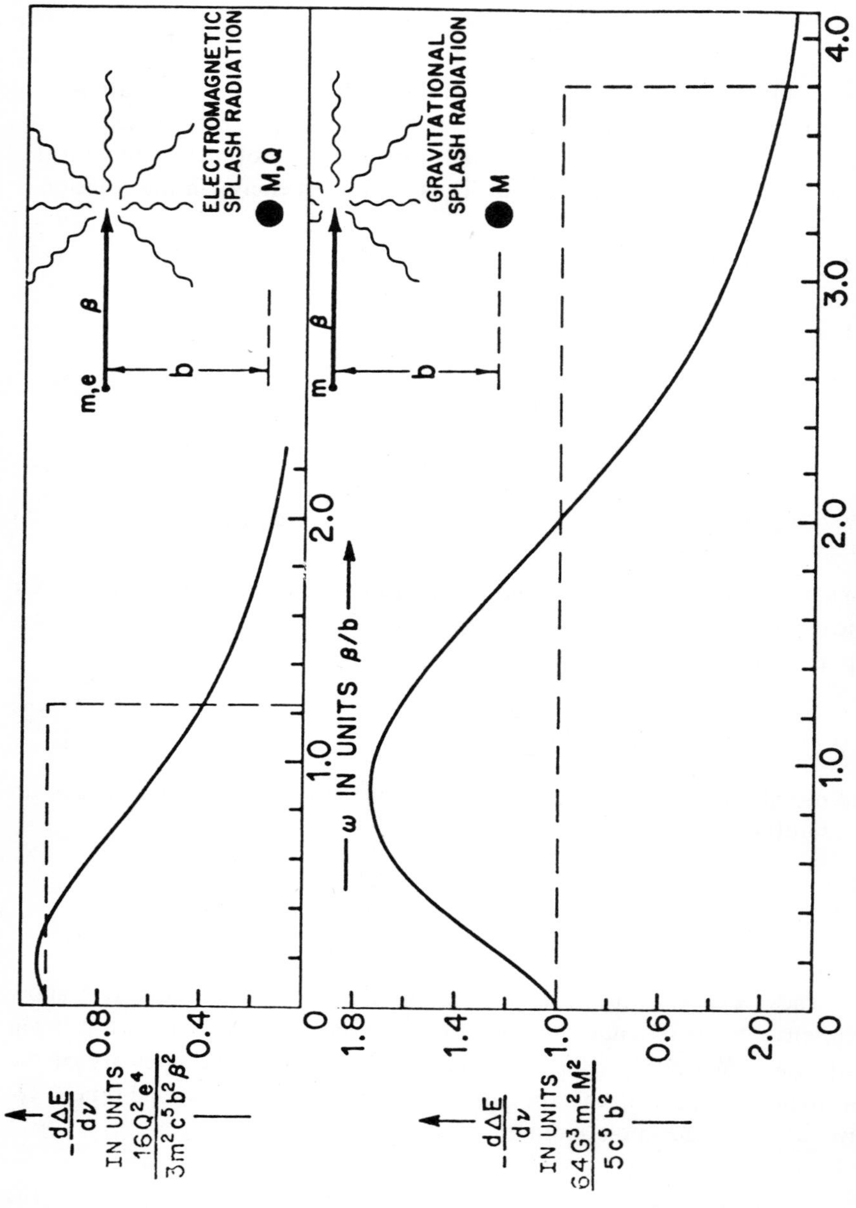
ELECTROMAGNETIC SPLASH RADIATION
m,e
β
b
M,Q
GRAVITATIONAL SPLASH RADIATION
m
M
$-\frac{d\Delta E}{d\nu}$ IN UNITS $\frac{16Q^2e^4}{3m^2c^5b^2\beta^2}$
$-\frac{d\Delta E}{d\nu}$ IN UNITS $\frac{64G^3m^2M^2}{5c^5b^2}$
ω IN UNITS β/b
0
0.4
0.8
1.0
2.0
1.0
1.4
1.8
0.6
2.0
3.0
4.0

Figure 2 – Electromagnetic and gravitational splash radiation compared. A particle of mass m, charge e, travelling at the non-relativistic speed $v = \beta c$, flies past a particle of much greater mass M, charge Q, originally at rest, at such a large impact parameter b that the motion is almost straight line motion. The frequency distribution of the electromagnetic energy radiated is given by the formula

$$-d\Delta E/d\nu = (4e^2/3c^3)\left\{\left|\int_{-\infty}^{+\infty} \ddot{x}(t)\, e^{i\omega t} dt\right|^2 + \left|\int_{-\infty}^{+\infty} \ddot{y}(t)\, e^{i\omega t} dt\right|^2\right\}$$

with $(\ddot{x}, \ddot{y}) = (eQ/m)(vt, b)/[(vt)^2 + b^2]^{3/2}$; or

$$-d\Delta E/d\nu = (16Q^2 e^4/3 m^2 c^5 b^2 \beta^2)\,[u^2 K_0^2(u) + u^2 K_1^2(u)]$$

with $u = \omega b/v$ (conventional units) or $u = \omega b/\beta$ (geometric units) (formula given by N. Bohr[1]). The function of u in square brackets reduces to 1 at low frequencies and goes to $\pi u e^{-2u}$ at high frequencies, and, when integrated with respect to u, gives $\pi^2/8 = 1.234$ ("cut-off point of equivalent flat spectrum", as illustrated in diagram) similarly, the distribution in frequency of the gravitational splash radiation is given by

$$-d\Delta E/d\nu = (2/45)\sum_{p,q}\left|\int_{-\infty}^{+\infty} \dddot{Q}^{pq}(t)\, e^{i\omega t}\, dt\right|^2$$

where in

$$\dddot{Q}^{pq} = (mM/r^3)(2\delta^{pq}\, r\dot{r} + 18(\dot{r}/r)\, r^p r^q - 12\,\dot{r}^p r^q - 12\, r^p \dot{r}^q)$$

we insert on the right the expressions for the unperturbed straight line motion of m (now neutral!) relative to M (also uncharged). The first factor in $\dddot{Q}^{pq}$ has the value $mM(b^2 + \beta^2 t^2)^{-3/2}$ and the second has the value $\beta^2 t$ times:

$$\begin{aligned}
xx &: \; -4\beta^2 t - 18\, b^2\beta^2\, t/(b^2 + \beta^2 t^2) \\
xy &: \; 6 b\beta - 18\, b^3\, \beta/(b^2 + \beta^2 t^2) \\
yy &: \; 2\beta^2 t + 18\, b^2\beta^2 t/(b^2 + \beta^2 t^2) \\
zz &: \; 2\beta^2 t
\end{aligned}$$

The Fourier integrals $\int_{-\infty}^{+\infty} \dddot{Q}^{pq}(t)\, e^{i\omega t}\, dt$, relevant to the distribution in angle as well as to the distribution in frequency, are given by 4 m M/b (no restriction on ratio of m to M) times:

$$\begin{aligned}
xx &: \; -i\,(2\,u\,K_0 + 3\,u^2\,K_1) \\
xy &: \; -3\,u\,K_1 - 3\,u^2\,K_0 \\
yy &: \; i\,u\,K_0 \\
zz &: \; i\,(u\,K_0 + 3\,u^2\,K_1)
\end{aligned}$$

Thus

$$-d\Delta E/d\nu = (64\, m^2 M^2/5\, b^2)\,[(u^2/3 + u^4)\, K_0^2 + 3\, u^3 K_0 K_1 + (u^2 + u^4)\, K_1^2]$$

(gravitational units; multiply by $G^3/c^5 = 1.23 \times 10^{-74}$ (erg/Hz) (cm^2/g^4) for conventional units). The factor in square brackets reduces to 1 for low $u = \omega b/\beta$, goes to $\pi u^3 e^{-2u}$ at high frequencies, and, when integrated with respect to u, gives $37\,\pi^2/96$ ("cut-off value for $\omega b/\beta$ for equivalent flat spectrum", as illustrated in the lower diagram).

where M is the mass of the center of attraction (still geometric units!).

The quantity in (14) that governs the emission has qualitatively the form of a "kinetic energy" – a directed kinetic energy, to be sure, but still a kinetic energy. The same is true in the other principal impulsive mechanisms for the production of gravitational radiation. Of such mechanisms, in addition to (1) the collisional impact just considered it is appropriate to look at (3) the sudden collapse of a star with white dwarf core (supernova and ensuing pursuit-and-plunge scenario) and (2) the capture of a star by a black hole (formed in the final stages of such a scenario).

The simplest case of (2) capture is that in which a "particle" (star or planet) of mass m falls straight into a standard Schwarzschild black hole (zero angular momentum). No full treatment of this important problem is yet available. However, F. Zerilli has given the foundation for such a treatment in his 1969 Ph.D. thesis[3]. He considered the incoming particle as making a small perturbation upon the background of the Schwarzschild geometry. He analyzed this perturbation of the geometry into tensorial spherical harmonics. For the radial factor in each harmonic he wrote down a second order differential equation, a wave equation in curved space. On the right-hand side of each such equation appears a source term. The source represents the driving effect of the incoming particle. It contains as key factor the expression

$$m\,\delta\,(r - r(t))$$

The radial factor in the perturbation of the metric ("amplitude of specific harmonic") is obtained by solving the radial equation. The character of the solution is vitally affected by the choice of boundary conditions.

In the present problem, as in almost every problem of radiation physics, the appropriate boundary condition at infinity is supplied by the requirement that there should be no incoming wave at infinity, only an outgoing wave (no "timing of sources at infinity in anticipation of the acceleration of the source"!) However the same kind of causality requirement imposes still another demand on the radial wave: it may transport energy towards the black hole, but it must not transport energy out of the black hole. One (1) determines the wave uniquely by solving the wave equation subject to these boundary conditions (2) evaluates the amplitude of this wave asymptotically at great distances (3) squares this amplitude (4) inserts this square into formulas given by Zerilli and thus (5) finds the intensity of the outgoing radiation in units of energy per unit of frequency and per unit of solid angle.

James Bardeen has pointed out that the black holes formed in nature will normally have the extreme or nearly the extreme possible angular momentum, $L(cm^2) = [m(cm)]^2$. He has also noted that a particle spiralling into a black hole of this so-called "extreme Kerr character" will give off 43 percent of its rest energy as gravitational radiation, whereas one spiralling into a Schwarzschild black hole will give off only 5.72 percent of its rest energy. Both circumstances give strong motivation to extend the analysis of Zerilli to the gravitational radiation given off when the incoming mass m, rather than spiralling in (extended opportunity to radiate) plunges in. Such a plunge will create a powerful pulse of gravitational radiation with its own characteristic pulse shape and its own characteristic polarization. The "signature" of this pulse can be expected someday to be a key factor in identifying and diagnosing this type of event.

In default of a detailed analysis, one can make a tentative rough estimate of the "plunge spectrum" based upon the following two simple but in principle contradictory idealizations: (1) the particle starts from rest at infinity and falls straight in according to the exact law for geodetic motion in the Schwarzschild geometry and (2) the particle radiates as if it were moving in flat space.

Denote the mass of the black hole by M, and express the z coordinate of the plunging particle (x=0, y=0; mass m) in terms of a dimensionless parameter η; thus,

$$z = 2M\eta^2$$

$$dt = -\frac{4M\eta^2\, d\eta}{1 - (1/\eta^2)}$$

$$t = 4M\left(\eta^3/3 + \eta - \frac{1}{2}\ln\frac{\eta + 1}{\eta - 1}\right) \qquad (15)$$

The particle starts at $t = -\infty$, $z = +\infty$, $\eta = +\infty$ and in infinite Schwarzschild coordinate time (far away observer) arrives at $t = +\infty$, $z = 2M$, $\eta = 1$. Its velocity and the derivatives of this velocity are

$$\frac{dz}{dt} = -\frac{1}{\eta}\left(1 - \frac{1}{\eta^2}\right)$$

$$\frac{d^2z}{dt^2} = -\frac{1}{4M\eta^4}\left(1 - \frac{1}{\eta^2}\right)\left(1 - \frac{3}{\eta^2}\right)$$

$$\frac{d^3z}{dt^3} = -\frac{1}{4M^2\eta^7}\left(1 - \frac{1}{\eta^2}\right)\left(1 - \frac{6}{\eta^2} + \frac{6}{\eta^4}\right) \qquad (16)$$

Turn now from this motion as calculated in curved spacetime to the formula for rate of gravitational radiation as idealized to a flat background geometry. For the quadrupole moment of the system we have

$$Q^{ab} = m(3\, x^a x^b - \delta^{ab}\, |x|^2) + M(3X^a X^b - \delta^{ab}\, |X|^2)\,; \tag{17}$$

for its third time rate of change

$$\dddot{Q}^{zz}(t) = 4m\,(3\,\ddot{z}\dot{z} + z\dddot{z})$$

(idealization m < M); and for the Fourier transform

$$\dddot{Q}^{zz}(\omega) = \frac{1}{\sqrt{2\pi}} \int_{-\infty}^{+\infty} \dddot{Q}^{zz}\, e^{+i\omega t}\, dt \tag{18}$$

For the remaining components of the quadrupole moment different from zero we have

$$Q^{xx} = Q^{yy} = -\frac{1}{2}\, Q^{zz} \tag{19}$$

The calculated intensity is

$$-\frac{d\Delta E}{d\nu} = -2\pi\, \frac{d\Delta E}{d\omega} = \frac{4}{45}\, \dddot{Q}^{pq}(\omega)\, \dddot{Q}^{pq\,*}(\omega)$$

$$= \frac{16m^2}{15} \left| \int_{-\infty}^{+\infty} (3\,\ddot{z}\dot{z} + z\dddot{z})\, e^{i\omega t}\, dt \right|^2 \tag{20}$$

Expressed in terms of the dimensionless parameter η this formula becomes

$$-\frac{d\Delta E}{d\nu} = \frac{16m^2}{15} \left| \int_1^{\infty} d\eta \left(\frac{3}{\eta^7} - \frac{1}{\eta^3} \right) \times \right.$$

$$\left. \exp i\, 4M\omega \left(\frac{\eta^3}{3} + \eta - \frac{1}{2} \ln \frac{\eta + 1}{\eta - 1} \right) \right|^2 \tag{21}$$

The intensity as given by this very rough and crude analysis (Figure 3) goes to zero as well at low frequencies as at high frequencies. The calculated intensity peaks at

$$\nu_{max} \simeq 0.024/M \text{ (geometric units)}$$

$$\simeq 4.9 \times 10^3\, \frac{M_\odot}{M}\, \text{Hz (cgs units)} \tag{22}$$

The calculated intensity at the peak is

$$\left(-\frac{d\Delta E}{d\nu} \right)_{peak} \simeq 0.047\; m^2 \text{ (geometric units)}$$

$$= 3.9 \times 10^{47}\ \text{(erg/Hz)} \left(\frac{m}{M_\odot} \right)^2 \text{ (cgs units)} \tag{23}$$

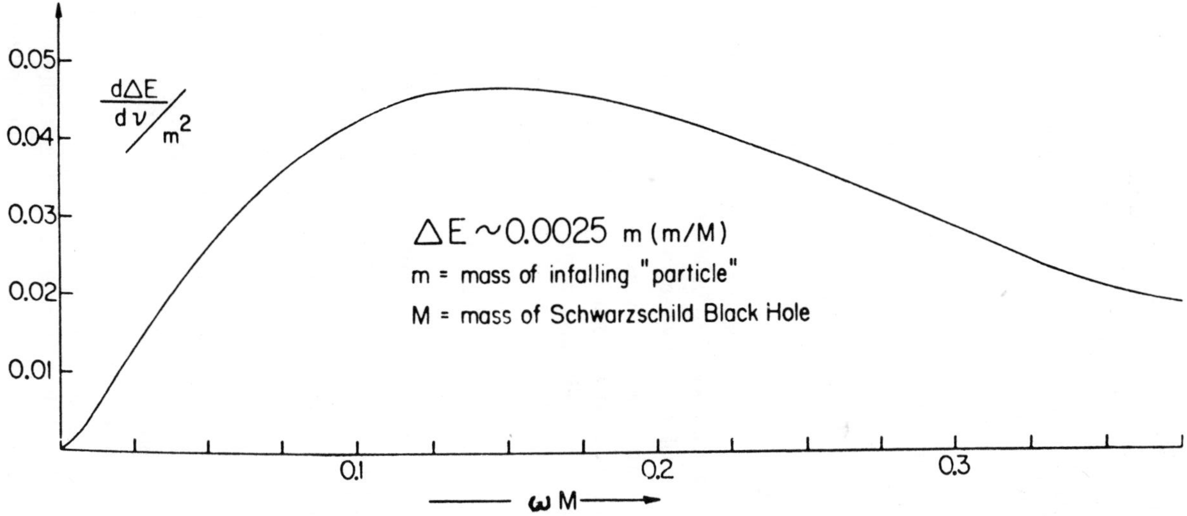

Figure 3 – Spectrum of plunge radiation as estimated from combination of two not quite consistent simplifications: (1) particle of mass m starts from rest at infinity and falls straight into standard Schwarzschild black hole of mass M; but (2) the gravitational radiation from m is calculated as if it were executing this motion towards the mass M ($\gg$ m) in flat space. Details and application in test.

The calculated total energy in the pulse is

$$-\Delta E = \frac{4m^2}{15M}\left(\frac{1}{7\cdot 9} - \frac{6}{11\cdot 13} + \frac{9}{15\cdot 17}\right) = 0.00246\,\frac{m^2}{M} \qquad (24)$$

For example let a neutron star with mass m = 0.604 km = 0.41 $M_\odot$ and a calculated radius[4] of 21 km fall into a non-rotating black hole of mass 50 $M_\odot$ (Schwarzschild radius 147 km). The foregoing rough estimates give

(energy of pulse)	$\sim 1.5 \times 10^{49}$	erg	$= 8.4 \times 10^{-6}\ M_\odot c^2$
(frequency of peak in spectrum)	~ 98	Hz	
(intensity at peak)	$\sim 6.2 \times 10^{46}$	erg/Hz	(25)

Tullio Regge suggested the adjective "Nimmersatt" as good characterizer of a black hole. That is the case here, where capture increases the mass from 50 $M_\odot$ to 50.4 $M_\odot$, and with it increases the appetite and cross section of the black hole. How big then can and will a black hole grow? Zel'dovich and Novikov have given reasons to believe[5] that a small black hole, if present in the earliest days of the universe, would have soaked up the primordial cosmic fireball radiation so rapidly as to acquire a mass far above the mass of a galaxy, perhaps even as high as 10^{17} $M_\odot$. A star of solar mass falling into such a black hole will give a pulse with the peak in its frequency spectrum at 1 cycle in 6×10^5 yr, too low to be detected by a simple bar of laboratory dimensions, and also too fantastically low in frequency even to be detected by the lowest mode of quadrupole vibration of the earth itself.

No convincing reason has ever been given to believe that such an early phase (and rapidly growing) black hole will ever have been formed in the beginning of the universe. Therefore it is of interest to see that the physics of a galactic nucleus offers a mechanism to produce a black hole of more modest mass, $\sim 10^7$ $M_\odot$ to $\sim 10^{10}$ $M_\odot$.

Standers and Spitzer[6] analyze by computer the evolution of a cluster of $\sim 10^8$ stars. The stars exchange energy with one another by gravitational interaction. Some gain energy and move out into a halo. Others lose energy and move together into a more compact configuration. Eventually the density of stars rises to the point where collisions occur. Gas is driven out in these collisions. It floats down near the bottom of the gravitational potential well, at the center of the star cluster. Out of this gas new stars form. The cycle repeats itself, but more quickly. Eventually the calculated compactification reaches the point where Sanders and Spitzer cut off the analysis. Pursued a little further, it could hardly have failed to result in the creation of a black hole with mass equal to some substantial fraction of the mass of the original star cluster. If such a process should have gone to the stage in our own galaxy where for example there is by now present a black hole of mass 10^8 $M_\odot$, then the fall of an additional solar mass into such an object at a distance of 8.2 kpc = 2.5×10^{19} cm from the earth will be expected to give a pulse of gravitational radiation with roughly these properties (with uncertainty arising as much as anything from question about the angular momentum of this black hole (Schwarzschild vs. Kerr; estimates made for Schwarzschild case)):

(energy of pulse)	$\sim 4.4 \times 10^{43}$	erg = 2.5×10^{-11} $M_\odot c^2$	
(frequency at peak in spectrum)	$\sim 4.9 \times 10^{-5}$	Hz	
(intensity at peak)	$\sim 3.9 \times 10^{47}$	erg/Hz	
(intensity at this frequency at earth)	$\sim 6.2 \times 10^{8}$	erg/cm² Hz	(26)

Collapse of a Star with White Dwarf Core; the "Pursuit - and - Plunge" Scenario.

Mechanism (3) for the production of powerful pulses of gravitational radiation starts, not with a black hole, but with a star of several solar masses. In the course of its normal astrophysical evolution[7,8] it has come to the point where it has developed, not only a dense core with roughly white dwarf densities, but also an incipient instability against gravitational collapse. Gravitational collapse gets under way, at first slowly, then with rapidly increasing speed. The collapse proceeds as indicated in considerable detail in the electronic computer analyses of Colgate and White[9] and May and White[10] and others, with one exception. The system to begin with, like most stars, has a non-trivial amount of angular velocity. As the collapse proceeds, the angular velocity goes up. For an example of one among many possibilities, consider what one might almost call a "maximal event": Collapse of a rotating star with core of mass $M \sim 6\ M_\odot$ (Figure 4).

The initial angular velocity in this example is such that the configuration after collapse is rotating very rapidly. Far from being spherical, it is forced by centrifugal effects to assume a pancake shape. In the example the thickness of this pancake is of the order of 16 km and its diameter roughly 100 km. Such a flattened configuration is unstable against fragmentation. The fragmentation will be expected to produce several neutron stars with or without accompanying fragments of planetary mass, and conceivably sometimes even already at the start one black hole.

The system of fragments whirls around, one mass in pursuit of another. Gravitational radiation of slowly varying period comes off, taking away

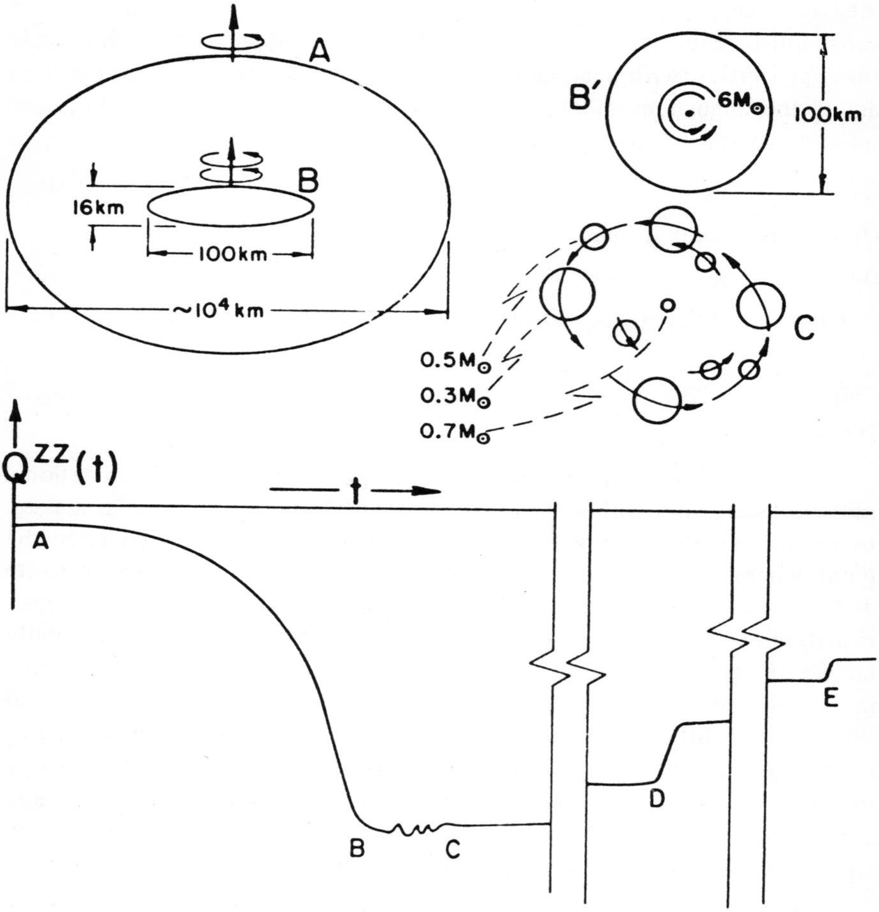

Figure 4 – A rotating star with white dwarf core (A) collapses to a pancake neutron star (B);it fragments (C); the fragments lose energy in periodic and splash gravitational radiation and recombine. The lower curve gives the schematic representation of the quadrupole moment as a function of time. Between B and C impulse radiation is created in the act of fragmentation which is not adequately described by the one indicated component of the quadrupole moment tensor. Between C and D multiply periodic radiation is given out until at D, two fragments have lost enough angular momentum so that they combine with a splash of gravitational radiation; similary at E, etc.

not only energy, but also angular momentum. In consequence the system becomes more compact. Inevitably one object plunges into coalescence with another, and from time to time this kind of scene repeats itself (time of minutes to days). Neutron stars or black holes or both grow in mass as the amalgamation proceeds. Thus the phases of emission of gravitational radiation with slowly varying frequency alternate with powerful pulses of amalgamation gravitational radiation ("plunges"; Figure 4). Few sequences of events can one easily imagine that give a stronger or more characteristic signature of what is going on! Moreover, the gravitational radiation from such an event penetrates out through the cloud of debris around the event unaffected in polarization or intensity. Few channels of communication from a far away event can be expected to send in such a characteristic signal as this pursuit - and - plunge scenario.

How often such a collapse-and-fragment, pursuit-and-plunge event happens in this galaxy is uncertain. The figure of Zwicky[11] for the occurrence of a supernova in our galaxy is one every 300 years; Shklovsky[12] estimates one every thirty years. There is nothing evident to exclude the possibility that events of "quiet collapse" and low optical luminosity – but of powerful gravitational wave signature – may be even more frequent.

Apart from the multiply periodic gravitational radiation of slowly changing frequency given out in the spin-down of the steeplechase of fragments, the truly characteristic parts of the signature are the pulses; (a) at the time of collapse (b) at the time of fragmentation (c) at the time two fragments plunge into coalescence.

A star collapsing gravitationally with spherical symmetry will not produce any gravitational radiation because no quadrupole moment or change of quadrupole moment is present in such event. In contrast, a star will have a quadrupole moment when it is in steady rotation, but this quadrupole moment will not change with time and no emission of gravitational radiation will result. However, the combination of rotation and collapse implies both quadrupole moment and time rate of change of this quadrupole moment, and therefore gravitational radiation (see graph of sudden change of quadrupole moment – negative to indicate flattening – in Figure 4). The estimated amount of gravitational energy emitted in the low frequency part of the spectrum during the collapse to the pancake configuration (0.2 m sec) is of the order $-\dfrac{d\Delta E}{d\nu} \sim 10^{48}$ erg/Hz. Assuming the source to be at a distance of 1000 pc away we have a flux at the earth's surface of $\sim 10^4$ ergs/cm^2 Hz.

(b) The rotating neutron pancake may not be circular. An oval pancake has a non-zero mass quadrupole moment turning periodically in time, and serves as an emitter in its own right, before fragmentation. It has long been known[13] that a rotating self gravitating fluid with an equation of state of the type $p = \rho^{\gamma}$ with $\gamma \gtrsim 2.2$ may have configurations of equilibrium with 3 unequal moments of inertia if it is rotating above a critical value of the angular velocity. (Figure 5). An eccentricity as small as $\epsilon = 5 \times 10^{-4}$ in the equatorial plane of a rotating neutron star is enough to give a powerful loss of energy by gravitational radiation.[14]

The fragmentation itself results in impulsive changes in quadrupole moment and associated pulses of gravitational radiation.

(c) After fragmentation the separated objects will gradually spiral in. One by one neutron stars will plunge into coalescence with each other or with a black hole circulating in the same company, each time giving off a pulse of gravitational radiation (cf. Eqs. 22-24).

No Obvious Nearby Sources of Strong Gravitational Radiation.

An earthquake might be imagined to be a significant source of gravitational radiation. Of course the seismic waves themselves can be excluded at the start. The detector of gravitational waves can be isolated from the seismic waves, both by proper mounting, and by any reasonable means of discrimination in time of arrival, so that a gravitational wave is clearly differentiated from a seismic wave. However one can easily estimate that the intensity of such a wave is absolutely negligible. Even a much more spectacular event like the impact of a very large meteorite (1 km^3) with the surface of the earth will emit a far too small amount of radiation to influence any conceivable detectors[2].

Detectors.

The normal modes of vibration of an aluminum bar (Figure 6) are not all active as detectors of gravitational waves. Changes in the quadrupole modes are associated only with the odd numbered modes. Any given mode can be described by its characteristic damping coefficient with respect to rate of loss of energy by gravitational radiation. The damping coefficient A_{grav} is obtained[2] by dividing the rate of loss of gravitational energy by the vibrational energy present in the bar

$$A_{grav} = \frac{-(dE/dt)_{grav}}{E} = \frac{64}{15} \frac{G}{c^5} \frac{M}{L^2} v^4 \qquad (27)$$

where M is the mass of the bar, L the length of the bar and v is the velocity of sound in the bar. We can describe the absorption cross section of bar for radiation of random polarization, averaged over all orientations of the

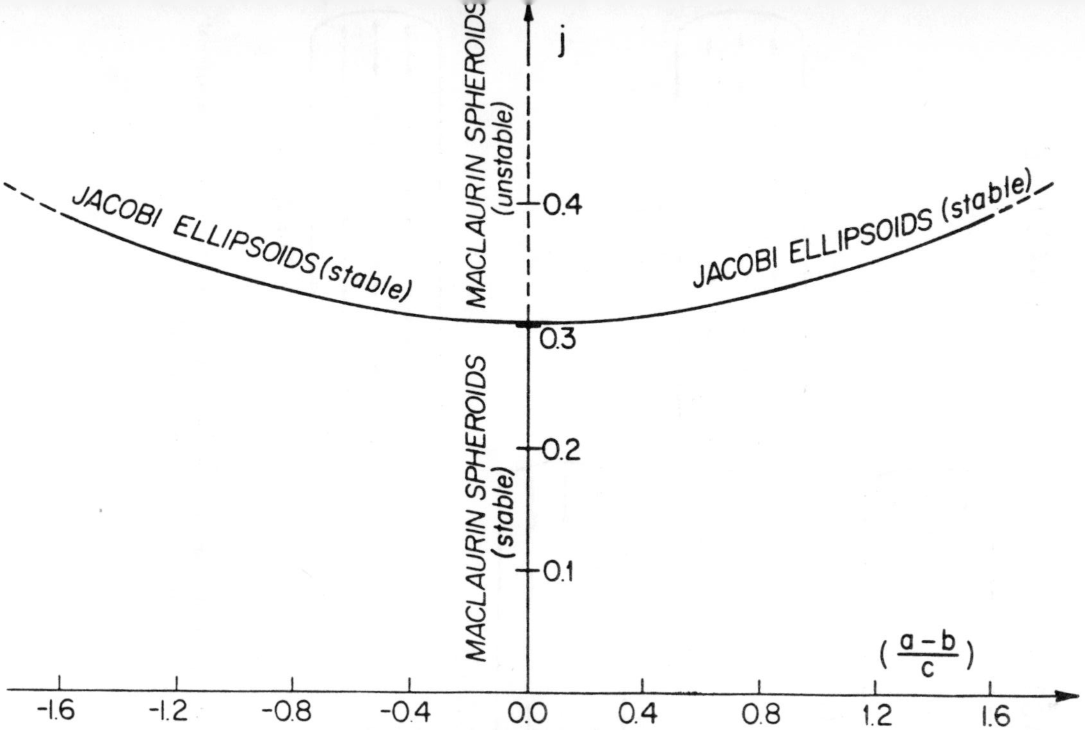

Figure 5 – Slowing down by gravitational radiation of a mass of incompressible fluid turning at a uniform angular velocity and with an angular momentum larger than the critical angular momentum $L = 0.303751\ L_0$, where $L_0 = (GM^3)^{1/2}\ (abc)^{1/6}$ Horizontal scale: difference of semiaxes of ellipsoid in equatorial plane divided by semiaxis along axis of rotation (zero when angular momentum is less than critical amount). Correlation between angular momentum and shape (cf. Jeans[13]) recomputed fresh for this curve.

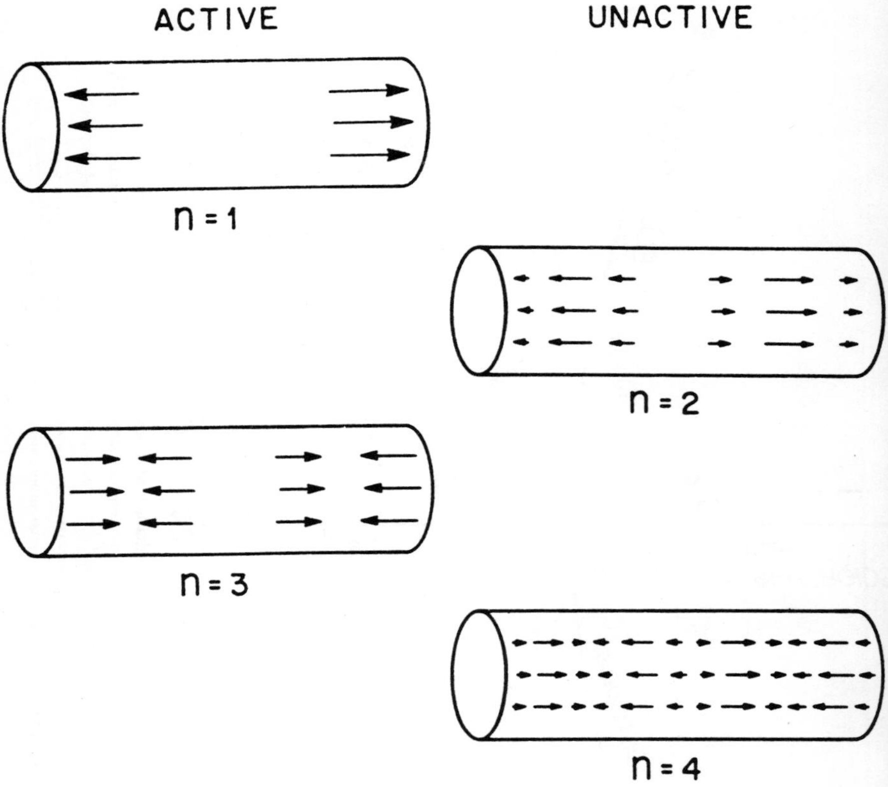

Figure 6 – Modes of vibration of a bar active and inactive with respect to emission of gravitational radiation as distinguished by odd or even character of index number n.

bar by the Breit-Wigner formula[2]

$$\sigma = \frac{\pi}{2} \lambdabar^2 \frac{A_{grav} \; A_{diss}}{(\omega - \omega_0)^2 + \frac{1}{4} (A_{grav} + A_{diss})^2} \tag{28}$$

where A_{grav} is the gravitational damping coefficient and the A_{diss} is the sum of coefficients of damping due to all dissipative modes other than gravitational radiation (mechanical friction, electrical resistance, signal read out, etc). The quantity $\lambdabar$ is the "reduced wave length", $\lambda/2\pi$. In any detector of gravitational radiation so far imagined the gravitational damping A_{grav} is many orders of magnitude smaller than the quantity A_{diss}. The relevant figure for our purpose is not this cross section itself. To work at resonance would require a too improbable degree of tuning. What counts for us is the integrated cross section. In the approximation $A_{diss} \gg A_{grav}$ the integrated cross section has a value which is independent of the details of the damping process, and primarily dependent upon the mass M of the bar:

$$\int_{resonance} \sigma(\nu)\, d\nu = (\pi/2)\, \lambdabar^2\, A_{grav} \doteq (32/15\pi)\, (G/c)\, (V^2/c^2)\, (M/n^2) \tag{29}$$

again for radiation incident with random polarization and from random directions. The odd-integer index n = 1, 3, 5, refers to the active modes of Figure 6.

When we apply the same reasoning to evaluate of the cross section of the earth for gravitational radiation of random polarization, we obtain with some simplifying assumptions[2] the integrated cross section

$$\int_{\substack{resonance\\random\\polarization}} \sigma(\nu)\, d\nu = (2\pi/25)\, (G/c)\, Ma^2/\lambdabar^2 = 4.7 \text{ cm}^2 \text{ Hz} \tag{30}$$

where $M = 5.98 \times 10^{27}$ g is the mass and a the radius of the earth and $\lambdabar = 1.55 \times 10^{13}$ cm is the reduced wave length of gravitational waves of 54 min period, the period of the lowest quadrupole mode of vibration of the earth. The noise power spectrum of normal mode of excitation (54 min) of the earth's surface during quiet periods as measured by a seismometer according to Weber and Larson[15] is given by 6.9×10^{-14} $(\text{cm/sec}^2)^2$/ /rad/sec). This mode of vibration is damped at a damping rate $\Delta\omega = \omega/Q \sim$ $\sim (2\pi/3240 \text{ sec})/400 = 4.86 \times 10^{-6}$ rad/sec. Therefore the power dissipated over resonance is

$$1.2 \times 10^9 \text{ erg/sec} \tag{31}$$

If we assume that all the energy that is dissipated is being fed into the mode by gravitational radiation we get a flux of 2.6×10^8 erg/cm^2 sec Hz required to keep the earth's vibration up to its present value. Assuming a radiation with a flat spectrum extending from zero frequency up to the frequency where we observe by way of the earth we obtain a total flux of 2.4×10^{-6} erg/cm^3 implying a mass density of 2.6×10^{27} g/cm^3 of effective gravitational radiation. This value is larger than the upper limit to the energy density (10^{-27} g/cm^3) compatible with homogeneous relativistic cosmology and maximum stretching of uncertainties in age and in the inverse Hubble constant.

Let us now consider the corresponding problem for Weber's aluminum bar. Eq. (29) with $M = 1.4 \times 10^6$ g and $v/c = 2.14 \times 10^{-5}$ gives for the integrated cross section for the fundamental mode $n = 1$ the result

$$\int_{\substack{\text{resonance}\\ \text{random polarization}\\ \text{random orientation}}} \sigma(\nu)\, d\nu = 1.0 \times 10^{-21}\ \text{cm}^2\, \text{H}_z \tag{32}$$

In the course of 81 days Weber's published work indicates that he found one event in which the Argonne detector and the Maryland one gave pulses in coincidence, each of the order of $5\ kT = 2 \times 10^{-13}$ erg. From this number follows an upper limit for the pulsed intensity of gravitational radiation. Dividing this amount of energy by the integrated cross section of the detector we obtain an upper limit for the strength of pulses coming in over that 81 day interval, namely

$$\sim 2 \times 10^8\ \text{erg/cm}^2\ \text{Hz in 81 days} \tag{33}$$

If one thinks of a spectrum flat from $\nu = 0$ up to the frequency of observation $\nu_0 = 1660$ cps he has an upper limit on the flux of $\sim 3 \times 10^{11}$ erg/cm^2. With one such pulse every 81 days (7×10^6 sec) this particular upper limit on the flux would correspond to $\sim 5 \times 10^4$ erg/cm^2 sec averaged over a long period of time. The equivalent upper limit on the effective density of mass energy in space, obtained by dividing the preceding figure by the cube of the speed of light, is

$$\rho_{\text{mass energy}} \sim 1.7 \times 10^{-27}\ \text{g/cm}^3$$

again more than appears reasonable on cosmological grounds. Weber has reported evidence that the events observed did not originate in:

a) a surge of voltage in the interstate power grid

b) a coupling of the recorders by way of a telephone line

c) solar flares

d) low frequency signals sent by powerful transmitters to submarines
e) lightning strokes
f) cosmic rays.

Weber has also conducted an experiment in which he put a time delay in one of the detectors. He found a ~10-fold smaller number of coincidences. It would be helpful to put time delays in each of the detectors and tune them so that the resulting time delay can be made the same or can be made different. This would definitely make clear that the difference in counting rates cannot be due to the fact that there was a different circuit in the two cases. It appears too early to assume that any real effect of gravitational waves has been demonstrated.

In summary, the amplitude of the quadrupole vibrations of the earth and of Weber's bar give for the first time upper limits on the strength of gravitational radiation coming in from space at 1 cycle/54 min. and at 1660 cycles/sec. Moreover, from what one knows about supernovae and about proximity of stars near the center of the galaxy it seems inescapable that from time to time an event must occur that gives off a powerful pulse of gravitational radiation with spectrum peaking at a frequency of the rough order of 10^3 Hz. The expected output of energy in a single event is a little different according as the event is the collapse of a dense star core to a neutron star pancake in a supernova event, or the union of one neutron star with another or with a black hole, but a figure of the order of $\sim 10^{51}$ erg, with a factor of 10 either way, seems representative. For a source located at the center of the galaxy (8 kpc or 24×10^{21} cm) this implies an integrated flux past the earth of $\sim 10^{51}$ erg/7×10^{45} cm^2 $\sim 10^5$ erg/cm^2, or $\sim 10^2$ erg/cm^2 Hz near 10^3 Hz. The cross-section, integrated over resonance, for a bar like Weber's being $\sim 10^{-21}$ cm^2 Hz, one expects an energy pickup from such a pulse of the order of $\sim 10^{-19}$ erg, compared to a value of k T at room temperature of 4×10^{-14} erg. Work is in progress at an increasing number of centers aimed at improved circuitry and lower temperatures and at the detection of such pulses.

REFERENCES

1. N. BOHR, *Roy. Soc. London, Proc.* **A84**, 395 (1910).
2. R. RUFFINI and J.A. WHEELER, "Relativistic Cosmology and Space Platforms", a chapter in *The Significance of Space Research for Fundamental Physics*, European Space Research Organization book Sp-52, Paris, 1971.
3. F. ZERILLI, Ph. D. Thesis, Princeton University 1969 (unpublished) and *Phys. Rev.* **D2**, 2141, 1970.

4. B.K. HARRISON, K.S. THORNE, M. WAKANO, J.A. WHEELER, *Gravitation Theory and Gravitational Collapse*, University of Chicago Press, Chicago 1965.

5. YA. B. ZEL'DOVICH and I.D. NOVIKOV, *Stars and Relativity*. translated by Eli Erlock, ed. by K.S. Thorne and W.D. Arnett, University of Chicago Press, 1971.

6. R.H. SANDERS, Ph. D. thesis, Princeton University, 1970, (unpublished); L. Spitzer, Jr., report at Semaine d'Etude on the structure and activity of galactic nuclei, Vatican City, April 1970, Pontificiae Academiae Scientiarum Scripta Varia 35, 1971.

7. B. PACZYNSKI, "Stellar Evolution from Main Sequence to White Dwarf or Carbon Ignition", *Acta Astronomica*, **20**, 47 (1970).

8. D. SUGIMOTO, "Mixing between Stellar Envelope and Core in Advanced Phases of Evolution. I", *Prog. Theor. Phys.*, **44**, 375 (1970).

9. S. COLGATE and R.H. WHITE, *Astrophys. J.*, **142**, 626 (1966).

10. M.M. MAY and R.H. WHITE, *Phys. Rev.* **141**, 1232 (1966).

11. F. ZWICKY, "Supernovae", a chapter in *Handbuch der Physik*, **51**, *Astrophysics II:* Stellar Structure, 766-785 S. Flügge, ed Springer, Berlin, 1958; also in *Stellar Structure*, p. 367, ed. L.H. Aller and D.B. Mc Laughlin, University of Chicago Press, Chicago, 1965.

12. J.S. SHKLOVSKY, *Supernova*, John Wiley-Interscience Publication, New York, 1968.

13. J.H. JEANS *Astronomy and Cosmology*, Cambridge University Press, Cambridge, 1919.

14. A. FERRARI and R. RUFFINI, *Ap. J. Letters* **158**, L71 (1969).

15. J. WEBER and J.W. LARSON, *J. Geophys. Res.* **41**, 6005 (1966).

A 3.2

Gravitational Radiation in the Presence of a Schwarzschild Black Hole. A Boundary Value Search (*).

M. Davis and R. Ruffini
Joseph Henry Laboratories, Princeton University - Princeton, N. J.

(ricevuto il 7 Ottobre 1971)

The analysis of the spectrum of the energy emitted in gravitational radiation by a particle falling into a Schwarzschild black hole was first analysed by Ruffini and Wheeler (1) in the framework of the linearized slow-motion approximation of general relativity. Zerilli (2) has shown that the tensorial-perturbation technique introduced by Regge and Wheeler (3) can be used in the analysis of this problem. Following this approximation we do not impose any restriction on the velocity of the incoming particle but we still assume that the perturbation introduced into the metric of the Schwarzschild black hole by the incoming particle be small ($m \ll M$). The radial functions of the tensorial perturbations are solutions of differential equations of either « magnetic » or « electric » parity. In the particular case where we analyse the motion of a particle falling on a radial geodesic path starting from rest at infinity only the « electric » parity need be considered (see Zerilli (2)). The radial function $R_l(r)$, which is a combination of Fourier-transformed metric perturbations of the l-th multipole then satisfies the following equation:

$$\frac{d^2 R_l}{dr^{*2}} + [\omega^2 - V_l(r)]R_l = S_l \tag{1}$$

with $r^* = r + 2M \ln(r/2M - 1)$. Here and in the following we are using geometrical units $G = c = 1$. $V_l(r)$ is an « effective » potential defined by

$$V_l(r) = \left(1 - \frac{2M}{r}\right) \frac{2\lambda^2(\lambda+1)r^3 + \sigma\lambda^2 Mr^2 + 18M^2 r + 18M^3}{r^3(\lambda r + 3M)}, \tag{2}$$

(*) Work supported in part by the National Science Foundation Grant GP7669.
(1) R. Ruffini and J. A. Wheeler: *Gravitational Radiation* in the *Cortona Symposium on Weak Interactions*, edited by L. Radicati (1971).
(2) F. J. Zerilli: *Phys. Rev. D*, **2**, 2141 (1970).
(3) T. Regge and J. A. Wheeler: *Phys. Rev.*, **108**, 1063 (1957).

here $\lambda = \frac{1}{2}(l-1)(l+2)$ and $S_l(r)$ is the l-pole component of the source of the wave

$$S_l = (S_l^1 + iS_l^2)\exp[i\omega T] \tag{3}$$

with

$$S_l^1 = -2m\frac{r-2M}{\lambda r + 3M}\left[\frac{\mathrm{d}T}{\mathrm{d}r}\frac{\sqrt{2}}{r} + \frac{2}{r-2M}\left(\frac{M}{r}\right)^{\frac{1}{2}}\right](2l+1)^{\frac{1}{2}}\,, \tag{4.1}$$

$$S_l^2 = -2m\frac{r-2M}{\lambda r + 3M}\left[\frac{\sqrt{2}(2l+1)^{\frac{1}{2}}}{r} - \frac{2\lambda}{\omega(\lambda r + 3m)}\right] \tag{4.2}$$

and

$$T(r) = -\frac{4}{3}\left(\frac{r}{2M}\right)^{\frac{3}{2}} - 4\left(\frac{r}{2M}\right)^{\frac{1}{2}} + 2\ln\left[\frac{(r/2M)^{\frac{1}{2}}+1}{(r/2M)^{\frac{1}{2}}-1}\right]. \tag{5}$$

The boundary conditions adopted in solving the problem have been such to obtain outgoing waves at infinity and ingoing waves at the black-hole surface. We have integrated eq. (1) by a direct numerical technique. Initial values are chosen to satisfy one boundary condition (*e.g.*, ingoing waves for $r^* \to -\infty$) and the equation integrated to the other boundary (*e.g.*, $r^* \to \infty$). The initial values are modified up to the point that both boundary conditions are identically satisfied.

Since eq. (1) has real and imaginary components, it is equivalent to two uncoupled second-order equations. The four resulting initial-value variables are reduced to two by the imposition of ingoing ($r^* \to -\infty$) or outgoing ($r^* \to +\infty$) waves at one of the boundaries. The two remaining variables may be thought of as the amplitude and the phase of the radiation function $R_l(r)$ at one point. In Fig. 1 we show some values of

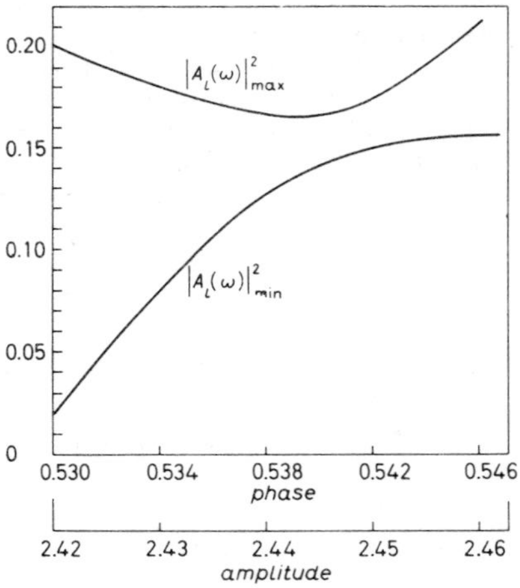

Fig. 1. – The energy emitted by a particle falling radially into a Schwarzschild black hole can be expressed by the equation $-\mathrm{d}E/\mathrm{d}\omega = (1/32\pi)\sum_l l(l+1)(l-1)(l+2)|A_l(\omega)|^2$. Asymptotically, when the proper boundary conditions are fulfilled (outgoing waves at infinity) we have $\mathrm{d}R_l/\mathrm{d}r \sim A_l(\omega)\exp[i\omega r^*]$. Here R_l has to fulfill eq. (1) in the text. For one and only one combination of the amplitude and the phase (here plotted on the abscissa) of the ingoing radiation, we obtain a purely outgoing radiation at infinity (*i.e.* $|A_l(\omega)|^2_{max} = |A_l(\omega)|^2_{min}$). In this Figure the noncrossing of the two curves is due to the contribution of the source which is still not negligible at the point where our asymptotic considerations apply ($r^* \sim 380$).

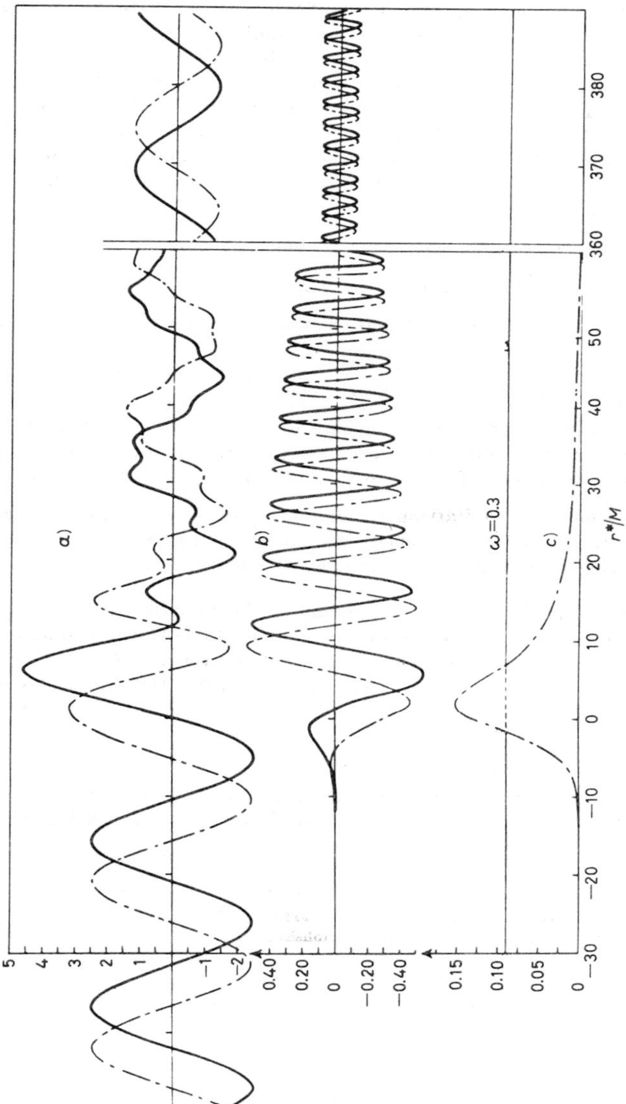

[R119]

Fig. 2. – For selected values of l and ω ($l = 2$ $\omega M = 0.3$) *a*) the radiation function, *b*) the source and *c*) the effective potential are plotted as a function of r^*. The outgoing ($r^* \to +\infty$) and ingoing ($r^* \to -\infty$) wave behavior of the radiation function as well as its behavior in the « near zone » region are here shown with particular clarity: —— real component, —·—·— imaginary component.

the parameters tried to fulfill the boundary conditions. All these computations refer to a value of $l = 2$ and a value of the frequency $\omega M = 0.3$. In Fig. 2 the main results of the computation are summarized. The radiation function, the source term of eq. (1) and the effective potential are plotted as a function of the radial coordinate r^*. For negative r^* the real component of the radiation function leads the imaginary component by 90° ($R \propto \exp[-i\omega r^*]$ purely ingoing radiation) whereas for large r^* the real component trails the imaginary component by 90° ($R \propto \exp[i\omega r^*]$ purely outgoing radiation). Because the source falls off only as $r^{*-\frac{1}{2}}$ for large r^* it is necessary to integrate to large distances. The slow decay of the source limits the numerical accuracy of our results to a few percent. Given the amplitude of the outgoing wave R one can compute the intensity of the outgoing radiation for that frequency and multipole. We are therefore able to give the complete structure of the radiation function in the three separate regions:

i) $r^* \to -\infty$ radiation ingoing into the black hole,

ii) $r^* \to +\infty$ radiation outgoing toward infinity,

iii) « near zone » where the contributions due to the source and to the potential are maximal,

From these preliminary results we can conclude:

1) at the frequency examined the amount of gravitational radiation as computed in this new approximation is substantially larger than the one predicted by the linearized slow-motion approximation of the theory of gravitation (Ruffini and Wheeler [1]);

2) a large amount of gravitational radiation appears to fall into the black hole and, at least at the frequency examined, the amount of this radiation appears to be substantially larger than the radiation escaping at infinity.

We have also examined the entire spectrum of the radiation as a function of the frequency by Green's function techniques [4]. The fact that we can compute both the amount of radiation ingoing into the black hole and outgoing at infinity gives a good hint in attacking the problem of the radiation reaction [5]. Details on this work will be presented elsewhere.

* * *

Preliminary results of this work were presented by one of us (R.R.) at the Copenhagen meeting GR6. A travel grant by the N.S.F. to one of us (R.R.) is gratefully acknowledged.

---

(4) M. Davis, W. H. Press, R. H. Price and R. Ruffini: to be published.
(5) M. Davis, R. Ruffini and J. Tiomno: to be published.

VOLUME 27, NUMBER 21 PHYSICAL REVIEW LETTERS 22 NOVEMBER 1971

Gravitational Radiation from a Particle Falling Radially into a Schwarzschild Black Hole*

Marc Davis and Remo Ruffini
Joseph Henry Laboratories, Princeton University, Princeton, New Jersey 08540

and

William H. Press† and Richard H. Price‡
Kellogg Radiation Laboratory, California Institute of Technology, Pasadena, California 91109
(Received 24 September 1971)

We have computed the spectrum and energy of gravitational radiation from a "point test particle" of mass m falling radially into a Schwarzschild black hole of mass $M \gg m$. The total energy radiated is about $0.0104mc^2(m/M)$, 4 to 6 times larger than previous estimates; the energy is distributed among multipoles according to the empirical law $E_{2^l\text{-pole}} \approx (0.44m^2c^2/M)e^{-2l}$; and the total spectrum peaks at an angular frequency $\omega = 0.32c^3/GM$.

In view of the possibility that Weber may have detected gravitational radiation,[1] detailed calculations of the gravitational radiation emitted by fully relativistic sources are of considerable interest. Three such calculations have been published in the past: waves from pulsating neutron stars, by Thorne[2]; waves from rotating neutron stars, by Ipser[3]; and waves from a physically unrealistic collapse problem (important for the points of principle treated), by de la Cruz, Chase, and Israel.[4] To these, this paper adds a fourth calculation: the waves emitted by a body falling radially into a nonrotating black hole. This calculation is particularly important for two reasons: (i) It is the first accurate calculation of the spectrum and energy radiated by any realistic black-hole process (though upper limits on the energy output have been derived by Hawking[5]); (ii) Weber's events involve such high fluxes that black holes are more attractive as sources than are neutron stars.

A first analysis of the radial-fall problem was done by Ruffini and Wheeler[6] with a simple idealization: The particle's motion is derived from the Schwarzschild metric, but its radiation is calculated using the flat-space linearized theory of gravity. This scheme yielded a total energy radiated of $0.00246mc^2(m/M)$ and a spectrum peaked at an angular frequency $0.15c^3/GM$. Zerilli,[7] using the formalism of Regge and Wheeler,[8] gave the mathematical foundations for a fully relativistic treatment of the problem. Unfortunately, Zerilli's equations are sufficiently complicated as to make a calculation of the energy release inaccessible to analytic means.

We have used Zerilli's equations (corrected for errors in the published form), and by numerical techniques we have (i) computed the wave form of gravitational radiation, (ii) evaluated the amplitude of this wave asymptotically at great distances, and (iii) used this amplitude to compute the outgoing wave intensity in units of energy per unit frequency per unit of solid angle.

Zerilli describes the 2^l-pole component of gravitational waves by a radial function $R_l(r)$ which is a combination of the Fourier transform of metric perturbations in the Regge-Wheeler formalism. The function $R_l(r)$ satisfies the remarkably simple Zerilli wave equation ($G = c = 1$)

$$d^2R_l/dr^{*2} + [\omega^2 - V_l(r)]R_l = S_l, \tag{1}$$

with

$$r^* = r + 2M\ln(r/2M - 1). \tag{2}$$

$V_l(r)$ is an "effective potential" defined by

$$V_l(r) = (1 - 2M/r)[2\lambda^2(\lambda+1)r^3 + 6\lambda^2Mr^2 + 18M^2r + 18M^3]/r^3(\lambda r + 3M). \tag{3}$$

Here, $\lambda = \frac{1}{2}(l-1)(l+2)$ and $S_l(r)$ is the 2^l-pole component of the source of the wave. We are interested in the particular case of a particle initially at infinity ($t = +\infty, r = +\infty$) and falling radially into a Schwarzschild black hole ($t = +\infty, r = 2M$). For this simple case the source may be written as

$$S_l(r) = \frac{4M}{\lambda r + 3M}(l+\tfrac{1}{2})^{1/2}\left(1 - \frac{2M}{r}\right)\left[\left(\frac{r}{2M}\right)^{1/2} - \frac{i2\lambda}{\omega(\lambda r + 3M)}\right]e^{i\omega T(r)}. \tag{4}$$

Here $t = T(r)$ describes the particle's radial trajectory giving the time as a function of radius along the

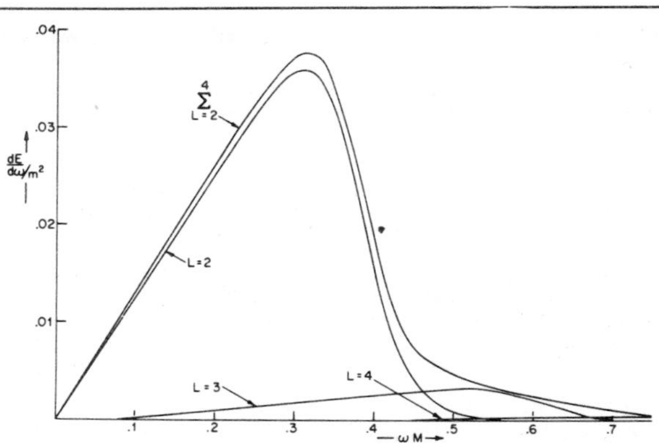

FIG. 1. Spectrum of gravitational radiation emitted by a test particle of mass m falling radially into a black hole of mass M (in geometrical units $c=G=1$).

geodesic

$$T(r) = -\frac{4}{3}\left(\frac{r}{2M}\right)^{3/2} - 4\left(\frac{r}{2M}\right)^{1/2} + 2\ln\left\{\left[\left(\frac{r}{2M}\right)^{1/2}+1\right]\left[\left(\frac{r}{2M}\right)^{1/2}-1\right]^{-1}\right\}. \tag{5}$$

The effect of gravitational radiation reaction on the particle's motion is therefore ignored. This is justified by the final result: The total energy radiated, of order m^2c^2/M, is negligible compared to the particle's final kinetic energy, of order mc^2. Equation (1) is solved with boundary conditions of purely outgoing waves at infinity and purely ingoing waves at the Schwarzschild radius:

$$R_l \to \begin{cases} A_l^{\text{out}}(\omega)\exp(i\omega r^*) \text{ as } r^* \to +\infty, \\ A_l^{\text{in}}(\omega)\exp(-i\omega r^*) \text{ as } r^* \to -\infty. \end{cases} \tag{6}$$

The energy spectrum is determined by Zerilli's formula,

$$\left(\frac{dE}{d\omega}\right)_{2^l\text{-pole}} = \frac{1}{32\pi}\frac{(l+2)!}{(l-2)!}\omega^2|A_l^{\text{out}}(\omega)|^2.$$

Two distinct methods were used to calculate $A_l^{\text{out}}(\omega)$: (i) direct integration of Eq. (1) with a numerical search technique to determine both the phase and the amplitude of the outgoing wave at infinity that would give a purely ingoing wave at the black-hole surface [details of this analysis done by two of us (M.D. and R.R.) will be published elsewhere]; (ii) integration by a Green's-function technique (see Zerilli[7]). This method allows the coefficient A_l^{out} to be computed directly as an integral involving the source term Eq. (4) and certain homogeneous solutions to Eq. (1).

All these calculations gave results in agreement within a few percent. The results are summarized in Figs. 1–3. The total energy radiated away in gravitational waves is

$$E_{\text{total}} \approx 0.0104mc^2(m/M). \tag{8}$$

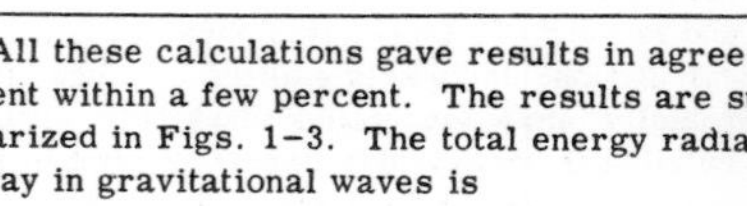

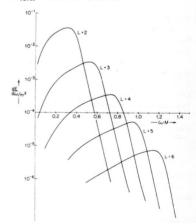

FIG. 2. Details of the spectrum of gravitational radiation integrated over all angles for the lowest five values of the multipoles.

This is about 6 times larger than Zerilli's estimate of the energy and 4 times larger than the estimate of Ruffini and Wheeler based on a purely linearized theory. The spectrum of the outgoing radiation is the superposition of a series of overlapping peaks, each peak corresponding to a certain multipole order l. Roughly 90% of the total energy is in quadrupole ($l=2$) radiation and 9% is in octupole ($l=3$). The total energy contributed by each multipole falls off quickly with l obeying the empirical relation (Fig. 3)

$$E_{2^l\text{-pole}} \approx (0.44m^2c^2/M)e^{-2l}. \tag{9}$$

The spectrum shown in Fig. 1 is for the energy integrated over all angles. An observer at a particular angle θ from the path of the particle's fall will see a slightly different spectrum because of the different angular dependence of the various 2^l-poles. For example, a pure 2^l-pole has the angular dependence

$$(dE/d\Omega)_{2^l\text{-pole}} = E_{2^l\text{-pole}}[(l-2)!/(l+2)!]\{2\partial_\theta{}^2 Y_0{}^l(\theta,\varphi) + l(l+1)Y_0{}^l(\theta,\varphi)\}^2. \tag{10}$$

As shown in Fig. 2, the energy contribution of progressively higher multipoles peaks at progressively higher angular frequencies, with the approximate relation

$$\omega(E_{2^l\text{-pole}}, \text{peak}) \approx \{c^2[V_l(r)]_{\max}\}^{1/2} \approx lc^3(27)^{-1/2}/GM \text{ for large } l. \tag{11}$$

Each energy peak may be interpreted as due to a train of gravitational waves produced by 2^l-pole normal-mode vibrations of the black hole which the in-falling body excites (see Press[9]). Averaging over angular factors and summing the various l's, one finds that the total spectrum is peaked at $\omega = 0.32c^3/GM$, and falls off at higher ω according to the empirical law

$$dE_{\text{total}}/d\omega \sim \exp(-9.9GM\omega/c^3) \tag{12}$$

Aside from the interesting details of our numerical results, the very fact that they are well behaved is important. Extrapolation of the flat-space linearized theory indicates that only a small fraction of a test body's rest mass $[\sim(m/M)mc^2]$ should be converted to wave energy during "fast" parts of its orbit (parts with durations $\sim GM/c^3$). It has been an open question whether this estimate holds in the region of strong fields very near the black hole. If the estimates were wrong, our results would have been divergent, with either increasing l or increasing ω. In fact, our results are strongly convergent.

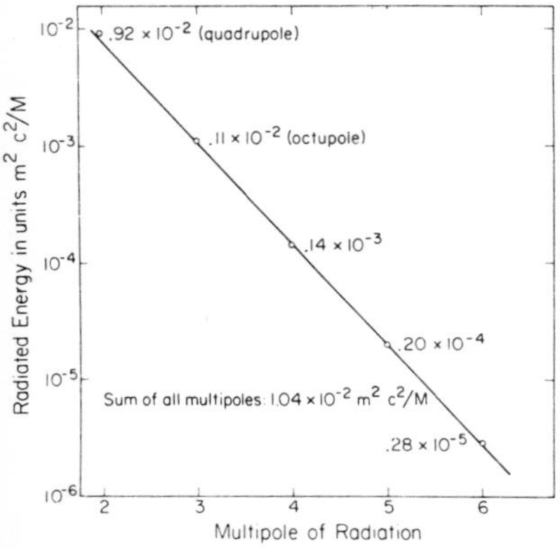

FIG. 3. Total energy radiated by each multipole. Quadrupole radiation contributes 90% of the total energy, and higher multipoles contribute progressively smaller amounts. The solid line is a plot of const e^{-2l} an empirical fit to the data.

The other side of the coin is equally important: Although our computation verifies the linearized theory's dimensional estimate, it shows that a completely relativistic treatment can give quantitative amounts of gravitational radiation substantially larger than the linearized theory would predict.[10]

This research was performed independently and simultaneously at Caltech and Princeton, using different integration techniques but arriving at identical results. We thank Kip S. Thorne and Jayme Tiomno for helpful suggestions.

---

*Work supported in part by the National Science Foundation under Grants No. GP-19887, No. GP-28027, No. GP-27304, and No. GP-30799X.

†Fannie and John Hertz Foundation Fellow.

‡Present address: Department of Physics, University of Utah, Salt Lake City, Utah 84112.

[1]J. Weber, Phys. Rev. Lett. 22, 1320 (1969), and 24, 276 (1970), and 25, 180 (1970).

[2]K. S. Thorne, Atrophys. J. 158, 1 (1969).

[3]J. R. Ipser, Astrophys. J. 166, 175 (1970).

[4]V. de la Cruz, J. E. Chase, and W. Israel, Phys. Rev. Lett. 24, 423 (1970).

[5]S. Hawking, to be published.

[6]R. Ruffini and J. A. Wheeler, in *Proceedings of the Cortona Symposium on Weak Interactions*, edited by L. Radicati (Accademia Nazionale Dei Lincei, Rome, 1971).

[7]F. J. Zerilli, Phys. Rev. D 2, 2141 (1970).

[8]T. Regge and J. A. Wheeler, Phys. Rev. 108, 1063 (1957).

[9]W. H. Press, to be published.

[10]M. Davis, R. Ruffini, and J. Tiomno, to be published, will give further details on the intensity and pattern of radiation for this problem and more general particle orbits.

A 3.4

Reprinted from:

PHYSICAL REVIEW D VOLUME 5, NUMBER 12 15 JUNE 1972

Pulses of Gravitational Radiation of a Particle Falling Radially into a Schwarzschild Black Hole*

Marc Davis, Remo Ruffini, and Jayme Tiomno†

Joseph Henry Laboratories, Princeton University, Princeton, New Jersey 08540

(Received 20 December 1971)

Using the Regge-Wheeler-Zerilli formalism of fully relativistic linear perturbations in the Schwarzschild metric, we analyze the radiation of a particle of mass m falling into a Schwarzschild black hole of mass $M \gg m$. The detailed shape of the energy pulse and of the tide-producing components of the Riemann tensor at large distances from the source are given, as well as the angular distribution of the radiation. Finally, analysis of the energy going down the hole indicates the existence of a divergence; implications of this divergence as a testing ground of the approximation used are examined.

In a recent series of investigations Zerilli,[1] Davis and Ruffini,[2] and Davis, Ruffini, Press, and Price[3] have analyzed the problem of a particle falling radially into a Schwarzschild black hole. In this paper this process is analyzed further. We are concerned with the features of the burst of the components of the Riemann tensor significant in the use of a detector and of the angular distribution of gravitational radiation. General and apparently contradictory considerations on the structure of a burst of gravitational radiation in black-hole physics were presented by Gibbons and Hawking[4] and by Press.[5] Some of the major features predicted in these two treatments are found indeed to be present in the detailed analysis of the physical example under consideration. An analysis for the radiation going into the hole is presented and its implications are examined.

We can expand[6] the perturbations $h_{\mu\nu}=g_{\mu\nu}-(g_{\mu\nu})_{\text{Schw}}$ of a Schwarzschild background in spherical harmonics of multipole orders l and m. In our case (a particle falling radially in along the z axis starting at rest from infinity) the "magnetic" and the $m\neq 0$ "electric" terms identically vanish (see Zerilli[1]). We have in this case for large values of the radial distance, in Zerilli's radiation gauge,

$$h_{\mu\nu}(t,r,\theta,\phi) \sim p_{\mu\nu}\sum_l R_l(r,t)\left(\frac{\partial^2}{\partial\theta^2}-\cot\theta\frac{\partial}{\partial\theta}\right)Y_{l0}(\theta,\phi)/2r \qquad (\mu,\nu=0,3). \quad (1)$$

Here $p_{\mu\nu}$ is the polarization tensor with the only nonvanishing components $p_{22}=r^2$ and $p_{33}=-r^2\sin^2\theta$. Thus the outgoing radiation will be totally polarized with the principal axes in the θ and ϕ directions. The function $R_l(r,t)$ satisfies the Zerilli equation which in Fourier-transformed form gives

$$\frac{d^2R_l(r,\omega)}{dr^{*2}}+[\omega^2-V_l(r)]R_l(r,\omega)=S_l(r,\omega). \quad (2)$$

Here $r^*=r+2M\ln(r/2M-1)$, $V_l(r)$ is the effective curvature potential, and $S_l(r,\omega)$ is the Fourier-transformed electric source term generated by the incoming particle expressed in tensor harmonics.[2,3] Equation (2) has been numerically integrated with the asymptotic boundary conditions

$$R_l(r,\omega)=\begin{cases}A_l^{\text{out}}(\omega)e^{i\omega r^*} & \text{as } r^*\to+\infty\\ A_l^{\text{in}}(\omega)e^{-i\omega r^*} & \text{as } r^*\to-\infty,\end{cases} \quad (3)$$

where A_l^{out} is given in the Green's function technique by

$$A_l^{\text{out}}(\omega)\propto\int_{-\infty}^{\infty}u_l(r^*,\omega)S_l(r^{*\prime},\omega)\,dr^{*\prime}, \quad (4a)$$

$$A_l^{\text{in}}(\omega)\propto\int_{-\infty}^{\infty}v_l(r^{*\prime},\omega)S_l(r^{*\prime},\omega)\,dr^{*\prime}. \quad (4b)$$

Here u_l (v_l) is the solution of the homogeneous equation obtained from (2) specifying a purely ingoing wave at $r^*=-\infty$ (outgoing at $r^*=+\infty$). By a further Fourier transformation we obtain the asymptotic expression

$$R_l^{\text{out}}(r^*,t)=\int_{-\infty}^{\infty}A_l^{\text{out}}(\omega)e^{i\omega(r^*-t)}\,d\omega. \quad (5)$$

The explicit results of this integration are given in Fig. 1(b) as a function of the retarded time $t-r^*$ for $l=2$.

The asymptotic expression of the tide-producing components of the Riemann tensor, which is what is measured by gravitational-wave detectors,[7] is easily obtained in the radiation region from

$$R_{\alpha\beta\gamma\delta} = \tfrac{1}{2}(h_{\alpha\delta,\beta\gamma} + h_{\beta\gamma,\alpha\delta} - h_{\alpha\gamma,\beta\delta} - h_{\beta\delta,\alpha\gamma}), \qquad (6)$$

where the comma (,) means ordinary derivative. If we assume the z axis is pointed along the line of propagation of the wave and the principal axes of polarization [see Eq. (1)] are pointed in the x and y directions, the only nonzero Newtonian tide-producing components of the Riemann tensor are, in our problem, $R^{y}{}_{0y0} = -R^{x}{}_{0x0}$, with

$$R^{y}{}_{0y0}(r^*, t) = \sum_{l} \ddot{R}_l(r^*, t) W_l(\theta, \phi)/2r . \qquad (7)$$

Here the dot indicates normal derivative with respect to time. The components of the Riemann tensor for selected l, without their angular dependence factor

$$W_l(\theta) = \left(\frac{\partial^2}{\partial\theta^2} - \cot\theta \frac{\partial}{\partial\theta}\right) Y_{l0}(\theta), \qquad (8)$$

and the factors $1/(8\pi)^{1/2} r$ are plotted in Fig. 1(c). Finally, we have also computed the outgoing energy flux from the stress-energy pseudotensor which becomes in the asymptotic region

$$t_{01} \sim \frac{1}{16\pi} \sum_{ll'} \dot{R}_l \dot{R}_{l'} W_l(\theta, \phi) W_{l'}(\theta, \phi)/4r^2 . \qquad (9)$$

In Fig. 1(d) we give the outgoing energy flux integrated over all directions for selected values of l. The interference between terms of different l is zero due to the orthogonality of the functions $W_l(\theta, \phi)$. As a check on our entire treatment we have verified that the total energy $\int_{-\infty}^{+\infty}(dE/dt)\,dt$ for every l agrees with the value $\int_{-\infty}^{\infty}(dE/d\omega)\,d\omega$ as given in Ref. 3 within 1%. From (7) we can compute the total flux per steradian; this quantity is plotted in Fig. 2. For pure quadrupole radiation the angular pattern of the radiation has a $\sin^4\theta$ dependence (z axis $\to \theta = 0$). Inclusion of higher multi-

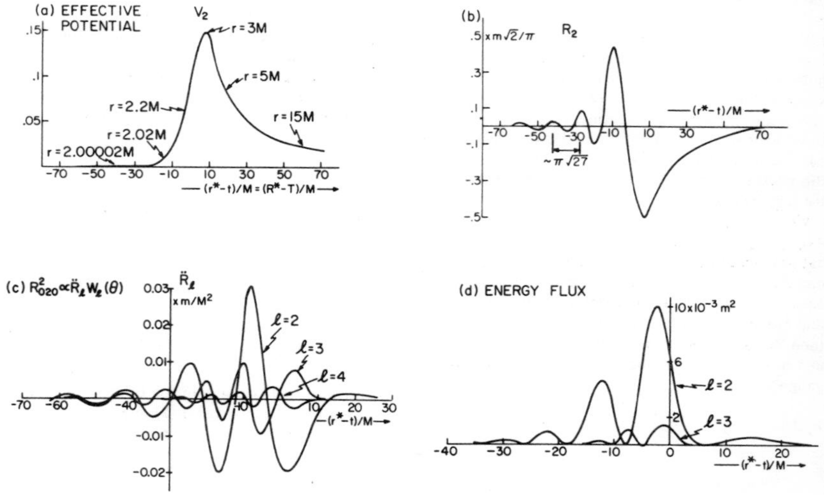

FIG. 1. Asymptotic behavior of the outgoing burst of gravitational radiation compared with the effective potential, as a function of the retarded time $(t - r^*)/M$. (a) Effective potential for $l = 2$ in units of M^2 as a function of the retarded time $(t - r^*)/M = (T - R^*)/M$. For selected points the value of the Schwarzschild coordinate r is also given. (b) Radial dependence of the outgoing field $R_l(r, t)$ as a function of the retarded time for $l = 2$. (c) $\ddot{R}_l(r^*, t)$ factors of the Riemann tensor components (see text) given as a function of the retarded time for $l = 2, 3, 4$. (d) Energy flux integrated over angles for $l = 2, 3$; the contributions of higher l are negligible.

poles introduces interference terms which tip the peak of the pattern forward by $7\frac{1}{2}°$. Figure 2 shows that there is no beaming of the radiation.

From the comparison of the different diagrams in Fig. 1 we can distinguish and characterize three different regions in the total energy flux:

(i) $5 \lesssim (r^* - t)/M \lesssim 30$, a precursor,

(ii) $-10 \lesssim (r^* - t)/M \lesssim 5$, a sharp burst,

(iii) $(r^* - t)/M \lesssim -10$, a ringing tail.

The precursor corresponds to the first part of the pulse as produced in the Ruffini-Wheeler approximation.[8] The sharp burst has a width $\sim 10M$ in agreement with the predictions on qualitative ground by Gibbons and Hawking[4] referring to any emission process taking place during the formation of (or capture by) a collapsed object. However, the present results do not support their suggestion that the "number of zeros" of the Riemann tensor could discriminate between sources of different origin since the ringing tail produces many zero crossings of the Riemann tensor. Finally, the oscillating tail has characteristic frequencies $\omega \sim l/\sqrt{27}$ which correspond to the ringing modes of the black hole found by Press.[5] It is interesting, however, that these ringing modes are energetically significant in this physical example only for low values of l, as is clear from Fig. 1(d).

A deeper insight in the three regions (i), (ii), and (iii) can be gained by the study of the effective potential plotted in Fig. 1 as a function of the retarded time $(t - r^*)/M$ referred to the observer at large distances. This is equal to $(T - R^*)/M$ as "seen" by the ingoing particle as the outgoing wave sweeps past it. R^* and T are the particle's position and Schwarzschild coordinate time computed from the geodesic trajectory (with $T = -\infty$ at $R^* = +\infty$, and $T = +\infty$ at $R^* = -\infty$). The implication is that for radiation "directly" emitted outward from the particle and not reflected, one can specify the radial position of the particle when it supposedly "emits" this radiation. Notice, for example, that the peak of the radiation flux occurs near retarded time $(r^* - t)/M = -2$, when the particle itself is at $2.3M$ (very near the horizon indeed), and just inside the peak of the curvature potential, which peaks at $r = 3M$.

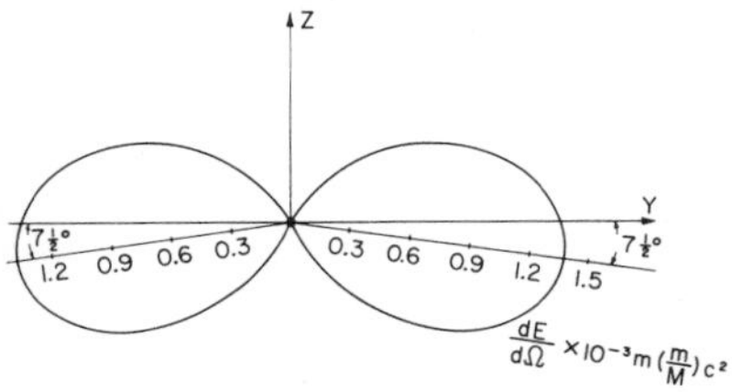

FIG. 2. Angular pattern of the radiation. The particle is supposed to fall down in the z direction which corresponds to $\theta = 0$.

We see that starting from large values of $(r^* - t)/M$ the field $R_l(r, t)$ builds up slowly and thus the energy emission (proportional to $\dot{R}_l^2$) and the Riemann tensor (proportional to $\ddot{R}_l$) gives rise to the very small "precursor" as the particle approaches the effective potential barrier. Concerning the emission of the main burst we have noticed in the evaluation of the integral (4a) for $A_l^{\text{out}}(\omega)$ starting from $r = 2M$ that the main contribution came from the interval $2.1 \lesssim r/M \lesssim 10$. The contributions beyond this point were oscillating, and very slowly damped, since the source decreases only as $r^{-1/2}$ for asymptotic distances. This averaging out of the contributions for large values of r is expected from the linearized theory of gravitation. Note that the ringing comes out after the particle is already inside the barrier. Here we cannot be seeing direct radiation from the particle because the driving source S_l is exponentially decaying, and the contributions to $A_l^{\text{out}}(\omega)$ for $2 \lesssim r/M \lesssim 2.1$ are very small. The wave emitted in this region for a given mode has a characteristic frequency as expected from the frequency spectrum calculated previously.[3] These facts suggest that part of the energy produced in the strong-burst region (ii) was stored in the "resonant cavity" of the geometry and then slowly released in the ringing modes.

We can now briefly summarize the main results of the analysis of the radiation going into the black hole. We have proceeded as follows: (1) Evaluate the amplitude A of the ingoing wave $R_l(r^*, \omega)$ for $r^* \to -\infty$ (purely ingoing waves), (2) solve for the scattering problem of Eq. (2) without source, imposing a purely ingoing wave of amplitude A at $r^* = -\infty$. As a consequence at $r^* = +\infty$ we have an ingoing wave with amplitude B and an outgoing wave with amplitude C. The energy flux going into the black hole is evaluated at $r^* = +\infty$, subtracting from the energy going in (proportional to $|B|^2$) the energy coming out (proportional to $|C|^2$). From the structure of the homogeneous Eq. (2) we also have $|A|^2 = |B|^2 - |C|^2$; therefore we can evaluate the energy going into the black hole simply by the same expression as used to calculate outgoing energy flux[1]:

$$\frac{dE}{d\omega} = \frac{1}{32\pi} \sum_l l(l-1)(l+1)(l+2)\omega^2 |A_l^{\text{in}}|^2 . \qquad (10)$$

The results of this analysis are given in Fig. 3. The spectral distribution for every multipole is

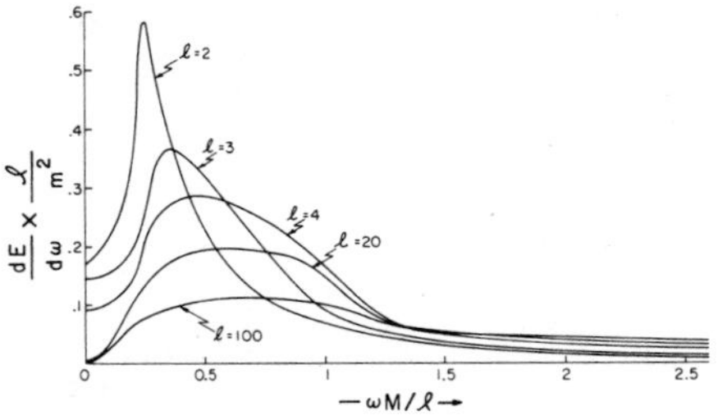

FIG. 3. Energy spectrum of the radiation going into the black hole for selected values of l. The total energy radiated per multipole is roughly constant and $\sim 0.25m^2/M$ for all considered l.

well behaved, and the total energy per multipole is roughly constant $\sim 0.25m^2/M$ (at least up to $l=100$). This is in contrast to the case of outgoing radiation, where the higher multipoles were exponentially damped.[3] It is plausible to assume that this behavior is indeed valid even for larger l. At first it would seem that the energy summed over all l diverges. This divergence results from the fact that we are treating the incoming object as a point particle. Under this assumption the Regge-Wheeler condition that the perturbation introduced by the particle be small by comparison with the background metric is no longer fulfilled. However, this circumstance is automatically eliminated as soon as a minimum size is assumed for the particle. We take as minimum size $2m$. From the preceding results we have seen that most of the radiation is emitted in the region near the horizon. This and the assumption of the minimum size $2m$ for the particle implies a cutoff in l given by $l_{\max} \simeq \frac{1}{2}\pi M/m$. The total energy going in is thus of the order

$$\frac{m^2}{4M}\frac{\pi M}{2m} \sim \frac{\pi m}{8}.$$

It is remarkable that the ingoing radiation does not depend, as does the outgoing radiation, on the ratio m/M, and that the amount of this radiation is a sizable fraction to the total rest mass of the ingoing particle, for an incident particle of minimum size.

Study of the numerical integration of $A_l^{\rm in}(\omega)$ shows that for calculations of ingoing radiation, one can neglect contributions beyond $r=5M$. Most of the ingoing radiation is generated inside the barrier, where $\omega^2 - V_l(r) < 0$, typically in the interval $2.01 \lesssim r/M \lesssim 3.5$. This large inward burst of energy is therefore generated in a finite Schwarzschild coordinate time interval; it occurs for $-20 < (r^* - t)/M < 15$ and vanishes as $(r^* - t)/M \to -\infty$. It occurs early enough that its reaction on the geodesic path of the incoming particle could affect the nature of the outgoing radiation, because the integral for $A_l^{\rm out}(\omega)$ has significant contributions beginning as $r \gtrsim 2.1M$. Further analysis may give a deeper understanding of this process, and details on exactly how the reaction of ingoing radiation affects the outgoing burst.

*Work partially supported by the National Science Foundation under Grant No. 30799X.

†At the Institute for Advanced Study, Princeton, N. J., when this work was initiated.

[1]F. Zerilli, Phys. Rev. D 2, 2141 (1970).

[2]M. Davis and R. Ruffini, Lett. Nuovo Cimento 2, 1165 (1971).

[3]M. Davis, R. Ruffini, W. Press, and R. Price, Phys. Rev. Letters 27, 1466 (1971).

[4]G. Gibbons and S. Hawking, Phys. Rev. D 4, 2191 (1971).

[5]W. Press, Astrophys. J. Letters 170, L105 (1971).

[6]T. Regge and J. A. Wheeler, Phys. Rev. 108, 1063 (1957).

[7]J. Weber, *General Relativity and Gravitational Waves* (Interscience, New York, 1961).

[8]R. Ruffini and J. A. Wheeler, in *Proceedings of the Cortona Symposium on Weak Interactions*, edited by L. Radicati (Accademia Nazionale die Lincei, Rome, 1971).

PHYSICAL REVIEW D VOLUME 7, NUMBER 4 15 FEBRUARY 1973

Gravitational Radiation from a Mass Projected into a Schwarzschild Black Hole*

Remo Ruffini
Joseph Henry Physical Laboratories, Princeton, New Jersey 08540
(Received 14 July 1972)

Gravitational radiation emitted by a particle projected with nonzero kinetic energy from infinite distance into a Schwarzschild black hole is examined. Direct comparison between a semirelativistic approach and the fully relativistic approach in the Regge-Wheeler-Zerilli formalism gives an insight into the nature of the results. Detailed spectral distributions are given. Contrary to the case in which the particle falls in with zero kinetic energy, the spectrum does not vanish any more at low frequencies and a considerably larger amount of radiation is emitted.

I. INTRODUCTION

In recent works Ruffini and Wheeler,[1] Davis and Ruffini,[2] Davis, Ruffini, Press, and Price,[3] and Davis, Ruffini, and Tiomno[4] have examined the details of a burst of gravitational radiation emitted by a particle falling radially into a Schwarzschild black hole. Different aspects of this problem were separately analyzed. In Ref. 1 two apparently contradictory assumptions are made: (a) The particle follows a geodetic in the curved Schwarzschild background, and (b) the system radiates as if it were in flat space, the energy radiated being simply proportional to the square of the third time derivative of the quadrupole moment of the system (see, e.g., Landau and Lifshitz[5]). The reason for adopting these approximations is simply explained: We can by an easy computational and analytic analysis put in evidence some of the main qualitative features of the spectrum of the radiation.

We leave to more detailed analysis the search for the exact numerical values, as well as the details of the polarization of the radiation and the study of the shape of the burst.

The subsequent analysis by Davis and Ruffini[2] and Davis, Ruffini, Press, and Price[3] was approached with the help of a different formalism and was mainly directed to analyze these details. Here the Regge-Wheeler formalism[6] as developed by Zerilli[7] is adopted. The spectrum is found to be qualitatively similar to the one given in Ref. 1, but the amount of radiation is approximately four times as large. Also, more details are given of the multipole distribution of the radiation. Nearly 90% of the total energy emitted is contained in the quadrupole mode ($l=2$), 9% in the octupole mode ($l=3$), and the remaining part in higher multipoles. In the subsequent paper by Davis, Ruffini, and Tiomno[4] the analysis is further generalized and the shape of the burst of the radiation is obtained for the first time.

In this paper it is our aim to examine the spectral distribution of the radiation emitted by a particle falling radially inward, starting its motion from infinite distance ($r=+\infty$) with a *finite* value of the kinetic energy. In the previous analysis we always supposed that the particle was falling inward starting *at rest* at infinity. That the spectrum of the radiation should indeed have completely new features can be deduced from the treatment given in Ref. 1. On the basis of these preliminary qualitative results we have developed the full Regge-Wheeler treatment. The spectrum, as expected using the approximation in Ref. 1, does not go to zero at low frequencies. A considerably larger amount of energy is emitted by the system consisting of the particle and the black hole. Detailed results of the total energy emitted as well as spectral and multipole distribution are given here.

II. LOW-FREQUENCY RADIATION IN SEMIRELATIVISTIC APPROACH

Following the formalism developed in Refs. 1 and 5, we have that the gravitational radiation emitted by a system with quadrupole moment Q_{rs} is given by

$$-\frac{dE}{dt}=\frac{1}{45}\dddot{Q}^{rs}\dddot{Q}_{rs}\ , \tag{1}$$

and the total amount of radiation emitted is given by

$$-\Delta E=\frac{1}{45}\int_{-\infty}^{+\infty}\dddot{Q}^{rs}(t)\dddot{Q}_{rs}(t)\,dt$$
$$=\frac{1}{45}\int_{-\infty}^{+\infty}\dddot{Q}^{rs}(\omega)\dddot{Q}_{rs}(\omega)\,d\omega\ . \tag{2}$$

Here and in the following we assume geometrical

units with $G=c=1$, and the dot represents differentiation with respect to time. In Eq. (2) $Q^{rs}(\omega)$ is the Fourier transform of the quadrupole moment, namely

$$\dddot{Q}^{rs}(\omega)=\frac{1}{\sqrt{2\pi}}\int_{-\infty}^{+\infty}\dddot{Q}^{rs}(t)e^{i\omega t}\,dt. \tag{3}$$

We are mainly concerned with the features of the spectrum at low frequencies. We then have

$$\lim_{\omega\to 0}\dddot{Q}^{rs}(\omega)=\frac{[\ddot{Q}^{rs}(t)]_{-\infty}^{+\infty}}{\sqrt{2\pi}}, \tag{4}$$

and for the emission of energy at low frequencies

$$\begin{aligned}-\frac{dE}{d\nu}&=-2\pi\frac{dE}{d\omega}\\ &=\tfrac{4}{45}\pi\dddot{Q}^{rs}(\omega\to 0)\,\dddot{Q}_{rs}(\omega\to 0)\\ &=\tfrac{2}{45}[\ddot{Q}^{rs}(t)]_{-\infty}^{+\infty}[\ddot{Q}_{rs}(t)]_{-\infty}^{+\infty}.\end{aligned} \tag{5}$$

Turning now from these general arguments to the case of a particle of mass m falling into a larger mass M, we have for the quadrupole moment of the system

$$Q^{rs}=m(3x^r x^s-\delta^{rs}|x|^2)+M(3X^rX^s-\delta^{rs}|X|^2), \tag{6}$$

where with $x^r(t)$ and $X^r(t)$ we indicate respectively the coordinates of the masses m and M. We then clearly have

$$\ddot{Q}^{rs}=m(3\ddot{x}^r x^s+6\dot{x}^r\dot{x}^s+3x^r\ddot{x}^s-2\delta^{rs}\ddot{x}^p x_p-2\delta^{rs}\dot{x}^p\dot{x}_p)$$
$$+\text{analogous term for the other mass.} \tag{7}$$

In the asymptotic regime ($r\to+\infty$), we also have

$$\ddot{x}^r=M(X^r-x^r)/r^3, \tag{8a}$$

$$\ddot{x}^s=m(X^s-x^s)/r^3. \tag{8b}$$

In the other asymptotic direction ($r\to 2m$) the particle, in the Ruffini-Wheeler[1] approximation, is "frozen" at $r\sim 2m$ in the final evolution of the implosion. From (7), (8a), and (8b) we can, therefore, immediately conclude that:

(a) Quite apart from the details of the implosion, uniquely from the asymptotic regimes, we can predict that the intensity of the gravitational radiation emitted has to vanish in the limit of low frequencies if the particle starts its implosion from rest. This effect was, indeed, clearly confirmed in the detailed analysis in Ref. 3.

(b) If the implosion takes place with a nonzero initial velocity ($\dot{x}^r\neq 0$) then the spectrum will not be mainly concentrated around frequencies ~$0.024/M$ (see Ref. 1) but will have sizable low-frequency components.

(c) The energy radiated at low frequencies should be proportional to the fourth power of the velocity, in the limit that the initial velocity of the particle at infinity is small.

From these general conclusions let us proceed to a detailed treatment.

III. RADIATION IN THE REGGE-WHEELER FORMALISM

The radiation as well as the particle of mass m are treated in the Regge-Wheeler formalism as a small perturbation of the given Schwarzschild background metric generated by the larger mass M:

$$g_{\mu\nu}=(g^0_{\mu\nu})_{\text{Schwarzschild}}+h_{\mu\nu}. \tag{9}$$

Using the symmetry properties of the background space the perturbations $h_{\mu\nu}$ have been expanded in tensorial spherical harmonics. The equations governing their radial dependence have been obtained by requiring that the metric $g_{\mu\nu}$ fulfill Einstein's equations with a source term given by the tensor energy-momentum of a pointlike particle of mass m.

Details of the treatment with associated gauge conditions are given by Zerilli in Ref. 7. For our purpose it is enough to know that the magnetic parity terms in the Zerilli formalism do not give any contribution in the case of radial fall. The solution for the electric parity terms reduces to the integration of equations governing the radial part $R_l(r)$ of the perturbation, the angular part being automatically described by the orthonormal set of spherical harmonics. We have

$$\frac{d^2R_l}{dr^{*2}}+[\omega^2-V_l(r)]R_l(r)=S_l(\omega,r). \tag{10}$$

Here

$$r^*=r+2M\ln(r/2M-1); \tag{11}$$

$$V_l(r)=\frac{(1-2M/r)[2\lambda^2(\lambda+1)r^3+6\lambda^2Mr^2+18M^2r\lambda+18M^3]}{r^3(\lambda r+3M)^2}, \tag{12}$$

λ being given by $\lambda=\frac{1}{2}(l-1)(l+2)$; and

$$S_l(\omega,r)=-\frac{8\pi(r-2M)}{r}\,\frac{\lambda r+3M}{\omega}\,\frac{d}{dr}\left[\frac{r(r-2M)}{(\lambda r+3M)^2}\,\frac{A_l^{(1)}(\omega,r)}{\sqrt{2}}\right]+\frac{r(r-2M)}{\lambda r+3M}A_l(\omega,r). \tag{13}$$

We have

$$A_{lm}(r,t)=m\,\frac{dt}{d\tau}\left(\frac{dR}{dt}\right)^2\frac{\delta(r-R(t))Y^*_{lm}(\Omega(t))}{(r-2m)^2}\,, \quad (14a)$$

$$A^{(1)}_{lm}(r,t)=\sqrt{2}\,im\frac{dt}{d\tau}\frac{dR}{dt}\,\frac{\delta(r-R(t))Y^*_{lm}(\Omega(t))}{r^2}\,, \quad (14b)$$

τ being the proper time of the inward-falling particle.

For a particle projected radially from infinity with nonzero initial velocity we have for the Fourier-transformed quantities (14a) and (14b)

$$A_l(\omega,r)=\left(\frac{m}{2\pi}\right)\left[l+\tfrac{1}{2}\left(\gamma^2-1+\frac{2m}{r}\right)\right]^{1/2}\frac{e^{i\omega T(r)}}{(r-2m)^2}\,, \quad (15a)$$

$$A^{(1)}_l(\omega,r)=-i\left(\frac{m}{2\pi}\right)\gamma(2l+1)^{1/2}\frac{e^{i\omega T(r)}}{r(r-2m)}\,, \quad (15b)$$

the quantity γ being

$$\gamma=\left(1-\frac{2m}{r}\right)\frac{dt}{d\tau}\,. \quad (16)$$

We have to integrate Eq. (10) with the usual boundary conditions of purely ingoing waves at the surface of the black hole and purely outgoing waves at infinity:

$$A_l(\omega)e^{i\omega r^*},\quad r^*\to+\infty \quad (17a)$$

$$B_l(\omega)e^{-i\omega r^*},\quad r^*\to-\infty\,. \quad (17b)$$

The integration of the equations has been carried out by using numerical Green's function techniques. More details will be given elsewhere (see Ref. 8).

IV. RESULTS AND CONCLUSIONS

In Fig. 1 we compare and contrast the results of the present analysis for $\gamma=1.2$ and $\gamma=1.4$ with the ones presented in Ref. 3 for $\gamma=1.0$.

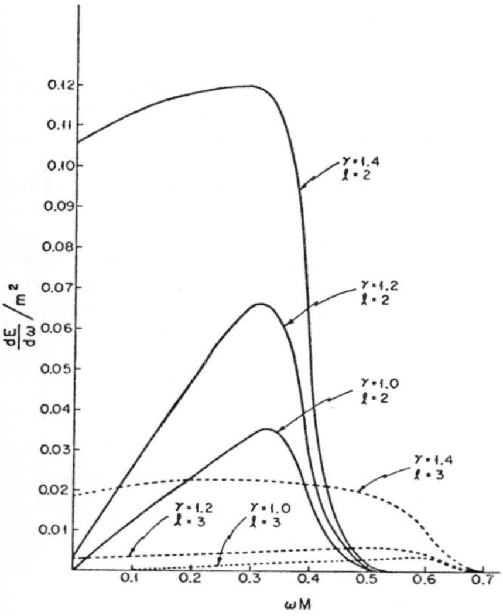

FIG. 1. Compared and contrasted are the spectra of the gravitational energy emitted by a particle radially projected into a Schwarzschild black hole. The solution for $\gamma=1.0$ refers to the case in which the particle starts at rest at infinity. The solid lines give the quadrupole radiation ($l=2$) for selected values of γ ($\gamma=1.0$, $\gamma=1.2$, $\gamma=1.4$). The dashed lines show the octupole radiation ($l=3$) for the corresponding selected values of γ.

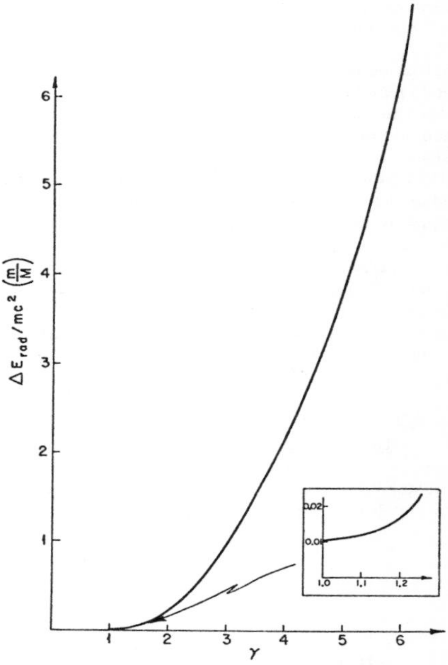

FIG. 2. Total energy radiated outward by a particle projected from infinite distance into a Schwarzschild black hole as a function of γ. The radiation for $\gamma\sim1$ is here magnified to allow a direct comparison with the treatment presented in Sec. III.

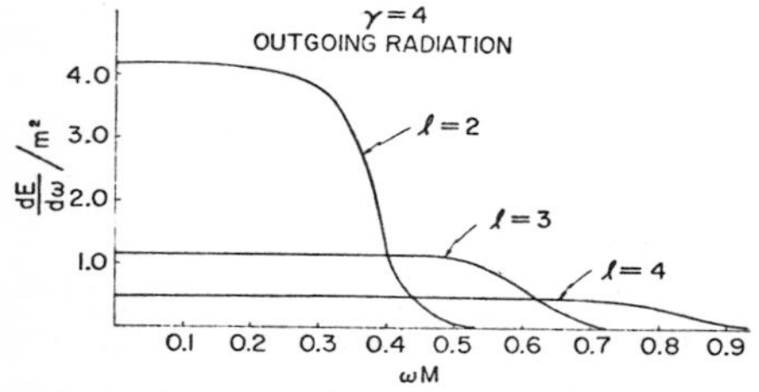

FIG. 3. Gravitational radiation emitted outward by a particle projected from infinity with $\gamma = 4$. The contribution due to lowest multipoles as well as the total energy emitted is clearly enhanced.

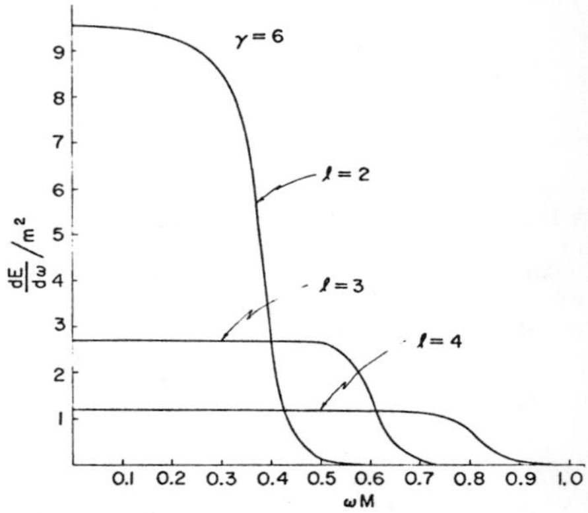

FIG. 4. Gravitational radiation emitted outward by a particle projected from infinity with $\gamma = 6$. The $l = 3$ mode is approximately one half of the $l = 2$; the $l = 4$ one fourth. The contribution due to the $l > 2$ modes is much larger than the one given in Ref. 3 for the case $\gamma = 1.0$.

As we expected already on the basis of the semi-relativistic approach presented in Sec. III, we can reach the following conclusions:

(1) The spectral distribution does not go to zero at low frequencies.

(2) The amount of energy increases substantially with γ, and at least for values of γ near unity the qualitative behavior suggested in Sec. III is, indeed, found to be correct (see Fig. 2).

(3) The contribution of higher-order multipoles becomes more and more significant for larger values of γ.

(4) To the sizable change in the amount of energy radiated there does not correspond a sizable change in the *maximum* frequency of the spectrum. The bulk of the radiation is still emitted at values of $\omega \le 0.7/M$.

These results are made even more transparent and enhanced by going to larger values of γ. In Figs. 3 and 4 we summarize the results, respectively, for $\gamma = 4$ and $\gamma = 6$. Finally, Fig. 2 summarizes the results for the total energy radiated summed over all multipoles for values of $\gamma > 1$. The solution for $\gamma < 1$ corresponding to the motion of a particle falling inward from rest at *finite* distance presents difficulties in handling the Fourier transforms properly. We do not, however, expect any new peculiar property in this regime apart from a reduction in the amount of radiated energy.

The violent dependence of the energy emitted on the value of the initial kinetic energy (see Fig. 2), while predictable on the ground of the work of Ruffini and Wheeler,[1] is of the greatest interest and importance for the understanding of the nature of gravitational radiation.

On the other hand, from a realistic astrophysical point of view we find it hard to conceive such a relativistic source. Details of the polarization and gravitational radiation going *into* the black hole will be presented elsewhere.[8]

ACKNOWLEDGMENT

It is a pleasure to acknowledge discussions with J. A. Wheeler on some aspects of this problem.

---

*Work partially supported by NSF Grant No. 30799X.

[1]R. Ruffini and J. A. Wheeler, in *The Significance of Space Research for Fundamental Physics*, edited by A. F. Moore and V. Hardy (European Space Research Organization, Paris, 1971).

[2]M. Davis and R. Ruffini, Lett. Nuovo Cimento 2, 1165 (1971).

[3]M. Davis, R. Ruffini, W. Press, and R. Price, Phys. Rev. Letters 27, 1466 (1971).

[4]M. Davis, R. Ruffini, and J. Tiomno, Phys. Rev. D 5, 2932 (1972)

[5]L. Landau and E. Lifshitz, *Théorie des Champs* (MIR, Moscow, 1970).

[6]T. Regge and J. A. Wheeler, Phys. Rev. 108, 1063

A 3.6

Can Synchrotron Gravitational Radiation Exist?*

Marc Davis and Remo Ruffini
Joseph Henry Physical Laboratories, Princeton, New Jersey 08540

and

Jayme Tiomno
Institute for Advanced Study, Princeton, New Jersey 08540

and

Frank Zerilli
University of North Caroline, Chapel Hill, North Carolina 27514
(Received 10 March 1972)

A complete relativistic analysis for gravitational radiation emitted by a particle in circular orbit around a Schwarzschild black hole is presented in the Regge-Wheeler formalism. For completeness and contrast we also analyze the electromagnetic and scalar radiation emitted by a suitably charged particle. The three radiation spectra are drastically different. We stress some important consequences and astrophysical implications.

It has been recently suggested by Misner,[1] Misner *et al.*,[2] and Campbell and Matzner[3] that in the emission process of gravitational radiation, high beaming due to synchrotron effects could take place in extremely relativistic regimes. Indeed the existence of this phenomenon would be of great importance for experimental detection of gravitational radiation. The required total energy corresponding to an observed event may be very much smaller than usually estimated if (in order of importance !) (a) the beaming effect exists; (b) a privileged plane of emission is found for the beamed radiation; and (c) the detector happens to be in that plane. The failure to fulfill one of these three circumstances would make the phenomenon largely uninteresting. In this Letter we address ourselves to condition (a). We analyze the gravitational radiation emitted by a particle moving in the field of a Schwarzschild black hole in stable ($r > 6M$) as well as unstable ($3M \leqslant r \leqslant 6M$) circular orbits (geometrical units, $G = c = 1$). To compare and contrast the results we also give the explicit analytic formulas and the energy fluxes for the cases of a charged particle emitting electromagnetic radiation (for details see Ruffini, Tiomno, and Vishveshwara,[4] Ruffini and Tiomno,[5] and Denardo, Ruffini, and Tiomno[6]) and a particle[7] emitting scalar radiation in the same orbit. As a biproduct of our results it will become evident that an extrapolation from the results obtained in the case of scalar radiation to the case of gravitational synchrotron radiation does not properly account for the complexity of the tensor field. The complete treatment and details of these works will be published in later papers.[5,6,8]

Before giving the main results of our treatment let us recall that the circular orbits between $3M \leqslant r \leqslant 6M$ are all *unstable* and therefore unphysical.[9] The captured star will in fact spiral in down to $r = 6M$ ($r = M$ in the case of a co-rotating star in the extreme Kerr geometry) in the family of stable circular orbits and then plunge in. However, these stable configurations are uninteresting from the point of view of beaming of the radiation because the velocity of the particle is too small to have any beaming effect.[10,11] We here consider the orbits $3M \leqslant r \leqslant 6M$ only to explore the more fundamental question of the physics of the emission of gravitational radiation in a case where the particle does indeed reach a velocity comparable to the velocity of light. We have to use a fully relativistic formalism in the sense that (a) it be valid even for $v/c \sim 1$; (b) it takes into proper account the contribution from the given background. Thus we use the Regge-Wheeler[12] formalism with the Green's function techniques as previously developed by Zerilli,[13] Davis and Ruffini,[14] Davis, Ruffini, Press, and Price,[15] and Davis, Ruffini, and Tiomno.[16]

The power emitted from a particle in a circular orbit around a Schwarzschild black hole is given in all the three cases (scalar, vector, tensor) by

$$P(\omega) = \sum_{l,m>0} \frac{\omega^2}{2\pi}\left(|R_{(M)}{}^{lm}|^2 + |R_{(E)}{}^{lm}|^2\right). \tag{1}$$

Here $\omega = m\omega_0$ is the frequency of the radiation, with $\omega_0 = (M/r_0{}^3)^{1/2}$, r_0 being the radial coordinate of the circular orbit. The functions $R_{(M)}{}^{lm}$ (nonexistent in the scalar case) have parity $(-1)^{l+1}$;

TABLE I. The magnetic and electric components of the power radiated from a particle in circular orbit are here given in the case of scalar, vector, and tensor radiation. M is the mass of the black hole, m_0 the mass of the particle, $\gamma = dt/ds = (1-3M/r_0)^{-1/2}$, $r^* = r + 2M\ln(r/2M - 1)$, and r_0 and r_0^* are the orbit's coordinates. W is the Wronskian of the two independent solutions $u(r)$ and $v(r)$, and $\omega_0 = (M/r_0^3)^{1/2}$.

	$(-1)^{\ell}$ electric	$(-1)^{\ell+1}$ magnetic
Scalar	$R^{\ell m}_{(E)}(n\omega_o, r^*) = 4\pi s Y^m_{\ell}(\frac{\pi}{2}, 0)\, \frac{u(r^*)v(r^*_o)}{\gamma r_o W}\, \delta^m_n$	none
Vector	$R^{\ell m}_{(E)}(n\omega_o, r^*) = \frac{4\pi q Y^m_{\ell}(\frac{\pi}{2}, 0)\delta^m_n}{[\ell(\ell+1)]^{\frac{1}{2}}\, W}\, u(r^*)\, \frac{d}{dr^*_o}\, v(r^*_o)$	$R^{\ell m}_{(M)}(n\omega_o, r^*) = -\, 4\pi q\omega_o Y^{m+1}_{\ell}(\frac{\pi}{2}, 0)\delta^m_n C^{\ell}_m\, \frac{u(r^*)v(r^*_o)}{W}$
Tensor	$R^{\ell m}_{(E)}(n\omega_o, r^*) = 4\pi m_o \gamma D^{\ell}_m Y^{\ell}_m(\frac{\pi}{2}, 0)\delta^m_n\, u(r^*)\, \frac{1}{W} \times$ $\times \left[\alpha(r_o)v(r^*_o) + \frac{1}{\lambda}\frac{d}{dr^*_o}\left(\beta(r_o)v(r^*_o)\right)\right]$ $\lambda = \frac{1}{2}(\ell-1)(\ell+2)$ $\beta(r_o) = \left(1 - \frac{2M}{r_o}\right)\left(1 + \frac{3M}{\lambda r_o}\right)^{-1}$ $D^{\ell}_m = ((\ell-1)(\ell+2))^{\frac{1}{2}}\,(\ell(\ell+1))^{-\frac{1}{2}}$ $\alpha(r_o) = \frac{r_o - 2M}{(r_o + \frac{3M}{\lambda})^2}\left(1 + \frac{1}{\lambda} + \frac{M}{\lambda r_o} - \frac{3M}{\lambda^2 r_o} + \frac{6M^2}{\lambda^2 r_o^2}\right) -$ $-\,(2m^2 - \ell(\ell+1))\,(\ell(\ell+1) - 2)^{-1}\,\omega_o^2 r_o$	$R^{\ell m}_{(M)}(n\omega_o, r^*) = 4\pi m_o \gamma C^{\ell}_m Y^{\ell}_{m+1}(\frac{\pi}{2}, 0)\omega_o\, \frac{\delta^m_n\, u(r^*)}{\sqrt{(\ell-1)(\ell+2)}\, W} \times$ $\times \frac{d}{dr^*_o}\,[r_o v(r^*_o)]$ $C^{\ell}_m = (\ell(\ell+1) - m(m+1))^{\frac{1}{2}}\,(\ell(\ell+1))^{-\frac{1}{2}}$

the functions $R_{(E)}{}^{lm}$ have parity $(-1)^l$. They are computed in the asymptotic region $r \to +\infty$. Their structure is different depending on whether they refer to the scalar, electromagnetic, or gravitational case as summarized in Table I. They are expressed in terms of the two radial functions $u(r_*)$ and $v(r_*)$ (purely outgoing wave at infinity and ingoing wave at the black-hole surface, respectively) which are solutions of Schrödinger-type equations:

$$d^2u/dr_*{}^2 + (\omega^2 - V_{\rm eff})u = 0. \tag{2}$$

Here $\omega = n\omega_0$, $r_* = r + 2M\ln(r/2M - 1)$, and $V_{\rm eff}(r)$ will depend on the particular field under examination. In the limit of high l, the potential approaches $(1 - 2M/r)l(l+1)/r^2$ for all three fields. We have integrated Eq. (2) numerically for the three cases.[17] The powers radiated at the orbit $r = 3.05M$ are given in Fig. 1. The difference in the three cases is manifest. Only for scalar radiation can we find something similar to what is usually called synchrotron radiation: *most* of the radiation concentrated at high multipoles. The contributions of lowest multipoles is significant in the case of electromagnetic radiation and very important in the case of gravitational radiation.

To complete the analysis of gravitational radiation from circular orbits, we have summarized some of the main results in Fig. 2. The energy emitted is here plotted again as a function of the harmonic index l for different circular orbits corresponding to different radii. In particular it is clear from these results that for the gravitational radiation (1) the beaming (high-l component) is insignificant for orbits up to the last stable circular orbit; (2) for orbits $3M \leq r \leq 6M$ we have indeed an enhancement of high-l components. *However*, the lower multipoles continue to give very substantial contributions.

The analysis of the physical reasons for the drastic differences in the three cases is under examination and will be presented elsewhere. It is evident that the "water-sprinkler effect" of the radiation arises because even zero-mass quanta in extremely relativistic orbits ($r \to 3M$)

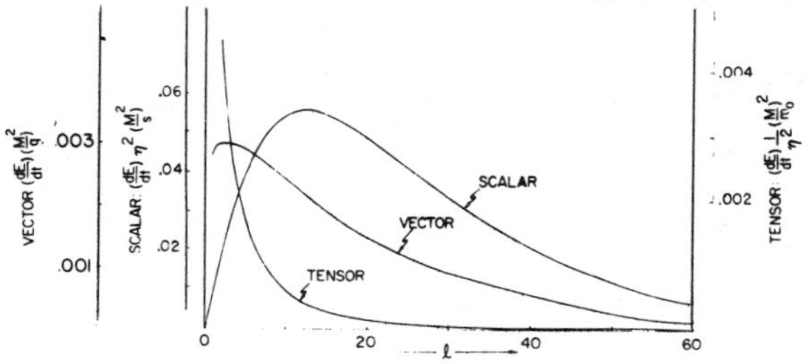

FIG. 1. Scalar, vector, and tensor radiation as functions of l compared and contrasted. The particle is assumed to be in a circular orbit at $r = 3.05M$ with $\eta^2 = (1 - 3M/r_0)^{-1} = 61$. Here, m is the mass, q the electric charge, and s the scalar charge of the particle. Since most of the radiation (~97%) comes from the $l = m = n$ modes, this plot is in fact a power spectrum.

are strongly influenced by the background geometry.[18] In this connection it has been proposed[18] that a particle *not* in a geodesic orbit but suitably propelled by nongravitational forces to acquire relativistic velocities should emit "synchrotron"-like gravitational radiation. We do not think however that this effect has any realistic astrophysics application. We are currently considering the corresponding effect in the case of electromagnetic radiation by analyzing the motion of a relativistic particle in a Reissner-Nordstrom background.[6] We summarize our results as follows: (a) No gravitational synchrotron radiation can be expected from stable circular orbits in Schwarzschild geometry (this result has been generalized to the stable Kerr orbits both co-rotating and counter-rotating[10]); (b) for unstable circular orbits a large amount of the radiation is still emitted at low multipoles, and thus at larger angles. In computing the effect of the enhancements of the highest multipoles we have to compensate for this defocusing effect.

These results raise serious questions about the effectiveness of the so-called gravitational

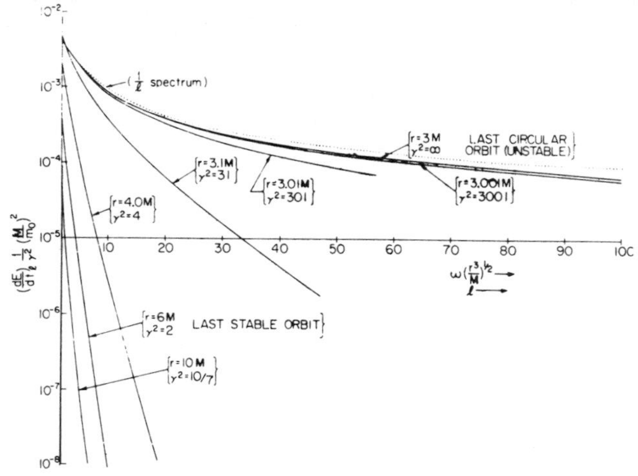

FIG. 2. The spectrum of gravitational radiation emitted by a particle in circular orbit around a black hole for selected values of the orbital radius. Beaming effects are totally negligible up to the last stable circular orbit ($6M$). When $r \rightarrow 3M$, higher multipoles are enhanced; however, the amount of energy radiated at lower multipoles is far from negligible.

synchrotron radiation mechanism in concentrating energy into a plane. We think, however, that quite apart from any discussion of gravitational synchrotron radiation a detailed analysis of the polarization, intensity, and angular distribution of the electromagnetic and gravitational radiation emitted by material falling into a collapsed object can greatly help in the understanding of the physical processes and in the design of the most descriminating detectors.

We are thankful to John A. Wheeler for encouragement in this work.

---

*Work partially supported by the National Science Foundation under Grant No. 30799X.

[1]C. W. Misner, Phys. Rev. Lett. 28, 994 (1972).

[2]C. W. Misner, R. A. Breuer, D. R. Brill, P. L. Chrzanovski, H. G. Hughes, III, and C. M. Pereira, Phys. Rev. Lett. 28, 998 (1972).

[3]G. A. Campbell and R. A. Matzner, to be published.

[4]R. Ruffini, J. Tiomno, C. V. Vishveshwara, Nuovo Cimento Lett. 3, 211 (1972).

[5]R. Riffini and J. Tiomno, to be published (derivation of equation in Table I for the electromagnetic case, with details of polarization and radiation features).

[6]G. Denardo, R. Ruffini, and J. Tiomno, to be published (radiation features for very highly bound orbits in the Reisner-Nordstrom field).

[7]Details of this problem can be found in Ref. 2. In Table I we have adapted the formulas of Misner *et al.* for the scalar case to our Green's function approach.

[8]M. Davis, R. Ruffini, and J. Tiomno, to be published (derivation of gravitational radiation equations given in Table I, and a detailed analysis of the polarization of the radiation).

[9]For analysis of geodesics in the Schwarzschild and Kerr cases see R. Ruffini and J. A. Wheeler, in *The Significance of Space Research for Fundamental Physics*, edited by A. F. Moore and V. Hardy (European Space Record Organization, Paris, 1971).

[10]R. Ruffini, Bull. Amer. Phys. Soc. 17, 449 (1972).

[11]Moreover, tidal effects have to be taken into account as shown and computed by Ruffini and Wheeler (see Ref. 9) and B. Mashoon (to be published). The validity of the "particle" approximation should be carefully analyzed for any astrophysical application!

[12]T. Regge and J. A. Wheeler, Phys. Rev. 108, 1063 (1957).

[13]F. Zerilli, Phys. Rev. D 2, 2141 (1970).

[14]M. Davis and R. Ruffini, Nuovo Cimento Lett. 2, 1165 (1971).

[15]M. Davis, R. Ruffini, W. Press, and R. Price, Phys. Rev. Lett. 27, 1466 (1971).

[16]M. Davis, R. Ruffini, and J. Tiomno, Phys. Rev. D 5 (to be published).

[17]It has been shown that for very large values of l it is possible to give analytic solutions of Eq. (2) which are the same in all three cases (see Ref. 2 and R. A. Breuer, P. L. Chrzanowski, H. G. Hughes, and C. W. Misner, to be published). This result is currently being used to give asymptotic formulas ($l > 100$) for the electromagnetic and gravitational case by R. A. Breuer, R. Ruffini, J. Tiomno, and C. V. Vishveshwara (to be published).

[18]R. Ruffini, unpublished.

[19]The spectrum of bursts of gravitational radiation emitted by a pair of particles in narrow elliptical orbits about their mutual center of gravity has been analyzed in the linearized approximation by Ruffini and Wheeler (Ref. 9). They point out a clear enhancement of high multipoles. Quite independently from the present work on synchrotron radiation, we are currently analyzing this interesting process in a completely relativistic approach.

A 3.7

Polarization of Gravitational and Electromagnetic Radiation from Circular Orbits in Schwarzschild Geometry*

Remo Ruffini

Joseph Henry Physical Laboratories, Princeton, New Jersey 08540

Frank Zerilli+

University of North Carolina, Chapel Hill, North Carolina 27514

ABSTRACT

We give explicit analytic formulae for the polarization of gravitational and electromagnetic radiation emitted by particles in circular ultrarelativistic (unbounded) orbits in the Schwarzschild geometry. Also the response of a gravitational wave detector to this radiation field is analyzed.

*Work partially supported by National Science Foundation Grant GP-30799X

+Present Address: Department of Astronomy, University of Washington, Seattle, Washington 98195

In a recent letter, gravitational, electromagnetic and scalar radiation were examined in the limit of ultrarelativistic (unbounded) orbits by Davis, Ruffini, Tiomno, Zerilli[1]. In that paper the main features of the power spectrum of the radiation were summarized; a further generalization of these results has been accomplished by Breuer, Ruffini, Tiomno, Vishveshwara[2]. These authors, using asymptotic formulae for the Greens function have been able to show that the main features of the spectrum, discovered in Ref. (1) are, indeed, valid even in the extreme relativistic regimes where the particle has an orbit infinitely near to 3M (M being the mass of the black hole). In this letter our main goal is to give detailed results of the polarization pattern both for tensor and vector radiation from these orbits. Some indication of the possible polarization limited to the gravitational case have been presented by Khalilov, Loskutov, Sokolov, and Ternov[3] in the framework of a semi-relativistic approach. In the present paper we give the explicit analytic formulae for the polarization which can be obtained from the relativistic Regge-Wheeler formalism. Our results are not in agreement with the ones presented in Ref. (3).

It was already pointed out in Ref. (1) that (a) no physical mechanism has yet been conceived to enhance the highest multipole radiation characteristic of ultrarelativistic orbits and (b) the gravitational and electromagnetic case are less effective than the scalar one in beaming the radiation. The fact that both electromagnetic and gravitational radiation are purely linearly polarized in the plane determined by the circular orbit of the source is the main conclusion of this paper which puts a "signature" to this radiation from ultrarelativistic orbits. The interest in the polarization of the radiation as a tool to infer the location of the source of graviational radiation has been, recently, emphasized by Douglas and Tyson[4].

In the Regge-Wheeler formalism the motion of the particle in the given background metric is described by a perturbation

$$g_{\alpha\beta} = (g_{\alpha\beta})_{schw} + h_{\alpha\beta} \qquad (1)$$

where the symmetric tensor $h_{\alpha\beta}$ is expanded in a complete set of tensor harmonic. The field equations for perturbations are immediately obtained by writing the Einstein equations for the perturbed metric, with the energy-momentum tensor of a test particle in a circular orbit[5]. The polarization of the radiation is completely determined by the analysis in the asymptotic region (far away from the source) of the first order corrections to the Riemann tensor of the background geometry.

$$R^{(1)}_{\alpha\beta\delta\sigma} = \frac{1}{2}(h_{\alpha\sigma,\beta\delta} + h_{\beta\delta,\alpha\sigma} - h_{\alpha\delta,\beta\sigma} - h_{\beta\sigma,\alpha\delta}) \qquad (2)$$

We have in fact, asymptotically

$$R_{joko} = -\frac{1}{2} h^{TT}_{jk,oo} \qquad (3)$$

Here as usual, we indicate by h^{TT} the transverse traceless part of the radiation field. The only non-vanishing components are

$$R_{2020} = -R_{3030} = \Sigma_{\ell m} \beta_{\ell m}(\omega) e^{i\omega r^*}\left(\frac{\partial^2 Y^{\ell m}}{\partial\theta^2} + \frac{\ell(\ell+1)}{2} Y^{\ell m}\right)/(r-2m) \qquad (4.1)$$

$$R_{2030} = R_{3020} = \Sigma_{\ell m} \gamma_{\ell m}(\omega) e^{i\omega r^*}\left(\frac{1}{\sin\theta} \frac{2}{\partial\theta}\left(\frac{\partial y^{\ell m}}{\partial\theta} - \cot \partial y^{\ell m}\right)/(r-2m)\right. \qquad (4.2)$$

here $r^* = r + 2M\ lg(r/2M - 1)$ and $\omega = m\omega_o, \omega_o$ being the angular velocity of the particle in circular orbit. It would be extremely difficult to give any compact expression for the polarization if all the terms in the sums (4.1) and (4.2) were equally important. It is possible, however, to introduce a drastic simplification: in the case of circular orbits only the electric [parity $(-1)^{\ell}$] perturbations with $\ell = m$ give a significant contribution. Moreover, the contribution of the magnetic terms or of the term with $\ell \neq m$ are largely negligible. These results have been explicitely pointed out by Davis, Ruffini, Tiomno and Zerilli(1) and confirmed also in the asymptotic regimes by Breuer, Ruffini, Tiomno, and Vishveshwara(2).

We have, therefore, for the transverse traceless component of the perturbation written in orthonormal frame the very simple expression

$$h_{22}^{TT} = - h_{33}^{TT} \sim e^{i\omega r^*} \alpha_{\ell\ell}\,\ell(\ell-1)(1+\cos^2\theta)P_{\ell}^{\ell}(\cos\theta)/(2\omega(r-2m)\sin^2\theta) \qquad (5.1$$

$$h_{32}^{TT} = h_{23}^{TT} \sim e^{i\omega r^*} \alpha_{\ell\ell}\ \ell(\ell-1)\ i\ \cos\theta\ P_{\ell}^{\ell}(\cos\theta)/(\omega(r-2m)\sin^2\theta) \qquad (5.2$$

where $\alpha_{\ell\ell} = [(2\ell+1)/(4\pi\ 2\ell!)]^{1/2}$. We can now immediately conclude from these expressions that the gravitational radiation is:

(a) linearly polarized in the plane of the orbit ($\theta = \pi/2$).

(b) circularly polarized on the axis of rotation ($\theta = 0$).

(c) elliptically polarized in between.

In the case of electromagnetic radiation, we expand the vector potential associated with the particle moving in the circular orbit on the curved background in terms of four-dimensional vector spherical harmonic[6]. Once again only the electric parity $(-1)^{\ell}$ terms give a significant contribution and the magnetic parity can be neglected. Also the term with $\ell \neq m$ can be neglected[1][2]. We then have for the vector potential

$$A_{\mu}(r,\theta,\phi,t) = \sum_{\ell=|m|} \begin{vmatrix} 0 \\ 0 \\ \dfrac{a^{\ell m}(r,t)}{\sin\theta} \dfrac{\partial y^{\ell m}}{\partial \phi} \\ - a^{\ell m}(r,t) \sin\theta \dfrac{\partial y^{\ell m}}{\partial \theta} \end{vmatrix} \tag{6}$$

where $a^{\ell m}(r,t)$ satisfies a Schrodinger like equation[2]. In an orthonormal tetrad we obtain for the non vanishing components of the electromagnetic field

$$E_{\theta} = \Sigma_{\ell}\, a^{\ell\ell}(r,t)_{,t}\ \ell[(2\ell-1)!!][(2\ell+1)/(2\ell!4\pi)]^{\frac{1}{2}} e^{im(\phi+\pi/2)}/r \tag{7.1}$$

$$E_{\phi} = \Sigma_{\ell}\, a^{\ell\ell}(r,t)_{,t}\ [(2\ell-1)!!][(2\ell!4\pi)]^{\frac{1}{2}} \cos\theta\, e^{im\phi}/r \tag{7.2}$$

and

$$H_{\theta} = \Sigma_{\ell}\, a^{\ell\ell}(r,t)_{,r}\ \ell[(2\ell-1)!!][(2\ell+1)/(2\ell!4\pi)]^{\frac{1}{2}} \cos\theta\, e^{im\phi}/r \tag{8.1}$$

$$H_{\phi} = \Sigma_{\ell} -a^{\ell\ell}(r,t)_{,r}\ \ell[(2\ell-1)!!][(2\ell+1)/(2\ell!4\pi)]^{\frac{1}{2}} e^{i(m\phi+\pi/2)}/r \tag{8.2}$$

We, therefore, immediately conclude that the electromagnetic radiation emitted by a particle in a circular orbit in a gravitational field is:

(a) linearly polarized in the plane of the orbit

(b) circularly polarized on the axis of rotation

(c) elliptically polarized in between.

It is of interest to notice here that the polarization formulae are not, speaking, functions of the energy of the orbit and are valid as well in the case of non-relativistic orbits. The condition to be fulfilled in our derivation (main contribution coming only from electric perturbations with $\ell = m$)

holds also outside the extreme relativistic regime[1] . The <u>average</u> polarization of the radiation is, however, strongly dependent on the orbital energy of the particle. The beaming effects and the consequent processing of the radiation in the equatorial plane are, in fact, a direct result of the contribution due to the highest multipoles which in turn are enhanced only under extreme relativistic circumstances. For different energy of the orbits we have different angular distribution of the radiation and consequently different average polarization.

We can finally study some of the implications of the fact that the polarization of the gravitational radiation is, indeed, on hundred per cent polarized in the plane of the orbit. If we assume the source to be located at the center of the galaxy we can examine the response of a detector oriented east-west at the surface of the earth (see Fig. 1). Using the formula given by Ruffini and Wheeler[7] in the case of randon polarization, we have for the response function of the detector

$$W(H,\delta) = (\cos^2 H - \sin^2\delta \sin^2 H)^2 + (\sin\delta \sin 2H)^2 \tag{9.1}$$

H being the hour angle and δ the declination of the source. For a completely polarized source we have

$$W(H,\delta) = [\cos^2 H - \sin^2 H \sin^2 \delta)\cos 2\alpha + \sin 2\alpha \sin\delta \sin 2H]^2 \tag{9.2}$$

α being, in our case, the angle between the plane of the orbit of the source and the meridian for which the source is at the zenith ($\alpha = -31.7°$ for a cource at the galactic center and plane of polarization coinciding with the plane of the galaxy). The response function of the detector is given in Fig. (2). Also in Fig. (2) is the direct comparison with a source also located at the galactic center, but emitting unpolarized radiation. These results agree with the analysis presented by Tyson and Douglas (apart from a minus sign in their angle ψ)[4]. Considerably different in intensity

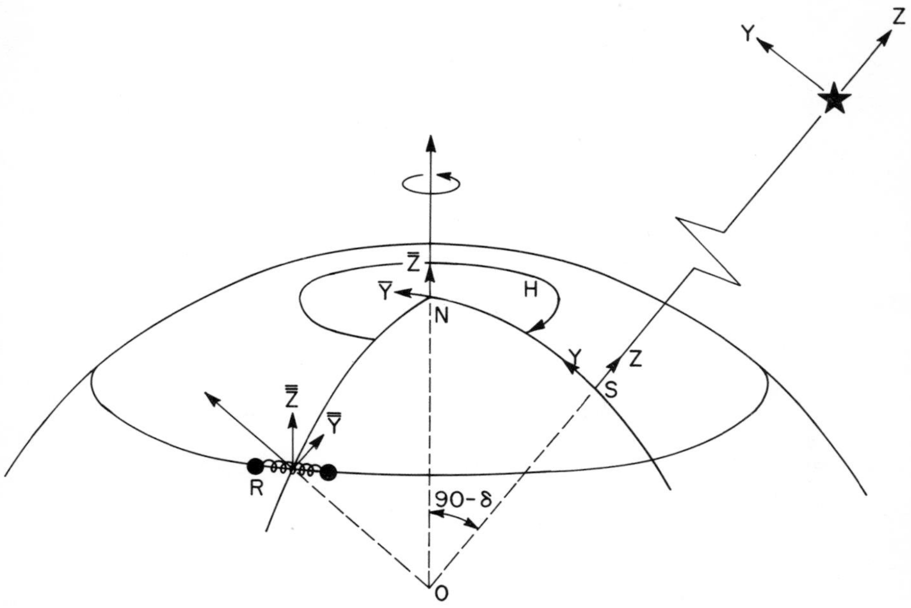

Fig. 1. (reproduced from R. Ruffini and J. A. Wheeler ref. 8, with their permission). An idealized detector of gravitational waves, R, on the surface of the Earth, is driven by a far away source. The coupling between source and detector is analyzed by transforming tensorial components in turn from the laboratory frame (double barred coordinates) to a frame at the North Pole (barred coordinates) and then to a frame at S. The angle α, in the text, is simply the angle between the plane of the source and the plane identified by the axis Y and Z.

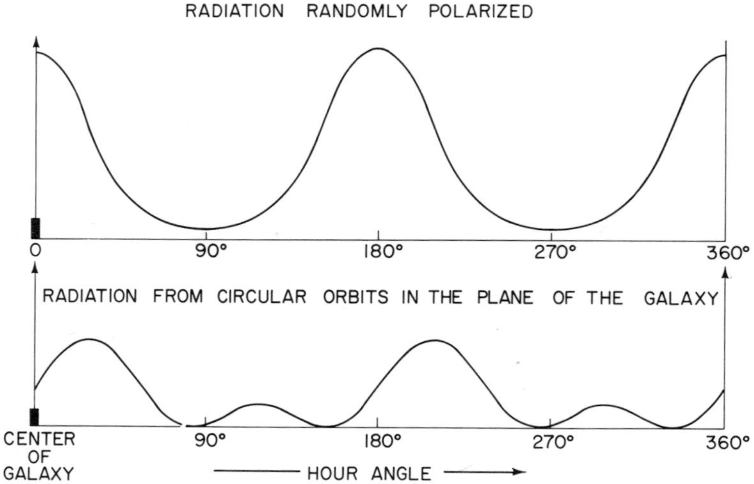

Fig. 2. Response of a detector of gravitational radiation, applied East-West as indicated in Fig. (1) to a source of gravitational radiation located at hour angle H and declination $\delta = -28^{o}.9$ (center of the galaxy). Compared and contrasted are the two examples in which the radiation is randomly polarized (upper part of the figure) and the radiation is fully polarized and originates from circular orbits in the plane of the galaxy (see text). The intensity as well as the peaks of the response function are clearly markedly different.

from the case of unpolarized radiation the peak of the response function of the detector <u>does</u> <u>not</u> take place now at hour angle 0° or 180°. This is of interest for a direct comparison with experimental results.

If we can compare these results with the one of a particle falling radially into a black hole[8] we can see what a large amount of information can be gained in the understanding of the sources of gravitational radiation just by the analysis of the shape and polarization of the pulses. It is, therefore, clear that, if gravitational radiation is indeed detected, a careful analysis of the polarization of the radiation by setting three detectors at different angles as suggested by Hamilton[9] and Fairbank[10] or just by changing the orientation of the detector as suggested by Tyson and Douglas[4] could greatly help in the indentification of the source of gravitational waves.

REFERENCES

1. M. Davis, R. Ruffini, J. Tiomno, F. Zerilli, Phys. Rev. Lett. 28, 1352 1972.

2. R. Breuer, R. Ruffini, J. Tiomno, C. V. Vishveshwara, Phys. Rev. Scheduled February 15.

3. V. R. Khalilov, Ju. M. Loskutov, A. A. Sokolov, I. M. Ternor, Phys. Lett. 42A, 43, 1972.

4. J. A. Tyson, D. H. Douglas, Phys. Rev. Lett. 28, 991, 1972.

5. F. Zerilli, Phys. Rev. 2, 2141, 1970.

6. R. Ruffini, J. Tiomno, C. V. Vishveshwara, Nuovo Cimento Lett. 3, 211, 1972.

7. R. Ruffini, J. A. Wheeler, Cosmology from Space Platform-ESRO Report SP. 52 - A. F. Moore and V. Hardy, ed.1971

8. M. Davis, R. Ruffini, J. Tiomno, Phys. Rev. - 5, 2932, 1972.

9. W. Hamilton - Invited talk at the A.P.S. Meeting Nov. 1970, New Orleans Bull. Am. Phys. Soc., Vol. 15, 1361, 1970.

10. W. Fairbank - Private communication.

PART IV Energetics of Black Holes

Introduction

The emphasis in Part I was on the unavoidability of the formation of black holes and on a critical approach to the key concept of critical mass against gravitational collapse. The essential role of X-ray sources in discriminating between neutron stars and black holes was also discussed. In Parts II and III the emphasis has been on black holes as "particle accelerators" and consequently as strong emitters of both gravitational and electromagnetic radiation. There the black hole acts only as a "catalyst" for the emission of radiation which takes place in its surrounding field. In this part the emphasis is on the black hole itself as a source of energy (Appendix 4.1 and 4.2) and on a strong field analysis of collapsed objects (Appendix 4.3).

In the previous two chapters our approach was based substantially on a perturbation analysis of the background geometry of a given black hole. This perturbation was usually driven by a test particle following a geodesic in the Schwarzschild or Reissner-Nordstrøm geometry. In Appendix 4.2 a different route is followed: we look for an exact two body solution. Using the approach due to Israel and Wilson[1] and to Perjes[2] an explicit form is given for the metric corresponding to two identical Kerr-Newman sources. They are not, of course, two black holes since

$$L^2/M^2 + Q^2 > M^2 \tag{1}$$

for each one of them and they should therefore be considered as two naked singularities in equilibrium under their mutual electromagnetic and gravitational forces. The spin of the two solutions are oppositely directed along the symmetry axis. First we have examined the coalescence of these two naked singularities. The problem of the coalescence of two black holes is one of the most complicated problems to be attacked. There are two different aspects to be solved: (a) changes in singularity structure and formation of a new horizon (b) emission of gravitational radiation due to the change of quadrupole moment of the system. In our case the system of two particles is in equilibrium (no matter what the distance between the particles is) under the two different effects of gravitational and electromagnetic interactions. The gravitational attraction is balanced by the Coulomb repulsion (the particles are assumed to have charge equal to mass); the repulsion due to the magnetic fields (the systems are indeed endowed with a magnetic field due to the presence of the spin and since they are rotating in opposite directions the interaction will be repulsive) is totally balanced by the presence of a *gravitational* analogue to the magnetic field which is attractive for oppositely directed spins and repulsive for aligned spins. We can therefore neglect in our case all the complications coming from the emission of gravitational radiation emitted during the coalescence and focus entirely on the problem of the analysis of the singularity's structure and the possible formation of a new global horizon. Since the two particles are counterrotating and Eq. (1) is identically satisfied we should expect that upon reducing their distance the system would approach an *extreme* Reissner-Nordstrøm solution with $Q = M$. This is not the case (see Appendix 4.3). More work is needed in this field to enlighten not least the main issue of the final stages of gravitational collapse from this "strong field approximation" point of view.

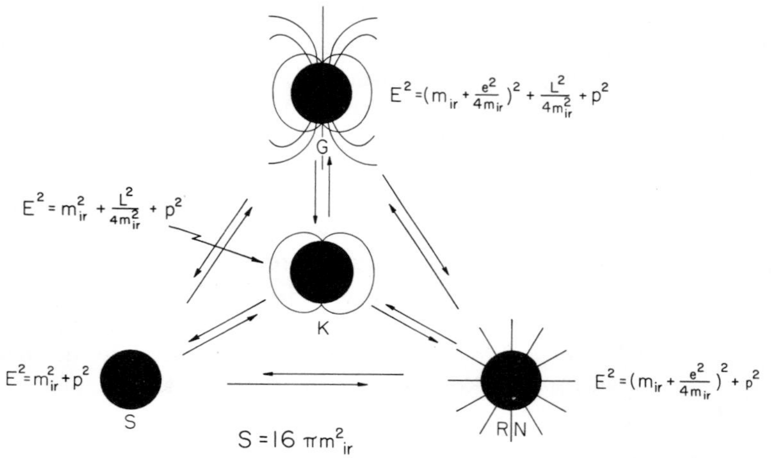

Figure 1. The entire family of black holes is here presented with the formulae determining their energy as a function of their characterizing parameters.

The perturbation approach analyzed in the last two chapters is of the maximum importance not only to compute realistically the process of the emission of radiation from test particles in extremely relativistic regimes but also as a theoretical tool to probe the *global* features of collapsed objects. The main idea here is to analyze the *reaction* of a black hole to the capture of infinitesimal and suitably directed "gedanken" projectiles! How difficult these projectiles are to be concretized in a realistic physical application is no matter here. Their "raison d'être" is only one of exploring and understanding, through a detailed mathematical analysis, the structure of black holes and obtaining formulae and relations valid "in se" and totally independent of the idealizations needed to obtain them. There is no better or more powerful example than the one summarized by the general formula governing the energetics of black holes (see Figure 1)

$$M^2 = \left(m_{ir} + \frac{Q^2}{4m_{ir}}\right)^2 + \frac{L^2}{4m_{ir}^2} \tag{2}$$

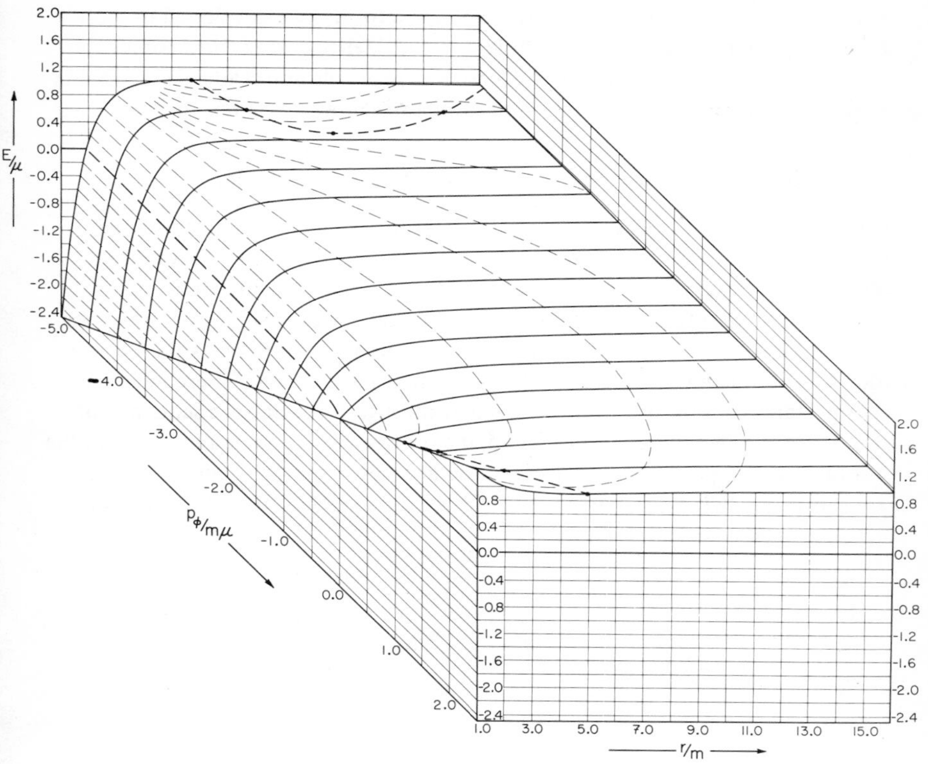

Figure 2. "Effective potential" experienced by a test particle moving in the equatorial plane of an extreme Kerr black hole $a/m = 1$. Here m is the mass of the black hole expressed in geometrical units $m = (G/c^2)m_{\text{conv}}$ and a is the angular momentum per unit of mass. We have considered values of angular momentum of the test particles ($p\phi/m\mu$) and distance r close to the critical region of transition from stable circular orbit (minimum in effective potential) to types of motion in which the capture of the test particle is immediate (no minimum in effective potential). In contrast to the case of the Schwarzschild black hole the behaviour of the effective potential is now considerably different depending on whether the angular momentum p_ϕ has positive or negative values. For negative angular momentum the last circular orbit occurs at $p\phi/m\mu = -22/(3\sqrt{3})$. Here μ is the mass of the test particle and r is the "Schwarzschild distance" (proper circumference/2π). The binding at this point is 3.78 per cent. For positive angular momentum we have circular orbits all the way down to $\hat{r} = r/m = 1$. The last circular orbit occurs at $\hat{p}_\phi = p_\phi/m\mu = 2\sqrt{3}$, $\hat{r} = 1$ with a value of the total energy $E = \mu/\sqrt{3}$. The binding energy is at this point ~42%. The effective potential is defined as the value of the energy which annuls the expression

$$\hat{E}^2(\hat{r}^3 + \hat{r} + 2) - 4\hat{E}\hat{p}_\phi - \hat{r}(\hat{r} - 1)^2 + (2 - \hat{r})\hat{p}_\phi^2 = 0$$

Of the two solutions of this equation we have here considered the one with the positive sign in front of the radical. The other solution (not shown) is easily obtained by noticing that the expression of the effective potential is unchanged with respect to a sinultaneous change of E in $-E$ and p_ϕ in $-p_\phi$. The potential energy surfaces for the two solutions ("positive" and "negative" energies) meet at the "knife edge" $\hat{r} - 1$, $\hat{E} = E/\mu = \hat{p}_\phi/2$. From Ruffini and Wheeler Ref. (5).

This formula was obtained in a form analogous to the one developed in section 2 for a Reissner-Nordstrøm black hole just considering a set of reversible "gedanken" transformations on the Kerr-Newman solution. Details are given in Appendix 4.1 and 4.2. But let us here summarize a few main results;

1 In What Sense Can Energy be Extracted from a Black Hole?

It has been pointed out[4] that energy, at least in principle, can be extracted from black holes. The simplest way to visualize this process is to look at the effective potential of an uncharged Kerr black hole (see Figure 2). Compare and contrast this effective potential with the one corresponding to a Reissner-Nordstrøm solution presented in Figure 2 of Part II. In both cases the positive root solutions can have negative values for their total energy. These negative energy states cannot, however, be reached through a continuous sequence of circular orbits. A splitting process and, *most essential*, a drastic reduction in the rest mass of the particles is needed. In the concrete example given in Figure 3 an initial particle with mass μ_0 and energy E_0 comes in from infinite distance

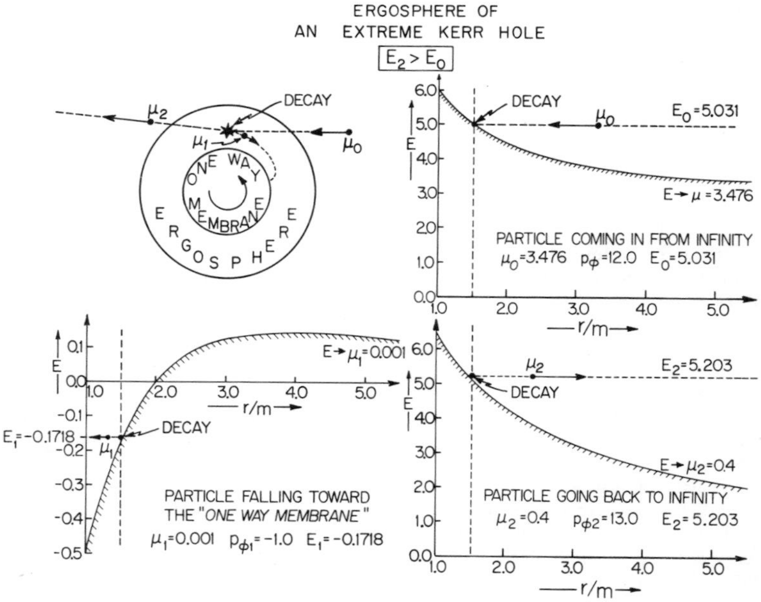

Figure 3. A particle of mass μ_0 coming from infinity with total energy E_0 and a positive value of the angular momentum p_ϕ can penetrate the ergosphere of an extreme Kerr hole and here decay into two particles. One particle of mass μ_1, negative value of the angular momentum $p_{\phi 1}$ and a negative value of the total energy E_1, falls towards and penetrates the one-way membrane. The second particle of mass μ_2, positive value of the angular momentum $p_{\phi 2}$ and a positive value of the total energy E_2, goes back to infinity. The remarkable new feature in this process is that the energy E_2 of the particle coming back to

and splits in the ergosphere of a Kerr black hole, one debris (of mass μ_1 and energy $E_1 < 0!$) is projected into the black hole, the other of mass μ_2 and energy E_2 comes out to infinite distance. We then have

$$E_2 > E_0 \tag{3}$$

but with

$$\mu_1 + \mu_2 \ll \mu_0. \tag{4}$$

Limitations on the efficiency of this process of energy extraction as a consequence of this drastic reduction in rest mass have been analyzed by B. Mashhoon. In the case of generalized Kerr solution both extraction of Coulomb and rotational energy are possible.

---

infinity is larger than the energy E_0 of the particle coming in. For simplicity we consider a decay process in the equatorial plane of a Kerr hole. The required conservation of the 4 momentum gives

$$E_0 = E_1 + E_2 \tag{a}$$

$$p_{\phi 0} = p_{\phi 1} + p_{\phi 2} \tag{b}$$

$$\begin{aligned}&[(E_0^2 - \mu_0^2)r^3 + 2\mu_0^2 mr^2 + (a^2E_0^2 - p_{\phi 0}^2 - a^2\mu_0^2)r + 2m(E_0a - p_{\phi 0})^2]^{1/2}\\ &= - [(E_1^2 - \mu_1^2)r^3 + 2\mu_1^2 mr^2 + (a^2E_1^2 - p_{\phi 1}^2 - a^2\mu_1^2)r + 2m(E_1a - p_{\phi 1})^2]^{1/2}\\ &\quad + [(E_2^2 - \mu_2^2)r^3 + 2\mu_2^2 mr^2 + (a^2E_2^2 - p_{\phi 2}^2 - a^2\mu_2^2)r + 2m(E_2a - p_{\phi 2})^2]^{1/2}\end{aligned} \tag{c}$$

Here a is the angular momentum J per unit of mass of the Kerr hole $a = J/m$. We have further simplified the problem by considering an extreme Kerr hole $a/m = 1$ and a decay process in which the incoming particle is at a turning point of its trajectory. Equations (a) and (b) are unaltered and Eq. (c) (conservation of radial momentum) simplifies to

$$\begin{aligned}&[(E_1^2 - \mu_1^2)\hat{r}^3 + 2\mu_1^2\hat{r}^2 + (E_1^2 - \hat{p}_{\phi 1}^2 - \mu_1^2)\hat{r} + 2(E_1 - \hat{p}_{\phi 1})^2]^{1/2} =\\ &[(E_2^2 - \mu_2^2)\hat{r}^3 + 2\mu_2^2\hat{r}^2 + (E_2^2 - \hat{p}_{\phi_2}^2 - \mu_2^2)\hat{r} + 2(E_2 - \hat{p}_{\phi 2})^2]^{1/2}\end{aligned} \tag{c'}$$

where $\hat{r} = r/m$ and $p_\phi = p_\phi/m$. The system of Eqs (a) and (b), (c') has been solved assuming $\hat{r} = 1.5$, $\mu_1 = 0.01$, $p_{\phi 1} = -1.0$, $\mu_2 = 0.40$, $p_{\phi 2} = 12.0$. An additional condition is the requirement that the total energies of the two particles E_1 and E_2 be larger than or equal to the respective turning point energies. One of the numerical solutions is given in this figure. On the upper left side a qualitative diagram shows the main feature of the decay process in the equatorial plane of the ergosphere of a Kerr hole. In the upper right side is the effective potential (energy required to reach r as a turning point) for the incoming particle. The effective potential is plotted at the lower left and at the lower right side for the particle falling toward the one-way membrane and for the particle going back to infinity. This "energy gain process" critically depends on the existence and on the size of the ergosphere which in turn depends upon the value a/m of the hole. In the equatorial plane we have for the ergosphere $1 + (1 - a^2/m^2)^{1/2} \leqslant r \leqslant 2$; when $a/m \to 0$ (Schwarzschild hole) the one-way membrane expands and coalesces with the infinite red shift surface, wiping out the ergosphere. The particle falling towards the one-way membrane will in general alter and reduce the ratio a/m of the black hole.
From Ruffini and Wheeler Ref. (9).

2 In What Sense Can We Speak of the Total Energy of a Black Hole?

The mass as sensed from infinity in a Kerr solution can be split into three major components as given by Eq. (1),

Total mass-energy = rest mass + coulomb energy + rotational energy.

The rest mass is just given by the irreducible mass m_{ir}. Both the Coulomb and the rotational energy can be augmented or depleted at will be the use of reversible or irreversible transformations (details in Appendix 4.1). Most important for us here is to evaluate through formula (1) how much energy can still be stored in rotational or Coulomb energy in a black hole. Starting from an extreme black hole with

$$a^2 + Q^2 = M^2 \tag{5}$$

Up to 50% of the total energy can be extracted in the limit $a = 0$ and 29% in the limit $Q = 0$ (here, as usual, we use geometrical units and $a = L/M$ where L is the angular momentum, Q the charge and M the mass of the black hole as sensed from infinity).

If we focus on a large amount of hydrogen ($M \gtrsim M_\odot$) and let this cloud undergo all the thermonuclear evolution of a star, we can extract in the entire life of the star (10^9–10^{10} years!) up to 1–2% of the total mass-energy of the initial system in the form of electromagnetic radiation. If the star finally collapses and coalesces into a black hole a much larger amount of energy can still be stored there, as given by formula (1) and, at least in principle, amounts of energy larger than the ones available from *any* other energy source can be extracted. This led[6] to considering black holes as "the largest storehouses of energy in the universe". Much work is currently going on to find realistic methods of extracting this enormous amount of energy.

3 In What Sense is m_{ir} Irreducible?

From formula (1) we can write down a formula for the variation in irreducible mass due to the capture of a given particle with energy $E = \delta m$ angular momentum $p_\phi = \delta L$ by a generalized Kerr-Newman black hole

$$4m_{ir}\delta m_{ir} = (E(r_+^2 + a^2) - ap_\phi - r_+ \, qQ/(M^2 - a^2 - Q^2)^{1/2} \tag{6}$$

with $r_+ = M + (M^2 - a^2 - Q^2)^{1/2}$.

From this formula and from the explicit expression for E it follows then

$$\delta m_{ir} \geqslant 0 \tag{7}$$

the equal sign being valid only at the "knife edge" or horizon r_+ where (reversible transformations!)

$$E = (a\,p_\phi + r_+ Qq)/(r_+^2 + a^2) \tag{8}$$

See Appendix 4.1.

4 "In What Sense must the Area of the Horizon Increase?"

If we compute the area of the horizon of a generalized Kerr black hole we have

$$S = 4\pi(r_+^2 + a^2) = 16\pi m_{ir}^2 \tag{9}$$

The area uniquely depends on the value of the irreducible mass. Since $\delta m_{ir} \geqslant 0$, we have that the area of a black hole is left unchanged by reversible transformations and must increase in irreversible transformations. The theorem of the increase of the black hole surface area was independently found in a different context by S. Hawking.[7]

5 "In What Sense Can We Speak of an Angular Velocity of a Black Hole?

In Appendix 4.3 we have given the explicit analytic expressions for the four velocity of an inertial observer momentarily at rest at a given value of the r and

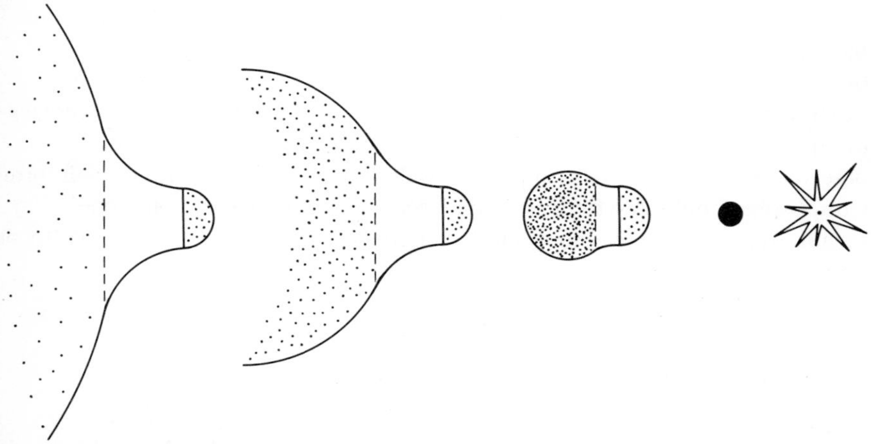

Figure 4. Final state of collapse of a black hole matched to a closed Universe.

θ coordinates. On the horizon, $r_+ = m + (m^2 - (a^2 + Q^2))^{1/2}$, the angular velocity of the inertial observer as measured from infinity, is given by the equation

$$\Omega = a/(r_+^2 + a^2) \tag{10}$$

Moreover, if we follow the geodesic motion of any particle falling into a generalized Kerr black hole, it will approach asymptotically with time a circular motion of constant latitude at the horizon. We then have $dr/d\phi = d\theta/d\phi = 0$, and the particle will have an angular velocity in the ϕ direction given by Eq. (10) or $d\phi/dt = \Omega$. This circumstance led us[6] to define Ω as the angular velocity of the black hole. J. A. Wheeler pointed out that an alternative definition of angular velocity could be obtained directly from Eq. (2) by requiring as in classical physics

$$\text{angular velocity} = \frac{\partial m}{\partial L}.$$

Indeed this definition of angular velocity coincides with the one given by Eq. (10) and greatly contributes to a deeper understanding of the meaning of Eq. (2).

6 What Happens to a Black Hole after It Has Been Formed?[6]

At its moment of formation, the black hole will most likely be a magnetic black hole and will lose both rotational and electromagnetic energy by accretion of matter and acceleration mechanisms, tending towards a Schwarzschild black hole. However, as seen from infinity, it takes an infinite time to form a black hole. In the Schwarzschild case as the collapsing star approaches the Schwarzschild radius, the time of a far away observer, in approximately flat space increases exponentially with respect to the time of an observer comoving with the surface of the star. Therefore, since the lifetime of the Universe is finite, an observer on the surface of the collapsing star will be able to calculate, tracing back the light signal received, that the Universe is rapidly passing the moment of maximum expansion and recontracting. However, when the radius of the Universe becomes of dimensions comparable to the radius of the collapsing star, no far away observer exists and the collapsing star as well as the rest of the Universe will form a three sphere and the system as a whole will collapse to one and only one singularity! The same reasoning applies, of course, if we have a system of collapsing stars.

References

1. Israel, W. and Wilson, G. (1972) *Journ. Mat. Phys.* **13**, 865.
2. Perjes, Z. (1971) *Phys. Rev. Lett.* **27**, 1668.
3. The existence of this gravitational spin interaction was first pointed out by O'Connel, R. F. see e.g. O'Connel, R. F. and Rasband, S. N. (1971) *Nature*–**PS232**, 193. See also Wilkins, D. Ph.D., Thesis, Stanford University, unpublished.
4. Floyd, R. and Penrose, R. (1971) *Nature* **229**, 193.
5. Ruffini, R. and Wheeler, J. A. (1970) *Cosmology from space platform* E.S.R.O. book Sp 52, Moore, A. F. and Hardy, V. eds. Paris.
6. Christodoulou, D. and Ruffini R. (1971) Gravity Research Foundation Essay, Third prize.
7. Hawking, S. (1971) *Phys. Rev. Letters* **26**, 1344.

A 4.1

Reprinted from:

PHYSICAL REVIEW D VOLUME 4, NUMBER 12 15 DECEMBER 1971

Reversible Transformations of a Charged Black Hole*

Demetrios Christodoulou
Joseph Henry Laboratories, Princeton University, Princeton, New Jersey 08540

and

Remo Ruffini
Joseph Henry Laboratories, Princeton University, Princeton, New Jersey 08540, and Institute for Advanced Study, Princeton, New Jersey 08540

(Received 1 March 1971; revised manuscript received 26 July 1971)

A formula is derived for the mass of a black hole as a function of its "irreducible mass," its angular momentum, and its charge. It is shown that 50% of the mass of an extreme charged black hole can be converted into energy as contrasted with 29% for an extreme rotating black hole.

The mass m of a rotating black hole can be increased and (Penrose[1]) decreased by the addition of a particle and so can its angular momentum L; but (Christodoulou[2]) there is no way whatsoever to decrease the irreducible mass m_{ir} in the equation

$$E^2 - p^2 = m^2 = m_{ir}{}^2 + L^2/4m_{ir}{}^2 \qquad (1)$$

for the mass of a black hole. The concept of reversible (m_{ir} unchanged) and irreversible transformations (m_{ir} increases), which was introduced and exploited by one of us to obtain this result, is extended here to the case where the object also has charge, to yield the following four conclusions:

(1) The rest mass of a black hole is given in

terms of its irreducible mass and its angular momentum L and charge e by the formula[3]

$$m^2 = (m_{ir} + e^2/4m_{ir})^2 + L^2/4m_{ir}^2 . \tag{2}$$

(2) Reversibility implies and demands zero separation between the "negative-root states" and "positive-root states" of the particle defined by a quadratic equation of the form

$$\alpha E^2 - 2\beta E + \gamma = 0 , \tag{3}$$

a requirement which is met and can only be met at the horizon itself.

(3) There exists a one-to-one connection between (a) the irreducible mass (as defined here and previously exclusively through the theory of reversible and irreversible transformations), and (b) the proper surface area S of the horizon (shown by Hawking[4] never to decrease),

$$S = 16\pi m_{ir}^2 . \tag{4}$$

(4) The innermost stable circular orbit is the simplest place for a black hole to hold a particle bound and ready for capture. This orbit lies just outside the horizon only when the black hole is an extreme Kerr-Newman[5] black hole in the sense

$$(L^2/m^2) + e^2 \ (= a^2 + e^2 \text{ in the notation of Kerr}) = m^2 \tag{5}$$

or parametrically

$$L = 2m_{ir}^2 \cos\chi , \quad e = \pm 2m_{ir}(\sin\chi)^{1/2} ,$$
$$m = 2^{1/2} m_{ir}(1 + \sin\chi)^{1/2} \tag{6}$$

(χ has any value from 0 to $\frac{1}{2}\pi$). The binding energies of a particle in this most-bound stable orbit are given by the formula

$$\frac{E}{\mu} = \frac{(2a^2 - m^2)\lambda + a(\lambda^2 m^2 + 4a^2 - m^2)^{1/2}}{4a^2 - m^2} , \tag{7}$$

where $\lambda = \epsilon e/\mu m$ (cf. Fig. 1). The transformation on the black hole affected by capture of a particle from such an orbit becomes reversible only when the charge-to-mass ratios of the particle (ϵ/μ) and the black hole (e/m) attain the limits $|\epsilon/\mu| \to \infty$ and parameter $\chi \to 0$ ($e/m \to 0$, $L \to m^2$) such that $\epsilon e/m\mu \to -\infty$. The binding of the particle in this orbit is 100% of its rest mass.

In black-hole physics one has reversibility without reversibility. As compared to such frictional processes as a brick sliding on a pavement, or an accelerated charged particle radiating, or a freely falling deformed droplet of molasses reverting to sphericity, it is difficult to name an act more impressively lacking in reversibility than capture of a particle by a black hole. Lost beyond recall is not a part of the mass-energy of the moving system but all of it; and not only mass-energy but also identity. The resulting black hole, like the original black hole, is characterized by three "independent determinants," mass, charge, and angular momentum, and by nothing else; all particularities (anomalies in higher multipole moments; also baryon number, lepton number, and strangeness) are erased, according to all available indications.[6-11] To reverse the change in a black hole brought about by the addition of a particle with a given rest mass, charge, and angular momentum (μ, ϵ, p_φ) one does not and cannot cause the black hole to reexpel the particle. Nor is there any such thing as a particle with a negative rest mass that one can add to cancel the first addition. Add instead (B in Fig. 1) a particle of the original rest mass μ but of charge $-\epsilon$ and angular momentum $-p_\varphi$. This addition restores the determinants m,

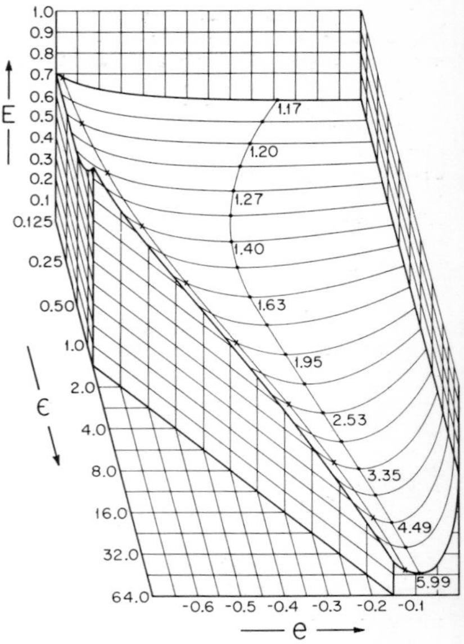

FIG. 1. Energy of a test particle of specified charge ϵ, corresponding to the circular orbit touching the one-way membrane of a black hole of specified charge e of the limiting configurations $a^2 + e^2 = m^2$, versus ϵ and e. Such orbits will be stable and will be the orbits of lowest energy if $E \le e(\mu/m)$ for $e\epsilon < 0$. The crossed curve is the boundary between stable and unstable orbits. The numbers on the energy minima correspond to the values of the angular momentum for the most-bound orbit.

e, L of the black hole to their original value when and only when positive- and negative-root states have zero separation, a condition that is fulfilled only at the horizon itself, $r = r_{\text{horizon}} = r_+ = m + (m^2 - a^2 - e^2)^{1/2}$. At the horizon the positive- and negative-root surfaces $E = E_\pm(p_\varphi, \epsilon)$ meet at a "knife edge" (cf. Fig. 18 in Ruffini and Wheeler[12]) and the two "hyperbolas" of Fig. 2 degenerate to the "straight line" (acquires one more dimension and becomes a plane when the charge as well as the angular momentum of the test particle is taken into account),

$$E = \frac{a p_\varphi + e\epsilon r_+}{a^2 + r_+^2}. \tag{8}$$

Details follow.

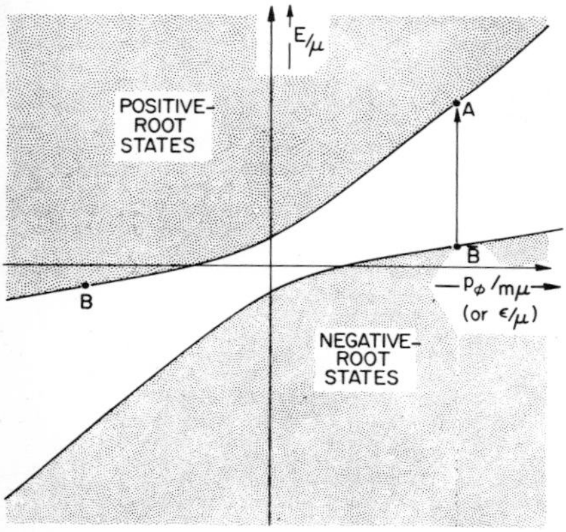

FIG. 2. Reversing the effect of having added to the black hole one particle (A) by the Penrose process of adding another particle (B) of the same rest mass but of opposite angular momentum and charge in a "positive-root negative-energy state." The diagram shows schematically the energy E of the particle (measured in the Lorentz frame tangent at $r \to \infty$) as a function of the charge ϵ of the particle and its angular momentum p_φ about the axis of the black hole (for simplicity one of these two dimensions has been suppressed in the frame). The particle under examination is in the equatorial plane of the black hole ($\theta = \frac{1}{2}\pi$) at a specified value of r. The energy lies on the indicated "positive-root" curve when p_r and p_θ are zero, otherwise in the dotted region above the curve. Addition of B is equivalent to subtraction of $\bar{B}$. Thus the combined effect of the capture of particles A and B is an increase in the mass of the black hole given by the vector $\bar{B}A$. This vector vanishes and reversibility is achieved when and only when the separation between positive-root states and negative-root states is zero [in this case the hyperbolas coalesce to the straight line given by Eq. (8)].

The Hamilton-Jacobi equation

$$g^{\alpha\beta}\left(\frac{\partial S}{\partial x^\alpha} + \epsilon A_\alpha\right)\left(\frac{\partial S}{\partial x^\beta} + \epsilon A_\beta\right) + \mu^2 = 0 \tag{9}$$

for the motion of the particle, separated and solved by Carter,[13] leads to the quadratic equation (3) for the energy with

$$\begin{aligned} \alpha &= r^4 + a^2(r^2 + 2mr - e^2), \\ \beta &= (2mr - e^2)ap_\varphi + \epsilon er(r^2 + a^2), \\ \gamma &= -(r^2 - 2mr + e^2)p_\varphi^2 + 2\epsilon earp_\varphi + \epsilon^2 e^2 r^2 \\ &\quad -(r^2 - 2mr + a^2 + e^2)(\mu^2 r^2 + Q) \\ &\quad -[(r^2 - 2mr + a^2 + e^2)p_r]^2, \end{aligned} \tag{10}$$

see Fig. 1. Here Q is a constant of the motion (generalization of the usual expression for the square of the angular momentum) related to the polar momentum p_θ at any angle θ by the equation

$$Q = \cos^2\theta\,[a^2(\mu^2 - E^2) + p_\varphi^2(\sin\theta)^{-2}] + p_\theta^2. \tag{11}$$

The positive- and negative-root solutions of (a) coincide only when the discriminant of this equation vanishes. This condition is satisfied only at the horizon where the energy is given by Eq. (8). The derivation of formula (1) assumes and demands that the in-falling particles make an effectively infinitesimal change in the properties of the black hole. Applying the laws of conservation of the three determinants and writing $E = dm$, $p_\varphi = dL$, $\epsilon = de$, we obtain the partial differential equation

$$dm(L, e) = \frac{(L/m)dL + r_+ e\,de}{r_+^2 + L^2/m^2}. \tag{12}$$

Integration gives Eq. (1), provided that the following condition is satisfied:

$$\frac{L^2}{4m_{\text{ir}}^2} + \frac{e^4}{16 m_{\text{ir}}^4} \leq 1. \tag{13}$$

For it to be possible to reverse the transformation, both the original particle (positive energy) and the added particle (negative energy) must arrive at the horizon with zero radial momentum. Otherwise there is a nonzero kinetic energy that is irretrievably lost.

When one turns from reversibility as a criterion for an interesting transformation to merely the ability to extract energy, it becomes important to specify under what conditions a positive-root state has negative energy. From Eq. (10) it follows that the region (outside the one-way membrane) where positive-root states of negative energy are available to a particle of specified rest mass μ, charge ϵ, and angular momentum p_φ extends to the surface

$$(r^2-2mr+e^2+a^2)[p_\varphi{}^2+\sin^2\theta(r^2+a^2\cos^2\theta)\mu^2] = \sin^2\theta(\epsilon er+ap_\varphi)^2. \quad (14)$$

When the transformation is reversible, the energy-extraction process has its maximum possible efficiency. Repetition of an energy-extraction process with maximum possible efficiency results in conversion into energy of 50% of the mass of an extreme charged black hole and 29% of that of an extreme rotating black hole. Thus, black holes appear to be the "largest storehouse of energy in the universe."[14]

ACKNOWLEDGMENT

We are indebted to John A. Wheeler for stimulating discussions and also for advice on the wording of this paper.

---

*Work supported by the National Science Foundation under Grant No. GP-30799X.

[1]R. Penrose, Nuovo Cimento, Rivista, Vol. 1, special issue, 1969.
[2]D. Christodoulou, Phys. Rev. Letters 25, 1596 (1970).
[3]D. Christodoulou and R. Ruffini, Bull. Am. Phys. Soc. 16, 34 (1971).
[4]S. Hawking, Phys. Rev. Letters 26, 1344 (1971).
[5]E. T. Newman, E. Couch, R. Chinnapared, A. Exton, A. Prakash, and R. Torrence, J. Math. Phys. 6, 918 (1965).
[6]J. Bekenstein, Bull. Am. Phys. Soc. 16, 34 (1971).
[7]B. Carter, Phys. Rev. Letters 26, 331 (1971).
[8]J. B. Hartle, Phys. Rev. D 1, 394 (1970).
[9]W. Israel, Phys. Rev. 164, 1776 (1967).
[10]C. Teitelboim, Bull. Am. Phys. Soc. 16, 35 (1971).
[11]J. A. Wheeler, in a chapter in the Mendeleev anniversary volume (Accademia delle Scienze di Torino, Torino, 1971).
[12]R. Ruffini and J. A. Wheeler, in *The Significance of Space Research for Fundamental Physics*, edited by A. F. Moore and V. Hardy (European Space Research Organization, Paris, 1970).
[13]B. Carter, Phys. Rev. 174, 1559 (1968).
[14]D. Christodoulou and R. Ruffini (unpublished).

A 4.2

ON THE ELECTRODYNAMICS OF COLLAPSED OBJECTS*

Demetrios Christodoulou

Department of Physics

California Institute of Technology

Pasadena, California 91109

and

Remo Ruffini

Joseph Henry Laboratories

Princeton University

Princeton, New Jersey 08540

* Work supported in part by National Science Foundation Grant No. 30799X to Princeton University.

ABSTRACT

The details of the magnetic and electric field to be expected in a collapsed object nonradiative at ∞ ("Black Hole") are here given. Also given are the formulae determining the maximum total energy extractable from a collapsed object and the definition of its angular velocity as seen from infinity. Physical meaning is given to the expression of the surface area. Typical order of magnitute of the preceding quantities for collapsed objects of different masses are here estimated.

Newman et. al. have shown that an exact solution to the Einstein equation exists characterized only by mass m, charge e and angular momentum L. Boyer and Lindquist have introduced a particularly advantageous coordinate system in which the metric takes the form

$$ds^2 = g_{\mu\nu}\, dx^\mu\, dx^\nu = \rho^2\ \Delta^{-1}\ dr^2 + \rho^2\ d\theta^2 + \rho^{-2}\ \sin^2\theta\ [a\ dt - (r^2+a^2) d\varphi]^2 - \rho^{-2}\ \Delta\ [dt - a\ \sin^2\theta\ d\varphi\ ]^2 \quad \text{(Greek indices varying from 0, 3; Latin index from 1, 3)} \tag{1}$$

with $\Delta = r^2 - 2mr + a^2 + e^2$ and $\rho^2 = r^2 + a^2\ \cos^2\theta$; $a = L/m$ is the angular momentum per unit mass. Here and in the following we use geometrical units, $m = Gm_{conv}/c^2$, $e = G^{1/2}\ e_{conv}/c^2$, $L = G\ L_{conv}/c^3$. The electromagnetic field form associated with this solution is

$$F_{\mu\nu} = \tfrac{1}{2}\ f_{\mu\nu}\ dx^\mu\ \Lambda\ dx^\nu = e\ \rho^{-4}(r^2 - a^2\ \cos^2\theta) dr\ \Lambda\ [dt - a\ \sin^2\theta\ d\varphi] - 2\ e\ \rho^{-4}\ a\ r\ \cos\theta\ \sin\theta\ d\theta\ \Lambda\ [a\ dt - (r^2 + a^2) d\varphi]. \tag{2}$$

It has been conjectured that this is the most general geometry left behind after a stellar object ~~object~~ has undergone complete gravitational collapse and has finally reached stationary conditions. We are here mainly interested in giving the explicit form of the electric and magnetic fields and the formulae determining the dragging of the inertial frames in the space surrounding the collapsed object. For this we limit our attention to the singularity free range of the coordinate, namely $r > r_+ = m + \sqrt{m^2 - e^2 - a^2}$, and we define the surface $r = r_+$ to be the surface of the black hole ($\Delta = 0$).

Consider a local inertial frame in the neighborhood of each circle r = const, θ_{const}. Pick at each circle an inertial frame with r and ϑ 4-velocity components momentarily zero, φ components corresponding to that of a particle of zero azimuthal angular momentum. As seen from the coordinate system (fixed) at infinity the angular velocity of this local Lorentz frame is

$$\omega = - g_{03}/g_{33} = \frac{a\ (2\ m\ r - e^2)}{(r^2 + a^2)^2 - a^2\ \Delta\ \sin^2\theta} \tag{3}$$

Christodoulou and Ruffini have shown that the total energy of a collapsed object described by the metric (1) can be written as

$$m^2 = (m_{ir} + e^2/4m_{ir})^2 + L^2/4m_{ir}{}^2 \tag{4}$$

Here m_{ir} is the irreducible mass (total energy of the collapsed object after all the rotational and electromagnetic energy has been extracted). We can associate therefore to the collapsed object the angular velocity

$$\Omega = \frac{\partial m}{\partial L} = \frac{L/4m_{ir}{}^2}{[(m_{ir} + e^2/4m_{ir})^2 + L^2/4m_{ir}{}^2]1/2} \tag{5}$$

A surprising result is that

$$\Omega = \lim_{r \to r^+} \omega = a/(r_+^2 + a^2) \tag{6}$$

which can be expressed in the following words: The angular velocity of a collapsing object and the angular velocity of dragging of the inertial frames on its surface tend to a common value (given by (5), (6)) in the limit in which the collapsing object tends to its final stationary configuration.

The Lorentz force experienced by a test charge q of mass μ at rest in a reference frame with four velocity u^β is

(3)

$$q\, F^{\alpha}{}_{\beta} u^{\beta} = \mu \frac{\delta u\alpha}{\delta\tau}. \qquad (7)$$

n complete analogy we define the force experienced by a fictitous est magnetic monopole p

$$-p * F^{\alpha}{}_{\beta} \mu^{\beta} = \mu \frac{\delta u^{\alpha}}{\delta\tau}. \qquad (8)$$

ere $\frac{\delta u^{\alpha}}{\delta\tau}$ gives the deviation of the particle trajectory from the eodesic $\delta u^{\alpha}/\delta\tau = du^{\alpha}/d\tau + \Gamma^{\alpha}_{\beta\gamma}\, u^{\beta}\, u^{\gamma} = 0$, and the symbol * denotes he duality operation:

$$* F_{\alpha\beta} = \tfrac{1}{2}(-g)^{\frac{1}{2}}\, [\alpha\ \beta\ \gamma\ \delta]\ F^{\gamma\delta}$$

ere, as usual, $[\alpha, \beta, \gamma, \delta]$ is the Levi-Civita indicator which changes ign on interchange of any two indices, and [0, 1, 2, 3] has the value +1. ssuming as reference frame the inertial system of reference previously efined, we obtain for the components of the four velocity of an ob-erver at rest with respect to that system of reference

$$u^{r} = 0 \qquad u^{\varphi} = \frac{a(-e^{2} + 2mr)}{\Delta^{\frac{1}{2}}\, \rho} \frac{1}{\sqrt{(r^{2} + a^{2})^{2} - a^{2}\, \Delta \sin^{2}\theta}}$$

$$u^{\theta} = 0 \qquad u^{t} = \frac{\sqrt{(r^{2} + a^{2})^{2} - a^{2}\, \Delta \sin^{2}\theta}}{\Delta^{\frac{1}{2}}\, \rho}. \qquad (9)$$

he components of the electric and magnetic field measured by such an bserver are immediately derived from (7) and (8) to be

$$D^{\mu} = F^{\mu}{}_{\beta} u^{\beta} \qquad B^{\mu} = - * F^{\mu}{}_{\beta} u^{\beta} \qquad (10)$$

r explicitly for the components of the electric field:

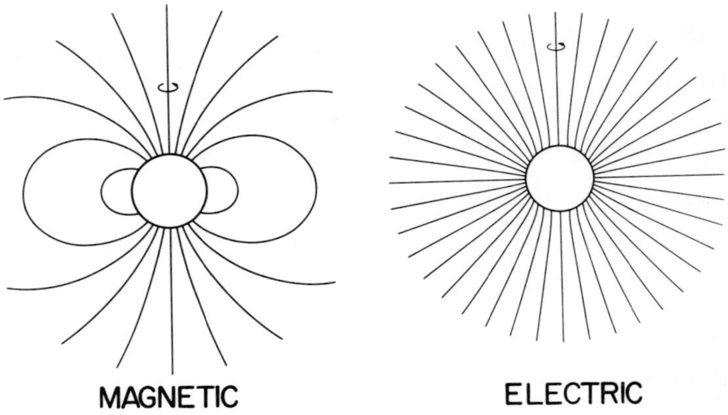

FIGURE 1 Magnetic and electric lines of force for a magnetic black hole

$$D^r = [e\,\rho^{-5}\,\Delta^{\frac{1}{2}}\,(r^2+a^2)\,(r^2-a^2\cos^2\theta)]\;[(r^2+a^2)^2-a^2\,\Delta\sin^2\theta]^{-\frac{1}{2}}$$

$$D^\theta = -[2\,e\,a^2\,\rho^{-5}\,\Delta^{\frac{1}{2}}\;r\cos\theta\sin\theta]\;[(r^2+a^2)^2-a^2\,\Delta\,\sin^2\theta]^{-\frac{1}{2}}$$

$$D^\varphi = 0 \tag{11}$$

and for the components of the magnetic field:

$$B^r = [2\,e\,a\;\rho^{-5}\,r\,(r^2+a^2)\cos\theta\,\Delta^{\frac{1}{2}}]\;[(r^2+a^2)^2-a^2\,\Delta\,\sin^2\theta]^{-\frac{1}{2}}$$

$$B^\theta = [e\,a\,\rho^{-5}\,\Delta^{\frac{1}{2}}\,(r^2-a^2\cos^2\theta)\sin\theta]\;[(r^2+a^2)^2-a^2\,\Delta\,\sin^2\theta]^{-\frac{1}{2}}$$

$$B^\varphi = 0 \tag{12}$$

We obtain for the lines of force of the field the following close form expressions

$$(r^2-a^2\cos^2\theta)\,(r^2+a^2)\,d\theta + 2\,a^2 r\cos\theta\sin\theta\,dr = 0 \qquad \text{electric} \tag{13.1}$$

$$2\,r\cos\theta\,(r^2+a^2)\,d\theta - (r^2-a^2\cos^2\theta)\sin\theta\,dr = 0 \qquad \text{magnetic} \tag{13.2}$$

In Fig. 1 the electric and magnetic lines of force are plotted. In Tab. 1 we give some typical value of the characteristic parameters of a collapsed object as a function of the mass.

It is interesting to analyze the limit of expressions (11) and (12) when terms of the order $\frac{1}{r^4}\,(|a| + |e| + m)^3$ are considered small and higher order terms neglected. To make the comparison with analogous result in flat space, we will consider the physical components: $A_i = \sqrt{A_i A^i}$. We have

$$D_r = \sqrt{D_r D^r} = \frac{e(r^2+a^2)(r^2-a^2\cos^2\theta}{\rho^4[(r^2+a^2)^2-a^2\Delta\,\sin^2\theta\;]^{\frac{1}{2}}} \sim \frac{e}{r^2} - \frac{1}{2}\frac{\mu a}{r^4}(3\cos^2\theta+1) - \frac{ea^2}{r^4}\left(\frac{\cos 2\theta+3}{4}\right) \tag{1 .1}$$

$$D_\theta = \sqrt{D_\theta D\theta} = -\frac{2\,e\,a^2 r\cos\theta\sin\theta\;\Delta^{\frac{1}{2}}}{\rho^4[(r^2+a^2)^2-a^2\Delta\,\sin^2\theta\;]^{\frac{1}{2}}} \sim -\frac{\mu a}{r^4}\sin(2\theta) \tag{1 .2}$$

$$D_\varphi = 0 \tag{1 .3}$$

and

$$B_r = \sqrt{B_r B^r} \sim \frac{2\,e\,a\,r\,(r^2 + a^2)\cos\theta}{[(r^2 + a^2)^2 - a^2\,\Delta\,\sin^2\,]^4} \sim \frac{2\,e\,a}{r^3}\cos\theta = \frac{2\mu}{r^3}\cos\theta \quad (1\ .1)$$

$$B_\theta = \sqrt{B_\theta B^\theta} \sim \frac{e\,a\,(r^2 - a^2\cos^2\theta)\sin\theta\,\Delta^{\frac{1}{2}}}{[(r^2 + a^2)^2 - a^2\,\Delta\,\sin\ \,]^4} \sim \frac{e\,a}{r^3}\sin\theta = \frac{\mu}{r^3}\sin\theta \quad (1\ .2)$$

$$B_\varphi = 0 \quad (1\ .3)$$

We have assumed in Eq. 1 and corresponding Eqs. 1 $\mu = e\,a$. Equation 17 gives us then the usual formula of a dipole field of magnetic moment μ. The first term in Eq. (1 .1) is clearly the Coulomb term due to the charge. The last term in 1 .1 is an additional component of the electromagnetic field seen from the particular observer we have chosen. J. Tiommo has shown that this last term disappears if we chose the particular set of observer introduced by Tamm (flat space background observers!). We can now give an important physical interpretation to the terms in Eq. 16 containing the magnetic moment μ. Recalling Eq. (6) and the expression of the area of a black hole as given by Christodoulou and Ruffini

$$S = 16\,\Pi\,m^2_{ir} \quad (1\ )$$

and

$$r_+^2 + a^2 = 4\,m^2_{ir} \quad (1\ )$$

we can write (16.2) in the following form

$$D_\theta = -\frac{\mu\,a}{r^4}\sin(2\theta) = -\frac{\mu}{r^4}\,\omega\,\frac{S}{4\Pi}\sin(2\theta)$$

and the corresponding expression in (1 .1). This is exactly the electric field generated by a classical aligned magnetic field at the surface S of a star rotating with angular velocity ω! (see e.g. Deutsch). This new relation greatly unifies and gives physical insight, both into the expression of the angular velocity here introduced and the expression of the surface area of a black hole.

TABLE I

Typical values of the parameters associated with an extreme collapsed object ($m^2 = e^2 + a^2$ and $e = a$) for selected values of the mass.

Mass	Initial radius of collapsed object $r=m+\sqrt{m^2-(a^2+e^2)}$ in cm	Minimum rotation period $\omega=a/(r_+^2 + a^2)$ in sec	Maximum strength of the magnetic field see Eq. 12 in text in Gauss	Maximum net charge (in electron charge)	Maximum energy extractable from the collapsed object in ergs	Maximum power assuming a lifetime of 10^8 years
$10\ M_\odot$	1.47×10^6	0.653×10^{-3}	1.05×10^{18}	7.59×10^{39}	0.689×10^{55}	2.19×10^{39}/
$10^8\ M_\odot$	1.47×10^{13}	0.653×10^4	1.05×10^{11}	7.59×10^{46}	0.689×10^{62}	2.19×10^{46}
$10^{12} M_\odot$	1.47×10^{17}	0.653×10^8	1.05×10^7	7.59×10^{50}	0.689×10^{66}	2.19×10^{50}

References

R. H. Boyer and R. W. Lindquist. Journ. Math. Phys. 8, 265, 1967.

D. Christodoulou and R. Ruffini. Phys. Rev. 4, 3552, 1971.

A. J. Deutsch. Annales d'Astrophysique. 1, 1, 1955.

L. Landau and E. Lifschitz. Theorie des Champs-third Edition. Edit. MIR. Moscow 1970.

E. T. Newman, E. Couch, R. Chinnapared, A. Exton, A. Prakash and R. Torrence. Journ. Math. Phys. 6, 918, 1965.

I. Tamm, J. Russ. Phys.-Chem. Soc. Phys. Div. 56, 2-3, 248, (1924).

J. Tiomno. Private Communication

A 4.3

Metric of Two Spinning Charged Sources in Equilibrium

Leonard Parker*†
University of Wisconsin-Milwaukee, Milwaukee, Wisconsin 53201

and

Remo Ruffini‡
Princeton Univeristy, Princeton, New Jersey 08540

and

Daniel Wilkins§
University of California, Santa Barbara, California 93107

Using the approach of Israel, Wilson and Perjes we give ne explicit form of the metric corresponding to two identical Kerr-ewman sources in equilibrium under their mutual electromagnetic and ravitational forces, with their spins oppositely oriented along a given xis. Symmetries, the complete analytic, extension, the limit of infinite eparation of the sources, and the two types of solution with vanishing eparation are discussed.

* * *

A class of static solutions of the Einstein-Maxwell equations corres-onding to an equilibrium distribution of charged sources was found in 1947 y Majumdar[1] and Papapetrou[2]. These solutions have recently been generalized o the stationary case (sources with spin) by Israel and Wilson[3], and erjes[4] (IWP). The first order set of partial differential equations char-cterizing the IWP solutions must be solved to obtain the metric explicitly.

One member of the class of IWP solutions is the Kerr-Newman solution with harge equal to mass (in units with G = c = 1) as has been proved in that ase by direct integration of the IWP differential equations[3,4]. It is of interest to find the solution for several Kerr-Newman sources. The result would provide a new explicit exact solution of the Einstein-Maxwell equations. It could, for example, serve as the basis for an analysis of solutions with slightly different parameters by means of perturbation or related techniques.

The problem of forces and torques between spinning objects has been extensively treated[5,6,7]. All these analyses, however, are based on the assumption that one of the bodies is a test particle whose own effect can be neglected. The solution we present in this paper allows for this interaction in the realistic case of a two body problem and provides a standard against which results obtained for test particles may be judged. It is interesting that the solution here presented demonstrates the balance (correct to all multipole orders) of the forces between two spinning sources.

In the present paper, we give the explicit stationary metric corresponding to two identical Kerr-Newmann sources in equilibrium under their mutual gravitational and electromagnetic forces, with their spins oppositely oriented along a given axis. We confirm that the result obtained has the expected properties. To our knowledge, this is the first explicit non-static two-body solution of the Einstein-Maxwell equation. It is of interest in connection with certain aspects of coalescence, such as changes in singularity structure and possible horizon formation, and the uniqueness of the final configuration, as well as in analyzing the force balance and the electromagnetic field, which can only be given explicitly in this highly relativistic regime now that the analytic expression for he metric is known.

The IWP class of metrics are written in the standard form

$$s^2 = -f^{-1}\gamma_{mn}\, dx^m dx^n + f(\omega_m dx^m + dt)^2 \tag{1}$$

here f, ω_m and γ_{mn} are independent of t. The line element $\gamma_{mn}\, dx^m dx^n$ orresponds to a Euclidean 3-space (in arbitrary coordinates). Any function U of the spatial coordinates which satisfies Laplace's equation in the flat -space generates an IWP metric by means of the equations

$$f = U^{-2} \tag{2}$$

nd

$$\varepsilon^{npq}\, \partial_p \omega_q = i\, \gamma^{\frac{1}{2}} f^{-1} \gamma^{nm}\, \partial_m \ln\, (U/U^*) \tag{3}$$

here U* is the complex conjugate of U.

he electromagnetic field follows from γ_{mn}, U and ω_p by means of the ex-ression given in Refs. 3 and 4.

The solution corresponding to two identical Kerr-Newman sources, with harge equal to mass, lying in equilibrium on the z-axis at ± b, each ith angular momentum per unit mass, a, pointing towards the origin, is enerated by [3,4]

$$U = 1 + \frac{m}{R_1} + \frac{m}{R_2} \tag{4}$$

ith

$$R_1{}^2 = x^2 + y^2 + (z - b - ia)^2 \tag{5}$$

nd

$$R_2{}^2 = x^2 + y^2 + (z + b + ia)^2 \tag{6}$$

n order to obtain the explicit form of the metric, one must solve Eq. (3) or ω_j. We use spherical coordinates

$$x + iy \equiv r \sin\theta \, e^{i\phi} , \quad z = r \cos\theta \qquad ($$

in terms of which

$$R_1^{\;2} = r^2 - 2\ell r \cos\theta + \ell^2, \qquad ($$

and

$$R_2^{\;2} = r^2 + 2\ell r \cos\theta + \ell^2, \qquad ($$

where

$$\ell = b + ia. \qquad ($$

The axial symmetry implies that the left side of Eq. (3) vanishes when the index $n = \phi$. Therefore, the coordinate t can be chosen such that ω_r and ω_θ vanish, since the addition of a scalar function of x^m to t chang ω_j by a gradient. Then Eq. (3) reduces to

$$\frac{\partial\omega_\phi}{\partial\theta} = - \text{Im} \{2mr^2 \sin\theta \, (1 + \frac{m}{R_1{*}} + \frac{m}{R_{2*}}) \cdot [r(\frac{1}{R_1{}^3} + \frac{1}{R_2{}^3}) - \ell\cos\theta \, (\frac{1}{R_1{}^3} - \frac{1}{R_2{}^3})]\} , \qquad ($$

and

$$\frac{\partial\omega_\phi}{\partial r} = \text{Im} \{2m\ell r \sin^2\theta \, (1 + \frac{m}{R_1{*}} + \frac{m}{R_2{*}}) \, (\frac{1}{R_1{}^3} - \frac{1}{R_2{}^3})\} , \qquad ($$

where Im denotes the imaginary part. One can check directly that (11) and (12) satisfy the integrability condidtion $\frac{\partial}{\partial r} (\frac{\partial}{\partial\theta} \omega_\phi) = \frac{\partial}{\partial\theta} (\frac{\partial}{\partial r}$ After a very lengthy calculation one obtains the integral of (11) in the f

$$\omega_\phi = - \text{Im} \{\frac{2m}{R_1} [\ell - r \cos\theta] + \frac{im^2(r^2 - \ell^2)}{2a(r^2 - |\ell|^2)} \frac{R_1{}^*}{R_1} + \frac{m^2}{2b} \frac{(r^2 - \ell^2)}{(r^2 + |\ell|^2)} \frac{R_1{}^*}{R_2} \} \qquad ($$

+ [all previous terms with $a \to - a$, $b \to - b$, where $\ell = b$

o additive function of r appears for reasons discussed further on. Note hat the branches of the square root functions R_1 and R_2 can be chosen in- ependently. Equations (1), (2), (4), and (13) give the explicit expression or the two-body metric in this case.

For the following discussion, it is convenient to introduce two redun- ant sets of oblate spheroidal coordinates, (r_1, θ_1, ϕ_1) and (r_2, θ_2, ϕ_2) entered at $z = +b$ and $z = -b$, respectively. Each point (x,y,z) is relat- d to the new coordinates as follows:

$$\begin{aligned} x + iy &= [(r_1 - m)^2 + a^2]^{\frac{1}{2}} \sin\theta_1 \, e^{i\phi} \\ z &= b + (r_1 - m)\cos\theta_1 \end{aligned} \qquad (14)$$

$$\begin{aligned} x + iy &= [(r_2 - m)^2 + a^2]^{\frac{1}{2}} \sin\theta_2 \, e^{i\phi} \\ z &= -b + (r_2 - m)\cos\theta_2 \, . \end{aligned}$$

Now R_1 and R_2 assume the simple forms

$$R_1 = r_1 - m - ia\cos\theta_1, \qquad (15)$$

nd

$$R_2 = r_2 - m + ia\cos\theta_2. \qquad (16)$$

eplacing $(r_j - m, \theta_j)$ by $(m - r_j, \pi - \theta_j)$ leaves (x,y,z) unchanged, but oes change the sign of R_j $(j = 1$ or $2)$, thus changing the branch of the quare root function.

In the new coordinates, (13) becomes

$$\begin{aligned} \omega_\phi = {} & \frac{(m^2 - 2mr_1)\, a \sin^2\theta_1}{(r_1 - m)^2 + a^2\cos^2\theta_1} - \frac{(m^2 - 2mr_2)\, a \sin^2\theta_2}{(r_2 - m)^2 + a^2\cos^2\theta_2} \\ & + m^2 a F \left[(2b)^{-1} (r^2 + |\ell|^2) H - (G + bH) \right] , \end{aligned} \qquad (17)$$

where

$$F = \{[(r_1 - m)^2 + a^2\cos^2 \theta_1]^{-1} - [(r_2 - m)^2 + a^2\cos^2\theta_2]^{-1}\} (r^2 + |\ell|^2)^-$$

$$G = (r_1 - m)(r_2 - m) + a^2\cos\theta_1 \cos\theta_2 \ , \qquad (1$$

$$H = (r_2 - m)\cos\theta_1 - (r_1 - m)\cos\theta_2 \ , \qquad (1$$

and

$$r^2 + |\ell|^2 = (r_1 - m)^2 + a^2(1 + \sin^2\theta_1) + 2b^2 + 2b(r_1-m)\cos\theta_1$$

$$= (r_2 - m)^2 + a^2 (1 + \sin^2\theta_2) + 2b^2 - 2b(r_2 - m)\cos\theta_2 \ . \qquad (1$$

Each of the first two terms in (17) is the ω_ϕ of a single Kerr-Newman source with $q = m$[3,4].

Equation (12) requires that on the symmetry axis, where $\sin\theta$ vanishes ω_ϕ is a constant, which must be set equal to zero so as to assure an asymptotically Minkowskian metric. The right side of Eq. (17) vanishes on the symmetry axis, so that it is correct as it stands, without an added function of r. To obtain that result, one notes that the first two terms of Eq. (17) clearly vanish on the symmetry axis. Moreover, if one solves fo r_1 in terms of r_2 by means of Eq. (14), and notes that $\cos^2\theta_2 = 1$ on the symmetry axis, then one finds that the factor in square brackets in Eq. (17) vanishes. Furthermore, one confirms by inspection that (17) vanishes as expected when $a = 0$.

There are four regions: I, in which $(r_1 - m)$ and $(r_2 - m)$ (equal, in coordinate free form, to $\mathrm{Re}R_1$ and $\mathrm{Re}R_2$, respectively) are positive, II, and II_2, in which they have opposite signs, and III, in which they are both negative. One would expect by symmetry (since the spins are oppositely directed) that, in region I and III, ω would vanish in the equatorial plane ($z = 0$) as well as when the two particles "coalesce" ($b = 0$). In the former case, Eq. (14) implies that $r_1 = r_2$ and $\cos\theta_1 = -\cos\theta_2$, from which it follows that ω_ϕ vanishes. In the latter case, it is necessary to consider the limit as b approaches zero, since b^{-1} appears in the third term of Eq. (17). For small b, one finds from Eq. (14) that

$$r_1 = r_2 + O(b),$$

$$\cos\theta_1 = \cos\theta_2 + O(b) \quad ,$$

$$F = O(b),$$

and

$$H = O(b),$$

whence $\omega_\phi = O(b)$

As a final check on the metric, fix (r_1, θ_1, ϕ_1) and let b become large, effectively carrying the second source off to infinity. One finds from Eq. (14) that in the above limit

$$|(r_2 - m)| = O(b),$$

$$\theta_2 = O(b),$$

and

$$r^2 + |\ell|^2 = 2b^2 + O(b),$$

so that all terms but the first in Eq. (17) vanish as b^{-1}, leaving

$$\omega_\phi = \frac{(m^2 - 2mr_1)\, a \sin^2\theta_1}{(r_1 - m)^2 + a^2 \cos^2\theta_1} \tag{19}$$

which is the ω_ϕ for a simple Kerr-Newman source with $q = m$, and angular mometum per unit mass, a, in the $-z$ direction. We now turn to a discussion of further properties of our metric.

Rings of infinite redshift occur where $g_{tt} = 0$, or $U = \infty$. At the first ring $r_1 - m = 0$ and $\theta_1 = \pi/2$, while at the second ring $r_2 - m = 0$ and $\theta_2 = \pi/2$. In order to get, say, from gregion I to II, one must change the sign of $r_1 - m$, which entails passing through the interior of the first ring.

Naked ring singularities occur at $U = 0$, where one component of the stree-energy tensor,

$$8\pi\, T_{tt} = (\gamma^{mn} \frac{\partial U}{\partial x^m} \frac{\partial U^*}{\partial x^n})\, |U|^{-6} \tag{20}$$

diverges[8]. Apart from exceptional cases such as $b = 0$ (see below) or $a = 0$, which is treated by Hartle and Hawking[9], the real and imaginary parts of $U = 0$ will give two independent equations for the two unknowns, e.g. r_1, θ_1, thus determining a set of rings. We shall also see in the following that T_{tt} may sometime diverge at a ring of infinite redshift, where $U = \infty$.

The circle r_1, θ_1, t = constant is a closed timelike line if

$$g_{\phi\phi} = f^{-1}\gamma_{\phi\phi} - f\,(\omega_\phi)^2 < 0. \tag{21}$$

This inequality is indeed usually satisfied near a ring singularity where $f = |U|^{-2}$ approaches $+\infty$, ω_ϕ does not vanish, and

$$\gamma_{\phi\phi} = [(r_1 - m)^2 + a^2]\sin^2\theta_1$$

is finite. In this paper we investigate the so-called electrovac solution in which free electromagnetic fields provide the stress-energy. It is customarily assumed that if corresponding exterior.solutions were found in nature, closed time-like lines and singularities not surrounded by event horizon would not occur thanks to an interior solution; that is to say the electrovac condition would not hold everywhere.

The problem of dynamical coalescence is a very difficult one, which may conveniently be split into two parts, namely (a) change in singularity stnucture and horizon formation, and (b) emission of gravitational radiation. Consideration of the quasi-stationary approach of our two sources through a sequence of equilibrium configurations, corresponding to successively smaller values of b, may throw light on part (a) of the more general problem of dynamical coalescence. The condition that the charge equal the mass is necessary to allow part (a) to be focused by the use of the electromagnetic field to produce equilibrium.

The Majumdar-Papapetrou geometry corresponding to many Reissner-Nordstrøm black holes, each with charge equal to mass has the metric[1,2,9]

$$ds^2 = U^{-2}\, dt^2 - U^2\, \gamma_{mn}\, dx^m\, dx^n, \tag{22}$$

where

$$U = 1 + \sum_i (m_i/r_i).$$

If the masses are equal for a system of N such black holes, the area of a spherical surface enclosing them satisfies $S \geq 4\pi(Nm)^2$. The equal sign is only possible in the limit that the horizon ($r_i = 0$) touch one another. If coalescence were to occur the horizon of the resulting black hole would have area $4\pi(Nm)^2$. Examples given by Brill and Lindquist[10] in the analysis of time-symmetric solutions of the Einstein-Maxwell equations suggest that the formation of a horizon may normally have to be accompanied by radiation of energy with consequent binding of the set of black holes. It is interesting that quasistatic coalescence of our two sources does not yield a Reissner-Nordstrøm black hole. Rather, one obtains the following results.

When b = 0, the two rings of infinite redshift coincide. It is then impossible to change the signs of $r_1 - m$ and $r_2 - m$ independently. There are two types of solution, according to whether the interior of the redshift ring connects regions I and III or II_1 and II_2. In the former case,

$$
\begin{aligned}
& r_1 - m = r_2 - m \ , \ \cos\theta_1 = \cos\theta_2 \\
& U = 1 + m\,[(R_1)^{-1} + (R\,*)^{-1}]
\end{aligned}
\tag{23}
$$

Since U is real-valued ω_ϕ vanishes. The solution is static. Expansion of the axially symmetric U in powers of the spherical radius defined by Eq. (7) shows that it differs from the Reissner-Nordstrøm form $U = 1 + 2m/r$ (for mass = ± charge = 2m) by higher multipoles which depend on even integral powers of a.

In this static solution, solving $U = 0$ for the radius of the singularities yields in general two roots:

$$
r_1 = \begin{cases} \sqrt{m^2 - a^2\cos^2\theta_1}\ , \text{ and} \\ \sqrt{m^2 - a^2\cos^2\theta_1}, \text{ if } a\cos\theta_1 \neq 0 \end{cases}
\tag{24}
$$

When $a\cos\theta_1 = 0$, the second root is spurious. Assuming $a \neq 0$, -a gap in the singularity's "surface" thus occurs at $r_1 - m = 0$, $\theta_1 = \pi/2$ the location of the ring of infinte redshift. Although the singularity has vanishing surface area according to (eqs.) (1) and (2), its structure is much more complicated than that of the point singularity in the Reissner-Nordstrøm solution.

For the other type of solution with $b = 0$,

$$
\begin{aligned}
& r_1 - m = m - r_2 \ , \ \cos\theta_1 = -\cos\theta_2 \ , \\
& U = 1 + 2iam\cos\theta_1\,[(r_1 - m)^2 + a^2\cos^2\theta_1]^{-1}
\end{aligned}
\tag{25}
$$

Expanding U in powers of r_1, and comparing with that for a single Kerr-Newman particle allows the mass and angular momentum to be read off directly[4]. The coefficient of $(r_1)^{-1}$ is the mass and the imaginary part of the coefficient of $-\cos\theta_1\ (r_1)^{-2}$ is the angular momentum (taken positive along the + z direction) In this case, the mass vanishes and the angular momentum is -2am. U does not vanish anywhere. But eq.(20) enables one to show that the redshift ring is a singularity: T_{tt} diverges as one approaches this ring in the equatorial plane from outside ($\theta_1 = \pi/2$, $r_1 - m \neq 0$). In eq. (17) the "interaction" term vanishes, leaving

$$\omega_\phi = \frac{4m(m-r_1)\ a \sin^2\theta_1}{(r_1-m)^2 + a^2\cos^2\theta_1} \tag{26}$$

The values of the electric and magnetic field at a point in space-time become well-defined only when a local observer is specified. It is natural to choose an observer with velocity $u^\phi = dx^\phi/ds$ along the symmetry directions $\partial/\partial\phi$, $\partial/\partial t$ of the field such that $g_{o\phi}$ vanishes with respect to his local co-moving frame[11]. This choice is clearly suitable for any axially symmetric static geometry; in the ergosphere of a Kerr-Newman black hole it yields, as desired, a timelike world-line. The non-zero components of u^ν are

$$u^t = dt/ds = (g_{tt} - g_{t\phi}^2/g_{\phi\phi})^{-\frac{1}{2}} \tag{27}$$

$$u^\phi = d\phi/ds = -(g_{t\phi}/g_{\phi\phi})(g_{tt} - g_{t\phi}^2/g_{\phi\phi})^{-\frac{1}{2}}$$

It follows that all components of u_ν vanish except for u_t. The electric and magnetic fields defined, respectively, by the Lorentz force on a unit electric and magnetic monopole moving with velocity u^ν are given by

$$E^\nu = F^{\nu t} u_t\ ,\quad B^\nu = {}^*F^{\nu t} u_t\ , \tag{28}$$

$$^*F^{\mu\nu} = (-g)^{-\frac{1}{2}}\, [\mu\nu\rho\sigma]\, F_{\rho\sigma}$$

Here the completely anti-symmetric symbol $[\mu\nu\rho\sigma]$ satisfies $[0\ 1\ 2\ 3] = +1$, it being understood that t corresponds to 0.

Those components of $F_{\mu\nu}$ not given in the following [3] can be readily derived from these:

$$F_{tn} = \partial_n A_t \ , \tag{29a}$$

$$F^{mn} = f\gamma^{-\frac{1}{2}} \varepsilon^{mnp} \partial_p \Phi \ , \tag{29b}$$

$$F^{nt} = \omega_m F^{mn} + F_{tm} \gamma^{mn} \ . \tag{29c}$$

The two potentials A_t and Φ are determined by

$$A_t + i\Phi = e^{i\alpha}(1 - U^{-1}) \ , \tag{30}$$

where the real parameter α effects a duality transformation.

Let a caret "^" over a quantity denote the value when $\alpha = 0$. From eq. (i),

$$\begin{aligned} A_t &= \cos\alpha\ \hat{A}_t - \sin\alpha\ \hat{\Phi} \\ \Phi &= \cos\alpha\ \hat{\Phi} + \sin\alpha\ \hat{A}_t \end{aligned} \tag{31}$$

For the two source solution described here, it is straightforward to prove that

$$\begin{aligned} E^{\mu} &= \cos\alpha\ \hat{E}^{\mu} + \sin\alpha \cdot \hat{B}^{\mu} \\ B^{\mu} &= \cos\alpha\ \hat{B}^{\mu} - \sin\alpha \cdot \hat{E}^{\mu} \end{aligned} \tag{32}$$

To show this start by using oblate spherical coordinates (t, r_1, θ_1, ϕ) for which $\omega_m = \partial_m \omega_\phi^{\phi}$ and γ_{mn} is diagonal. Then from eqs. (28) and (29) $E^{\phi} = B^{\phi} = E^{t} = B^{t} = 0$. Relations of the form (32) can be derived for the r_1, θ_1 components. The same linear relation holds trivially for the vanishing ϕ and t components, and being tensorial (32) must then be true in arbitrary coordinates.

The static b = 0 solution goes asymptotically to the Reissner-Nordstrøm solution for large radii. It has no magnetic field unless a duality transformation is performed, in which case it will acquire a magnetic monopole moment.

The stationary b = 0 solution is more unusual. Although the mass and charge vanish, it possesses angular momentum and , for $\alpha \neq 0$, a magnetic dipole moment. Expanding in powers of r_1, one obtains by eq. (25) and eq. (30) with $\alpha = 0$,

$$\hat{A}_t = (2a^2m^2 \cos^2\theta_1)\, r_1^{-4} + 0\,(r_1^{-5})\,, \qquad (33)$$
$$\hat{\Phi} = (am \cos\theta_1)\, r_1^{-2} + 0(r_1^{-3})$$

The resulting electromagnetic fields are

$$\hat{E}^{r_1} = -4m^2a^2\,(1 + \cos^2{}_1)\, r_1^{-5} + 0(r_1^{-6}),$$
$$\hat{E}^{\theta_1} = (4m^2a^2 \sin\theta_1 \cos\theta_1)\, r_1^{-6} + 0(r_1^{-7})\,, \qquad (34)$$
$$\hat{B}^{r_1} = (2am \cos\theta_1)\, r_1^{-3} + 0(r_1^{-4})\,,$$
$$\hat{B}^{\theta_1} = (am \sin\theta_1)\, r_1^{-4} + 0(r_1^{-5})\,.$$

To compare these with the results of classical electromagetism, one must project at each point onto an orthonormal pair of vectors transported with the preferred observer and directed along the r_1, θ_1 coordinate axes. At large radii, this effectively multiples the θ_1-components by r_1. The magnetic field is a dipole field with moment am directed along the +z direction. If, however, the fields are expressed in terms of r_2, θ_2 coordinates a must be changed to -a, and the dipole moment changes sign. The electric field does not have the requisite angular dependence to be an octopole field, which may be derived by taking the gradient of $U \sim (5\cos^3\theta - 3\cos\theta)\, r^{-4}$. Mathematically, this arises both because $\hat{A}_\xi$ is not octopolar[12] and because thanks to $\omega_\phi \neq 0$, $F^{\phi n}$ contributes to F^{nt} in eq. (29c).

The table I summarizes the two types of solution with b = 0.

TABLE I

Two Types of Solution with b = 0

Type	Coordinate Relations	Mass	Angular Momentum	Charge	Magnetic Dipo Moment
				($\alpha = 0$)	
1. Static	$r_1 - m = r_2 - m$ $\theta_1 = \theta_2$	2m	0	2m	0
2. Stationary	$r_1 - m = m - r_2$ $\theta_1 = \pi - \theta_2$	0	$\mp 2am\hat{z}$*	0	$\mp am\hat{z}$*

*Upper sign refers to r_1, θ_1 coordinates, lower to r_2, θ_2. $\hat{z}$ denotes unit vector in +z-direction.

Acknowledgments

We would like to thank R. A. Kobishe for checking many of the calculations.

References and Footnotes

*Reseach supported by NSF Grant GP-19432, and by the University of Wisconsin-Milwaukee, Graduate School.

†On leave to Department of Physics, Princeton University, Princeton, New Jersey 08540 for the current year

‡Partially supported by NSF Grant GP-30799X.

§Work supported in part by NASA Grant 05-020-019.

[1] S. D. Majumdar, Phys. Rev. 72 , 390 (1947).

[2] A. Papapetrou, Proc. Roy. Irish Accad. A51, 191 (1947).

[3] W. Israel and G. A. Wilson, "A Class of Stationary Electromagnetic Vacuum Fields" (preprint, 1971), Jrl. of Math. Phys. 1972.

[4] Z. Perjes, Phys. Rev. Letters 27, 1668 (1971).

[5] A. Papapetrou, Proc. Roy. Soc., 209, 248 (1951).

[6] L. I. Schiff, Proc. Nat. Acad. Sci. 46, 871 (1960).

[7] D. C. Wilkins, Ann. Phys. 61, 277 (1970).

[8] The two electromagnetic invariants $F_{\mu\nu}F^{\mu\nu}$ and $*F_{\mu\nu}F^{\mu\nu}$ (where $*F_{\mu\nu} = \frac{1}{2}(-g)^{\frac{1}{2}} [\mu\nu\rho\sigma]\, F^{\rho\sigma}$) can also be used to test for singularities but they are not invariant under duality transformations, which mix the electric and magnetic fields in such a way as to preserve Maxwell's equations in vacuum. The stress-energy tensor and the metric are invariant under such transformations.

[9] J. B. Hartle and S. W. Hawking, "Solutions of the Einstein-Maxwell Equations with Many Black Holes" (preprint, 1971) have found no event horizons in the non-static IWP solutions.

[10] D. R. Brill and R. W. Lindquist, Phys. Rev. 131, 471 (1963).

[11] D. Christodoulou and R. Ruffini, "On the Electrodynamics of Collapsed Objects" (unpublished preprint, 1972).
R. S. Hanni, "Slicing the Electromagnetic Field Around a Black Hole" (unpublished junior paper, Princeton, 1972).

[12] In general, from eq. (30), the complex electromagnetic potential $A_t + i\Phi$ does not satisfy Laplace's equation with respect to the flat background.

ACKNOWLEDGEMENTS

The author thanks the editors and publishers for permission to reproduce the appendices, as follows:

Appendix	Source
A 1.1	The Astrophysical Journal, 163 (1971), The University of Chicago
A 2.2	Lettere al Nuovo Cimento, 3 (1972)
A 2.3	Physics Letters, 41B (1972), North-Holland Publishing Company, Amsterdam
A 2.6	Astrophysical Letters, 13 (1973), Gordon and Breach, London
A 3.1	Reprinted from Atti del Convegno Internazionale sul Tema: The Astrophysical Aspects of the Weak Interactions (Cortona "Il Palazzone", 10-12 Giuno 1970). Quaderno N. 157. Roma, Accademia Nazionale dei Lincei, 1971
A 3.2	Lettere al Nuovo Cimento, 2 (1971)
A 3.3	Physical Review Letters, 27 (1971)
A 3.4	Physical Review D, 5 (1972)
A 3.5	Physical Review D, 7 (1973)
A 3.6	Physical Review Letters, 28 (1972)
A 4.1	Physical Review D, 4 (1971)

Table of Contents

Contributors vi
Preface (English) vii
Préface (Français) ix
List of Participants xi

S. W. HAWKING

The Event Horizon 1

Introduction 5
1 Spherically Symmetric Collapse 6
2 Nonspherical Collapse 11
3 Conformal Infinity 13
4 Causality Relations 16
5 The Focusing Effect 20
6 Predictability 24
7 Black Holes 29
8 The Final State of Black Holes 37
9 Applications 44
A Energy Limits, 44
B Perturbations of Black Holes, 46
C Time Symmetric Black Holes, 51
References 54

B. CARTER

Black Hole Equilibrium States 57

Part I Analytic and Geometric Properties of the Kerr Solutions
1 Introduction 61
2 Spheres and Pseudo-Spheres 62
3 Derivation of the Spherical Vacuum Solutions 68
4 Maximal Extensions of the Spherical Solutions 74
5 Derivation of the Kerr Solution and its Generalizations . . 89
6 Maximal Extensions of the Generalized Kerr Solutions . . 103
7 The Domains of Outer Communication 112
8 Integration of the Geodesic Equations 117

Part II General Theory of Stationary Black Hole States

1 Introduction 125
2 The Domain of Outer Communications and the Global Horizon . 133
3 Axisymmetry and the Canonical Killing Vectors 136
4 Ergosurfaces, Rotosurfaces and the Horizon 140
5 Properties of Killing Horizons 146
6 Stationarity, Staticity, and the Hawking–Lichnerowicz Theorem 151
7 Stationary-Axisymmetry, Circularity, and the Papapetrou Theorem 159
8 The Four Laws of Black Hole Mechanics 166
9 Generalized Smarr Formula and the General Mass Variation Formula 177
10 Boundary Conditions for the Vacuum Black Hole Problem . 185
11 Differential Equation Systems for the Vacuum Black Hole Problem 197
12 The Pure Vacuum No-Hair Theorem 205
13 Unsolved Problems 210

J. M. BARDEEN

Timelike and Null Geodesics in the Kerr Metric 215
I Introduction 219
II Orbits in the Equatorial Plane 221
III Photon Orbits 227
A The Equations, 228
B The Apparent Shape of the Black Hole, 229
C The Throat of the Extreme Kerr Metric, 233
D Geometrical Optics, 236
Bibliography 239

J. M. BARDEEN

Rapidly Rotating Stars, Disks, and Black Holes 241

Introduction 245
I The Equations 246
II Boundary Conditions 249
III The Angular Momentum and Mass 253
IV The Structure of the Matter Configuration 261
V Stability 266
VI Uniformly Rotating Disks 271
Bibliography 289

H. Gursky

Observations of Galactic X-Ray Sources 291

I Introduction 295
II Observational Techniques 297
A Proportional Counters, 297
B Mechanical Collimators, 300
C Considerations on Statistics, 301
1 Analysis of spectral data, 301
2 X-ray pulsations, 304
3 Association of X-ray sources with known objects, 305
D Description of the X-ray Satellite UHURU, 306
III Distribution of Sources in the Galaxy 308
A Longitude Distribution, 310
B Maximum Luminosity, 311
C Minimum Luminosity, 311
D Spectral Content of the Radiation, 314
E Effect of Anisotropic X-ray Emission, 314
F Summary, 315
IV Properties of Individual X-ray Sources 315
A Scorpius X-1, 316
B Cygnus X-2, 319
C Cygnus X-1, 320
D Centaurus X-3, 323
E Hercules X-1, 325
F Other X-ray Sources of Interest, 330
G Summary, 330
Bibliography 335
Appendix 337
A Galactic X-ray Sources, 337
B Supernova Remnants, 340
C Transient Sources, 340
D Magellanic Cloud Sources, 341

I. D. Novikov and K. S. Thorne

Astrophysics of Black Holes 343

1 Introductory Remarks 347
2.1 Thermal Bremsstrahlung ("Free–Free Radiation" From a Plasma) 347
2.2 Free-Bound Radiation 355
2.3 Thermal Cyclotron and Synchrotron Radiation . . 357

2.4 Electron Scattering of Radiation 360
2.5 Hydrodynamics and Thermodynamics 362
2.6 Radiative Transfer 370
2.7 Shock Waves 379
2.8 Turbulence 383
2.9 Reconnection of Magnetic Field Lines 384

3 The Origin of Stellar Black Holes 386

4 Black Holes in the Interstellar Medium 389
4.1 Accretion of Noninteracting Particles onto a Nonmoving Black Hole 389
4.2 Adiabatic, Hydrodynamic Accretion onto a Nonmoving Black Hole 390
4.3 Thermal Bremsstrahlung from the Accreting Gas . . 396
4.4 Influence of Magnetic Fields and Synchrotron Radiation . 397
4.5 Interaction of Outflowing Radiation with the Gas . . 399
4.6 Validity of the Hydrodynamical Approximation . . 401
4.7 Gas Flow Near the Horizon 402
4.8 Accretion onto a Moving Hole 404
4.9 Optical Appearance of Hole:Summary 407

5 Black Holes in Binary Star Systems and in the Nuclei of Galaxies 408
5.1 Introduction 408
5.2 Accretion in Binary Systems: The General Picture . . 409
5.3 Accretion in Galactic Nuclei: The General Picture . . 420
5.4 Properties of the Kerr Metric Relevant to Accreting Disks . 422
5.5 Relativistic Model for Disk: Underlying Assumptions . 424
5.6 Equations of Radial Structure 427
5.7 Equations of Vertical Structure 431
5.8 Approximate Version of Vertical Structure . . . 434
5.9 Explicit Models for Disk 435
5.10 Spectrum of Radiation from Disk 438
5.11 Heating of the Outer Region by X-rays from the Inner Region 443
5.12 Fluctuations on the Steady State Model 443
5.13 Supercritical Accretion 444
5.14 Comparison with Observations 445

6 White Holes and Black Holes of Cosmological Origin . . 445
6.1 White Holes, Grey Holes, and Black Holes . . . 445
6.2 The Growth of Cosmological Holes by Accretion . . 446
6.3 Limit on the Number of Baryons in Cosmological Holes . 447
6.4 Caution 448

References 448

R. RUFFINI

On the Energetics of Black Holes 451

Part I On the Critical Mass of a Neutron Star

1.1 Introduction 455
1.2 Special Relativity and the Concept of Critical Mass . . . 457
1.3 Nuclear Interaction and the Value of the Critical Mass . . 466
1.4 Critical Mass Without any Detailed Knowledge of the Equation of State at Supranuclear Densities 482
1.5 The Critical Mass and the Late Stage of Evolution of Stars . 483
1.6 X-ray Sources and the Determination of the Critical Mass . 488
References 495
Appendix 1.1 Hagedorn Equation of State in Neutron Stars
C. Rhoades and R. Ruffini R1
Appendix 1.2 On the Masses of X-ray Sources
R. Leach and R. Ruffini R7
Appendix 1.3 On the Maximum Mass of a Neutron Star
C. Rhoades and R. Ruffini R19

Part II Electromagnetic Radiation in Static Geometries

2.1 Introduction 497
2.2 Timelike Geodesics in Schwarzschild and Reissner–Nordstrøm Geometries 499
2.3 Electromagnetic Perturbations in a Static Geometry . . 508
2.4 Electromagnetic Radiation from a Particle Falling Radially into a Schwarzschild or Reissner–Nordstrøm Black Hole . . 511
2.5 Electromagnetic Radiation from a Particle in Circular Orbits in a Schwarzschild or a Reissner–Nordstrøm Geometry . . 516
2.6 In what Sense can we Speak of an Object Endowed with a Net Charge in Astrophysical Problems? 525
References 527
Appendix 2.1 On the Energetics of Reissner–Nordstrøm Geometries
G. Denardo and R. Ruffini R33
Appendix 2.2 Electromagnetic Field of a Particle Moving in a Spherically Symmetric Black-Hole Background
R. Ruffini, J. Tiomno and C. Vishveshwara . . . R45
Appendix 2.3 Fully Relativistic Treatment of the Brehmstrahlung Radiation from a Charge Falling in a Strong Gravitational Field
R. Ruffini. R51

Appendix 2.4 Lines of Force of a Point Charge near a Schwarzschild Black Hole
R. Hanni and R. Ruffini R57
Appendix 2.5 Ultrarelativistic Electromagnetic Radiation in Static Geometries
R. Ruffini and F. Zerilli R75
Appendix 2.6 On a Magnetized Rotating Sphere
R. Ruffini and A. Treves R89

Part III Gravitational Radiation in Schwarzschild Geometries

Introduction 528
References 538
Appendix 3.1 Gravitational Radiation
R. Ruffini and J. A. Wheeler R95
Appendix 3.2 Gravitational Radiation in the Presence of a Schwarzschild Black Hole. A Boundary Value Search
M. Davis and R. Ruffini R117
Appendix 3.3 Gravitational Radiation from a Particle Falling Radially into a Schwarzschild Black Hole
M. Davis, R. Ruffini, W. Press and R. H. Price . . R121
Appendix 3.4 Pulses of Gravitational Radiation of a Particle Falling Radially into a Schwarzschild Black Hole
M. Davis, R. Ruffini and J. Tiomno R125
Appendix 3.5 Gravitation Radiation from a Mass Projected into a Schwarzschild Black Hole
R. Ruffini R129
Appendix 3.6 Can Synchrotron Gravitational Radiation Exist?
M. Davis, R. Ruffini, J. Tiomno and F. Zerilli . . R133
Appendix 3.7 Polarization of Gravitational and Electromagnetic Radiation from Circular Orbits in Schwarzschild Geometry
R. Ruffini and F. Zerilli R137

Part IV Energetics of Black Holes

Introduction 539
References 546
Appendix 4.1 Reversible Transformations of a Charged Black Hole
D. Christodoulou and R. Ruffini R147
Appendix 4.2 On the Electrodynamics of Collapsed Objects
D. Christodoulou and R. Ruffini R151
Appendix 4.3 Metric of Two Spinning Charged Sources in Equilibrium
L. Parker, R. Ruffini and D. Wilkins R161

Table of Contents 547

Printed in Great Britain by T. & A. Constable Ltd., Edinburgh

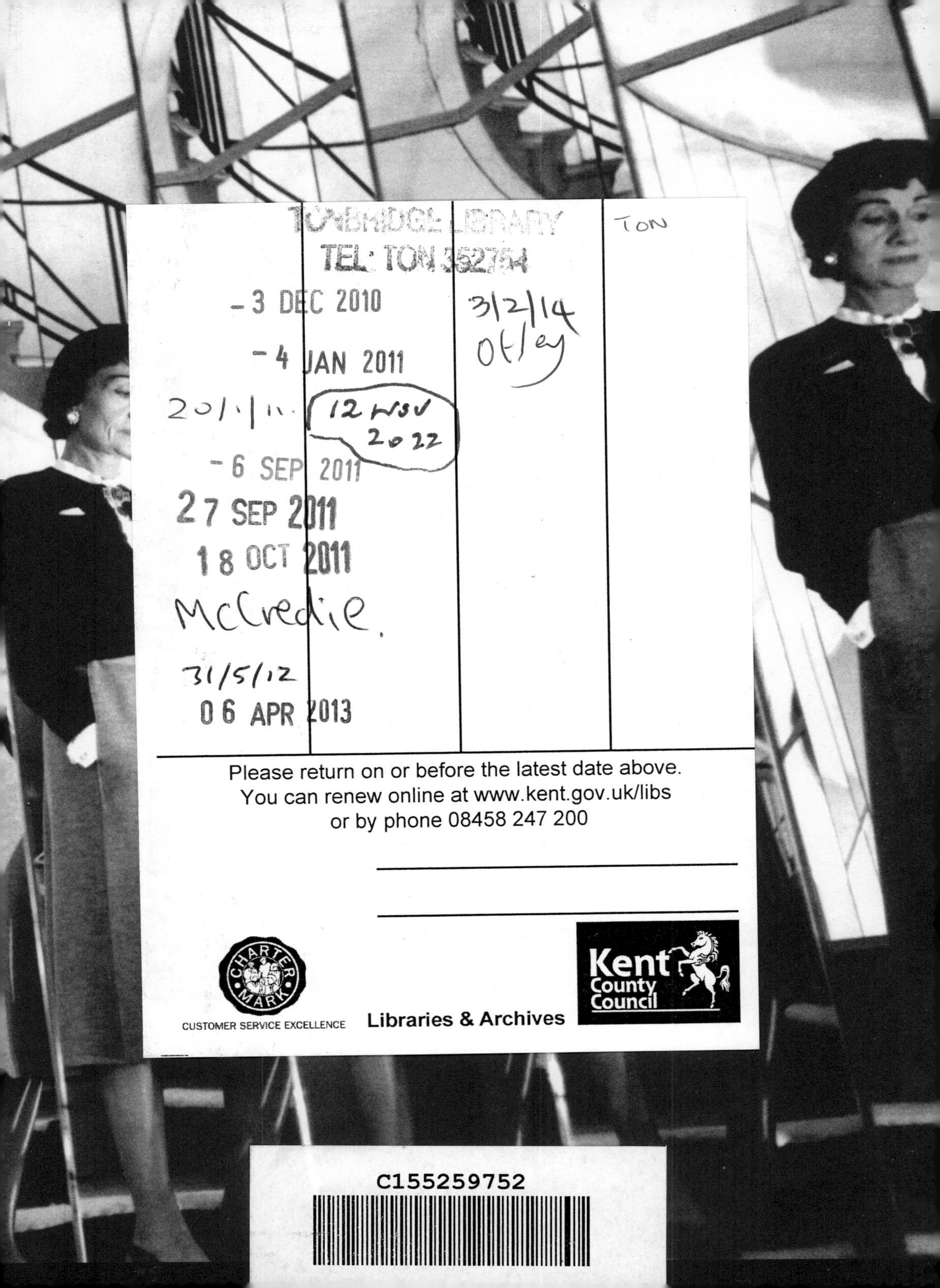

Chanel on the balcony of her suite at the Ritz. Roger Schall, 1937.